Encapsulated Catalysts

Encapsulated Catalysts

Edited by

Samahe Sadjadi

Iran Polymer and Petrochemical Institute, Tehran, Iran

ELSEVIER

ACADEMIC PRESS

An imprint of Elsevier

Academic Press is an imprint of Elsevier
125 London Wall, London EC2Y 5AS, United Kingdom
525 B Street, Suite 1800, San Diego, CA 92101-4495, United States
50 Hampshire Street, 5th Floor, Cambridge, MA 02139, United States
The Boulevard, Langford Lane, Kidlington, Oxford OX5 1GB, United Kingdom

Library of Congress Cataloging-in-Publication Data
A catalog record for this book is available from the Library of Congress

British Library Cataloguing-in-Publication Data
A catalogue record for this book is available from the British Library

ISBN: 978-0-12-803836-9

Publisher: John Fedor
Acquisition Editor: Emily M. McCloskey
Editorial Project Manager: Jill Cetel
Production Project Manager: Maria Bernard
Cover Designer: Matthew Limbert

Typeset by SPi Global, India

Dedication

I dedicate this book to my beloved father, who has supported me in everything I have ever done or gone through. He has constantly amazed me with his support, inspiration, and invaluable advice.

Contents

Contributors

Mariam Attoui, University of Bordeaux, Pessac Cedex, France

Claudia Bernal, Universidad de La Serena, La Serena, Chile

Marilena Carbone, University of Rome Tor Vergata, Rome, Italy

Ummuhan Cimenler, University of South Florida, Tampa, FL, United States

Esmail Doustkhah, University of Maragheh, Maragheh, Iran

David Farrusseng, Univ Lyon, Université Claude Bernard Lyon 1, Villeurbanne, France

Claire Jahier, University of Bordeaux, Pessac Cedex, France

Raksh V. Jasra, Reliance Industries Limited, Vadodara, India

Harjinder Kaur, Gujarat University, Ahmedabad, India

Fei Ke, Anhui Agricultural University, Hefei, China

Farooq-Ahmad Khan, Lakehead University, Thunder Bay, ON, Canada

John N. Kuhn, University of South Florida, Tampa, FL, United States

Junqiu Liu, Jilin University, Changchun, China

Monica Mesa, Universidad de Antioquia, Medellín, Colombia

Sylvain Nlate, University of Bordeaux, Pessac Cedex, France

Tiezheng Pan, Jilin University, Changchun, China

Sadegh Rostamnia, University of Maragheh, Maragheh, Iran

Samahe Sadjadi, Iran Polymer and Petrochemical Institute, Tehran, Iran

Sumeet K. Sharma, Reliance Industries Limited, Vadodara, India

Bing Sun, University of Neuchâtel, Neuchâtel, Switzerland

Georg Süss-Fink, University of Neuchâtel, Neuchâtel, Switzerland

Pietro Tagliatesta, University of Rome Tor Vergata, Rome, Italy

Bruno Therrien, University of Neuchâtel, Neuchâtel, Switzerland

Alain Tuel, Univ Lyon, Université Claude Bernard Lyon 1, Villeurbanne, France

Shuai Wang, Huazhong University of Science and Technology, Wuhan, China

Lixin Wu, Jilin University, Changchun, China

Zheye Zhang, Huazhong University of Science and Technology, Wuhan, China

Junfa Zhu, University of Science and Technology of China, Hefei, China

About the Authors

Junqiu Liu received his MSc in chemistry in 1990 at Jilin University and his PhD in macromolecular chemistry from the State Key Laboratory of Supramolecular Structure and Materials, College of Chemistry, Jilin University in 1999 under the supervision of Professor Jiacong Shen. Following his doctoral studies, he was a Humboldt Fellow with Professor Günter Wulff in Heinrich-Heine University, Germany. In 2003 he joined the faculty of the State Key Laboratory of Supramolecular Structure and Materials at Jilin University as a professor of chemistry. His main research interests include design of supramolecular catalysts and artificial enzymes, biomimetic microsystems, and biofunctional supramolecules.

Tiezheng Pan was born in Jilin, China, in 1989. He received his BSc degrees from both the College of Chemistry and College of Life Science, Jilin University in 2012. He is currently a PhD graduate student with Professor Junqiu Liu at the State Key Laboratory of Supramolecular Structure and Materials, College of Chemistry, Jilin University.

Lixin Wu obtained his BS in chemistry at Heilongjiang University (China) in 1982 and received his PhD degree in 1993 at Jilin University (China) on the synthetic bilayer membranes. He then undertook postdoctoral research at Changchun Institute of Physics (CAS, China) in 1995 and Hokkaido University in Japan (JSPS) in 1996 on photophysical and photochemical process of self-assemblies in solution systems. He went to the University of Hong Kong as a visiting scholar in 1997–98, studying the synthesis and photochromism of metal complexes. Dr. Wu became a full professor of chemistry at Jilin University in 2000 and the vice director of State Key Laboratory of Supramolecular Structure and Materials in the same year. His research interest concentrates on the self-assemblies and nano-structures of polyoxometalate-based hybrid supramolecular complexes, organic-inorganic polymer liquid crystal, colloid and interface phenomena on solid surface, supramolecular catalysis, and bio-related functionalization of hybrid assemblies.

Chapter 1

Organically Encapsulated Polyoxometalate Catalysts: Supramolecular Composition and Synergistic Catalysis

Lixin Wu
Jilin University, Changchun, China

Contents

1 INTRODUCTION OF POLYOXOMETALATES

Polyoxometalates (POMs) in chemistry refer to a class of nanosized inorganic polyanionic clusters (from less than one to several nanometer) in which transition metal ions are linked by corner- and edge-shared oxyanions to build a large, closed-framework entity. In an early definition, POMs are those mainly comprised of the metal ions belonging to the early transition metals (groups 5 and 6) in high oxidized states with their electron configuration of d(0) or d(1). Some typical examples include vanadium(V), niobium(V), tantalum(V), molybdenum(VI), and tungsten(VI). In the present stage, the POM family has been expanded to the metal-oxide clusters consisting of wider transition metal ions such as iron, cobalt, nickel, manganese, and even actinide series such as uranium ion and so forth.

POMs as a kind of inorganic compound have been known for over 190 years since the first member, ammonium phosphomolybdate containing

Encapsulated Catalysts. http://dx.doi.org/10.1016/B978-0-12-803836-9.00001-8

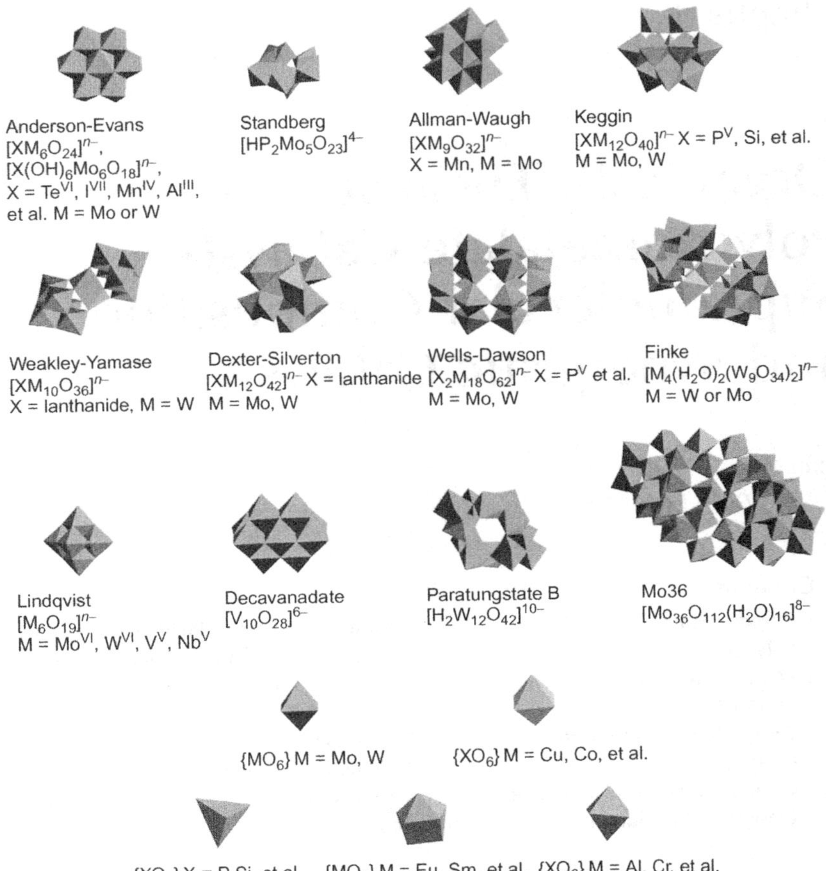

FIG. 1.1 Typical structural framework motifs of hetero- and homo-polyoxometalates.

the chemical composition of $(NH_4)_3[PMo_{12}O_{40}]$, was first discovered in 1826 [1], although its chemical structure was not identified until the isostructuralism of phosphotungstate was determined by Keggin in 1934 [2]. With the development of POM chemistry, the POMs' topologic morphologies are summarized and divided into several types (Fig. 1.1). The most fundamental type in general chemical structure ($XM_{12}O_{40}{}^{n-}$) with the ratio of heteroatom to metal ion at 1:12 is called Keggin structure. In the following discovery, other representative structure motifs such as Wells-Dawson ($X_2M_{18}O_{62}{}^{n-}$) with the ratio of 2:18, Anderson-Evans ($XM_6O_{24}{}^{n-}$) with the ratio of 1:6, Lindqvist ($M_6O_{19}{}^{n-}$), Allman-Waugh ($XM_9O_{32}{}^{n-}$), Dexter-Silverton ($XM_{12}O_{42}{}^{n-}$), Weakley-Yamase ($XM_{10}O_{36}{}^{n-}$), Standberg ($X_2M_5O_{(21\sim23)}{}^{n-}$), and Finke structure $[M_4(XW_9)_2{}^{n-}$ or $M_4(X_2W_{15})_2{}^{n-}]$, were found successively, where "X" denotes the heteroatom, "M" represents the metal ion, and "n" means the number of charges in the chemical formula. In addition to those well-known motifs, isomerism, lacunary structure, and crosslinked POMs connected by fundamental clusters via corner-shared and/or edge-shared oxyanions and/or substituted metal ions in diverse shapes have been created recently. In particular, giant clusters with metal ions numbering in the hundreds, including wheel-shaped molybdenum blue anions, spherical keplerates, cryptates, and clathrates, exploited by Müller and coworkers, represent the new generation of POMs' structure chemistry.

Beside the progress of topological motifs, the chemical compositions of POMs are also extended and the metal ion substitution has been conducted to enrich the structures and frameworks [3,4]. Organic modifications through covalent bond further expand the POM family to hybrid organic/inorganic materials [5–7]. Accompanying the development of chemical structure and cluster topology, the functionalization and applied materials of POMs also received wide interest. The structural variety based on highly symmetrical core assemblies brings the moderation of charge and electronic structure as well as the change of solution behavior. The optimization of electronic structures affords mixed-valence species, intramolecular electron transfer, ring current, electron-storage, and release properties. Some specific POMs are applied in analytical chemistry, catalysis, and photocatalysis. In solid-state, POMs are utilized as electronic and protonic conductors [8–10]. In medicine, POMs could be used for antitumor, antiviral, and antiretroviral activity [11–15]. Some important results on disease-related protein inhibitors, photoredox of water to generate hydrogen and oxygen, photo- and electro-chromic materials and devices, and single-molecular magnets have been reported over recent years [16–26].

In comparison to the diverse potential applications in multidisciplinary fields, the practically realizable application of POMs is the catalytic property. Because of their strong acidity when bearing protons as the counterions, POMs can be used as a catalyst for the synthesis and hydrolysis of esters. Because the redox of metal ions in the cluster occurs from their highest oxidation state to a lower oxidation state, the catalytic redox property does not break the closed framework of clusters.

Although interesting phenomena found on POMs depict a tempting foreground, several challenges remain. Due to the sensitivity of POMs regarding pH, their catalytic reactions are normally conducted in a restricted condition to retain the stability of POMs' framework. POMs as polyanionic inorganic clusters are not soluble in weak polar solvents, meaning that the direct utilization as the catalysts in non-polar organic media often does not afford satisfactory efficiency. Although there are several methods to transfer POMs from aqueous solution to organic phase, the dispersion of the POM cored catalysts should always be censured to avoid a worse specific surface. To overcome these disadvantages, many effective methods in conducting the catalytic functions are adopted by using the intrinsic features of POMs. By embedding at the lower domains of alkyl chain surface on the inner channel of porous materials, POMs conduct the high effective hydrolysis of esters. Through a joint sol-gel reaction, POMs have been embedded in the silica matrix. Some POMs that are grafted by organic groups via covalent bonds can be anchored on a solid surface for catalysis. Among those reported approaches, the most common method in dispersing POMs in various environments for the purpose of catalysis is the encapsulation with organic components bearing counterions driven by the electrostatic interaction. Because most POMs possess negative charges, cationic organic molecules are employed. This supramolecular route has several important advantages in comparison to other methods, such as:

- It is a general way suitable for all POMs regardless of structural morphology, chemical composition, organic modification, and the amount of charges.
- The main property of POMs in catalysis is well maintained because the encapsulation just uses the ionic replacement.
- The POMs encapsulated with organic cations can be suitably modulated for incorporating into different systems where individual POMs are not compatible.
- The Encapsulation can both protect the structural framework of POMs from the decomposition and preserve the sustainable catalytic functions.

As POMs have been widely used as catalysts, and as several representative reviews and books have summarized the corresponding progress in the field [27–31], we shall focus here only on the catalytic functions of encapsulated systems regarding the clusters over recent years, which have not yet been dealt with in detail. Some independent examples are introduced to demonstrate the extension of the method. In addition to the encapsulated POM catalysts, some relevant results may be included in the chapter for the purposes of ensuring the wholeness of the content.

2 POMs AS INORGANIC POLYANIONIC CLUSTER CATALYSTS

Due to the structural framework, chemical composition and electronic properties, POMs display diverse catalytic properties, and can perform as models for fundamental chemistry in catalysis. Among them, two well-known aspects are acid and oxidation catalysis. Through the prolongation of the catalytic properties, the clusters also exhibit catalytic properties of inoxidative polymerizations [32,33], coupling of aromatics [34–36], and addition reaction [37], as well as reductions [38–40]. When POMs exist in the form of strong polybasic acids, as depicted by following equation, the robust catalytic reactions of the type of solid acids can be carried out as those conducted by other strong Brønsted acids [41–43].

$$H_nPOM = nH^+ + [POM]^{n-}$$

$$\text{Substrate} \xrightarrow{H^+} \text{Product}$$

In principle, most acid-catalyzed reactions can be accomplished by POMs in the form of heteropoly acid. For POMs substituted by transition metals with

Lewis acid characteristic, such as Zr and rare earth ions, Lewis acid catalyzed reactions can be conducted [44–47]. Because the basicity of the polyanion increases when the cluster appears in incomplete topologic frameworks, the lacunary POMs normally show a tendency to become protonated and perform base catalyzed catalysts [48–50]. In comparison to the general strong acid such as hydrochloric and sulfuric acid, one of the main advantages is that POMs can be used in both homogeneous and heterogeneous acid catalysis because of the capability of POMs loading in solid supports such as silica gels or polymer matrixes [51,52].

When POMs are used as oxidation catalysts, the active sites move from protons to surface oxygen atoms and metal ions locating at the surface of cluster frameworks. Because the metal ions in their high oxidation state bind one or two terminal oxygens, the active oxygen atoms deriving from oxidants, such as hydrogen peroxide, organic peroxide alcohols, or acids, and even air, insert into the M=O double bond (Fig. 1.2), forming the POM peroxide that catalyzes the oxidation reaction of substrates. After the accomplishment of oxidation, the metal ions on the surface of POMs lose the active oxygen atom and return to the initial state. With the utilization of POM catalysts, alcohols are dehydrogenated to generate unsaturated bonds [53–56], while aldehydes are oxidized into the corresponding carboxylic acids [57–59]. Alkanes and olefins can also be oxidized to the corresponding oxides or epoxyl products [60–65].

In some cases, catalytic reactions can be carried out under light radiation and/or electrochemistry. With light irradiation, the electron of oxygen atom jumps to the excited state and moves to the *d* orbital of a metal ion, forming intramolecular electron transition, that is, O → M ligand to metal charge transfer (LMCT). The metal atom can obtain electrons from the external sacrificial agent, forming a reduced POM. In other words, the metal ions in the highest oxidized state are reduced by photo-irradiation in the presence of the reductant agent (Fig. 1.3). When mixing with a substrate, a POM at a lower oxidized state

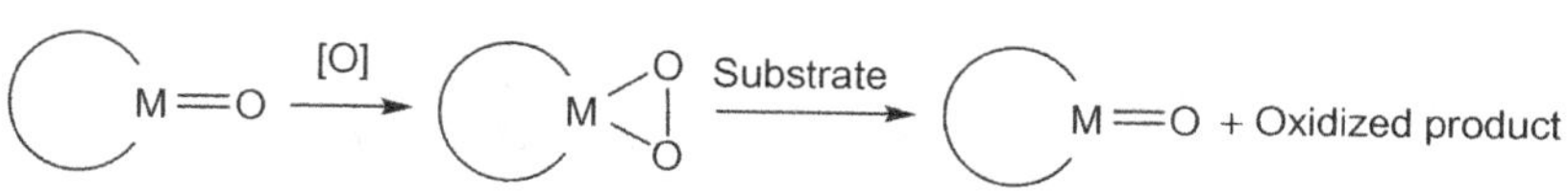

FIG. 1.2 The schematic drawing of terminal oxygen atom of POMs inserted by an activated oxygen from oxidant and then oxidizing substrate with activated oxygen atom.

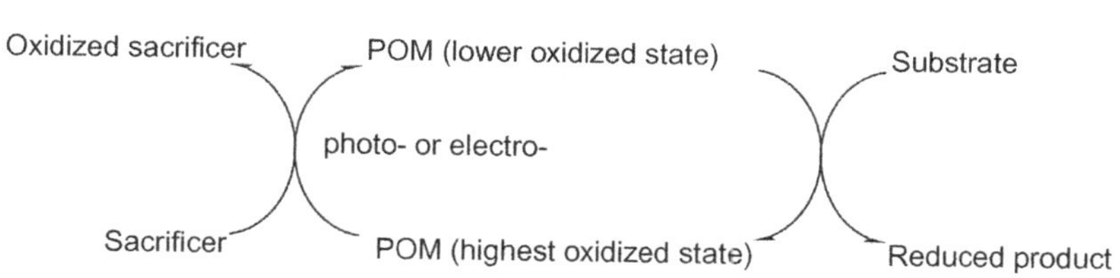

FIG. 1.3 The schematic drawing of POM catalysis for photo- or electro-stimulated reduction and reoxidation in the presence of sacrificer and oxidant substrates.

performs as reductant and releases the reduced electron to it, realizing a redox cycle. The catalytic activity has been demonstrated to source from the contribution of both thermal and photochemical redox of organic and inorganic molecules [66–69]. If the sacrificer is used as the substrate, the light-stimulated reaction becomes the catalytic oxidation reaction of POMs. Following a similar process, in one of the very hot research fields, POMs as the homogeneous phase catalyst have been used for the dehydrogenation for H_2 and O_2 evolution in water-splitting processes [70–77].

3 ENCAPSULATIONS OF INORGANIC CLUSTERS

Although POMs exhibit the robust catalytic properties, some critical factors still affect their utilization in different catalytic environments: for instance, the framework stability, the solubility as the homogeneous catalyst, the dispersion as the pure solid catalyst, and the ability to incorporate into the solid carriers as the heterogeneous catalyst. Considering the possible influence for the catalytic activity, the direct covalent modification to POMs is not usually thought to be a preferable approach, although the method is indeed effective in some special cases [78,79]. In fact, the covalent modification of organic groups is less used in POM catalyst. This is because few POMs are suitable for the purpose. Instead, the modification to the POMs through intermolecular interaction becomes more superable, since in this way, the basic electronic configuration and the framework of POM clusters can be maintained to the utmost extent. In comparison to charge transfer interaction, hydrogen bond, π-π interaction, and other intermolecular interaction, the multiple negative charges on surface are the most important characteristics of POMs, and the ionic interaction with these charges represents the common route, which shows no modification limitation for all POMs acting as the catalysts. By using organic cations to replace the counterions of POMs, the newly formed organic-inorganic salts become mostly insoluble in water, but soluble in polar and/or non-polar organic solvents. The substitution of counterions with organic cations offers POMs the homogenous property, in contrast to the naked POMs in organic solvents. Actually, in the initial stage, the phase transfer of POMs from water to organic media was accomplished by using tetrabutylammonium cation as the counterions [80]. Currently, several types of surfactants and organic cations, such as the single-chain, double-chain, and multiple chain cationic surfactants, cationic dendrimers, ionic liquids, cationic oligomers, and block copolymers, as well as some biomolecules and artificial peptides, have been developed for the modification of POMs. The cationic heads of organic molecules are generally the quaternary ammonium, pyridinium, and pronated amino groups [81–83].

Different from alkaline metal ions, the organic cations can cover the full surface of POMs, depending on the size of organic cations and inorganic polyanionic clusters as well as the number of charges and charge neutralization. When the organic cations bearing at least one long alkyl chain, such as dodecyl, tetradecyl, cetyl, and octadecyl groups, the surface of clusters could be well covered due to the chain flexibility and disordered conformation, forming the surfactant encapsulated POM (SEPs) complexes. More significantly, the procedure for the preparation is very simple, and is applicable for a wide range of purposes. Because of the higher solubility, most preparation routes of encapsulated POMs involve dispersion in water. For organic components having strong polarity, they are also dissolved in an aqueous solution. Mixing the two solutions or adding one solution to another gives rise to the precipitate due to the charge neutralization causing insolubility in water. By washing the crude product with water to remove counterionic salt, one can obtain the SEPs, where the inorganic cluster locates at the center and the organic cation's cover the surface of the polyanionic core via electrostatic interaction (Fig. 1.4). The SEPs prepared from this route normally dissolve in weak polar to polar organic solvents such as chloroform, dichloromethane, DMSO, DMF, and so forth.

Less polar surfactants, such as the ammonium cations bearing two or more long alkyl chains for the encapsulation of POMs, do not dissolve in water but become compatible with weak polar solvents, and the phase transfer method has to be used. In this case, by stirring the two incompatible solutions that contain hydrophobic organic cations and hydrophilic inorganic polyanions according to the charge ratio for some time at room temperature or encountering a gentle heating, the inorganic component is transferred into organic phase. After solvent evaporation and drying, the SEP complexes are obtained with high purity because the counterion of organic cation moves to the aqueous solution,

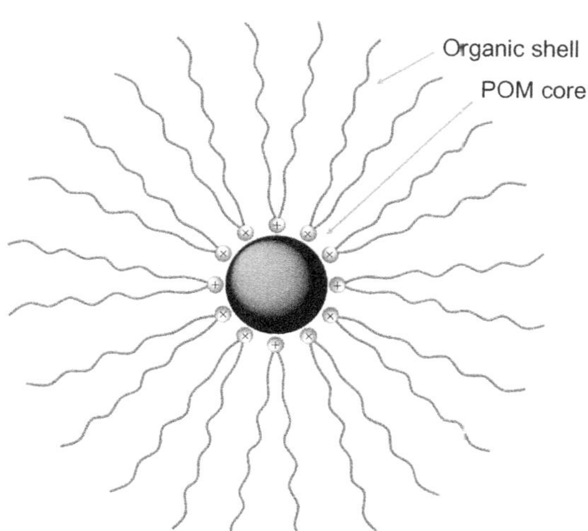

FIG. 1.4 The schematic drawing of surfactant encapsulated polyoxometalate complex in core-shell structure.

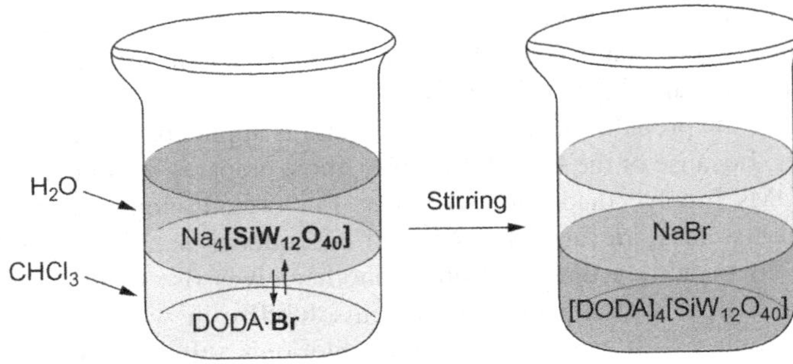

FIG. 1.5 The schematic route of the preparation of SEP complexes by mixing the POM aqueous solution and the DODA·Br in chloroform and encountering a stirring or gentle heating, which gives rise to the polyanions going into the organic phase while the bromide moves to the aqueous phase.

forming salt with the counterions of POMs (Fig. 1.5). There is still a method to prepare SEP complexes without using solvents, although it is rarely used, in which the solids of the two components are mixed and ground until they interact with each other. The extraction of the solid powder mixture with organic solvents and then washing the solutions with water give the target products. The method is useful for those POM components that are instable in water. The chemical composition of the prepared SEPs can be well identified by NMR, FT-IR and MS spectra, XPS, and elemental and thermogravimetric analysis. In most cases, the prepared SEP complexes can be found in a fully charged neutralized state, while in some other cases, the counterions of POMs could not be substituted completely. In addition to phase transfer, another important property deriving from organic cationic molecules is that surface encapsulation enhances structural stability of POMs during the catalytic process. For example, $(NH_4)_{28}$ $[MO_{154}\,(NO)_{14}\,O_{448}H_{14}\,(H_2O)_{70}]$ represents a new type of giant POM clusters bearing applicable properties. However, it could not be used in a harsh environment because of its structural instability. After the surface encapsulation with cationic surfactant dioctadecyldimethylammonium (DODA), the POMs in the formed complex raise their tolerance for external condition and could be used for Lewis and Brønsted acid catalysis [84].

The surfactant encapsulation is apparently not the only approach for the catalytic reactions of POMs in organic phase. Following the ionic interaction, ionic liquid, linear/branched polymers have also been employed for the encapsulation of POMs. The ionic modification of POMs by taking synzyme concept makes the catalytic oxidation applicable for water-insoluble molecules such as alcohols in an aqueous solution. A crosslinked polyethyleneimine with epichlorohydrinammonia for connection of alkylamine can be used to accommodate POMs through electrostatic interaction (Fig. 1.6) [85]. In contrast to the SEP catalysts dissolving in the organic phase, the polymer-encapsulated POM hydrogel matrix performs the oxidation for secondary alcohols such as those lipophilic substrates from 2-pentanol to 2-hexadecanol to ketones in the presence of hydrogen peroxide in an aqueous solution [86,87].

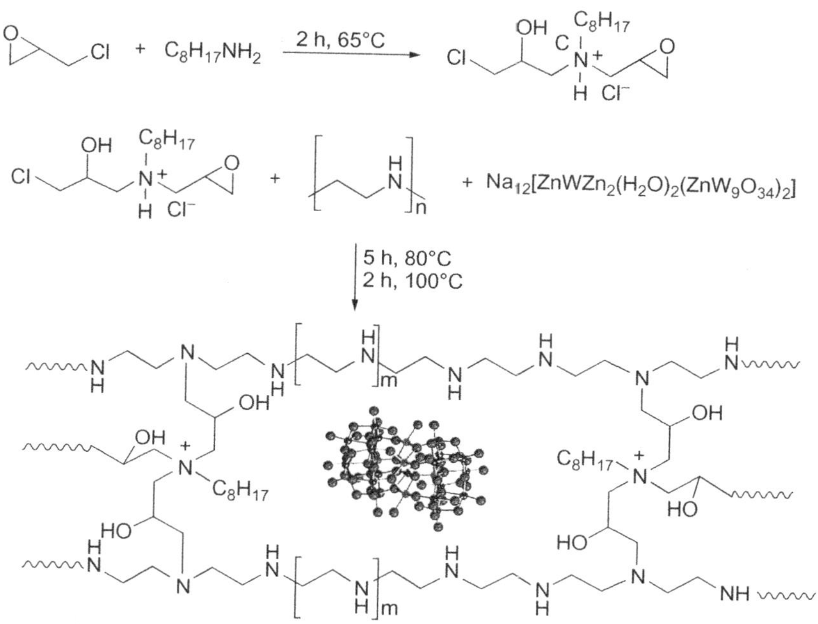

FIG. 1.6 Preparation of crosslinked polyethyleneimine and for the encapsulation for $[ZnWZn_2(H_2O)_2(ZnW_9O_{34})_2]^{12-}$, where for clarity, the network structures simplified to 2D representation and polyethyleneimine is simplified as a linear polymer. *(Reproduced from A. Haimov, R. Neumann, An example of lipophiloselectivity: the preferred oxidation, in water, of hydrophobic 2-alkanols catalyzed by a cross-linked polyethyleneimine polyoxometalate catalyst assembly, J. Am. Chem. Soc. 128 (2006) 15697–15700, with permission of the American Chemical Society. Copyright 2006.)*

4 CATALYSIS OF ENCAPSULATED POM COMPLEXES IN SOLUTION

In the initial stage of encapsulated POM catalysts, organic cations such as tetrabutylammonium (TBA) were just used to replace the counterions of POMs. The reduced polarity of the formed organic cations encapsulating complexes makes inorganic clusters no longer soluble in water but in weak polar solvents, which extends the application of POMs as a homogenous catalyst for water-insoluble substrates in high efficiency. However, the surface covering degree surrounding the inorganic clusters was not evaluated systematically. By the analysis of the surface area of POMs and the lateral size of cationic head of organic components, the coverage could be determined quite precisely and expressed via the following equation:

$$\mathrm{Ra} = n[\mathrm{La}] - [\mathrm{Sa}]$$

where Ra is the residual area, n denotes the charge number of a POM, Sa represents the POM's surface area, and La is the lateral area of one surfactant molecule at the cationic head. When Ra is positive, it means that the charges of the

POM could not be fully neutralized because the surface area on the POM is not large enough for accommodation of an equivalent charge number of organic cations. When the Ra value is zero, the surface of the POM is just fully encapsulated by the organic cations. When Ra is negative, it implies that the surface area around the POM could not be covered entirely even if the POM's charges are completely combined by the organic cations. Since the active center for catalysis locates at the surface of POMs, the diffusion and transportation of substrates to the POMs across the covering organic layer are prerequisite. Thus, controlling the surface coverage and building suitable slits for the substrates and product molecules become important. However, the blocking of the organic covering layer on POMs was dealt with much more carefully, comparing with those mostly used POMs; the organic cations are also quite limited and the total surface area of surfactants is usually less than that of POMs. In addition, the surface coverage can be easily adjusted by changing the chemical structure of the cationic head.

The early research work on the catalysis of cetylpyridinium cation encapsulated POMs for the oxidation of alkene, alcohol, thioether, etc., demonstrated the possibility of highly efficient homogeneous catalysis [88–91]. Following pioneering results, the POMs encapsulated by other surfactant molecules bearing a diverse cationic head were applied for catalytic purposes [92–94]. In recent years, some new catalytic reactions have been developed by using the prior properties. Systematic reviews summarized the general progress. Here, the stimuli-response catalytic reaction and the automatic separation of catalyst are focused in the homogeneous catalysis of SEPs.

For a typical example, Xi and his coworkers used hexadecyl pyridinium to encapsulate a heteropolyoxotungstate $[PO_4(WO_3)_4]^{3-}$ cluster to prepare an incompletely encapsulated complex. The catalyst is used for the catalytic oxidation of propene and olefin in the mixed solution of oil/water. For a fresh catalyst, very high conversion and selectivity (both reaching and/or excessing 90%) are found during the epoxidation of propylene at 65°C in 5 h. Interestingly, the complex is insoluble in organic solvents such as dichloromethane, but becomes soluble when H_2O_2 (30%) or its aqueous solution is added into the reaction solution. This catalyst also brings about another interesting property: the full exhaustion of hydrogen peroxide can be used for triggering the separation of the catalyst from the reaction mixture by forming precipitate automatically. After centrifugation, the catalyst is recovered and reused in the next reaction [95]. The authors thought that the peroxidation to the inorganic cluster made the encapsulated complex exist in a well-dispersed state while after the catalysis, the reduced complex catalysts linked each other due to the coordination bridging of terminal oxygen atoms to tungsten ion, giving rise to the cross-linking of the catalyst and yielding insoluble precipitate.

Neumann and his coworkers adopted a different strategy to perform a modulated phase transfer of encapsulated catalyst. They used polyfluorinated quaternary ammonium $[CF_3(CF_2)_7(CH_2)_3]_3N^+CH_3(RFN^+)$ to encapsulate polyoxometalate anions $[WZnM_2(H_2O)_2(ZnW_9O_{34})_2]^{12-}$ (M = Mn(II), Zn(II)). The obtained complexes $(RFN^+)_{12}[WZnM_2(H_2O)_2(ZnW_9O_{34})_2]$ performs multiphasic catalysis for alcohol and alkenol oxidation, and alkene epoxidation by the presence of aqueous H_2O_2 oxidant and absence of fluorous solvent (by aqueous H_2O_2 oxidant in the presence and absence of fluorous solvents) (Fig. 1.7) [96]. In both cases, the catalyst, hydrogen peroxide and the substrate are not compatible with each other and exist in separated states. However, upon heating, the substrate and catalyst become compatible with each other in the mixed organic phase because of the improved solubility of fluorous solvent and the encapsulation of the catalyst by the fluorous surfactant, so that the catalytic oxidation of aliphatic alcohols, alkenols, and alkenes can be carried out efficiently. After the reaction finishes and the system cools down, the fluorous phase and fluorous covered catalyst separate automatically from the organic phase with products. Due to this feature, the catalyst could be recycled.

Neumann's group also investigated the heterogeneous polyoxometalate catalysts for oxidation reactions [97]. They used a sandwiched cluster to be encapsulated by branched tripodal organic polyammonium salt, tris [2-(trimethylammonium)ethyl]-1,3,5-benzenetricarboxylate or 1,3,5-tris[4-(N, N,N-trimethylammoniumethylcarboxyl)phenyl] benzene trications. A three-dimensional perforated coral-shaped amorphous aggregation with the organic cations surrounding polyoxometalate polyanions is confirmed due to the irregular shape of the organic cations. Because of the interaction between complexes, they do not exist independently and the disordered packing of the encapsulated catalysts affords a mesoporous structure with an average pore diameter of 36 Å. The heterogeneous catalysts show effective and selective catalysis for the epoxidation of allylic alcohols and oxidation of secondary alcohols to ketones in the presence of hydrogen peroxide.

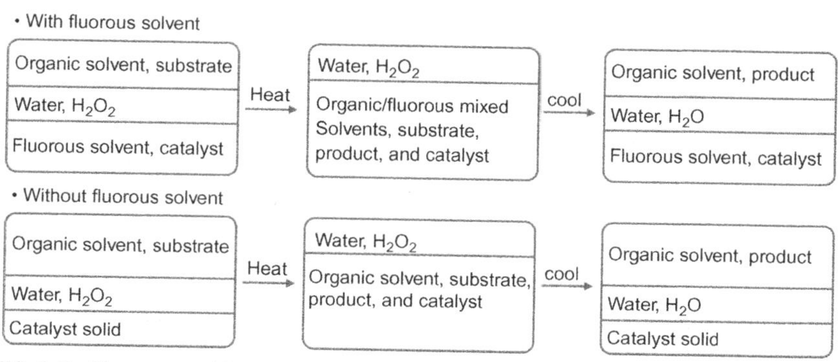

FIG. 1.7 The process of fluorous SEPs catalyzing the oxidation of substrates in the mixture solution of hydrocarbon and fluorous solvents. *(Reproduced from G. Maayan, R.H. Fish, R. Neumann, Polyfluorinated quaternary ammonium salts of polyoxometalate anions: fluorous biphasic oxidation catalysis with and without fluorous solvents, Org. Lett. 5 (2003) 3547–3550, with permission of the American Chemical Society. Copyright 2003.)*

Due to the solubility of organic molecules and the restriction of preparation condition such as the co-deposition method, the incomplete encapsulation phenomenon is not occasional. By analyzing the surface occupation of organic cations on POMs, the hydrophilic residue area always needs to be filled with hydrophilic part of other POMs driven by the interfacial energy in organic phase. With this driving force, the polarity of the complexes inevitably makes them aggregate in the solution. Wu's group employed the change of amphiphilic property of SEPs to control the catalytic reaction process and the photo irradiation to realize the phase transfer of the catalyst between water and organic phase as well as the recycled utilization [98]. Reversible *trans-/cis*-isomerization of azobenzene (Azo) group under photo-irradiation is normally accompanied by spatial and polarity changes [99]. With this property, the photo-sensitive group was grafted to the hydrophobic terminals of dihexyldimethylammonium bromide. The Azo-contained cationic surfactant encapsulated $K_{12.5}Na_{1.5}$ $[NaP_5W_{30}O_{110}]$ complex displays acute photo-responsivity in organic phase. At the initial stage (Azo groups in *trans*-form), the sample in toluene/THF mixture solution exists at the individual state due to the similar surface polarity of the complex with the mixture solvent. After encountering a UV light irradiation (such as 365 nm) to the solution, Azo groups surrounding POMs change to *cis*-form. Because the *cis*-conformation has a higher polarity than the *trans*-conformation, the increased polarity induces the aggregation of the complex. When the sample solution is irradiated with visible light (such as 450 nm), the *cis*-conformation returns to a *trans*-state, and the aggregation decomposes into a reversible individual state. In a polar solution such as water/THF mixture, the opposite behavior occurs. Interestingly, the reversible change of aggregation and dispersion of the photo-sensitive SEP can be applied to modulate the catalytic activity for an oxidation reaction. As shown in Fig. 1.8, for the individual complex in toluene, its catalysis for methyl phenyl sulfide in toluene/THF solution yields a higher conversion, while for the aggregated state after the UV light irradiation, lower conversion is found because most of the catalytic surface of POMs has been embedded in the aggregations. After a visible light irradiation, the higher conversion recovers. The alternate UV and visible light irradiations dominate the reaction going fast and decelerating. The plot of the same oxidation catalyzed by the complex versus the time indicates the modulation of light irradiation.

The polar change of the Azo modified SEP catalyst, which is derived from conformation transformation, can be also used for the light controlled phase transfer catalysis [100]. The encapsulation of Azo terminated dihexyldimethylammonium bromide to Na_{12} $[WZn_3(H_2O)_2[ZnW_9O_{34})_2]$ gives a complex with chemical formula of $[\text{Azo-surfactant}]_9Na_3[WZn_3(H_2O)_2(ZnW_9O_{34})_2]$ through a two-phase procedure. In a toluene solution of prepared complex, oxidant cumyl hydroperoxide (CHP), and phenothiazine (PH), a useful intermediate in organic synthesis is added as the substrate to carry out the homogeneous catalysis at 30°C. After the accomplishment of the homogeneous reaction in 4 h,

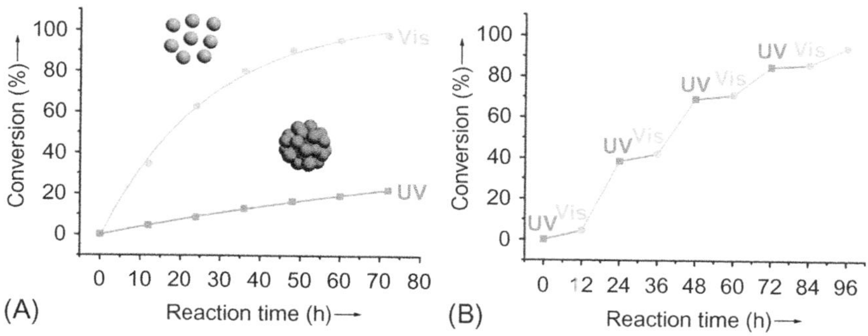

FIG. 1.8 The plots of the oxidation conversion of methylphenyl sulfide in toluene/THF solution versus the time (A) catalyzed by dispersed and aggregated SEP catalyst prepared upon irradiation of visible and UV light, respectively, and (B) upon alternate UV and visible light irradiation, where the insets indicate existing states of Azo-SEP catalyst. *(Reproduced from Y. Yang, L. Yue, H.L. Li, E. Maher, Y.G. Li, Y.Z. Wang, et al., Photo-responsive self-assembly of an azobenzene-ended surfactant-encapsulated polyoxometalate complex for modulating catalytic reactions, Small 8 (2012) 3105–3110, with permission of WILEY-VCH Verlag GmbH & Co. KGaA, Weinheim. Copyright 2012.)*

the product *S,S*-dioxide (PHDO) and the reduced oxidant dimethylphenylcarbinol (DMPC) remain in the toluene while the catalyst is transferred into the added water by a simple UV light irradiation. By mixing another port of fresh reaction toluene solution with the catalyst aqueous solution and experiencing a visible light irradiation, the catalyst transfers automatically back to the organic phase for catalysis. Five cycles later, more than 88% of the catalyst remains. Following the same idea, other POMs such as Keplerate-type $(NH_4)_{72}\{(Mo)Mo_5O_{21}(H_2O)_6\}_{12}\{Mo_2O_4(SO_4)\}_{30}$($Mo_{132}$) and disk-like $K_{12.5}Na_{1.5}[NaP_5W_{30}O_{110}]$ (P_5W_{30}) are used for the smart photo-responsive catalysis, and similar homogeneous catalysis and heterogeneous separation are achieved.

By carefully selecting the reaction media, SEPs could be used as the temperature-controlled phase transfer catalyst for the catalytic oxidation of pyridine and alcohols. Several representative results have been summarized in a mini review recently by Wang et al. [29,101,102]. When ionic liquid molecules encapsulate POM clusters, the formed POM complex catalysts perform the catalytic oxidation in ionic liquid for alcohols [103].

5 CATALYSIS OF ENCAPSULATED POM COMPLEXES IN SELF-ASSEMBLIES

Actually, most organically covered POM complexes do not exist in individual states, instead, they are mostly in an aggregated state despite being in polar and nonpolar solvents because of their amphiphilicity. Therefore, the aggregation of POM complexes in Neumann's work is apparently not an occasional example

[104,105]. Systematic investigation of the phenomenon demonstrates that there is a hydrophobic/hydrophilic separation between the surface covering organic cations and the ball like POMs driven by the interfacial energy. On the other hand, the electrostatic interaction provides a favorable condition because of its irrelevance to the preferential orientation between the organic and inorganic components, which gives rise to the ordered self-assembly of SEPs in organic solvents. Kurth et al. reveal that for the giant clusters, the surfactant DODA cations disperse on POMs' surface with a disordered covering and the complexes form a tight packing structure similar to rigid balls [106]. Wu et al. demonstrate the phase separation structure when small clusters bearing lower surface charges are encapsulated by the same surfactant [107]. The detailed characterization for the self-assembly of SEPs indicates the onion-like reverse bilayer structure of surfactant components with POMs locating at the middle of the reverse bilayer [108].

This special assembly structure is used as both a catalyst and the template for the preparation of metal nanoparticles in certain morphologies. In the presence of a sacrificer, the POMs in SEP complexes can be reduced by light irradiation, forming heteropoly blue. The reduced SEPs are often used for general reduction and preparation of metal nanoparticles by adding noble metal ions into the reaction solution.

For protonated dioctadecylamine encapsulated POMs such as $[PMoW_{12}O_{40}]^{3-}$ complex, the assembly with inversed vesicular assembled structure maintains well even when encountering a light irradiation in organic phase, in which the POM clusters have been reduced to the lower oxidized state such as $W^{6+} \rightarrow W^{5+}$. With this property, the photocatalytic reduction of some metal ions can be performed. Wu's group used the encapsulated catalyst assembly to realize the preparation of metal nanoparticles. The self-assembly of the catalyst is obtained by simply dissolving the sample in methanol. Here the solvent plays a key role for the template in the formation of nanoparticles with specific morphology. On one hand, the polar solvent helps to increase the structural stability; on the other hand, it serves as the sacrificer for the reduction of the POMs used under the light irradiation [109,110]. Dynamic light scattering and TEM imaging reveal the formation of assemblies in sizes of ca. 200 nm. X-ray diffraction (XRD) data confirm that the spherical assembly has a layered sub-structure with a thickness of ca. 2.7 nm and a larger lateral area of alkyl chains than in the regular packing. The observed layer spacing is much smaller than that of the ideal length estimated from the alkyl chains in full *trans* form and the diameter of POMs, indicating the partial disorder and the interdigitation of alkyl chains between two adjacent layers. The partial disorder of alkyl chains is useful for the entry and exit of the substrate across the organic coating. In contrast to those naked POM catalysts affording the formation of spherical-shaped nanoparticles, the self-assemblies of the encapsulated catalysts at the reduced state produce flower-like nanoparticles due to the confined environment for the growth and assembly of nanoparticles when an $HAuCl_4$ alcohol solution is added dropwise to the reaction system. With an increased time for light

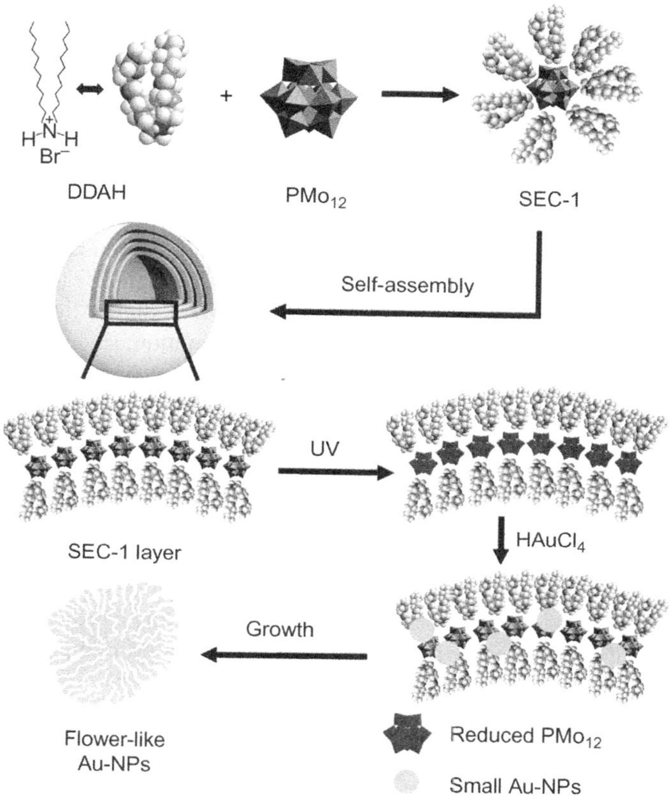

FIG. 1.9 Schematic presentation of preparation of POM catalyst and its self-assembly as the template for the catalytic reduction of gold ions into patterned nanoparticles. *(Reproduced from H.L. Li, Y. Yang, Y.Z. Wang, W. Li, L.H. Bi, L.X. Wu, In situ fabrication of flower-like gold nanoparticles in surfactant-polyoxometalate-hybrid spherical assemblies. Chem. Commun. 46 (2010) 3750–3752, with permission of the Royal Society of Chemistry. Copyright 2010.)*

irradiation, the reduced metal atoms deposit surrounding the assemblies and finally form the nanostructure, as shown in Fig. 1.9 [111].

Following a similar approach, Wang and his coworkers used a series single chain cationic surfactants, dodecyltrimethylammonium (DTMA), tetradecyl trimethylammonium (TTMA), hexadecyl trimethylammonium (HTMA), and octadecyl trimethylammonium (OTMA) to encapsulate phosphotungstate and then prepared SEP catalysts, $(DTMA)_3PW_{12}O_{40}$, $(TTMA)_3PW_{12}O_{40}$, $(HTMA)_3PW_{12}O_{40}$, and $OTMA)_3 PW_{12}O_{40}$ [112]. In contrast to those spherical self-assemblies of POM complexes in chloroform (or dichloromethane) and their mixture solution with methanol, these complexes form semi-tubular and fiber-like assemblies in the mixture of butanone and butan-1-ol (v/v, 2:1), although the reverse bilayer sub-structure remains the same. From the antisymmetric and symmetric stretching bands of CH_2 appearing at 2922 and 2851 cm^{-1}, it could be deduced that the ordered packing of alkyl chains exists in the inner bilayer. The measured bilayer spacing is 2.6–2.7 nm for the two short chain surfactants and the layer spacing is ca. 3.4–3.8 nm for two long chain surfactants (Fig. 1.10). Considering the ideal length of these surfactants and the ordered assembly of POM complexes, the interdigitation among alkyl chains between adjacent bilayers seems inevitable. The authors did not evaluate the influence of organic layer spacing for the catalytic efficiency of substrate dibenzothiophene, but from the HPLC data, one can see that the oxidation of POM assembly for the substrate is irrelevant to the alkyl chain length and the structural morphology of assemblies because the conversions in a similar time scale are at the similar level. However, this is obviously not common, since the catalytic reactions are generally involved in the reaction temperature and the molecular size of the substrates.

A solvent is very important for both the catalytic reaction and the morphology of self-assemblies constructed by POM complex catalysts. By encapsulating the phosphotungstate with dioctadecyldimethylammonium (DODA), the formed encapsulated catalyst aggregates into conical self-assembly in the mixture of chloroform and *n*-butanol-1, which shows the catalytic properties to dibenzothiophene (DBT), diphenyl sulfide (DPS), and dimethyl sulfide (DMS), as many similar encapsulated POMs performed [113]. In comparison to the scale of substrates used, the assemblies of the POM complex catalysts

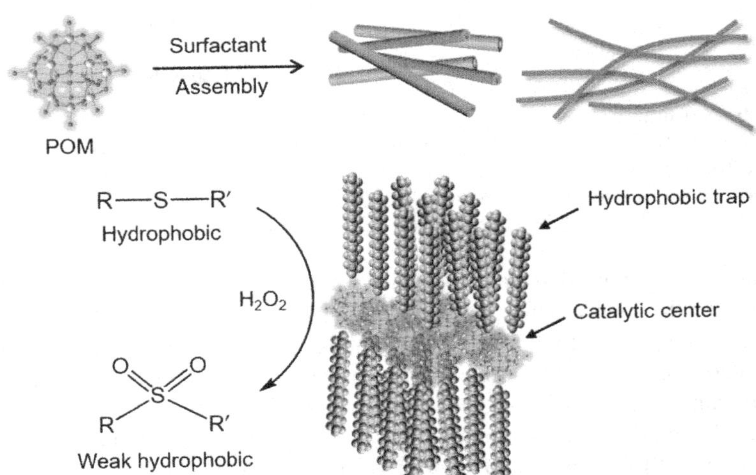

FIG. 1.10 Schematic illustrations for (A) preparation of POM semitube and wire like assemblies and (B) oxidation of sulfides to sulfones. *(Reproduced from A. Nisar, J. Zhuang, X. Wang, Construction of amphiphilic polyoxometalate mesostructures as a highly efficient desulfurization catalyst, Adv. Mater. 23 (2011) 1130–1135, with permission of WILEY-VCH Verlag GmbH & Co. KGaA, Weinheim. Copyright 2011.)*

possess bigger sizes. Due to the similar catalytic center around a POM's surface, the difference of the catalytic efficiency is mainly derived from the effective surface area and the substrates entering and exiting the organic covering layer. The incorporation of Fe_3O_4 nanoparticles stabilized by oleic amine into the SEP self-assemblies is found to increase the catalytic activity. Considering that the size (7–8 nm) of Fe_3O_4 nanoparticles is much larger than the POM complex, the hybridization should play a role in decrease of the order of assemblies so that the substrate could penetrate into organic coating conveniently.

As a special example, Liu and Wei prepared a small cluster $[V_6O_{19}]^{2-}$ that is covalently grafted with two hexadecyl groups at opposite positions. Its TBA covered complex affords the self-assemblies with reverse bilayer structure similar to the abovementioned sub-structure. The belt-like assemblies are stable in acetonitrile and display catalytic property for the oxidation of sulfide in the presence of hydroperoxide, although the activity does not seem very high [114].

6 CATALYSIS OF ENCAPSULATED POMs COMPLEXES AT SOLUTION INTERFACE

When the number of surfactant molecules is larger, the formed SEP complexes can be regarded as the inverse micelles bearing a POM core that replaces all counterions of the surfactant and water molecules. Because of the electrostatic interaction, the micelle like complex becomes stable in organic phase due to the shared counterionic cluster. However, in response to the chemical environment, the complex itself undergoes a polarity separation between the POM core and the surfactant. Thus, the phase separated POM complexes can accomplish partial functions of surfactants. In many cases, the biphasic catalytic reactions are needed and the amphiphilic POM complex catalysts become useful for a highly efficient catalysis. In the early stage, Neumann et al. prepared SEPs by covering the $PO_4[W(O)(O_2)_2]_4^{3-}$ core with series organic cations, tetrahexylammonium (THA), trioctylmethylammonium (TOMA), (−)-dodecyl-*N*-methylephedrinium (DME), and hexadecyltrimethyl ammonium (HTMA). When 1,2-dichloroethane solution of these SEP catalysts in the presence of 30% H_2O_2 and the substrate 1-octene is stirred at room temperature, the inverse emulsion (water in oil) is obtained. The catalytic oxidation of the substrate into 1-octene oxide takes place quickly. But the conversion of the product varies against the change of surfactant molecules around the POM and the best value is about 57%. Clearly, it is not very high. As mentioned above, in addition to the POM's catalytic center itself, many factors play roles in the reaction activity of the SEP catalytic system, and the ability of the SEP catalysts for stabilizing the emulsion by locating at the oil and water interface is another very important factor. Therefore, the selection of surfactant for interfacial catalysis in biphasic system becomes crucial [115].

To raise the catalytic activity toward industrial application, with the same route, the $[PW_{12}O_{40}]^{3-}$ is encapsulated with DODA and it is expected to be fixed at the oil and water interface more efficiently. To develop an effective approach for oxidative desulfurization of diesel, Li et al. made a detailed investigation for the inverse emulsion solution [116]. By dissolving H_2O_2 in aqueous solution (30%) and the SEP catalyst in the organic phase, the metastable emulsion forms under stirring. In ca. 80 min at 30°C, the model substrate 4,6-dimethyldibenzothiophene, refractory sulfide in hydrodesulfurization processing, could be fully oxidized in a turnover number (TON) higher than 300. By demulsification and sedimentation of the product, the used catalyst is separated and used again (Fig. 1.11). In the case of the same POM cluster, when a different surfactant such as hexadecyl pyridinium $(n\text{-}C_{16}H_{33})C_5H_5N^+$ is employed, the formed emulsion becomes too stable to be broken so that the separation of the catalyst becomes difficult for recycling. Thus, the suitable surface activity of the SEP catalysts is important for both catalysis and the separation from reaction system.

Not only surfactant, the change of POMs affects the surface activity and the catalysis of SEP at liquid-liquid interface. When a lacunary Keggin POM,

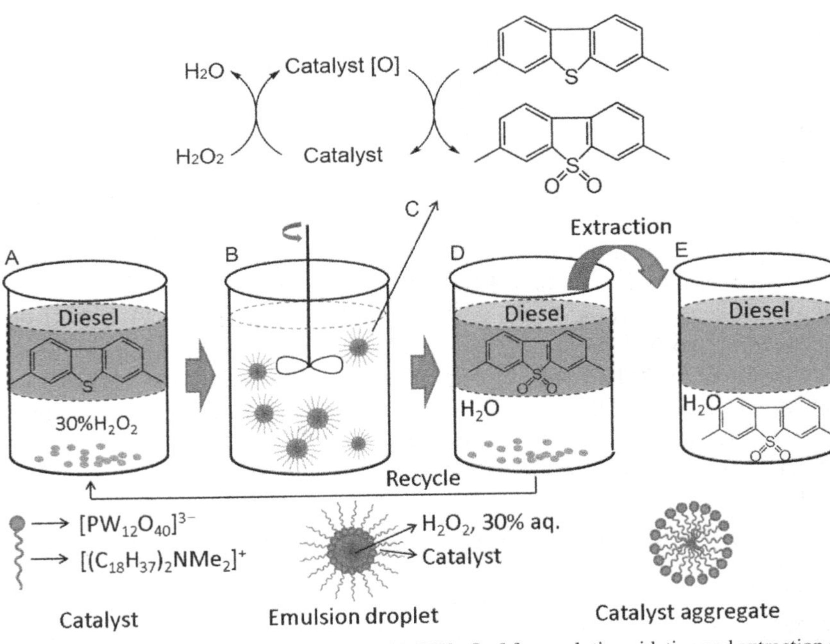

FIG. 1.11 The schematic process of $[DODA]_3[PW_{12}O_{40}]$ for catalytic oxidation and extraction of sulfur-containing molecules present in diesel (A) before oxidation, (B, C) during oxidation of sulfur-containing molecules in emulsion droplets, (D) after oxidation, and (E) in extraction of reaction with a polar solvent. *(Reproduced from C. Li, Z.X. Jiang, J.B.Gao, Y.X. Yang, S.J. Wang, F.P. Tian, et al., Ultra-deep desulfurization of diesel: oxidation with a recoverable catalyst assembled in emulsion, Chem. Eur. J. 10 (2004) 2277–2280, with permission of WILEY-VCH Verlag GmbH & Co. KGaA, Weinheim. Copyright 2004.)*

$[PW_{10}O_{36}]^{7-}$, is encapsulated by octadecyl trimethylammonium unsaturatedly, the formed SEP complex $[C_{18}H_{37}N(CH_3)_3]_4[H_2NaPW_{10}O_{36}]$ shows a much better catalytic activity than $[DODA]_4[H_2NaPW_{10}O_{36}]$ and $[DODA]_3[PW_{12}O_{40}]$ (Fig. 1.12) [117]. These results further demonstrate the importance of the synergistic property of both organic and inorganic components for catalysis at the interface.

In the catalytic system of SEP complexes, H_2O_2 is the most often used but not the only oxidant. Cheap and environment friendly oxidant such as air is also considered. But air is normally difficult to be used directly. With the assistance of aldehyde such as isobutyl aldehyde, in the presence of molecular oxygen, the SEP$[C_{18}H_{37}N(CH_3)_3]_5[PV_2Mo_{10}O_{40}]$ can perform a catalytic oxidation for both aldehyde to peracid and sulfide to sulfoxide and sulfone in the emulsion system with acetonitrile [118]. Interestingly, instead of the POM catalytic center, the cationic head surrounding a POM can also be the catalytic reaction site for the catalysis of cross-aldol reaction in an emulsion system, where the POM core collects the cations, yielding a synergistic effect [119,120].

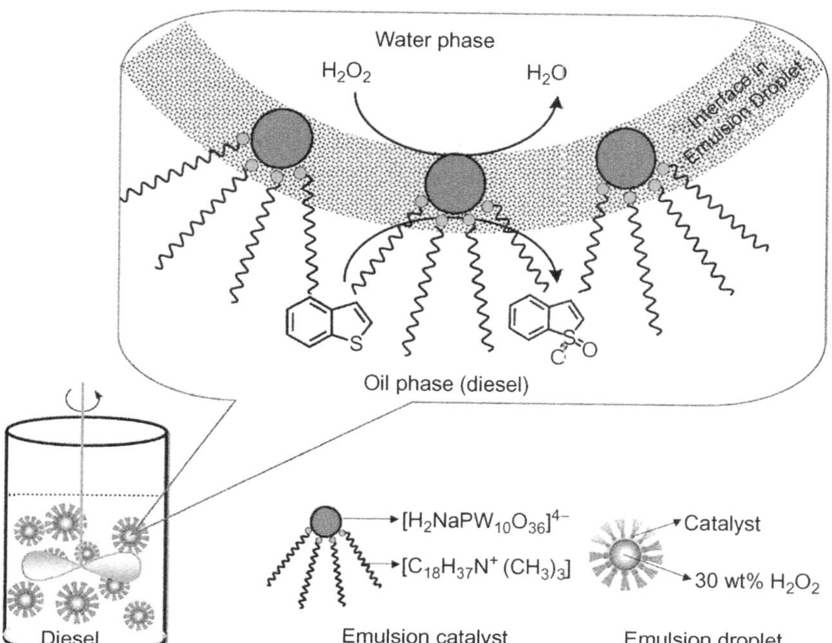

FIG. 1.12 The catalytic oxidation of SEP complexes for sulfides such as benzothiophene in water/oil emulsion droplets. *(Reproduced from H.Y. Lü, J.B. Gao, Z.X. Jiang, F. Jing, Y.X. Yang, G. Wang, et al., Ultra-deep desulfurization of diesel by selective oxidation with $[C_{18}H_{37}N(CH_3)_3]_4[H_2NaPW_{10}O_{36}]$ catalyst assembled in emulsion droplets, J. Catal. 239 (2006) 369–375, with permission of Elsevier Inc. Copyright, 2006.)*

In some special cases, the naked POMs mix with cationic micelle bearing anionic substrate, yielding electrostatic immobilization of substrate and POM catalyst at the surface, which affords enhanced oxidation efficiency in water [121].

7 CATALYSIS OF ENCAPSULATED POMs COMPLEXES IN SOLID MATRIX

The organically encapsulated POM complexes as homogeneous and/or phase transfer catalysts exhibit highly catalytic effect for compatible substrates in solutions, but their separation from the reaction system often involves complicated procedures. Like general catalysts applied in industry, therefore, developing applicable methods that maintain catalytic activity such as those occurring in homogeneous catalysis while render the catalyst easily removable from the reaction is highly desired. By simply incorporating POMs into silica-bearing channels and other solid surfaces, the catalytic reactions are carried out and the POM catalysts loaded on/in solid matrixes are reused after the reaction through a quick filtration [122–124].

The covering of ionic liquid bola-form cations bridged by polyethylene glycol (PEG) onto $Na_7PW_{11}O_{39}$ via electrostatic interaction gives an ionic liquid encapsulated POM complex catalyst. Because of the excess positive charges on the catalyst surface, the formed complex can further be incorporated into the polymer matrix, sodium carboxymethyl cellulose, a negatively charged polyelectrolyte through co-precipitation. With this polymer encapsulated POM catalyst, the epoxidation of *cis*-cyclooctene by H_2O_2 in ethyl estate can be carried out and the catalyst is easily separated by filtration for further catalytic recycling after cooling down the reaction [125].

Leng, Wang et al. simplified the above method. They prepared a suitably cross-linked cationic copolymer through the polymerization of an amine-functionalized ionic liquid [3-aminoethyl-1-vinylimidazolium]Br·HBr([AVIM]Br·HBr) and a dicationic ionic liquid [1,1′-(butane-1,4-diyl)-bis(3-vinylimidazolium)]Br_2 ([BVIM]Br_2) monomer. By mixing the cationic polymer matrix with polyanionic cluster $H_3PW_{12}O_{40}$, the polyelectrolyte encapsulated POM composite is obtained via co-precipitation and stabilized by an electrostatic interaction (Fig. 1.13). In an optimized condition, with the addition of H_2O_2 aqueous solution, the cyclooctene can be oxidized into epoxyl product with a conversion of 98.5% at 70°C within 10 min in the solid catalyst dispersed in acetonitrile. The composite catalyst displays similar catalytic activity for other alkenes. The recovery could be realized through a quick filtration. In contrast, for naked POMs, the conversion becomes very low [126]. By using different polymers that are copolymerized from amidazol ionic liquid derivative monomers, the hybrid matrix encapsulated $[PW_{12}O_{40}]^{3-}$ catalyzes the oxidation of benzyl alcohol in the presence of H_2O_2 with high conversion and selectivity [127].

Wu et al. adopted a different approach by using surface modified SEPs to realize the covalent incorporation of POMs in silica matrix. They grafted

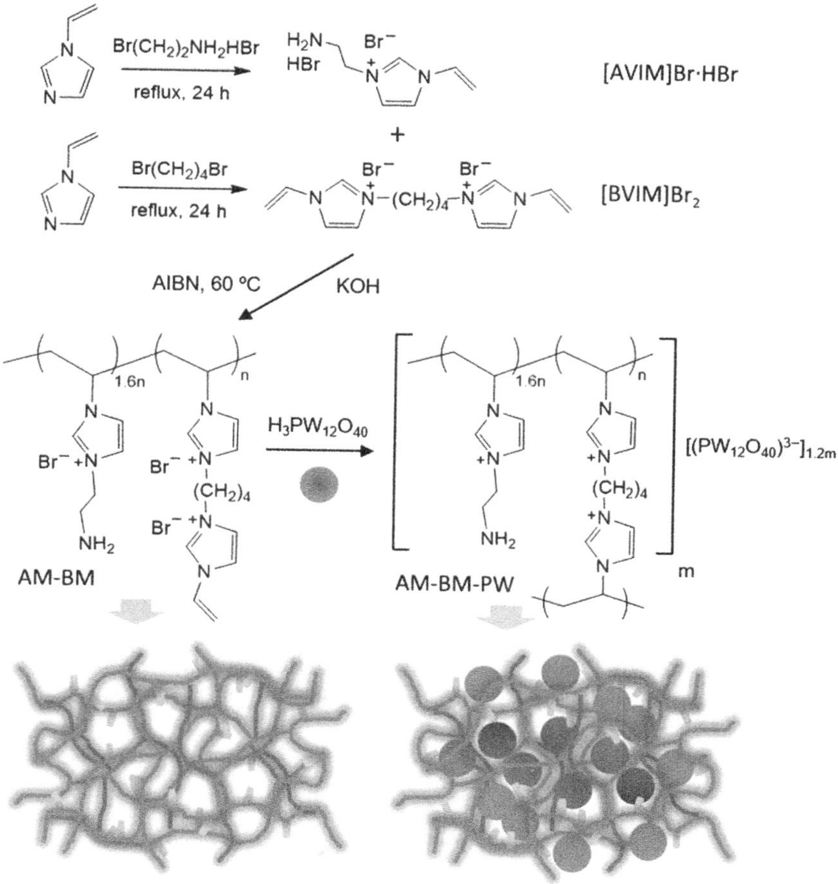

FIG. 1.13 Synthesis of the polymer ionic liquid and the preparation of polymer encapsulated POM composite, where the copolymer derives from polymerization of the prepared monomers. *(Reproduced from Y. Leng, W.J. Zhang, J. Wang, P.P. Jiang, A novel heteropolyanion-based amino-containing cross-linked ionic copolymer catalyst for epoxidation of alkenes with H_2O_2, Appl. Catal. A Gen. 445–446 (2012) 306–311, with permission of Elsevier B. V. Copyright, 2012.)*

hydroxyl group at the hydrophobic terminals of double alkyl chain surfactants. Thus, the formed SEPs bear hydroxyl groups at the outside surface due to the encapsulation for POMs. Interestingly, the incorporated OH groups weaken the self-assembly on one hand, which is favorable for the dispersion of SEPs in a solid matrix during the subsequent grafting procedure. The active OH groups on the periphery of the SEP complex participate in the hydrolysis of organic silane through a sol-gel reaction and finally connect to the silica matrix. As a result, the complex disperses in the silica. Because of the disordered porosity in silica gel, the embedded SEPs can be used for the catalytic oxidation of thioethers. Typically, a di(11-hydroxyundecyl)dimethylammonium was used to-encapsulate the $[PW_{12}O_{40}]^{3-}$ cluster via the aco-deposition method in an aqueous solution, forming a SEP-bearing hydroxyl group surface. By mixing the complex with tetraethoxylsilane in the mixture solution of ethanol and water, the SEP covalently grafted silica solid is obtained after stirring and encountering a drying process, as seen in Fig. 1.14 [128].

Thus the obtained catalyst is found to display high activity compared to the homogenous catalyst. Although it is difficult to make an exact evaluation for the measured pore size due to the limitation of the method and the Brunauer–Emmett–Teller (BET) surface area 365.9 $m^2 g^{-1}$ (total pore volume of 1.05 $cm^3 g^{-1}$) is not very high, the SEP catalyst anchored in silica displays much higher efficiency of H_2O_2 usage and TOF, and lower reaction temperature and time for the oxidation of dibenzothiophene (DBT), as seen in Table 1.1. This method is also suitable for the catalytic oxidation of other thio-substrates, olefins, and alcohols, but the TOF and the efficiency of the oxidant are subjected to the substrates used and reaction conditions.

Following the same method, the cationic head with chirality has been used to substitute the general ammonium head group. Because of the introduced chiral environment surrounding the surface of general POMs, the SEP catalyst obtains a supramolecular chirality and is used to the resolution of racemic alcohols via a catalytic oxidation reaction. When one of the methyl groups in the above-used surfactant di(11-hydroxyundecyl)dimethylammonium is replaced by (*R*,*S*)1-phenylethyl, the POMs in the silica matrix present a chiral surface. The doping

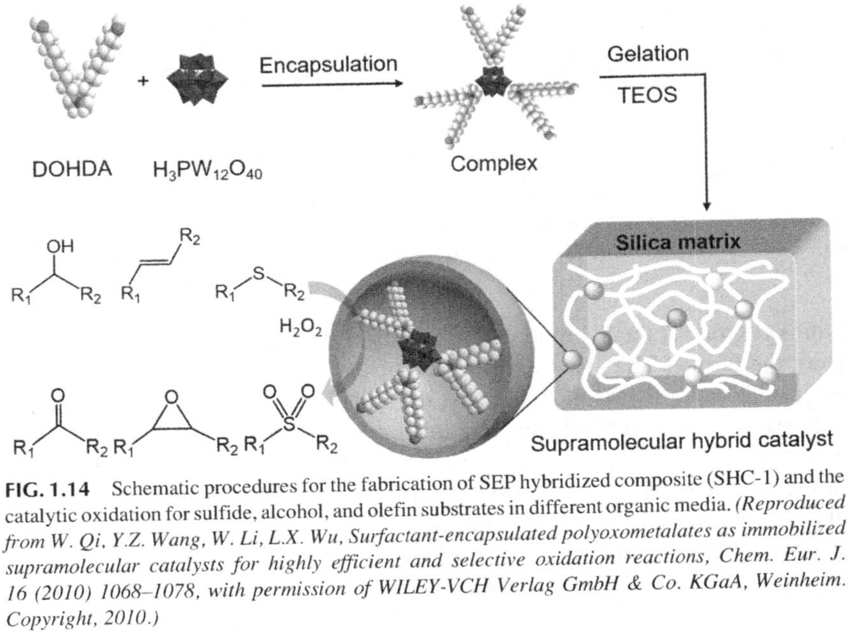

FIG. 1.14 Schematic procedures for the fabrication of SEP hybridized composite (SHC-1) and the catalytic oxidation for sulfide, alcohol, and olefin substrates in different organic media. *(Reproduced from W. Qi, Y.Z. Wang, W. Li, L.X. Wu, Surfactant-encapsulated polyoxometalates as immobilized supramolecular catalysts for highly efficient and selective oxidation reactions, Chem. Eur. J. 16 (2010) 1068–1078, with permission of WILEY-VCH Verlag GmbH & Co. KGaA, Weinheim. Copyright, 2010.)*

TABLE 1.1 Comparison of Catalytic Activities of POM-1-Based Catalysts for the Oxidation of DBT

Reaction System	T (°C)[a]	t (min)[b]	k (min^{-1})[c]	TOF (min^{-1})	C (%)[d]	E (%)[e]
SEP/SiO_2 (heterogeneous)	40	42	0.111	9.52	100	80
Bare $H_3PW_{12}O_{40}$ [20] (homogeneous)	50	60	0.078	1.72	100	1.2
Phase transfer [9] (homogeneous)	50	25	0.160	0.07	100	30
Microemulsion [14] (heterogeneous)	60	90	0.013	5.32	100	66.7

[a]*T is the reaction temperature.*
[b]*t is the reaction time.*
[c]*k is the reaction rate constant.*
[d]*C is the conversion.*
[e]*E is the efficiency of H_2O_2 usage.*
(Reproduced from W. Qi, Y.Z. Wang, W. Li, L.X. Wu, Surfactant-encapsulated polyoxometalates as immobilized supramolecular catalysts for highly efficient and selective oxidation reactions, Chem. Eur. J. 16 (2010) 1068–1078, with permission of WILEY-VCH Verlag GmbH & Co. KGaA, Weinheim. Copyright, 2010.)

capacity of the SEP catalysts can then be modulated by controlling the ratio during the hydrolyzation reaction of TEOS. For example, the loading content of encapsulated Na_{12} $[WZn_3(H_2O)_2(ZnW_9O_{34})_2]$ catalyst in silica can reach over 80% (Table 1.2). With the increase of catalyst in silica, the BET surface area decreases quickly, and for the higher doping content, the surface area becomes undetectable. However, the catalysis of the species with higher doping content shows higher activity, which demonstrates that the reaction mainly occurs on the surface of the ground carrier grains. In comparison to the homogeneous reaction, the anchored catalyst performs a much higher selectivity for a chiral product by oxidizing one enantiomer to the ketone. The catalytic ability of the obtained catalyst has a dependence on the substrates used. Considering the limited surface area, the size of substrates should be one of the important factors [129].

Such a method can also be applied for the photo catalytic reduction of metal ions. The sol-gel coating that SEP complex is incorporated on quartz plate is applied for the in situ catalytic reduction of metal ions. The coating firstly encounters a light irradiation to reduce the W atom to a low oxidized state. By dipping the coated plate into a silver or gold ion solution, the metal substrates are then automatically reduced to the metal nanoparticles while the W atom is oxidized back to the highest oxidation state [130]. Clearly, toward solid catalysts, the in situ photo catalysis can be realized in solutions bearing other substrates.

TABLE 1.2 Summary of the Catalytic Activities of CSHCs With Different CSEP Loading Contents for Different Aromatic Alcohols Under the Optimized Reaction Conditions[a]

Entry	Catalyst	Substrate	*t* (h)	Conv.[b] (%)	ee[c]
1	*R*-CSHC-62	1-Phenylpropanol	72	60.1	3.4 (*R*)
2	*R*-CSHC-62	Benzoin	72	85.6	60.0 (*R*)
3	*S*-CSHC-58	Benzoin	72	80.6	51.6 (*S*)
4	*R*-CSHC-86	1-Phenylethanol	48	86.4	55.6 (*R*)
5	*R*-CSHC-86	Benzoin	72	92.1	89.4 (*R*)

[a]*All reactions are carried out in acetonitrile (1 mL) with racemic substrate (0.2 mmol), catalyst (1 μmol), 30%* H_2O_2 *(0.5 mmol) at 50°C.*
[b]*The conversions are determined by HPLC based on the crude reaction mixture.*
[c]*ee values are determined by HPLC analysis on a chiralcel OD-H column and the absolute configuration is determined by comparison of the HPLC results with the data in the literature [24].*
(Reproduced from L. Shi, Y.Z. Wang, B. Li, L.X. Wu, Polyoxometalate complexes for oxidative kinetic resolution of secondary alcohols: unique effects of chiral environment, immobilization and aggregation, Dalton Trans. 43 (2014), 9177–9188, with permission of the Royal Society of Chemistry. Copyright, 2014.)

8 CONCLUSION

In summary, POMs represent a class of nano-sized inorganic polyanionic catalysts bearing well-defined topologic architecture. Transition metal ions in the clusters are linked by oxyanions to build a large, closed-framework entity. In comparison to the diverse potential applications, the catalytic property of POMs is realized in practical applications. To develop new applications for catalytic purposes, the most common approach in employing the intrinsic features of POMs in various environments is the encapsulation with organic components bearing counterions driven by the electrostatic interaction. Due to the reduced polarity of the encapsulated POM catalysts, the POM complexes can be extended as the homogenous catalyst for water-insoluble substrates in high efficiency. Moreover, most organically covered POM complexes are actually in an assembled state despite being in polar and nonpolar solvents, because of their amphiphilicity. The special assembly structure can be used as both a catalyst and the template for the preparation of metal nanoparticles in certain morphologies. In addition, for the emulsion system, the phase-separated POM complexes can both accomplish partial functions of surfactants and perform the biphasic catalytic reactions to a high degree of efficiency. In addition, the synergistic property of both organic and inorganic components for catalysis at the interface is also important. On the other hand, by simply incorporating POMs into solid matrixes, such as silica, ionic liquid, polymer, and so forth, catalytic reactions can be carried out and the POM catalysts loaded on/in solid matrixes can be reused after the reaction through a separated procedure. In brief, the encapsulated POM catalysts are stable, effective, and designable, which makes them outstanding candidates for industrial catalysis. From the current development of encapsulated POM catalysts, it can be envisioned that more precise and smart catalysts will be developed for special catalytic reactions in a restricted system such as on a chip or in a nanotube. Because of the easily organic modification on coverings, the formed POM complexes could also be used for controlled catalytic reactions in biomaterials through a stimuli-response process.

REFERENCES

[1] J. Berzerius, The preparation of the phosphomolybdate ion $[PMo_{12}O_{40}]^{3-}$, Pogg. Ann. 6 (1826) 369–371.

[2] J.F. Keggin, The structure and formula of 12-phosphotungstic acid, Proc. R. Soc. London, Ser. A 144 (1934) 75–100.

[3] J.W. Zhao, H.P. Jia, J. Zhang, S.T. Zheng, G.Y. Yang, A combination of lacunary polyoxometalates and high-nuclear transition-metal clusters under hydrothermal conditions. Part II: from doublecluster, dimer, and tetramer to three-dimensional frameworks, Chem. Eur. J. 13 (2007) 10030–10045.

[4] R. Al-Oweini, A. Sartorel, B.S. Bassil, M. Natali, S. Berardi, F. Scandola, U. Kortz, M. Bonchio, Photocatalytic water oxidation by a mixed-valent $Mn_3^{III}Mn^{IV}O_3$ manganese oxo core that mimics the natural oxygen-evolving center, Angew. Chem. Int. Ed. 53 (2014) 11182–11185.

[5] A. Proust, B. Matt, R. Villanneau, G. Guillemot, Functionalization and post-functionalization: a step towards polyoxometalate-based materials, Chem. Soc. Rev. 41 (2012) 7605–7622.

[6] A. Dolbecq, E. Dumas, C.R. Mayer, P. Mialane, Hybrid organic-inorganic polyoxometalate compounds: from structural diversity to applications, Chem. Rev. 110 (2010) 6009–6048.

[7] A. Blazevic, A. Rompel, The Anderson–Evans polyoxometalate: From inorganic building blocks via hybrid organic-inorganic structures to tomorrow's "Bio-POM", Coord. Chem. Rev. 307 (2016) 42–64.

[8] E.B. Wang, Introduction of Polyoxometalates, Northeast Normal University Press, Changchun, 2009.

[9] J.M. Poblet, X. López, C. Bo, Ab initio and DFT modelling of complex materials: towards the understanding of electronic and magnetic properties of polyoxometalates, Chem. Soc. Rev. 32 (2003) 297–308.

[10] T. Yamase, Photo- and electrochromism of polyoxometalates and related materials, Chem. Rev. 98 (1998) 307–325.

[11] S. Shigeta, S. Mori, E. Kodama, J. Kodama, K. Takahashi, T. Yamase, Broad spectrum anti-RNA virus activities of titanium and vanadium substituted polyoxotungstates, Antiviral Res. 58 (2003) 265–271.

[12] M. Inoue, T. Suzuki, Y. Fujita, M. Oda, N. Matsumoto, T. Yamase, Enhancement of antibacterial activity of b-lactam antibiotics by $[P_2W_{18}O_{62}]^{6-}$, $[SiMo_{12}O_{40}]^{4-}$, and $[PTi_2W_{10}O_{40}]^{7}$ against methicillin-resistant and vancomycin-resistant Staphylococcus aureus, J. Inorg. Biochem. 100 (2006) 1225–1233.

[13] M. Inoue, K. Segawa, S. Matsunaga, S. Matsumoto, M. Oda, T. Yamase, Antibacterial activity of highly negative charged polyoxotungstates, and, and Keggin-structural polyoxotungstates $K_{27}[KAs_4W_{40}O_{140}]$ and $K_{18}[KSb_9W_{21}O_{86}]$ against Helicobacter pylori, J. Inorg. Biochem. 99 (2005) 1023–1031.

[14] H. Yanagie, A. Ogata, S. Mitsui, T. Hisa, T. Yamase, M. Eriguchi, Anticancer activity of polyoxomolybdate, Biomed. Pharmacother. 60 (2006) 349–352.

[15] T. Yamase, Anti-tumor, -viral, and -bacterial activities of polyoxometalates for realizing an inorganic drug, J. Mater. Chem. 15 (2005) 4773–4782.

[16] K. Dan, T. Yamase, Prevention of the interaction between HVEM, herpes virus entry mediator, and gD, HSV envelope protein, by a Keggin polyoxotungstate, PM-19, Biomed. Pharmacother. 60 (2006) 169–173.

[17] Y.V. Geletii, B. Botar, P. Kögerler, D.A. Hillesheim, D.G. Musaev, C.L. Hill, An all-inorganic, stable, and highly active, tetraruthenium homogeneous catalyst for water, oxidation, Angew. Chem. Int. Ed. 47 (2008) 3896–3899.

[18] Q. Yin, J.F. Tan, C. Besson, Y.V. Geletii, D.G. Musaev, A.E. Kuznetsov, Z. Luo, K.I. Hardcastle, C.L. Hill, A fast soluble carbon-free molecular water oxidation catalyst based on abundant metals, Science 328 (2010) 342–345.

[19] Y.V. Geletii, Z. Huang, Y. Hou, D.G. Musaev, T. Lian, C.L. Hill, Homogeneous light-driven water oxidation catalyzed by a tetraruthenium complex with all inorganic ligands, J. Am. Chem. Soc. 131 (2009) 7522–7523.

[20] T. Rüther, V.M. Hultgren, B.P. Timko, A.M. Bond, W.R. Jackson, A.G. Wedd, Electrochemical investigation of photooxidation processes promoted by sulfo-polyoxometalates: coupling of photochemical and electrochemical processes into an effective catalytic cycle, J. Am. Chem. Soc. 125 (2003) 10133–10143.

[21] A. Hiskia, A. Mylonas, E. Papaconstantinou, Comparison of the photoredox properties of polyoxometalates and semiconducting particles, Chem. Soc. Rev. 30 (2001) 62–69.

[22] L.C. Palilis, M. Vasilopoulou, D.G. Georgiadou, P. Argitis, A water soluble inorganic molecular oxide as a novel efficient electron injection layer for hybrid light-emitting diodes (HyLEDs), Org. Electron. 11 (2010) 887–894.

[23] Y.B. Yang, L. Xu, F.Y. Li, X. Du, Z.X. Sun, Enhanced photovoltaic response by incorporating polyoxometalate into a phthalocyanine-sensitized electrode, J. Mater. Chem. 20 (2010) 10835–10840.

[24] S. Zeror, J. Collin, J.-C. Fiaud, L.A. Zouioueche, A recyclable multi-substrates catalytic system for enantioselective reduction of ketones in water, J. Mol. Catal. A: Chem. 256 (2006) 85–89.

[25] J.D. Compain, P. Mialane, Iron polyoxometalate single-molecule magnets, Angew. Chem. Int. Ed. 48 (2009) 3077–3081.

[26] M. Ibrahim, Y. Lan, B.S. Bassil, Y. Xiang, A. Suchopar, A.K. Powell, U. Kortz, Hexadecacobalt(II)-containing polyoxometalate-based single-molecule magnet, Angew. Chem. Int. Ed. 50 (2011) 4708–4711.

[27] A. Nisar, X. Wang, Surfactant-encapsulated polyoxometalate building blocks: controlled assembly and their catalytic properties, Dalton Trans. 41 (2012) 9832–9845.

[28] K. Zhu, D.H. Wang, J. Liu, Self-assembled materials for catalysis, Nano Res. 2 (2009) 1–29.

[29] Y. Zhou, Z.J. Guo, W. Hou, Q. Wang, J. Wang, Polyoxometalate-based phase transfer catalysis for liquid–solid organic reactions: a review, Catal. Sci. Technol. 5 (2015) 4324–4335.

[30] S.S. Wang, G.Y. Yang, Recent advances in polyoxometalate-catalyzed reactions, Chem. Rev. 115 (2015) 4893–4962.

[31] Y. Zhou, G.J. Chen, Z.Y. Long, J. Wang, Recent advances in polyoxometalate-based heterogeneous catalytic materials for liquid-phase organic transformations, RSC Adv. 4 (2014) 42092–42113.

[32] Y. Lan, E.B. Wang, Y.H. Song, Y.L. Song, Z.H. Kang, L. Xu, Z. Li, An effective layer-by-layer adsorption and polymerization method to the fabrication of polyoxometalate-polypyrrole nanoparticle ultrathin films, Polymer 47 (2006) 1480–1485.

[33] A. Aoshima, S. Tonomura, S. Yamamatsu, New synthetic route of polyoxytetramethyleneglycol by use of heteropolyacids as catalyst, Polym. Adv. Technol. 2 (1990) 127–132.

[34] T. Yokota, M. Tani, S. Sakaguchi, Y. Ishii, Direct coupling of benzene with olefin catalyzed by $Pd(OAc)_2$ combined with heteropolyoxometalate under dioxygen, J. Am. Chem. Soc. 125 (2003) 1476–1477.

[35] M. Tani, S. Sakaguchi, Y. Ishii, $Pd(OAc)_2$-catalyzed oxidative coupling reaction of benzenes with olefins in the presence of molybdovanadophosphoric acid under atmospheric dioxygen and air, J. Org. Chem. 69 (2004) 1221–1226.

[36] T. Yamada, S. Sakaguchi, Y. Ishii, Oxidative coupling of benzenes with α,β-unsaturated aldehydes by the $Pd(OAc)_2$/molybdovanadophosphoric acid/O_2 system, J. Org. Chem. 70 (2005) 5471–5474.

[37] K. Kamata, Y. Nakagawa, K. Yamaguchi, N. Mizuno, 1,3-Dipolar cycloaddition of organic azides to alkynes by adicopper-substituted silicotungstate, J. Am. Chem. Soc. 130 (2008) 15304–15310.

[38] S.S.R. Putluru, S. Mossin, A. Riisager, R. Fehrmann, Heteropoly acid promoted Cu and Fe catalysts for the selective catalytic, Catal. Today 176 (2011) 292–297.

[39] S.S.R. Putluru, A.D. Jensen, A. Riisager, R. Fehrmann, Heteropoly acid promoted V_2O_5/TiO_2 catalysts for NO abatement with ammonia in alkali containing flue gases, Catal. Sci. Technol. 1 (2011) 631–637.

[40] O. Branytska, L.J.W. Shimonb, R. Neumann, Tandem pinacol coupling-rearrangement of aromatic aldehydes with hydrogen catalyzed by a combination of a platinum complex and a polyoxometalate, Chem. Commun. (2007) 3957–3959.

[41] J. Kaur, K. Griffin, B. Harrison, I.V. Kozhevnikov, Friedel–Crafts acylation catalysed by heteropoly acids, J. Catal. 208 (2002) 448–455.

[42] J. Kaur, I.V. Kozhevnikov, Efficient acylation of toluene and anisole with aliphatic carboxylic acids catalysed by heteropoly salt $Cs_{2.5}H_{0.5}PW_{12}O_{40}$, Chem. Commun. (2002) 2508–2509.

[43] A.M. Alsalme, P.V. Wiper, Y.Z. Khimyak, E.F. Kozhevnikova, I.V. Kozhevnikov, Solid acid catalysts based on $H_3PW_{12}O_{40}$ heteropoly acid: acid and catalytic properties at a gas–solid interface, J. Catal. 276 (2010) 181–189.

[44] C. Boglio, K. Micoine, P. Rémy, B. Hasenknopf, S. Thorimbert, E. Lacôte, M. Malacria, C. Afonso, J.C. Tabet, Increased Lewis acidity in hafnium-substituted polyoxotungstates, Chem. Eur. J. 13 (2007) 5426–5432.

[45] N. Dupré, P. Rémy, K. Micoine, C. Boglio, S. Thorimbert, E. Lacôte, B. Hasenknopf, M. Malacria, Chemoselective catalysis with organosoluble Lewis acidic polyoxotungstates, Chem. Eur. J. 16 (2010) 7256–7264.

[46] E. Derat, E. Lacôte, B. Hasenknopf, S. Thorimbert, M. Malacria, Water dissociation on α1-hafnium and ytterbium substituted Dawson polyoxotungstates: a density functional theory study, J. Phys. Chem. A 112 (2008) 13002–13005.

[47] S. Vanhaecht, G. Absillis, T.N. Parac-Vogt, Hydrolysis of DNA model substrates catalyzed by metal-substituted Wells–Dawson polyoxometalates, Dalton Trans. 41 (2012) 10028–10034.

[48] A. Yoshida, M. Yoshimura, K. Uehara, S. Hikichi, N. Mizuno, Formation of S-shaped disilicoicosatungstate and efficient Baeyer–Villiger oxidation with hydrogen peroxide, Angew. Chem. Int. Ed. 45 (2006) 1956–1960.

[49] A. Yoshida, S. Hikichi, N. Mizuno, Acid–base catalyses by dimeric disilicoicosatungstates and divacant γ-Keggin-type silicodecatungstate parent: reactivity of the polyoxometalate compounds controlled by step-by-step protonation of lacunary W = O sites, J. Organomet. Chem. 692 (2007) 455–459.

[50] K. Sugahara, T. Kimura, K. Kamata, K. Yamaguchi, N. Mizuno, A highly negatively charged γ-Keggin germanodecatungstate efficient for Knoevenagel condensation, Chem. Commun. 48 (2012) 8422–8424.

[51] E.F. Kozhevnikova, J. Quartararo, I.V. Kozhevnikov, Fries rearrangement of aryl esters catalysed by heteropoly acid, Appl. Catal. A. Gen. 245 (2003) 69–78.

[52] E.F. Kozhevnikova, E. Rafiee, I.V. Kozhevnikov, Fries rearrangement of aryl esters catalysed by heteropoly acid: catalyst regeneration and reuse, Appl. Cataly. A. Gen. 260 (2004) 25–34.

[53] A.M. Khenkin, L.J.W. Shimon, R. Neumann, Preparation and characterization of new ruthenium and osmium containing polyoxometalates, $[M(DMSO)_3Mo_7O_{24}]^{4-}$ (M Ru(II), Os (II)), and their use as catalysts for the aerobic oxidation of alcohols, Inorg. Chem. 42 (2003) 3331–3339.

[54] V. Kogan, Z. Aizenshtat, R. Neumann, Preferential catalytic hydrogenation of aromatic compounds versus versusketones with a palladium substituted polyoxometalate as pre-catalyst, New J. Chem. 26 (2002) 272–274.

[55] H. Benaissa, P.N. Davey, Y.Z. Khimyak, I.V. Kozhevnikov, Heteropoly compounds as catalysts for hydrogenation of propanoic acid, J. Catal. 253 (2008) 244–252.

[56] T. Okuhara, N. Mizuno, M. Misono, Catalytic chemistry of heteropoly compounds, Adv. Catal. 41 (1996) 113–252.

[57] O.A. Kholdeeva, M.P. Vanina, M.N. Timofeeva, R.I. Maksimovskaya, T.A. Trubitsina, M.S. Melgunov, E.B. Burgina, J. Mrowiec-Bialon, A.B. Jarzebski, C.L. Hill, Co-containing polyoxometalate-based heterogeneous catalysts for the selective aerobic oxidation of aldehydes under ambient conditions, J. Catal. 226 (2004) 363–371.

[58] O.A. Kholdeeva, M.N. Timofeeva, G.M. Maksimov, R.I. Maksimovskaya, W.A. Neiwert, C.L. Hill, Aerobic oxidation of formaldehyde mediated by a Ce-containing polyoxometalate under mild conditions, Inorg. Chem. 44 (2005) 666–672.

[59] W.W. Guo, Z. Luo, H.J. Lv, C.L. Hill, Aerobic oxidation of formaldehyde catalyzed by Polyvanadotungstates, ACS Catal. 4 (2014) 1154–1161.

[60] I.C.M.S. Santos, F.A.A. Paz, M.M.Q. Simões, M.G.P.M.S. Neves, J.A.S. Cavaleiro, J. Klinowski, A.M.V. Cavaleiro, Catalytic homogeneous oxyfunctionalization with hydrogen peroxide in the presence of a peroxotungstate, Appl. Catal. A. Gen. 351 (2008) 166–173.

[61] I. Bar-Nahum, A.M. Khenkin, R. Neumann, Mild, aqueous, aerobic, catalytic oxidation of methane to methanol and acetaldehyde catalyzed by a supported bipyrimidinylplatinum-polyoxometalate hybrid compound, J. Am. Chem. Soc. 126 (2004) 10236–10237.

[62] A.M. Khenkin, R. Neumann, Redirection of oxidation reactions by apolyoxomolybdate: oxydehydrogenation instead of oxygenation of alkanes with *tert*-butylhydroperoxide in acetic acid, J. Am. Chem. Soc. 123 (2001) 6437–6438.

[63] S.L. Linguito, X. Zhang, M. Padmanabhan, A.V. Biradar, T.T. Xu, T.J. Emge, T. Asefa, J. Li, New polyoxomolybdate compounds synthesized in situ using ionic liquid 1-butyl-3-methylimidazoliumtetrafluoroborate as green solvent, New J. Chem. 37 (2013) 2894–2901.

[64] A.M. Khenkin, R. Neumann, Oxygen transfer from sulfoxides: selective oxidation of alcohols catalyzed by polyoxomolybdates, J. Org. Chem. 67 (2002) 7075–7079.

[65] L.B. Ni, J. Patscheider, K.K. Baldridge, G.R. Patzke, New perspectives on polyoxometalate catalysts: alcohol oxidation with Zn/Sb-polyoxotungstates, Chem. Eur. J. 18 (2012) 13293–13298.

[66] D. Schaming, C. Allain, R. Farha, M. Goldmann, S. Lobstein, A. Giraudeau, B. Hasenknopf, L. Ruhlmann, Synthesis and photocatalytic properties of mixed polyoxometalate-porphyrin copolymers obtained from Anderson-type polyoxomolybdates, Langmuir 26 (2010) 5101–5109.

[67] D. Schaming, R. Farha, H.L. Xu, M. Goldmann, L. Ruhlmann, Formation and photocatalytic properties of nanocomposite films containing both tetracobalt Dawson-derived sandwich polyanions and tetracationic porphyrins, Langmuir 27 (2011) 132–143.

[68] D. Schaming, C. Costa-Coquelard, S. Sorgues, L. Ruhlmann, I. Lampre, Photocatalytic reduction of Ag_2SO_4 by electrostatic complexes formed by tetracationic zinc porphyrins and tetracobalt Dawson-derived sandwich polyanion, Appl. Catal. A. Gen. 373 (2010) 160–167.

[69] I. Ahmed, X.X. Wang, N. Boualili, H.L. Xu, R. Farha, Photocatalytic synthesis of silver dendrites using electrostatic hybrid films of porphyrin-polyoxometalate, Appl. Catal. A. Gen. 447–448 (2012) 89–99.

[70] A.R. Howells, A. Sankarraj, C. Shannon, A diruthenium-substituted polyoxometalate as an electrocatalyst for oxygen generation, J. Am. Chem. Soc. 126 (2004) 12258–12259.

[71] Y.V. Geletii, B. Botar, P. Kögerler, D.A. Hillesheim, D.G. Musaev, C.L. Hill, An all-inorganic, stable, and highly active tetraruthenium homogeneous catalyst for water oxidation, Angew. Chem. Int. Ed. 120 (2008) 3960–3963.

[72] A. Sartorel, M. Carraro, G. Scorrano, R. Zorzi, S. Geremia, N.D. McDaniel, S. Bernhard, M. Bonchio, Polyoxometalate embedding of a tetraruthenium (IV)-oxo-coreby template-directed metalation of $[\gamma\text{-SiW}_{10}O_{36}]_8$: a totally inorganic oxygen-evolving catalyst, J. Am. Chem. Soc. 130 (2008) 5006–5007.

[73] M. Murakami, D.C. Hong, T. Suenobu, S. Yamaguchi, T. Ogura, S. Fukuzumi, Catalytic mechanism of water oxidation with single-site ruthenium-heteropolytungstate complexes, J. Am. Chem. Soc. 133 (2011) 11605–11613.

[74] Y.X. Li, C.X. Ren, S.Q. Peng, G.X. Lu, S.B. Li, Effect of preparation method of colloidal platinum on performance of 12-tungstosilicate for photocatalytic hydrogen generation, J. Mol. Catal. A Chem. 246 (2006) 212–217.

[75] X. Liu, Y.X. Li, S.Q. Peng, G.X. Lu, S.B. Li, Photocatalytic hydrogen evolution under visible light irradiation by the polyoxometalateα- $[AlSiW_{11}(H_2O)O_{39}]^{5-}$ Eosin Y system, Int. J. Hydrog. Energy 37 (2012) 12150–12157.

[76] H.J. Lv, J. Song, H.M. Zhu, Y.V. Geletii, J. Bacsa, C.C. Zhao, T.Q. Lian, D.G. Musaev, C.L. Hill, Visible-light-driven hydrogen evolution from water using a noble-metal-free polyoxometalate catalyst, J. Catal. 307 (2013) 48–54.

[77] H.J. Lv, W.W. Guo, K.F. Wu, Z.Y. Chen, J. Bacsa, D.G. Musaev, Y.V. Geletii, S.M. Lauinger, T.Q. Lian, C.L. Hill, A noble-metal-free, tetra-nickel polyoxotungstate catalyst forefficient photocatalytic hydrogen evolution, J. Am. Chem. Soc. 135 (2014) 14015–14018.

[78] P.C. Yin, J. Wang, Z.C. Xiao, P.F. Wu, Y.G. Wei, T.B. Liu, Polyoxometalate–organic hybrid molecules as amphiphilic emulsion catalysts for deep desulfurization, Chem. Eur. J. 18 (2012) 9174–9178.

[79] Y. Xiao, D. Chen, N. Ma, Z.Y. Hou, M.B. Hu, C.H. Wang, W. Wang, Covalent immobilization of a polyoxometalate in a porous polymer matrix: a heterogeneous catalyst towards sustainability, RSC Adv. 3 (2013) 21544–21551.

[80] C. Venturello, E. Alneri, M. Ricci, A new, effective catalytic system for epoxidation of olefins by hydrogen peroxide under phase-transfer conditions, J. Org. Chem. 48 (1983) 3831–3833.

[81] A. Nisar, X.X. Xu, S.L. Shen, S. Hu, X. Wang, Noble metal nanocrystal-incorporated fullerene-like polyoxometalate based microspheres, Adv. Funct. Mater. 19 (2009) 860–865.

[82] W.F. Bu, S. Uchida, N. Mizuno, Micelles and vesicles formed by polyoxometalate–block copolymer composites, Angew. Chem. Int. Ed. 121 (2009) 8431–8434.

[83] H.B. Wei, J.L. Zhang, N. Shi, Y. Liu, B. Zhang, J. Zhang, X.H. Wan, A recyclable polyoxometalate-based supramolecular chemosensor for efficient detection of carbon dioxide, Chem. Sci. 6 (2015) 7201–7205.

[84] S.I. Noro, R. Tsunashima, Y. Kamiya, K. Uemura, H. Kita, L. Cronin, T. Akutagawa, T. Nakamura, Adsorption and catalytic properties of the inner nanospace of a gigantic ring-shaped polyoxometalate cluster, Angew. Chem. Int. Ed. 121 (2009) 8859–8862.

[85] A. Haimov, R. Neumann, An example of lipophiloselectivity: the preferred oxidation, in water, of hydrophobic 2-alkanols catalyzed by a cross-linked polyethyleneimine polyoxometalate catalyst assembly, J. Am. Chem. Soc. 128 (2006) 15697–15700.

[86] A. Haimov, H. Cohen, R. Neumann, Alkylated polyethyleneimine/polyoxometalate synzymes as catalysts for the oxidation of hydrophobic substrates in water with hydrogen peroxide, J. Am. Chem. Soc. 126 (2004) 11762–11763.

[87] N.V. Maksimchuk, M.N. Timofeeva, M.S. Melgunov, A.N. Shmakov, Y.A. Chesalov, D.N. Dybtsev, V.P. Fedin, O.A. Kholdeeva, Heterogeneous selective oxidation catalysts based on coordination polymerMIL-101 and transition metal-substituted polyoxometalates, J. Catal. 257 (2008) 315–323.

[88] Y. Matoba, H. Inoue, J.I. Akagi, T. Okabayashi, Y. Ishii, M. Ogawa, Epoxidation of allylic alcohols with hydrogen peroxide catalyzed by $[PMO_{12}O_{40}]^{3-}[C_5H_5N^+(CH_2)_{15}CH_3]_3$, Synth. Commun. 14 (1984) 865–873.

[89] C. Venturello, R. D'Aloisio, Quaternary ammonium tetrakis(diperoxotungsto)phosphates(3-) as a new class of catalysts for efficient alkene epoxidation with hydrogen peroxide, J. Org. Chem. 53 (1988) 1553–1557.

[90] Y. Ishii, K. Yamawaki, T. Yoshida, T. Ura, M. Ogawa, Oxidation of olefins and alcohols by peroxo-molybdenum complex derived from tris(cetylpyridinium) 12-molybdophosphate and hydrogen peroxide, J. Org. Chem. 52 (1987) 1868–1870.

[91] Y. Ishii, K. Yamawaki, T. Ura, H. Yamada, T. Yoshida, M. Ogawa, Hydrogen peroxide oxidation catalyzed by heteropoly acids combined with cetylpyridinium chloride: epoxidation of olefins and allylic alcohols, ketonization of alcohols and diols, and oxidative cleavage of 1,2-diols and olefins, J. Org. Chem. 53 (1988) 3581–3593.

[92] Y. Nakagawa, K. Kamata, M. Kotani, K. Yamaguchi, N. Mizuno, Polyoxovanadometalate-catalyzed selective epoxidation of alkenes with hydrogen peroxide, Angew. Chem. Int. Ed. 44 (2005) 5136–5141.

[93] N. Mizuno, C. Nozaki, I. Kiyoto, M. Misono, Highly efficient utilization of hydrogen peroxide for selective oxygenation of alkanes catalyzed by diiron-substituted polyoxometalate precursor, J. Am. Chem. Soc. 120 (1998) 9267–9272.

[94] K. Kamata, M. Kotani, K. Yamaguchi, S. Hikichi, N. Mizuno, Olefin epoxidation with hydrogen peroxide catalyzed by Lacunary Polyoxometalate $[\gamma\text{-}SiW_{10}O_{34}(H_2O)_2]^{4-}$, Chem. Eur. J. 13 (2007) 639–648.

[95] Z.W. Xi, N. Zhou, Y. Sun, K.L. Li, Reaction-controlled phase-transfer catalysis for propylene epoxidation to propylene oxide, Science 292 (2001) 1139–1141.
[96] G. Maayan, R.H. Fish, R. Neumann, Polyfluorinated quaternary ammonium salts of polyoxometalate anions: fluorous biphasic oxidation catalysis with and without fluorous solvents, Org. Lett. 5 (2003) 3547–3550.
[97] M.V. Vasylyev, R. Neumann, New heterogeneous polyoxometalate based mesoporous catalysts for hydrogen peroxide mediated oxidation reactions, J. Am. Chem. Soc. 126 (2004) 884–890.
[98] Y. Yang, L. Yue, H.L. Li, E. Maher, Y.G. Li, Y.Z. Wang, L.X. Wu, W.W.W. Yam, Photoresponsive self-assembly of an azobenzene-ended surfactant-encapsulated polyoxometalate complex for modulating catalytic reactions, Small 8 (2012) 3105–3110.
[99] G.S. Kumar, Photochemistry of azobenzene-containing polymers, Chem. Rev. 89 (1989) 1915–1925.
[100] Y. Yang, B. Zhang, Y.Z. Wang, L. Yue, W. Li, L.X. Wu, A photo-driven polyoxometalate complex shuttle and its homogeneous catalysis and heterogeneous separation, J. Am. Chem. Soc. 135 (2013) 14500–14503.
[101] Y. Ding, W. Zhao, The oxidation of pyridine and alcohol using the Keggin-type lacunary polytungstophosphate as a temperature-controlled phase transfer catalyst, J. Mol. Catal. A Chem. 337 (2011) 45–51.
[102] J. Rieger, T. Antoun, S.H. Lee, M. Chenal, G. Pembouong, J.L. Haye, I. Azcarate, B. Hasenknopf, E. Lacôte, Synthesis and characterization of a thermoresponsive polyoxometalate–polymer hybrid, Chem. Eur. J. 18 (2012) 3355–3361.
[103] X.X. Duan, G. Sun, Z. Sun, J.X. Li, S.T. Wang, X.H. Wang, S.W. Li, Z.J. Jiang, A heteropolyacid-based ionic liquid as a thermoregulated and environmentally friendly catalyst in esterification reaction under microwave assistance, Catal. Commun. 42 (2013) 125–128.
[104] G. Maayan, R. Popovitz-Biro, R. Neumann, Micelle directed synthesis of polyoxometalate nanoparticles and their improved catalytic activity for the aerobic oxidation of sulfides, J. Am. Chem. Soc. 128 (2006) 4968–4969.
[105] M. Vasylyev, R. Popovitz-Biro, L.J.W. Shimon, R. Neumann, Inorganic–organic hybrid materials based on Keggin type polyoxometalates and organic polyammonium cations, J. Mol. Struct. 656 (2003) 27–35.
[106] D.G. Kurth, P. Lehmann, D. Volkmer, H. Cölfen, M.J. Koop, A. Müller, A.D. Chesne, Surfactant-encapsulated clusters (SECs): $(DODA)_{20}(NH_4)[H_3Mo_{57}V_6(NO)_6O_{183}(H_2O)_{18}]$, a case study, Chem. Eur. J. 6 (2000) 385–393.
[107] W.F. Bu, H.L. Fan, L.X. Wu, X.L. Hou, C.W. Hu, G. Zhang, X. Zhang, Surfactant-encapsulated polyoxoanion: structural characterization of its Langmuir films and Langmuir-Blodgett films, Langmuir 18 (2002) 6398–6403.
[108] H.L. Li, H. Sun, W. Qi, M. Xu, L.X. Wu, Onionlike hybrid assemblies based on surfactant-encapsulated polyoxometalates, Angew. Chem. Int. Ed. 46 (2007) 1300–1303.
[109] A. Troupis, A. Hiskia, E. Papaconstantinou, Synthesis of metal nanoparticles by using polyoxometalates as photocatalysts and stabilizers, Angew. Chem. Int. Ed. 41 (2002) 1911–1914.
[110] T. Yamase, Photo- and electro-chromism of polyoxometalates and related materials, Chem. Rev. 98 (1998) 307–325.
[111] H.L. Li, Y. Yang, Y.Z. Wang, W. Li, L.H. Bi, L.X. Wu, In situ fabrication of flower-like gold nanoparticles in surfactant-polyoxometalate-hybrid spherical assemblies, Chem. Commun. 46 (2010) 3750–3752.
[112] A. Nisar, J. Zhuang, X. Wang, Construction of amphiphilic polyoxometalate mesostructures as a highly efficient desulfurization catalyst, Adv. Mater. 23 (2011) 1130–1135.

[113] A. Nisar, Y. Lu, J. Zhuang, X. Wang, Polyoxometalate nanocone nanoreactors: magnetic manipulation and enhanced catalytic performance, Angew. Chem. 123 (2011) 3245–3250.

[114] P.C. Yin, A. Bayaguud, P. Cheng, F. Haso, L. Hu, J. Wang, D. Vezenov, R.E. Winans, J. Hao, T. Li, Y.G. Wei, T.B. Liu, Spontaneous stepwise self-assembly of a polyoxometalate–organic hybrid into catalytically active one-dimensional anisotropic structures, Chem. Eur. J. 20 (2014) 9589–9595.

[115] R. Neumann, A.M. Khenkin, Peroxometalate catalyzed oxidations with hydrogen peroxide in biphasic reaction media: reactions in inverse emulsions, J. Org. Chem. 59 (1994) 7577–7579.

[116] C. Li, Z.X. Jiang, J.B. Gao, Y.X. Yang, S.J. Wang, F.P. Tian, F.X. Sun, X.P. Sun, P.L. Ying, C.R. Han, Ultra-deep desulfurization of diesel: oxidation with a recoverable catalyst assembled in emulsion, Chem. Eur. J. 10 (2004) 2277–2280.

[117] H.Y. Lü, J.B. Gao, Z.X. Jiang, F. Jing, Y.X. Yang, G. Wang, C. Li, Ultra-deep desulfurization of diesel by selective oxidation with $[C_{18}H_{37}N(CH_3)_3]_4[H_2NaPW_{10}O_{36}]$ catalyst assembled in emulsion droplets, J. Catal. 239 (2006) 369–375.

[118] H.Y. Lü, J.B. Gao, Z.X. Jiang, F. Jing, Y.X. Yang, G. Wang, C. Li, Oxidative desulfurization of dibenzothiophene with molecular oxygen using emulsion catalysis, Chem. Commun. (2007) 150–152.

[119] Q. Gao, Y. Liu, S.M. Lu, J. Li, C. Li, Recyclable enamine catalysts for asymmetric direct cross-aldol reaction of aldehydes in emulsion media, Green Chem. 13 (2011) 1983–1985.

[120] J.Y. Li, S.Z. Luo, J.P. Cheng, Chiral primary-tertiary diamine catalysts derived from natural amino acids for *syn*-aldol reactions of hydroxy ketones, J. Org. Chem. 74 (2009) 1747–1750.

[121] L. Schue, P.M. Jean-Baptiste, Y.Y. Du, H. Jintoku, H. Ihara, R. Oda, S. Nlate, Electrostatic immobilization of substrate and polyoxotungstate catalyst at the surface of micelles or enhanced reaction efficiency in water, Catal. Commun. 59 (2015) 65–68.

[122] M.R. Farsani, B. Yadollahi, Synthesis, characterization and catalytic performance of a Fe polyoxometalate/silicacomposite in the oxidation of alcohols with hydrogen peroxide, J. Mol. Catal. A: Chem. 392 (2014) 8–15.

[123] H.W. Zheng, Z. Sun, X.L. Chen, Q. Zhao, X.H. Wang, Z.J. Jiang, A micro reaction-controlled phase-transfer catalyst for oxidative desulfurization based on polyoxometalate modified silica, Appl. Catal. A. Gen. 467 (2013) 26–32.

[124] M. Maldotti, A. Molinari, G. Varani, M. Lenarda, L. Storaro, F. Bigi, R. Maggi, A. Mazzacani, G. Sartori, Immobilization of $(n\text{-}Bu_4N)_4W_{10}O_{32}$ on mesoporous MCM-41 and amorphous silicas for photocatalytic oxidation of cycloalkanes with molecular oxygen, J. Catal. 209 (2002) 210–216.

[125] L. Hua, J.Z. Chen, C. Chen, W.W. Zhu, Y.Y. Yu, R. Zhang, L. Guo, B.N. Song, H.M. Gan, Z.S. Hou, Immobilization of polyoxometalate-based ionic liquid on carboxymethyl cellulose for epoxidation of olefins, New J. Chem. 38 (2014) 3953–3959.

[126] Y. Leng, W.J. Zhang, J. Wang, P.P. Jiang, A novel heteropolyanion-based amino-containing cross-linked ionic copolymer catalyst for epoxidation of alkenes with H_2O_2, Appl. Catal. A. Gen. 445–446 (2012) 306–311.

[127] Y. Leng, J. Wang, P.P. Jiang, Amino-containing cross-linked ionic copolymer-anchored heteropoly acid for solvent-free oxidation of benzyl alcohol with H_2O_2, Catal. Commun. 27 (2012) 101–104.

[128] W. Qi, Y.Z. Wang, W. Li, L.X. Wu, Surfactant-encapsulated polyoxometalates as immobilized supramolecular catalysts for highly efficient and selective oxidation reactions, Chem. Eur. J. 16 (2010) 1068–1078.

[129] L. Shi, Y.Z. Wang, B. Li, L.X. Wu, Polyoxometalate complexes for oxidative kinetic resolution of secondary alcohols: unique effects of chiral environment, immobilization and aggregation, Dalton Trans. 43 (2014) 9177–9188.

[130] W. Qi, H.L. Li, L.X. Wu, Stable photochromism and controllable reduction properties of surfactant-encapsulated polyoxometalate/silica hybrid films, J. Phys. Chem. B 112 (2008) 8257–8263.

Chapter 2

Encapsulation of Polyoxometalate Catalysts Into Dendrimers: One Step in Making Stable and Recoverable Homogeneous DENDRI-POM Catalysts for Oxidation Reactions

Sylvain Nlate, Claire Jahier and Mariam Attoui
University of Bordeaux, Pessac Cedex, France

Contents

1 INTRODUCTION

Polyoxometalates (POMs) [1] are discrete, anionic metal-oxide clusters with a diverse range of properties that render them attractive candidates for potential applications in a variety of fields such as catalysis, medicine, analytical chemistry, and material science [2]. Owing to their unique metal-oxide cage structure and reversible electron transformation abilities [3], POMs have been extensively investigated as catalysts in a variety of oxidation reactions [4].

However, in order to study the activity of POMs as homogenous catalysts in common organic solvents and due to the poor solubility of most alkali-POM salts in organic media, organic salts of POM catalysts are prepared.

Encapsulated Catalysts. http://dx.doi.org/10.1016/B978-0-12-803836-9.00002-X

The coupling of POMs with organic supports can be achieved in various ways, depending on the structure of both components. Noncovalent interactions are mostly used to confine POMs with the support. Since POMs are negatively charged, the first strategy to couple them on support relies on the counterion's exchange with positively charged materials, leading to electrostatic interactions between both components [5]. Hence, to overcome the poor solubility problem of alkali-POM salts in organic solvents, a variety of positively charged organic moieties have been used as countercations of polyanions [6], resulting in a wide range of hybrids, soluble in organic media. Among the organic cations, the use of cationic dendrimers constitutes an excellent contribution to green chemistry because it focuses on the development of efficient and recoverable dendritic polyoxometalate (DENDRI-POM) catalysts able to perform organic transformations without any significant loss in activity. Dendrimers and metallodendrimers are generating much attention due to their applications in various areas [7]. The increasing use of dendrimers in catalysis, pioneered by van Leeuwen [7b, d], is an emerging field because the size of these macromolecules allows their easy recovery, an essential feature for reaction efficiency, economy, and environmental concern. Dendritic structures are also attractive because they can modify the properties of polyanions, such as their solubility, catalytic efficiency, stability, and recycling capacity [8]. Thus, by using an appropriate dendritic organic countercation, a variety of organic-inorganic hybrids with specific properties can be prepared. In this context, various metal catalysts have been encapsulated into dendritic structures, using the convergent synthetic approach, in which bifunctional dendrons are associated with the catalyst. Several reviews on the topic have already been reported and showed numerous applications of metallodendritic materials [7]. In contrast, dendritic polyoxometalate hybrids are still underrepresented, in spite of their potential use as recoverable catalysts in the transformation of organic substrates [5c]. The fixation of POM catalysts onto dendrimers is a very promising route that allows the coupling of molecular control and mechanistic knowledge, with the catalytic efficiency of the POM unit. An increasing effort has been made in the design of new POM-based catalysts, and a variety of DENDRI-POM hybrids are now reported [5,8]. For all the systems studied so far, it was clearly established that the properties of the anionic POMs are closely related to the structure of their counteranions.

For several years, only a few examples of DENDRI-POM hybrids have been reported for oxidation transformations, and most of them are reported from our group [5c]. Before our first report, Newkome and Hill reported the synthesis of DENDRI-POM hybrids built by covalent grafting of an $[HP_2V_3W_{15}O_{59}]^{5-}$ polyanion to a tetra-armed core, they were used as recoverable catalysts in the oxidation of tetrahydrothiophene to its sulfoxide, using *t*BuOOH and H_2O_2 as oxidants [8a]. Two years later, DENDRI-POM hybrids based on the encapsulation of a heteropolytungstate cluster in a shell of dendritic surfactants were reported by Volkmer and coworkers [8b]. The objective of this chapter is to present an overview of DENDRI-POM hybrid catalysts in which POM units

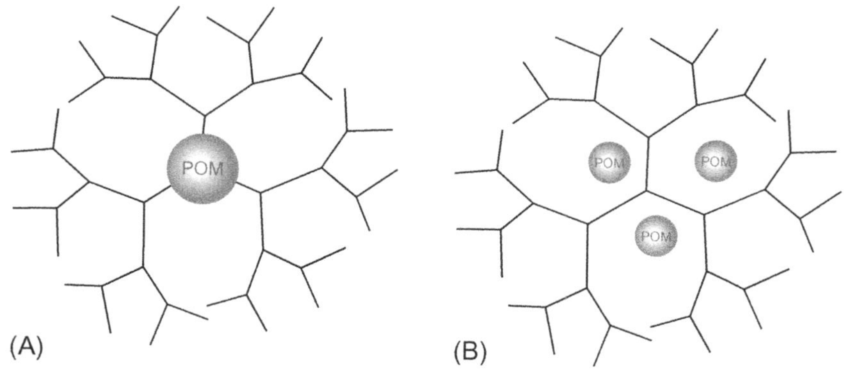

FIG. 2.1 POM-cored dendrimer (A) and DENDRI-POM hybrid with POM units encapsulated within the cavities of the dendrimer (B). *(Reprinted from S. Nlate, C. Jahier, Eur. J. Inorg. Chem. (2013) 1606–1619. Copyright 2013, with permission of John Wiley and Sons and Copyright Clearence Center.)*

are encapsulated into the dendritic structure. DENDRI-POM hybrids in which POM units are located at the periphery of the dendrimers are not considered in this chapter. We will focus on the synthesis of encapsulated DENDRI-POM hybrids and the investigation of their properties. The effects of the structure of the POM anion and the dendritic cation on the properties of the DENDRI-POM hybrid will be highlighted. We will particularly focus on pioneering work, with a brief summary of recently published work in this field. Two families of DENDRI-POM hybrids will be considered in this chapter. The first one involves DENDRI-POM in which the POM unit is located at the core (Fig. 2.1A), and the second family involves hybrids with POM units encapsulated within the cavities of the dendrimers (Fig. 2.1B).

2 DENDRI-POM HYBRIDS IN WHICH POM UNITS ARE ENCAPSULATED INTO THE DENDRIMERS

Given that the structures of POMs are very sensitive to the environmental pH values, their incorporation into dendrimers represents a promising strategy to protect them against external acidic or basic attacks, contrary to their homologues with POM species located at the periphery, which decompose slowly. Thus, the encapsulation of POM units into the dendritic structures provides an effective protection to POMs by enhancing their stability against the ambient environment, one of the most important features for their catalytic properties. The immobilization of POMs onto a support or into a matrix can be achieved in various ways, depending on the type of support. The main approach for the encapsulation of POM units into dendrimers is based on the electrostatic assembly of cationic dendrons or dendrimers with anionic POM species [8]. Surface modification of POM anions via electrostatic interactions with suitable cations, rather than covalent tethering, is a very attractive strategy because it prevents the need for time-consuming and often ligand functionalization. The encapsulation of catalytically active POMs within a shell of cationic dendrimer may help to tune the properties of the POM catalyst. It is known from phase-transfer experiments that the organic structure of the cation not only influences its ability to transfer an anion from the aqueous to the organic phase, but also strongly affects the rate of the reaction in the organic phase. For several years, based on the report of Volkmer, many groups have investigated the encapsulation of POM units into dendrimers [8d–m] and the effects of the dendritic wedges on the properties of polyanions. Families of chiral and achiral POM-cored dendrimers were prepared by electrostatic coupling of POM units with chiral and achiral dendritic cations, and they were tested as recoverable catalysts in the oxidation of alkenes, alcohols, and sulfides. These transformations have clearly highlighted a close relationship between the properties of the catalytically active POM species and the structure of the dendritic countercations. Thus, varying the structure of the dendritic cation can enable the modulation of the properties of POMs for specific applications.

2.1 Achiral DENDRI-POM Hybrids

2.1.1 DENDRI-POMs With POM Units Encapsulated Into the Cavities of Dendrimers

The encapsulation of POMs into the cavities of functionalized dendrimers opens the possibility of stable DENDRI-POMs bearing more than one POM unit in their structures. The strategy to introduce POM units into the cavities of dendrimers is similar to the dendritic host-guest catalyst stabilization by hydrogen bonding and ionic interactions studied by van Leeuwen and coworkers [7b,d]. The 36-armed DENDRI-POM hybrid bearing two-POM units were prepared in two main steps, namely, (1) the synthesis of the 36-armed dendrimers bearing 6-internal amino groups, and (2) the electrostatic encapsulation of POMs into the 36-armed dendrimers.

Synthesis of 36-Armed Dendrimers Bearing 6-Internal Amino Groups:

36-armed dendrimers **3a** and **3b** containing 6 internal amino groups were synthesized from a hexa-amine dendritic core **1** with benzylbromide allyl-terminated dendrons **2a** and the *p*-tolyl-terminated dendron **2b**, respectively (Fig. 2.2). Attempts to quaternarize amino groups of **3a** and **3b** with bromide dendrons **2a-b** failed.

Encapsulation of POM Units Into 36-Armed Dendrimers:

36-armed dendrimer **3b** containing six internal amino groups were used to introduce anionic POM units into their cavities through electrostatic interaction in acidic mediums (Fig. 2.3). The 36-armed dendrimer **3b** reacts with the

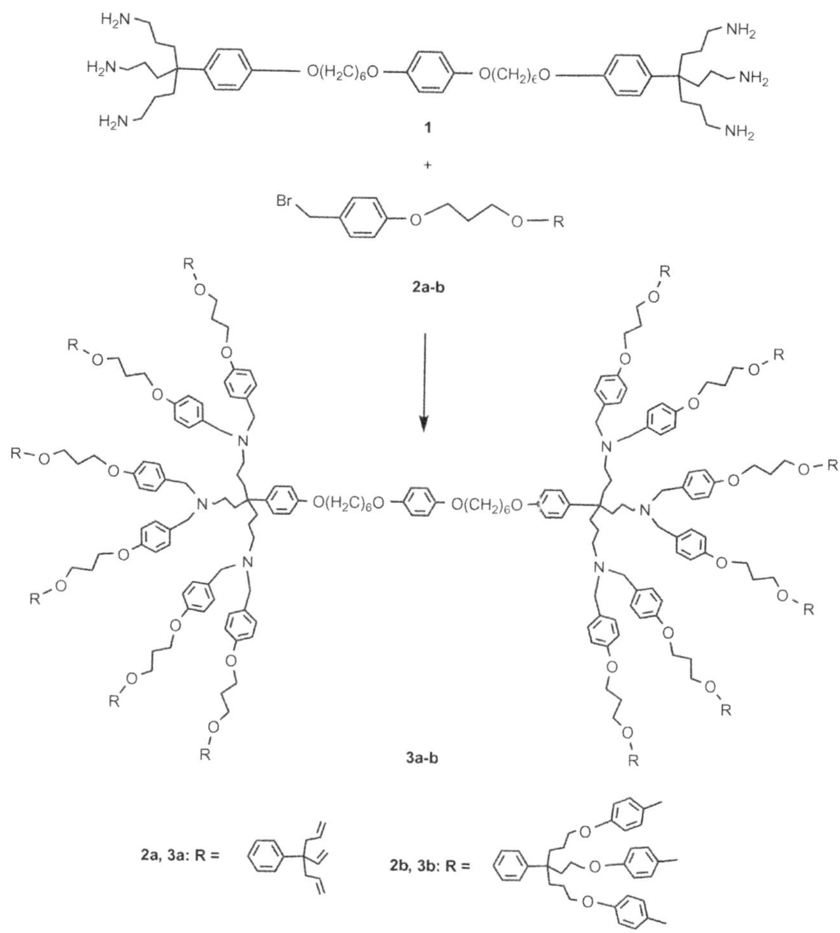

FIG. 2.2 36-Armed amino dendrimer **3a-b**. *(Reprinted from C. Jahier, L. Plault, S. Nlate, Isr. J. Chem. 49 (2009) 109–118. Copyright 2009, with permission of John Wiley and Sons and Copyright Clearence Center.)*

Venturello tri-anion $[PO_4\{WO(O_2)_2\}_4]^{3-}$ to result in the corresponding 36-armed *p*-tolyl-terminated DENDRI-POM **4b** in which two POM units are encapsulated into the dendritic structure through electrostatic interaction with internal ammonium groups [8i]. Attempts to prepare the 36-allyl-terminated DENDRI-POM by reacting the amino dendrimer **3a** with the POM anion failed, because of the slow epoxidation of the 36-peripheral olefinic groups of **3a**, which decomposed the POM catalyst after a long reaction time in an acidic medium. In contrast to dendrimers with nonoxidizable end groups such as **3b**, the encapsulation of POMs into dendritic structure **3a** bearing oxidizable

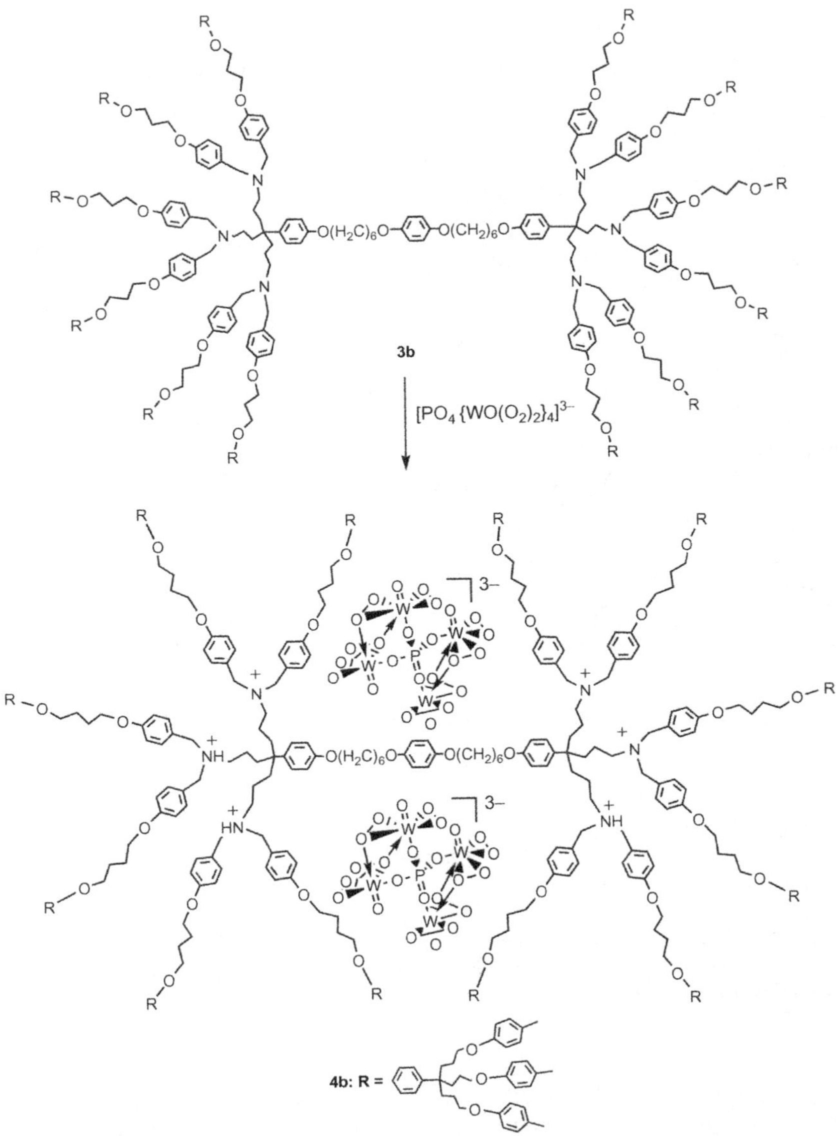

FIG. 2.3 36-Armed dendrimer encapsulating two POM units **4b**. *(Reprinted from C. Jahier, L. Plault, S. Nlate, Isr. J. Chem. 49 (2009) 109–118. Copyright 2009, with permission of John Wiley and Sons and Copyright Clearence Center.)*

alkene groups gives unsatisfactory results. Indeed, the POM unit oxidizes the peripheral olefinic end groups at the surface of dendrimers, which prevents its efficient incorporation into the cavities of the dendrimer. Thus, an appropriate elaboration of dendritic structures is crucial for efficient encapsulation of POM catalysts.

Catalytic Evaluation of 36-Armed DENDRI-POM Hybrid 4b in the Oxidation Reaction:

The catalytic performance of the POM-encapsulated dendrimer **4b** was tested in the oxidation of cyclooctene **5** with H_2O_2. These studies indicated that DENDRI-POM **4b** oxidizes cyclooctene to the corresponding epoxide **6** with 100% conversion after 3 h (Fig. 2.4). The properties of the catalyst **4b** are retained under the protection of the dendritic wedge.

Since the structures of POMs are very sensitive to the environmental pH range, their incorporation into dendritic structures retained their properties by protecting them against external acidic or basic attacks. This family of compounds is very promising for multifunctional systems, which can be used as dual encapsulation and release systems. Interestingly, the protection of POM by its encapsulation into the dendritic structure enhances its stability and recovery abilities without significantly affecting the catalytic properties. Thus, the microenvironment around the catalytic species determines its properties.

2.1.2 *Polyoxometalate With a Single POM Anion Encapsulated Into the Dendritic Core*

Since the pioneering work by Volkmer and coworkers [5a], POM-cored dendrimers have attracted much attention, due to the potential applications of DENDRI-POM materials. Thus, exploring new DENDRI-POM hybrids remains a challenging topic for chemists. Various allyl and *n*-propyl-derived ammonium dendrons were prepared and assembled to POM anions by anion exchange, giving soluble DENDRI-POM hybrids in which a single POM anion is located at the core of compact shell of dendrons. They have been used as recoverable catalysts in the oxidative transformation of sulfides, alkenes, and alcohols.

Preparation of Allyl and *n*-Propyl-Terminated Dendritic Structures Able to Encapsulate POM Anions:

Due to the affinity of multibranched dendrons for a rapid growth of dendrimers as well as the functionalization of molecular and nanoscale materials, a variety of useful methods have been reported for the synthesis of dendrimers and dendrons [9]. Among them, a fast and scalable procedure has been proposed for the

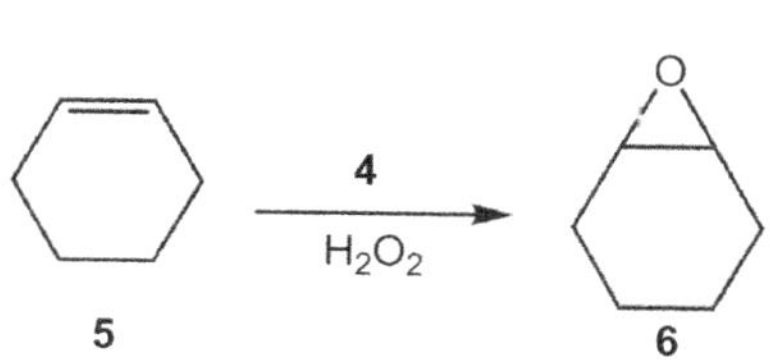

FIG. 2.4 Oxidation of cyclohexene **5** with the 36-armed DENDRI-POM hybrid **4b**.

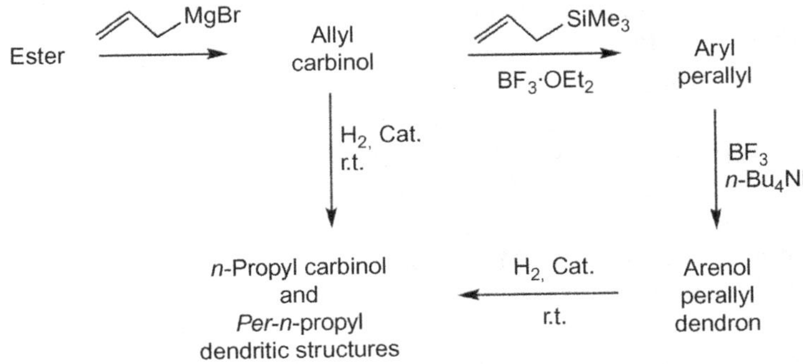

FIG. 2.5 Strategy to prepare polyallyl and poly-*n*-propyl dendrons and dendrimers. *(Adapted from S. Nlate, L. Plault, F.-X. Felpin, D. Astruc, Adv. Synth. Catal. 350 (2008) 1419–1424.)*

synthesis of aryl polyallyl and aryl poly-*n*-propyl dendrons and dendrimers (Fig. 2.5) [10]. In contrast to the organometallic method reported in ref. [9], this new procedure consists of a direct allylation of esters with a Grignard reagent that results in the corresponding allylcarbinol compounds. A subsequent allylation of the allylcarbinol intermediate with allyltrimethylsilane gives the perallylated compound. A catalytic hydrogenation of the latter gives the corresponding *n*-propyl derivative [10].

The efficiency of this procedure has been evaluated in the preparation of tripodal dendrons 4-triallylmethylphenol **10** and 4-tri-*n*-propylmethylphenol **11** (Fig. 2.6). These dendrons are interesting building blocks for a fast synthesis of giant dendrons and dendrimers [10].

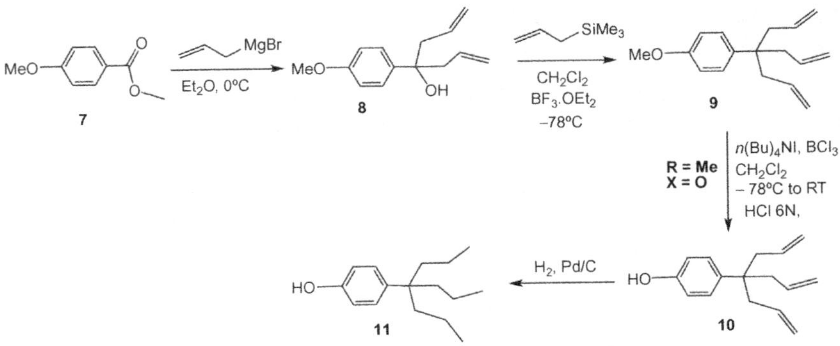

FIG. 2.6 Synthesis of 4-triallylmethylphenol and 4-tri-*n*-propylmethylphenol dendrons **6** and **7**. *(Reprinted from S. Nlate, L. Plault, F.-X. Felpin, D. Astruc, Adv. Synth. Catal. 350 (2008) 1419–1424. Copyright 2008, with permission of John Wiley and Sons and Copyright Clearence Center.)*

DENDRI-POM Hybrids Based on the Venturello Polyanion:

Various DENDRI-POM with the Venturello anion $[PO_4\{WO(O_2)_2\}_4]^{3-}$ located at the core were synthesized by the electrostatic coupling approach. The Venturello tri-anion is generally used as a model catalyst because of its easy preparation and catalytic efficiency in the oxidation of organic substrates.

DENDRI-POMs Bearing Epoxy, n-Propyl and Aryl Sulfide Peripheral Groups: In order to evaluate the effects of dendritic structures on the properties (solubility, stability, catalytic efficiency, and selectivity) of POM catalysts, various dendrons were prepared and coupled with the Venturello POM tri-anion. Families of 9- and 27-armed DENDRI-POM hybrids bearing respectively epoxy **13a**, *n*-propyl **13b** and aryl sulfide **13c** peripheral groups on the branch termini and a catalytically active POM unit at the core have been prepared from dendrons **12a-c** and evaluated in oxidation reactions (Fig. 2.7).

It appears that the dendritic wedge around the POM increases its stability. The corresponding DENDRI-POM hybrid can thus be stored and handled without any special precaution. In contrast, with nondendritic POM hybrids and DENDRI-POMs in which POM units are located at the periphery, the stability decreased significantly, showing a positive effect of the dendritic structure on the catalytic species, and thus the advantage of the encapsulation of POM catalysts into dendritic structures. Oxidation transformations of selected alkenes, sulfides, and alcohols with hydrogen peroxide were used as model reactions to evaluate the catalytic efficiency of these DENDRI-POM hybrids as shown in Fig. 2.8.

A study of countercation effects indicated that the dendritic structure increases the stability of the POM species and facilitate the recovery and recyclability of the catalyst up to the eighth cycle. On the other hand, the increased bulkiness around the POM catalyst slowed down the reaction kinetics, highlighting a negative dendritic effect on the reaction kinetics. Within the 9-armed POM-cored dendrimer series, the reaction kinetics was sensitive to the nature of the peripheral end groups of POM hybrids. Indeed, compared to epoxy- and aryl sulfide-terminated DENDRI-POMs the 9-armed *n*-propyl-derivative was the most active catalyst. A comparison between the 9-armed and 27-armed DENDRI-POM series shows a negative dendritic effect on the reaction rate. This negative effect is probably caused by the increased bulk around the POM center that makes more difficult the access of substrates molecules to the catalytically active species. Furthermore, it is known that the increased bulkiness of the cation reduces the electrophilicity of the peroxopolyoxo species [11]. If the ions are bulky, the interionic distance increases, and this decreases the interionic interaction. In the case of the 27-armed POM-cored dendrimers, the bulkiness of the dendritic cation might be responsible for the large distance between the ammonium function of the dendron and the anionic POM. Within the 9-armed DENDRI-POM series, in comparison to the *n*-propyl termini DENDRI-POMs, the lower electrophilicity of the aryl sulfide and epoxy

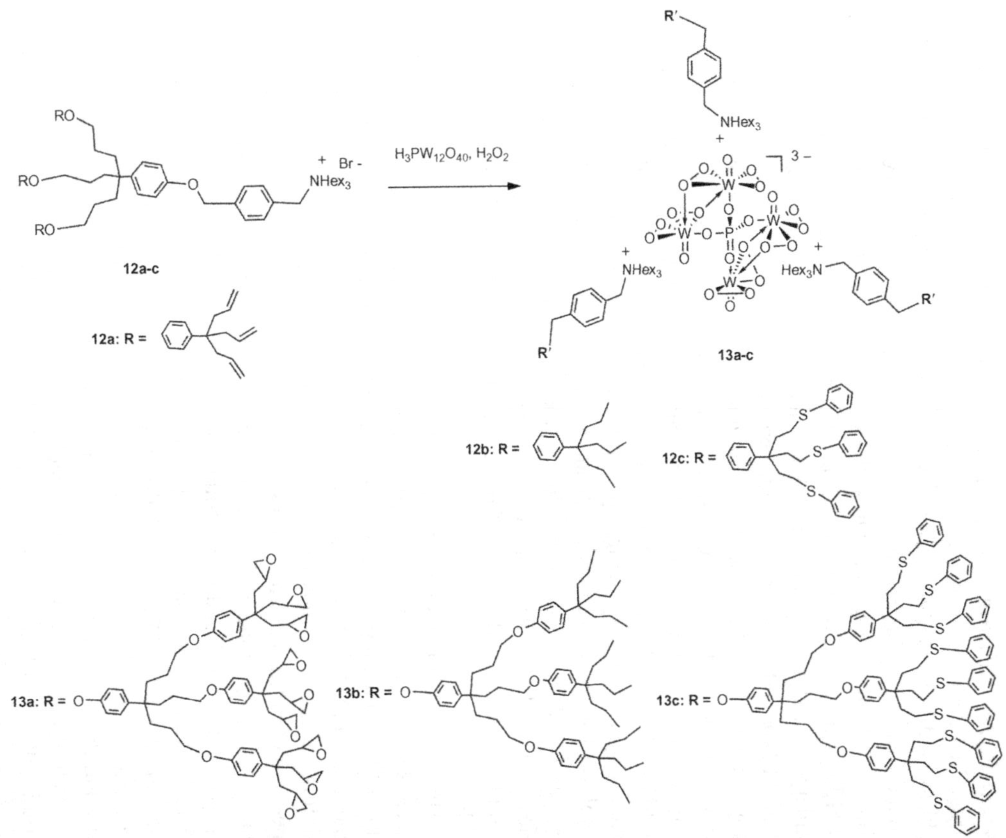

FIG. 2.7 Synthesis of 27-armed DENDRI-POM hybrids **13a-c** for oxidation reactions. *(Reprinted from S. Nlate, L. Plault, D. Astruc, Chem. Eur. J. 12 (2006) 903–914. Copyright 2006, with permission of John Wiley and Sons and Copyright Clearence Center.)*

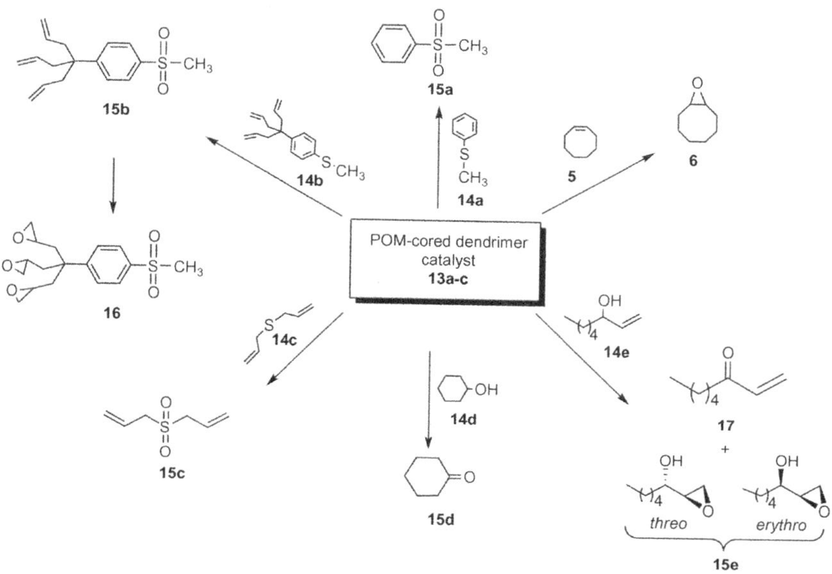

FIG. 2.8 Oxidation of sulfides, alkenes, and alcohols by DENDRI-POM hybrids **13a-c**. *(Reprinted from S. Nlate, L. Plault, D. Astruc, Chem. Eur. J. 12 (2006) 903–914. Copyright 2006, with permission of John Wiley and Sons and Copyright Clearence Center.)*

end groups might be responsible for the lower reactivity of the epoxy and aryl sulfide-based DENDRI-POM hybrids. After the reaction, the DENDRI-POM catalyst is easily recovered and recycled without a discernible loss of activity, contrary to the nondendritic counterpart. A significant advantage of the dendritic structures around the POM is that they increase its stability and enable its recovery and recyclability, contrary to the nondendritic counterparts that decomposed slowly in solution.

POM-Cored Dendrimers Based on 1,1′-Binaphthyl-2-Methylpyridine Ligand

Preparation of 1,1′-Binaphthyl-2-Methylpyridine-Based DENDRI-POMs: A mixture of binaphthylamino-containing ligands **17**, **18**, and **19** reacts with the Venturello tri-anion to give the corresponding POM hybrids **20**, **21**, and **22** in which the POM anion is encapsulated into the binaphthylamino ligands, in their racemic form (Fig. 2.9) [12].

Structures of these DENDRI-POMs were examined in solution by IR and NMR spectroscopy as well as elemental analysis. Particular focus was devoted to their ^{31}P and ^{183}WNMR spectroscopy, because these techniques offer evidence for the presence of the intact polyanion in the corresponding hybrids. Although the expected ^{31}PNMR signal was observed for all the DENDRI-POM hybrids studied, no signal was observed in their ^{183}WNMR spectrum.

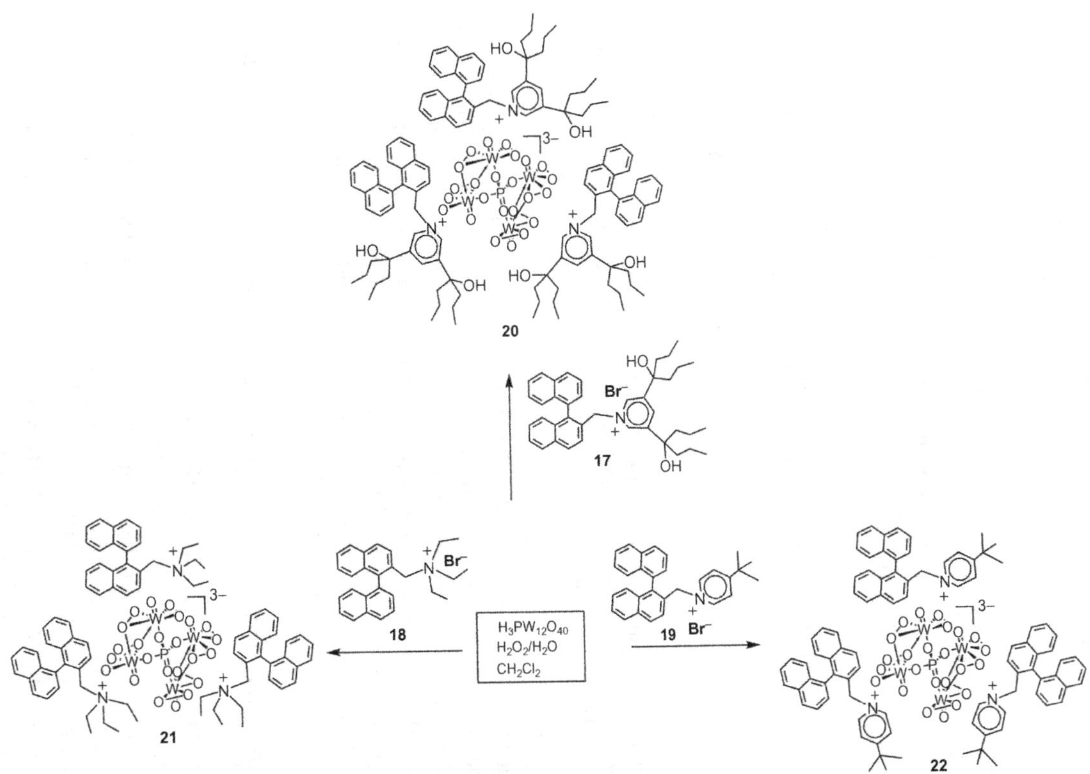

FIG. 2.9 1,1′-binaphtyl-2-methylammonium-based DENDRI-POM for oxidation reactions. *(Reprinted from C. Jahier, F.-X. Felpin, C. Meliet, F. Agbossou-Niedercorn, J.-C. Hierso, S. Nlate, Eur. J. Inorg. Chem. (2009) 5148–5155. Copyright 2009, with permission of John Wiley and Sons and Copyright Clearence Center.)*

Evaluation of 1,1′-Binaphthyl-2-Methylpyridine DENDRI-POMs in Oxidation Reactions: The performances of the 1,1′-binaphthyl-based POM salts in catalysis were tested in the oxidation of cyclooctene **5**, thioanisole **14a**, cyclohexanol **14d**, and styrene **23**. In contrast to styrene **23**, POM hybrids **20**, **21**, and **22** oxidize alcohols and sulfides to the corresponding epoxides, sulfoxides, and sulfones. The oxidation of thioanisole **14a** presumably gives, in a first step, sulfoxide **15**, which is in turn oxidized into sulfone **15a** either through a POM-catalyzed process or directly by the action of hydrogen peroxide (Fig. 2.10).

Diminishing the oxidant/substrate ratio down to 1.1 increases the selectivity to sulfoxide. It appears that DENDRI-POM hybrids oxidized sulfides with a better selectivity than their nondendritic counterparts. Interestingly, the use of enantiopure forms of 1,1′-binaphthyl-2-methyl-based ligands is potentially one approach in which enantiopure 1,1′-binaphthyl-2-methylpyridine-based DENDRI-POMs for asymmetric transformations can be obtained.

DENDRI-POM Hybrids Based on Polyanions Other Than the Venturello Species:

Volkmer and coworkers have reported the synthesis of the DENDRI-POM hybrid $(C_{52}H_{60}NO_{12})_{12}[(Mn(H_2O)_3(SbW_9O_{33})_2]$ **26** obtained by electrostatic encapsulation of the polyanion $[(Mn(H_2O))_3(SbW_9O_{33})_2]^{12-}$ within a shell of

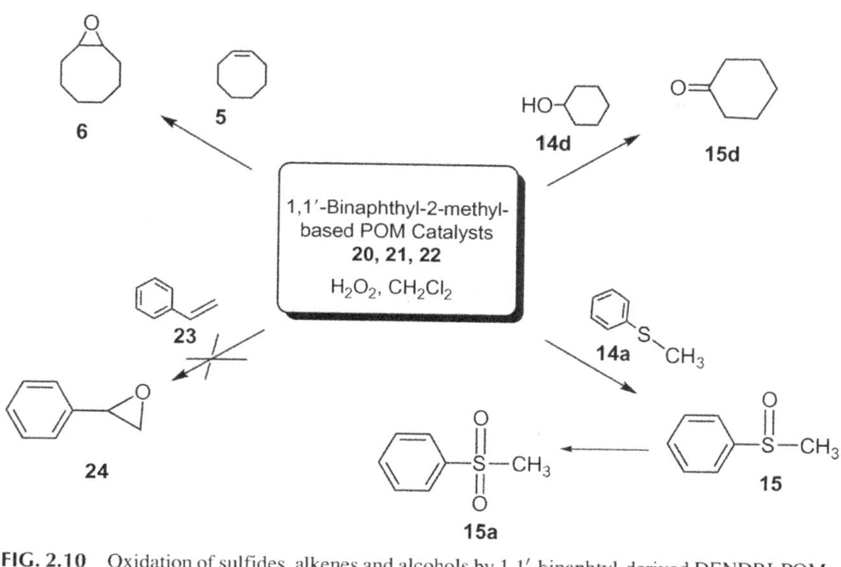

FIG. 2.10 Oxidation of sulfides, alkenes and alcohols by 1,1′-binaphtyl-derived DENDRI-POMs. *(Reprinted from C. Jahier, F.-X. Felpin, C. Meliet, F. Agbossou-Niedercorn, J.-C. Hierso, S. Nlate, Eur. J. Inorg. Chem. (2009) 5148–5155. Copyright 2009, with permission of John Wiley and Sons and Copyright Clearence Center.)*

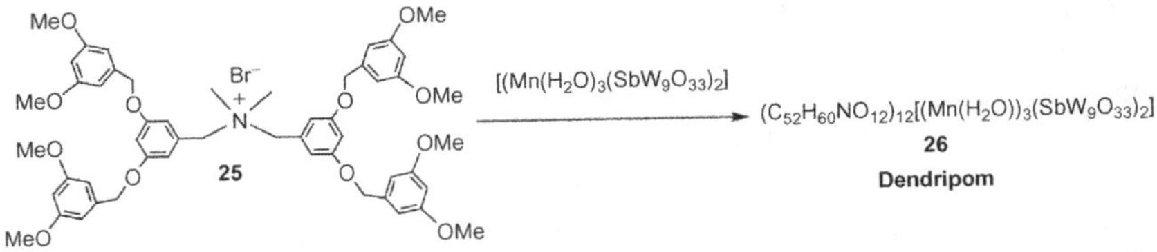

FIG. 2.11 Sandwich-type transition metal-substituted heteropolytungstate DENDRI-POM for oxidation reactions. *(Adapted from D. Volkmer, B. Bredenkötter, J. Tellenbröker, P. Kögerler, D.G. Kurth, P. Lehmann, H. Schnablegger, D. Schwahn, M. Piepenbrink, B. Krebs, J. Am. Chem. Soc. 124 (2002) 10489–10496, with permission of the American Chemical Society.)*

a cationic dendron **25** as illustrated in Fig. 2.11. This sandwich-type transition metal-substituted heteropolytungstate was chosen because of its excellent efficiency in the catalytic epoxidation of dienes. The structure and properties of the POM unit located at the core of **26** were investigated using analytical and theoretical methods [8b].

Wu and coworkers have reported a series of DENDRI-POM hybrids **28a-d** by encapsulating four polyanions, $[SiW_{12}O_{40}]^{4-}$, $[BW_{12}O_{40}]^{5-}$, $[KSiW_{11}O_{39}]^{7-}$ and $[Na_2P_2W_{15}O_{56}]^{10-}$ into the shell forms with the amphiphilic benzyl dendron **27** (Fig. 2.12.) [13].

With the aim to develop a robust POM-containing hybrid gelator, Wang and coworkers have covalently grafted two poly(urethane amide) dendrons containing four alkyl chains to the modified Anderson-type POM cluster [14]. The structure of this hybrid was confirmed by ESI-MS analysis.

DENDRI-POM Hybrids Based on Zirconium-Peroxotungstosilicate-Anion: As a part of our ongoing research in the field of DENDRI-POMs, we have prepared a series of POM-cored dendrimers by the coupling of zirconium-peroxotungstosilicate polyanion $[Zr_2(O_2)_2(SiW_{11}O_{39})_2]^{12-}$ [15] with various dendritic ammonium salts [8k]. The synthesis of the 36-(4-*tert*-butylphenyl) DENDRI-POM hybrid **30** from benzylbromide salt **29** is shown in Fig. 2.13.

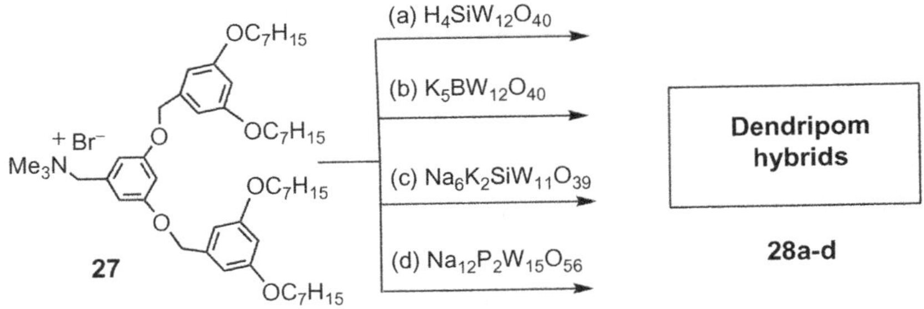

FIG. 2.12 DENDRI-POM hybrids based on amphiphilic benzyl ammonium dendron **27** [13]. *(Adapted from Y. Yang, Y. Wang, H. Li, W. Li, L. Wu, Chem. Eur. J. 16 (2010) 8062–8071, with permission of Wiley.)*

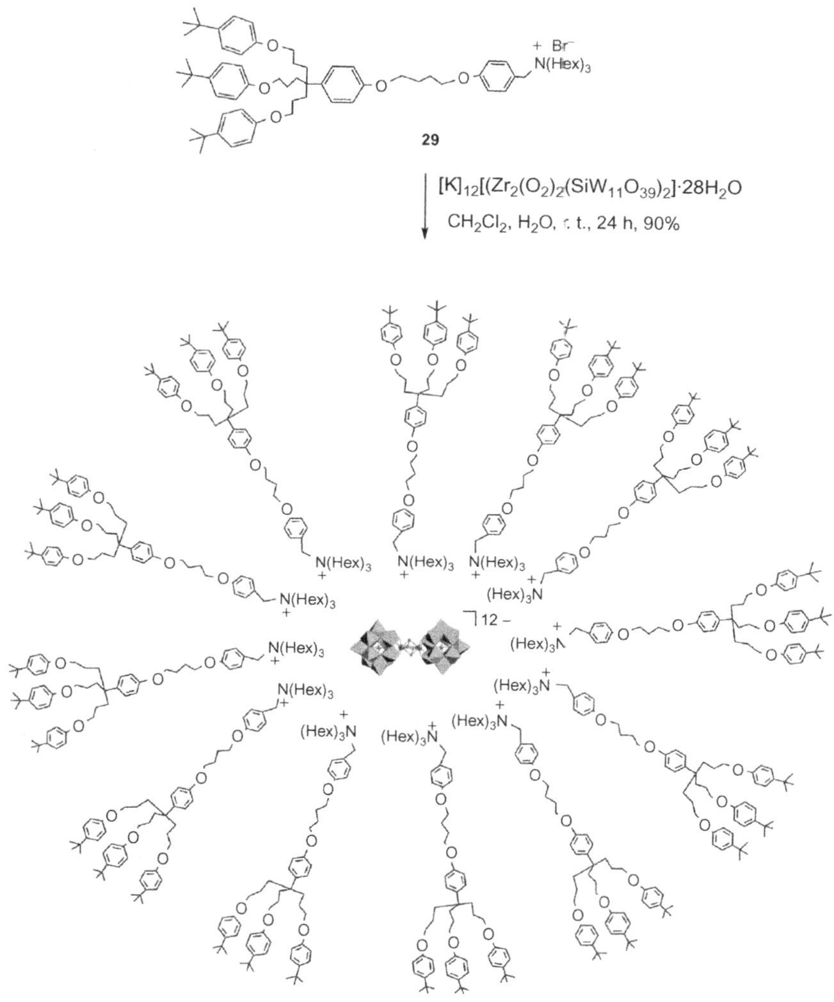

30

FIG. 2.13 36-armed DENDRI-POM hybrid based on zirconium-peroxotungstosilicate anion. *(Reprinted from C. Jahier, S.S. Mal, U. Kortz, S. Nlate, Eur. J. Inorg. Chem. (2010) 1559–1566. Copyright 2010, with permission of John Wiley and Sons and Copyright Clearence Center.)*

Contrary to the potassium salt of $[Zr_2(O_2)_2(SiW_{11}O_{39})_2]^{12-}$, DENDRI-POM hybrids are soluble in common organic solvents, an interesting feature for their use in homogeneous catalysis. These DENDRI-POMs are stable and efficient catalysts in the oxidation of sulfides. They can be recovered and reused. Their catalytic properties are also influenced by the structure of the dendrimer around the POM species. In contrast to the Venturello trianion, zirconium-peroxotungstosilicate-based DENDRI-POMs do not oxidize alkenes and alcohols.

DENDRI-POM Hybrids Based on Keggin, Dawson, and Preyssler Anions: DENDRI-POM hybrids **31**, **32**, and **33** based on Keggin $[PW_{12}O_{40}]^{3-}$, Dawson $[P_2W_{18}O_{62}]^{6-}$, and Preyssler $[NaP_5W_{30}O_{110}]^{14-}$ polyanions, respectively, have been synthesized by electrostatic coupling of the corresponding anion with a diallyl carbinol benzyl ammonium dendrons **3** as shown in Fig. 2.14 [8n].

The structures of these DENDRI-POM hybrids were examined in solution by IR, NMR spectroscopy and elemental analysis. Particular focus was devoted to ^{31}P and ^{183}WNMR spectroscopy, because these techniques offer evidence for the presence of the intact polyanion in corresponding hybrids. ^{183}WNMR spectroscopy was used to identify the polyanions in the DENDRI-POM hybrids and to obtain information about the interaction between the dendritic structure and the polyanion. Contrary to Dawson- and Preyssler-based DENDRI-POM hybrids, the ^{183}WNMR spectrum of the Keggin-based DENDRI-POM **31** shows one peak at $\delta = -88.6$ ppm in [D$_6$]-acetone (Fig. 2.15), indicating that all tungsten atoms are equivalent in this molecule. This result was the first successful example of a ^{183}WNMR experiment in which a DENDRI-POM hybrid shows a clear-cut signal. The small chemical shift variation in the ^{183}WNMR spectrum of **31** *versus* that of the sodium precursor ($Na_3[PW_{12}O_{40}]$) indicates no structural change of the polyanion, but rather it reflects the substitution of the sodium cation by a dendritic cation, as well as solvent effects.

Concerning the ^{31}PNMR experiments, the expected signal was obtained for all the DENDRI-POMs studied. We have tried diligently to crystallize all the studied DENDRI-POM hybrids, but we did not succeed in obtaining single crystals suitable for XRD studies. In addition, some ^{29}Si and ^{183}WNMR spectroscopic data would be a good line of evidence for the presence of the polyanion in all these systems. Unfortunately, all attempts to perform ^{29}Si and ^{183}WNMR spectroscopy in most of the POM hybrids with dendritic structures were futile. Indeed, because of the low receptivity of Si and W in dilute systems, it is difficult to observe any signal in ^{29}Si and ^{183}WNMR spectra of DENDRI-POMs. Contrary to DENDRIPOMs **32** and **33**, we have observed the corresponding signals in the ^{29}Si and ^{183}WNMR spectra of POM hybrid **31**. The catalytic efficiency of POM-based dendrimers **31**, **32**, **and 33** was evaluated in the oxidation of methyl phenyl sulfide and cyclooctene with H_2O_2. These hybrids catalyzed the oxidation of methyl phenyl sulfide to the corresponding sulfoxide and sulfone, but they were inert toward alkenes in similar conditions. Catalyst recovery experiments showed that, after two cycles of oxidation reactions, the activity of the Keggin-based DENDRI-POMs **31** were not changed, whereas that of the Dawson-based hybrid **32** decreased dramatically after the first cycle.

2.2 Optically Active POM-Encapsulated Dendrimers

The quest for a new catalytic and highly enantioselective process is currently one of the most intensively studied areas in chemical synthesis [16]. In this

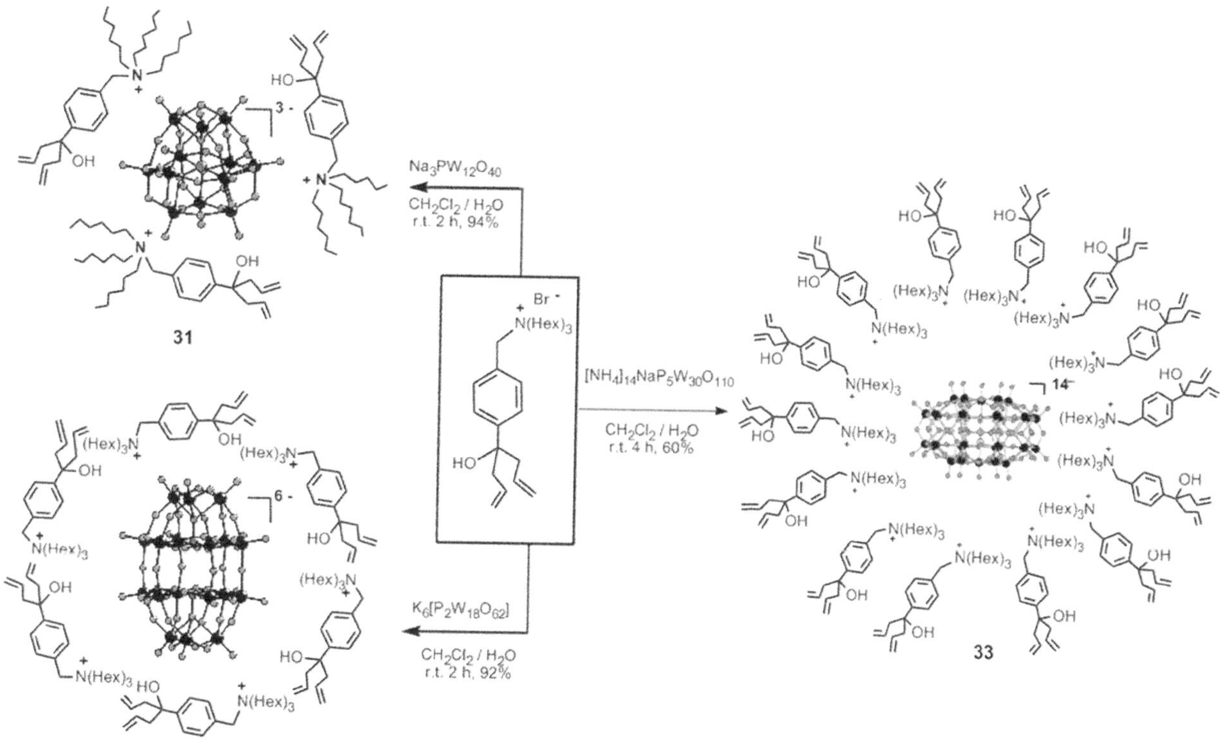

FIG. 2.14 DENDRI-POM hybrids based on the Keggin, Dawson, and Preyssler polyanions. *(Reprinted from C. Jahier, S.S. Mal, R. Al-Oweini, U. Kortz, S. Nlate, Polyhedron 57 (2013) 57–63. Copyright 2013, Elsevier.)*

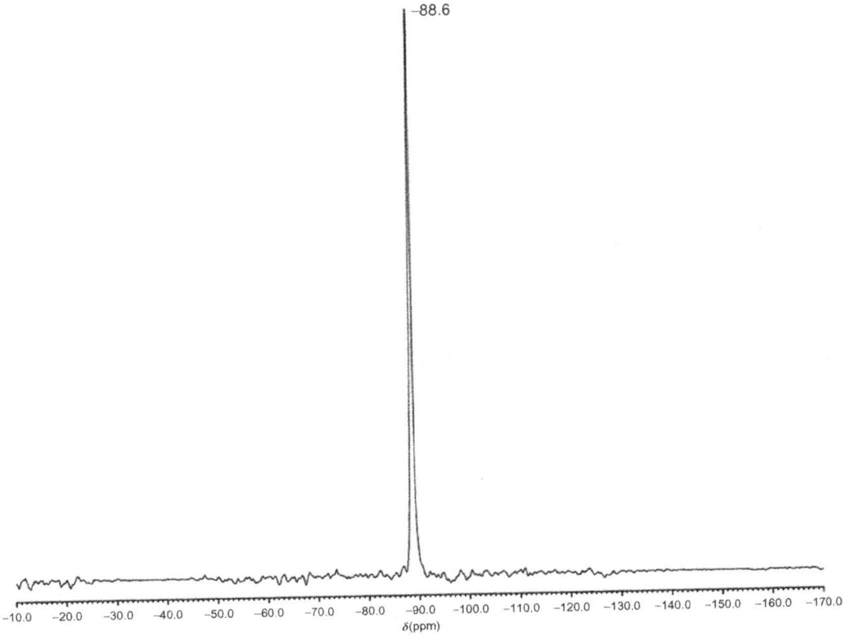

FIG. 2.15 ^{183}WNMR spectrum of the Keggin-based DENDRI-POM hybrid **31**. *(Reprinted from C. Jahier, S.S. Mal, R. Al-Oweini, U. Kortz, S. Nlate, Polyhedron 57 (2013) 57–63. Copyright 2013, Elsevier.)*

context, chiral polyoxometalates have become a topic of recent interest owing to their potential applications in medicine and stereoselective catalysis [17]. Among numerous applications of POM, catalysis is by far the most studied area, owing to the enormous versatility that POMs offer in the clean synthesis of fine chemicals. Chiral POM hybrids for asymmetric catalysis are still underrepresented [18], especially chiral DENDRI-POM hybrids. To date, numerous inorganic chiral POMs have been developed using two main synthetic approaches: the use of chiral species (organic moieties or metal complexes) as chirality transfer agents [18] and the strategy based on a spontaneous resolution upon crystallization in the absence of any chiral auxiliary that yields conglomerates [19]. However, resolution of the enantiomeric structure is frequently complicated by racemization, so the spontaneous resolution of enantiopure conglomerates was successfully achieved in only a few cases.

Chiral DENDRI-POM hybrids have been synthesized and their chiroptical properties investigated in the enantioselective oxidation of organic substrates [20]. Chiral organic structures were directly attached to the achiral POM moieties by electrostatic interaction, yielding enantiopure DENDRI-POMs on a large scale. Two series of DENDRI-POM hybrids prepared from enantiopure 1-phenylethylamine and L-proline were respectively considered for these studies.

2.2.1 Optically Active DENDRI-POM Hybrids Based on 1-Phenylethylamine

Enantiopure DENDRI-POM Hybrids Prepared From Monopodal Dendrons:

Enantiopure 12-armed and 24-armed DENDRI-POM hybrids **36** and **37** were obtained by ionic coupling of enantiopures 4-armed and 8-armed *n*-propyl dendritic structures **34** and **35**, with the Venturello trianion $\{PO_4[WO(O_2)_2]_4\}^{3-}$, respectively (Fig. 2.16) [8j,o]. The optical and chiroptical properties of these POM-cored dendrimers have been demonstrated in a solution by UV.vis spectroscopy, circular dichroïsm (CD), vibrational circular dichroïm (VCD) spectroscopy, and fluorometry, as well as in the asymmetric oxidation of thioanisole **14a** as a model reaction.

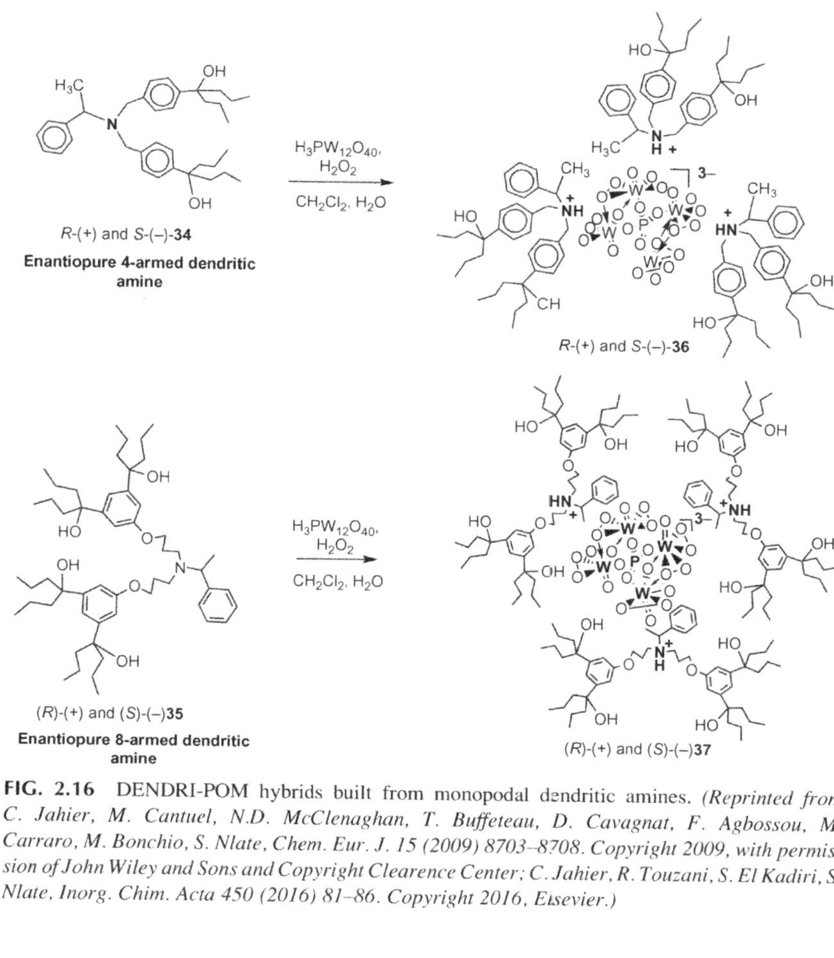

FIG. 2.16 DENDRI-POM hybrids built from monopodal dendritic amines. *(Reprinted from C. Jahier, M. Cantuel, N.D. McClenaghan, T. Buffeteau, D. Cavagnat, F. Agbossou, M. Carraro, M. Bonchio, S. Nlate, Chem. Eur. J. 15 (2009) 8703–8708. Copyright 2009, with permission of John Wiley and Sons and Copyright Clearence Center; C. Jahier, R. Touzani, S. El Kadiri, S. Nlate, Inorg. Chim. Acta 450 (2016) 81–86. Copyright 2016, Elsevier.)*

Circular Dichroism (CD) Experiments of DENDRI-POM Hybrids: The CD spectra of the DENDRI-POM *R*-(+)-**36** and *S*-(−)-**36** (Fig. 2.17A) as well as the ammonium chloride of *R*-(+)-**34**.HCl and *S*-(−)-**34**.HCl (Fig. 2.17B) are mirror images of one another, indicating the enantiopurity of the corresponding compounds. The spectra of DENDRI-POM hybrids *R*-(+)-**36** and *S*-(−)-**36** show strong and weak Cotton effects with maxima at 222 and 298 nm; these values are comparable to those obtained for oxygen-to-tungsten charge transfer bands of Keggin polyoxoanions [18k,21]. The CD spectra of the ammonium chloride *R*-(+)-**34**.HCl and *S*-(−)-**34**.HCl (Fig. 2.17B) exhibit strong and weak Cotton effects with maxima at 200 and 228 nm.

Vibrational Absorption and Circular Dichroism (CVD) Analyses of DENDRI-POMs: VCD spectroscopy is a well-defined method to obtain information about the chirality of the sample, and in particular for determining the absolute configuration and solution conformation of chiral molecules. Thus, VCD has been used to obtain information about the interaction between ligands and POM units. The IR spectra of ammonium chloride of ligand **34** and dendritic POM hybrids **36** in a CH_2Cl_2 solution are reported in Fig. 2.18. Most of the bands for the DENDRI-POM hybrids are related to the corresponding ligand, whereas broad bands are observed around 1075 and 945 cm^{-1} for the POM unit. These spectral contributions are assigned to stretching vibrations of the P—O and W═O bonds, respectively. The interaction between the ligand and the POM unit also induces spectral modifications to several bands of the ligand. Indeed, the broad band associated with the stretching vibration of the NH^+ group shifts from 2400 cm^{-1} for the ligand alone to 2900 cm^{-1} for the DENDRI-POM **36**. A smaller interaction of the NH^+ group with the POM (with respect to the chloride anion of the ligand) explains this blueshift.

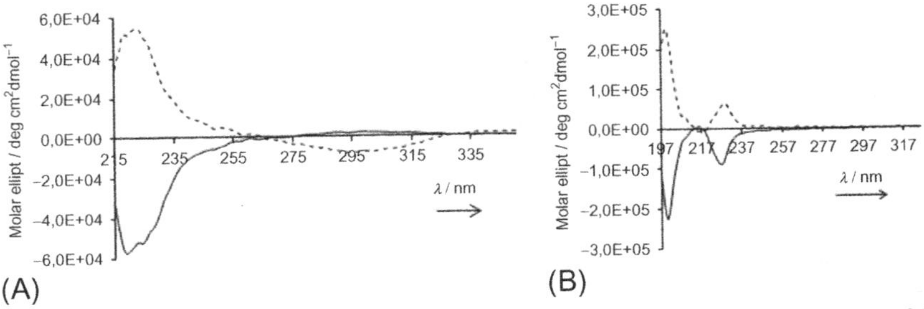

FIG. 2.17 CD spectra in CH_3CN of (A): monopodal enantiopure DENDRI-POMs *R*-(+)-**36** and *S*-(−)-**36** ($c = 2.3 \times 10^{-5}$M), and (B): spectra of ammonium chloride *R*-(+)-**34**·HCl and *S*-(−)-**34**·HCl ($c = 1.0 \times 10^{-5}$M). *(Reprinted from C. Jahier, M. Cantuel, N.D. McClenaghan, T. Buffeteau, D. Cavagnat, F. Agbossou, M. Carraro, M. Bonchio, S. Nlate, Chem. Eur. J. 15 (2009) 8703–8708. Copyright 2009, with permission of John Wiley and Sons and Copyright Clearence Center.)*

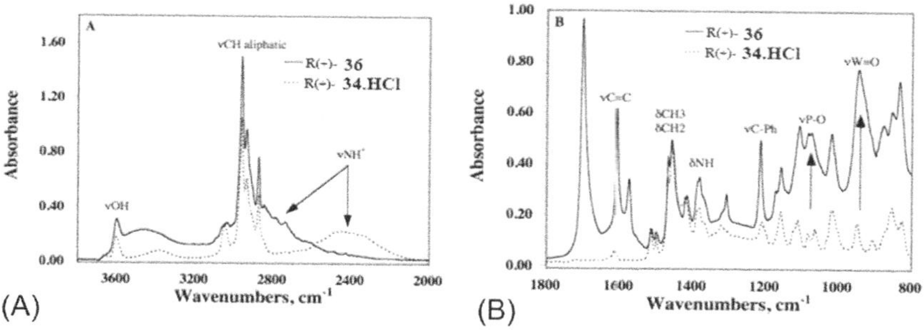

FIG. 2.18 IR spectra of R-(+)-**34**·HCl and R-(+)-**36** in the: (A) 3800–2000 cm^{-1} range and (B): 1800–800 cm^{-1} range under the following experimental conditions: Co-addition of 50 scans, 4 cm^{-1} resolution, concentration of 150 mm for R-(+)-**34**·HCl and 50 mm for R-(+)-**36** in CH_2Cl_2, path length 200 μm. *(Reprinted from C. Jahier, M. Cantuel, N.D. McClenaghan, T. Buffeteau, D. Cavagnat, F. Agbossou, M. Carraro, M. Bonchio, S. Nlate, Chem. Eur. J. 15 (2009) 8703–8708. Copyright 2009, with permission of John Wiley and Sons and Copyright Clearence Center.)*

The VCD spectra of R-(+)-**36** and S-(−)-**36** in a $CDCl_3$ solution are shown in Fig. 2.19. The two VCD spectra are perfect mirror images, which is indicative of the identical optical purity of the two enantiomers. The most important contributions in the VCD spectra come from vibrational modes related to the ligand. In particular, the VCD bands at 1468 and 1452 cm^{-1} are associated with the

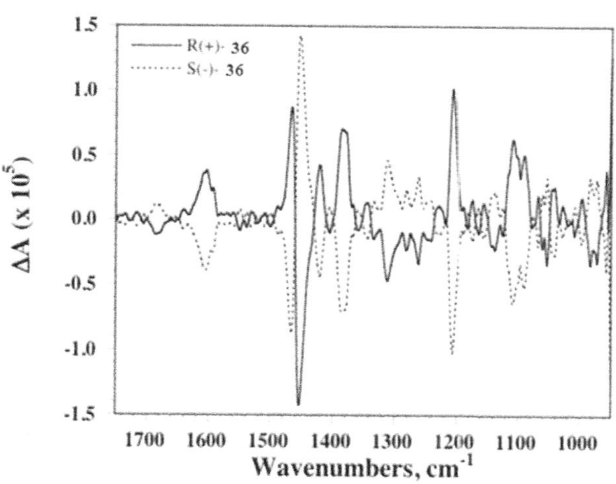

FIG. 2.19 Experimental VCD spectra of R-(+)-**36** and S-(−)-**36** in the 1750–950 cm^{-1} region under the following experimental conditions: 36,000 scans (12 h acquisition time), 4 cm^{-1} resolution, concentration of 0.6 m in $CDCl_3$, path length (250 μm). *(Reprinted from C. Jahier, M. Cantuel, N.D. McClenaghan, T. Buffeteau, D. Cavagnat, F. Agbossou, M. Carraro, M. Bonchio, S. Nlate, Chem. Eur. J. 15 (2009) 8703–8708. Copyright 2009, with permission of John Wiley and Sons and Copyright Clearence Center.)*

CH_2 and CH_3 bending vibrations of the propyl groups. Two other VCD bands with significant intensities are located at 1383 and 1206 cm^{-1}. They are assigned to the C—NH bending and C—Ph stretching vibrations, respectively. The investigation of chirality transfer from the ligand to the POM unit is not straightforward. Indeed, induced chirality on achiral molecules gives rise to very low VCD signals and is generally associated with the modified VCD of the chiral inducer.

Electronic Absorption and Emission of DENDRI-POM Hybrids: The electronic absorption spectra of (a) DENDRI-POM hybrid *S*-(−)-**36**, (b) tetra-*n*-propyl dicarbinol amine *S*-(−)-**34**.HCl in CH_3CN, and (c) the nondendritic POM-Arquad ($[\{n\text{-}C_{18}H_{37}\ (75\%)+n\text{-}\{(C_{16}H_{33}\ (25\%)\}_2NMe_2]_3^+[PO_4\{WO(O_2)_2\}_4]^{3-}$) in which Arquad$=[\{n\text{-}C_{18}H_{37}\ (75\%)+n\text{-}\{(C_{16}H_{33}\ (25\%)\}_2NMe_2]$ are shown in Fig. 2.20. Absorption in CH_2Cl_2 and CH_3CN were not simply a summation of absorption bands associated with constituent POM and dendritic chromophores. An additional absorption band between 280 and 350 nm is observed with respect to free dendron salts **34**·HCl and the nondendritic POM-Arquad. The latter absorption is consistent with significant ground-state interactions with at least one chiral auxiliary, even at micromolar concentrations. This interaction

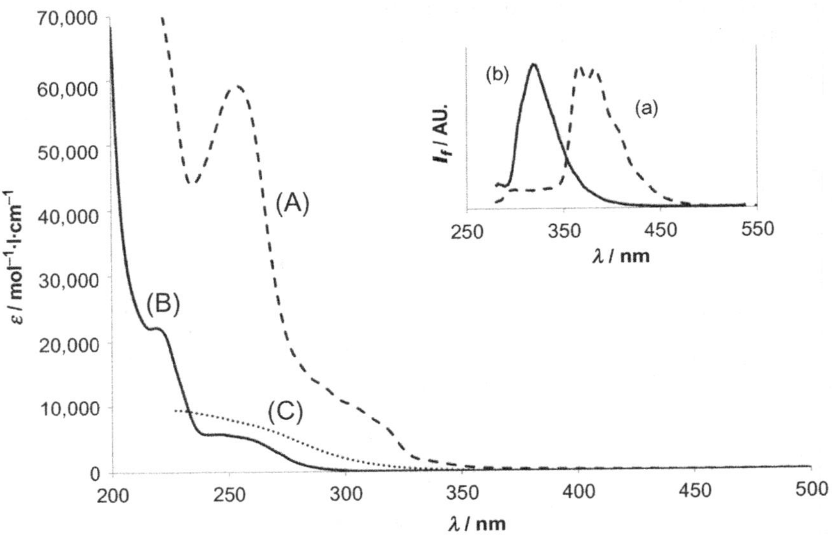

FIG. 2.20 Electronic absorption spectra of (A) DENDRI-POM *S*-(−)-**36**, (B) tetra-*n*-propyl dicarbinol amine *S*-(−)-**34**.HCl in CH_3CN, and (C) POM-Arquad in CH_2Cl_2. Inset: luminescence spectrum of (a) *S*-(−)-**36**, and (b) *S*-(−)-**34**.HCl in CH_3CN. *(Reprinted from C. Jahier, M. Cantuel, N.D. McClenaghan, T. Buffeteau, D. Cavagnat, F. Agbossou, M. Carraro, M. Bonchio, S. Nlate, Chem. Eur. J. 15 (2009) 8703–8708. Copyright 2009, with permission of John Wiley and Sons and Copyright Clearence Center.)*

appears unaltered upon adding aliquots of water, including doing so in simulated catalytic conditions. However, increasing the concentration of water disfavored the formation of the ionic complex.

The fluorescence emission and excitation spectra show that structured POM-Arquad emission was slightly blueshifted when it was with dendritic wedges ($\Delta\omega = c.1000\ cm^{-1}$), whereas blue ammonium ligand **34**·HCl fluorescence ($\Phi_F = 0.03$, $\lambda_{max} = 320$ nm) appeared to be quenched or partly quenched owing to the close proximity of the POM. All of these results are consistent with the ground-state (and excited-state) complex comprising the anionic POM and at least one cationic dendritic wedge, highlighting a close interaction.

Evaluation of Enantiopure DENDRI-POM Hybrids in an Oxidation Reaction: To evaluate the catalytic efficiency of enantiopure 12-armed DENDRI-POM hybrids (*S*)-(−)- and (*R*)-(+)-**36** as well as 24-armed DENDRI-POM hybrids (*S*)-(−) and (*R*)-(+)-**37** in an asymmetric reaction, thioanisole **14a** was used as a model substrate and oxidized with H_2O_2 (35%) in a biphasic mixture of water and $CDCl_3$. DENDRI-POM hybrids **36** and **37** selectively oxidize thioanisole **14a** to the corresponding chiral sulfoxide **15a** with up to 14% and 10% enantiomeric excess (*ee*) respectively [8j,o]. The selectivity to chiral sulfoxide and the *ee* values increased by decreasing the reaction temperature, depending on the structure of the catalyst, the nature of the substrate, and reaction conditions. Compared to enantiopure 12-armed *n*-propyl DENDRI-POM **36**, the oxidation of thioanisole **14a** with 24-armed *n*-propyl DENDRI-POM **37** indicated a significant decrease in the reaction kinetics and the enantioselectivity, whereas the selectivity to the chiral sulfoxide is improved. Thus, increasing the bulkiness around the POM catalyst with larger *n*-propyl-based dendritic structures decreased the reaction kinetics and the enantioselectivity, when the selectivity to chiral sulfoxide is improved. This result indicates that the encapsulation of POM catalysts into a larger dendritic structure promotes the selectivity to the chiral sulfoxide, but it decreases the reaction rate and the chiral environment around the POM. This result is probably due to the decreased interaction between dendritic cations and the anionic POM. The negative effect on the reaction rate may arise from the fact that the increased bulk around the POM center caused by the 24-peripheral-*n*-propyl groups in POM hybrid **37** prevents the easy access of substrate molecules to the catalytic POM center and thus slows down the reaction rate. It is important to note that all the reactions performed in the absence of the POM catalyst did not proceed when carried out under similar conditions. Furthermore, all the studied DENDRI-POM hybrids are more enantioselective than their nondendritic counterparts under similar conditions, highlighting a positive dendritic effect on the enantioselectivity. It is important to note that all the DENDRI-POM hybrids based on studied 1-phenylethylamine did not oxidize aromatic alkenes under similar conditions.

Enantiopure Tripodal DENDRI-POM Hybrids:

Enantiopure tripodal *n*-propyl-based DENDRI-POM hybrids *R*-(+)-**39** and *S*-(−)-**39** were built up by ionic assemblies of the trianionic POM $\{PO_4[WO(O_2)_2]_4\}^{3-}$ with enantiopure *n*-propyl tripodal dendritic amines *R*-(+)-and *S*-(−)-**38** (Fig. 2.21). In this case, the triply-charged anionic POM is encapsulated by a single dendritic structure bearing three ammonium groups The chiroptical properties of the tripodal DENDRI-POMs *R*-(+)-**39** and *S*-(−)-**39** have been demonstrated in a solution by UV-vis spectroscopy, circular dichroïsm (CD), vibrational circular dichroïm (VCD) spectroscopy, and fluorometry. The enantioselectivity was evaluated in the oxidation of thioanisole **14a** as a model reaction.

Circular Dichroïsm (CD) Experiments of DENDRI-POM Hybrids: The CD spectra of enantiomers *R*-(−)-**39** and *S*-(+)-**39** (Fig. 2.22A) are mirror images of one another, confirming their enantiopurity. These spectra show strong and weak Cotton effects with maxima at 199, 224, and 302 nm, whereas the CD spectra of the ammonium chloride counterparts *R*-(−)-**38**.3HCl and *S*-(+)-**38**.3HCl (Fig. 2.22B) exhibit strong and weak Cotton effects with maxima at 225 nm. The values obtained for DENDRI-POMs *R*-(−)-**39** and *S*-(+)-**39** are comparable to those obtained for oxygen-to-tungsten charge-transfer bands of Keggin polyoxoanions [18k,21].

Vibrational Absorption and Circular Dichroïsm (CVD) Analyses: The infrared spectra of the ammonium chloride tripodal ligand **38**.3HCl and the corresponding tripodal DENDRI-POM hybrids **39** in CH_2Cl_2 solution are reported in Fig. 2.23.

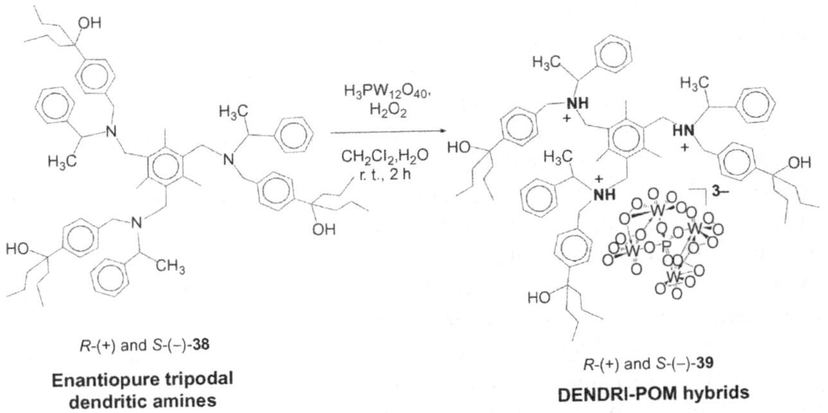

FIG. 2.21 DENDRI-POM hybrids built from tripodal dendritic amines. *(Reprinted from C. Jahier, M.-F. Coustou, M. Cantuel, N.D. McClenaghan, T. Buffeteau, D. Cavagnat, M. Carraro, S. Nlate, Eur. J. Inorg. Chem. (2011) 727–738. Copyright 2011, with permission of John Wiley and Sons and Copyright Clearence Center.)*

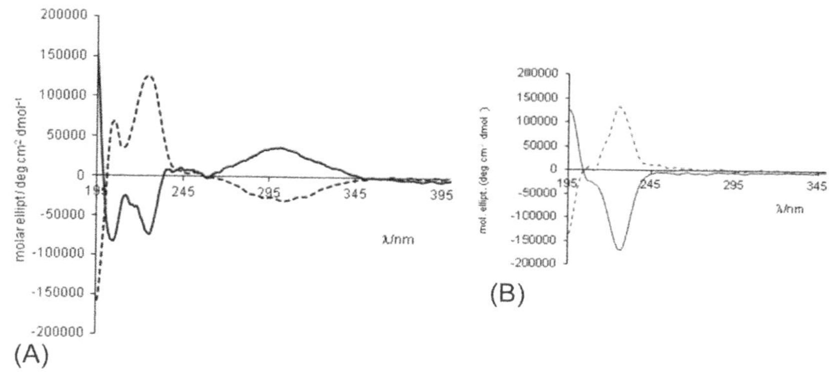

FIG. 2.22 CD spectra in CH_3CN of (A) enantiopure DENDRI-POMs *R*-(−)-**39** and *S*-(+)-**39** ($c = 5.0 \times 10^{-6}$M) and (B): spectra of ammonium chloride *R*-(−)-**38**·3HCl and *S*-(+)-**38**·3HCl ($c = 2.8 \times 10^{-6}$M)]. Ellipt = ellipticity. *(Reprinted from C. Jahier, M.-F. Coustou, M. Cantuel, N.D. McClenaghan, T. Buffeteau, D. Cavagnat, M. Carraro, S. Nlate, Eur. J. Inorg. Chem. (2011) 727–738. Copyright 2011, with permission of John Wiley and Sons and Copyright Clearence Center.)*

Most of the bands of the DENDRI-POM complex are related to the tripodal ligand, whereas broad bands of the POM unit are observed around 1075 cm^{-1} and at 945 cm^{-1}. These spectral contributions are assigned to stretching vibrations of P—O and W=O bonds, respectively. The interaction between the tripodal ligand and the POM unit induces spectral modifications for several ligand bands. Indeed, the broad band associated with the stretching vibration of the NH^+ group shifts from 2590 cm^{-1} for the ligand alone to 2900 cm^{-1} for the POM complex. A lower interaction of the NH^+ group with the POM (with respect to the chloride anion for the ligand) explains this blue shift.

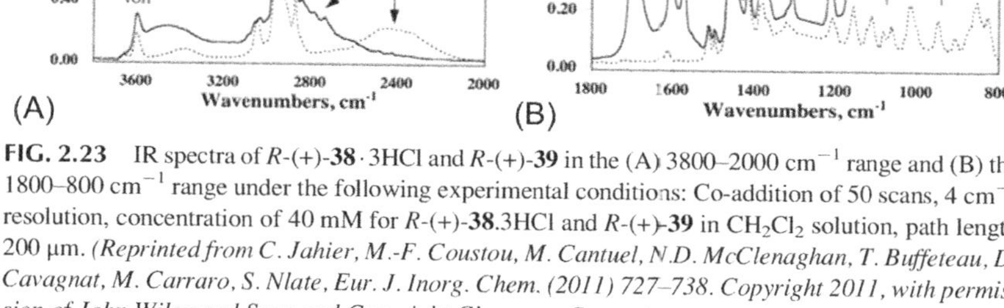

FIG. 2.23 IR spectra of *R*-(+)-**38**·3HCl and *R*-(+)-**39** in the (A) 3800–2000 cm^{-1} range and (B) the 1800–800 cm^{-1} range under the following experimental conditions: Co-addition of 50 scans, 4 cm^{-1} resolution, concentration of 40 mM for *R*-(+)-**38**.3HCl and *R*-(+)-**39** in CH_2Cl_2 solution, path length 200 μm. *(Reprinted from C. Jahier, M.-F. Coustou, M. Cantuel, N.D. McClenaghan, T. Buffeteau, D. Cavagnat, M. Carraro, S. Nlate, Eur. J. Inorg. Chem. (2011) 727–738. Copyright 2011, with permission of John Wiley and Sons and Copyright Clearence Center.)*

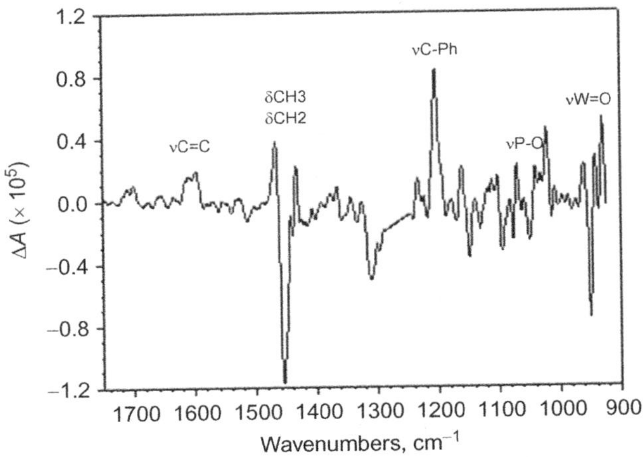

FIG. 2.24 VCD spectrum of *R*-(+)-**39** in CH_2Cl_2 solution. *(Reprinted from C. Jahier, M.F. Coustou, M. Cantuel, N.D. McClenaghan, T. Buffeteau, D. Cavagnat, M. Carraro, S. Nlate, Eur. J. Inorg. Chem. (2011) 727–738. Copyright 2011, with permission of John Wiley and Sons and Copyright Clearence Center.)*

A second important spectral modification upon complexation is the enhancement of the bands associated with the aromatic C═C stretching vibrations above 1500 cm^{-1} due to charge transfer between the POM unit and aromatic rings. The VCD spectra of *R*(+)-**39** in a CH_2Cl_2 solution is shown in Fig. 2.24.

The most important contributions in the VCD spectrum of the tripodal DENDRI-POM hybrid come from vibrational modes related to the ligand. In particular, the VCD bands at 1467 and 1454 cm^{-1} are associated with the CH_2 and CH_3 bending vibrations of the propyl groups. Likewise, the VCD band located at 1204 cm^{-1} is assigned to the chiral C—Ph stretching vibrations. A negative VCD band is observed around 1050 cm^{-1} for the P—O bonds whereas a positive-negative-positive pattern is detected in the 960–920 cm^{-1} spectral range for the W═O bonds. The intensities of the latter bands are significantly higher than those observed for the P—O bonds, indicating that the W═O mode seems to be a good probe to follow the chirality transfer from the ligand to the POM cluster.

Electronic Absorption and Emission Studies: The electronic absorption spectra of POM-Arquad and the tripodal ammonium salt **38**.3HCl as well as the corresponding dendritic-POM **39** are shown in Fig. 2.25.

For the chloride salt **38**.3HCl, the intense high energy band at 220 nm is ascribed to peripheral aromatics while the electronically isolated central aromatic lies at 260 nm. Concerning the DENDRI-POM **39**, significantly different absorption is observed compared to the parent POM and tripodal ligand in CH_2Cl_2 and CH_3CN. The absorption spectrum of tripod-POM **39** is altered upon dilution of an acetonitrile solution, or on adding water, indicating a decreased

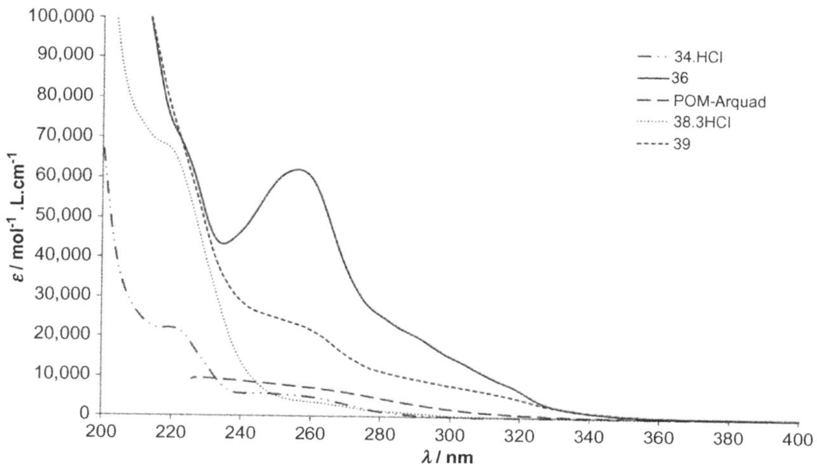

FIG. 2.25 Electronic absorption spectra of the hydrochloride salt of the monopodal ligand **34**.HCl, the corresponding DENDRI-POM **36**, the POM-Arquad in CH_2Cl_2, the hydrochloride salts of the tripodal dendrimers *R*-(+)-**38**·3HCl, and the corresponding cendritic POM **39** in CH_3CN. ($c=10^{-5}$). *(Reprinted from C. Jahier, M.-F. Coustou, M. Cantuel, N.D. McClenaghan, T. Buffeteau, D. Cavagnat, M. Carraro, S. Nlate, Eur. J. Inorg. Chem. (2011) 727–738. Copyright 2011, with permission of John Wiley and Sons and Copyright Clearence Center.)*

interaction/dissociation, consistent with the proposed interaction. Fluorescence emission and excitation spectra showed that the structured POM-Arquad emission is slightly blue-shifted when elaborated with dendritic wedges ($\Delta\omega$ = c.1000 cm^{-1}) while blue ammonium ligand **38**.3HCl fluorescence ($\Phi_F = 0.004$, $\lambda_{max} = 303$ nm) appears partly quenched due to the close proximity of the POM in **39**, although no sensitized emission arising from the POM part is detected. These results are consistent with a ground-state (and excited-state) complex comprising the anionic POM and at least one cationic dendritic wedge, necessitating close interaction. The weaker charge transfer interaction in the POM-tripod species may be due to differing charge densities within the dendron brick **38**.3HCl compared to the monopodal ligand **34**.HCl. Indeed, the protonation of the amino groups can have a dramatic effect on the neighboring aromatics.

Oxidation Catalysis With Enantiopure Tripodal DENDRI-POM Hybrids: DENDRI-POMs *R*-(−)-**39** and *S*-(+)-**39** oxidize methyl phenyl sulfide **14a**, methyl *p*-tolyl sulfide **40**, and 2-chloroethyl phenyl sulfide **41** to the corresponding methyl phenyl sulfoxide **15**, methyl *p*-tolyl sulfoxide **42**, and 2-chloroethyl phenyl sulfoxide **43** with up to 13% enantiomeric excess. A small amount of sulfones **15a**, **42a**, and **43a** were also obtained respectively (Fig. 2.26).

In all cases, the selectivity to sulfoxide and the enantiomeric excess are sensitive to the reaction temperature and the nature of the substrate. The methyl

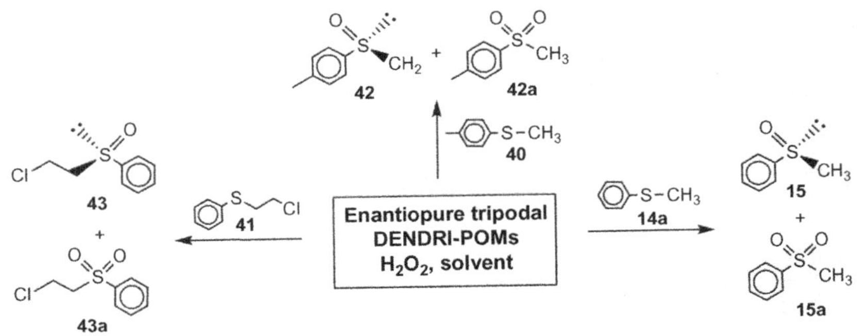

FIG. 2.26 Oxidation of sulfides by enantiopure tripodal DENDRI-POM **39**. *(Reprinted from C. Jahier, M.-F. Coustou, M. Cantuel, N.D. McClenaghan, T. Buffeteau, D. Cavagnat, M. Carraro, S. Nlate, Eur. J. Inorg. Chem. (2011) 727–738. Copyright 2011, with permission of John Wiley and Sons and Copyright Clearence Center.)*

p-tolyl sulfide **40** substituted with an electron donating group (CH_3) accelerated the oxidation reaction compared to methyl phenyl sulfide and 2-chloroethyl phenyl sulfide **41**. The best enantioselectivity was obtained in the oxidation of methyl phenyl sulfide **14a** with DENDRI-POM **39**. The results obtained are consistent with an electrophilic oxygen transfer mechanism. Overall, compared to reactions performed at low temperature (−50°C), the reaction kinetics were extremely fast at 35°C, whereas the selectivity and enantioselectivity decreased. DENDRI-POM catalysts were more enantioselective than their non-dendritic counterparts under similar conditions, highlighting a slight dendritic effect on the *ee*. Furthermore, increasing the catalyst concentration increases the oxidation kinetics without significantly affecting the *ee* values.

Catalyst Recovery and Reuse of Enantiopure DENDRI-POM Catalysts:

Up to four catalytic cycles were performed at 35°C using the oxidation of thioanisole as a model reaction. The catalyst was recovered by precipitation after each catalytic cycle and analyzed by ^{1}H and ^{31}PNMR spectroscopy before a new catalytic reaction. Monopodal DENDRI-POM hybrids based on 1-phenylethylamine oxidize thioanisole **14a** to the corresponding sulfoxide with up to 8% *ee*, indicating no discernible decrease in their activity, selectivity, and enantioselectivity, contrary to tripodal counterparts that slowly decomposed in the reaction mixture. However, if the reaction is performed at −50°C with tripodal DENDRI-POM hybrids, no change was observed in their activity, selectivity, and enantioselectivity over three cycles of reaction.

2.2.2 *L-Proline-Derived DENDRI-POM Hybrids*

Two enantiopure DENDRI-POM salts were prepared by electrostatic assembly of L-proline-derived dendron **44** with the Venturello trianion $[PO_4\{WO(O_2)_2\}_4]^{3-}$

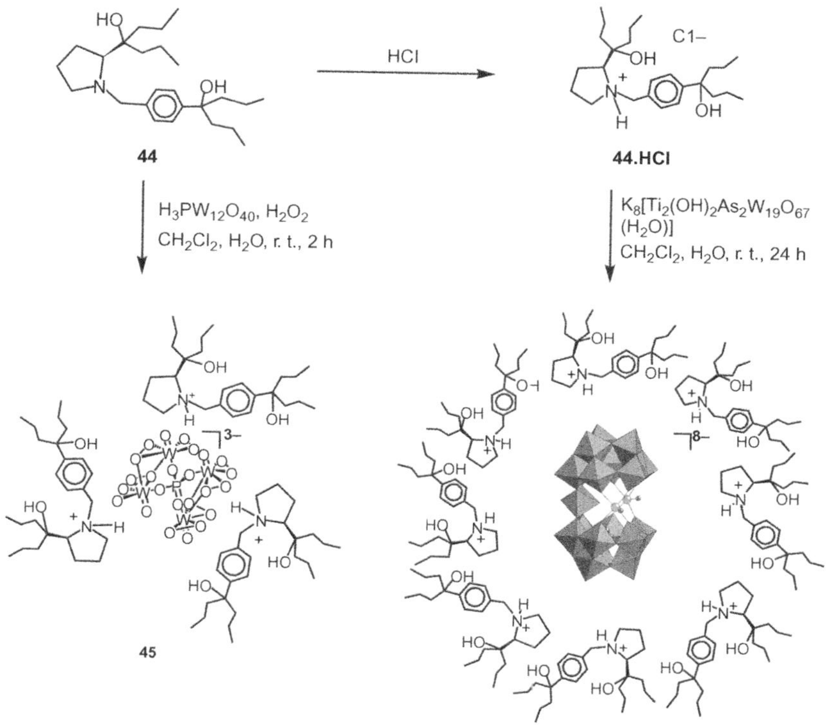

FIG. 2.27 L-Proline-derived DENDRI-POM hybrids. *(Reprinted from C. Jahier, S. Nlate, Eur. J. Inorg. Chem. (2012) 833–840. Copyright 2012, with permission of John Wiley and Sons and Copyright Clearence Center.)*

resulting in the 12-armed *n*-propyl DENDRI-POM hybrid **45**, and the coupling of chloride salt **44.HCl** with the dititanium(IV) containing 19-tungstodiarsenate(III) $[Ti_2(OH)_2AsW_{19}O_{67}(H_2O)]^{8-}$ results in the 32-armed *n*-propyl DENDRI-POM hybrid **46** (Fig. 2.27).

The structure of DENDRI-POM **46** in which the titanium polyanion $[Ti_2(OH)_2(As_2W_{19}O_{67}(H_2O)]^{8-}$ is surrounded by eight tetra-*n*-propyl pyrrolidine cations was confirmed by NMR spectroscopy, IR, and elemental analysis.

Evaluation of L-Proline-Derived DENDRI-POMs in Oxidation Reactions:

The catalytic efficiency of L-Proline-derived DENDRI-POMs has been evaluated in the oxidation of thioanisole **14a** and selected aromatic alkenes, in the presence of H_2O_2 (Fig. 2.28). DENDRI-POM **45** based on the Venturello trianion oxidizes thioanisole **14a** to the corresponding sulfoxide **15** with up to 4% *ee*, with a small amount of sulfone **15a**. Whereas the reactivity of **45** toward thioanisole **14a** decreased at low −50°C, no significant change was observed in

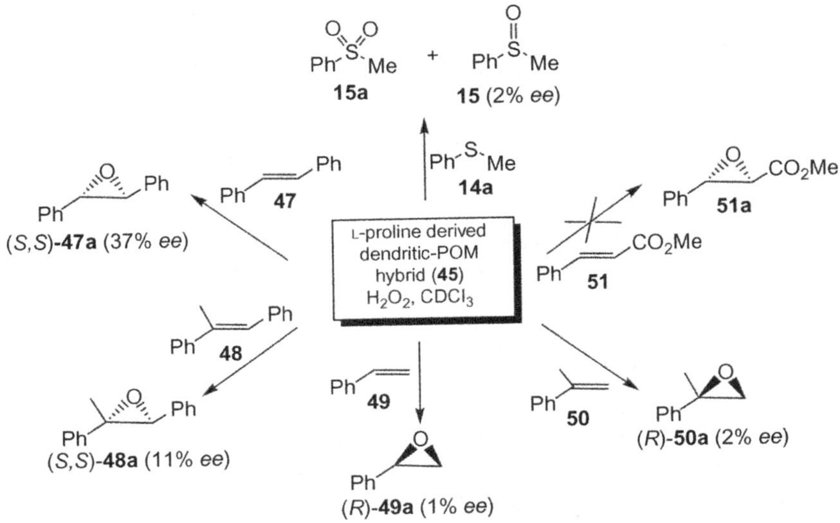

FIG. 2.28 Catalytic oxidation of thioanisole and selected alkenes by L-proline derived DENDRI-POM **45** with H_2O_2. *(Reprinted from C. Jahier, S. Nlate, Eur. J. Inorg. Chem. (2012) 833–840. Copyright 2012, with permission of John Wiley and Sons and Copyright Clearence Center.)*

the *ee* values. These results contrast with those obtained with DENDRI-POMs built from 1-phenylethylamine-based dendrons. The latter oxidized **14a** with up to 14% *ee*, and the *ee* values are sensitive on the reaction temperature. No significant difference was observed in the oxidation of sulfide **14a**, with hybrid **45** bearing the Venturello polyanion *versus* POM hybrid **46** composed with the dititanium-containing polyanion.

DENDRI-POM hybrids **45** and **46** were also evaluated in the asymmetric oxidation of selected aromatic alkenes **47–51** showed in Fig. 2.28. DENDRI-POM hybrid **45** oxidizes these aromatic alkenes to the corresponding *trans*-stilbene oxide **47a**, α-*trans*-methylstilbene oxide **48a**, phenyloxirane **49a**, and 2- methyl-2-phenyloxirane **50a**, if the reaction was performed at 50°C. Enantiomeric excesses of 37% and 11% were respectively obtained in the oxidation of **47** and **48**, with POM **45**. The enantioselectivity was not dependent on the reaction temperature, contrary to the oxidation of sulfides. Alkenes **49** and **50** were inert at room temperature. Significant substrate conversion was obtained by heating the reaction mixture to 50°C, with unfortunately no significant enantioselectivity (up to 2%). Alkene **51** bearing a withdrawing substituent is not oxidized under these conditions, and the epoxide **51a** was not obtained, even after modifications of reaction parameters. These results indicate that the reaction kinetics and the enantioselectivity are controlled by the electronic structure and the bulkiness of the alkene substrates. Contrary to the hybrid **45**, the dititanium polyanion-based hybrid **46** is inert toward oxidation of alkenes in similar conditions. These results contrast to those of Kholdeeva and Kortz, in which alkenes were oxidized with the alkyl ammonium salt of $[Ti_2(OH)_2(As_2W_{19}O_{67}(H_2O)]^{8-}$, confirming that the properties of POM anions are closely dependent on the nature of the associated cations [22].

Effect of Solvents, Catalysts, and Oxidant Concentration on Oxidation Reactions:

The catalytic oxidation of *trans*-stilbene **47** with DENDRI-POM **45** was performed in various solvents (chloroform, dichloromethane, toluene, acetonitrile, acetone, and tetrahydrofuran), in the presence of H_2O_2 at 50°C. The result obtained shows a dramatic influence of solvents' polarities on the reaction kinetics and the enantioselectivity. In these conditions, the *trans*-stilbene **47** was not oxidized in CH_3CN, acetone, and THF, whereas in $CDCl_3$, CH_2Cl_2, and toluene, the epoxide **47a** was obtained with *ee* values ranging from 19% to 37%. The reaction seemed favorable in less polar and noncoordinating solvents. It becomes apparent that out of all solvents evaluated in the oxidation of sulfides and alkenes with POM **45**, the reaction in $CDCl_3$ was found to be accelerated to some extent compared with that in CH_2Cl_2 and toluene, and the *ee* was increased in CH_2Cl_2 and toluene to 19% and 28%, respectively; it increased to 37% in $CDCl_3$ [8m]. The influence of the catalyst concentration showed that increasing the catalyst concentration increased the reaction kinetics without affecting the *ee* values. By varying the oxidant/substrate ratio, no significant change was observed in the reaction kinetics and the enantioselectivity.

Recovery and Reuse of Enantiopure DENDRI-POM Hybrids:

In order to evaluate the recovery and reuse abilities of L-proline-based DENDRI-POM **45**, four catalytic cycles were performed in the oxidation of *trans*-stilbene **47**. At room temperature, *trans*-stilbene **47** was oxidized to the corresponding sulfoxide without any discernible loss in activity and enantioselectivity, whereas at 50°C, the conversion and the yield of the recovered catalyst decreased significantly after the first cycle, due to its slow decomposition in the reaction mixture. However, *ee* values were not affected in these conditions.

2.2.3 Other Examples of Reported Chiral Organic-Inorganic POM-Based Systems

Chiral organic cations have been used by Wu and coworkers to encapsulate a catalytically active polyoanion $Na_{12}[WZn_3(H_2O)_2(ZnW_9O_{34})_2]$ through electrostatic interaction. The obtained chiral POM hybrids self-assembled into spherical supramolecular nanostructures with a diameter of c.100 nm, efficient catalysts in the oxidation of sulfides with up to 72% *ee* [23].

Two dendritic poly(urethane amines) (PUAs) have been grafted to a modified Anderson-type POM cluster $(Bu_4N^+)_3[(MnMo_6O_{18})^{3-}[(OCH_2)_3CNH_2]_2]$

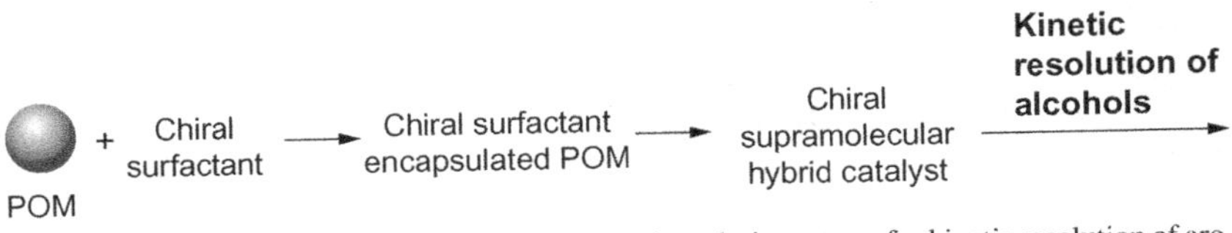

FIG. 2.29 Schematic representation of POM-based catalytic systems for kinetic resolution of aromatic alcohols reported by Wu. *(Adapted from L. Shi, Y. Wang, B. Li, L. Wu, Dalton Trans. 43 (2014) 9177–9188, with permission of the Royal Society of Chemistry.)*

through covalent bonds by Wang and coworkers, with the aim to develop robust POM-containing hybrid organogels of self-assembled hybrid nanofiber [24].

Wu and coworkers have used chiral surfactants bearing two long alkyl chains with hydroxyl groups at their terminal for the encapsulation of catalytically active POMs as illustrated in Fig. 2.29. The obtained chiral surfactant-encapsulated POM was covalently immobilized into a silica matrix via a sol-gel process to result in chiral supramolecular hybrid catalysts (CSHCs) [25]. Theses catalysts are efficient in the oxidative kinetic resolution of aromatic secondary alcohols with up to 89% *ee*. The covalent grafting of the chiral POM complex onto silica matrix seems to be one of the factors that promote enantioselectivity with these processes.

3 CONCLUSIONS

In the past decade, the concept of green chemistry and sustainable development has become a strategic focus in both industrial and academic research. Thus, stabilization of a catalyst through its encapsulation into matrixes has quickly become more common. In this context, DENDRI-POM hybrids represent an increasingly important class of environmentally benign catalysts for oxidation reactions. It has been demonstrated that the encapsulation of POM catalysts into dendritic structures has an important effect on the properties of the POM, especially its stability. The synthesis of DENDRI-POM hybrids by electrostatic coupling of POM anions and dendritic cations has been largely studied. The encapsulation of a POM unit into the dendritic structure increases its stability and efficiency toward oxidation reactions. Within the encapsulated DENDRI-POM hybrids studied, POM-cored dendrimers are the most stable. These results indicate that the properties of POM catalysts can be tuned by modifying their local environment. For catalytic systems reported so far, better *ee* values have been obtained with nondendritic systems by varying the structure of the organic cation around the POM catalyst, or by immobilizing chiral complexes into inorganic supports. Thus, new encapsulated DENDRI-POM catalysts have to be developed with the aim to improve the enantioselectivity through strong interactions between the POM unit and chiral dendritic structures.

ACKNOWLEDGMENTS

We acknowledge the Agence National de la Recherche (France), the University of Bordeaux (France), the Centre National de la Recherche Scientifique (France), and the European Cooperation in Science and Technology (COST D40 action) for supporting the research described in this chapter. We would like to thank Professor Ulrich Kortz of Bremen University for the generous donation of polyoxometalate salts.

REFERENCES

[1] (a) M.T. Pope, A. Müller, Angew. Chem. Int. Ed. Engl. 30 (1991) 34–48. (b) J.T. Rhule, C.L. Hill, D.A. Judd, R.F. Schinazi, Chem. Rev. 98 (1998) 327–358. (c) D.E. Katsoulis, Chem. Rev. 98 (1998) 359–388. (d) M.T. Pope, A. Müller (Eds.), Polyoxometalate Chemistry: From Topology via Self-Assembly to Applications, Kluwer Academic, Dordrecht, 2001.

[2] (a) C.L. Hill, Special thematic issues, polyoxometalates, Chem. Rev. 98 (1998) 1–389. (b) A. Proust, R. Thouvenot, P. Gouzerh, Chem. Commun. 16 (2008) 1837–1852. (c) D.L. Long, E. Burkholder, L. Cronin, Chem. Soc. Rev. 36 (2007) 105–121. (d) D. Long, R. Tsunashima, L. Cronin, Angew. Chem. Int. Ed. 49 (2010) 1736–1758. (e) P. Mialane, A. Dolbecq, F. Sécheresse, Chem. Commun. 33 (2006) 3477–3485. (f) A. Dolbecq, E. Dumas, C.R. Mayer, P. Mialane, Chem. Rev. 110 (2010) 6009–6048.

[3] (a) D.A. Judd, J.H. Nettles, N. Nevins, J.P. Snyder, D.C. Liotta, J. Tang, J. Ermolieff, R.F. Schinazi, C.L. Hill, J. Am. Chem. Soc. 123 (2001) 886–897. (b) S. Shigeta, S. Mori, E. Kodama, J. Kodama, K. Takahashi, T. Yamase, Antivir. Res. 58 (2003) 265–271. (c) X. Wang, J. Liu, M.T. Pope, Dalton Trans. 5 (2003) 957–960. (d) Y. Tajima, Microbiol. Immunol. 47 (2003) 207–212.

[4] (a) M. Sadakane, E. Steckhan, Chem. Rev. 98 (1998) 219–238. (b) C. Fleming, D.-L. Long, N. McMillan, J. Johnston, N. Bovet, V. Dhanak, N. Gadegaard, P. Kögerler, L. Cronin, M. Kadodwala, Nat. Nanotechnol. 3 (2008) 229–233.

[5] (a) D.G. Kurth, P. Lehmann, D. Volkmer, H. Cölfen, M.J. Koop, A. Müller, A.Du. Chesne, Chem. Eur. J. 6 (2000) 385–393. (b) Y. Yang, Y. Wang, H. Li, W. Li, L. Wu, Chem. Eur. J. 16 (2010) 8062–8071. (c) S. Nlate, C. Jahier, Eur. J. Inorg. Chem. (2013) 1606–1619.

[6] (a) N. Mizuno, K. Yamaguchi, K. Kamata, Coord. Chem. Rev. 249 (2005) 1944–1956. (b) W. Zhu, H. Li, X. Jiang, Y. Yan, J. Lu, L. He, J. Xia, Green Chem. 10 (2008) 641–646.

[7] (a) D. Astruc, F. Chardac, Chem. Rev. 101 (2001) 2991–3023. (b) G.E. Oosterom, J.N.H. Reek, P.C.J. Kramer, P.W.N.M. van Leeuwen, Angew. Chem. Int. Ed. 40 (2001) 1828–1849. (c) R. Kreiter, A. Kleij, R.J.M. Klein Gebbink, G. van Koten, Top. Curr. Chem. 217 (2001) 163–199. (d) R. van Heerbeek, P.C.J. Kamer, P.W.N.M. van Leeuwen, J.N.H. Reek, Chem. Rev. 102 (2002) 3717–3756. (e) L.J. Twyman, A.S.H. King, I.K. Martin, Chem. Soc. Rev. 31 (2002) 69–82.

[8] (a) H. Zeng, G.R. Newkome, C.L. Hill, Angew. Chem. Int. Ed. 39 (2000) 1771–1774. (b) D. Volkmer, B. Bredenkötter, J. Tellenbröker, P. Kögerler, D.G. Kurth, P. Lehmann, H. Schnablegger, D. Schwahn, M. Piepenbrink, B. Krebs, J. Am. Chem. Soc. 124 (2002) 10489–10496. (c) M.C. Rogers, B. Adisa, D.A. Bruce, Catal. Lett. 98 (2004) 29–36. (d) L. Plault, A. Hauseler, S. Nlate, D. Astruc, J. Ruiz, S. Gatard, R. Neumann, Angew. Chem. Int. Ed. 43 (2004) 2924–2928. (e) M.V. Vasylyev, D. Astruc, R. Neumann, Adv. Synth. Catal. 347 (2005) 39–44. (f) S. Nlate, D. Astruc, R. Neumann, Adv. Synth. Catal. 346 (2004) 1445–1448. (g) S. Nlate, L. Plault, D. Astruc, Chem. Eur. J. 12 (2006) 903–914. (h) S. Nlate, L. Plault, D. Astruc, New J. Chem. 31 (2007) 1264–1274. (i) C. Jahier,

L. Plault, S. Nlate, Isr. J. Chem. 49 (2009) 109–118. (j) C. Jahier, M. Cantuel, N.D. McClenaghan, T. Buffeteau, D. Cavagnat, F. Agbossou, M. Carraro, M. Bonchio, S. Nlate, Chem. Eur. J. 15 (2009) 8703–8708. (k) C. Jahier, S.S. Mal, U. Kortz, S. Nlate, Eur. J. Inorg. Chem. (2010) 1559–1566. (l) C. Jahier, M.-F. Coustou, M. Cantuel, N.D. McClenaghan, T. Buffeteau, D. Cavagnat, M. Carraro, S. Nlate, Eur. J. Inorg. Chem. (2011) 727–738. (m) C. Jahier, S. Nlate, Eur. J. Inorg. Chem. (2012) 833–840. (n) C. Jahier, S.S. Mal, R. Al-Oweini, U. Kortz, S. Nlate, Polyhedron 57 (2013) 57–63. (o) C. Jahier, R. Touzani, S. El Kadiri, S. Nlate, Inorg. Chim. Acta 450 (2016) 81–86.

[9] (a) V. Sartor, L. Djakovitch, J.-L. Fillaut, F. Moulines, F. Neveu, V. Marvaud, J. Guittard, J.-C. Blais, D. Astruc, Am. Chem. Soc. 121 (1999) 2929–2930. (b) V. Sartor, S. Nlate, J.-L. Fillaut, L. Djzakovitch, F. Moulines, V. Marvaud, F. Neveu, J.-C. Blais New, J. Chem. Theory Comput. 24 (2000) 351–370.

[10] S. Nlate, L. Plault, F.-X. Felpin, D. Astruc, Adv. Synth. Catal. 350 (2008) 1419–1424.

[11] F.P. Ballistreri, A. Bazzo, G.A. Tomaselli, R.M. Toscano, J. Org. Chem. 57 (1992) 7074–7077.

[12] C. Jahier, F.-X. Felpin, C. Meliet, F. Agbossou-Niedercorn, J.-C. Hierso, S. Nlate, Eur. J. Inorg. Chem. (2009) 5148–5155.

[13] Y. Yang, Y. Wang, H. Li, W. Li, L. Wu, Chem. Eur. J. 16 (2010) 8062–8071.

[14] B. Liu, J. Yang, M. Yang, Y. Wang, N. Xia, Z. Zhang, P. Zheng, W. Wang, I. Lieberwirth, C. Kübel, Soft Matter 7 (2011) 2317–2320.

[15] (a) S.S. Mal, N.H. Nsouli, M. Carraro, A. Sartorel, G. Scorrano, H. Oelrich, L. Walder, M. Bonchio, U. Kortz, Inorg. Chem. 49 (2010) 7–9. (b) B.S. Bassil, S.S. Mal, M.H. Dickman, U. Kortz, H. Oelrich, L. Walder, J. Am. Chem. Soc. 130 (2008) 6696–6697.

[16] S. Bräse, F. Lauterwasser, R.E. Ziegert, Adv. Synth. Catal. 346 (2003) 869–929.

[17] (a) D.A.J. Judd, H. Nettles, N. Nevins, J.P. Snyder, D.C. Liotta, J. Tang, J. Ermolieff, R.F. Schinazi, C.L. Hill, J. Am. Chem. Soc. 123 (2001) 886–897. (b) A. Müller, F. Peters, M.T. Pope, D. Gattesschi, Chem. Rev. 98 (1998) 239–271. (c) B. Hasenknopf, K. Micoine, E. Lacôte, S. Thorimbert, M. Malacria, R. Thouvenot, Eur. J. Inorg. Chem. 32 (2008) 5001–5013 (references cited therein). (d) M. Carraro, A. Sartorel, G. Scorrano, C. Maccato, M.H. Dickman, U. Kortz, M. Bonchio, Angew. Chem. 120 (2008) 7385–7389. Angew. Chem. Int. Ed. 47 (2008) 7275–7279. (e) C. Boglio, B. Hasenknopf, G. Lenoble, P. Rémy, P. Gouzerh, S. Thorimbert, E. Lacôte, M. Malacria, R. Thouvenot, Chem. Eur. J. 14 (2008) 1532–1540.

[18] (a) M. Inoue, T. Yamase, Bull. Chem. Soc. Jpn. 68 (1995) 3055–3063. (b) F.B. Xin, M.T. Pope, J. Am. Chem. Soc. 118 (1996) 7731–7736. (c) M. Inoue, T. Yamase, Bull. Chem. Soc. Jpn. 69 (1996) 2863–2868. (d) D.-L. Long, P. Kögerler, L.J. Farrugia, L. Cronin, Chem. Asian. J. 1 (2006) 352–357. (e) D.-L. Long, E. Burkholder, L. Cronin, Chem. Soc. Rev. 36 (2007) 105–121. (f) C. Streb, D.-L. Long, L. Cronin, Chem. Commun. (2007) 471–473. (g) U. Kortz, M.G. Savelieff, F.Y.A. Ghali, L.M. Khalil, S.A. Maalouf, D.I. Sinno, Angew. Chem. 114 (2002) 4246–4249. Angew. Chem. Int. Ed. 41 (2002) 4070–4073. (h) X.K. Fang, T.M. Anderson, C.L. Hill, Angew. Chem. 117 (2005) 3606–3610. Angew. Chem. Int. Ed. 44 (2005) 3540–3544. (i) X.K. Fang, T.M. Anderson, Y. Hou, C.L. Hill, Chem. Commun. (2005) 5044–5046. (j) L.M. Zheng, T. Whitfield, X. Wang, A.J. Jacobson, Angew. Chem. 112 (2000) 4702–4705. Angew. Chem. Int. Ed. 39 (2000) 4528–4531. (k) H.Y. An, E.B. Wang, D.R. Xiao, Y.G. Li, Z.M. Su, L. Xu, Angew. Chem. 118 (2006) 918–922. Angew. Chem. Int. Ed. 45(2006) 904–908. (l) H.Q. Tan, Y.G. Li, Z.M. Zhang, C. Qin, X.L. Wang, E.B. Wang, Z.M. Su, J. Am. Chem. Soc. 129 (2007) 10066–10067.

[19] (a) Y. Hou, X. Fang, C.L. Hill, Chem. Eur. J. 13 (2007) 9442–9447. (b) Y.Q. Lan, S.L. Li, M. Su, K.Z. Shao, J.F. Ma, X.L. Wang, E.B. Wang, Z.M. Su, Chem. Commun. (2008)

58–60. (c) Y.Q. Lan, S.L. Li, X.L. Wang, K.Z. Shao, D.Y. Du, Z.M. Su, E.B. Wang, Chem. Eur. J. 14 (2008) 9999–10006.
[20] M. Carraro, G. Modugno, A. Sartorel, G. Scorrano, M. Bonchio, Eur. J. Inorg. Chem. (2009) 5164–5174.
[21] A. Tézé, J. Canny, L. Gurban, R. Thouvenot, H. Hervé, Inorg. Chem. 35 (1996) 1001–1005.
[22] (a) F. Hussain, B.S. Bassil, U. Kortz, O.A. Kholdeeva, M.N. Timofeeva, P. De Oliveira, B. Keita, L. Nadjo, Chem. Eur. J. 13 (2007) 4733–4742. (b) O.A. Kholdeeva, B.G. Donoeva, T.A. Trubitsina, G. Al-Kadamany, U. Kortz, Eur. J. Inorg. Chem. 34 (2009) 5134–5141. (c) B.G. Donoeva, T.A. Trubitsina, N.S. Antonova, J.J. Carbo, J.M. Poblet, G. Al-Kadamany, U. Kortz, O.A. Kholdeeva, Eur. J. Inorg. Chem. 34 (2010) 5312–5317. (d) N.S. Antonova, J.J. Carbo, U. Kortz, O.A. Kholdeeva, J.M. Poblet, J. Am. Chem. Soc. 132 (2010) 7488–7497.
[23] Y. Wang, H. Li, W. Qi, Y. Yang, Y. Yan, B. Li, L. Wu, J. Mater. Chem. 22 (2012) 9181–9188.
[24] B. Liu, J. Yang, M. Yang, Y. Wang, N. Xia, Z. Zhang, P. Zheng, W. Wang, I. Liberwirth, C. Kübel, Soft Matter (2011) 2317–2320.
[25] L. Shi, Y. Wang, B. Li, L. Wu, Dalton Trans. 43 (2014) 9177–9188.

FURTHER READING

(a) T.A. Sullens, R.A. Jensen, T.Y. Shvareva, T.E. Albrecht-Schmitt, J. Am. Chem. Soc. 126 (2004) 2676–2677. (b) S. Bareyt, S. Piligkos, B. Hasenknopf, P. Gouzerh, E. Lacôte, S. Thorimbert, M. Malacria, Angew. Chem. 115 (2003) 3526–3528. Angew. Chem. Int. Ed. 42 (2003) 3404–3406. (c) M. Sadakane, M.H. Dickman, M.T. Pope, Inorg. Chem. 40 (2001) 2715–2719. (d) V. Soghomonian, Q. Chen, R.C. Haushalter, J. Zubieta, C.J. O'Connor, Science 259 (1993) 1596–1599. (e) H.Y. An, D.R. Xiao, E.B. Wang, Y.G. Li, L. Xu, New J. Chem. 29 (2005) 667–672.

Chapter 3

Catalysts Encapsulated in Nanostructured Carbon Systems

Zheye Zhang and Shuai Wang
Huazhong University of Science and Technology, Wuhan, China

Contents

1 INTRODUCTION

Catalysts are substances that can accelerate chemical reactions or improve reaction selectivity without consuming themselves [1,2]. Currently, catalysts are widely used in organic synthesis, environmental protection, and energy related systems [3–8], which are of direct relevance to many aspects of our daily life. Since most catalysis is carried out by solids and the reaction occurs only when the reactants are adsorbed onto the catalyst surface, the total surface area of the solid active phase has a substantial effect on the reaction rate. With the emergence of "nanotechnology," great efforts have been made in the field of catalysis to engineer the structural details of catalysts, including their size, shape, and surface properties, needed to properly tune their catalytic performance [9–13]. Designed nanostructures of transition metals (TMs) with well-defined sizes and shapes, for instance, have already been recognized as promising catalysts with enhanced selectivity [14–17]. However, the high surface energy of metal nanoparticles (NPs) makes them easily agglomerated or shape-changed during catalytic reactions, which results in the dramatic decrease of their activity and selectivity [18–20]. For example, AuNPs dispersed on high-surface-area supports have been shown to

Encapsulated Catalysts. http://dx.doi.org/10.1016/B978-0-12-803836-9.00003-1

be quite active in many low-temperature reactions, including carbon monoxide and hydrocarbon oxidations, but they also lose their activity over time as they sinter into larger particles [21]. Therefore, it is highly desirable to use new approaches to prevent such agglomeration, for instance, partially encapsulating the dispersed metal NPs with additional porous materials [22,23].

An alternative approach for the protection of catalytically active NPs is to develop them into core-shell nanostructures[24–28]. These nanostructures are NPs encapsulated and protected by an outer shell that isolates the NPs and prevents their migration and coalescence during the catalytic reactions. Efficient core-shell catalysts require porous shells that allow free access of chemical species from the outside to the surface of nanocatalysts. For this purpose, great effort has been made in developing various porous shell materials, including polymers[29–31], carbons [32–36], mesoporous silica ($mSiO_2$) [37–39], and metal oxides[40–43]. Among these encapsulating materials, carbon-based materials have attracted tremendous attention as a result of the integration of their excellent conductivity for electron capture and transport, the large specific surface area, and their unique interaction with catalyst particles [32–36,44–49]. Furthermore, the carbon shells are anticipated to provide an intriguing confinement environment for metal catalysts and catalytic reactions [49,50]. Hence, we believe there is a need to summarize recent developments in catalysts encapsulated in nanostructured carbon systems. In this chapter, we review the recent advancements on the synthesis and applications of catalysts encapsulated in nanostructured carbon systems. In particular, the role of the nanostructured carbon in boosting the catalytic performance is highlighted.

2 SYNTHESIS AND PROPERTIES OF THE CATALYSTS

Carbon plays a key role in the earth's civilizations. It can form strong bonds in different forms. In the last three decades, a lot of new carbon nanomaterials were discovered, including fullerene, carbon nanotube (CNT), graphene, amorphous porous carbon, and so on. These carbon nanomaterials exhibit promising applications in the field of catalysis. Metal NPs encapsulated in carbon cages are promising highly active nanostructures with high chemical and thermal stability, because the encapsulating shell protects the metal particle from atmospheric-oxygen-induced degradation and agglomeration of neighboring NPs. In this section, we provide a brief survey of the synthesis and properties of three widely explored nanostructured carbon (i.e., CNT, graphene, and amorphous porous carbon) encapsulated catalysts.

2.1 Catalysts Encapsulated in CNT

CNT is a representative one-dimensional structure allotrope of carbon. Since the discovery in 1991 [51], CNT has been a superstar in nanomaterials by attracting a great deal of interest in both academy and industry due to its remarkable properties and potential applications in a wide variety of fields. According to the number of layers, single-walled CNT (SWCNT), double-walled CNT (DWCNT), and multiwalled CNT (MWCNT) are distinguished. Their high electron and thermal conductivity, large surface area, and functionalizable surfaces have attracted wide interest for catalytic applications [52,53]. For example, nitrogen-doped CNT was reported as a metal-free catalyst for the aerobic oxidation of cyclohexane in the liquid phase [54]. As a catalyst support, CNT was widely studied as an alternative for dispersion of TMs on their exterior walls for various catalytic reactions [55–58]. However, catalysis utilizing the interior surface of CNT has been less explored, although theoretical studies predicted that chemical reactions may be influenced inside such an intriguing confined environment [59,60].

The efforts to fill the CNT channels started shortly after their discovery and several techniques were developed including in situ filling during arc discharge, vapor deposition of volatile complexes, and wet chemistry methods [61–65]. Among these synthetic methods, wet chemistry methods are simple and most versatile for catalysis since they are applicable to most metals. Incipient wetness impregnation techniques rely on the filling of oxidized CNTs by capillarity with a solution containing a metal salt [66–70] or preformed NPs [71–73]. X-ray photoelectron spectroscopy [66] and electron tomography (3D TEM) [67] studies have indicated that the fraction of metal present in the inner cavity of CNTs is between 15% and 50%. However, the main drawback of these methods is that a significant amount of precursor is lost during the synthesis. To overcome this problem, Serp and coworkers reported a simple and efficient method for the selective confinement of NPs in the inner cavity of CNTs, and the effect of spatial confinement of bimetallic PtRu NPs on their catalytic performance for the selective hydrogenation of cinnamaldehyde. The strategy developed to selectively confine NPs inside CNTs was based on surface chemistry (Fig. 3.1)

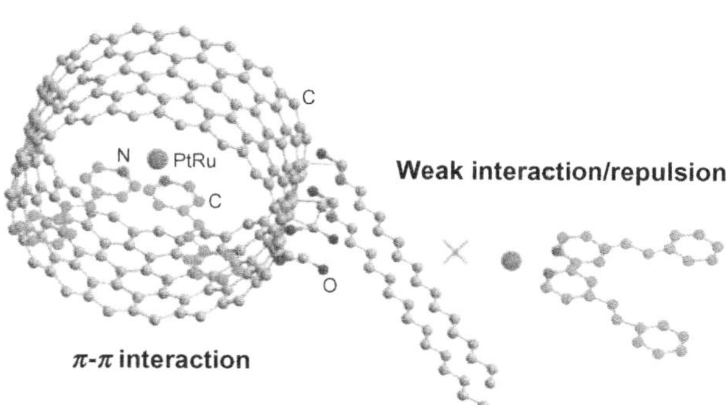

FIG. 3.1 Strategy adopted to drive NPs into CNTs. *(Reproduced from E. Castillejos, P. J. Debouttiere, L. Roiban, A. Solhy, V. Martinez, Y. Kihn, O. Ersen, K. Philippot, B. Chaudret, P. Serp, Angew. Chem. Int. Ed. 48 (2009) 2529–2533, with the permission of Wiley.)*

[64]. The first step consists of the controlled preparation by an organometallic route of NPs stabilized by a ligand possessing two functionalities: one has an affinity for the surface of the NPs and the other presents an affinity for the CNT graphene layers (π-π interaction). The second step is the functionalization of the CNT external surface to introduce different surface species that should induce weak interaction or repulsion between the NPs and CNT external surface. According to this method, 5–23 wt% PtRu@CNT samples could be successfully prepared. PtRu NPs of approximately 2 nm were uniformly confined in segmented CNTs of 40 nm in internal diameter.

A general method has also been developed for the deposition of metal NPs selectively either inside or outside of CNTs (Fig. 3.2). This method was based on the difference in the liquid/solid interface energies of an aqueous and an organic solution with the CNT surface [65]. Because of their lipophilic character, the organic solvent better wets the surface of the nanotubes compared to water and penetrates into the inner volume. The precise control of the volume of each phase allows filling the CNT with the organic phase and covering its outer surface with the aqueous one. Hence, metal NPs can be put with high selectivity either inside or outside the CNT, just by choosing in which solvent

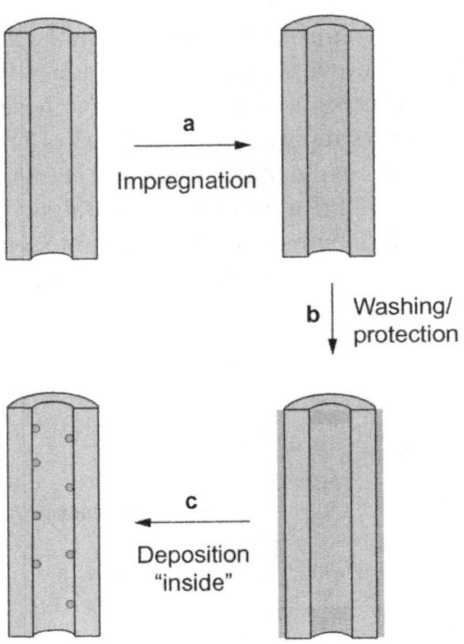

FIG. 3.2 Schematic view of a longitudinal cross-section of a CNT during the different steps for the selective deposition of NPs inside CNTs: (a) impregnation with the ethanolic solution containing the metal precursor; (b) impregnation with distilled water to wash and protect the outer surface from metal deposition; (c) after drying, calcination, and reduction, metal particles are decorating the inner surface of the CNT with a high selectivity. *(Reproduced from J.P. Tessonnier, O. Ersen, G. Weinberg, C. Pham-Huu, D.S. Su, R. Schlogl, ACS Nano 3 (2009) 2081–2089, with the permission of the American Chemical Society.)*

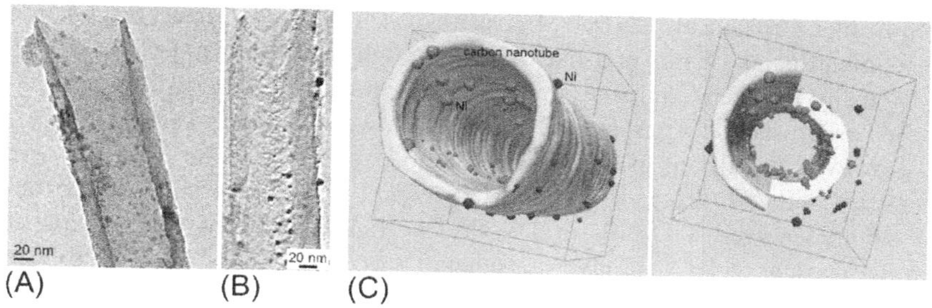

FIG. 3.3 (A) Typical 2D TEM image of the sample "metal inside," used to reconstruct its volume. (B) Longitudinal section through the reconstructed volume. (C) Modeling of the reconstructed volume: carbon nanotube, Ni particles inside the tube, Ni particles on the external surface. *(Reproduced from J.P. Tessonnier, O. Ersen, G. Weinberg, C. Pham-Huu, D.S. Su, R. Schlogl, ACS Nano 3 (2009) 2081–2089, with the permission of the American Chemical Society.)*

the metal precursor is dissolved. 3D TEM also validated the hypothesis of a high selectivity inside of the particles, as shown in Fig. 3.3. However, the weight loading of the filled NPs is generally low compared to the overall weight of the CNTs; thus it hindered their subsequent use in the field of catalysis. In this case, Baaziz et al. reported the development of a new liquid-phase filling of the CNT channel with magnetic cobalt-based NPs of uniform size and shape along with an extremely high loading, in other words, up to 60 wt% [74]. The Co NPs casted inside the CNTs were obtained by introducing the CNTs in an octadecene suspension containing a Co stearate complex and oleic acid as surfactant and by submitting this mixture to a thermal decomposition at 318°C.

Other approaches of dispersing metal NPs inside the CNT channels were mainly developed by Bao's group. As shown in Fig. 3.4, the freshly produced MWCNTs were opened at the ends and shortened via a controlled oxidation process catalyzed by Ag NPs. Metal NPs were uniformly distributed inside MWCNTs using ultrasonic treatment and extended stirring to aid the capillary forces [75]. Following the same method, Ru [76], Fe [77], and Rh NPs [49] have been successfully dispersed within MWCNTs too, demonstrating that this can be applied for many TMs. A facile and scalable method was also developed to

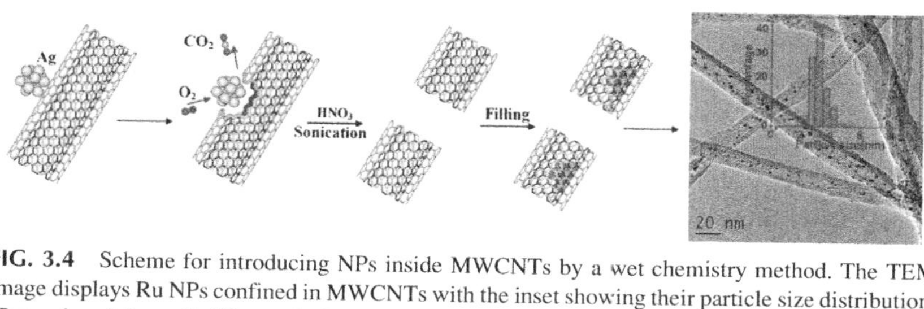

FIG. 3.4 Scheme for introducing NPs inside MWCNTs by a wet chemistry method. The TEM image displays Ru NPs confined in MWCNTs with the inset showing their particle size distribution. *(Reproduced from C. Wang, S. Guo, X. Pan, W. Chen, X. Bao, J. Mater. Chem. 18 (2008) 5782–5786, with the permission of the Royal Society of Chemistry.)*

confine metal NPs inside CNTs by Bao et al. [47]. They encapsulated Fe NPs into the compartments of pea-pod like CNTs through one-step synthesis at 350°C in N_2 using ferrocene and sodium azide as the precursors. Fe particles outside of the carbon shells were removed by thorough washing in an acid solution.

However, the filling of DWCNTs was not feasible with the described wet chemistry method. Instead, using $TiCl_4$ vapor, they successfully dispersed subnanometer-sized TiO_2NPs in DWCNTs. Nanotubes were first evacuated before exposure to $TiCl_4$. Anhydrous ethanol was used to allow slow hydrolysis of $TiCl_4$ under Ar atmosphere. Subsequently, the sample was treated with HF to remove titanium species on the outer walls and washed with deionized water. These NPs exhibited a rather good stability since no obvious aggregation was observed when the sample was heated up to 500°C [78]. To develop SWCNT-encapsulated metal NPs, Laasonen et al. reported a one-step floating catalyst (aerosol) chemical vapor deposition (CVD) synthesis process [79]. An aerosolized feedstock solution of ferrocene and thiophene in toluene was introduced into a high-temperature reactor to produce Fe catalyst NPs. Ethylene (C_2H_4) was utilized as a carbon source for the CNT synthesis because hydrocarbons have shown a high reaction yield, which is a key factor for industrial applications. Carbon dioxide was used as an etching agent that can remove amorphous carbon coated on the surfaces of Fe catalyst NPs. It can also prevent the formation of cementite (Fe_3C), which is believed to be an inactive phase for the growth of nanotubes.

2.2 Catalysts Encapsulated in Graphene

Graphene, a single layer of carbon atoms bonded together in a honeycomb crystal lattice, is the fundamental structure in the carbon world. It was discovered by Novoselov and Geim with a micromechanical exfoliation method in 2004 [80]. Similar to CNTs, graphene has also been used as an effective catalyst or as a support of other catalysts [81–84]. However, compared with other carbon nanomaterials, graphene has recently attracted the most intensive attention [85–87], which is mainly due to its several intrinsic advantages. First, graphene has an ultrahigh theoretical specific surface area of 2600 $m^2\ g^{-1}$, being twice that of SWCNTs and much higher than those of most carbon black and activated carbon (AC). These characteristics make graphene highly desirable for supporting or encapsulating catalysts [88–90]. Moreover, the locally conjugated structure endows graphene with enhanced adsorption capacities towards substrates in the catalytic reaction. Second, the excellent electron mobility of graphene facilitates the electron transfer during the catalytic reactions, improving its catalytic activity [91,92]. Third, graphene also has high thermal, chemical, and electrochemical stabilities, which can improve the lifetime of catalysts [93].

Single-atomic-layer graphene was reported to serve as a passivation layer to prevent some metal (Cu, Ni, etc.) from oxidization due to its diffusion suppression and chemical inertness to many oxidizing gas [94]. Furthermore, the work function (potential energy surface) of graphene can shift up and down from

0.15 to 1 eV on different metal substrates (Ni, Co, etc.), but the electronic structure of the graphene-coated metal is still dominated by the original metal [95,96]. On this basis, graphene is a perfect coating material to stabilize nanocatalysts without many changes to their initial catalytic properties, and it may even improve their performance especially under harsh conditions. In this part, we mainly report the synthesis of catalysts encapsulated in single or multilayer graphene nanosheets.

Although intensive works have been devoted to encapsulating catalysts into CNTs, however, the carbon shells previously used for these catalysts are too thick and usually consist of multilayer graphitic carbon or composites thereof, which may significantly reduce the catalytic activity because the electronic structure of the outermost carbon layer is only modulated by the electron transferred from the encapsulated metal core when the shell consists of no more than three to four carbon layers [97,98]. Therefore, it is essential to synthesize carbon-encapsulated catalysts with a controllable number of graphene layers, especially for catalysts with less than three layers of graphene. Recalling the graphene preparation, two major methods include the top-down exfoliation techniques and bottom-up, for example, CVD methods [80,99,100]. For instance, Chen and coworkers developed a novel 3D nanoporous catalyst by encapsulating RuO_2NPs using nanoporous N-doped graphene to stabilize the RuO_2NPs, as shown in Fig. 3.5. First, N-doped graphene was uniformly grown

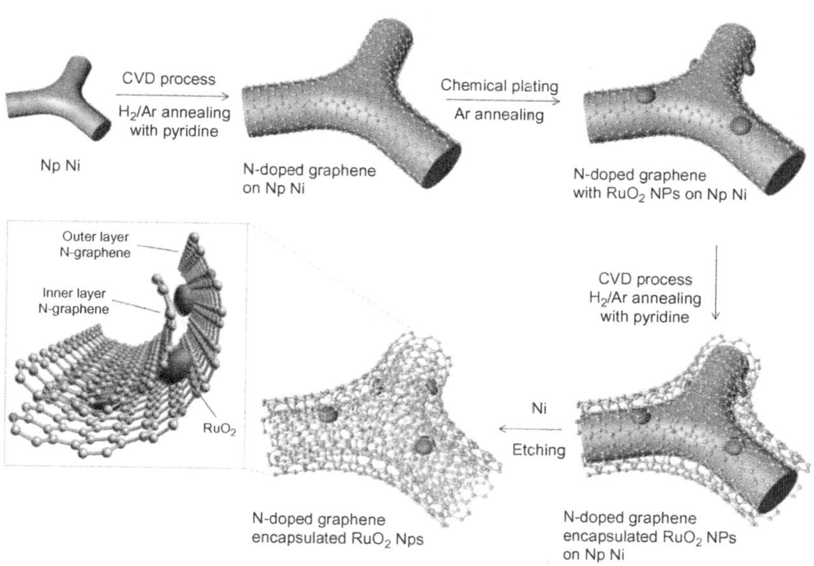

FIG. 3.5 Schematic illustration of the preparation process of nanoporous N-doped graphene with encapsulated RuO_2NPs, and the schematic of the structure of nanoporous N-doped graphene layers and RuO_2NPs. *(Reproduced from X. Guo, P. Liu, J. Han, Y. Ito, A. Hirata, T. Fujita, M. Chen, Adv. Mater. 27 (2015) 6137–6143, with the permission of Wiley.)*

on the nanoporous Ni ligament surface through the CVD method. Second, RuO_2NPs were chemically plated on the internal surface of nanoporous N-doped graphene using an $RuCl_3$-based organic solution as the Ru source and then the sample was annealed in an Ar gas atmosphere for the formation of RuO_2NPs. Third, the RuO_2 loaded nanoporous graphene with nanoporous Ni substrate was used as the template to grow one more layer N-doped graphene by CVD with the same growth condition as the first step. Finally, freestanding nanoporous N-doped graphene with encapsulated RuO_2NPs was exfoliated by etching away the Ni template in 2 M HCl solution, in which the RuO_2 was stabilized by the inner and outer layers of N-doped graphene [89]. Tan et al. also reported the synthesis of graphene-encapsulated Pt nanocrystal through a simple CVD method [101]. However, the CVD method requires a high temperature above 800°C to obtain graphene on Cu or Ni metal substrates, and these nanometal particles easily surface-melt and grow up at this temperature [99,100,102]. Therefore, a completely new synthetic route is needed to obtain graphene-coated TM NPs. Huang et al. reported a one-step simple route to prepare in situ graphene-encapsulated Fe NPs (Fe@G) via an arc-discharge method [90]. The iron NPs with 5–15 nm diameter are encapsulated by two to five layers of graphene (Fig. 3.6). The graphene layer outsides could protect the inner Fe NPs from oxidizing in air or etching by acid. However, metal NPs@graphene fabricated by the arc discharge method usually had large sizes and a broad distribution of particle sizes, in some cases empty carbon cages and CNTs were accompanied. To overcome this problem, Nagatsu et al. developed a simple one-step synthesis method of graphene encapsulated Ni (Ni@G) NPs using a long pulse laser to ablate an Ni target in methane or a mixture of methane and helium at room temperature [103]. By this method, a high yield of core-shell structure Ni@GNPs with a typical size of 5 nm and size distribution of 3–10 nm was fabricated.

Very recently, 3D TMs and their alloys encapsulated in graphene have been emerging as promising candidates for highly efficient catalysis. Bao and coworkers reported a bottom-up method for the preparation of ultrathin graphene spheres with only one to three graphene layers that encapsulate CoNi nanoalloy (CoNi@NC) electrocatalysts by using Co^{2+}, Ni^{2+}, and ethylene diaminetetraacetate anions ($EDTA^{4-}$) as precursors [45]. The graphene layers prevented the corrosion of 3D TMs from a harsh environment and simultaneously promote the catalytic reaction on a carbon surface due to the electron transfer from encapsulated 3D TMs. However, the electron transfer significantly relies on the work function of different metals and their interaction strength with graphene [95,96]. Therefore, it is imperative to develop a universal method to achieve the encapsulation of all kinds of metals. To this end, Bao's group introduced a facile and universal method for the preparation of single layer graphene encapsulating uniform 3D TMs, including Fe, Co, Ni, and their alloy NPs via a CVD process [48]. As depicted in Fig. 3.7, metal precursors were firstly filled into the channels of $mSiO_2$(SBA-15) by an impregnation method in methanol solution, and then the samples were heated to 700°C under H_2, followed by

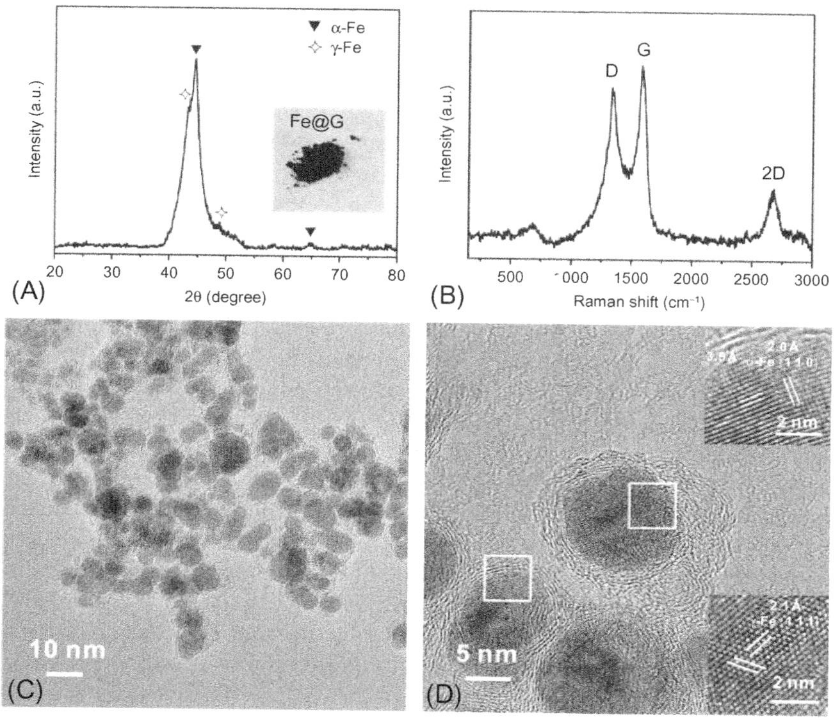

FIG. 3.6 (A) XRD pattern and (B) Raman spectrum of the Fe@G sample. (C, D) TEM and HRTEM images of the Fe@G sample. The insets are the enlarged HRTEM images of the nanostructures marked by *squares*. *(Reproduced from C. Wang, P. Zhai, Z. Zhang, Y. Zhou, J. Ju, Z. Shi, D. Ma, R.P.S. Han, F. Huang, Part. Part. Syst. Charact. 32 (2015) 29–34, with permission of Wiley.)*

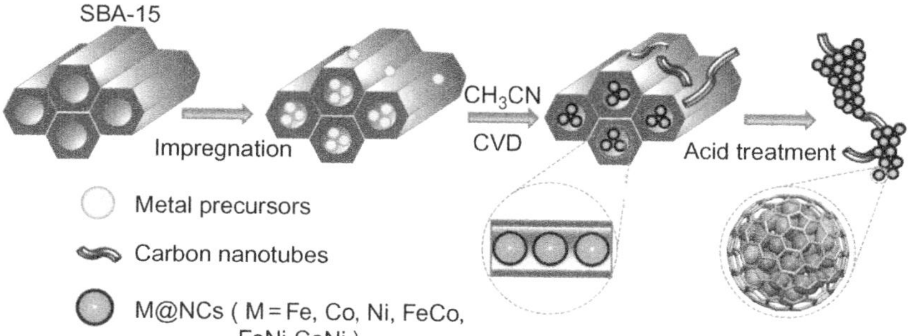

FIG. 3.7 Schematic illustration of the synthesis process of M@NCs from metal-containing precursors and SBA-15. *(Reproduced from X. Cui, P. Ren, D. Deng, J. Deng, X. Bao, Energy Environ. Sci. 9 (2016) 123–129, with the permission of the Royal Society of Chemistry.)*

CVD of the carbon source (acetonitrile) on the metal NPs formed at a high temperature. Due to the confinement of the SBA-15 channels, the metal precursors can firstly form uniform particles under high temperature, and single layer graphene will be formed between the metal NPs surface and the pore walls of SBA-15. Subsequently, a single-layer graphene will grow on the surface of the metal NPs in the channels of SBA-15. The final single-layer graphene encapsulating metal NPs on the interlaced CNTs can be obtained after etching SBA-15 and the bare metals with an HF solution. Through this method, a series of single-layer graphene encapsulating 3D TMs (M@NCs) can be easily prepared.

However, nearly all previous studies used complex and expensive CVD methods or multistep synthesis processes, hindering the large scale commercialization. Chen et al. adopted a facile synthesis of an FeCo alloy encapsulated in highly nitrogen-doped graphene layers by one-step annealing of $Fe_3[Co(CN)_6]_2$ NPs [104]. As depicted in Fig. 3.8, $Fe_3[Co(CN)_6]_2$ spheres were synthesized and used as metal-organic framework (MOF) precursors for further annealing under a nitrogen flow without adding any other carbon sources. During the annealing process, Fe and Co atoms from the precursor will form FeCo alloy NPs under high temperatures, meanwhile some CN^- group linkers will serve as both carbon and nitrogen sources for the formation of nitrogen-doped graphene layers outside the alloy particles. The in situ formed graphene layers will avoid the agglomeration of FeCo NPs to provide more active sites for catalysis. Moreover, the nitrogen-doped graphene shells will also protect the alloy core from corrosion so as to enhance the durability of the catalyst. Other metals or alloy materials coated with a graphene shell can be also obtained by this process via simply changing the metal ion centers or ligands of the MOF precursor.

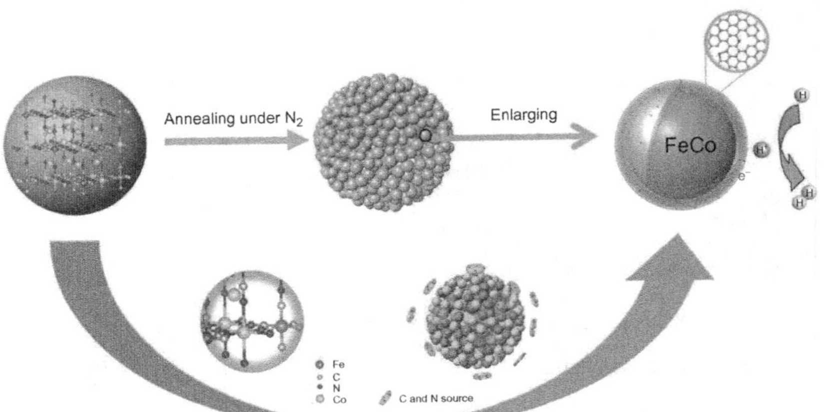

FIG. 3.8 The synthetic route and model of the FeCo alloys encapsulated in nitrogen-doped graphene layers. *(Reproduced from Y. Yang, Z. Lun, G. Xia, F. Zheng, M. He, Q. Chen, Energy Environ. Sci. 8 (2015) 3563–3571, with the permission of the Royal Society of Chemistry.)*

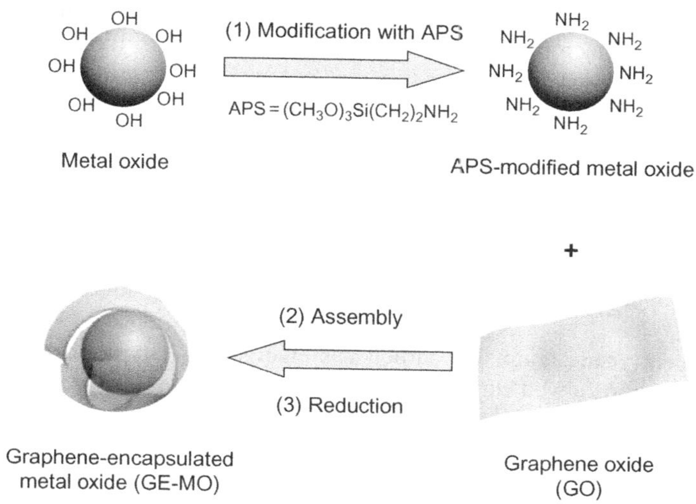

FIG. 3.9 Schematic illustration of the synthesis process of graphene-encapsulated metal oxide. *(Reproduced from S. Yang, X. Feng, S. Ivanovici, K. Mullen, Angew. Chem. Int. Ed. 49 (2010) 8408–8411, with the permission of Wiley.)*

Other approaches of encapsulating NPs by graphene mainly include electrostatic interactions [105,106]. For example, Yang et al. developed a novel strategy for the fabrication of graphene-encapsulated metal oxide by coassembly between negatively charged graphene oxide and positively charged oxide NPs [105]. The process is driven by the mutual electrostatic interactions of the two species, and it is followed by chemical reduction. For a typical synthesis (Fig. 3.9), metal oxide NPs were first modified by surface grafting of aminopropyltrimethoxysilane (APS) to render the oxide surface positively charged. The modified oxide NPs were then assembled with negatively charged graphene oxide by electrostatic interactions. Under optimal conditions, almost all the graphene oxide and modified oxide NPs coassembled to leave a transparent aqueous solution. Finally, the resulting aggregates were chemically reduced with hydrazine to result in graphene-encapsulated metal oxide NPs. This simple and low-cost assembly protocol provided a new pathway for the large-scale production of various graphene-encapsulated catalysts.

2.3 Catalysts Encapsulated in Amorphous Porous Carbon

Porous carbon materials have received a great deal of attention due to their many applications, including gas separation, water purification, catalyst supports, and electrodes for electrochemical double layer capacitors and fuel cells [107–112]. According to the International Union of Pure and Applied Chemistry (IUPAC) recommendation, porous carbon materials can be classified into three types based on their pore sizes: microporous <2 nm, 2 nm $<$ mesoporous <50 nm, and macroporous >50 nm. As the other major form of carbon, amorphous porous carbon is primarily made of graphitic sheetlets with no regular order, and they are characterized by strong covalent bonding in the basal plane. AC is one of the most common amorphous carbons. Modern manufacturing processes basically involve three main steps, namely, raw material preparation, low-temperature carbonization, and activation. The raw materials are carbonaceous matter such as wood, peat, coals, coconut shell, fruit nuts, and other biomass-derived sources. The usefulness of AC mainly derives from its high micropore and mesopore volumes and the resulting high surface area [113,114].

In recent years, templated carbon has emerged as a type of amorphous porous carbon with well controlled pore size and structures. It was synthesized by using hard and soft templating approaches [115,116] and it possesses extremely high surface areas (as high as 4100 m^2g^{-1}) [117] and large pore volumes (as large as 2.25 cm^3 g^{-1}), which makes it an extremely promising candidate for applications such as catalyst supports and adsorbents [118–120]. The synthesis and structural characterization of porous carbon supported catalysts have been well summarized in some recent authoritative review articles [107,109,121,122], and it will not be thoroughly discussed here. Moreover, it has been pointed out in some reports that these supported systems often induce leaching of the catalyst into the reaction solution and aggregation of NPs to form larger ones after reaction runs. Hence, in this section, we focus on the preparation and synthesis of amorphous porous carbon-encapsulated catalysts.

A typical strategy that has been widely investigated is the use of microporous and mesoporous carbon materials as hosts for the immobilization of various metal or metal oxide NPs. In these systems, NPs with uniform particle sizes can be encapsulated in a specific location due to the presence of designated anchoring sites inside well-defined pore systems. Among various encapsulating approaches, the creation of a core-shell structure represents a fascinating technique to stabilize particle cores. A promising structure for a catalyst in a core-shell approach is core NPs encapsulated by a hollow porous carbon shell whose size is larger than that of the core particle. This "rattle-type" nanostructure enables stabilization of the core NPs without any ligands even under severe reaction conditions in a catalytic system. Ikeda et al. demonstrated a ligand-free PtNP 2.2 nm in diameter as a typical catalyst that is encapsulated in a hollow porous carbon shell 30–40 nm in diameter (Pt@hmC) [123]. The fabrication procedure was based on the successive coating of a SiO_2 layer and a $mSiO_2$ layer on the surface of poly(*N*-vinyl-2-pyrrolidone) (PVP) stabilized Pt NPs, subsequent growth of a carbon shell inside the $mSiO_2$ shell using resol-type phenol-formaldehyde resin as precursor, and removal of the siliceous components by treatment with aqueous HF (Fig. 3.10). The rattle like nanostructure as-obtained is effective for preventing aggregation of Pt core particles. Moreover, the void space between the Pt core and porous carbon shell is suitable as a

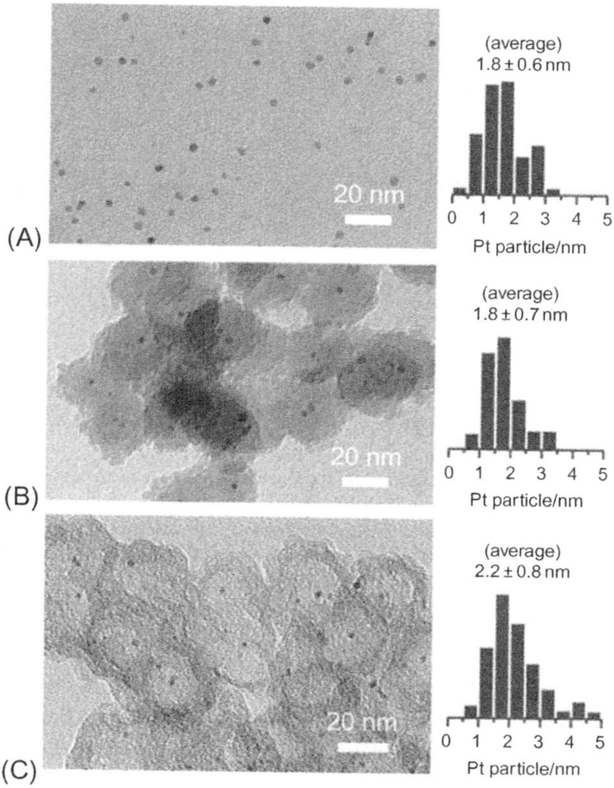

FIG. 3.10 TEM images and Pt particle size distributions of (A) Pt-PVP, (B) Pt@SiO_2-mSiO_2, and (C) Pt@hmC. *(Reproduced from S. Ikeda, S. Ishino, T. Harada, N. Okamoto, T. Sakata, H. Mori, S. Kuwabata, T. Torimoto, M. Matsumura, Angew. Chem. Int. Ed. 45 (2006) 7063–7066, with the permission of Wiley.)*

catalytic reaction site. This basic concept of the system could also be applicable to other metal or metal oxide NPs encapsulated in hollow porous carbon shell. For instance, Rh NPs encapsulated in a hollow porous carbon shell were successfully prepared using the similar method [124]. The specific characteristics of the composite and their effect on catalytic activity have been discussed.

Carbon precursors have an important effect on the preparation and final physical and chemical properties of the resulting carbon framework [125]. In addition to a good carbon yield, simple and effective uniform coating is another important issue in the templated synthesis of carbon materials. To this end, surface-initiated atom transfer radical polymerization from SiO_2 NPs and mSiO_2 has been used to make uniform polyacrylonitrile coating for the preparation of nanoporous carbon [126,127]. Another approach of the carbonization of phenol/formaldehyde resin (PFR) has been accepted as the most effective

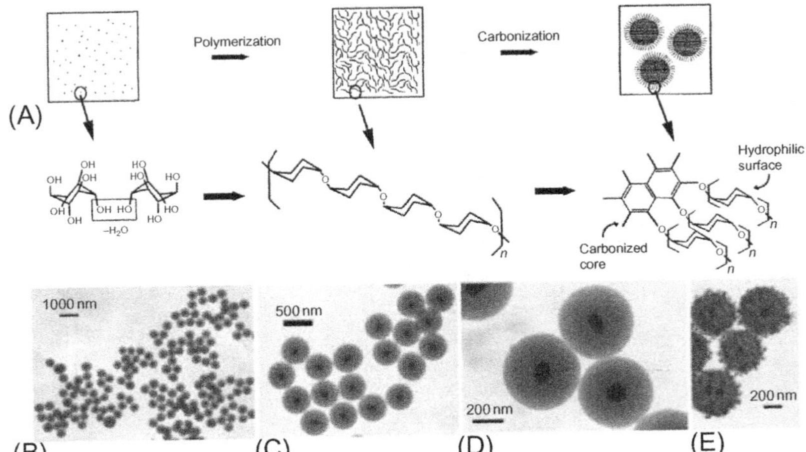

FIG. 3.11 (A) Schematic growth model for carbon spheres using glucose as a carbon source. TEM images of (B) Au-cored carbon spheres from hydrothermal reduction and the encapsulation method, (C, D) silver-cored carbon spheres from encapsulation of silver nanoparticle seeds, and (E) layered structure with a silver core, a platinum shell, and a carbon interlayer, formed by seed-based encapsulation followed by the reflux method. *(Reproduced from X. Sun, Y. Li, Angew. Chem. Int. Ed. 43 (2004) 597–601, with the permission of Wiley.)*

method for preparing porous carbon, due to several advantages of PFR, including facile preparation, high thermal stability, and easy conversion to carbon materials [128–130]. However, it would be desirable to use nontoxic carbon resources instead of the strongly carcinogenic phenol/formaldehyde. White et al. have developed a novel hydrothermal carbonization method to synthesize carbon capsules using biomass as a carbon source and polymer latex as a template [131]. Li et al. reported that monodisperse colloidal carbon microspheres were prepared from aqueous glucose solutions by hydrothermal synthesis, and encapsulated metal NPs were also obtained by this procedure (Fig. 3.11) [132]. This synthesis is convenient, and the as-formed colloidal spheres inherit functional groups from the starting material and have reactive surfaces, which facilitate loading with metal NPs and make it possible to integrate differently sized metal NPs in layered structures. Furthermore, the synthetic strategy can be transferred to other systems such as oxides, chalcogenides, and nitrides for the formation of hybrid, hollow, or porous structures. Inspired by this method, Zeng et al. reported the hydrothermal followed by carbonization synthesis of molybdenum oxide nanocrystals encapsulated within a highly porous carbonaceous matrix with a large surface area [133]. Using Au@SiO_2 as a template and sugar as a carbon source, Xu et al. reported the synthesis of carbon foam-encapsulated Au nanocatalysts with interconnected pores in the carbon shells and highly dispersed AuNPs confined in the cavities [134]. The Au NPs are separated from each other by permeable carbon shells, which efficiently prevent aggregation of the AuNPs even under high temperature conditions. In all cases, it is still challenging but desirable to explore new polymer analogues that feature low toxicity, excellent thermal stability, and high monodispersity within a framework to prepare porous carbon for encapsulating catalysts.

Dopamine, a biomolecule that contains catechol and amine functional groups, can self-polymerize at alkaline pH values and spontaneously deposit polydopamine conformal films on virtually any surface [135]. The thickness of such a conformal coating can be precisely controlled with a resolution of approximately 1 nm. Considering the structural similarity of polydopamine to phenolic resins, Dai and coworkers developed a versatile and facile method to prepare hollow carbon spheres (HCS) and yolk-shell carbon nanocomposite catalysts using dopamine as the carbon source (Fig. 3.12) [136]. The essence of their method lies in the exploitation of (1) the conformal nature of polydopamine coatings and the (2) high carbonization yield of polydopamine. Importantly, it is the first time that dopamine has been utilized in the construction of carbon-based nanomaterials. Inspired by this method, Lu et al. reported the hollow porous carbon shell-confined Fe NPs through polymerization of dopamine on SiO_2 nanospheres in the presence of $FeCl_3$, followed by carbonization and etching of the SiO_2 templates [137]. However, this approach requires a multistep synthesis procedure and also uses highly toxic HF. To overcome this problem, Lu and coworkers presented for the first time a simple yet versatile strategy for the large-scale manufacture of monodisperse and size-controlled N-doped carbon submicrometer spheres [138]. In this approach, dopamine was selected as the carbon resource due to its biocompatibility and self-polymerization capability at alkaline conditions. Moreover, $Fe@Fe_3C$

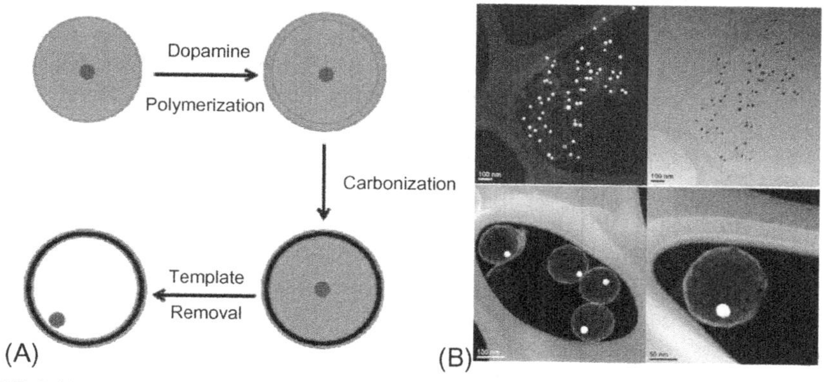

FIG. 3.12 (A) Schematic illustration of the synthesis Au@C nanorattle structure using $Au@SiO_2$ as a template and dopamine as a carbon source. (B) Scanning transmission electron microscopic (STEM) images of Au@C yolk-shell nanocomposites. *(Reproduced from R. Liu, S.M. Mahurin, C. Li, R.R. Unocic, J.C. Idrobo, H. Gao, S.J. Pennycook, S. Dai, Angew. Chem. Int. Ed. 50 (2011) 6799–6802, with the permission of Wiley.)*

catalysts encapsulated in N-doped porous carbon spheres were also successfully prepared. This feasible strategy presents new opportunities for tailoring the structure, composition, and size of porous carbon spheres for encapsulating various nanocatalysts.

Porous carbon-encapsulated catalysts can also be prepared by the ultrasonic spray pyrolysis method [139]. In a typical synthesis of porous carbon-encapsulated AuNPs, ultrasound is applied to generate an aerosol of a precursor solution ($HAuCl_4$, sucrose and $NaNO_3$) that is swept through a heated zone (e.g., 700°C) by an inert gas (Ar) with a residence time of 10 s. The temperature is set high enough to decompose $HAuCl_4$ into gold and sucrose into carbon in the presence of $NaNO_3$ as catalyst and pore formation agent. Matsumura and coworkers developed a simple and unique photocatalysis method for obtaining a carbon-metal catalyst composite, that is, Pt NPs incorporated in microporous hollow carbon nanospheres, through the photocatalytic reaction of TiO_2NPs in deaerated aqueous media [140]. In this method, TiO_2NPs not only act as molds of hollow carbon but also induce simultaneous deposition of Pt NPs and phenolic polymers by photocatalytic reduction and oxidation of a Pt (IV) precursor and phenol, respectively. Our group also reported novel double-shelled reduced graphene oxide@Pd@carbon (RGO@Pd@C) hollow spheres with reduced graphene oxide as the inner shell and amorphous porous carbon as the outer shell [44]. Ultrafine Pd NPs grew well on the surface of the inner shell by spontaneous redox reaction between graphene oxide and $PdCl_4^{2-}$. Moreover, the outer porous carbon shell combined with the inner shell could confine Pd NPs in it, which effectively prevent the aggregation and leaching of Pd NPs and increase their catalytic active area, therefore enhancing their stability and catalytic performance.

Although considerable efforts have been made in developing porous carbon-encapsulated metal nanocatalysts, preparing bimetallic NPs (containing noble metals and TMs) directly encapsulated in porous carbon, while maintaining a small particle size, has proved to be difficult. This is due to the fact that the formation of intermetallic phases generally needs high temperatures, which easily leads to the aggregation of bimetallic NPs [46]. Therefore, often special synthetic methods are required to control the size of carbon-encapsulated alloy particles and at the same time achieve good alloying. Schüth et al. reported a simple method to prepare PtCo bimetallic nanocatalysts with diameters of 3.6 ± 0.7 nm in HCS (Fig. 3.13) [46]. First, platinum NPs with a uniform particle size of 1.2 ± 0.3 nm were synthesized and encapsulated in hollow polymer spheres in one step (Pt@HPS). Subsequently, cobalt ions were introduced into the polymer shells through ion-exchange with the abundant carboxylate groups (Pt@HPS-Co^{2+}). In the third step, the polymer product of Pt@HPS-Co^{2+} was treated at 500°C under an H_2/Ar (5%/95%) atmosphere, leading to the formation of PtCo bimetallic NPs encapsulated inside the HCS (PtCo@HCS). Interestingly, the surface area of only the hollow carbon shells of Pt@HCS and PtCo@HCS could be calculated to be 575 and 712 $m^2\ g^{-1}$, respectively. This illustrated that introducing Co indeed helps in generating pores in the hollow

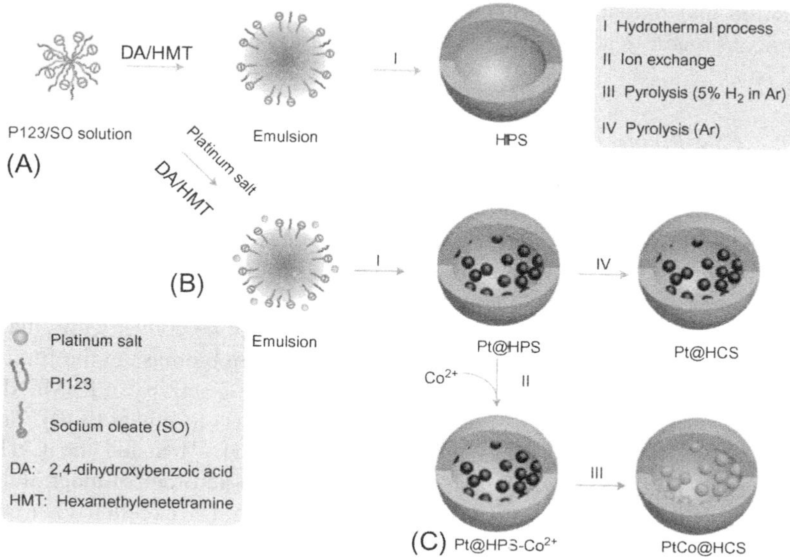

FIG. 3.13 Scheme for the synthesis of three different materials. (A) The synthesis of uniform HPS. (B) The synthesis of Pt@HCS. (C) The synthesis of PtCo@HCS. *(Reproduced from G. Wang, J. Hilgert, F. Richter, F. Wang, H. Bongard, B. Spliethoff, C. Weidenthaler, F. Schüth, Nat. Mater. 13 (2014) 293–300, with the permission of Nature Publishing Group.)*

carbon shells. Furthermore, the synthesis methodology described here was also broadly applicable to many other bimetallic nanocatalysts, including Cu_3Pd@HCS, $CoPd_2$@HCS, and Cu_3Pt@HCS.

3 APPLICATIONS

As discussed above, nanocatalysts (i.e., metal or metal oxide NPs) can be encapsulated in various nanostructured carbon systems (i.e., CNT, graphene, and amorphous porous carbon). These catalysts combine the electrical, mechanical, optical, and thermal properties of nanostructured carbon materials with the abundant functions of the other catalysts. Moreover, the confinement effects may influence the catalytic performance of these carbon-encapsulated catalysts. As a result, they have been widely explored for catalyzing chemical, electrochemical and photocatalytic reactions. In this part, the above three catalytic applications of nanostructured carbon-encapsulated catalysts are briefly summarized.

3.1 Catalysts for Chemical Reactions

Nanostructured carbon-based catalysts perform a range of chemical transformations, including gas-phase conversion reactions, liquid-phase hydrogenation, reduction and oxidation reactions, and so on. This section summarizes the chemical/organic reactions catalyzed by nanostructured carbon-encapsulated catalysts.

3.1.1 Gas-Phase Conversion Reactions

With soaring oil prices and dwindling resources, ethanol and other alternative energy sources have moved into the spotlight in recent years as clean, sustainable, and transportable fuel alternatives [141]. Syngas (a mixture of CO and H_2) conversion over Rh-based catalysts to ethanol [142] offers a good alternative because other C_2 oxygenate by-products (e.g., acetaldehyde and acetic acid) can be readily hydrogenated to ethanol, and syngas can be conveniently manufactured from natural gas and coal at present and from biomass in the future. Bao et al. reported a striking enhancement of the catalytic activity of RhMnNPs confined inside CNTs for the conversion of CO and H_2 to ethanol [49]. The overall formation rate of ethanol (30.0 mol mol^{-1}Rh h^{-1}) inside the CNTs exceeds that on the outside of the CNTs by more than an order of magnitude, although the latter is much more accessible (Fig. 3.14A). The excellent catalytic performance can be attributed to the confinement of the NPs within the CNTs, leading to the peculiar interaction of the interior nanotube surface with the metal particles, which benefits the dissociation of CO (Fig. 3.14B). Furthermore, theoretical studies on noncatalytic gas-phase reactions have prefigured that confinement within small channels could increase the density of reactants [143], hence creating a locally higher pressure, which will favor syngas conversion to C_2 oxygenates in the present case.

Confinement of Fe NPs inside CNT channels (Fe-in) was also found to enhance its catalytic activity for syngas conversion reaction with respect to the outside catalyst (Fe-out) [144]. TEM analysis indicated that over 70% of iron

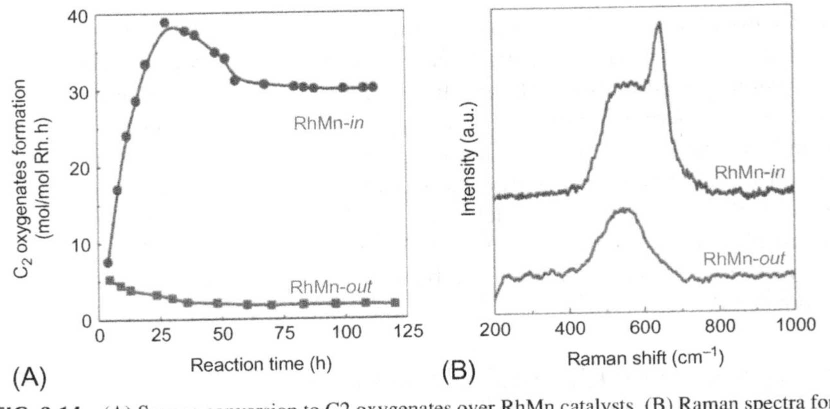

FIG. 3.14 (A) Syngas conversion to C2 oxygenates over RhMn catalysts. (B) Raman spectra for RhMn catalysts upon in situ CO adsorption. *(Reproduced from X. Pan, Z. Fan, W. Chen, Y. Ding, H. Luo, X. Bao, Nat. Mater. 6 (2007) 507–511, with the permission of Nature Publishing Group.)*

particles of Fe-in were distributed inside CNT channels, while almost all particles of Fe-out were on the outside. The syngas conversion reaction results showed that CO conversion was almost 1.5 times, and the yield of C_{5+} hydrocarbons was twice as high as those of Fe-out. Furthermore, the yield was 6 times higher than that over the XC-72 carbon black-supported Fe catalyst. A better reducibility and an enhanced catalytic performance of CNT-encapsulated Fe nanocatalysts were observed by Dalai and coworkers [145]. Both the Fe-in and Fe-out catalysts exhibited a similar initial CO conversion at 2 MPa and space velocity of 2 Lg^{-1} h^{-1}. However, Fe-in was much more stable because CO conversion was 6%–10% higher than that over Fe-out in the temperature range of 265–285°C, and selectivity toward C_{5+} hydrocarbon products was higher by 13% after 125 h time on stream.

Apart from CNT-encapsulated catalysts, graphene-encapsulated highly stable Fe NPs (Fe@G) were reported by Huang and coworkers [90]. This room temperature-stable sample exhibits excellent catalytic performance in the syngas conversion reaction. Specifically, it can deliver a stable CO conversion (≈75%) and a high yield on light olefins (C_2-C_4, 746 mg g_{cat}^{-1} h^{-1}) and C_{5+} hydrocarbon (706 mg g_{cat}^{-1} h^{-1}) at 553 K, which is more than three times that of bare Fe NPs (Fig. 3.15). The graphene shell was proposed to be responsible for the olefins preference and chain growing enhancing for heavy hydrocarbon products. Furthermore, the Fe@G shows no apparent changes on NP size and morphology and the graphene shell still survives after the catalytic reaction. Similar results were also observed in porous carbon-encapsulated Fe nanocatalysts [146]. The syngas conversion reaction showed that CO conversion and C_{5+} hydrocarbon selectivity enhanced to 95% and 70%, respectively, compared with AC-supported Fe NPs (CO conversion: 17 %; C_{5+} hydrocarbon selectivity: 10%).

Carbon-encapsulated catalysts can also be applied in many other gas-phase conversion reactions. For example, it has been reported that NH_3 decomposition

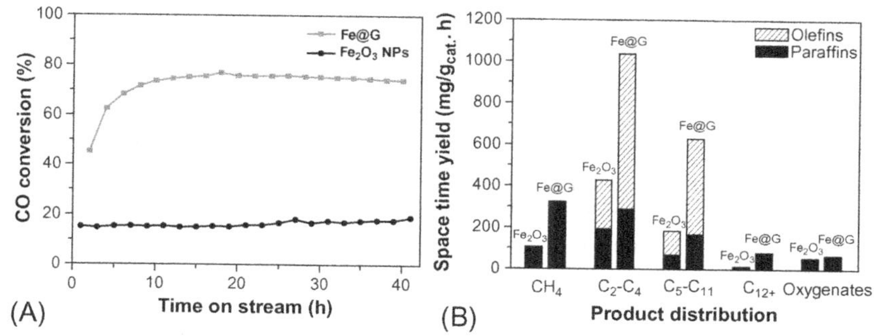

FIG. 3.15 (A) CO conversion vs. time on stream. (B) Space time yield of catalytic products. *(Reproduced from C. Wang, P. Zhai, Z. Zhang, Y. Zhou, J. Ju, Z. Shi, D. Ma, R.P.S. Han, F. Huang, Part. Part. Syst. Charact. 32 (2015) 29–34, with permission of Wiley.)*

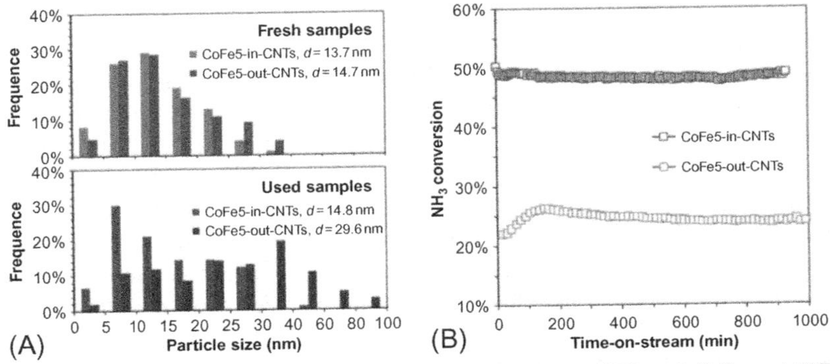

FIG. 3.16 Characterization and activity evaluation of $CoFe_5$-in-CNTs and $CoFe_5$-out-CNTs: (A) particle size distributions and (B) comparison of catalytic performance. *(Reproduced from J. Zhang, J.O. Muller, W. Zheng, D. Wang, D.S. Su, R. Schlogl, Nano Lett. 8 (2008) 2738–2743, with the permission of American Chemical Society.)*

benefits from confinement of the bimetal $CoFe_5$ inside CNTs ($CoFe_5$-in) [147]. To demonstrate the advantages of the $CoFe_5$-in catalyst, $CoFe_5$-out was prepared for comparison. The two fresh catalysts exhibited a similar particle size distribution (Fig. 3.16A) and the same $CoFe_2O_4$(311) planes. However, NH_3 conversion over $CoFe_5$-out was roughly half of that over $CoFe_5$-in during ~16 h time on stream (Fig. 3.16B). The superior thermal stability of the inside particles was proposed to be responsible for their higher conversion (Fig. 3.16A). Galeano et al. reported the synthesis of porous carbon-encapsulated Au NPs for catalytic oxidation of CO [148]. The half conversion temperature obtained for the encapsulated catalyst was much lower than for the carbon supported Au NPs. The excellent catalytic activity could be attributed to the intrinsic activity of carbon protected Au NPs and the porous carbon shell without mass-transfer limitations in the reaction tested.

3.1.2 Liquid-Phase Hydrogenation and Reduction Reactions

The liquid-phase hydrogenation reaction is one of organic transformations, which is a highly attractive route to a renewable liquid fuel. This reaction has been frequently carried out in carbon-encapsulated catalysts. Rh-based catalysts were first introduced on open CNTs for liquid-phase hydroformylation of propylene, which exhibited a turnover frequency (TOF) of 0.10 s^{-1} and a molar ratio of the normal/branched products $n/i = 9$ in comparison to 0.06 s^{-1} (TOF) and $n/i = 6$ over the catalysts on closed CNTs [149]. The higher activity and regioselectivity of the open catalyst were suggested to be attributed to the Rh species inside CNTs and its surface consisting of six-membered C-rings. Confinement of RhNPs inside hollow porous carbon shell was also found to benefit hydrogenation of *t*-butylbenzene in water with a much higher yield (97%) for *t*-butylcyclohexane compared with an AC-supported Rh catalyst (71%) and porous Al_2O_3 supported Rh catalyst (36%) [124]. Benefiting from the unique rattle-like structure, this heterogeneous catalyst was also active for the hydrogenation of benzoic acid, an electron-poor aromatic compound, and a heterocyclic compound of 3-hydroxypyridine to form cyclohexanecarboxylic acid and 3-hydroxypiperidine, respectively, indicating applicability for relatively inactive unsaturated compounds.

Encapsulating Pt NPs inside porous carbon shell has been demonstrated to have a high level of catalytic activity for the hydrogenation of nitrobenzene and olefins in comparison with PVP protected Pt NPs and AC supported Pt NPs (Fig. 3.17), which suggests that there is no significant effect of the porous carbon shell on the mass transfer of such a bulky substrate [123]. Similar results were observed in hollow porous carbon-encapsulated Pt catalysts for the catalytic hydrogenation of olefins [140]. Pd NPs were introduced inside MWCNTs for cinnamaldehyde hydrogenation, which exhibited a faster hydrogenation rate and a much higher selectivity (90%) to hydrocinnamaldehyde compared with an AC supported Pd catalyst [150]. Confinement of Pd NPs inside MWCNTs was also found to benefit hydrogenation of benzene with a TOF value twice as high as that over zeolite Y and AC supported Pd catalysts [151]. Serp and coworkers reported an enhancement of the catalytic activity of PtRuNPs confined inside CNTs for the selective hydrogenation of cinnamaldehyde to cinnamyl alcohol [65]. The TOF value and the selectivity toward cinnamyl alcohol was much higher than that of an unsupported PtRu catalyst. A higher selectivity toward cinnamyl alcohol was also observed over a CNT-confined Pt catalyst compared with the outside Pt catalyst [152].

The reduction of 4-nitrophenol (4-NP) by $NaBH_4$ has been considered a standard benchmark reaction for nanometal catalysis [153]. For example, Au NPs confined inside porous carbon sphere exhibited a superior catalytic performance for the reduction of 4-NP [139]. The reaction rate constant in terms of surface Au atoms is about ten times higher than that of the SiO_2 [38], ZrO_2 [154], and Fe_3O_4 [155] supported Au catalysts. The high catalytic activity could be ascribed to the high porosity and amphiphilicity of the carbon support. Likewise, Dai and coworkers reported an enhanced catalytic reduction of 4-NP over an Au@C yolk-shell catalyst [136]. The reduction of 4-NP into 4-aminophenol (4-AP) was completely finished in 5 min. The fast reaction kinetics compared to the Au@SiO_2 [38] and Au@polymer [156] nanorattle structures are believed to come from a combination of Au@C nanorattle structure characteristics: the facile diffusion of 4-NP through the porous carbon layer sheet within a 4 nm ultrathin shell, the large free reaction voids inside the carbon sphere, and highly dispersed AuNPs in each capsule. Moreover, the catalysts were still highly active after five successive cycles of reactions, with conversion close to 100% within 5 min reaction periods. Similar results were also observed in nitrogen-doped graphene-encapsulated FeCo nanocatalysts [157]. Due to the intrinsic magnetic character of FeCo nanocrystals, the reusable catalysis became very easy through a magnet near the vessel. A much better catalytic

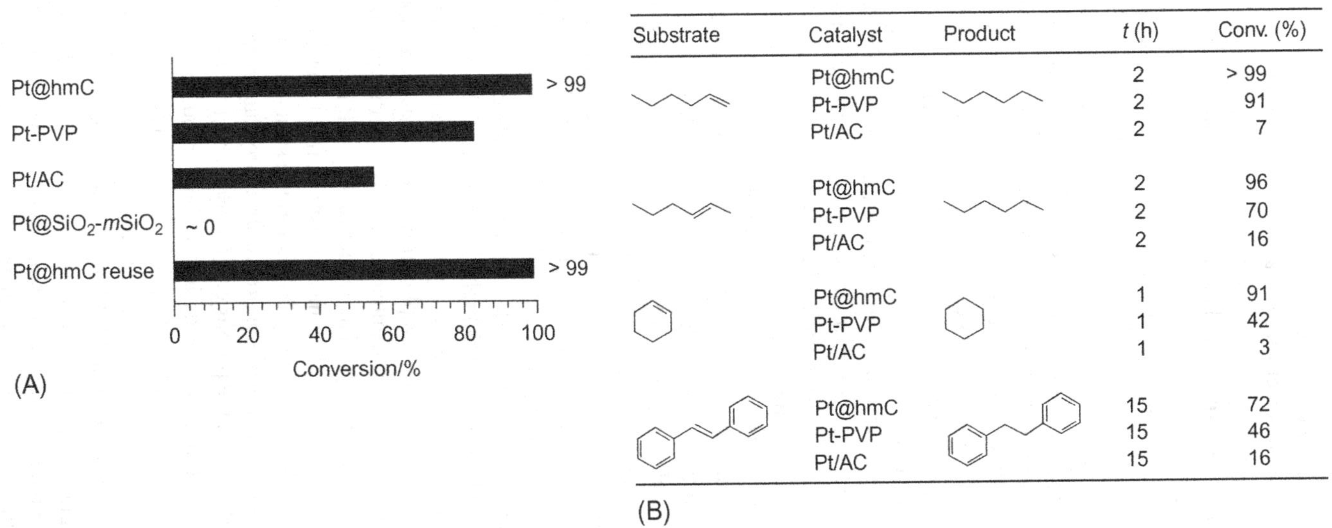

FIG. 3.17 Liquid-phase hydrogenation of (A) nitrobenzene and (B) various olefins by various Pt catalysts. *(Reproduced from S. Ikeda, S. Ishino, T. Harada, N. Okamoto, T. Sakata, H. Mori, S. Kuwabata, T. Torimoto, M. Matsumura, Angew. Chem. Int. Ed. 45 (2006) 7063–7066, with the permission of Wiley.)*

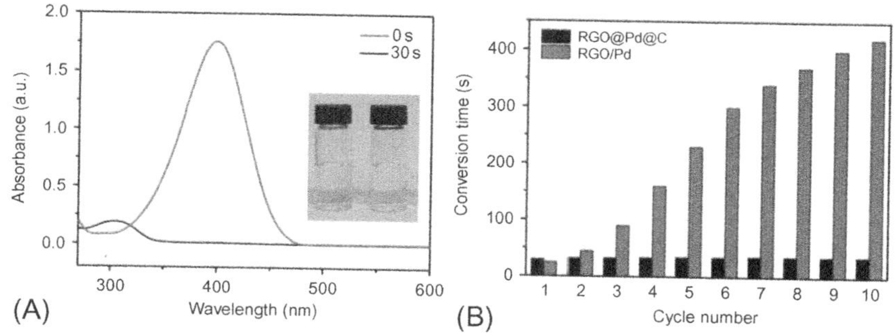

FIG. 3.18 (A) UV/vis spectra of 4-NP reduction reaction in the absence and presence of RGO@Pd@C catalyst. Insert: color changes of the conversion of 4-NP. (B) The stability of the catalyst by performing the same reduction reaction with 10 cycles. *(Reproduced from Z.Y. Zhang, F. Xiao, J. Xi, T. Sun, S. Xiao, H. Wang, S. Wang, Y.Q. Liu, Sci. Rep. 4 (2014) 4053, with the permission of Nature Publishing Group.)*

performance for ultrafine Pd NPs confined in double-shelled graphene@carbon hollow spheres for the same reaction was reported by our group [44]. With the aid of our prepared catalysts, the reduction reaction of 4-NP to 4-AP with $NaBH_4$ as a reducing agent can be finished within only 30 s (Fig. 3.18A); even the content of Pd is as low as 0.28%. To demonstrate the catalytic stability of RGO@Pd@C catalysts, RGO supported Pd (RGO/Pd) nanocatalysts were prepared for comparison. Obviously, RGO@Pd@C catalysts exhibited good stability after 10 cycles of reactions, with 100% conversion within 40 s reaction periods. However, the reaction time increased from 25 s to 420 s under the same condition when RGO/Pd was used as the catalyst (Fig. 3.18B).

3.1.3 Liquid-Phase Oxidation and Specific Reactions

Liquid-phase oxidation of alcohols into the corresponding carbonyl compounds and carboxylic acids is one of the fundamental and important reactions in organic synthesis. Nanocarbon-encapsulated catalysts have also been widely applied in these reactions. For example, Matsumura and coworkers reported a core-shell structure consisted of a Pd NP core and a hollow porous carbon shell (Pd@hmC) for the oxidation of alcohols in the liquid phase [158]. When compared to AC supported Pd catalysts, Pd@hmC showed a high level of catalytic activity for oxidation of benzyl alcohol into benzaldehyde using atmospheric pressure of O_2 as an oxidant. The Pd@hmC catalysts also exhibited a high level of catalytic activity for aerobic oxidations of other primary benzylic and allylic alcohols into corresponding aldehydes (Fig. 3.19). Furthermore, the specific core-shell structure of Pd@hmC resulted in possible regeneration of catalytic activity for these aerobic oxidations by a high-temperature heat treatment of the sample recovered after catalytic reactions. Carbon foam-confined Au (Au@CF) nanostructures have been successfully used in a glycerol oxidation

Entry	substrate	Catalyst	Time/h	TOF ^{b}/h^{-1}
1	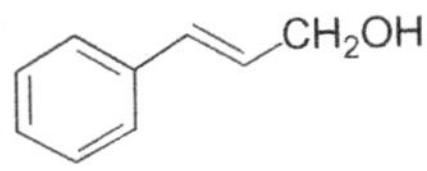	Pd@hmC	5	560
2		Pd/AC	5	150
3	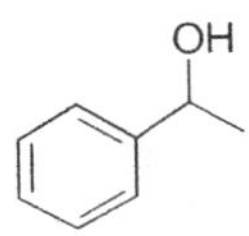	Pd@hmC	15	150
4		Pd/AC	15	50
5	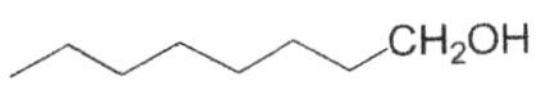	Pd@hmC	24	–
6		Pd/AC	24	–
7	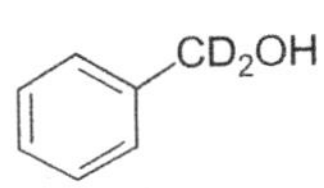	Pd@hmC	2	340
8		Pd/AC	2	200
9	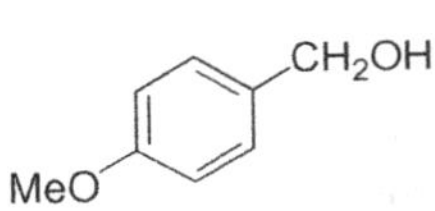	Pd@hmC	1	4470
10		Pd/AC	1	2500
11	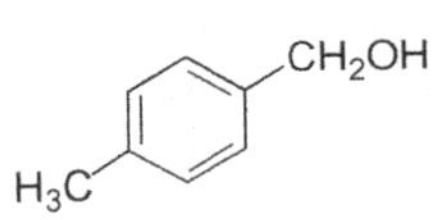	Pd@hmC	1	3370
12		Pd/AC	1	2150
13	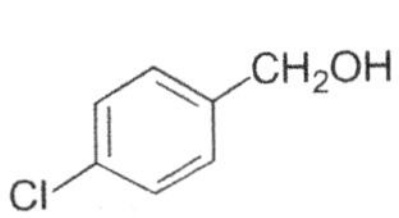	Pd@hmC	5	540
14		Pd/AC	5	220

FIG. 3.19 Aerobic oxidation activity of various alcohols over Pd@hmC and Pd/AC catalysts. *(Reproduced from T. Harada, S. Ikeda, F. Hashimoto, T. Sakata, K. Ikeue, T. Torimoto, M. Matsumura, Langmuir 26 (2010) 17720–17725, with the permission of American Chemical Society.)*

reaction [134]. The outside deposited Au/CF was used as a reference sample. In the oxidation of glycerol, the Au/CF and Au@CF show similar curves of conversion as a function of reaction time. The conversion of glycerol increases with the reaction time. Glycerol was nearly completely converted after 4 h for both Au/CF and Au@CF. These suggest that AuNPs in Au@CF can be accessible to the substrate comparable to that in Au/CF. Besides, the activity of AuNPs in Au@CF after thermal treatment could be retained; however, Au/CF could not be retained.

In addition to these well-known organic reaction, Zeng et al. reported Mo-embedded mesoporous carbon microspheres for a Friedel-Crafts alkylation

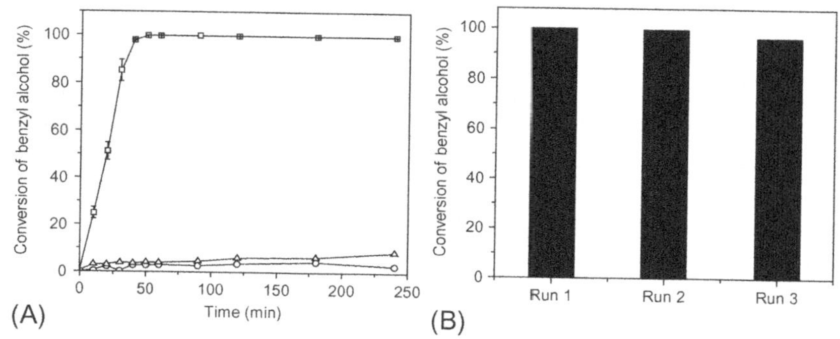

FIG. 3.20 (A) Conversion of benzyl alcohol in the benzylation of toluene catalyzed by Mo@C-1 (○), Mo@C-2 (△), and Mo@C-3 (□). (B) Catalytic stability tests of a Mo@C-3 catalyst. *(Reproduced from J. Dou, H.C. Zeng, J. Phys. Chem. C 116 (2012) 7767−7775, with the permission of the American Chemical Society.)*

reaction [133]. A100% conversion of benzyl alcohol was achieved within 1 h at a reaction temperature of 110°C (Fig. 3.20A). The observed high activity was attributed to the mesoporous structure of carbon microspheres, allowing free traveling of reactants and products between the embedded α-MoO_3 and liquid medium. The high oxidation state Mo^{6+} is responsible for the active sites, since the MoO_2/Mo_2C embedded catalyst is inactive. Along with its high activity, the α-MoO_3 embedded mesoporous carbon catalyst was also highly robust and could be regenerated after use (Fig. 3.20B). For a hydrogenolysis reaction, a hollow porous carbon nanosphere-encapsulated PtCo bimetallic nanocatalyst has been prepared for the production of 2,5-dimethylfuran (DMF) by hydrogenolysis of 5-hydroxymethylfurfural (HMF) [46]. Due to the synergistic effect from ultrafine encapsulated PtCo alloy NPs and mesoporous carbon shell, the conversion of HMF over PtCo catalysts was 100% within 10 min and the yield to DMF reached 98% after 2 h, which substantially exceeds the best supported catalysts reported in the literature. For a C-C coupling reaction, Wang et al. firstly reported the Pd-embedded 3D graphene architecture as a fixed-bed reactor for the Heck reaction [159]. K_2CO_3 was chosen as the base in the reaction in which iodobenzene and methyl acrylate were reacted in the presence of the 3D cylinder (0.1 mol% Pd) as the catalyst. Both the selectivity and conversion were 100% (below the detection limit of GC-MS), which is much higher than the results obtained when the Pd catalyst is loaded onto the other supports [160]. Likewise, Vinu and coworkers prepared a mesoporous carbon nitride-encapsulated Au (Au@MCN) nanocatalyst for a three-component coupling reaction [161]. As a result, the obtained Au@MCN could be used as a highly active, selective, and recyclable heterogeneous catalyst for coupling benzaldehyde, piperidine, and phenylacetylene for the synthesis of propargylamine.

3.2 Catalysts for Electrochemical Reactions

Nanostructured carbon-encapsulated catalysts have attracted increasing interest in the past decade. Benefiting from their remarkable electrochemical catalytic properties derived from chemical compositions and synergetic effects of multi-functionalities, these carbon-encapsulated nanocatalysts play a significant role in various electrochemical reactions. This section mainly summarizes their electrochemical applications including fuel cells, electrochemical water splitting, and electrochemical sensors.

3.2.1 *Fuel Cells*

A fuel cell is an important device for energy conversion and storage, which generates electricity via the reaction between a fuel and an oxidant in its electrodes [162–166]. The typical fuel cells, especially low-temperature fuel cells such as H_2-O_2 fuel cells or direct alcohol fuel cells (DAFCs) can electrochemically convert hydrogen into water or alcohol into water and carbon dioxide. The main reactions in a fuel cell include the oxygen reduction on cathode and the fuel oxidation on anode. As is widely known, the oxygen reduction reaction (ORR) on a cathode plays a critical role in determining the performance of a fuel cell. Pt and Pt alloys have been accepted as the most active and effective cathodic catalysts for ORR [167]. However, a number of challenges remain for the Pt-based catalysts, such as low durability, easy poisoning from methanol crossover, high cost, and so on. To resolve these problems, low-Pt or non-Pt and metal-free electrocatalysts with low-cost, high electrocatalytic activity, and stability have been extensively studied as efficient ORR electro catalysts in recent years [168–178].

Nonprecious metal NPs encapsulated in nanocarbons are promising highly active nanostructures with high chemical and thermal stability, because the encapsulating shell protects the metal NPs from atmospheric-oxygen-induced degradation and agglomeration of neighboring NPs. Thus, these new structures enable technological applications of metal NPs as the naked ones are not stable enough in ambient air [47,179–182]. Recently, Wu et al. reported an encouraging catalyst synthesized from large-surface-area carbon, polyaniline, Fe, and Co at a high temperature (PANI-FeCo-C) for ORR catalysis in H_2-O_2 fuel cell [179]. The FeCo NPs were to a large extent encapsulated in onion-like carbon nanoshells. As a result, the PANI-FeCo-C catalysts exhibited a high catalytic activity (excellent four-electron selectivity, hydrogen peroxide yield $<1.0\%$) and stability (a 700-h fuel-cell performance test at a constant cell voltage of 0.4 V) that approaches the performance of Pt-based systems at the fuel cell cathode. However, developing highly active nonprecious metal catalysts that outperform the most active Pt-based catalysts is still a challenge. A new type of N-doped CNT-encapsulated Fe catalyst has been found to possess the highest ORR activity and stability in alkaline media of any nonprecious metal catalysts [180]. As shown in Fig. 3.21, when used at a sufficiently high loading, this

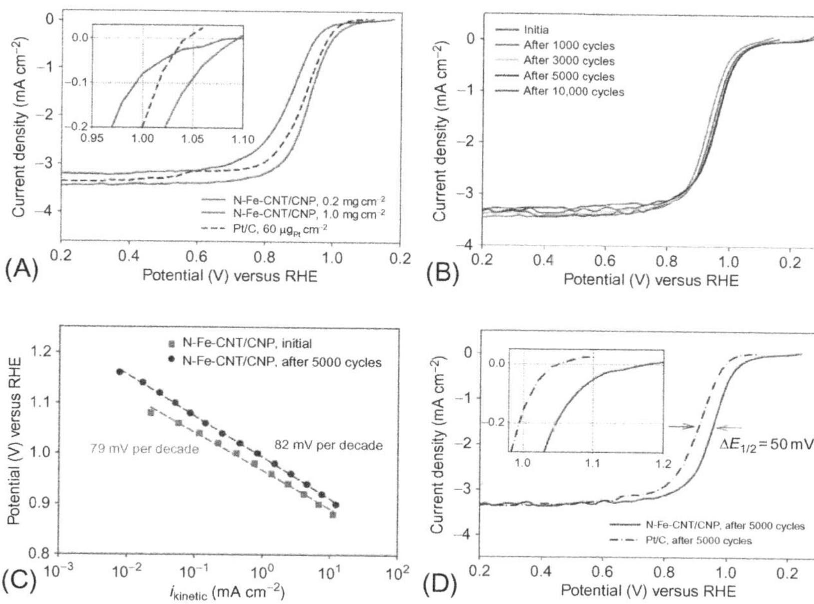

FIG. 3.21 (A) ORR polarization plots before a cycling durability test. (B) ORR polarization plots measured during cycling durability in O_2-saturated 0.1 M NaOH. (C) Tafel plots for the N-doped CNT-encapsulated Fe catalyst before and after 5000 cycles in an O_2-saturated electrolyte. (D) ORR polarization plots after 5000 cycles in an O_2-saturated electrolyte. *(Reproduced from H.T. Chung, J.H. Won, P. Zelenay, Nat. Commun. 4 (2013) 1922, with the permission of Nature Publishing Group.)*

catalyst exhibited a more negative onset potential and excellent cycling durability, which outperforms the most active Pt-based catalysts. Likewise Yu et al. reported a novel nonprecious metal catalyst based on Fe_3C NPs encapsulated in mesoporous Fe-N-doped carbon nanofibers; the electrocatalyst shows outstanding ORR activity closely comparable to the Pt/C catalyst in alkaline media, and there is good ORR activity in acidic media [181]. The superior catalytic activity could be attributed to multiple highly effective intrinsic active sites (i.e., Fe-N_x/C and Fe_3C@graphitic carbon), high specific surface area (425 $m^2\ g^{-1}$), and mesoporous structure, as well as highly graphitic carbon fibrous networks. A meso/macroporous nitrogen-doped carbon architecture of Fe_3C encapsulated in graphitic layers was also prepared as a nonprecious ORR catalyst [182]. Outstanding activity and stability toward the ORR in both acidic and alkaline solutions was observed due to the unique confined structure and chemical composition.

Although various nanostrucrued carbon-encapsulated catalysts have been prepared for ORR catalysis, the nature of the active ORR catalytic sites was still not clear. To this end, Bao and coworker intended to design active and stable nonprecious metal ORR catalysts with features of coordinatively unsaturated metal sites or low valence state metals physically isolated from the acid environment to avoid leaching, which, however, should not impede the catalytic properties of metals. They encapsulated Fe NPs into the compartments of pea-pod like CNTs as a well-defined model for understanding the nature of Fe-based cathode catalysts in ORR catalysis [47]. In this system, direct contact of metal particles with harsh environments including acid medium, oxygen, and sulfur contaminations is avoided. However, this protection does not impede the activation of O_2 and the catalyst has a rather high activity and long-term stability in ORR catalysis even in the presence of SO_2 poison. The DFT calculations indicate that the catalytic activity could arise from the electron transfer from Fe particles to the CNTs leading to a decreased local work function on the carbon surface. Interestingly, doping N into the carbon lattice further increases the projected density of states (DOS) near Fermi level and reduces the local work function, and experiments show a significantly enhanced ORR activity (Fig. 3.22) [47]. This work provides a new concept in designing efficient and durable electrocatalysts, as well as heterogeneous catalysts under harsh reaction conditions.

On the other hand, methanol, with its low molecular weight, simplest structure, high energy density, large scale of production from renewable sources, and easy storage and transportation, has been considered as one of the ideal combustible materials for direct methanol fuel cells (DMFCs) [183]. The development of highly active electrocatalysts for the oxidation of the methanol molecule is one of the vital needs for practical applications of DMFCs. Due to the low activity of nonnoble catalysts, considerable efforts have been devoted to the optimization of the existing Pt-based nanocatalysts [184–187]. However, according to the dual-pathway mechanism, CO is a kind of poisoning intermediate species, which can largely reduce the catalytic activity of Pt-based catalysts. Integrating the unique properties of nanostructured carbon and Pt offers new opportunities to reduce the CO poisoning effect in DMFCs. Recently, various nanostructured carbon supported noble metal (i.e., Pt or its alloy) nanocatalysts for DMFCs have been reported [188–192]. However, maintaining efficient contact between Pt and nanocarbons, while at the same time maximizing catalytic activity, remains a key challenge. To address this problem, Tan et al. reported a nitrogen-doped graphene-encapsulated ultrafine Pt nanocrystal (N-GPN) as an efficient DMFC catalyst [101]. As shown in Fig. 3.23A, the electrochemical surface area of Pt was determined to be around 112.7 $m^{-2}\ g^{-1}$ for the N-GPN catalyst, which is much higher than that of either the commercial J.M. catalyst or pristine GPN. Fig. 3.23B shows the electrocatalytic performance of the Pt nanocrystal for methanol oxidation. Compared with GPN and J.M. Pt/.C, a significant enhancement of the peak current as well as an obvious negative shift of the peak potential and the onset potential of methanol oxidation were observed on N-GPN. The excellent electrocatalytic activity was mainly due to the synergistic effect from ultrafine encapsulated Pt NPs and the N-doped graphene shell, resulting in the large electrochemical active surface

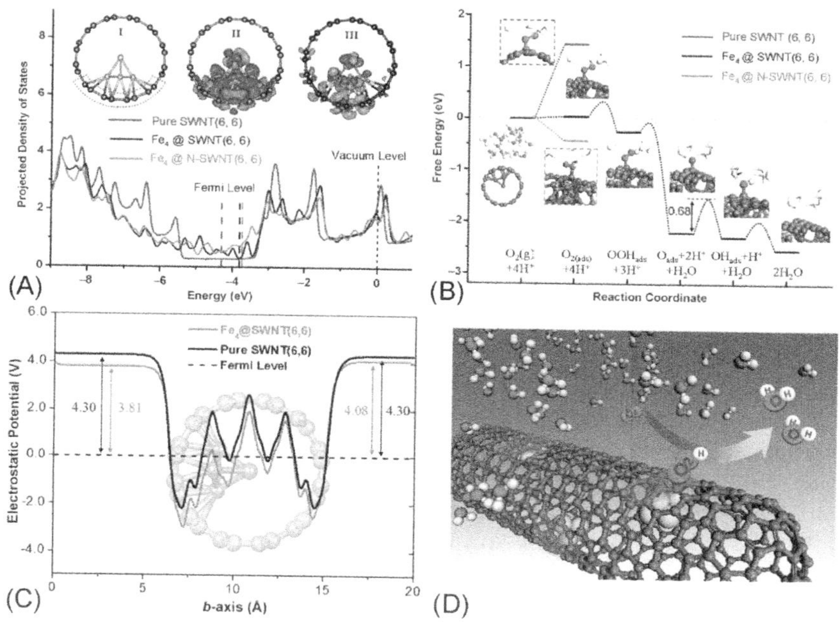

FIG. 3.22 (A) Projected DOS of the p orbitals of C atoms bonded to Fe_4 in Fe_4@SWNT and Fe_4@N-SWNTcompared with that in pure SWNT. The vacuum level is aligned at 0 eV. Inserted plot I and plot II show the optimized structure of Fe_4@SWNT and its in difference charge density, and plot III corresponds to that of Fe_4@N-SWNT. (B) Free-energy diagram of ORR on Fe_4@SWNT in water (*black line*) under experimental conditions (onset potential = 0.6 V vs. normal hydrogen electrode (NHE), pH 0). (C) The electrostatic potential profiles averaged on the plane perpendicular to the *b*-axis as a function of the *b*-axis of the supercell of Fe_4@SWNT and pure SWNT, respectively. The structure of Fe_4@SWNT is shown in the background. (D) A schematic representation of the ORR process at the surface of Fe_4@SWNT. *(Reproduced from D. Deng, L. Yu, X. Chen, G. Wang, L. Jin, X. Pan, J. Deng, G. Sun, X. Bao, Angew. Chem. Int. Ed. 52 (2013) 371–375, with permission of Wiley.)*

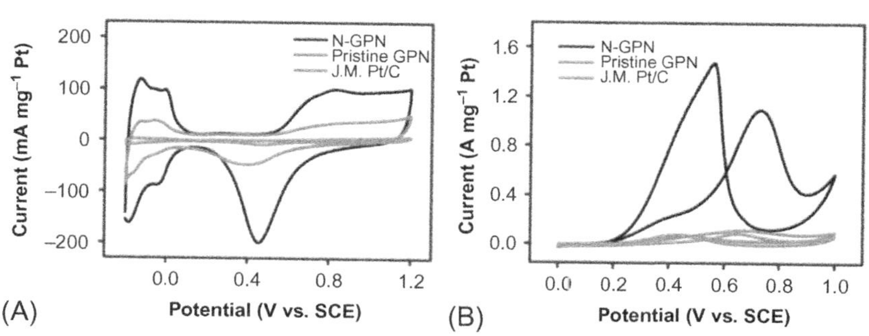

FIG. 3.23 Cyclic voltammograms of the N-GPN, GPN, and commercial J.M. Pt/C catalysts in (A) nitrogen-saturated 0.5 M H_2SO_4 and (B) N_2-saturated 0.5 M H_2SO_4 in 1.0 M methanol at a scan rate of 50 mV s^{-1}. *(Reproduced from D. Ding, Z. Song, Z. Cheng, W. Liu, X. Nie, X. Bian, Z. Chen, W. Tan, J. Mater. Chem. A 2 (2014) 472–477, with the permission of the Royal Society of Chemistry.)*

area of Pt and efficient electron transport path. This study clearly indicates that graphene encapsulation could enhance the catalytic performance of catalysts for methanol oxidation. Besides Pt, nanoalloy electrocatalysts have also been studied for methanol oxidation. Fisher et al. reported the synthesis of PtRu alloy confined inside CNTs and studied their electrochemical properties toward methanol electrooxidation in acid electrolyte [193]. The current density of methanol oxidation at PtRu catalysts is ~20 times higher than at bulk Pt electrodes. This enhancement of current density is due to both the high surface area and the high electrocatalytic activity of nanoclusters dispersed inside the CNT.

3.2.2 *Electrochemical Water Splitting*

Among the numerous energy conversion technologies available, electrochemical water-splitting provides a fascinating way of converting electrical energy into a renewable energy resource [194,195]. However, to overcome the large water-splitting overpotential (commercial electrolyzers generally work at a cell voltage of 1.8–2.0 V, which is much higher than the theoretical minimum value of 1.23 V) [196], efficient electrocatalysts for cathodic hydrogen evolution reactions (HER) and anodic oxygen evolution reactions (OER) must be developed. Currently, the state-of-the-art electrocatalysts for splitting water are Pt group metals and Ru- or Ir-based compounds for HER and OER [197,198], respectively, but their applications are limited by the high cost and scarcity. For HER, great effort has been made to develop low-cost and stable nonnoble metal-based electrocatalysts with commensurate performance [199–206]. Inspired by carbon-encapsulated nonprecious metal catalysts used for ORR catalysis, Bao et al. developed highly efficient HER electrocatalysts by using N-doped CNTs (NCNTs) encapsulating 3D TM Fe, Co, and FeCo alloy NPs [207]. To study the role of doped N atoms for HER, they introduced NH_3 during the preparation of electrocatalysts to increase the concentration of N. These corresponding catalysts are denoted as Fe@NCNTs-NH, Co@NCNTs-NH, and FeCo@NCNTs-NH. Remarkably, FeCo@NCNTs-NH showed a very high HER activity with an onset overpotential of only 70 mV vs. the reversible hydrogen electrode (vs. RHE), which is among the smallest overpotential for nonprecious metal HER catalysts in acidic electrolytes reported so far (Fig. 3.24A). These results indicate that the introduction of N in CNT-encapsulated metal catalysts can significantly enhance their HER activity. Furthermore, the polarization curve of FeCo@NCNTs after 10,000 cycles retained an almost similar performance to the initial test (Fig. 3.24B), demonstrating its long-term durability for HER in acidic media. DFT calculations (Fig. 3.24C and D) indicate that the adsorption free energy of the H atom on CNTs can be decreased by the enclosed metal cluster and N-dopants, implying that the introduction of metal and N in these catalysts can synergistically enhance the HER activity in accordance with the experimental results. These findings provide a new route in the design of efficient and durable nonprecious-metal

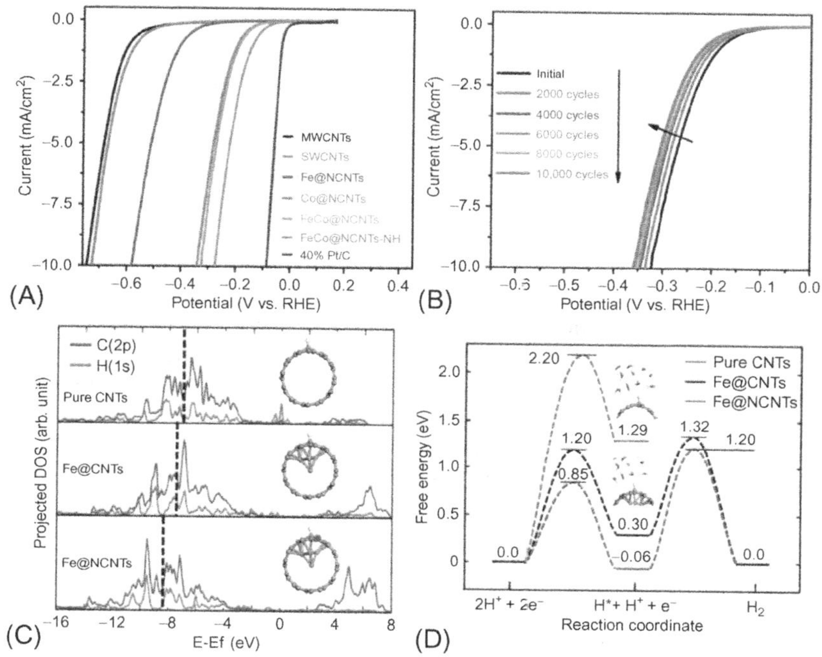

FIG. 3.24 (A) Polarization curves of Fe@NCNTs, Co@NCNTs,FeCo@NCNTs, FeCo@NCNTs-NH along with MWCNTs, SWCNTs, and 40% Pt/C for comparison. (B) Durability measurement of FeCo@NCNTs: polarization curves recorded initially and after every 2000 CV sweeps between +0.77 and 0.18 V (vs. RHE) at 100 mV s^{-1}. All the polarization curves were performed in 0.1 M H_2SO_4 at a scan rate of 2 mV s^{-1}. (C) Comparison of projected density of states (DOS) of H(1 s) and its bonded C(2p) when H is adsorbed on the surface of pristine CNTs, Fe@CNTs, and Fe@NCNTs. The *dashed lines* present the center of the occupied band. (D) The free-energy profiles of the Heyrovsky route for pristine CNTs, Fe@CNTs, and Fe@NCNTs. *(Reproduced from J. Deng, P. Ren, D. Deng, L. Yu, F. Yang, X. Bao, Energy Environ. Sci. 7 (2014) 1919–1923, with the permission of the Royal Society of Chemistry.)*

electrocatalysts for HER. Likewise Asefa and coworkers reported the synthesis of Co-embedded N-rich CNTs (Co-NRCNTs) that can serve as highly active electrocatalysts for HER under acidic, neutral, or basic media [208]. The enhanced catalytic activity could be ascribed to cooperative and/or synergistic catalytic effects between the N dopants and carbon-encapsulated Co NPs in Co-NRCNTs. Similar results have also been observed in N-doped multilayer graphene-encapsulated FeCo nanoalloy [104] and N-doped carbon-wrapped Co NPs on N-doped graphene nanosheets [209,210] for the same reaction.

Considering that the carbon shells previously used for these catalysts are too thick and usually consist of multilayer graphitic carbon or composites thereof, which may significantly reduce the catalytic activity, the synthesis of carbon-encapsulated nonprecious catalysts with few graphene layers will be important for the development of nonprecious catalysts with superior HER activity. Based on this point of view, Bao et al. also reported a hierarchical architecture that consists of ultrathin graphene shells (only 1–3 layers) that encapsulate a uniform CoNi nanoalloy (CoNi@NC), which provides a well-defined model for elucidating the role of ultrathin carbon-encapsulated metal catalysts in the HER from theoretical calculations to the real system [45]. The prepared CoNi@NC with optimized calcination temperature displayed superior HER activity with an onset overpotential of only approximately 30 mV and an overpotential of 224 mV at a current density of 10 $mAcm^{-2}$. DFT calculations indicate that the superior HER performance originates from the modulation of the electron density and the electronic potential distribution at the graphene surface by a penetrating electron from the CoNi core. To further reduce the carbon layers, a single-shell carbon-encapsulated Fe nanocatalyst (SCEFC) was reported by Laasonen and coworkers [79]. To obtain the desired catalytic activity for HER, they focused on the growth of SCEFC decorated on SWNT. Fig. 3.25A shows typical polarization curves of HER on the pristine SWNT, SCEFC/SWNT, and Pt/C electrodes. The SCEFC/SWNT material exhibited

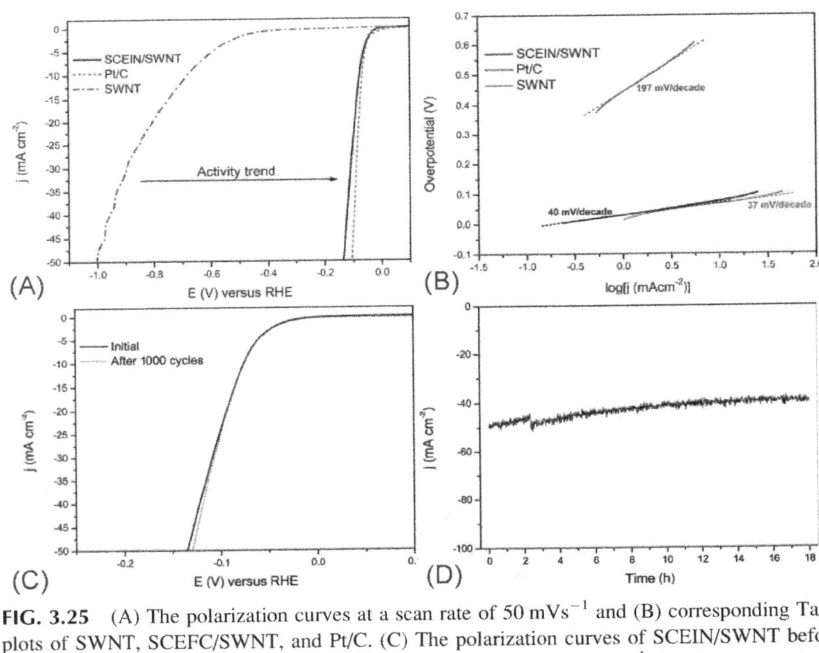

FIG. 3.25 (A) The polarization curves at a scan rate of 50 mVs^{-1} and (B) corresponding Tafel plots of SWNT, SCEFC/SWNT, and Pt/C. (C) The polarization curves of SCEIN/SWNT before the cycling durability test and after 1000 cycles at a scan rate of 100 mVs^{-1}. (D) Time dependence of the current density obtained for SCEFC/SWNT at a static potential of −135 mV for 18 h. All these tests were conducted in a 0.5 MH_2SO_4 acidic solution. *(Reproduced from M. Tavakkoli, T. Kallio, O. Reynaud, A.G. Nasibulin, C. Johans, J. Sainio, H. Jiang, E.I. Kauppinen, K. Laasonen, Angew. Chem. Int. Ed. 54 (2015) 4535–4538, with the permission of Wiley.)*

an onset potential of about 0 V vs. RHE, which is very close to the thermodynamic potential of HER. The low Tafel slope of 40 $mVdec^{-1}$ for SCEFC/SWNT (Fig. 3.25B) showed that this material catalyzed the HER through a Volmer-Heyrovsky mechanism in which a fast discharge step (Volmer reaction, $H_3O^+ + e^- \rightarrow H_{ads} + H_2O$) is followed by an electrochemical desorption step (Heyrovsky reaction, $H_{ads} + H_3O^+ + e^- \rightarrow H_2 + H_2O$) and the desorption of hydrogen is the rate-limiting step. Furthermore, long-term stability tests (Fig. 3.25C and D) indicated that SCEIN/SWNT was a stable catalyst for HER. All these results demonstrated that SCEFC/SWNT could act as a highly active and durable nonprecious catalyst for HER in an acidic electrolyte.

Besides nonprecious metal-based nanocatalysts, 2D MoS_2 has also been intensively studied for HER. For example, Zhu et al. reported a facile synthesis of S-rich small and single-layered MoS_2 nanoplates embedded in N-doped carbon nanofibers (MoS_2-NCNFs) [211]. It was shown that the S-rich MoS_2-NCNFs hybrid nanomaterial exhibited extraordinary HER activity, with a very low onset potential of 30 mV and a small Tafel slope of 38 mV per decade. Graphitic carbon layers acted as a channel for transferring electrons to MoS_2 in the MoS_2-NCNFs hybrid; this, together with the abundant catalytic activity of the MoS_2 nanoplates and pyridinic N and graphitic N, leads to excellent HER activity.

OER at anode is the most sluggish step in overall water splitting because of its high overpotential and slow kinetics. Precious Ir and Ru oxides have been tested as state-of-the-art catalysts for OER. The main disadvantages in using nonprecious metals in place of these precious metal catalysts for OER are their low efficiency and poor stability, which lead to the urgent demand for new concepts and strategies to solve these issues. To this end, Bao et al. reported a single layer graphene-encapsulated nonprecious metal NP (metal@NCs) as a high-performance electrocatalyst for OER [48]. As shown in Fig. 3.26A, the onset potential of the FeNi@NC sample is only 1.44 V vs. RHE, and the potentials at 10 and 20 mA cm^{-2} were 1.51 V vs. RHE and 1.54 V vs. RHE, respectively, which is even better than that of commercial IrO_2. The Tafel slope of 70 mV dec^{-1} for FeNi@NC is comparable to the value of 63 mV dec^{-1} of IrO_2 (Fig. 3.26B), showing that the FeNi@NC sample possessed a similar ability to drive the OER process compared to IrO_2 at low overpotential. Fig. 3.26C and D demonstrated that FeNi@NC had a much superior electrocatalytic stability compared with commercial IrO_2 catalysts. DFT calculations indicated that the high OER performance originates from the single layer graphene promoting the electron transfer from the encapsulated metals to the graphene surface, which optimized the electronic structure of the graphene surface and thereby triggers the OER activity on the inert graphene surface [48]. A porous carbon coated Ni-P nanoplate derived from MOFs has also been synthesized for OER catalysis [212]. As a result, the as-derived Ni-P catalysts manifested excellent electrocatalytic activity (deliver a current density of 10 mA cm^{-2} at 1.53 V) for OER superior to NiO and $Ni(OH)_2$ porous nanoplates. The outstanding electrocatalytic activity was mainly due to the

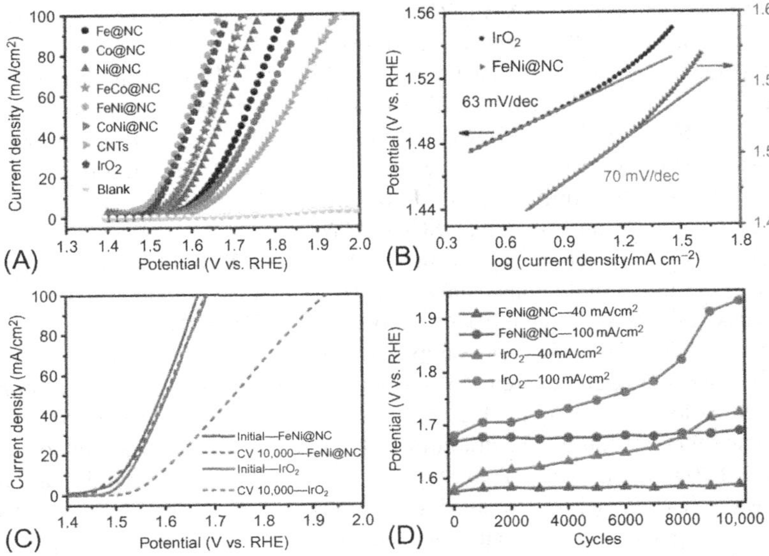

FIG. 3.26 (A) OER polarization curves for metal@NCs in comparison with CNTs and IrO_2 with the same mass loading. (B) Tafel plots for FeNi@NC and IrO_2. (C) Durability test of FeNi@NC in an alkaline electrolyte in contrast to IrO_2. (D) The potential changes in current densities at 40 and 100 mA cm^{-2} for FeNi@NC and IrO_2 during the durability test. All these tests were conducted in a 1.0 M NaOH base solution. *(Reproduced from X. Cui, P. Ren, D. Deng, J. Deng, X. Bao, Energy Environ. Sci. 9 (2016) 123–129, with the permission of the Royal Society of Chemistry.)*

cooperative effect from amorphous porous carbon and Ni-P nanoplates, resulting in the excellent conductivity and charge transfer capability, and the rapid diffusion of reactants in the electrolyte.

3.2.3 Electrochemical Sensors

Carbon-encapsulated electrocatalysts have also been intensively used in electrochemical sensors. The catalysts can enhance the electron transfer rate, electrochemical stability, and selectivity of the sensing electrodes. Electrodes modified with a carbon-encapsulated electrocatalyst have been used for detecting various inorganic or organic electroactive compounds, such as glucose, dopamine (DA), ascorbic acid (AA), hydrogen peroxide (H_2O_2), and so on [213–221].

The determination of H_2O_2 is of practical importance in chemical, biological, clinical, and many other fields [222,223]. A sensitive H_2O_2 sensor is also useful for enzyme-based biosensing because H_2O_2 is the product of most oxidase enzyme reactions [224]. You et al. reported a typical carbon confined Pt (6.5%) film nanocatalyst for electrochemical detection of H_2O_2 [213]. The film electrode was extremely active for the electro-oxidation of H_2O_2, which exhibited a much lower oxidation peak potential (0.5 V vs. Ag/AgCl) than that of a Pt bulk electrode (0.67 V vs. Ag/AgCl). An Au-embedded porous graphene

(AuEPG) thin film-based H_2O_2 sensor is another example; the response of AuEPG film sensor exhibited good linearity in the concentration range of 0.5 μM to 4.9 mM ($R = 0.998$) and high sensitivity of 75.9 μAmM^{-1} at a working potential of 0.5 V [214]. Besides noble metal-based electrocatalysts, Chen et al. reported a H_2O_2 sensor based on the hybrid nanomaterial of nonprecious Cu_2O nanocubes wrapped by graphene nanosheets (Cu_2O/GNs) [215]. The graphene coating was found to be able to effectively improve the electrochemical sensitivity and cycling stability of the fabricated sensor. As shown in Fig. 3.27A, it is obvious that the cathodic current increases with the increasing concentration of H_2O_2 and both Cu_2O and Cu_2O/GNs exhibited a fast response time for H_2O_2 (<7 s), indicating a sensitive and rapid response to H_2O_2 reduction. From the fitting lines (Fig. 3.27B), the linear current responses can be obtained in the H_2O_2 concentration ranges between 0.6–7.8 mM (correlation coefficient $R = 0.9970$; Cu_2O) and 0.3–7.8 mM ($R = 0.9989$; Cu_2O/GNs). The limits of detection for Cu_2O/GNs were calculated to be 20.8 mM in the linear concentration range, which is much lower than that of Cu_2O (64.4 mM). Moreover, Fig. 3.27C demonstrated excellent electrochemical stability after 1500 volt ammetric cycles for Cu_2O/GNs (5.4% loss) compared with Cu_2O (18.6% loss).

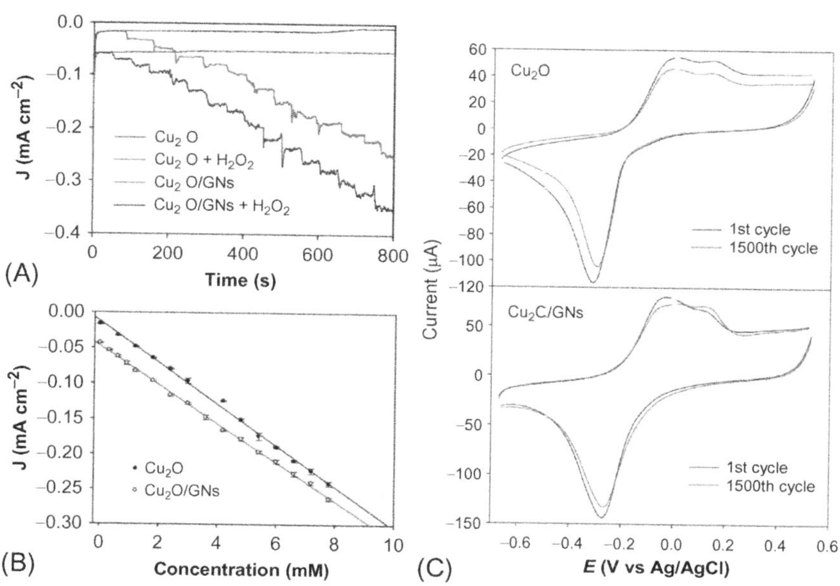

FIG. 3.27 (A) Steady-state current-time responses of the Cu_2O nanocubes and Cu_2O/GNs upon successive addition of H_2O_2 in N_2-saturated 0.1 M PBS (pH 7.4) under an applied potential of 0.4 V (vs. Ag/AgCl). (B) The corresponding calibration curves. (C) The 1st and the 1500th cyclic voltammetry curves obtained on Cu_2O nanocubes and Cu_2O/GNs with a scan rate of 0.1 V s^{-1}. *(Reproduced from M. Liu, R. Liu, W. Chen, Biosens. Bioelectron. 45 (2013) 206–212, with the permission of Elsevier.)*

Glucose determination has also attracted considerable interest in the diagnosis and management of diabetes mellitus as well as monitoring and control of food processes. Ferricyanide-confined CNT nanocomposite has been prepared as electrocatalysts for glucose determination [216]. Due to the confinement effect, the biosensor exhibited good linearity in the concentration range of 0.05–0.5 mM with a low detection limit of 0.01 mM. To further improve the electrocatalytic performance, Liu et al. reported a Ni-embedded carbon nanofiber (NiCNF)-based glucose sensor [217]. A higher stability and lower detection limit of 1 μM with wide linear range from 2 μM to 2.5 mM ($R = 0.9997$) could be obtained for NiCNF compared with Ni bulk electrode. This could be attributed to the high electrocatalytic activity of the Ni NPs confined in the CNF with a large electroactive surface area. Similar results could also be observed in graphite-like carbon film-encapsulated Ni electrode for glucose determination [218]. Recently, a new type nanohybrid of few layer graphene-encapsulated Au NP (AuNP-FLG) nanohybrid has been developed for electrochemical detection of glucose effectively [219]. The AuNP-FLG electrode exhibited a strong characteristic peak towards oxidation of glucose; however, bare ITO and FLG/ITO electrodes did not show any signal (Fig. 3.28A). The peak potential of glucose oxidation does not change, even with the increase in scan rate, suggesting a high electron transfer between the redox reaction center and the AuNP-FLG electrode (Fig. 3.28B). Furthermore, the AuNP-FLG sensor displayed good linearity in the concentration range of 6 μM to 28.5 mM ($R = 0.9937$) with a high sensitivity of 0.195 μA mM^{-1} cm^{-2} (Fig. 3.28C). The low limit of detection was found to be 1 μM. Particularly, the sensor also showed high selectivity for the detection of glucose in the presence of AA, acetaminophen (AP), and uric acid (UA) (Fig. 3.28D).

Other electrochemical sensors, for example, a DA sensor based on fullerene-coated Au electrode was reported by Sharma and coworkers [220]. As a result, the modified electrode not only exhibited strong catalytic activity towards the oxidation of dopamine with an ultralow detection limit of 0.26 nM and a high sensitivity of 0.4275 μA μM^{-1} in the presence of high concentration AA (0.1 mM), but also separated their voltammetric responses in each other. A graphitized mesoporous carbon-encapsulated Au (Au@GMC) electrocatalyst has been synthesized for efficient detection of human prostate-specific antigen (PSA) [221]. Differential pulse voltammetry tests demonstrated that the current changes were proportional to PSA concentration in the range from 0.25 to 200 ng mL^{-1}. The detection limit was found to be as low as 0.25 ng mL^{-1}.

3.3 Catalysts for Photocatalytic Reactions

Over the past decade, various semiconductors have been prepared as photocatalysts in the field of solar energy conversion and environmental protection [225–232]. However, the rapid recombination rate of photogenerated electron-hole pairs within the photocatalysts greatly decreases their catalytic

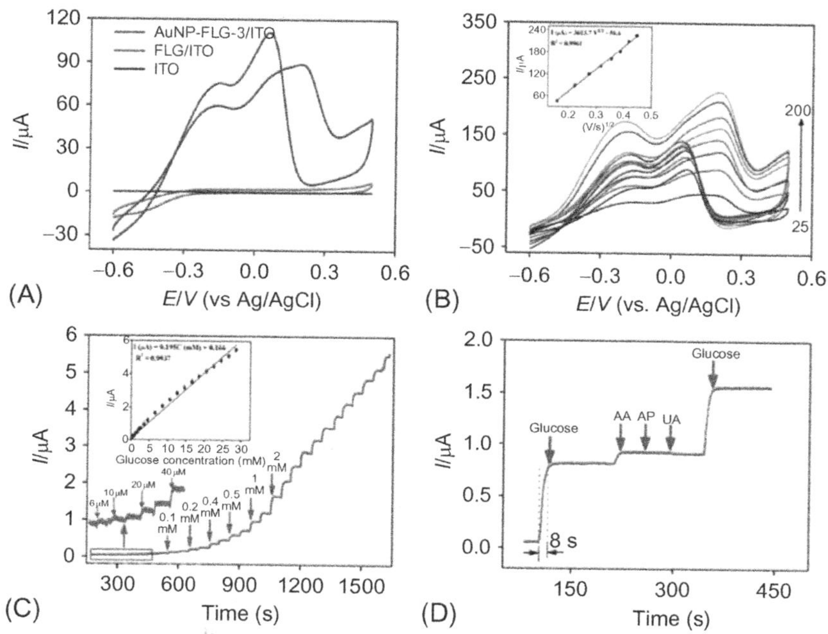

FIG. 3.28 (A) CVs of bare ITO, FLG/ITO and AuNP-FLG-3/ITO electrodes in 0.1 M NaOH containing 2 mM glucose at scan rate 50 mVs^{-1}. (B) CVs of AuNP-FLG-3/ITO electrode at various scan rates (inset: the plot of peak currents vs. square root of the scan rate). (C) Amperometric (*i-t*) curve of AuNP-FLG-3/ITO electrode with injection of different glucose concentrations at 0.0 V and regular intervals of 50 s (inset: plots of concentration of glucose vs. current). (D) Effect of interferences: 0.1 mM AA, 0.05 mM AP, and 0.1 mM UA. *(Reproduced from T. D. Thanh, J. Balamurugan, J.Y. Hwang, N.H. Kim, J.H. Lee, Carbon 98 (2016) 90–98, with the permission of Elsevier.)*

activities. Moreover, some semiconductors like metal sulfides were highly prone to suffer photocorrosion in aqueous media containing oxygen under light irradiation [233–235]. The photocorrosion effect made these semiconductors very unstable as a photocatalyst and greatly obstructed its practical application. To address these issues, nanostructured carbon-encapsulated semiconductors have been explored as a new class of photocatalysts. This section mainly summarizes their photocatalytic applications including photodegradation of organic compounds, photocatalytic water splitting, and some special photocatalytic conversion reactions.

3.3.1 Photodegradation of Organic Compounds

Serious environmental pollution has been considered as one of the major challenges in the twenty-first century. Among various technologies available today, heterogeneous photocatalytic degradation based on semiconductors making use of renewable solar energy was regarded as one of the most promising strategies for solving the environmental issues, and it has thus received much attention during recent decades [236–240]. Semiconducting TiO_2 as a well-known photocatalyst has been extensively studied. Combining the properties of both components, nanostructured carbon-encapsulated TiO_2 composites have been considered as effective photocatalysts for photodegradation of organic compounds [106,241,242]. For example, the CNT-confined TiO_2 photocatalyst exhibited improved visible-light activity in the degradation of methylene blue (MB) relative to the outside TiO_2 and commercial P25 [241]. Mechanism studies revealed that it was Ti^{3+} or oxygen vacancies generated from the modified electronic structure of TiO_2 via the confinement inside CNTs that accounts for the extension of the photocatalytic activity from the UV to the visible-light region. Lee et al. studied the performance of graphene-encapsulated TiO_2 NPs as photocatalysts for decomposing organic pollutants [106]. The rate constant of MB degradation with graphene-encapsulated TiO_2 catalyst was measured to be of 3.41×10^{-2} min^{-1}. This value was about one order of magnitude higher than that of a hydrothermally produced graphene-TiO_2 (4.73×10^{-3} min^{-1}) or anatase TiO_2 (3.28×10^{-3} min^{-1}) (Fig. 3.29A). The superior performance of the graphene-encapsulated TiO_2 catalyst was attributed to its narrower bandgap of 2.8 eV than that of TiO_2 (3.2 eV). The energy level difference between TiO_2 and graphene was changed and allowed electrons to flow from the excited MB to the conduction band of TiO_2 via graphene (Fig. 3.29B). Surface hybridization of TiO_2 with graphite-like carbon layers (TiO_2@GCLs) of a few molecular layers thickness as a photocatalyst was reported by Zhu and coworkers [242]. After 10 min irradiation with UV light about 85% of formaldehyde was degraded in the presence of TiO_2@GCLs, while the complete degradation level could be reached in approximately 20 min. Moreover, the catalyst exhibited the highest photocatalytic activity, which is about two times higher than that of Degussa P25 TiO_2. The electronic contact between semiconductor and graphite-like carbon results in an efficient separation of electron-hole pairs, which minimizes the energy-dissipating electron-hole recombination and subsequently leads to higher photocatalytic activity under UV irradiation.

Apart from carbon-encapsulated TiO_2, other semiconductor hybrids, such as carbon-encapsulated SnO_2 [243], ZnO [244,245], WO_3 [246], CdS [247,248], Ag@AgCl [249], and ZnO/Co_3O_4 [250] photocatalysts have also been explored. For instance, CNT-encapsulated CdS was used as a stable and efficient photocatalyst for the decomposition of MB under visible light irradiation [248]. It was found that the composites of CdS and CNTs showed much higher degradation activity compared with the pure CdS. The CdS-in-CNTs had a 2.5 times higher photocatalytic activity that pure CdS, which could degrade 84% of MB within 2 h (Fig. 3.30A).The reaction rate constants *k* of CdS-in-CNTs and pure CdS are 0.015 and 0.006 min^{-1}, respectively (Fig. 3.30A inset). Furthermore, after four cycles of testing, the photocatalytic activity of CdS-out-CNTs

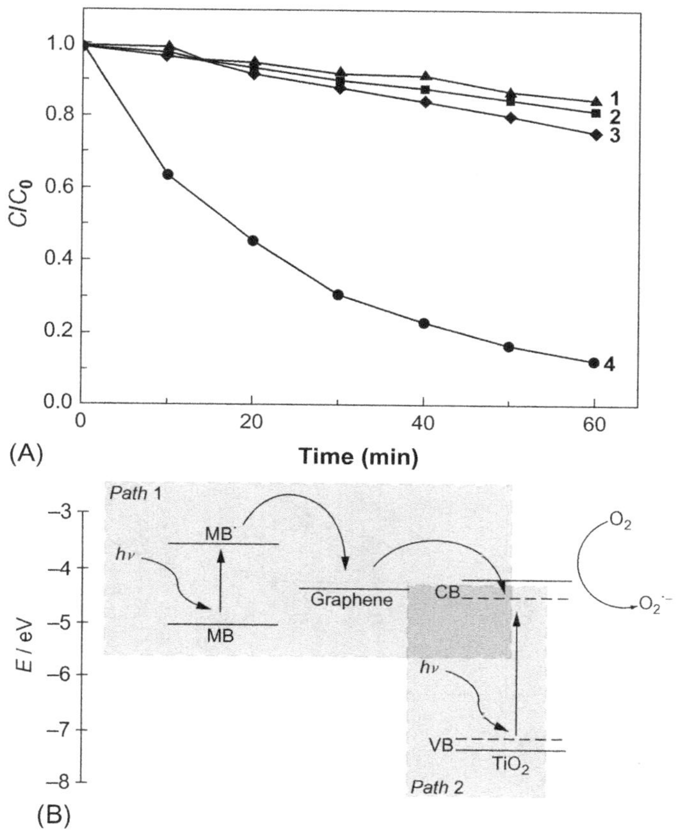

FIG. 3.29 (A) Photodegradation of MB under visible light (λ > 420 nm) by (1) P25, (2) bare anatase TiO_2 NPs, (3) graphene-TiO_2 NPs (two-step hydrothermal), and (4) graphene-TiO_2 NPs. The weight ratio of graphene to TiO_2 in the graphene-TiO_2 hybrid materials was 0.02: 1. (B) Suggested mechanism for the photocatalytic degradation of MB by graphene-encapsulated anatase TiO_2 NPs under visible light. *(Reproduced from J.S. Lee, K.H. You, C.B. Park, Adv. Mater. 24 (2012) 1084–1088, with the permission of Wiley.)*

significantly decreased, while CdS-in-CNTs maintained good, indicating that CdS-in-CNTs are photostable and not photocorroded (Fig. 3.30B). The mechanism analysis suggests that the spatial synergy of the CdS and powerful adsorptivity of CNTs are the primary causes for photocorrosion inhibition. Graphite-like encapsulated ZnO showed superior photocatalytic activity and stability under extreme pH conditions for the degradation of MB [245]. The presence of carbon layers could significantly suppress the photocorrosion of ZnO under the UV light irradiation, as well as the susceptibility of ZnO to facile dissolution at extreme pH values. Graphene-encapsulated WO_3 also exhibited much higher photocatalytic performance in degrading rhodamine

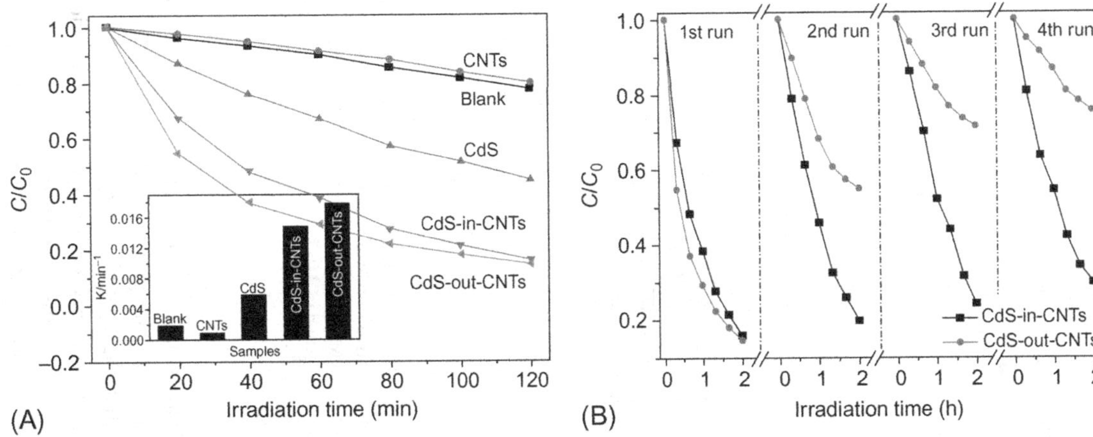

FIG. 3.30 (A) Photocatalytic degradation of MB over the as-prepared samples under visible light irradiation ($\lambda > 420$ nm). (B) Recyclable photocatalytic degradation of MB as a function of the irradiation period over the CdS-in-CNTs and CdS-out-CNTs under visible light irradiation ($\lambda > 420$ nm). The data were displayed as the normalized temporal MB concentration changes (C/C_0). *(Reproduced from X. Cui, Y. Wang, G. Jiang, Z. Zhao, C. Xu, A. Duan, J. Liu, Y. Wei, W. Bai, J. Mater. Chem. A 2 (2014) 20939–20946, with the permission of the Royal Society of Chemistry.)*

B and phenol under simulated solar light irradiation than the bare WO_3 [246]. Another photocatalyst, graphene-encapsulated ZnO/Co_3O_4(ZnO/Co_3O_4@G) core-shell hybrid, showed photocatalytic ability toward degradation of 2,4-dichlorophenol (2,4-DCP) in aqueous phase [250]. About 91% degradation of 2,4-DCP (20 mgL^{-1}) was achieved in only 150 min of visible light irradiation in the presence of 1.0 gL^{-1} of the ZnO/Co_3O_4@G catalyst at pH 5.0. The improvement in photocatalytic activity of the ZnO/Co_3O_4@G hybrid could be ascribed by an increase in contact surface between metal oxide core and graphene shell more effectively than the traditional graphene-metal oxide composites.

3.3.2 *Photocatalytic Water Splitting*

Photocatalytic water splitting to form H_2 and O_2 has attracted considerable attention as a potential means of renewable energy production. Over the past decades, various semiconductors have been prepared for photocatalytic splitting of water [251,252,197,253–255]. To improve their photocatalytic performance, recently, nanostructured carbon was used to encapsulate semiconductors for increasing their charge separation efficiencies, absorptions of sunlight, and stability [256–259]. For example, carbon-encapsulated TiO_2 has been prepared as an excellent catalyst for photodecomposing water. The coated carbon in the carbon-encapsulated TiO_2 nanotube (TNT@C) films is partly graphitized and the films have a strong photo response in the visible region. The photocatalytic tests show that the TNT@C films reveal higher photocatalytic activity for splitting $Na_2S+Na_2SO_3$ solution into H_2, compared with the TNT films [256]. A better photocatalytic H_2 generation performance could be achieved by Pt modified TiO_2-embedded CNFs [257]. The Pt photodeposition enhances the photocatalytic activity of the material due to increased electron transfer and reduced electron hole recombination.

Besides TiO_2-based photocatalysts, Ho and coworkers reported a series of ZnS/carbon photocatalysts by manipulating the forms of carbon nanomaterials from carbon dots (CDs) to a thin carbon shell on ZnS nanospheres [258]. As shown in Fig. 3.31A, under visible light irradiation, the carbon encapsulated ZnS core-shell structure showed the highest hydrogen rate of 969.6 $\mu mol\ g^{-1}\ h^{-1}$ that was 4.9 times over bare ZnS (198.0 $\mu mol\ g^{-1}\ h^{-1}$) and 1.5–4.2 times over ZnS/CDs nanocomposites (from 229.3 to 630.6 $\mu mol\ g^{-1}\ h^{-1}$). Furthermore, the photocurrent density of bare ZnS nanospheres is ca. 10 $nA\ cm^{-2}$, which is enhanced to 28 $nA\ cm^{-2}$ for the ZnS/CDs nanocomposite and 69 $nA\ cm^{-2}$ for the carbon encapsulated ZnS core-shell structure (Fig. 3.31B). The enhanced hydrogen rate for carbon-encapsulated ZnS photocatalyst should be associated with the UCPL properties and better charge transport performances of the thin carbon layer. To combat the photocorrosion problem of unstable semiconductors, carbon-coated Cu_2O nanowire arrays were synthesized as efficient photocatalysts for H_2 generation upon visible light

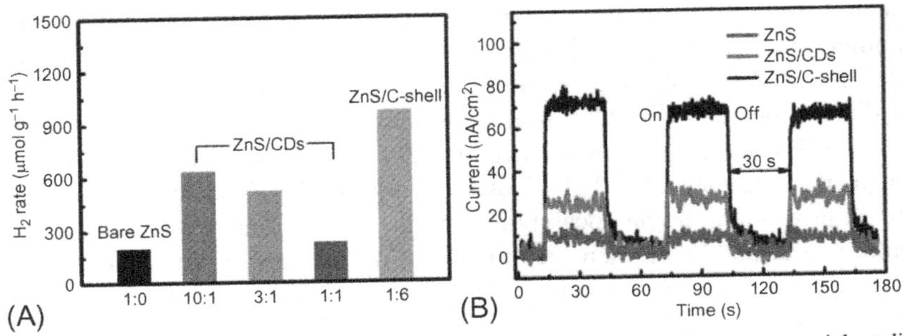

FIG. 3.31 (A) H_2 production from different samples of ZnS and ZnS/carbon nanomaterials at different weight ratios. (B) Photocurrent responses of the photoelectrochemical cells using ZnS nanospheres, ZnS/CD nanomaterials, and ZnS/carbon core-shell nanostructures as working electrodes for three 30 s light-on/off cycles. *(Reproduced from J. Wang, Y. Lim, G. W. Ho, Nanoscale 6 (2014) 9673–9680, with the permission of the Royal Society of Chemistry.)*

irradiation [259]. The carbon layer not only facilitated the electron transfer from Cu_2O to the electrolyte but also protected the Cu_2O nanowires, resulting in an improved photocurrent density and photostability.

3.3.3 Special Photocatalytic Conversions

Photocatalytic conversion of CO_2, which uses solar light energy to recycle CO_2 into fuels and chemicals, provides a promising and straightforward strategy to simultaneously reduce the atmospheric CO_2 level and fulfil the future energy demand. Recently, nanocarbon/semiconductor hybrids have been considered as ideal photocatalysts for CO_2 conversion into renewable fuels due to the synergistic effect from each component [260–263]. To increase the contact surface between semiconductor and nanocarbon, Zou et al. reported a novel graphene-encapsulated $Ti_{0.91}O_2$ hollow sphere structure for photocatalytic CO_2 conversion reaction [264]. As shown in Fig. 3.32A, a nine times and five times increase of the photocatalytic activity of graphene-encapsulated $Ti_{0.91}O_2$ hollow spheres relative to commercial P25 TiO_2 and pure $Ti_{0.91}O_2$ hollow spheres, respectively, was confirmed with photoreduction of CO_2 into renewable fuels (CO and CH_4). Furthermore, a decrease in the photoluminescence (PL) intensity of $Ti_{0.91}O_2$ hollow spheres after graphene encapsulation indicated the depression of electron-hole recommendation (Fig. 3.32B). Thus, the sufficiently compact stacking of ultrathin $Ti_{0.91}O_2$ nanosheets with graphene shell enables the photogenerated electron to quickly transfer from $Ti_{0.91}O_2$ nanosheets to graphene to enhance the lifetime of the charge carriers, leading to the improvement of the efficiency of the photocatalytic CO_2 conversion process.

Other photocatalytic conversion reactions, for instance, reduced graphene oxide wrapped Cu_2O supported on C_3N_4 (Cu_2O-rGO-C_3N_4), have been prepared for the efficient visible light-driven reduction of 4-NP to 4-AP [265]. The Cu_2O-rGO-C_3N_4 catalyst showed the highest photocatalytic activity compared with Cu_2O-rGO, Cu_2O-C_3N_4, and bare Cu_2O and C_3N_4 catalysts

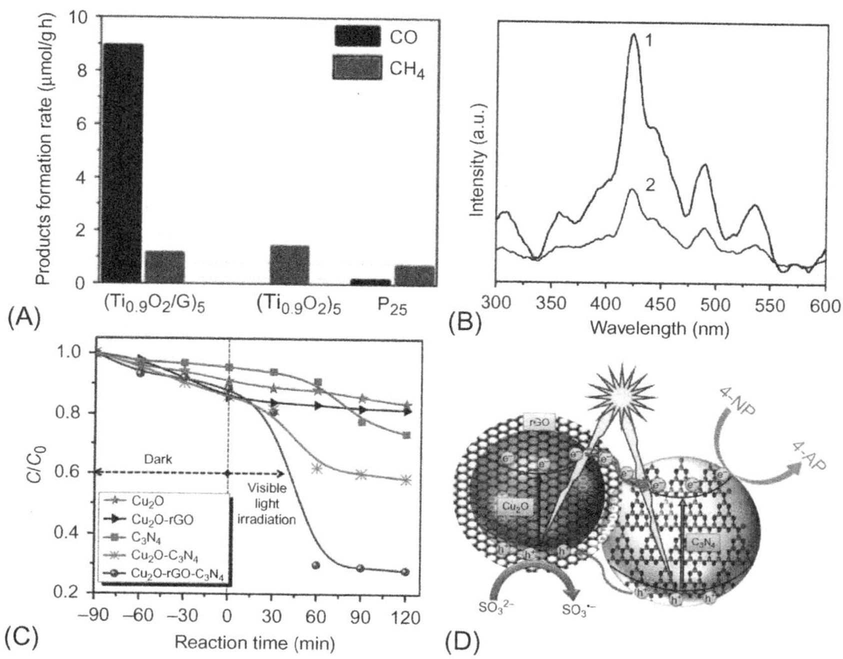

FIG. 3.32 (A) Comparison of the average product formation rates. (B) PL emission spectra of (1) $(Ti_{0.91}O_2)_5$ hollow spheres and (2) $(G\text{-}Ti_{0.91}O_2)_5$ hollow spheres. (C) Photocatalytic reduction of 4-NP to 4-AP in the presence of different photocatalysts. (D) Plausible mechanism for the photocatalytic reduction of 4-NP to 4-AP using Cu_2O-rGO-C_3N_4 photocatalyst in the presence of Na_2SO_3. *(A and B) Reproduced from W. Tu, Y. Zhou, Q. Liu, Z. Tian, J. Gao, X. Chen, H. Zhang, J. Liu, Z. Zou, Adv. Funct. Mater. 22 (2012) 1215–1221, with the permission of Wiley; (C and D) Reproduced from S.G. Babu, R. Vinoth, P.S. Narayana, D. Bahnemann, B. Neppolian, APL Mater. 3 (2015) 104415, with the permission of the American Institute of Physics.)*

(Fig. 3.32C). This can be well explained by the fact that the recombination rate was completely prevented in the Cu_2O-rGO-C_3N_4 photocatalyst, as evident by their PL studies [265]. The suggested photocatalytic mechanism for the photocatalytic reduction of 4-NP over Cu_2O-rGO-C_3N_4 catalyst in the presence of Na_2SO_3 under visible light irradiation is shown in Fig. 3.32D. When the rGO wrapped Cu_2O deposited on the surface of C_3N_4, the band structures were modified to reach the equilibrium between the Fermi levels of C_3N_4 and Cu_2O. In other words, the presence of rGO facilitated this carrier transport, and the holes were utilized for the conversion of $SO_3{}^{2-}$ to $SO_3{}^{-}$ thereby increasing the photocatalytic activity.

4 SUMMARY AND PERSPECTIVE

In summary, three widely explored nanostructured carbon (i.e., CNT, graphene, and amorphousporous carbon) encapsulated catalysts have been prepared and developed for various catalytic reactions including chemical, electrochemical, and photochemical catalysis. The superior properties of carbon-encapsulated catalysts are mainly ascribed to excellent conductivity for electron capture and transport, large specific surface area, sufficient contact surface, and their unique interaction with catalyst particles. For chemical catalysis such as gas-phase conversion reactions, CNT-encapsulated TM catalysts are widely studied as alternative materials to replace the commercial noble metal catalysts due to their confinement effects. The carbon-encapsulated nanocatalysts also demonstrate great potential in electrochemical catalysis. For instance, benefiting from the unique electron penetration effects, CNT and graphene-encapsulated 3D TMs have been widely used as electrocatalysts to enhance their HER performance in acidic media. In addition, to increase the photocatalytic activities and overcome the photocorrosion effect, nanostructured carbon-encapsulated semiconductors have also been explored as a new class of photocatalysts.

Although inspiring results have been made in recent years, the following challenges or issues still remain. First, the performance of carbon-encapsulated nanocatalysts strongly depends on their structure and composition. The relationship between the intrinsic catalytic activities of catalysts and their conjugated domains, defective sites, and functional groups should be clearly revealed in future studies. Second, rational designs of carbon-encapsulated composite structures with appropriate components, spatial arrangements, and interface engineering are essential. Third, a new type of carbon precursor with a good carbon or doped carbon yield and simple and effective uniform coating is needed for developing high-performance carbon-encapsulated nanocatalysts. Fourth, ideal carbon-encapsulated nanocatalysts should be made from single-layer carbon shell; thus, facile and effective approaches are desirable for the synthesis of single-layer carbon-encapsulated catalysts. Lastly, it is also important to produce highly active and stable nanostructured carbon-encapsulated catalysts with low costs to realize their practical applications on an industrial scale. This chapter represents a new and promising platform for the development of nanostructured carbon-encapsulated catalysts and will attract broad interest in the field of catalysis to further push it forward.

ACKNOWLEDGMENTS

This work was financially supported by the National Natural Science Foundation of China (Project No. 51173055 and No. 21305048) and the National Program on Key Basic Research Project (973 Program, Grant No. 2013CBA01600).

REFERENCES

[1] P. Ball, Designing the Molecular World, Princeton University Press, Princeton, 1994.
[2] C.-J. Li, L. Chen, Chem. Soc. Rev. 35 (2006) 68–82.
[3] T.Y. Lee, J. Suh, Chem. Soc. Rev. 38 (2009) 1949–1957.

[4] W.A. Herrmann, Angew. Chem. Int. Ed. 41 (2002) 1290–1309.
[5] P. Wasserscheid, W. Keim, Angew. Chem. Int. Ed. 39 (2000) 3772–3789.
[6] M. Kimura, J. Synth. Org. Chem., Jpn. 70 (2012) 216–226.
[7] X. Zhao, M. Yin, L. Ma, L. Liang, C. Liu, J. Liao, T. Lu, W. Xing, Energy Environ. Sci. 4 (2011) 2736–2753.
[8] P. Du, R. Eisenberg, Energy Environ. Sci. 5 (2012) 6012–6021.
[9] Y. Xiong, B.J. Wiley, Y. Xia, Angew. Chem. Int. Ed. 46 (2007) 7157–7159.
[10] A.R. Tao, S. Habas, P. Yang, Small 4 (2008) 310–325.
[11] M. Albiter, R. Crooks, F. Zaera, J. Phys. Chem. Lett. 1 (2010) 38–40.
[12] N. Pernicone, CATTECH 7 (2003) 196–204.
[13] G.A. Somorjai, F. Tao, J.Y. Park, Top. Catal. 47 (2008) 1–14.
[14] A.T. Bell, Science 299 (2003) 1688–1691.
[15] I. Lee, R. Morales, M.A. Albiter, F. Zaera, Proc. Natl. Acad. Sci. U. S. A. 105 (2008) 15241–15246.
[16] I. Lee, F. Delbecq, R. Morales, M.A. Albiter, F. Zaera, Nat. Mater. 8 (2009) 132–138.
[17] J. Grunes, J. Zhu, G.A. Somorjai, Chem. Commun. (2003) 2257–2260.
[18] C.H. Bartholomew, Appl. Catal., A 212 (2001) 17–60.
[19] R. Narayanan, M.A. El-Sayed, J. Am. Chem. Soc. 125 (2003) 8340–8347.
[20] R. Narayanan, M.A. El-Sayed, J. Phys. Chem. B 109 (2005) 12663–12676.
[21] G.J. Hutchings, M. Haruta, Appl. Catal., A 291 (2005) 2–5.
[22] Q. Zhang, I. Lee, J. Ge, F. Zaera, Y. Yin, Adv. Funct. Mater. 20 (2010) 2201–2214.
[23] J. Lu, B. Fu, M.C. Kung, G. Xiao, J.W. Elam, H.H. Kung, P.C. Stair, Science 335 (2012) 1205–1208.
[24] Q. Zhang, T. Zhang, J. Ge, Y. Yin, Nano Lett. 8 (2008) 2867–2871.
[25] W. Schartl, Nanoscale 2 (2010) 829–843.
[26] Q. Zhang, I. Lee, J.B. Joo, F. Zaera, Y. Yin, Acc. Chem. Res. 46 (2013) 1816–1824.
[27] Q. Zhang, W. Wang, J. Goebl, Y. Yin, Nano Today 4 (2009) 494–507.
[28] Q. Zhang, J. Ge, J. Goebl, Y. Hu, Z. Lu, Y. Yin, Nano Res. 2 (2009) 583–591.
[29] Y. Mei, Y. Lu, F. Polzer, M. Ballauff, M. Drechsler, Chem. Mater. 19 (2007) 1062–1069.
[30] G.L. Li, H. Möhwald, D.G. Shchukin, Chem. Soc. Rev. 42 (2013) 3628–3646.
[31] G.L. Li, C.A. Tai, K.G. Neoh, E.T. Kang, X. Yang, Polym. Chem. 2 (2011) 1368–1374.
[32] Z. Wen, J. Liu, J. Li, Adv. Mater. 20 (2008) 743–747.
[33] P. Zhang, C. Shao, Z. Zhang, M. Zhang, J. Mu, Z. Guo, Y. Liu, Nanoscale 3 (2011) 2943–2949.
[34] Y. Wan, Y.L. Min, S.H. Yu, Langmuir 24 (2008) 5024–5028.
[35] A.H. Lu, W.C. Li, N. Matoussevitch, B. Spliethoff, H. Bönnemann, F. Schüth, Chem. Commun. 1 (2005) 98–100.
[36] S. Shanmugam, A. Gabashvili, D.S. Jacob, J.C. Yu, A. Gedanken, Chem. Mater. 18 (2006) 2275–2282.
[37] S.H. Joo, J.Y. Park, C.K. Tsung, Y. Yamada, P. Yang, G.A. Somorjai, Nat. Mater. 8 (2009) 126–131.
[38] J. Lee, J.C. Park, H. Song, Adv. Mater. 20 (2008) 1523–1528.
[39] L.M. Liz-Marzán, M. Giersig, P. Mulvaney, Langmuir 12 (1996) 4329–4335.
[40] Z. Zhuang, W. Sheng, Y. Yan, Adv. Mater. 26 (2014) 3950–3955.
[41] G. Li, Z. Tang, Nanoscale 6 (2014) 3995–4011.
[42] T. Hirakawa, P.V. Kamat, J. Am. Chem. Soc. 127 (2005) 3928–3934.
[43] K. Maeda, K. Teramura, D. Lu, N. Saito, Y. Inoue, K. Domen, Angew. Chem. Int. Ed. 118 (2006) 7970–7973.
[44] Z.Y. Zhang, F. Xiao, J. Xi, T. Sun, S. Xiao, H. Wang, S. Wang, Y.Q. Liu, Sci. Rep. 4 (2014) 4053.

[45] J. Deng, P. Ren, D. Deng, X. Bao, Angew. Chem. Int. Ed. 54 (2015) 2100–2104.
[46] G. Wang, J. Hilgert, F. Richter, F. Wang, H. Bongard, B. Spliethoff, C. Weidenthaler, F. Schüth, Nat. Mater. 13 (2014) 293–300.
[47] D. Deng, L. Yu, X. Chen, G. Wang, L. Jin, X. Pan, J. Deng, G. Sun, X. Bao, Angew. Chem. Int. Ed. 52 (2013) 371–375.
[48] X. Cui, P. Ren, D. Deng, J. Deng, X. Bao, Energy Environ. Sci. 9 (2016) 123–129.
[49] X. Pan, Z. Fan, W. Chen, Y. Ding, H. Luo, X. Bao, Nat. Mater. 6 (2007) 507–511.
[50] X. Pan, X. Bao, Acc. Chem. Res. 44 (2011) 553–562.
[51] S. Iijima, Nature 354 (1991) 56–58.
[52] K.P. De Jong, J.W. Geus, Catal. Rev. 42 (2000) 481–510.
[53] P. Serp, M. Corrias, P. Kalck, Appl. Catal. A Gen. 253 (2003) 337–358.
[54] H. Yu, F. Peng, J. Tan, X. Hu, H. Wang, J. Yang, W. Zheng, Angew. Chem. Int. Ed. 50 (2011) 4064–4068.
[55] V. Lordi, N. Yao, J. Wei, Chem. Mater. 13 (2001) 733–737.
[56] B. Yoon, C.M. Wai, J. Am. Chem. Soc. 127 (2005) 17174–17175.
[57] Y. Mu, H. Liang, J. Hu, L. Jiang, L. Wan, J. Phys. Chem. B 10 (2005) 22212–22216.
[58] G.G. Wildgoose, C.E. Banks, R.G. Compton, Small 2 (2006) 182–193.
[59] X. Pan, X. Bao, Chem. Commun. 47 (2008) 6271–6281.
[60] P. Serp, E. Castillejos, ChemCatChem 2 (2010) 41–47.
[61] P.M. Ajayan, T.W. Ebbesen, T. Ichihashi, S. Iijima, K. Tanigaki, H. Hiura, Nature 362 (1993) 522–525.
[62] C. Guerret-Piecourt, Y. Lebouar, A. Loiseau, H. Pascard, Nature 372 (1994) 761–765.
[63] D. Ugarte, A. Chatelain, W.A. deHeer, Science 274 (1996) 1897–1899.
[64] E. Castillejos, P.J. Debouttiere, L. Roiban, A. Solhy, V. Martinez, Y. Kihn, O. Ersen, K. Philippot, B. Chaudret, P. Serp, Angew. Chem. Int. Ed. 48 (2009) 2529–2533.
[65] J.P. Tessonnier, O. Ersen, G. Weinberg, C. Pham-Huu, D.S. Su, R. Schlogl, ACS Nano 3 (2009) 2081–2089.
[66] J.-P. Tessonnier, L. Pesant, G. Ehret, M.J. Ledoux, C. PhamHuu, Appl. Catal., A 288 (2005) 203–210.
[67] F. Winter, G. Leendert Bezemer, C. vand der Spek, J.D. Meeldijk, A.J. van Dillen, J.W. Geus, K.P. de Jong, Carbon 43 (2005) 327–332.
[68] O. Ersen, J. Werckmann, M. Houlle, M.J. Ledoux, C. PhamHuu, Nano Lett. 7 (2007) 1898–1907.
[69] H.-Q. Wu, X.-W. Wei, M.-W. Shao, J.-S. Gu, M.-Z. Qu, J. Mater. Chem. 12 (2002) 1919–1921.
[70] J. Zhang, Y.-S. Hu, J.-P. Tessonnier, G. Weinberg, J. Maier, R. Schlogl, D.S. Su, Adv. Mater. 20 (2008) 1450–1455.
[71] B.M. Kim, S. Qian, H.H. Bau, Nano Lett. 5 (2005) 873–878.
[72] A.V. Bazilevsky, K. Sun, A.L. Yarin, C.M. Megaridis, J. Mater. Chem. 18 (2008) 696–702.
[73] Y. Sun, Y. Yao, C.G. Yan, Y. Han, M. Shen, ACS Nano 4 (2010) 2129–2141.
[74] W. Baaziz, S. Begin-Colin, B.P. Pichon, I. Florea, O. Ersen, S. Zafeiratos, R. Barbosa, D. Begin, C. Pham-Huu, Chem. Mater. 24 (2012) 1549–1551.
[75] C. Wang, S. Guo, X. Pan, W. Chen, X. Bao, J. Mater. Chem. 18 (2008) 5782–5786.
[76] S. Guo, X. Pan, H. Gao, Z. Yang, J. Zhao, X. Bao, Chem. Eur. J. 16 (2010) 5379–5384.
[77] W. Chen, X. Pan, X. Bao, J. Am. Chem. Soc. 129 (2007) 7421–7426.
[78] H. Zhang, X. Pan, J.J. Liu, W. Qian, F. Wei, Y. Huang, X. Bao, ChemSusChem 4 (2011) 975–980.
[79] M. Tavakkoli, T. Kallio, O. Reynaud, A.G. Nasibulin, C. Johans, J. Sainio, H. Jiang, E.I. Kauppinen, K. Laasonen, Angew. Chem. Int. Ed. 54 (2015) 4535–4538.

[80] K.S. Novoselov, A.K. Geim, S. Morozov, D. Jiang, Y. Zhang, S. Dubonos, I. Grigorieva, A. Firsov, Science 306 (2004) 666–669.
[81] E. Antolini, Appl. Catal., A 365 (2009) 1–24.
[82] R.J. White, R. Luque, V.L. Budarin, J.H. Clark, D.J. Macquarrie, Chem. Soc. Rev. 38 (2009) 481–494.
[83] Y.J. Song, K.G. Qu, C. Zhao, J.S. Ren, X.G. Qu, Adv. Mater. 22 (2010) 2206–2210.
[84] P. Serp, M. Corrias, P. Kalck, Appl. Catal., A 253 (2003) 337–358.
[85] M.J. Allen, V.C. Tung, R.B. Kaner, Chem. Rev. 110 (2010) 132–145.
[86] K.P. Loh, Q. Bao, P.K. Ang, J. Yang, J. Mater. Chem. 20 (2010) 2277–2289.
[87] C. Huang, C. Li, G.Q. Shi, Energy Environ. Sci. 5 (2012) 8848–8868.
[88] X. Huang, X.Y. Qi, F. Boey, H. Zhang, Chem. Soc. Rev. 41 (2012) 666–686.
[89] X. Guo, P. Liu, J. Han, Y. Ito, A. Hirata, T. Fujita, M. Chen, Adv. Mater. 27 (2015) 6137–6143.
[90] C. Wang, P. Zhai, Z. Zhang, Y. Zhou, J. Ju, Z. Shi, D. Ma, R.P.S. Han, F. Huang, Part. Part. Syst. Charact. 32 (2015) 29–34.
[91] K. Krishnamoorthy, R. Mohan, S.J. Kim, Appl. Phys. Lett. 98 (2011) 244101–2441013.
[92] P.D. Tran, L.H. Wong, J. Barber, J.S.C. Loo, Energy Environ. Sci. 5 (2012) 5902–5918.
[93] Y. Zhu, S. Murali, W. Cai, X. Li, J.W. Suk, J.R. Potts, R.S. Ruoff, Adv. Mater. 22 (2010) 3906–3924.
[94] S. Chen, L. Brown, M. Levendorf, W. Cai, S. Ju, J. Edgeworth, X. Li, C.W. Magnuson, A. Velamakanni, R.D. Piner, J. Kang, J. Park, R.S. Ruoff, ACS Nano 5 (2011) 1321–1327.
[95] B. Wang, M. Caffio, C. Bromley, H. Früchtl, R. Schaub, ACS Nano 4 (2010) 5773–5782.
[96] T. Olsen, J. Yan, J.J. Mortensen, K.S. Thygesen, Phys. Rev. Lett. 107 (2011) 156401.
[97] F. Guinea, Phys. Rev. B 75 (2007) 235433.
[98] H.A. Chen, C.L. Hsin, Y.T. Huang, M.L. Tang, S. Dhuey, S. Cabrini, W.W. Wu, S.R. Leone, J. Phys. Chem. C 117 (2013) 22211–22217.
[99] X. Li, W. Cai, J. An, S. Kim, J. Nah, D. Yang, R. Piner, A. Velamakanni, I. Jung, E. Tutuc, S.K. Banerjee, L. Colombo, R.S. Ruoff, Science 324 (2009) 1312–1314.
[100] H. Bi, F. Huang, W. Zhao, X. Lü, J. Chen, T. Lin, D. Wan, X. Xie, M. Jiang, Carbon 50 (2012) 2703–2709.
[101] D. Ding, Z. Song, Z. Cheng, W. Liu, X. Nie, X. Bian, Z. Chen, W. Tan, J. Mater. Chem. A 2 (2014) 472–477.
[102] D. Wan, T. Lin, H. Bi, F. Huang, X. Xie, I. Chen, M. Jiang, Adv. Funct. Mater. 22 (2012) 1033–1039.
[103] Q. Ou, T. Tanaka, M. Mesko, A. Ogino, M. Nagatsu, Diam. Relat. Mater. 17 (2008) 664–668.
[104] Y. Yang, Z. Lun, G. Xia, F. Zheng, M. He, Q. Chen, Energy Environ. Sci. 8 (2015) 3563–3571.
[105] S. Yang, X. Feng, S. Ivanovici, K. Mullen, Angew. Chem. Int. Ed. 49 (2010) 8408–8411.
[106] J.S. Lee, K.H. You, C.B. Park, Adv. Mater. 24 (2012) 1084–1088.
[107] J. Lee, J. Kim, T. Hyeon, Adv. Mater. 18 (2006) 2073–2094.
[108] C. Vix-Guterl, E. Frackowiak, K. Jurewicz, M. Friebe, J. Parmentier, F. Béguin, Carbon 43 (2005) 1293–1302.
[109] A. Stein, Z. Wang, M.A. Fierke, Adv. Mater. 21 (2009) 265–293.
[110] S. Subramoney, Adv. Mater. 10 (1998) 1157–1171.
[111] W.C. Choi, S.I. Woo, M.K. Jeon, J.M. Sohn, M.R. Kim, H.J. Jeon, Adv. Mater. 17 (2005) 446–451.
[112] J.B. Joo, P. Kim, W. Kim, J. Kim, J. Yi, Catal. Today 111 (2006) 171–175.
[113] J.S. Mattson, H.B.J. Mark, Activated Carbon, Decker, New York, NY, 1971.
[114] R.C. Bansal, J.B. Donnet, F. Stoeckli, Active Carbon, Decker, New York, NY, 1988.

[115] A.A. Zakhidov, R.H. Baughman, Z. Iqbal, C.X. Cui, I. Khayrullin, S.O. Dantas, I. Marti, V.G. Ralchenko, Science 282 (1998) 897–901.

[116] K. Matsuoka, Y. Yamagishi, T. Yamazaki, N. Setoyama, A. Tomita, T. Kyotani, Carbon 43 (2005) 876–879.

[117] B.Z. Fang, J.H. Kim, M. Kim, J.S. Yu, Chem. Mater. 21 (2009) 789–796.

[118] M.L. Lin, M.Y. Lo, C.Y. Mou, Catal. Today 160 (2011) 109–115.

[119] B. Liu, S. Creager, Electrochim. Acta 55 (2009) 2721–2726.

[120] S. Vengatesan, H.J. Kim, S.K. Kim, I.H. Oh, S.Y. Lee, E. Cho, H.Y. Ha, T.H. Lim, Electrochim. Acta 54 (2008) 856–861.

[121] Y. Yang, K. Chiang, N. Burke, Catal. Today 178 (2011) 197–205.

[122] C. Liang, Z. Li, S. Dai, Angew. Chem. Int. Ed. 47 (2008) 3696–3717.

[123] S. Ikeda, S. Ishino, T. Harada, N. Okamoto, T. Sakata, H. Mori, S. Kuwabata, T. Torimoto, M. Matsumura, Angew. Chem. Int. Ed. 45 (2006) 7063–7066.

[124] T. Harada, S. Ikeda, Y.H. Ng, T. Sakata, H. Mori, T. Torimoto, M. Matsumura, Adv. Funct. Mater. 18 (2008) 2190–2196.

[125] S.Y. Li, Y.R. Liang, D.C. Wu, R.W. Fu, Carbon 48 (2010) 839–843.

[126] C. Tang, L. Bombalski, M. Kruk, M. Jaroniec, K. Matyjaszewski, T. Kowalewski, Adv. Mater. 20 (2008) 1516–1522.

[127] M. Kruk, B. Dufour, E.B. Celer, T. Kowalewski, M. Jaroniec, K. Matyjaszewski, J. Phys. Chem. B 109 (2005) 9216–9225.

[128] J. Liu, S.Z. Qiao, H. Liu, J. Chen, A. Orpe, D. Zhao, G.Q. Lu, Angew. Chem. Int. Ed. 50 (2011) 5947–5951.

[129] S. Wang, W. Li, G. Hao, Y. Hao, Q. Sun, X. Zhang, A.H. Lu, J. Am. Chem. Soc. 133 (2011) 15304–15307.

[130] A.H. Lu, G. Hao, Q. Sun, Angew. Chem. Int. Ed. 50 (2011) 9023–9025.

[131] R.J. White, K. Tauer, M. Antonietti, M.-M. Titirici, J. Am. Chem. Soc. 132 (2010) 17360–17363.

[132] X. Sun, Y. Li, Angew. Chem. Int. Ed. 43 (2004) 597–601.

[133] J. Dou, H.C. Zeng, J. Phys. Chem. C 116 (2012) 7767–7775.

[134] M. Wang, J. Ma, C. Chen, F. Lu, Z. Du, J. Cai, J. Xu, Chem. Commun. 48 (2012) 10404–10406.

[135] H. Lee, S.H. Dellatore, W.M. Miller, P.H. Messersmith, Science 318 (2007) 426–430.

[136] R. Liu, S.M. Mahurin, C. Li, R.R. Unocic, J.C. Idrobo, H. Gao, S.J. Pennycook, S. Dai, Angew. Chem. Int. Ed. 50 (2011) 6799–6802.

[137] D. Zhou, L. Yang, L. Yu, J. Kong, X. Yao, W. Liu, Z. Xua, X. Lu, Nanoscale 7 (2015) 1501–1509.

[138] K. Ai, Y. Liu, C. Ruan, L. Lu, G. Lu, Adv. Mater. 25 (2013) 998–1003.

[139] J. Guo, K.S. Suslick, Chem. Commun. 48 (2012) 11094–11096.

[140] Y.H. Ng, S. Ikeda, T. Harada, S. Higashida, T. Sakata, H. Mori, M. Matsumura, Adv. Mater. 19 (2007) 597–601.

[141] A.E. Farrell, R.J. Plevin, B.T. Turner, A.D. Jones, M. O'Hare, D.M. Kammen, Science 311 (2006) 506–508.

[142] M.M. Bhasin, G.L. O'Connor, Procede de preparation selective de derives hydrocarbones oxygenes a deux atomes de carbone, Belgian Patent 824822, 1975.

[143] E.E. Santiso, A.M. George, C.H. Turner, M.K. Kostov, K.E. Gubbins, M. Buongiorno-Nardelli, M. Sliwinska-Bartkowiak, Appl. Surf. Sci. 252 (2005) 766–777.

[144] W. Chen, Z. Fan, X. Pan, X. Bao, J. Am. Chem. Soc. 130 (2008) 9414–9419.

[145] R.M.M. Abbaslou, A. Tavassoli, J. Soltan, A.K. Dalai, Appl. Catal. A Gen. 367 (2009) 47–52.

[146] Q. Yan, C. Wan, J. Liu, J. Gao, F. Yu, J. Zhang, Z. Cai. Green Chem. 15 (2013) 1631–1640.
[147] J. Zhang, J.O. Muller, W. Zheng, D. Wang, D.S. Su, R. Schlogl, Nano Lett. 8 (2008) 2738–2743.
[148] C. Galeano, R. Guttel, M. Paul, P. Arnal, A. Lu, F. Schuth, Chem. Eur. J. 17 (2011) 8434–8439.
[149] Y. Zhang, H. Zhang, G. Lin, P. Chen, Y. Yuan, K.R. Tsai, Appl. Catal. A Gen. 187 (1999) 213–224.
[150] J.P. Tessonnier, L. Pesant, G. Ehret, M.J. Ledoux, C. Pham-Huu, Appl. Catal. A Gen. 288 (2005) 203–210.
[151] A.M. Zhang, J.L. Dong, Q.H. Xu, H.K. Rhee, X.L. Li, Catal. Today 935 (2004) 347–352.
[152] H. Ma, L. Wang, L. Chen, C. Dong, W. Yu, T. Huang, Y. Qian, Catal. Commun. 8 (2007) 452–456.
[153] Y. Deng, Y. Cai, Z. Sun, J. Liu, C. Liu, J. Wei, W. Li, C. Liu, Y. Wang, D. Zhao, J. Am. Chem. Soc. 132 (2010) 8466–8473.
[154] X. Huang, C. Guo, J. Zuo, N. Zheng, G.D. Stucky, Small 5 (2009) 361–365.
[155] F. Lin, R. Doong, J. Phys. Chem. C 115 (2011) 6591–6598.
[156] H. Zhang, X. Yang, Polym. Chem. 1 (2010) 670–677.
[157] L. Hu, R. Zhang, L. Wei, F. Zhanga, Q. Chen, Nanoscale 7 (2015) 450–454.
[158] T. Harada, S. Ikeda, F. Hashimoto, T. Sakata, K. Ikeue, T. Torimoto, M. Matsumura, Langmuir 26 (2010) 17720–17725.
[159] Z. Tang, S. Shen, J. Zhuang, X. Wang, Angew. Chem. Int. Ed. 49 (2010) 4603–4607.
[160] X. Fan, M.G. Manchon, K. Wilson, S. Tennison, A. Kozynchenko, A.A. Lapkin, P.K. Plucinski, J. Catal. 267 (2009) 114–120.
[161] K.K.R. Datta, B.V. Reddy, K. Ariga, A. Vinu, Angew. Chem. Int. Ed. 49 (2010) 5961–5965.
[162] L. Carrette, K.A. Friedrich, U. Stimming, Fuel Cells 1 (2001) 5–39.
[163] F. Cheng, J. Chen, Chem. Soc. Rev. 41 (2012) 2172–2192.
[164] Y. Sun, Q. Wu, G. Shi, Energy Environ. Sci. 4 (2011) 1113–1132.
[165] D.A.C. Brownson, D.K. Kampouris, C.E. Banks, J. Power, Sources 196 (2011) 4873–4885.
[166] M. Liu, R. Zhang, W. Chen, Chem. Rev. 114 (2014) 5117–5160.
[167] U.A. Paulus, T.J. Schmidt, H.A. Gasteiger, R.J. Behm, J. Electroanal. Chem. 95 (2001) 134–145.
[168] A.C. Chen, P. Holt-Hindle, Chem. Rev. 110 (2010) 3767–3804.
[169] K.P. Gong, F. Du, Z.H. Xia, M. Durstock, L.M. Dai, Science 323 (2009) 760–764.
[170] H.B. Wang, T. Maiyalagan, X. Wang, ACS Catal. 2 (2012) 781–794.
[171] J.P. Paraknowitsch, A. Thomas, Energy Environ. Sci. 6 (2013) 2839–2855.
[172] W. Xia, A. Mahmood, Z. Liang, R. Zou, S. Guo, Angew. Chem. Int. Ed. 55 (2016) 2650–2676.
[173] Y. Nie, L. Li, Z. Wei, Chem. Soc. Rev. 44 (2015) 2168–2201.
[174] J. Masa, W. Xia, M. Muhler, W. Schuhmann, Angew. Chem. Int. Ed. 54 (2015) 10102–10120.
[175] W. Wang, B. Lei, S. Guo, Adv. Energy Mater. 6 (2016) 1–16. 1600236.
[176] L. Dai, Y. Xue, L. Qu, H.J. Choi, J.B. Baek, Chem. Rev. 115 (2015) 4823–4892.
[177] W. Wei, H. Liang, K. Parvez, X. Zhuang, X. Feng, K. Mullen, Angew. Chem. Int. Ed. 53 (2014) 1570–1574.
[178] Y. Liang, Y. Li, H. Wang, J. Zhou, J. Wang, T. Regier, H. Dai, Nat. Mater. 10 (2011) 780–786.
[179] G. Wu, K.L. More, C.M. Johnston, P. Zelenay, Science 332 (2011) 443–447.
[180] H.T. Chung, J.H. Won, P. Zelenay, Nat. Commun. 4 (2013) 1922.
[181] Z. Wu, X. Xu, B. Hu, H. Liang, Y. Lin, L. Chen, S. Yu, Angew. Chem. Int. Ed. 54 (2015) 8179–8183.

[182] M. Xiao, J. Zhu, L. Feng, C. Liu, W. Xing, Adv. Mater. 27 (2015) 2521–2527.
[183] A. Hamnett, Catal. Today 38 (1997) 445–457.
[184] W.C. Choi, J.D. Kim, S.I. Woo, Catal. Today 74 (2002) 235–240.
[185] J.H. Choi, K.W. Park, I.S. Park, W.H. Nam, Y.E. Sung, Electrochim. Acta 50 (2004) 787–790.
[186] W. Li, W. Zhou, H. Li, Z. Zhou, B. Zhou, G. Sun, Q. Xin, Electrochim. Acta 49 (2004) 1045–1055.
[187] W.H. Lizcano-Valbuena, V.A. Paganin, C.A. Leite, F. Galembeck, E.R. Gonzalez, Electrochim. Acta 48 (2003) 3869–3878.
[188] Y.G. Zhou, J.J. Chen, F.B. Wang, Z.H. Sheng, X.H. Xia, Chem. Commun. 46 (2010) 5951–5953.
[189] T. Maiyalagan, X.C. Dong, P. Chen, X. Wang, J. Mater. Chem. 22 (2012) 5286–5290.
[190] E. Yoo, T. Okata, T. Akita, M. Kohyama, J. Nakamura, I. Honma, Nano Lett. 9 (2009) 2255–2259.
[191] Y.C. Xin, J.G. Liu, Y. Zhou, W.M. Liu, J.A. Gao, Y. Xie, Y. Yin, Z.G. Zou, J. Power, Sources 196 (2011) 1012–1018.
[192] K. Kakaei, M. Zhiani, J. Power Sources 225 (2013) 356–363.
[193] G. Che, B.B. Lakshmi, C.R. Martin, E.R. Fisher, Langmuir 15 (1999) 750–758.
[194] J.A. Turner, Science 305 (2004) 972–974.
[195] T.E. Mallouk, Nat. Chem. 5 (2013) 362–363.
[196] K. Zeng, D. Zhang, Prog. Energy Combust. Sci. 36 (2010) 307–326.
[197] M.G. Walter, E.L. Warren, J.R. McKone, S.W. Boettcher, Q. Mi, E.A. Santori, N.S. Lewis, Chem. Rev. 110 (2010) 6446–6473.
[198] Y. Lee, J. Suntivich, K.J. May, E.E. Perry, Y. Shao-Horn, J. Phys. Chem. Lett. 3 (2012) 399–404.
[199] Y. Zheng, Y. Jiao, Y. Zhu, L.H. Li, Y. Han, Y. Chen, A. Du, M. Jaroniec, S.Z. Qiao, Nat. Commun. 5 (2014) 3783.
[200] M. Gao, J.X. Liang, Y. Zheng, Y. Xu, J. Jiang, Q. Gao, J. Li, S. Yu, Nat. Commun. 6 (2015) 5982.
[201] H. Fei, J. Dong, M.J. Arellano-Jimenez, G. Ye, N.D. Kim, E.L.G. Samuel, Z. Peng, Z. Zhu, F. Qin, J. Bao, M.J. Yacaman, P.M. Ajayan, D. Chen, J.M. Tour, Nat. Commun. 6 (2015) 8668.
[202] Y. Li, H. Wang, L. Xie, Y. Liang, G. Hong, H. Dai, J. Am. Chem. Soc. 133 (2011) 7296–7299.
[203] J. Tian, Q. Liu, N. Cheng, A.M. Asiri, X. Sun, Angew. Chem. Int. Ed. 53 (2014) 9577–9581.
[204] Z. Xing, Q. Liu, A.M. Asiri, X. Sun, Adv. Mater. 26 (2014) 5702–5707.
[205] J. Miao, F. Xiao, H.B. Yang, S.Y. Khoo, J. Chen, Z. Fan, Y. Hsu, H.M. Chen, H. Zhang, B. Liu, Sci. Adv. 1 (2015). e1500259.
[206] X. Long, G. Li, Z. Wang, H. Zhu, T. Zhang, S. Xiao, W. Guo, S. Yang, J. Am. Chem. Soc. 137 (2015) 11900–11903.
[207] J. Deng, P. Ren, D. Deng, L. Yu, F. Yang, X. Bao, Energy Environ. Sci. 7 (2014) 1919–1923.
[208] X. Zou, X. Huang, A. Goswami, R. Silva, B.R. Sathe, E. Mikmekova, T. Asefa, Angew. Chem. Int. Ed. 53 (2014) 4372–4376.
[209] W. Zhou, J. Zhou, Y. Zhou, J. Lu, K. Zhou, L. Yang, Z. Tang, L. Li, S. Chen, Chem. Mater. 27 (2015) 2026–2032.
[210] H. Fei, Y. Yang, Z. Peng, G. Ruan, Q. Zhong, L. Li, E.L.G. Samuel, J.M. Tour, ACS Appl. Mater. Interfaces 7 (2015) 8083–8087.
[211] Y. Zhang, L. Zuo, Y. Huang, L. Zhang, F. Lai, W. Fan, T. Liu, ACS Sustain. Chem. Eng. 3 (2015) 3140–3148.

[212] X. Yu, Y. Feng, B. Guan, X. Lou, U. Paik, Energy Environ. Sci. 9 (2016) 1246–1250.
[213] T. You, O. Niwa, M. Tomita, S. Hirono, Anal. Chem. 75 (2003) 2080–2085.
[214] Q. Xi, X. Chen, D.G. Evans, W. Yang, Langmuir 28 (2012) 9885–9892.
[215] M. Liu, R. Liu, W. Chen, Biosens. Bioelectron. 45 (2013) 206–212.
[216] L. Xiang, Z. Zhang, P. Yu, J. Zhang, L. Su, T. Ohsaka, L. Mao, Anal. Chem. 80 (2008) 6587–6593.
[217] Y. Liu, H. Teng, H. Hou, T. You, Biosens. Bioelectron. 24 (2009) 3329–3334.
[218] T. You, O. Niwa, Z. Chen, K. Hayashi, M. Tomita, S. Hirono, Anal. Chem. 75 (2003) 5191–5196.
[219] T.D. Thanh, J. Balamurugan, J.Y. Hwang, N.H. Kim, J.H. Lee, Carbon 98 (2016) 90–98.
[220] R.N. Goyal, V.K. Gupta, N. Bachheti, R.A. Sharma, Electroanalysis 20 (2008) 757–764.
[221] B. Liu, L. Lu, E. Hua, S. Jiang, G. Xie, Microchim. Acta 178 (2012) 163–170.
[222] J. Zhou, C. Liao, L. Zhang, Q. Wang, Y. Tian, Anal. Chem. 86 (2014) 4395–4401.
[223] S. Thiagarajan, T.H. Tsai, S.M. Chen, Int. J. Electrochem. Sci. 6 (2011) 2235–2245.
[224] J.M. You, Y.N. Jeong, M.S. Ahmed, S.K. Kim, H.C. Choi, S. Jeon, Biosens. Bioelectron. 26 (2011) 2287–2291.
[225] S. Linic, P. Christopher, D.B. Ingram, Nat. Mater. 10 (2011) 911–921.
[226] T.T. Xu, Q.Q. Qiao, Energy Environ. Sci. 4 (2011) 2700–2720.
[227] K.Q. Peng, S.T. Lee, Adv. Mater. 23 (2011) 198–215.
[228] S.C. Warren, E. Thimsen, Energy Environ. Sci. 5 (2012) 5133–5146.
[229] A. Mishra, P. Bauerle, Angew. Chem. Int. Ed. 51 (2012) 2020–2067.
[230] S. Chattopadhyay, L.C. Chen, K.H. Chen, NPG Asia Mater. 3 (2011) 74–81.
[231] K. Sivula, F. Le Formal, M. Gratzel, ChemSusChem 4 (2011) 432–449.
[232] M.I. Litter, Appl. Catal., B 23 (1999) 89–114.
[233] J.F. Reber, M. Rusek, J. Phys. Chem. 90 (1986) 824–834.
[234] D.N. Ke, S.L. Liu, K. Dai, J.P. Zhou, L.N. Zhang, T.Y. Peng, J. Phys. Chem. C 113 (2009) 16021–16026.
[235] A. Kudo, Y. Miseki, Chem. Soc. Rev. 38 (2009) 253–278.
[236] X.B. Chen, S.H. Shen, L.J. Guo, S.S. Mao, Chem. Rev. 110 (2010) 6503–6570.
[237] M.Q. Yang, N. Zhang, M. Pagliaro, Y.J. Xu, Chem. Soc. Rev. 43 (2014) 8240–8254.
[238] N. Zhang, Y.H. Zhang, Y.J. Xu, Nanoscale 4 (2012) 5792–5813.
[239] L. Zhao, X. Chen, X. Wang, Y. Zhang, W. Wei, Y. Sun, M. Antonietti, M.-M. Titirici, Adv. Mater. 22 (2010) 3317–3321.
[240] Q. Xiang, J. Yu, M. Jaroniec, Chem. Soc. Rev. 41 (2011) 785–796.
[241] W. Chen, Z. Fan, B. Zhang, G. Ma, K. Takanabe, X. Zhang, Z. Lai, J. Am. Chem. Soc. 133 (2011) 14896–14899.
[242] L.W. Zhang, H.B. Fu, Y.F. Zhu, Adv. Funct. Mater. 18 (2008) 2180–2189.
[243] H. Shen, X. Zhao, L. Duan, R. Liu, H. Wu, T. Hou, X. Jiang, H. Gao, Appl. Surf. Sci. 391 (2017) 627–634.
[244] Y. Guo, H. Wang, C. He, L. Qiu, X. Cao, Langmuir 25 (2009) 4678–4684.
[245] L. Zhang, H. Cheng, R. Zong, Y. Zhu, J. Phys. Chem. C 113 (2009) 2368–2374.
[246] Y. Li, Z. Tang, J. Zhang, Z. Zhang, Appl. Catal. A Gen. 522 (2016) 90–100.
[247] Y. Hu, X. Gao, L. Yu, Y. Wang, J. Ning, S. Xu, X. Lou, Angew. Chem. Int. Ed. 52 (2013) 5636–5639.
[248] X. Cui, Y. Wang, G. Jiang, Z. Zhao, C. Xu, A. Duan, J. Liu, Y. Wei, W. Bai, J. Mater. Chem. A 2 (2014) 20939–20946.
[249] Y. Min, G. He, Q. Xu, Y. Chen, J. Mater. Chem. A 2 (2014) 1294–1301.
[250] M. Rakibuddin, R. Ananthakrishnan, Photochem. Photobiol. Sci. 15 (2016) 86–98.

[251] Y. Tachibana, L. Vayssieres, J.R. Durrant, Nat. Photonics 6 (2012) 511–518.
[252] F.E. Osterloh, Chem. Soc. Rev. 42 (2013) 2294–2320.
[253] X. Wang, G. Liu, Z.-G. Chen, F. Li, L. Wang, G.Q. Lu, H.M. Cheng, Chem. Commun. 23 (2009) 3452–3454.
[254] S.K. Parayil, H.S. Kibombo, C.-M. Wu, R. Peng, J. Baltrusaitis, R.T. Koodali, Int. J. Hydrog. Energy 37 (2012) 8257–8267.
[255] H.J. Yun, H. Lee, N.D. Kim, D.M. Lee, S. Yu, J. Yi, ACS Nano 5 (2011) 4084–4090.
[256] F. Jia, Z. Yao, Z. Jiang, C.X. Li, Catal. Commun. 12 (2011) 497–501.
[257] M. Kim, A. Razzaq, Y.K. Kim, S. Kim, S. In, RSC Adv. 4 (2014) 51286–51293.
[258] J. Wang, Y. Lim, G.W. Ho, Nanoscale 6 (2014) 9673–9680.
[259] Z. Zhang, R. Dua, L. Zhang, H. Zhu, H. Zhang, P. Wang, ACS Nano 7 (2013) 1709–1717.
[260] H. Sun, S. Wang, Energy Fuel 28 (2014) 22–36.
[261] K. Mori, H. Yamashita, M. Anpo, RSC Adv. 2 (2012) 3165–3172.
[262] S. Navalon, A. Dhakshinamoorthy, M. Alvaro, H. Garcia, ChemSusChem 6 (2013) 562–577.
[263] W.Q. Fan, Q.H. Zhang, Y. Wang, Phys. Chem. Chem. Phys. 15 (2013) 2632–2649.
[264] W. Tu, Y. Zhou, Q. Liu, Z. Tian, J. Gao, X. Chen, H. Zhang, J. Liu, Z. Zou, Adv. Funct. Mater. 22 (2012) 1215–1221.
[265] S.G. Babu, R. Vinoth, P.S. Narayana, D. Bahnemann, B. Neppolian, Appl. Phys. Lett. 3 (2015) 104415.

Chapter 4

Polymer Resins as Nanoreactors for the Synthesis of Nanoparticles and Their Catalytic Application in C-C Coupling

Harjinder Kaur
Gujarat University, Ahmedabad, India

Contents

1 INTRODUCTION

Metal catalyzed carbon-carbon coupling reactions are one of the most practical methodologies that are routinely used to synthesize a wide variety of fine

Encapsulated Catalysts. http://dx.doi.org/10.1016/B978-0-12-803836-9.00004-3

chemicals, pharmaceuticals, materials etc. The 2010 Nobel Prize in Chemistry being given to Richard Heck, Ei-chi Negeshi, and Akira Suzuki is monumental proof of the usefulness of these transformations and a tribute to the revolution they have brought to the way organic transformations are conceptualized [1–4]. These coupling reactions have undergone tremendous change since their discovery in the early 20th century [5]. Generally, the reactions are catalyzed by palladium salts (2–10 mol%) or complexes in the presence of phosphine ligands in organic solvents. The reaction conditions are highly unfriendly to the environment because of the toxicity of palladium catalysts and phosphine ligands. Owing to its toxic nature, separation of the palladium catalyst from products, especially pharmaceuticals, is necessary. Many practical and economic protocols have been developed to enhance their industrialization. One of the main objectives is heterogenization of the catalyst onto insoluble inorganic or organic supports to facilitate their separation from the reaction mixture [6]. Since the advent of nanotechnology, which brought the concept of systematic synthesis of metal nanoparticles, there has been a diametric change in the amount of catalyst required [7–9]. Reactions are being carried out in the absence of toxic phosphine ligands, and organic solvents with high boiling points are replaced by water or benign solvents with low boiling point with very high turnover frequencies (TOF). Though the utility of palladium deposits as palladium black in Heck coupling reactions was previously reported [10], Beller et al. in 1996 first reported the use of palladium nanoparticles in the Heck arylation of styrene [11]. Colloidal palladium nanoparticles synthesized for this reaction were stabilized by tetraoctylammonium bromide. In the same year Reetz and coworkers independently reported the preparation of palladium colloid by electrochemical reduction [12]. They prepared highly stable Pd (0) nanoparticles stabilized by propylene carbonate. The particles were even stable under Heck reaction conditions (140–150°C). In 1997, Antoniette and coworkers reported preparation of (polystyrene)–*b*-(poly-4-vinyl pyridine) block polymer micelles for the synthesis of palladium nanoparticles and used them to study C-C coupling reactions [13]. Since then palladium nanoparticles have been synthesized on a large number of supports such as metal oxides [14], dendrimers [15], CNT [16], graphene [17], polymers [18] etc., and their catalytic activity in various C-C coupling reactions is reported. Although palladium is the most commonly metal used in C-C coupling reactions, a few reports involving Au [19–22], Ni [23], Pd-Cu [24], and Pd-Ni [25] (bimetallic) are also found in the literature.

From the plethora of reports available in the literature, the role of nanoparticles in developing environment-friendly protocols for C-C coupling reactions is unquestionable. However, these highly efficient catalytic species require protection or capping to prevent their agglomeration. Besides, efficient separation of colloidal metals from the reaction mixture is also important in terms of product purity and recyclability of the catalyst. Therefore, supporting nanoparticles onto solid supports in order to enhance their stability and ease of separation is equally essential [26,27].

The use of polymer resins in organic synthesis has changed and evolved dramatically over the past half century [28]. Cross-linked polymer resins, especially those derived from polystyrene and polyacrylic acid, were developed as ion exchangers for the purification of water. However, over the period of time researchers started using these highly porous materials in organic synthesis, initially to remove metal ions used in homogeneous catalysis and later for supporting reagents. Their popularity in organic synthesis increased tremendously from an early use in peptide synthesis as solid support (Merrifield resin) to a more recent use as stabilizer of reagents, ligands, metal ions, etc. [29,30]. The pores of resins facilitate transport of reagents and materials through them. Polarity of these resins can be changed by changing the functional groups to suit the reaction conditions. Hence, along with inorganic porous supports they also gained the attention of researchers for supporting nanoparticles and especially for solution-phase organic synthesis. A number of resins comprising cross-linked polystyrene (PS), polyacrylamide, polymethyl methacrylate (PMMA), poly vinyl pyridine (PVP), polyurea (PU), etc. have been used as supports for nanoparticles. Many of these are available in bead, pellet, or membrane forms that make their recovery and recycling very easy. A review by Biffis and Kralik [31] covers the use of polymers as a medium for catalytic support. In this chapter, we discuss the development of nanoparticle synthesis by using these resins as stabilizing agents and their subsequent applications to carbon-carbon coupling reactions such as Heck, Suzuki, Hiyama, and Sonogashira. Our focus is only on insoluble polymeric supports used for the heterogenization of catalysts. Soluble linear polymers such as PVP [18], PVA [32], and biopolymer [33] are excluded.

Polymer resins are insoluble cross-linked polymer chains that are classified as gel-type or macroreticular-type depending on the type of porosities. Both can be synthesized with different amounts of cross-linking agents [34,35]. Macroreticular or macroporous resins are synthesized in the presence of porogenic components that impart permanent micrometer-sized pores to these resins. On keeping with appropriate solvents they swell up, and besides the permanent micropores, they also develop nanoporosities due to extension of polymer chains. These nanoporosities help in stabilizing nanoparticles, whereas micropores help in transporting the material through the resin particles (Fig. 4.1).

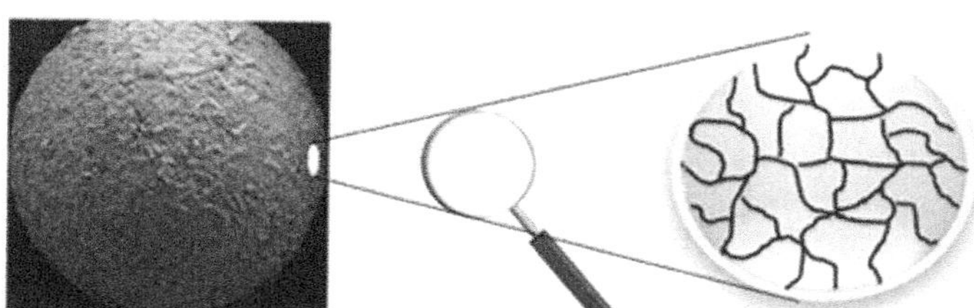

FIG. 4.1 Microscopic structure of a polymer resin bead.

Gel-type resins on the other hand do not have any permanent pores but develop porosities only on swelling. The size of the pores depends both the degree of cross-linking, type of solvent, and the nature of the polymer. Swelling in gel-type resins can be very large, resulting in a jelly-like mass. Swelling of macro-reticular resins is comparatively less than that of a gel-type resin. Both types of resins are commercially available in the form of small beads with diameters in the sub mm range or in powders. Gel-type resins can also be synthesized easily in the laboratory [28]. Though, commercially not as common as zeollites, these cross-linked resins in bead form are highly suitable for supporting catalysts especially for liquid-phase chemical reactions. In fact, as far as separation from solution phase is concerned, they are better than silicas, carbons, and zeolites, which are available in powder form only. They are also well suited for batch as well as continuous flow reactors.

Depending upon the functionalities in a polymer resin, nanoparticles can be supported on the exterior of a resin bead, interior of the bead, or on both (Fig. 4.2). Some of the reasons polymer resins are attractive for supporting nanoparticles are as follows:

1. Resins with ion-exchange groups are capable of sorbing metal ions in high concentrations and nanoparticles can be synthesized by simple reduction of sorbed metal species. That makes the synthesis very simple.
2. The nanoparticles in the resins are in close vicinity of acidic and basic groups that can provide additional catalytic sites. Especially noteworthy is the synthesis of the methyl isobutyl ketone by condensation of acetone and dihydrogen, in which the condensation step was catalyzed by the sulfonic acid group present in the resin and the hydrogenation step was promoted by the catalytic action of Pd (0) embedded in the resin [36].

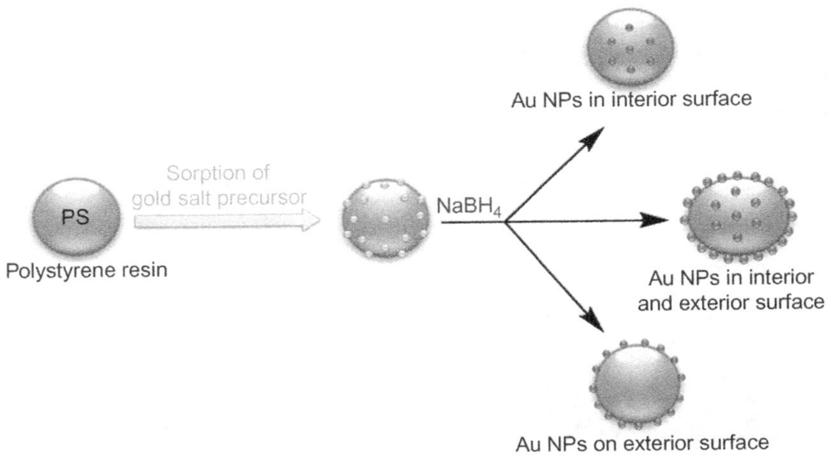

FIG. 4.2 Supporting of metal nanoparticles on polymer resins.

3. The resins can be conveniently made hydrophilic or lipophilic to suit the nature of the reagents required for the chemical reactions. This can have beneficial consequences on the selectivity of the catalyst.
4. The resins, especially polystyrene-based ones, are thermally stable up to 250°C and are ideal for solution-phase reactions. Nonfunctional resins are also chemically inert and do not interfere with the reaction conditions.
5. Their easy filterability in the lab makes them well suited to fulfill the needs of purification, reusability, and sustainability.

In general, these supports are highly versatile and can be easily modified as per reaction conditions and requirements. Mechanically, resin beads are slightly fragile and may disintegrate under hard stirring.

2 GENERAL METHODS FOR NANOPARTICLE SYNTHESIS AND CHARACTERIZATION

General methods reported for the synthesis of metal nanoparticles for porous materials can also be applied to the synthesis of supported nanoparticles [37]. Swelling of a resin in a good solvent is important to accomplish incorporation of metal into the polymer matrix and its uniform distribution. The percentage of swelling of the resin depends on the degree of cross-linking used and the nature of functional groups. For example, an ion-exchange polystyrene resin with a polar group will swell even in water, whereas a simple nonfunctional polystyrene will swell up only in organic solvents. Hence, a judicious choice of solvent is important. Resins should be thoroughly washed to remove any salts before starting the synthesis of nanoparticles. Three general methods of nanoparticle synthesis on polymeric supports can be classified as follows:

2.1 In Situ Synthesis of Nanoparticles in the Presence of Resin

This method is commonly known as the sorption reduction method (Fig. 4.3). This involves introduction of a metal precursor into the support by equilibrating the resin with the precursor solution followed by reduction. Most of the polymer

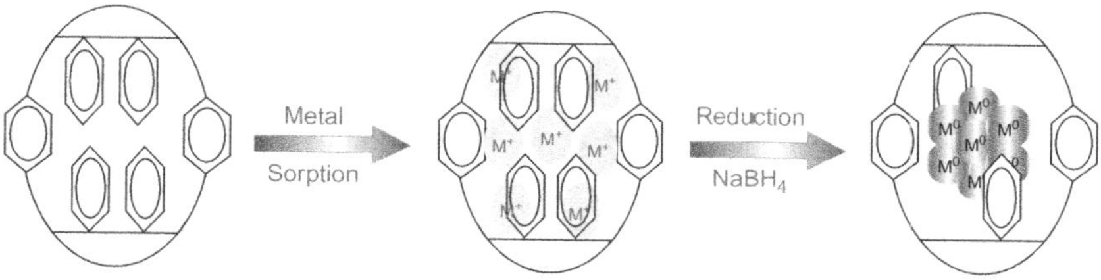

FIG. 4.3 Sorption reduction method for nanoparticle synthesis. *(Adapted from D. Shah, H. Kaur, A general sustainable catalyst for microwave enhanced carbon-carbon coupling reactions, J. Mol. Catal. A Chem. 424 (2016) 171–180, with permission of Elsevier.)*

resins having acidic, basic, or chelating groups can absorb metal precursors through ion-exchange or chelation. Metal uptake is generally quantitative and catalyst loading can be controlled by using an appropriate concentration of a metal precursor. If an appropriate solvent is chosen that can swell a resin, metal ions can penetrate inside the resin bead and uniform loading of the metal nanoparticles inside the resin is obtained. If a solvent is chosen that cannot swell the resin beads then nanoparticles are formed close to the surface of the resin beads.

In the synthesis of nanoparticle supported on nonfunctional polystyrene resins like HPS and Amberlite XAD-4, the resin beads were swollen using solvents like THF or ethanol before equilibrating them with the metal ion solution in the same solvent [38,39]. The resin particles were filtered off and an aqueous solution of sodium borohydride was passed through them. On reduction, nanoparticles were generated whose size was restricted by the resin pores. The metal loading on nonfunctional resin is comparatively low as compared to ion exchange resins. Here, passive diffusion of metal into micropores of the resin is solely controlled by the concentration of the precursor ion, although some complexing with the polystyrene rings has also been indicated [40]. The size of the nanoparticles may or may not be affected by the concentration of metal ions. In case of HPS, which has a highly cross-linked rigid structure, it has been reported that size of nanoparticles was completely controlled by resin pores. When the HPS batch with 2-nm pores was impregnated with a THF solution of H_2PtCl_6, dried, and reduced with H_2 for 3 h at room temperature, resin was impregnated with platinum nanoparticles of 1.3 nm. The closeness of the nanoparticles' size to pore volume obtained in HPS is really interesting [38]. In the case of Amberlite XAD-4, 5 g of resin was equilibrated with 25 mL of 1.0 mmol of palladium acetate solution prepared in ethanol at 10°C. After 1 h, excess solution was filtered off and 10 mL of cold aqueous $NaBH_4$ solution was passed through the beads. The colorless resin beads turned gray on impregnation with palladium nanoparticles of 5–10 nm [39].

Macroreticular resins have both micro and nano porosities. If a high concentration of metal precursor is used, then large nanoparticles are formed in the permanent pores of the polymer framework. In a pioneering experiment by Hanson et al., Amberlyst 15 (average pore size 20–60 nm) was used to synthesize nanoclusters. Palladium, platinum, silver, and nickel nanoparticles were synthesized by the sorption reduction method. The metal precursors were reduced by dihydrogen or hydrazine. In addition to the 2-nm size nanoparticles, they also observed large nanoparticles ranging from 10 to 30 nm. This data supported the conclusion that small nanoparticles are formed in nanopores after swelling of resin and large nanoparticles are formed in the micropores or permanent pores [41]. But the size of the nanoparticles also depends on the concentration of the metal ion solution. The distribution of the nanoparticles inside the resin matrix is also affected by the nature of the reducing agent. Nanoparticles were better dispersed in the resin matrix when reduced with dihydrogen. A correlation between the pore size distribution by size exclusion chromatography and the distribution of metal clusters size by TEM was carried out by Corain et al. for the copolymer of *n*-dodecyl methacrylate and vinyl pyridine cross-linked with ethylene glycol dimethacrylate (Copoly-[DOMA-VPY-EDMA]) [42].

2.2 In Situ Polymerization in the Presence of Nanoparticles

This method is more complicated because it involves synthesis of a polymer in the presence of preformed nanoparticles or a metal precursor. In this technique a metal precursor is included at the time of polymerization. It may get converted to nanoparticles during polymerization by decomposition, or else it can be reduced afterward. Though not very popular, this method was used to prepare EnCat's first commercial catalyst containing palladium nanoparticles [43]. Polymerization of isocyanate monomers was carried out in the presence of palladium acetate. After polymerization, the palladium precursor was reduced by using dihydrogen or formic acid. This method is much more suitable for condensation polymers. Polymerization processes catalyzed by free radical initiators are difficult to sustain in the presence of transition metals (Pd) with multiple oxidation states. A greater amount of the initiator AIBN (4% compared to usual 1%) was used to synthesize polyvinyl pyrrolidone cross-linked with diethyl phthalate in the presence of $PdCl_2$. X-ray photoelectron microscopy (XPS) studies confirmed that palladium species were reduced to Pd (0) during the polymerization process [44].

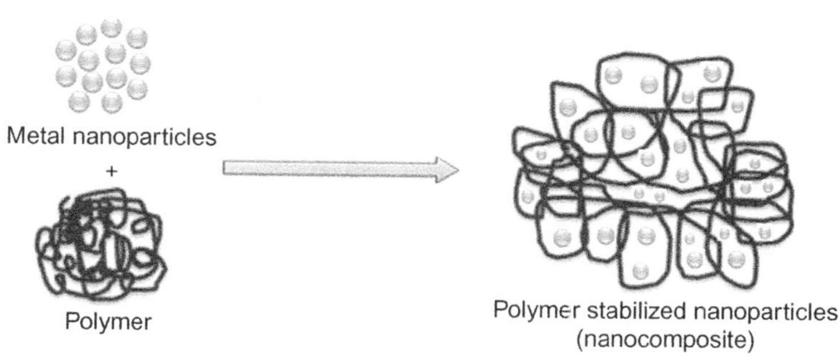

FIG. 4.4 Ex situ method for nanoparticle synthesis on polymer supports.

2.3 Ex Situ Synthesis

This involves synthesizing nanoparticles of known sizes by already established techniques and supporting them onto the polymers. The method is more suitable for linear polymers (Fig. 4.4). However, crosslinked polymer resins has also been used in few reports. Tarashaker Pal et al. synthesized gold nanoparticles by the well-known Frens method [45]. The synthesized nanoparticle solution was passed through an activated anion exchange resin column at a flow rate of 2 mL/min. As the nanoparticles were exchanged with resin beads their color changed from yellow to black.

Characterization of nanoparticles supported on polymeric resin can be done using methods similar to those used for other solid supports. Besides the usual methods used for characterization of polymers, some additional methods are required for characterization of nanoparticles. Microscopy is a very important tool for understanding the morphology and size of nanoparticles. Scanning electron microscopy (SEM) can be used for determining the presence of nanoparticles on the surface of polymer beads. In nonfunctional polymer resins such as AmberliteXAD-4, no agglomerations were observed on the surface when observed using SEM (Fig. 4.5), indicating that nanoparticles are formed inside the polymer beads only [22]. However, in case of AuNPs@PAMAM modified XAD-4 nanoparticles were also observed at the surface of beads [46].

Transmission Electron Microscopy (TEM) is especially very useful for determining the size and shape of the nanoparticles. As polymer resins do not disperse in solvents, preparation of the sample requires special effort. If a polymer is soft it can be sliced using a microtome and a section can be observed under TEM. Hard resins can be milled and a drop of alcoholic suspension is then placed on a carbon coated grid (Fig. 4.6). After drying the solvent the grid can be used for microscopy [47]. HRTEM attached to EDX can also be used for determining the metal loading. However, these values are area specific and may not give exact loading of metal if it is not uniformly distributed throughout the polymer matrix.

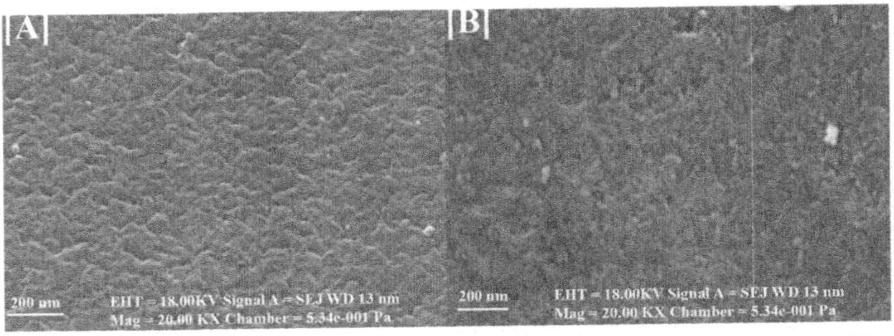

FIG. 4.5 SEM images (A) XAD-4 resin (B) Au NPs@ XAD-4-$G_{3.0}$ PAMAM dendrimer. *(Reproduced from A. Sharma, D. Shah, H. Kaur, Gold nanoparticles supported on dendrimer@ resin for the efficient oxidation of styrene using elemental oxygen, RSC Adv. 5 (2015) 42935–42941, with permission from The Royal Society of Chemistry.)*

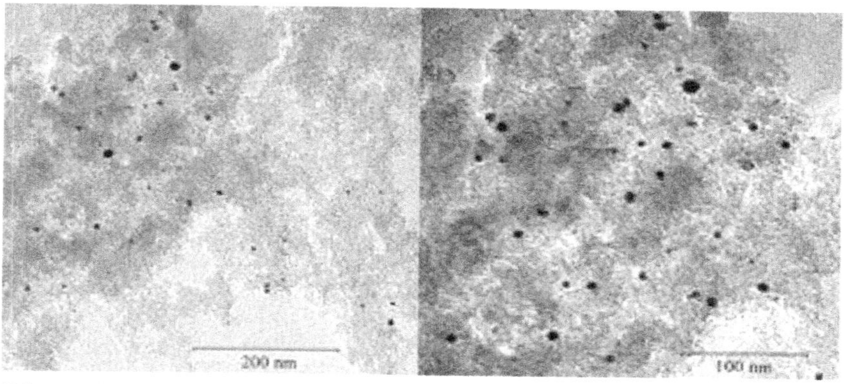

FIG. 4.6 TEM pictures of AuNPs @ Amberlite XAD-4 taken after milling. *(Reproduced from D. Shah, H. Kaur, Resin-trapped gold nanoparticles: an efficient catalyst for reduction of nitro compounds and Suzuki-Miyaura coupling, J. Mol. Catal. A Chem. 381 (2014) 70–76, with permission of Elsevier.)*

Powder XRD is another very useful technique for the characterization of resin supported nanoparticles. It is primarily used for phase identification of a material and provides information on unit cell dimensions. Due to the amorphous nature of polymer they generally give a signal below $2\theta = 30$, whereas the peaks for metal crystallite appear above this (Fig. 4.7). Scherrer's equation can be used to calculate the size of nanoparticles from the width of the peaks [48]. If the nanoparticles are very small in size (<5 nm) there is a considerable broadening of peaks. Sometimes this may lead to a difference in calculated size from that observed using TEM.

The surface plasmon of nanoparticles can be observed by UV-visible spectroscopy or by reflectance. AuNPs exhibit a strong surface plasmon resonance (SPR) absorption band that is absent in the spectrum of the bulk metal. This band is dependent on the size, shape, and aggregation of AuNPs. In the case of XAD-4 stabilized AuNPs, an SPR band was observed at 535 nm, which is the characteristic of the gold nanoparticles (Fig. 4.8). Mie theory can be used to determine the size of the nanoparticles from the wavelength of the SPR band. However, a shift in the wavelength can occur due to the stabilization effect of the ligand and π electron cloud of the benzene ring in resins [49].

Brunauer-Emmett-Teller (BET) method is an important technique for the measurement of the surface area of a catalyst. The analysis was carried out by nitrogen multilayer adsorption measured as a function of relative pressure using a fully automated BET analyzer. The technique encompasses external area and pore area evaluations to determine the total specific surface area in m^2/g yielding important information about surface porosity and particle size in many applications. In resin supported nanoparticles accessibility to the

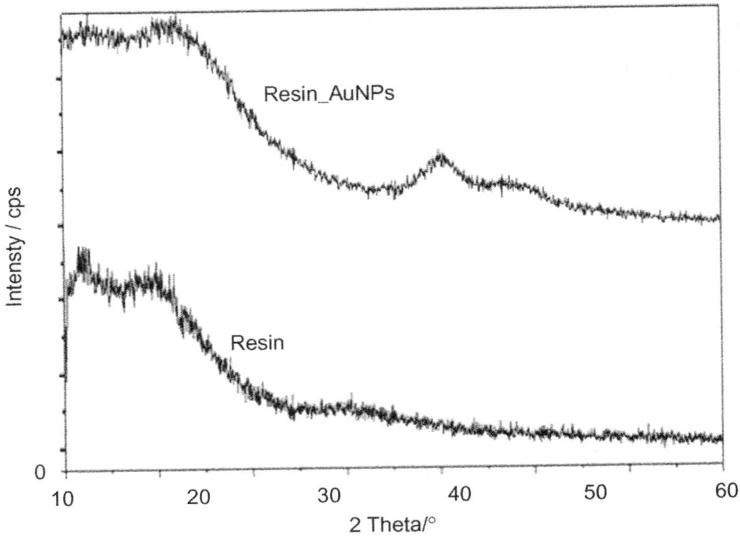

FIG. 4.7 XRD pattern of the polystyrene resin, resin supported AuNPs. *(Reproduced from D. Shah, H. Kaur, Resin-trapped gold nanoparticles: an efficient catalyst for reduction of nitro compounds and Suzuki-Miyaura coupling, J. Mol. Catal. A Chem. 381 (2014) 70–76, with permission from Elsevier.)*

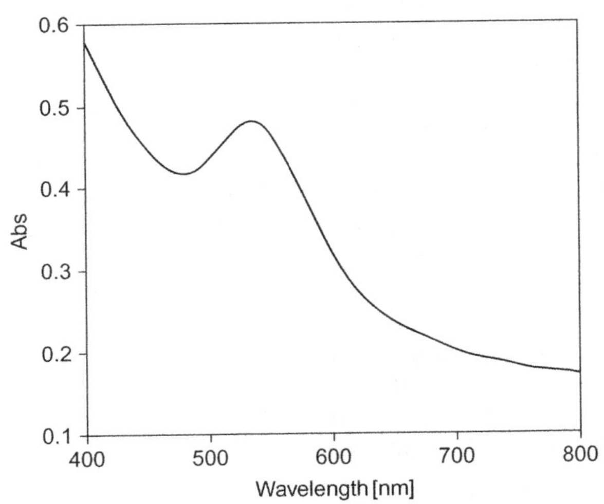

FIG. 4.8 UV-visible spectrum of AuNPs@Resin taken after milling. *(Reproduced from D. Shah, H. Kaur, Resin-trapped gold nanoparticles: an efficient catalyst for reduction of nitro compounds and Suzuki-Miyaura coupling, J. Mol. Catal. A Chem. 381 (2014) 70–76, with permission from Elsevier.)*

catalyst is highly dependent on the pore size. Hence, BET analysis for determination of the surface area is of prime importance for newly synthesized polymers [50].

3 CARBON-CARBON COUPLING REACTIONS CATALYZED BY RESIN-SUPPORTED METAL NANOPARTICLES

In the last century the heterogenization of catalyst was popular only for bulk industrial processes. However, growing environmental needs have urged academia as well as small pharmaceutical industries to adopt cleaner technologies. With the advent of nanocatalysts, the support has gained much attention not only for stabilization of the particle but also to solve the complex issue of nanoparticle separation from the reaction mixture.

3.1 Nanoparticles Supported on Cross-Linked Polystyrene Resins

Hansen et al. were first to note the ability of cross-linked polystyrene resins to act as templates for the synthesis of nanoparticles [41]. Cross-linked polystyrene resins (PS) are commercially available with different brand names and are mostly prepared by copolymerization of styrene with divinyl benzene (DVB). The porosity of the surface area of these resins depends upon the degree of cross-linking. They are commercially available with different % of DVB both in the form of nonfunctional hydrophobic types or functionalized (Fig. 4.9) hydrophilic types (R = sulfonic, carboxylic, amino, ammonium, etc.). They can also be functionalized with various chelating groups. Small nanoparticles formed in the polymer network are stabilized by the steric and electrostatic interactions of benzene rings or due to covalent and noncovalent interactions with various functional groups.

Hyper cross-linked polystyrene resins (HPS)—with their unique topologies, high surface, and nanostructure cavities—first drew the attention of researchers for the synthesis of platinum nanoparticles with controlled particle size. They

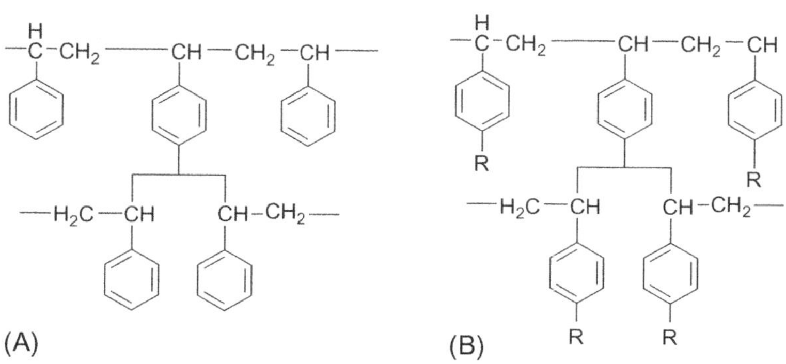

(A) (B)

FIG. 4.9 Structure of cross-linked polystyrene. (A) Nonfunctional; (B) R is any functional group.

are PS and PS-DVB resins that are 100% cross-linked by methylene bridges. These bridges are formed by treatment of polystyrene with ethylene dichloride and possess a highly rigid nanoporous framework with a very high surface area (1000 m^2 g^{-1}). The impregnation of the resin with platinum nanoparticles was reported by a simple sorption reduction method. TEM images revealed very narrow particle size distribution with a mean diameter of 1.4 nm. However, the catalyst showed poor activity owing to the fully nanoporous nature of the resin, which limited the transport of the material through the resin [51].

Sajiki et al. in 2008 reported DIAION HP20, another commercially available polystyrene resin that is used as a synthetic adsorbent, possessing a very high surface area (590 m^2 g^{-1}) and fine pore structure (30 nm of radius) for the preparation of supported palladium nanoparticles. Catalyst with concentrations of palladium varying from 0.5 to 10 mol% were synthesized and used to catalyze a ligand-free Suzuki Miyaura cross-coupling reaction. The authors were of the opinion that the catalyst can replace Pd/C in various reactions. The catalyst was separated by simple filtration and reused up to fourth run without any significant loss in the catalytic activity. Furthermore, no leached Pd species was detected in the filtrate after the removal of the catalyst by inductively coupled plasma (ICP) assay [40].

Kaur et al. synthesized Pd and AuNPs supported on polystyrene resin and used them for a number of C-C coupling reactions (Fig. 4.10) under microwave

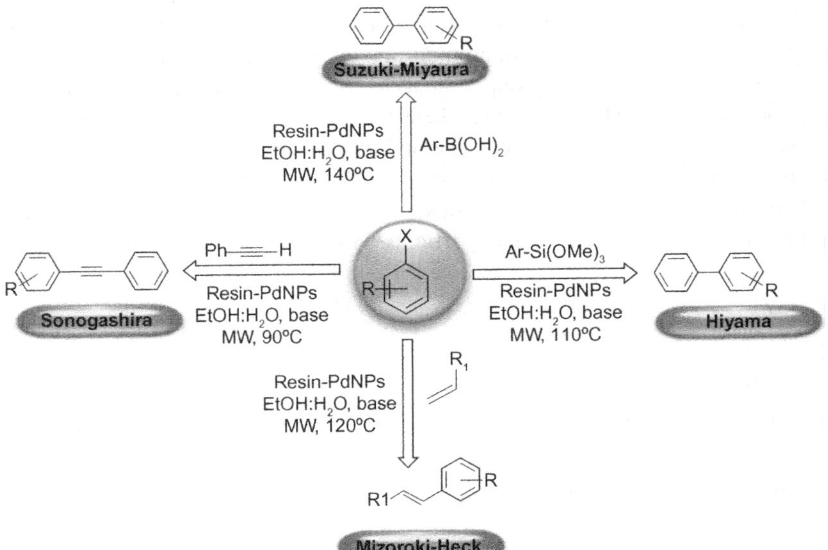

FIG. 4.10 Coupling reactions catalyzed by PdNPs@Amberlite XAD-4 under microwave heating. X = I, Br, Cl, R = –H, –OCH_3, –CH_3, –$COCH_3$, –CHO, –NO_2, –$NHCOCH_3$ R′ = –$COOCH_3$, Phenyl. *(Adapted from D. Shah, H. Kaur, A general sustainable catalyst for microwave enhanced carbon-carbon coupling reactions, J. Mol. Catal. A Chem. 424 (2016) 171–180, with permission of Elsevier.)*

heating [22,38,47,52]. Microwave-assisted heating in organic synthesis using a special synthesizer has become an integral part of most chemical laboratories owing to unprecedented acceleration in the reaction rates and improved yields. [53]. Leadbeater was the first to report a ligand-free Heck reaction of bromobenzene and styrene in water using TBAB as a surfactant under microwave heating at 170°C [54]. In our lab we synthesized palladium and gold nanoparticles on Amberlite XAD-4, a commercial polystyrene cross-linked macroporous resin. This resin is widely used as solid-phase support in organic synthesis. Nanoparticles were synthesized by the sorption reduction method. The metal sorption was governed by rules of diffusion only and it was kept low to prevent overheating of the resin during the microwave heating. At low metal concentrations, nanoparticles of less than 10 nm were obtained easily. SEM image of the resin beads did not show any particles or agglomerations on the surface. A number of C-C coupling reactions were studied using these resins in CEM microwave reactors using closed vessels with excellent conversions and yields. Ethanol/water mixture in various proportions was found suitable for all the reactions. An additional feature of microwave heating is its ability to heat solvents far above their boiling points using closed vessels. Thus, we could eliminate the need for high boiling solvents to reach high temperatures and were able to carry out reaction a at temperatures varying from 90°C to 140°C using solvents with low boiling points.

Some of the other highlights of the developed protocols were simplicity, wide functional group tolerance, easy separation, and recyclability of the catalyst. By using microwave heating, reaction times were reduced tremendously as shown in Fig. 4.11. No side products were observed and in most cases products were purified by simple crystallization [55]. While the presence of ligands/functional groups on a resin can enhance the stability of nanoparticles, strong ligands can reduce the catalytic activity [8]. The nanoparticles stabilized onto

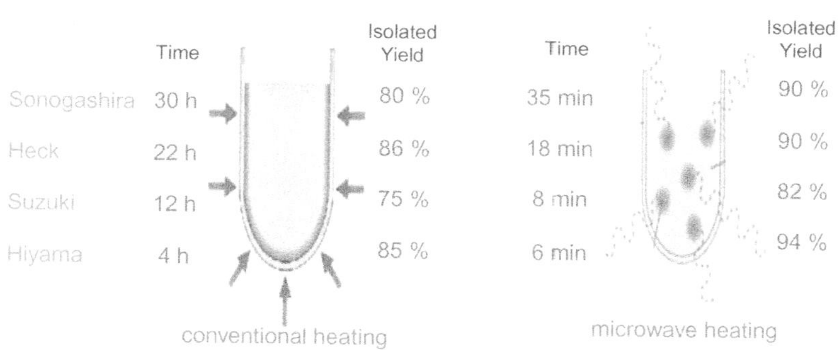

FIG. 4.11 A comparison of reaction time and yield under microwave heating with conventional heating. *(Reproduced from D. Shah, H. Kaur, A general sustainable catalyst for microwave enhanced carbon-carbon coupling reactions, J. Mol. Catal. A Chem. 424 (2016) 171–180, with permission of Elsevier.)*

nonfunctional resin by us were highly active, which was reflected by their high turnover number (TON) compared to other supported palladium nanoparticles. The presence of permanent pores and the hydrophobic nature or the resin creates a favorable mass transfer of the materials inside the resin matrix, which allows the reactants to come in contact with the catalyst, leading to high reaction rates. At the same time volatility of the products throws them out of the vicinity of hot metal nanoparticles embedded in resin, giving TON of 10^3 or more which is very high compared to other reported catalysts.

3.2 Carbon-Carbon Coupling Reactions Catalyzed by Nanoparticles Supported on Commercial Ion-Exchange Resins

Ion-exchange resins are generally cross-linked polystyrene-co-divinylbenzene or polyacrylate-co-divinylbenzene resins that are functionalized with acidic or basic groups (Fig. 4.12). The resins are known as cation exchange resins if they contain sulfonic, carboxylic, and phenolic groups; they are known as anion exchange resins if the functional group is a quaternary ammonium or ammonium group. Here the strong electrostatic field force of the charged ions and physical encapsulation by a polymer network can stabilize nanoparticles. Nanoparticles of a variety of metals such as Pd, Ag, Au, etc. have been synthesized by the sorption reduction method. The metal loading is dependent on the ion-exchange capacity of the resin, which in turn depends upon with size and charge on the metal ion. Charged functional groups on the resin and pores help in stabilizing the nanoparticles. Though a large number of catalytic applications of metal nanoparticles supported on ion-exchange resins are reported in literature, in the present review we are restricting our discussion to carbon-carbon coupling reactions only. For more details, readers can refer to a recent review by Liguori, Marrodan, and Barbaro [35]. Resins with basic groups are more suitable because coupling reactions are mostly carried out in the presence of a base.

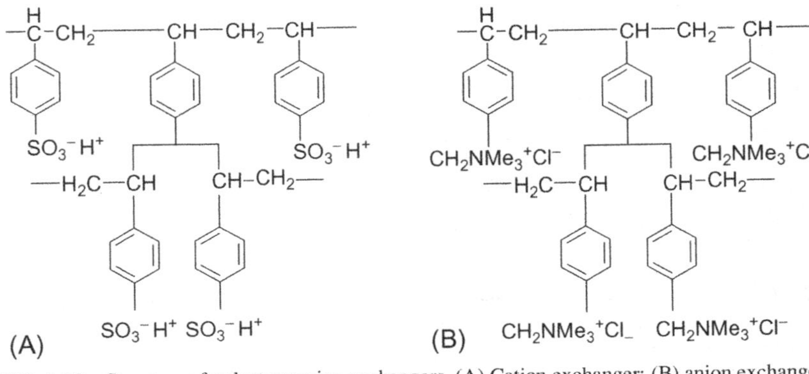

FIG. 4.12 Structure of polystyrene ion exchangers. (A) Cation exchanger; (B) anion exchangers.

Reetz and coworkers have reported a Suzuki reaction with palladium nanoparticles stabilized with $R_4N^+X^-$. Taking a cue from his work ionic polymers with a quaternary ammonium group were used to synthesize PdNPs and used to catalyze Suzuki reaction. Liang et al. used a metal organic chemical vapor deposition (MOCVD) approach to load Amberlite IRA-900(OH), a basic ion-exchanger, with a volatile palladium precursor $Pd(C_3H_5)(C_5H_5)$. The precursor was reduced by passing H_2 and hydrocarbon residue was removed by vacuum evacuation. Palladium nanoparticles of an average size of 2.6 nm were reported, whereas a pore size of 20 nm was reported from N_2 adsorption. It was reported that vapor deposition methods resulted in better dispersion of nanoparticles in the resin matrix. Ethanol/water in a 1:1 ratio was used as a solvent. A 99% conversion of iodobenzene was obtained at 70°C with a TOF of 2486 h^{-1}. The catalyst was used five times with a slight loss in activity. They also mentioned that when the chloride form of the ion-exchanger was replaced with an OH^- form, it lost its catalytic activity [56].

In another report, core shell Pd/Co nanoparticles supported on a sulfate ion-exchange resin were reported by an ex situ method. The bimetallic nanoparticles were present near the surface of the resin. They were found to be active for the cross-coupling of 4-bromo acetophenone with phenyl boronic acid in DMF/water using K_2CO_3 as a base [57]. Macronet MN100, a hyper cross-linked PS-DVB resin functionalized with tertiary amine groups, was used to synthesize palladium nanoparticles by the sorption reduction method. $PdCl_2$ was used as a precursor and reduced with sodium formate and sodium hydroxide in a solution. The catalyst was used to study Suzuki reaction. The catalyst was shown to have good efficiency for aryl bromides and chlorides using Cs_2CO_3 as a base [58].

Though Suzuki and Heck are the most common reactions explored by polymer stabilized nanoparticles, Sengupta et al. explored the feasibility of a Sonogashira reaction using bimetallic Pd/Cu nanoparticles. Cationic macroporous Amberlite resin with formate as a counteranion was used to support bimetallic Pd/Cu nanoparticles. Co-impregnatiom method was used to synthesize well dispersed nanoparticles with average size of 4.9 nm. XRD studies indicated that these nanoparticles were composed of three phases: Pd, PdO, and CuO. The synthesized catalyst exhibited high catalytic activity for Sonogashira cross-coupling reaction between aryl iodide and alkynes in CH_3CN at 80°C. Hot filtration and catalyst poisoning tests were carried out to prove heterogeneity of the catalyst. No homocoupled products were observed in the reaction mixture [59].

3.3 Carbon-Carbon Coupling Reactions Catalyzed by Nanoparticles Supported on Copolymers

In commercial resins, metal-complexing or ion-binding functional groups are usually highly polar moieties such as quaternary ammonium groups, sulfonic acid groups, etc., whereas most of the reactants used in C-C coupling reactions are hydrophobic in nature. To maintain a balance between metal complexing species that can enhance loading as well as lipophilicity, many copolymers have been used to stabilize palladium nanoparticles.

TentaGel, a commercially available copolymer of polystyrene and PVA, was used to immobilize palladium nanoparticles by Bradley et al. Amino methylated TentaGel was treated with a palladium acetate solution in toluene at 80°C and after sorption it was reduced by passing a solution of hydrazine hydrate in methanol. The resulting resin was cross-linked with succinyl chloride, which increased the stability of the catalyst. The average size of the nanoparticles was found to be 7.4 nm from TEM analysis. Suzuki coupling of aryl boronic acid with aryl bromides and chlorides was attempted using 10 mol% of the catalyst with K_2CO_3 as a base. Quantitative results were obtained only in the case of aryl bromides. The catalyst was recycled six times without any decrease in activity [60]. QuadraPure, a chelating commercial resin with an affinity for palladium, was impregnated with palladium nanoparticles by a method similar to the one mentioned above. Extra cross-linking steps were carried out to enhance the stability of the catalyst. Resin showed consistent good activity for Suzuki coupling of aryl boronic acids with various aryl halides up to ten cycles. ICP-AES studies of filtrate showed only 0.21 ppm of palladium in the filtrate [61]. Later on, this group further cross-linked the hydrogel with succinyl chloride before reducing Pd (II) to Pd (0). BET analysis showed a decrease in the pore size of the hydrogel. This catalyst showed better activity than the one prepared earlier [62] Biffis and coworkers reported palladium nanoparticles stabilized on cross-linked functional organic resins containing tertiary amino, cyano and pyridyl groups. Immobilization of the palladium precursor was carried out by chemical reduction and metal vapor deposition [63].

Core-shell microspheres of poly[styrene-*co*-2-(acetoacetoxy)ethyl methacrylate-*co*-methyl acrylic acid] [PS-co-PAEMA-co-PMAA] were synthesized by Zheng and Zhang in which PMAA forms the shell, whereas polystyrene and PAEMA segments form the core. The pH responsive microspheres were used as scaffolds to stabilize palladium nanoparticles of about 3 nm in size by sorption reduction method. The catalyst was used to study Suzuki and Heck coupling using water or DMF/water mixture as solvent and K_2CO_3 as a base. However, recyclability of the resin was only good in water. The highlight of the catalytic system was its easy separation by changing the pH of the solution. Recycling of the catalyst was studied by extracting the product in CH_2Cl_2 and using an aqueous layer for the next run. A 13% loss in the conversion was reported by the third cycle [64].

Palladium nanoparticles were also stabilized on a copolymer made up of 4-styrene sulfonic acid and maleic (PSSA-co-MA) using potassium tetrachloropalladate as precursor. An aqueous solution of ammonia borane was used as a reducing agent. Palladium nanoparticles with an average size of 3.4 nm thus were formed, and used to study the Suzuki cross-coupling reaction in water at

100°C. The catalyst showed excellent activity for iodo and bromo derivatives only. TOF of 1980 was reported for coupling of phenyl boronic acid with 4-bromo acetophenone in water using 1 mol% of catalyst and *t*-butoxide as a base. The catalyst retained 86% of its catalytic activity by the end of sixth cycle [65].

Sivudu et al. synthesized reusable poly (*N*-isopropylacrylamide-co-potassium methacrylate) hydrogel with different ratios of comonomers and stabilized palladium nanoparticles by the sorption reduction method using aqueous palladium chloride as a precursor and hydrazine hydrate as a reducing agent (Fig. 4.13). Methacrylate helped both in sorption of palladium ions and stabilization of nanoparticles. The swelling properties of the gel were studied in detail, and it was reported that the efficiency of the gel was highly dependent on the swelling that enhanced accessibility to the catalyst embedded inside the gel. It was used to examine a cross-coupling reaction between arylboronic acid and aryl halides in aqueous media. The catalyst showed recyclability up to six runs [66]. In another report glycidyl methacrylate (GMA) was copolymerized with styrene and cross-linked with DVB. The produced copolymer (Fig. 4.14)

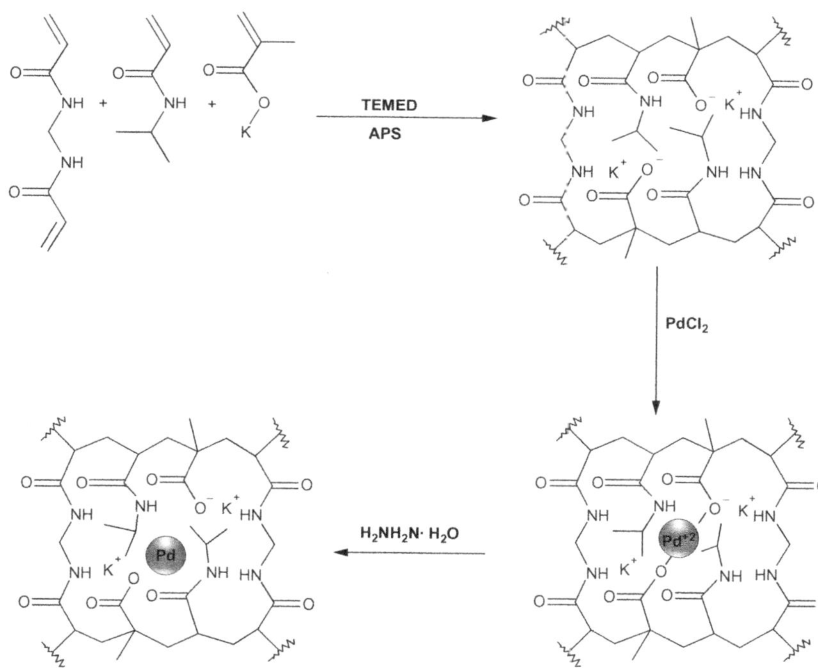

FIG. 4.13 Schematic representation of synthesis of hydrogel and nanoparticle stabilization. *(Redrawn from R.S. Sivudu, N.M. Reddy, M.N. Prasad, K.M. Raju, Y.M. Mohan, J.S. Yadav, et al., Highly efficient and reusable hydrogel-supported nano-palladium catalyst: evaluation for Suzuki–Miyaura reaction in water, J. Mol. Catal. A Chem. 295 (2008) 10–17, with permission from Elsevier.)*

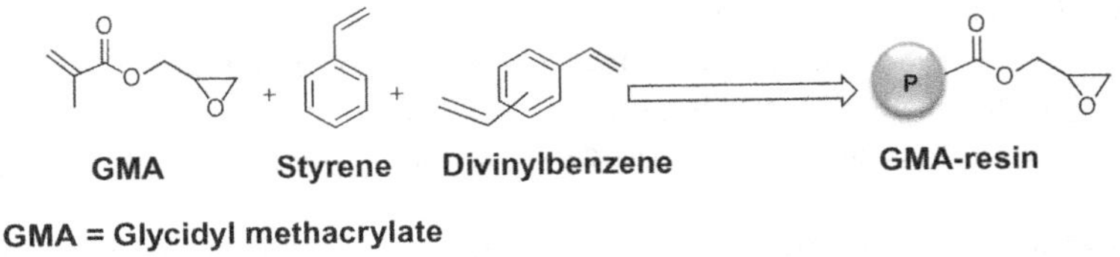

FIG. 4.14 Synthesis of GMA-co-PS-co-DVB polymer. *(Reproduced from A.M. Trzeciak, E. Mieczynska, J.J. Ziolkowski, W. Bukowski, A. Bukowska, J. Noworol, et al., Palladium(0) nanoparticles encapsulated in diamine-modified glycidyl methacrylate polymer (GMA-CHDA) applied as catalyst of Suzuki–Miyaura cross-coupling reaction, New J. Chem. 32 (2008) 1124–1130, with permission from The Royal Society of Chemistry.)*

was modified with ethylene diamine and used to support palladium nanoparticles. Study of its swelling in different solvents was carried out and on this basis DMF was chosen as a solvent for the metal impregnation step. Authors investigated the order of events in the synthesis of nanoparticles. They impregnated the resin first with a reducing agent that was hydrazine hydrate followed by a solution of $PdCl_2$ in DMF. This reduced the chloride content in the resin and influenced the distribution in size of nanoparticles. The different catalysts were used for studying Suzuki coupling in different solvents and best results were reported when ethylene glycol or isopropanol water mixture was used as the solvent and Cs_2CO_3 or KOH as a base. Recycling studies showed a substantial decrease in the activity by the third cycle. They also checked for the presence of catalytic species in the aqueous phase after extraction of the product and reported that it was not catalytically active [67].

Dendrimers are a type of polymeric support known for preparing nanoparticles with narrow size distribution. To facilitate their separation from the reaction mixture, they have been attached to various insoluble supports. Sreekumar et al. attached PAMAM dendrimers to chloromethyl polystyrene (1% DVB). The gel-type polymer was first converted to amine form and then a third generation dendrimer was grown, through Michael addition of methyl acrylate followed by the addition of ethylenediamine. Palladium nanoparticles were synthesized on this support and used to catalyze Suzuki reaction between 4-iodobenzene and phenylboronic acid in dioxane water mixture at 100°C [68]. In another example provided by Murugan et al. different generations of PAMAM dendrimers, 1st, 2nd and 3rd, were grown on chloromethylated polystyrene resin (2% DVB) and were used to stabilize palladium nanoparticles. The catalyst was used to catalyze Suzuki reaction in a water-ethanol medium. They reported that the activity of the catalyst increased slightly with an increase in the generation of the dendrimer [69].

In another study Bukowska et al. tried two different approaches to attach dendrimers on a GMA resin (glycidyl methacrylate, styrene, and divinyl benzene resin). First, they tried direct functionalization of epoxy resin with preformed PAMAM dendrimers; second, dendrimers were grown onto a resin in a stepwise approach (Fig. 4.15). Both resins were impregnated with palladium

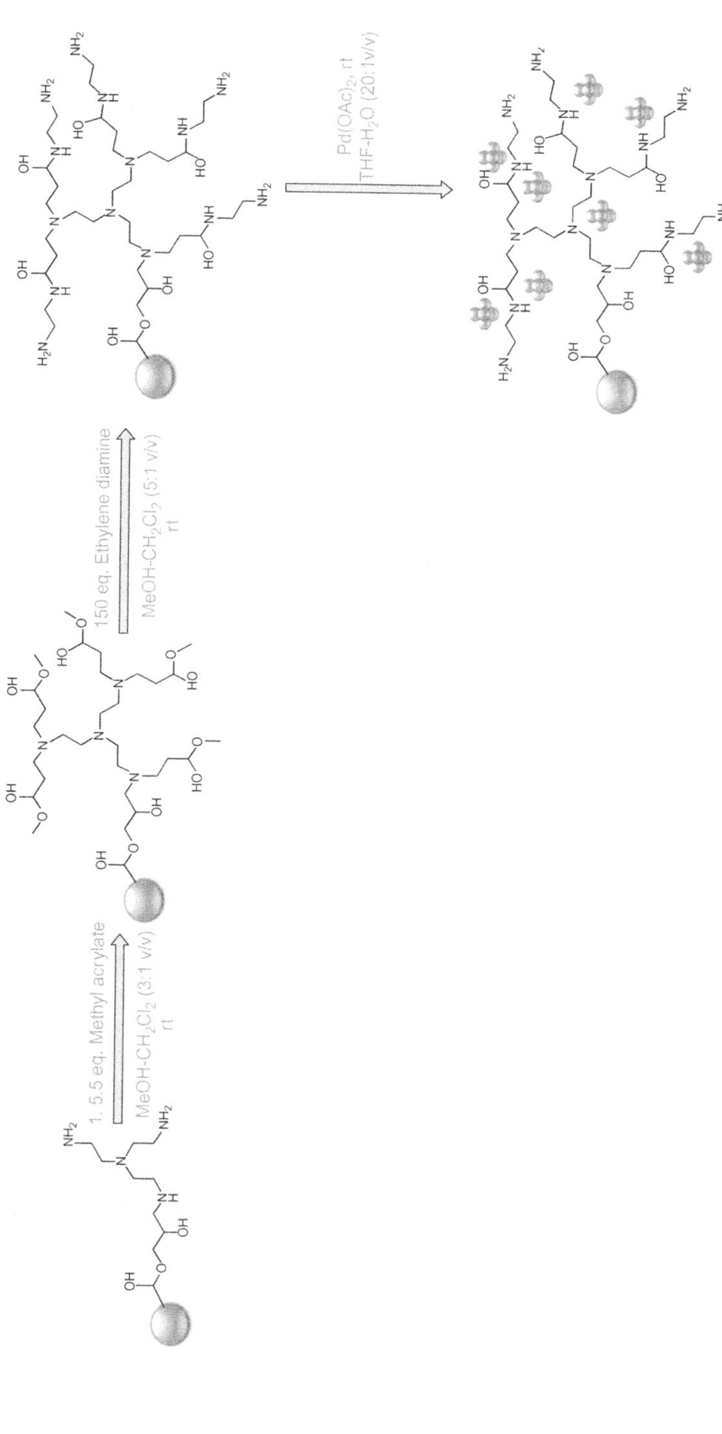

FIG. 4.15 Synthesis of the resin supported dendrimer and encapsulation of nanoparticles. *(Redrawn from A. Bukowska, W. Bukowski, K. Bestera, S. Flagaa, Linkage of the PAMAM type dendrimer with the gel type resin based on glycidyl methacrylate terpolymer as a method of preparation of the polymer support for the recyclable palladium catalyst for Suzuki–Miyaura cross-coupling reactions, RSC Adv. 5 (2015) 49036–49044, with permission from The Royal Society of Chemistry.)*

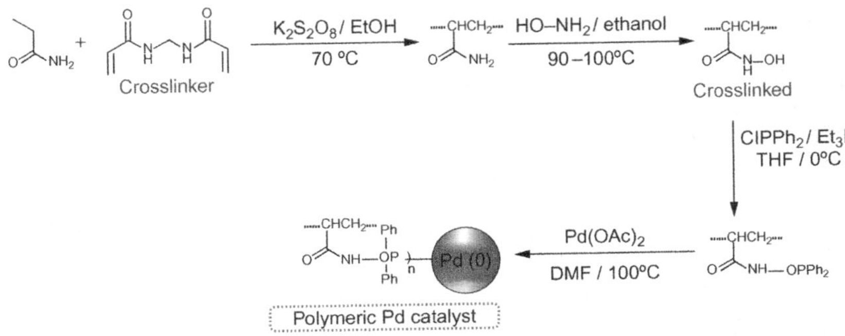

FIG. 4.16 Synthesis of cross-linked polyacrylamide resin with phosphinite ligand and encapsulation of nanoparticles. *(Redrawn from B. Tamami, S. Ghasemi, Palladium nanoparticles supported on modified crosslinked polyacrylamide containing phosphinite ligand: a novel and efficient heterogeneous catalyst for carbon–carbon cross-coupling reactions, J. Mol. Catal A Chem. 322 (2010) 98–105, with permission from Elsevier.)*

nanoparticles, and it was reported that the second resin absorbed more palladium compared to the first. This also resulted in higher catalytic activity of the second resin towards Suzuki reaction that was carried out in an isopropyl alcohol-water mixture. [70] The catalyst showed high recyclability, and its catalytic activity was retained up to 10th cycle.

Triphenyl phosphine is used as a ligand in most Suzuki reactions catalyzed by a homogeneous palladium catalyst. Tamami et al. modified cross-linked polyacrylamide with a phosphinite ligand (Fig. 4.16) and then treated the resin with a palladium acetate solution in DMF. It was visualized that palladium was first complexed with a phosphinite ligand then it was reduced to Pd (0). TEM images revealed palladium nanoparticles in the range of 18–30 nm. Catalytic activity of the resin for both Suzuki and Heck reactions was explored. Using 0.2 mol.% of the catalyst both reactions could be compelled in 1 h with iodobenzene, but considerably greater amount of time was required for bromo and chloro arenes. All the reactions were reported in DMF media and the catalytic activity in water was quite low. The catalyst was recycled up to 10 cycles, and it was observed that the TOF decreased from 475 to 158 by the end of last cycle. Hot filtration test indicated heterogeneous nature of the reaction [71]. A summary of polymer stabilized nanoparticles used for C-C coupling reactions is given in Table 4.1.

4 RECOVERY AND RECYCLING OF THE RESIN SUPPORTED NANOPARTICLES

One of the biggest advantages of polymer supported nanoparticles is their ease of separation from the reaction mixture. They can be removed from the reaction mixture by simple filtration and reused after washing. Polymer resins have

TABLE 4.1 Summary of Polymer Resin Supported Metal Nanoparticles for the C-C Coupling Reaction

No.	Support	Preparation Method	Catalytic Reaction	Ref.
1	Polystyrene	Sorption reduction	Heck, Suzuki, Sonogashira, Mizoroki-Heck coupling	[22,39,55]
2	Hyper cross-linked polystyrene	Adsorption reduction	Suzuki coupling	[58]
3	Polyionic resin	Deposition	Heck, Suzuki-Miyaura, Sonogashira coupling	[72]
4	Poly(amidoamine) dendrimer (cross-linked aminomethyl polystyrene)	Ligand exchange method	Suzuki coupling	[68]
5	Amine functionalized resin	Adsorption reduction method	Suzuki coupling	[73]
6	Anion exchange resin	Sorption reduction method	Suzuki coupling	[74]
7	Tertiary amino-functionalized resin	Adsorption reduction method	Suzuki coupling	[75]
8	ILs@ divinylbenzene	Adsorption reduction method	Suzuki coupling	[76]
9	Macroporous GMA resin	Adsorption reduction method	Heck coupling	[77]
10	Functional resin	Sorption reduction	Heck coupling	[64]
11	Diphenylphosphinomethyl polystyrene	Sorption reduction	Hiyama coupling	[78]
12	Cross-linked polyacryl amide	Sorption reduction	Heck, Suzuki	[71]
13	PSSA-co-MA	Adsorption reduction method	Suzuki-Miyaura	[65]

Continued

TABLE 4.1 Summary of Polymer Resin Supported Metal Nanoparticles for the C-C Coupling Reaction—cont'd

No.	Support	Preparation Method	Catalytic Reaction	Ref.
14	GMA-CHDA	Impregnation	Suzuki-Miyaura	[79]
15	g-SILLPs@ PS-DVB resin	Adsorption reduction method	Heck coupling	[80]
16	Cross-linked hyperbranched poly(phenylacetylene)	Deposition	Suzuki, Heck	[81]
17	Polystyrene-based HypoGel resins	Sorption reduction	Suzuki	[62]
18	Ion-exchange resin	Chemical vapor deposition (CVD)	Suzuki-Miyaura	[56]

a tendency to sorb reactants and products, so they need to be washed thoroughly with hot water, steam, or low boiling organic solvent before reuse. Tarashanker Pal et al. devised a novel method for retrieving gold nanoparticles from an ion-exchange resin, which was washing the resin with a surfactant (Fig. 4.17). The resin-bound nanoparticles were kept in contact with a concentrated solution of CPC. Gradually the solution became pink and resin beads were recovered for another experiment [45].

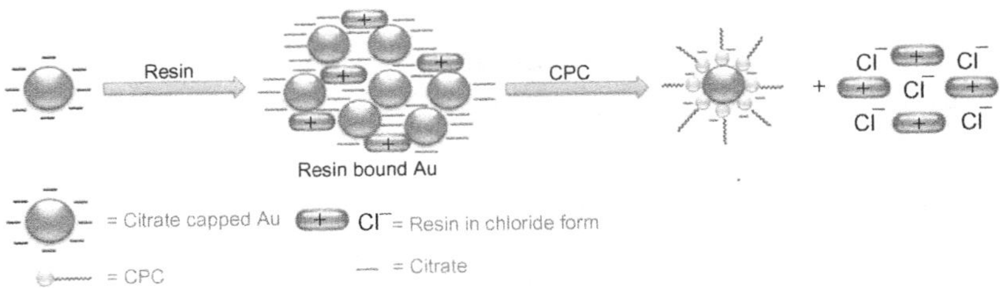

FIG. 4.17 Impregnation of AuNPs on ion-exchange resin and its retrieval. *(Redrawn with permission from S. Praharaj, S. Nath, S.K. Ghosh, S. Kundu, T. Pal, Immobilization and recovery of Au nanoparticles from anion exchange resin: resin-bound nanoparticle matrix as a catalyst for the reduction of 4-nitrophenol, Langmuir 20 (2004) 9889, Copyright 2004, American Chemical Society.)*

In most of the work reported above, metal nanoparticles supported on the resins have been recycled easily up to five or six cycles without any significant loss of the activity. In some cases activity up to 10th cycle was also reported [71]. Post-reaction TEM analysis in a few cases showed some increase in the average size of the nanoparticle [75]. Once the catalyst loses activity, expensive metals can be recovered from the polymer by incinerating the polymer in a furnace. The polymer gets oxidized leaving the metal residue behind. This is another point where organic supports are better compared to inorganic ones.

5 MECHANISM OF C-C COUPLING ON POLYMER SUPPORTED NANOPARTICLES

Understanding the mechanism of a catalytic reaction is important because it helps in recognizing active sites and designing better catalysts. Nanoparticle catalysis is at the interface of homogeneous and heterogeneous catalysis. In the case of carbon-carbon coupling reactions, whether the mechanism is truly homogeneous or heterogeneous is an ongoing debate, and the issue is still not resolved. What is irrevocably proven is that the activity of the catalyst depends upon the size of the nanoparticles, and smaller particles being more active lead to higher TON numbers. There are two main mechanistic approaches (Fig. 4.18) reported in the literature:

1. Active catalytic species leach out from the surface of nanoparticles and catalyze the reaction in a solution phase. The actual catalytic process is homogeneous in nature.
2. Catalysis occurs at the defect sites on the metal surface where highly unsaturated metal atoms are present, and it is heterogeneous in nature.

Most of the data reported in this regard is from Heck [82] and Suzuki reactions [83] carried out with colloidal nanoparticles. Many scientists are of the opinion

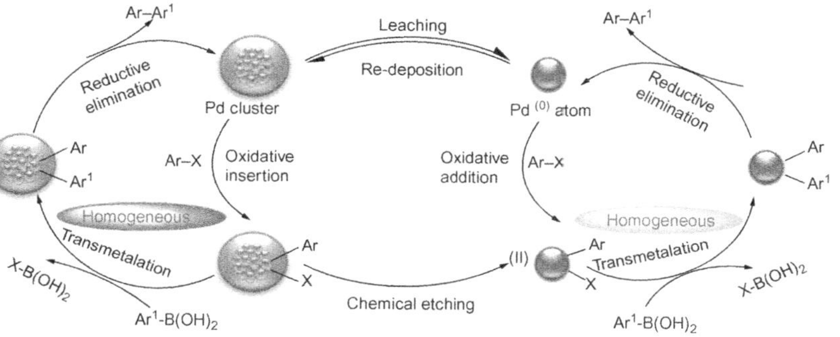

FIG. 4.18 Mechanism of nanoparticle catalyzed Suzuki coupling. *(Redrawn with permission from M.P. Lorenzo, Palladium nanoparticles as efficient catalysts for Suzuki cross-coupling reactions, J. Phys. Chem. Lett. 3 (2012) 167–174, Copyright 2012, American Chemical Society.)*

that nanoparticles act as reservoirs for active catalytic species that leach out from the surface and pass into the solution phase [84]. In the case of palladium nanoparticles supported on polystyrene resins, ICP-AES analysis of the reaction mixture hardly showed any leaching [39,50]. It is generally argued that the absence of leaching is not a proof for heterogeneous mechanism because it occurs only in the presence of aryl halides and leached out species are redeposited on the nanoparticle surface at the end of the reaction. Thus, the mechanism is actually pseudohomogeneous [85]. Results of the hot filtration tests from many polymer-stabilized palladium nanoparticles are negative, prompting the researchers to favor the heterogeneous nature of the catalysis. The data available in the literature is certainly not conclusive because leaching of metal species depends on a number of other factors like the nature of support, nature of the solvent, temperature, use of surfactants, stabilizing agents, etc. Leaching is also dependent on how firmly the nanoparticles are fixed on the support. Bradley et al. performed a three-phase test to ascertain the nature of the catalysis on a Suzuki reaction and concluded a heterogeneous mechanism. The latest development in this field involves the results of x-ray absorption spectroscopy published by Faulamp and Lee. In situ monitoring showed that the coordination environment of nanoparticles remains unchanged during the reaction. This rules out the homogeneous nature of the catalysis [86]. In the field of catalysis, poisoning tests are also widely applied to ascertain the nature of catalysis. Sengupta et al. performed a poisoning test in the case of a Sonogashira reaction using Hg (0) and CS_2 as poisoning agents. On the basis of these tests the authors concluded a heterogeneous nature of the catalyst supported on a microporous ion-exchange resin [59]. Studies of resin supported nanoparticles mainly favor the heterogeneous nature of the catalysis. Whatever the nature of the catalysis is, the efficiency and green nature of nanoparticles catalyzed C-C coupling reactions is indisputable.

6 CONCLUSION

Carbon-carbon coupling reactions are widely applicable in the synthesis of industrially important compounds due to mild reaction conditions associated with them, high efficiency, and wide functional group tolerance. With the usage of nanoparticles as catalysts, they have become even more attractive because nanoparticles are highly efficient owing to their high surface area to volume ratios. Besides this, nanoparticles are recyclable, reactions are being reported at low temperatures, without phosphine ligands and organic solvents with high boiling points are no longer essential.

The role of polymer resin as support is especially noteworthy as dispersed nanoparticles can be synthesized in one step without using extra stabilizing agents. In most polymeric supports nanoparticles are stabilized by noncovalent electrostatic forces or by steric hindrance. In the absence of strong ligating agents their catalytic activity is even greater. In addition to the above mentioned benefits they facilitate recovery and recycling of the catalyst, makes workup and product purification simple. It is also noteworthy that most of the resins are commercially available and leaching of the metal from these resins is negligible. Two commercially available polymer-impregnated palladium nanoparticles are already in the market. One of them is an ion-exchanger-based catalyst from Bayer [87] and another is NanoCat, which was mentioned earlier. Challenges still lie in understanding the accessibility of catalytic sites and designing catalysts with wide applications.

ACKNOWLEDGMENTS

The author is grateful to Nanomission DST for the financial support to carry out this work and to Inflibnet Gandhinagar for providing e-resources. The author would also like to thank Gujarat University as well as Anuj Sharma for his help in preparing the figures.

REFERENCES

[1] J. Tsuji, Modern Palladium Catalysis. Palladium Reagents and Catalysts: New Perspectives for the 21st Century, Wiley, Chichester, 2004.

[2] N. Miyaura, Y. Yanagi, A. Suzuki, The palladium-catalyzed cross-coupling reaction of phenylboronic acid with haloarenes in the presence of bases, Synth. Commun. 11 (1981) 513–519.

[3] F.R. Heck, P.J. Nolley, Palladium-catalyzed vinylic hydrogen substitution reactions with aryl, benzyl, and styryl halides, J. Org. Chem. 37 (1972) 2320–2322.

[4] Y. Hatanaka, T. Hiyama, Cross-coupling of organosilanes with organic halides mediated by a palladium catalyst and tris(diethylamino)sulfonium difluorotrimethylsilicate, J. Org. Chem. 53 (1988) 918–920.

[5] C.C. Carin, S. Johansson, O.K. Kitching, T.J. Colacot, V. Snieckus, Palladium-catalyzed cross-coupling: a historical contextual perspective to the 2010 Nobel prize, Angew. Chem. Int. Ed. 51 (2012) 5062–5085.

[6] N.E. Leadbeater, M. Marco, Preparation of polymer-supported ligands and metal complexes for use in catalysis, Chem. Rev. 102 (2002) 3217–3274.

[7] L. Yin, J. Liebscher, Carbon – carbon coupling reactions catalyzed by heterogeneous palladium catalysts, Chem. Rev. 107 (2007) 133–173.

[8] D. Astruc, Nanoparticles and Catalysis, Wiley-VCH Verlag Gmbh & Co., KGaA, Weinheim, 2008.

[9] A. Zapt, M. Beller, Fine chemical synthesis with homogeneous palladium catalysts: examples status and trends, Top. Catal. 19 (2002) 101–109.

[10] J.P. Nolley. Jr., R.F. Heck, Palladium-catalyzed vinylic hydrogen substitution reactions with aryl, benzyl, and styryl halides, J. Org. Chem. 37 (1972) 2320.

[11] M. Beller, H. Fischera, K. Kuhleina, C.P. Reisinger, W.A. Herrmann, First palladium-catalyzed heck reactions with efficient colloidal catalyst systems, J. Organomet. Chem. 520 (1996) 257–259.

[12] M.T. Reetz, G. Lohmer, Propylene carbonate stabilized nanostructured palladium clusters as catalysts in Heck reactions, Chem. Commun. 1921–1922 (1996).

[13] M. Antonietti, F. Gröhn, J. Hartmann, L. Bronstein, Nonclassical shapes of noble-metal colloids by synthesis in microgel nanoreactors, Angew. Chem. Int. Ed. Engl. 36 (1997) 2080–2083.

[14] A. Barau, V. Budarin, A. Caragheorgheopol, R. Luque, D. Macquarrie, A. Prelle, et al., A simple and efficient route to active and dispersed silica supported palladium nanoparticles, Catal. Lett. 124 (2008) 204–214.

[15] K.R. Gopidas, J.K. Whitesell, M.A. Fox, Synthesis, characterization, and catalytic applications of a palladium-nanoparticle-cored dendrimer, Nano Lett. 3 (2003) 1757–1760.

[16] X. Chen, Y. Hou, H. Wang, Y. Cao, J. He, Facile deposition of Pd nanoparticles on carbon nanotube microparticles and their catalytic activity for Suzuki coupling reactions, J. Phys. Chem. C 112 (2008) 8172–8176.

[17] G. Scheuermann, L. Rumi, P. Steurer, W. Bannwarth, R. lhaupt, Palladium nanoparticles on graphite oxide and its functionalized graphene derivatives as highly active catalysts for the Suzuki – Miyaura coupling reaction, J. Am. Chem. Soc. 131 (2009) 8262–8270.

[18] Y. Li, X.M. Hong, D.M. Collard, M.A. El-Sayed, Suzuki cross-coupling reactions catalyzed by palladium nanoparticles in aqueous solution, Org. Lett. 2 (2000) 2385–2388.

[19] J. Han, Y. Liu, R. Guo, Facile synthesis of highly stable gold nanoparticles and their unexpected excellent catalytic activity for Suzuki-Miyaura cross-coupling reaction in water, J. Am. Chem. Soc. 131 (2009) 2060–2061.

[20] Y. Li, X. Fan, J. Qi, J. Ji, S. Wang, G. Zhang, et al., Gold nanoparticles–graphene hybrids as active catalysts for Suzuki reaction, Mater. Res. Bull. 45 (2010) 1413–1418.

[21] D.K. Dumbre, P.N. Yadav, S.K. Bhargava, V.R. Choudhary, Suzuki–Miyaura cross-coupling reaction between aryl halides and phenylboronic acids over gold nano-particles supported on MgO (or CaO) and other metal oxides, J. Catal. 301 (2013) 134–140.

[22] D. Shah, H. Kaur, Resin-trapped gold nanoparticles: an efficient catalyst for reduction of nitro compounds and Suzuki-Miyaura coupling, J. Mol. Catal. A Chem. 381 (2014) 70–76.

[23] L. Wu, J. Ling, Z.-Q. Wu, A highly active and recyclable catalyst: phosphine dendrimer-stabilized nickel nanoparticles for the Suzuki coupling reaction, Adv. Synth. Catal. (2011) 1452–1456.

[24] M.B. Thathagar, J. Beckers, G. Rothenberg, Palladium-free and ligand-free Sonogashira cross-coupling, Green Chem. 6 (2004) 215–218.

[25] L.D. Pachon, M.B. Thathagar, F. Hartl, G. Rothemberg, Palladium-coated nickel nanoclusters: new Hiyama cross-coupling catalysts, Phys. Chem. Chem. Phys. 8 (2006) 151–157.

[26] A. Molnar, Efficient, selective, and recyclable palladium catalysts in carbon – carbon coupling reactions, Chem. Rev. 111 (2011) 2251–2320.

[27] S. Paul, M.M. Islam, M. Manirul, Suzuki–Miyaura reaction by heterogeneously supported Pd in water: recent studies, RSC Adv. 5 (2015) 42193–42221.

[28] D.C. Sherrington, Preparation, structure and morphology of polymer supports, Chem. Commun. (1998) 2275–2286.

[29] H. Widdecke, in: D.C. Sherrington, P. Hodge (Eds.), Synthesis and Separations Using Functional Polymers, Wiley, Chichester, 1988, p. 149.

[30] R.B. Merrifield, Solid phase peptide synthesis. I. The synthesis of a tetrapeptide, J. Am. Chem. Soc. 85 (1963) 2149.

[31] M. Kralik, A. Biffis, Catalysis by metal nanoparticles supported on functional organic polymers, J. Mol. Catal. A Chem. 177 (2001) 113–138.

[32] C. Luo, Y. Zhang, Y. Wang, Palladium nanoparticles in poly (ethyleneglycol): the efficient and recyclable catalyst for Heck reaction, J. Mol. Catal. A Chem. 229 (2005) 7–12.

[33] S.S. Yi, D.H. Lee, E. Sin, Y.S. Lee, Chitosan-supported palladium (0) catalyst for microwave-prompted Suzuki cross-coupling reaction in water, Tetrahedron Lett. 48 (2007) 6771.

[34] M. Kralik, B. Corain, M. Zecca, Catalysis by metal nanoparticles supported on functionalized polymers, Chem. Pap. 54 (2000) 254–264.

[35] F. Liguori, C. Moreno-Marrodan, P. Barbaro, Metal nanoparticles immobilized on ion-exchange resins: a versatile and effective catalyst platform for sustainable chemistry, Chin. J. Catal. 36 (2015) 1157–1169.

[36] B. Corain, M. Kralik, Dispersing metal nanoclusters inside functional synthetic resins: scope and catalytic prospects, J. Mol. Catal. A Chem. 159 (2000) 153–162.

[37] R.J. White, R. Luque, V.L. Budarin, J.H. Clark, D.J. Macquarrie, Supported metal nanoparticles on porous materials. Methods and applications, Chem. Soc. Rev. 38 (2009) 481–494.

[38] S.N. Sidorov, L.M. Bronstein, V.A. Davankov, M.P. Tsyurupa, S.P. Solodovnikov, P.M. Valetsky, Cobalt nanoparticle formation in the pores of hyper-cross-linked polystyrene: control of nanoparticle growth and morphology, Chem. Mater. 11 (1999) 3210.

[39] H. Kaur, D. Shah, U. Pal, Resin encapsulated palladium nanoparticles: an efficient and robust catalyst for microwave enhanced Suzuki–Miyaura coupling, Catal. Commun. 12 (2011) 1384–1388.

[40] Y. Monguchi, Y. Fujita, K. Endo, S. Takao, M. Yoshimura, Y. Takagi, et al., A highly active heterogeneous palladium catalyst supported on a synthetic adsorbent, Chem. Eur. J. 15 (2009) 834–837.

[41] D.L. Hanson, J.R. Katzer, B.C. Gates, G.C.A. Schuit, H.F. Harnesberger, Supported metal catalysts from cation exchange resins, J. Catal. 30 (1974) 204–215.

[42] B. Corain, K. Jerabek, P. Centomo, P. Canton, Generation of size-controlled Pd0 nanoclusters inside nanoporous domains of gel-type resins: diverse and convergent evidence that supports a strategy of template-controlled synthesis, Angew. Chem. Int. Ed. 43 (2004) 959–962.

[43] D. Pears, S.C. Smith, Polyurea-encapsulated palladium catalysts: the development and application of a new and versatile immobilized homogeneous catalyst technology, Aldrichim. Acta 38 (2005) 23–33.

[44] S. Huang, B. He, One-step preparation of polymer-supported colloidal palladium catalysts and their catalytic properties. Stability and catalytic properties in olefin hydrogenation, React. Polym. 23 (1994) 11–18.

[45] S. Praharaj, S. Nath, S.K. Ghosh, S. Kundu, T. Pal, Immobilization and recovery of Au nanoparticles from anion exchange resin: resin-bound nanoparticle matrix as a catalyst for the reduction of 4-nitrophenol, Langmuir 20 (2004) 9889.

[46] A. Sharma, D. Shah, H. Kaur, Gold nanoparticles supported on dendrimer@ resin for the efficient oxidation of styrene using elemental oxygen, RSC Adv. 5 (2015) 42935–42941.

[47] B.D. Cullity, S.R. Stock, Elements of X-Ray Diffraction, third ed., Prentice-Hall, Inc, USA, 2001.

[48] D. Shah, H. Kaur, Supported gold nanoparticle catalyzed cross-coupling of alkoxysilanes and aryl halides, Curr. Catal. 4 (2015) 224–230.

[49] Y. Mei, G. Sharma, Y. Lu, M. Ballauff, High catalytic activity of platinum nanoparticles immobilized on spherical polyelectrolyte brushes, Langmuir 21 (2005) 12229–12234.

[50] Y. Yabe, T. Maegawa, Y. Monguchi, H. Sajiki, Palladium on charcoal-catalyzed ligand-free stille coupling, Tetrahedron 66 (2010) 8654–8660.

[51] L.M. Bronstein, G. Goerigk, M. Kostylev, M. Pink, I.A. Khotina, P.M. Valetsky, et al., Structure and catalytic properties of Pt-modified hyper-cross-linked polystyrene exhibiting hierarchical porosity, J. Phys. Chem. B 108 (2004) 18234–18242.

[52] D. Shah, H. Kaur, Macroporous resin impregnated palladium nanoparticles: catalyst for a microwave-assisted green Hiyama reaction, J. Mol. Catal. A Chem. 359 (2012) 69–73.

[53] C.O. Kappe, Controlled microwave heating in modern organic synthesis, Angew. Chem. Int. Ed. 43 (2004) 6250–6284.

[54] N.E. Leadbeater, M. Marco, Rapid and amenable Suzuki coupling reaction in water using microwave and conventional heating, J. Org. Chem. 68 (2003) 888–892.

[55] D. Shah, H. Kaur, A general sustainable catalyst for microwave enhanced carbon-carbon coupling reactions, J. Mol. Catal. A Chem. 424 (2016) 171–180.

[56] M. Zhang, M. Jiang, C. Liang, Palladium supported on an ion exchange resin for the Suzuki-Miyaura reaction, Chin. J. Catal. 34 (2013) 2161–2166.

[57] A. Alonso, A. Shafir, J. Macanas, A. Vallribera, M. Munoz, D. Muraviev, Recyclable polymer-stabilized nanocatalysts with enhanced accessibility for reactants, Catal. Today 193 (2012) 200.

[58] S.E. Lyubimov, A.A. Vasilev, A.A. Korlyukov, M.M. Ilyin, S.A. Pisarev, V.V. Matveev, et al., Palladium-containing hypercrosslinked polystyrene as an easy to prepare catalyst for Suzuki reaction in water and organic solvents, React. Funct. Polym. 69 (2009) 755–758.

[59] D. Sengupta, J. Saha, G. De, B. Basu, Pd/Cu bimetallic nanoparticles embedded in macroporous ion-exchange resins: an excellent heterogeneous catalyst for the Sonogashira reaction, J. Mater. Chem. A 2 (2014) 3986.

[60] J.K. Cho, R. Najman, T.W. Dean, O. Ichihara, C. Muller, M. Bradley, Captured and cross-linked palladium nanoparticles, J. Am. Chem. Soc. 128 (2006) 6276–6277.

[61] K.H. Liew, P.L. Lou, J.C. Juan, M.A. Yarmo, R.M. Yusop, QuadraPure-supported palladium nanocatalysts for microwave-promoted Suzuki cross-coupling reaction under aerobic condition, Sci. World J. 2014 (2014) 796196.

[62] K.H. Liew, W.Z. Samad, N. Nordin, P.L. Loh, J.C. Juan, M.A. Yarmo, et al., Preparation and characterization of HypoGel-supported Pd nanocatalysts for Suzuki reaction under mild conditions, Chin. J. Catal. 36 (2015) 771–777.

[63] A.M. Caporusso, P. Innocenti, L.A. Aronica, G. Vitulli, R. Gallina, A. Biffis, et al., Functional resins in palladium catalysis: promising materials for Heck reaction in aprotic polar solvents, J. Catal. 234 (2005) 1–13.

[64] P. Zheng, W. Zhang, Synthesis of efficient and reusable palladium catalyst supported on pH-responsive colloid and its application to Suzuki and Heck reactions in water, J. Catal. 250 (2007) 324–330.

[65] O. Metina, F. Durap, M. Aydemir, S. Ozkara, PVP-stabilized palladium nanoparticles electrochemically obtained as effective catalysts in aqueous medium Suzuki–Miyaura reaction, J. Mol. Catal. A Chem. 337 (2011) 39–44.

[66] R.S. Sivudu, N.M. Reddy, M.N. Prasad, K.M. Raju, Y.M. Mohan, J.S. Yadav, et al., Highly efficient and reusable hydrogel-supported nano-palladium catalyst: evaluation for Suzuki–Miyaura reaction in water, J. Mol. Catal. A Chem. 295 (2008) 10–17.

[67] A.M. Trzeciak, E. Mieczynska, J.J. Ziolkowski, W. Bukowski, A. Bukowska, J. Noworol, et al., Palladium(0) nanoparticles encapsulated in diamine-modified glycidyl methacrylate polymer (GMA-CHDA) applied as catalyst of Suzuki–Miyaura cross-coupling reaction, New J. Chem. 32 (2008) 1124–1130.

[68] G.R. Krishnan, K. Sreekumar, Synthesis and characterization of polystyrene supported catalytically active poly(amidoamine) dendrimer-palladium nanoparticles conjugates, Soft Matter 8 (2010) 114–129.

[69] E. Murugan, J.N. Jebranjitham, A. Usha, Synthesis of polymer-supported dendritic palladium nanoparticle catalysts for Suzuki coupling reaction, Appl. Nanosci. 2 (2012) 211–222.

[70] A. Bukowska, W. Bukowski, K. Bestera, S. Flagaa, Linkage of the PAMAM type dendrimer with the gel type resin based on glycidyl methacrylate terpolymer as a method of preparation of the polymer support for the recyclable palladium catalyst for Suzuki–Miyaura cross-coupling reactions, RSC Adv. 5 (2015) 49036–49044.

[71] B. Tamami, S. Ghasemi, Palladium nanoparticles supported on modified crosslinked polyacrylamide containing phosphinite ligand: a novel and efficient heterogeneous catalyst for carbon–carbon cross-coupling reactions, J. Mol. Catal. A Chem. 322 (2010) 98–105.

[72] B. Basu, S. Das, P. Das, B. Mandal, D. Banerjee, F. Almqvist, Palladium supported on a Poly-ionic Resin as an efficient, ligand-free, and recyclable catalyst for heck, Suzuki-Miyaura, and Sonogashira reactions, Synthesis 7 (2009) 1137–1146.

[73] K. Bester, A. Bukowska, W. Bukowski, Palladium catalysts supported on amine-functionalized glycidyl methacrylate gel type terpolymers. Synthesis, characteristics and study of catalytic activity in Suzuki–Miyaura reactions, J. Mol. Catal. A Chem. 378 (2013) 124–134.

[74] K.A. De Castro, H. Rhee, Resin-immobilized palladium nanoparticle catalysts for Suzuki–Miyaura cross-coupling reaction in aqueous media, J. Incl. Phenom. Macrocycl. Chem. 82 (2015) 13–24.

[75] Y. Monguchi, T. Ichikawa, M. Netsu, T. Hattori, T. Mizusaki, Y. Sawama, et al., Tertiary-amino-functionalized resin-supported palladium catalyst for the heterogeneous Suzuki–Miyaura reaction of aryl chlorides, Synlett 26 (2015) 2014–2018.

[76] Y. Yu, T. Hu, X. Chen, K. Xu, J. Zhang, J. Huang, Pd nanoparticles on a porous ionic copolymer: a highly active and recyclable catalyst for Suzuki–Miyaura reaction under air in water, Chem. Commun. 47 (2011) 3592–3594.

[77] D. Yuan, B. Huang, Macroporous P (GMA–DVB–TRIM) microspheres supported diethylenetriamine palladium complex: an efficient and recyclable catalyst for Heck reactions, Catal. Commun. 18 (2012) 126–131.

[78] C. Diebold, A. Derible, J.M. Becht, C. Le Drian, A simple and efficient reusable polystyrene-supported palladium catalyst for Hiyama cross-coupling, Tetrahedron 69 (2013) 264–267.

[79] T. Borkowski, A.M. Trzeciak, W. Bukowski, A. Bukowska, W. Tylus, L. Kepinskid, Palladium (0) nanoparticles formed in situ in the Suzuki–Miyaura reaction: the effect of a palladium (II) precursor, Appl. Catal. A Gen. 378 (2010) 83–89.

[80] M.I. Burguete, E.G. Verdugo, I.G. Villar, F. Gelat, P. Licence, S.V. Luis, et al., Pd catalysts immobilized onto gel-supported ionic liquid-like phases (g-SILLPs): a remarkable effect of the nature of the support, J. Catal. 269 (2010) 150–160.

[81] Z. Dong, Z. Ye, Heterogeneous palladium catalyst constructed with cross-linked hyper-branched poly(phenylacetylene) as polymer support: a reusable highly active ppm-level catalyst for multiple cross-coupling reactions, Appl. Catal. A Gen. 489 (2015) 61–71.

[82] J.G. De Varies, A unifying mechanism for all high-temperature Heck reactions. The role of palladium colloids and anionic species, Dalton Trans. 421–429 (2006).

[83] R. Narayanan, C. Tabor, M.A. El-Sayed, Can the observed changes in the size or shape of a colloidal nanocatalyst reveal the nanocatalysis mechanism type: homogeneous or heterogeneous? Top. Catal. 48 (2008) 60–74.

[84] A. Biffis, M. Zecca, M. Basato, Palladium metal catalysts in heck C–C coupling reactions, J. Mol. Catal. A Chem. 173 (2001) 249.

[85] M.P. Lorenzo, Palladium nanoparticles as efficient catalysts for suzuki cross-coupling reactions, J. Phys. Chem. Lett. 3 (2012) 167–174.

[86] A.F. Lee, P.J. Ellis, J.S. Fairlamb, K. Wilson, A.F. Lee, Surface catalysed Suzuki–Miyaura cross-coupling by Pd nanoparticles: an operando XAS study, Dalton Trans. 39 (2010) 10473–10482.

[87] A.G. Bayer, Bayer Polymer-Based Catalysts. Properties and Applications, Liverkusen, 1989.

Chapter 5

Catalysis With Polymer-Incarcerated Metal(0) Clusters

Samahe Sadjadi
Iran Polymer and Petrochemical Institute, Tehran, Iran

Contents

1 INTRODUCTION TO POLYMER MICROENCAPSULATED AND POLYMER INCARCERATED CATALYSTS

Microencapsulation, the process of entrapping a substance in a coating, has been widely used in diverse research fields such as drug delivery and catalysis [1,2]. In 1998, the microencapsulation method was applied for the immobilization of a catalyst onto polymers [3–5]. In this process the catalyst would be wrapped by thin films of polymers that contain aromatic rings (mostly polystyrene derivatives) [6] and at the same time immobilized by interaction between π electrons of the aromatic moiety, which are used as a polymer backbone, and vacant orbitals of the catalysts (Fig. 5.1). This class of heterogeneous catalysts is referred to as microencapsulated catalysts. Up to now, various microencapsulated catalysts [7] such as microencapsulated palladium [8], microencapsulated osmium tetraoxide [9], microencapsulated bismuth trifluoromethanesulfonate [10], microencapsulated scandium trifluoromethanesulfonate [4], microencapsulated $VO(acac)_2$ [11], microencapsulated $Cu(acac)_2$ [12], microencapsulated

Encapsulated Catalysts. http://dx.doi.org/10.1016/B978-0-12-803836-9.00005-5

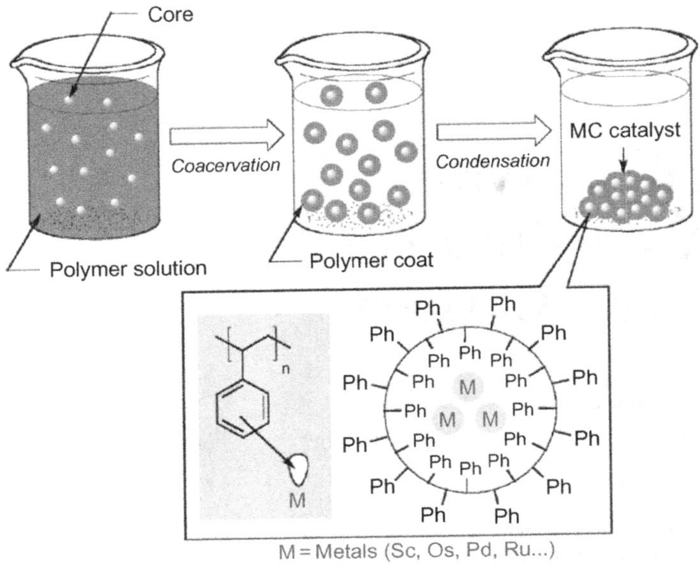

FIG. 5.1 Microencapsulation technique. *(Reprinted from R. Akiyama, S. Kobayashi, "Microencapsulated" and related catalysts for organic chemistry and organic synthesis, Chem. Rev. 109 (2009) 594–642 with permission of the American Chemical Society.)*

metathesis catalyst [13], and microencapsulated methylrhenium trioxide [14] have been developed and employed for catalyzing various chemical reactions.

It is worth noting that microencapsulation of transition metal catalyst can also be achieved by another strategy. In this method, an in situ interfacial polymerization is exploited for formation of microcapsules [3]. This process involves the dispersion of a solution of polyfunctional monomers or oligomers (mostly isocyanates) and the reagent to be encapsulated in a suitable organic solvent into an aqueous mixture containing surfactants. The resultant oil-in-water microemulsion is allowed to cure, during which time polymerization occurs at the oil-water interface [1]. The obtained microencapsules are soluble neither in organic solvents nor in aqueous ones and are robust without degradation under a normal reaction condition. The permeability and size of these microcapsules and the coordinating properties of the polymer matrix may be tuned by varying the nature of monomers/oligomers, other reagents, and conditions used in the encapsulation procedure [3]. As mentioned above, these microcapsules are based on polyurea and named Encat. Heretofore, various catalysts such as polyurea-encapsulated osmium tetroxide (Os EnCat), polyurea-encapsulated Pd, Pd EnCat, and Pd^0 EnCat have been developed and their utility for various organic reactions has been proved [1,15–17]. This microencapsulation method has been comprehensively reviewed by Kobayashi et al. [3]. This topic is beyond the scope of this chapter and readers are encouraged to study this reference for more details.

The promising features of microencapsulated catalysts are their heterogeneous nature, simple recovery, reusability without significant loss of activity and general ability. However, as in the microencapsulation process the polymer support must be dissolved in an organic solvent, the obtained catalysts were dissolved or swelled in some cases, and leaching of metals from supports occurred [3]. To circumvent this problem, a novel immobilization process, polymer incarceration (PI), was introduced by Kobayashi's research group [18]. In this method catalyst immobilization is based on two procedures: microencapsulation and cross-linking [19]. To prepare PI catalysts, a catalyst is introduced to a polymer that was dissolved in a suitable solvent. After cooling, formed microcapsules are washed and dried to afford a microencapsulated catalyst. It has already been shown that the catalyst locates not only inside the capsules, but also on the surface or close to the surface of the capsules. The microcapsules formed are then cross-linked to result in the desired catalyst [3].

Polymer incarceration is proved to be an effective method for controlling/preventing metal leaching. In most polymer incarcerated catalysts no or negligible metal leaching was observed. Furthermore, catalyst activity of polymer-incarcerated catalysts in most cases was comparable with the unsupported counterparts and they sometimes even exhibited higher activity. Another advantage of these catalysts is their facile and high efficient recovery and reusability [2]. Additionally, the catalytic performance of these catalysts can be tuned by various strategies such as modification/functionalization of the polymer backbone, using a second support such as carbon black (CB), and/or designing bimetallic systems. Although polymer-incarcerated catalysts are relatively new, the promising features of this new class of immobilized catalyst resulted in many advances in the field of catalysis. Up to now various polymer-incarcerated catalysts such as polymer-incarcerated metal(0) (nano)cluster catalysts have been developed mostly by the Kobayashi research group [20–27]. In the case of metal(0) (nano)clusters, microencapsulation keeps metal(0) nanoclusters stable and prevents them from aggregation without deactivating, while cross-linking heterogenizes the catalyst and prevents metal(0) (nano)clusters from leaching [18]. Kobayashi's research group provided an excellent review of various microencapsulated and polymer-incarcerated catalysts for organic chemistry and synthesis in 2009 [3]. They also published another review article in 2010 on the polymer-incarcerated metal(0) cluster catalyst [18].

Considering the diversity of the polymer microencapsulated and polymer incarcerated catalyst, we limit our discussion in this chapter to polymer-incarcerated metal(0) cluster catalysts for greater depth and cover examples of polymer-incarcerated mono metallic and bimetallic catalysts and their utility for organic reactions. This chapter is organized based on the types of the metal clusters incarcerated onto polymer and tries to provide more insight into this topic by discussing the catalyst preparation procedure, metal leaching, catalyst reusability, and optimal reaction conditions.

2 POLYMER-INCARCERATED PD (0) NANOPARTICLES

2.1 Suzuki-Miyaura Coupling

Kobayashi et al. introduced a variety of polymer incarcerated (PI) Pd catalysts and investigated their utilities in diverse organic reactions [28,29]. In this regard this research group developed a heterogeneous catalyst for the Suzuki-Miyaura coupling reaction based on Pd catalysts immobilized into polystyrene-based polymers [30]. Despite high catalytic activity, reusability, and circumventing Pd leaching, catalyst filtration resulted in loss of a phosphine ligand in each recycling process. Therefore, using an additional phosphine ligand was inevitable. To overcome this issue, a novel catalyst was synthesized by using $Pd(PPh_3)_4$ and a series of copolymers containing a PPh_2 group (Scheme 5.1) [31]. $HSiCl_3$ was employed to reduce the phosphine oxides formed during the microencapsulation and cross-linking of polymer chains. A comparison of phosphinated and nonphosphinated PI Pd catalyst in the Suzuki-Miyaura reaction (Scheme 5.2) proved the superior catalytic activity of the former. In the case of a nonphosphinated PI Pd catalyst, even the addition of triphenylphosphine did not lead to comparable catalytic activity to a phosphinated PI Pd catalyst. The optimum ratio of the diphenylphosphino groups in the polymer to Pd atoms was obtained as 2:1. Phosphine groups in the polymer backbone promoted the catalytic activity and suppressed Pd leaching. It is worth noting that the catalyst was reusable and could catalyze several reaction cycles without the loss of activity.

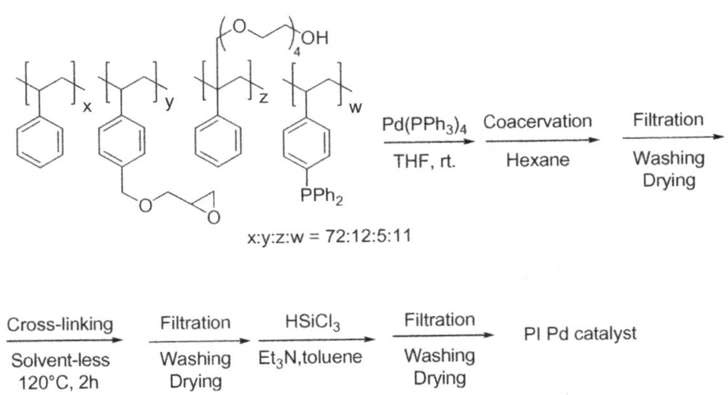

SCHEME 5.1 Preparation of catalyst. *(Reprinted from R. Nishio, M. Sugiura, S. Kobayashi, Novel polymer incarcerated palladium with phosphinated polymers: active catalyst for Suzuki-Miyaura coupling without external phosphines, Org. Lett. 7 (2005) 4831–4834 with permission of the American Chemical Society.)*

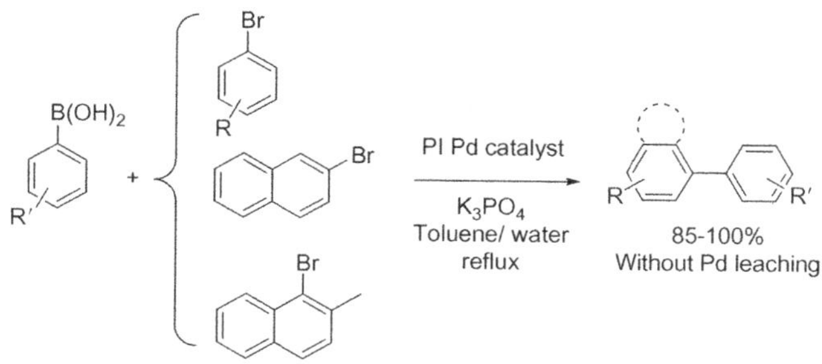

R = 2-Me, 4-*t*Bu, 4-COMe, 4-OH, 4-OMe, 4-CN
R′ = H, 2-Me

SCHEME 5.2 Phosphinated PI Pd-Catalyzed Suzuki-Miyaura coupling of various substrates without an external phosphine ligand. *(Adapted from R. Nishio, M. Sugiura, S. Kobayashi, Novel polymer incarcerated palladium with phosphinated polymers: active catalyst for Suzuki-Miyaura coupling without external phosphines, Org. Lett. 7 (2005) 4831–4834 with permission of the American Chemical Society and S. Sadjadi, M.M. Heravi, Pd(0) encapsulated nanocatalysts as superior catalytic systems for Pd-catalyzed organic transformations, RSC. Adv. 6 (2016) 88588–88624 with permission of the Royal Society of Chemistry.)*

2.2 Heterocycle Synthesis

Polymer-incarcerated Pd and Sc catalysts were employed by Kobayashi et al. for synthesis of Acridone derivatives [32]. The catalyst was synthesized by using a polystyrene-based copolymer and $Pd(PPh_3)_4$ (Scheme 5.3). The authors designed a reaction sequence including amination under Pd catalysis to afford *N*-phenyl anthranilic acid ester followed by its hydrolysis by an excess amount

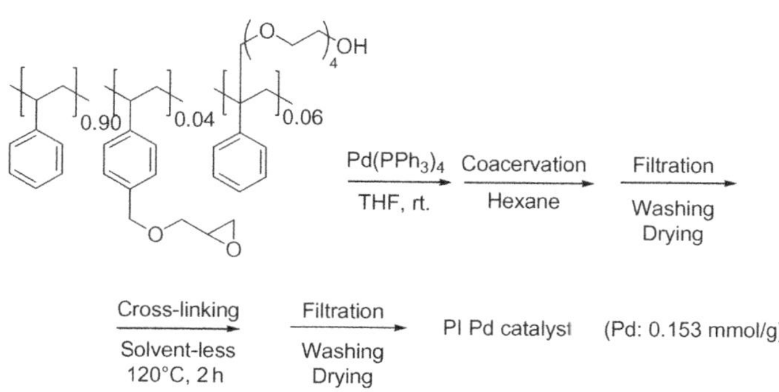

SCHEME 5.3 Preparation of the polymer-incarcerated (PI) Pd catalyst. *(Reprinted from R. Nishio, et al., Synthesis of acridone derivatives using polymer-supported palladium and scandium catalysts, J. Comb. Chem. 8 (2006) 459–461 with permission of the American Chemical Society, and S. Sadjadi, M.M. Heravi, Pd(0) encapsulated nanocatalysts as superior catalytic systems for Pd-catalyzed organic transformations, RSC. Adv. 6 (2016) 88588–88624 with permission of the Royal Society of Chemistry.)*

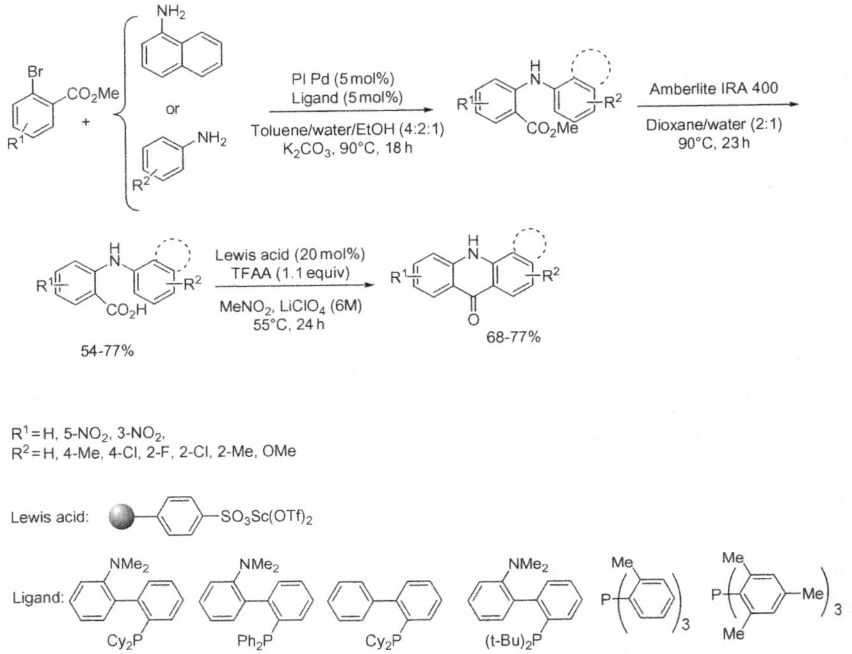

SCHEME 5.4 Synthesis of acridone derivatives. *(Adapted from R. Nishio, et al., Synthesis of acridone derivatives using polymer-supported palladium and scandium catalysts, J. Comb. Chem. 8 (2006) 459–461 with permission of the American Chemical Society, and S. Sadjadi, M.M. Heravi, Pd(0) encapsulated nanocatalysts as superior catalytic systems for Pd-catalyzed organic transformations, RSC. Adv. 6 (2016) 88588–88624 with permission of the Royal Society of Chemistry.)*

of basic resin Amberlite IRA 400 (OH) and formation of corresponding *N*-phenyl anthranilic acid. The latter tolerated intramolecular Friedel-Crafts acylation in $MeNO_2$-$LiClO_4$ solution in the presence of trifluoroacetic (TFAA) anhydride and polymer-supported scandium catalyst to furnish the desired product (Scheme 5.4). The authors investigated the amination step precisely by using a model reaction (aniline and aryl bromide). To suppress Pd leaching and promote catalytic activity, the mixture of toluene and water was used as reaction media. Using a polymer incarcerated catalyst, high yields were obtained for both aryl iodides and bromides while the reaction of aryl chlorides proceeded in moderate yields. Besides primary amines, the secondary amine could also be used. However, their reaction resulted in lower yields and Pd leaching.

2.3 Hydrogenation

Kobayashi et al. developed a novel hydrogenation catalyst based on polysilane incarcerated Pd on Al_2O_3 (PI Pd/PSi on Al_2O_3) [33]. This novel catalyst could circumvent the drawbacks of polysilane supported Pd catalysts (Pd/PSi) such as

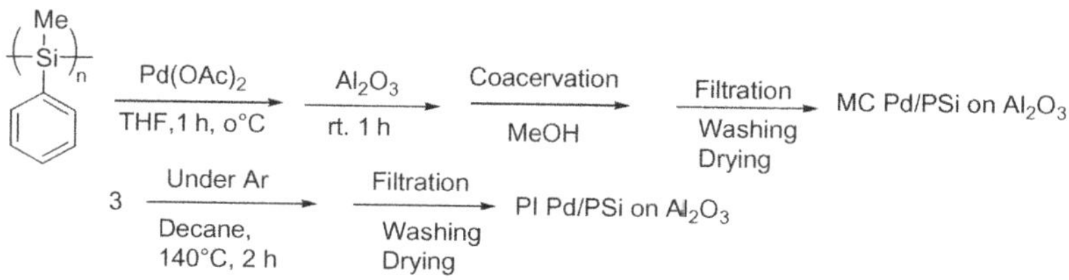

SCHEME 5.5 Catalyst preparation. *(Reprinted from H. Oyamada, et al., Development of polysilane-supported palladium/alumina hybrid catalysts and their application to hydrogenation, Org. Biomol. Chem. 6 (2008) 61–65 with permission of the Royal Society of Chemistry.)*

Pd leaching and solubility in some organic solvents. The schematic synthetic procedure of the catalyst is illustrated in Scheme 5.5. The authors also examined other supports such as TiO_2, SiO_2, and ZrO_2. Poor catalyst recovery was obtained in the case of silica. Additionally, the yields and qualities of catalysts prepared from other oxides were not comparable to alumina. The catalyst synthesis condition was also optimized by tuning temperature and reaction time. Moreover, the catalytic activity of the catalyst was investigated by using various substrates in the presence and absence of solvents (Table 5.1). As shown in Table 5.1, a broad range of substrates could be efficiently hydrogenated. A low amount of required catalyst as well as negligible Pd leaching were the merits of this procedure.

3 POLYMER-INCARCERATED Au

3.1 Oxidation

3.1.1 Oxidation of Alcohols to Carbonyl

Knowing the fact that gold nanoclusters can have weak interactions with aromatic molecules, Kobayashi et al. employed polymer containing alcohol and epoxide functionalities for preparing a polymer-incarcerated Au nanocluster, PI Au [34]. Two methods (A and B) were used for synthesis of PI-Au (Scheme 5.6). The catalytic activities of two obtained PI-Au, PI-Au 1(a-c), and PI-Au 2 were investigated for aerobic alcohol oxidation at ambient temperature. Optimization of reaction variables was achieved by using a model reaction. Studying PI-Au 1(a-c) catalysts, a converse relationship was found between catalyst loading and catalytic activity. This observation was attributed to the fact that lower catalyst loading resulted in smaller clusters. Furthermore, it was established that the presence of a base was essential for reaction. Comparison of the catalytic activities of the catalysts prepared by method B, PI-Au 2, and method A, PI-Au 1, showed superior catalytic performance of the former. It is worth noting that the PI-Au 2 efficiently catalyzed the reaction even in air.

TABLE 5.1 Hydrogenation of Various Substrates Using PI Pd/PSi on Al_2O_3 as a Catalyst

Substrate	Solvent	Time/h	Conc./mol L^{-1}	Product	Yield (%)	Leaching (%)
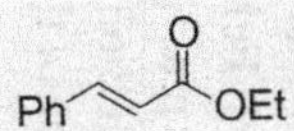	EtOH	1	0.33	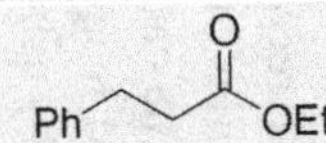	>99.9	<0.22
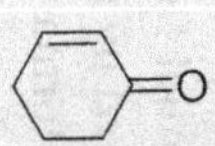	EtOH	4	1.0	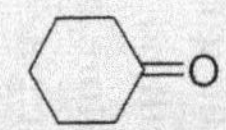	Quant.	<0.30
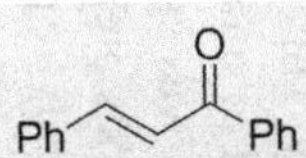	EtOH	24	1.0	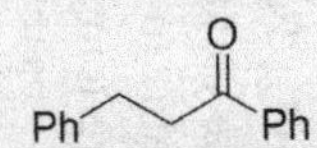 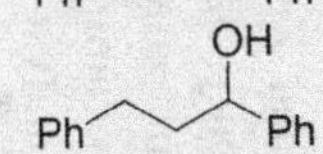	91 (6)	0.99
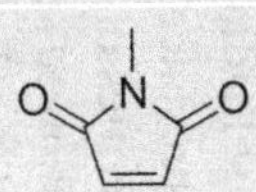	EtOH	1	0.50	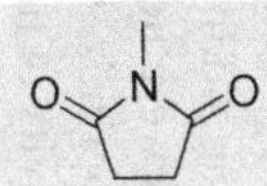	97	0.48
	EtOH	3.5	1.0	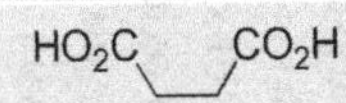	Quant.	0.94
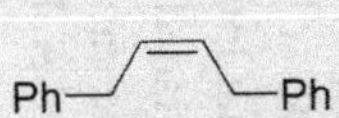	CH_2Cl_2	24	0.50	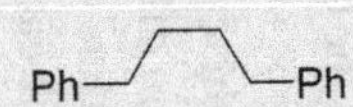	98	3.6
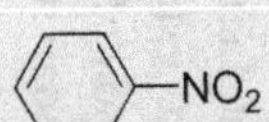	EtOH Hexane	8 1.5	1.0 0.33	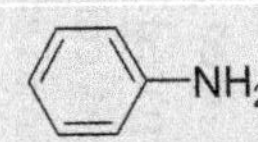	>99.9 >99.9	0.34 2.28

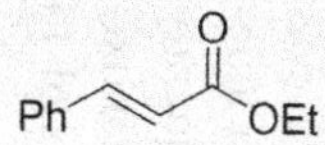	Neat-RT	8	0.05 mol%	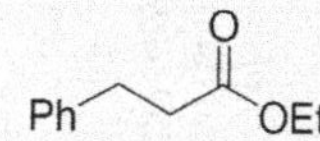	>99.9	<0.22
	Neat, 50°C	16	0.05 mol%		Quant.	<0.1
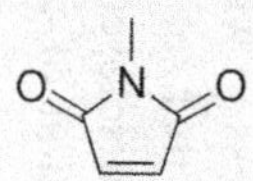	Neat, 50°C	1	0.05 mol%	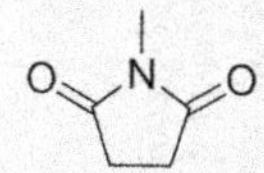	58	–
	Neat-RT	1	0.2 mol%		79	–
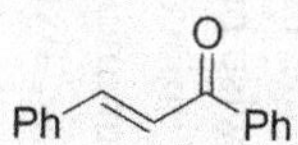	Neat, 50°C	24	0.05 mol%	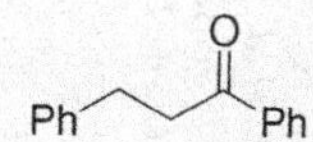 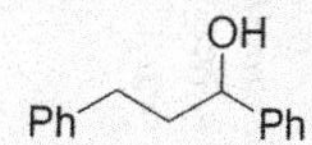	80(trace)	–
	Neat, 50°C	24	0.2 mol%		95(5)	0.12

Reprinted from H. Oyamada, et al., Development of polysilane-supported palladium/alumina hybrid catalysts and their application to hydrogenation, Org. Biomol. Chem. 6 (2008) 61–65 with permission of the Royal Society of Chemistry, and S. Sadjadi, M.M. Heravi, Pd(0) encapsulated nanocatalysts as superior catalytic systems for Pd-catalyzed organic transformations, RSC. Adv. 6 (2016) 88588–88624 with permission of the Royal Society of Chemistry.

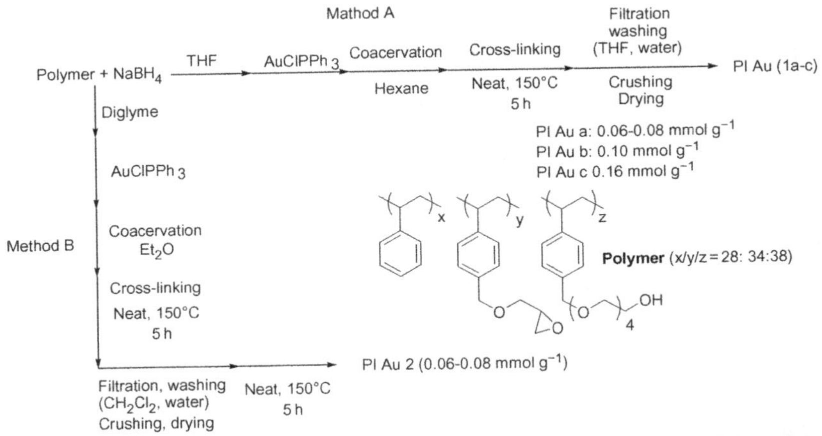

SCHEME 5.6 Catalyst synthesis. *(Adapted from Miyamura, H., et al., Aerobic oxidation of alcohols at room temperature and atmospheric conditions catalyzed by reusable gold nanoclusters stabilized by the benzene rings of polystyrene derivatives, Angew. Chem. 119 (2007) 4229–4232 with permission of Wiley-VCH Verlag.)*

Additionally, this catalyst could promote oxidation of alcohols under neat and base-free conditions. The reaction scope was studied by using various alcohols (Table 5.2). Beside high catalytic activity, the reusability of the catalyst and lack of leaching were other advantages of this catalyst.

3.1.2 Oxidation of Alcohols to Methyl Esters

Au nanoclusters were incarcerated in polymer and used as an efficient and reusable catalyst, PI-Au, for aerobic oxidation of alcohols to corresponding methyl esters under a mild reaction condition (atmospheric pressure and room temperature) [35]. The reaction variables were optimized by performing the model reaction, *p*-methylbenzyl alcohol oxidation, in the presence of various amounts of base and solvents. A 0.5 equivalent of K_2CO_3 and the mixture of methanol/water were selected as the best base and solvent respectively. The scope of the procedure was studied by using various alcohols including aliphatic, allylic, and aromatic alcohols with electron withdrawing and electron-donating functional groups (Table 5.3). As shown in this table, aromatic alcohols with various functional groups could be oxidized efficiently under PI-Au catalysis. However, oxidation of aliphatic alcohol did not proceed efficiently. In the case of allyl alcohol, further Michael addition of methanol to the ester occurred. It was claimed that esterification occurred through a stepwise process in which aldehydes, formed from alcohol oxidation, generated hemiacetal that subsequently oxidized to produce methyl ester. The kinetic measurements disclosed that oxidation of hemiacetal to ester was much faster than oxidation of alcohol to aldehyde.

TABLE 5.2 Oxidation of Various Alcohols

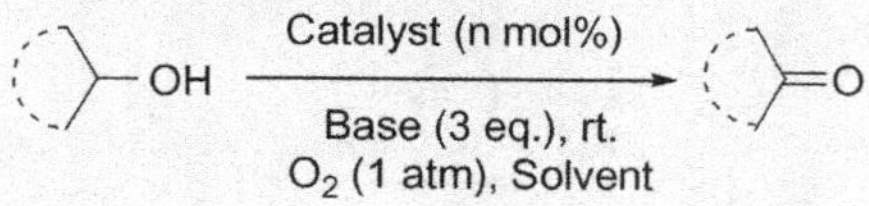

Substrate	*n*	Base	Time (h)	Solvent	Yield (%)
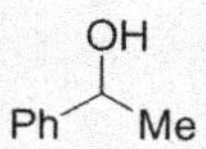	1	K_2CO_3	5	[a]BTF/Water 1:1	> 99
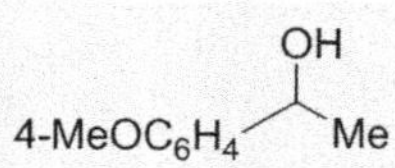	1	K_2CO_3	5	BTF/Water 1:1	> 99
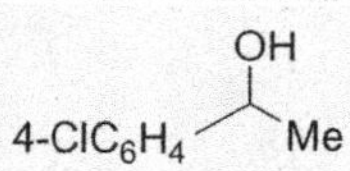	1	K_2CO_3	5	BTF/Water 1:1	91
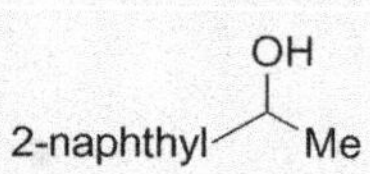	1	K_2CO_3	5	BTF/Water 1:1	> 99
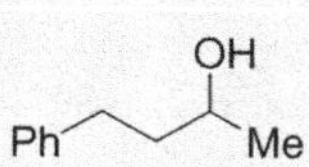	1	K_2CO_3	30	BTF/Water 1:1	95
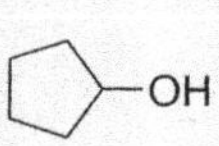	1	KOH	30	BTF/Water 1:1	> 99
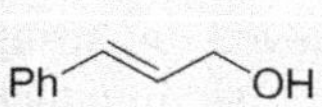	1	K_2CO_3	12	BTF/Water 1:1	64
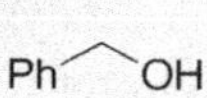	1	Cs_2CO_3	15	Diglyme	79
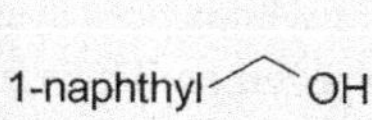	2	K_2CO_3	5	BTF/Water 1:1	78
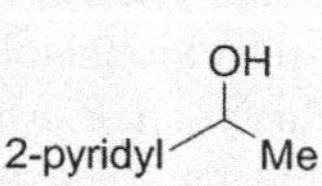	3	K_2CO_3	24	BTF/Water 1:1	> 99
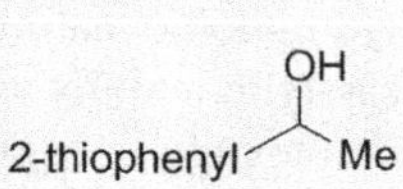	1	K_2CO_3	24	BTF/Water 1:1	94

[a]*BTF: benzotrifluoride.*

Reprinted from Miyamura, H., et al., Aerobic oxidation of alcohols at room temperature and atmospheric conditions catalyzed by reusable gold nanoclusters stabilized by the benzene rings of polystyrene derivatives, Angew. Chem. 119 (2007) 4229–4232 with permission of Wiley-VCH Verlag

TABLE 5.3 Aerobic Esterifications of Various Substrates Under Catalysis of a PI Au Catalyst.

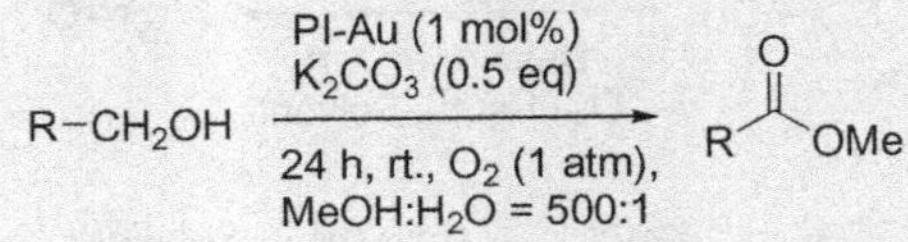

R	Conversion (%)[a]	Yield (%)[a]
Ph	>99	95
4-$MeOC_6H_4$	>99	>99
4-BrC_6H_4	>99	92
4-NO_2 C_6H_4	95	60
1-Naphthyl	>99	95
2-Pyridyl	>99	>99
Ph-CH=CH–	>99	69
$PhCH_2CH_2CH_2$–	12	10
CH_2=CH–	>99	60[b]

[a]*Determined by GC.*
[b]*Methyl 3-methoxypropionate was obtained.*
Reprinted from H. Miyamura, T. Yasukawa, S. Kobayashi, Aerobic oxidative esterification of alcohols catalyzed by polymer-incarcerated gold nanoclusters under ambient conditions, Green Chem. 12 (2010) 776–778 with permission of the Royal Society of Chemistry.

3.1.3 *Oxidation of Secondary Alcohols*

The oxidation of secondary alcohols was successfully achieved by using polymer-incarcerated, carbon-stabilized Au nanoclusters, PI/CB-Au, in which Au nanoclusters were stabilized by both CB and polymer [36]. The catalyst was synthesized via two synthetic methods (Scheme 5.7). Synthesis of catalyst "PI/CB-Au a" was carried out through a sequential process in which gold was loaded onto CB and subsequently isolated and combined with polymer. The catalyst "PI/CBAu b" was synthesized by a one-pot procedure. The catalytic activities of both catalysts were studied in a model reaction (($\pm$)-1-phenylethanol) and compared with that of a CB-supported gold nanocluster, CB-Au. It was found that the oxidation reaction did not proceed in the presence of CB-Au. This observation was attributed to the highly hydrophobic nature of CB. Additionally, the large diameter and resulting low catalytic activity of "PI/CB-Au a" was ascribed to its synthetic procedure that promotes aggregation (Table 5.4). Using a one-pot procedure for the synthesis of "PI/CB-Au b" resulted in smaller Au nanocluster diameter and high catalytic activity. The effect of the loading amount on the catalytic activity was studied by using various loadings (Table 5.4). A converse relationship was observed between

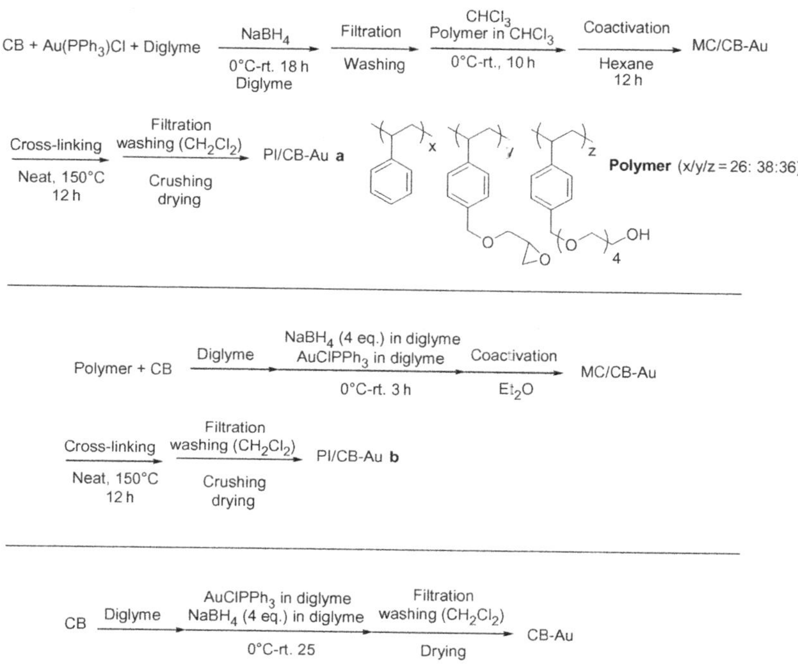

SCHEME 5.7 Catalyst preparation. *(Reprinted from S. Kobayashi, H. Miyamura, Polymer-incarcerated metal(0) cluster catalysts, Chem. Rec. 10 (2010) 271–290 with the permission of Wiley-VCH Verlag.)*

TABLE 5.4 Characterization of Catalysts CB-Au and PI/CB-Au a and b

Catalyst	MC/CB-Au Diameter (nm)	PI/CB-Au Diameter (nm)	PI/CB-Au Loading $(mmol\ g)^{-1}$	Oxidation Yield (%)
CB-Au	1–2	–	0.43	0
PI/CB-Au a	3.8	10.9	0.17	9
PI/CB-Au b	2.7	2.9	0.25	>99
PI/CB-Au b	2.7	5.1	0.40	90
PI/CB-Au b	2.2	3.8	0.52	91
PI/CB-Au b	1.8	3.8	0.60	86

Reprinted from C. Lucchesi, et al., Aerobic oxidation of alcohols under mild conditions catalyzed by novel polymer-incarcerated, carbon-stabilized gold nanoclusters, Adv. Synth. Catal. 350 (2008) 1996–2000 with the permission of Wiley-VCH Verlag.

the incorporation ratio and loading amounts. It was found that the loading amount of 0.60 mmol/g could be reached and successfully employed in the aerobic oxidation of the model reaction with a slight loss of catalytic activity. The substrate scope of reaction was investigated by performing oxidation of various alcohols including allylic, aromatic, and aliphatic alcohols (Table 5.5), among which aliphatic samples oxidation led to the lower yields.

In all of the above discussed samples, the copolymers used for incarceration were prepared from synthesized monomer and possessed hydroxyl and epoxy functionalities that served as internal cross-linkers. An alternative simplified approach is the use of the external cross-linkers. In this regard, Kobayashi et al. disclosed preparation of homopolymer-incarcerated gold nanocluster (2–3 nm) by employing (4-vinylphenyl) methanol-derived homopolymer as the support and sodium bis(2-methoxyethoxy)aluminum hydride (Red-Al). The latter served both as an inter-crosslinking agent to generate aluminates from the hydroxyl groups of polymer and as a reductant to reduce Au salt to an Au nanocluster. Various loadings of Au nanoclusters (0.08–0.48) could be achieved without agglomeration of Au nanoclusters. The catalytic activity of the obtained homopolymer incarcerated Au nanoclusters were confirmed for diverse aerobic oxidation reactions including oxidation of primary and secondary alcohols to corresponding aldehydes and ketones, amine to an imine, and hydroquinone to a quinone. Moreover, the novel catalyst could promote aerobic oxidative methyl esterification from an alcohol in methanol [22].

4 POLYMER-INCARCERATED Pt

Using styrene copolymer and Pt sources including platinum dichloride cyclooctadiene complex, $PtCl_2(COD)$, and $H_2PtCl_6{\cdot}6H_2O$, Kobayashi et al. reported synthesis of polymer incarcerated Pt nanocluster, PIPt, with catalytic activity for hydrogenation reaction [37]. The synthetic procedure of PIPt is represented in Scheme 5.8. It was found that the reducing agent and solvent could dramatically influence the process in terms of PIPt yield, Pt loading, and reproducibility. Among various tested reducing agents, H_2, hydrazine, $NaBH_4$, and silanes ($HSiEt_3$ and $HSi(OEt)_3$), $HSiEt_3$ exhibited the best results. In the case of hydrazine and H_2, no immobilized Pt nanoclusters were obtained. This observation was assigned to fast reduction and Pt precipitation. Upon using $NaBH_4$, however, cross-linking did not proceed. The solvent choice differed according to the Pt source. CH_2Cl_2 that could solubilize $PtCl_2(COD)$ was recognized as the best solvent for the synthesis of PIPt by using $PtCl_2(COD)$, while THF and hexane were used in the case of $H_2PtCl_6{\cdot}6H_2O$.

To elucidate the catalytic activity of the novel catalysts, hydrogenation of nitrobenzene to aniline was selected as a model reaction and performed in THF and under 1 atm H_2 at ambient temperature. The comparison of catalytic performance of PIPt obtained from $PtCl_2(COD)$ with PIPt formed from $Pt(PPh_3)_4$ and PIPd proved the higher activity of the former (Table 5.6).

TABLE 5.5 Oxidation of Various Secondary Alcohols Catalyzed by PI/CB-Au 1b

Substrate	Time (h)	Yield (%)	Substrate	Time (h)	Yield (%)
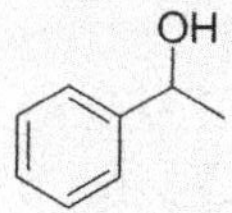	3	>99	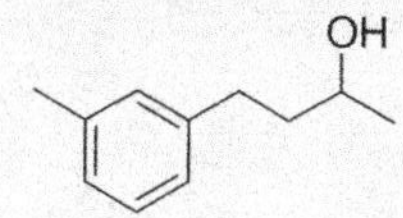	48	99
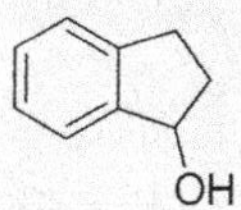	3	>99	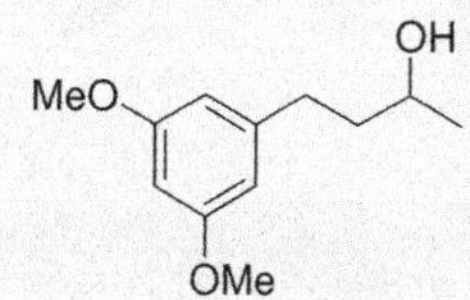	48	86
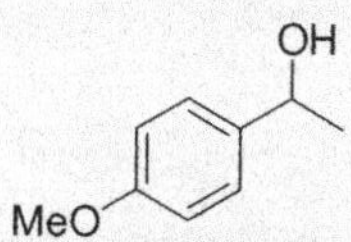	6	>99	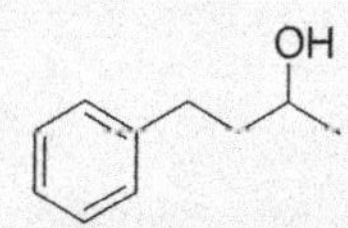	48	82
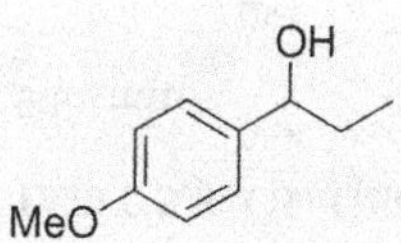	24	>99	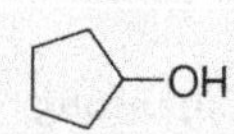	48	57

Continued

TABLE 5.5 Oxidation of Various Secondary Alcohols Catalyzed by PI/CB-Au 1b—cont'd

Substrate	Time (h)	Yield (%)	Substrate	Time (h)	Yield (%)
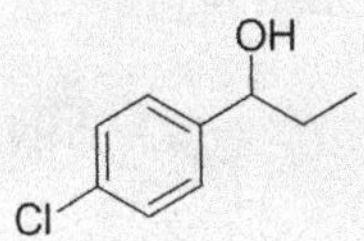	48	>99	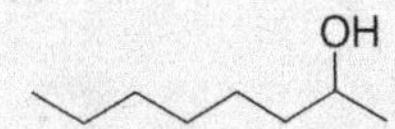	48	62
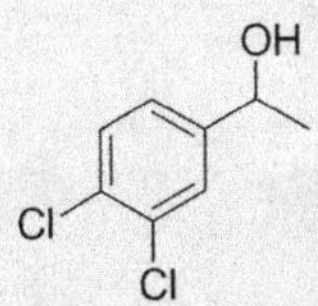	48	95	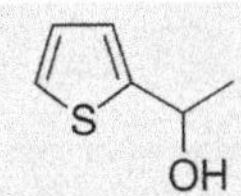	48	99
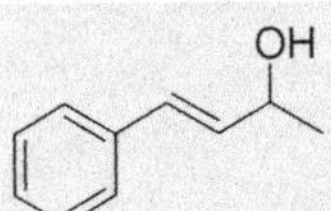	24	>99	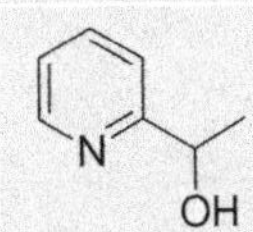	48	58

Reprinted from C. Lucchesi, et al., Aerobic oxidation of alcohols under mild conditions catalyzed by novel polymer-incarcerated, carbon-stabilized gold nanoclusters, Adv. Synth. Catal. 350 (2008) 1996–2000 with the permission of Wiley-VCH Verlag.

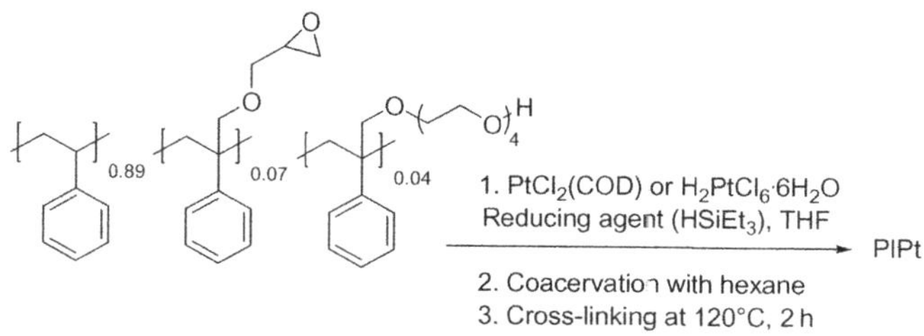

Using $H_2PtCl_6{\cdot}6H_2O$ as Pt source led to: Cluster size of 2–3 nm,
Pt loading of 0.286 mmol g^{-1}

Using $PtCl_2(COD)$ as Pt source led to:cluster size of less than 1 nm.
Pt loading of 0.142 mmol g^{-1}

SCHEME 5.8 Catalyst preparation. *(Reprinted from Y. Miyazaki, H. Hagio, S. Kobayashi, Practical access to the polymer incarcerated platinum (PI Pt) catalyst and its application to hydrogenation, Org. Biomol. Chem. 4 (2006) 2529–2531 with permission of the Royal Society of Chemistry.)*

TABLE 5.6 Hydrogenation of Nitrobenzene-Catalyzed by PI Catalysts

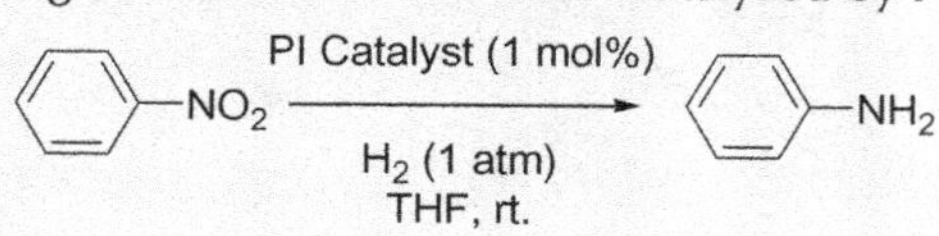

PI Catalyst	Time (h)	Yield (%)	Leaching (%)
PI Pt from $PtCl_2(COD)$	1	Quant.	Not detected
PI Pt from $Pt(PPh_3)_4$	6	Quant.	19
PIPd	3	Quant.	Not detected

Reprinted from Y. Miyazaki, H. Hagio, S. Kobayashi, Practical access to the polymer incarcerated platinum (PI Pt) catalyst and its application to hydrogenation, Org. Biomol. Chem. 4 (2006) 2529–2531 with permission of the Royal Society of Chemistry.

The catalyst showed high catalytic activity for hydrogenation of various substrates (Table 5.7) including alkenes, alkynes, sterically demanding C=C double bonds (Entry 5), heterocycles (Entry 6), and multifunctionalized substrates at room temperature (Entery 7–10) (Table 5.7). As obvious in this table, the sterically demanding substrate and heterocyclic substrate (Entry 5 and 6 respectively) required more reaction time. In the case of multifunctional substrates (Entery 7–10), selective hydrogenation of alkene, alkyne, and nitro groups was observed.

The utility of polymer-incarcerated Pt catalyst was also disclosed in a hydrosilylation reaction [38]. The procedure for the preparation of the catalyst from copolymer and $Pt(PPh_3)_4$ is illustrated in Scheme 5.9. The elemental analysis proved the presence of P atoms in the form of triphenylphosphine oxide in the PI Pt catalyst, which was plausibly due to the high stability of Pt-P. To study

TABLE 5.7 Hydrogenation of Various Substrates Catalyzed by PI Pt

Entry	Substrate	Product	Yield (%)	T (h)	Solvent	Catalyst (mol%)	H_2Pressure (atm)	Pt Source
1			Quant.	1	THF	1	1	$PtCl_2(COD)$
2	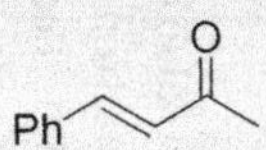	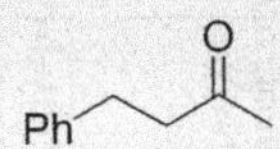 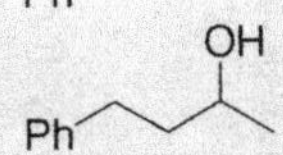	90 10	1	THF	1	1	$PtCl_2(COD)$
3	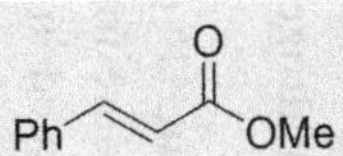	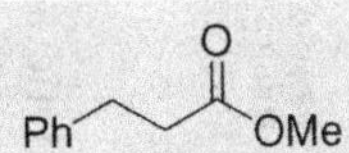	Quant.	1	THF	1	1	$PtCl_2(COD)$
4	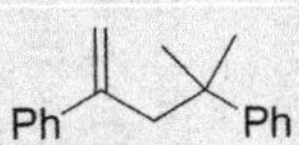	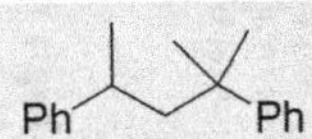	Quant.	1	THF	1	1	$PtCl_2(COD)$
5	Cholestol	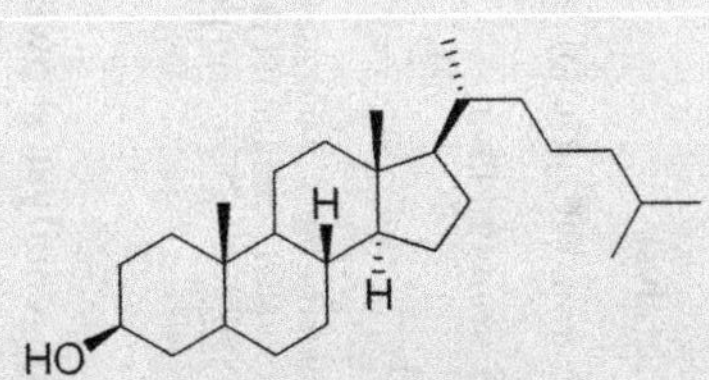	80	24	CH_2Cl_2	5	1	$H_2PtCl_6{\cdot}6H_2O$

6	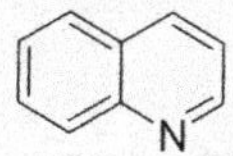	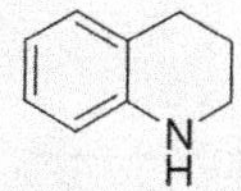	82	80	THF	1	1	PtCl$_2$(COD)
7	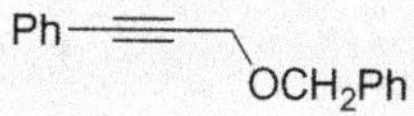		Quant.	1	THF	1	1	PtCl$_2$(COD)
8	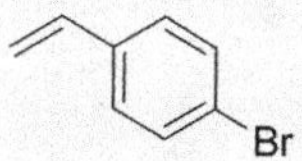	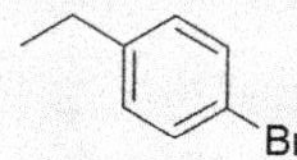	95	1	THF	1	1	PtCl$_2$(COD)
9	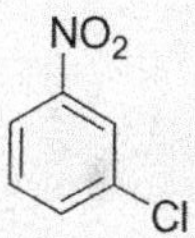	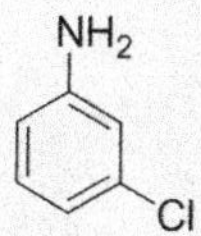	73	1	THF	1	1	PtCl$_2$(COD)
10	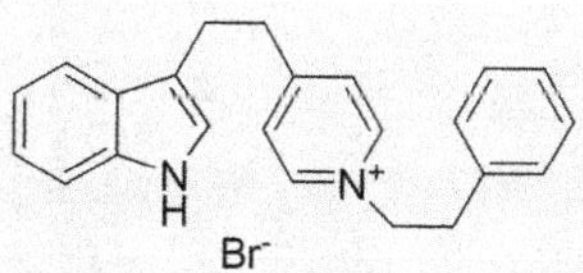	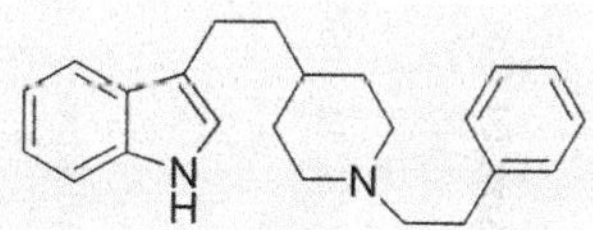	95	1	MeOH-THF	5	5	$H_2PtCl_6 \cdot 6H_2O$

Adapted from Y. Miyazaki, H. Hagio, S. Kobayashi, Practical access to the polymer incarcerated platinum (PI Pt) catalyst and its application to hydrogenation, Org. Biomol. Chem. 4 (2006) 2529–2531 with permission of the Royal Society of Chemistry.

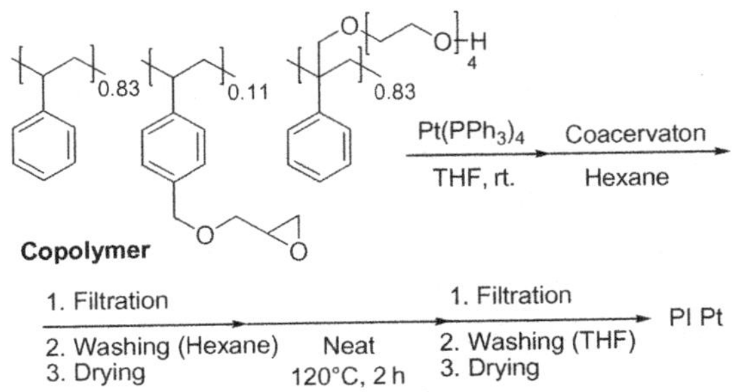

SCHEME 5.9 Catalyst preparation. *(Reprinted from H. Hagio, M. Sugiura, S. Kobayashi, Immobilization of a platinum catalyst using the polymer incarcerated (PI) method and application to catalytic reactions, Synlett (5) (2005) 813–816 with permission of Thieme.)*

the catalytic activity of the PI Pt catalyst, a reaction of pentamethyldisiloxane and 4-phenyl-1-butene was performed in the presence of a catalyst in various solvents. The results indicated that Pt leaching and reusability depended on the solvent. The 1:2 mixture of hexane-water was recognized as the best choice of the solvent in which the quantitative yield was obtained and no Pt leaching observed. In this reaction condition, the recovery and reusability of the catalyst without loss of activity for 5 runs was reported. Although the PI Pt catalyst showed a comparable catalytic activity to Pt/C (Table 5.8), superior catalytic activity was obtained compared to $Pt(PPh_3)_4$ and PtO_2. Furthermore, the presence of PPh_3 as an additive had a detrimental effect on the catalytic activity,

TABLE 5.8 Comparison of Catalytic Activity

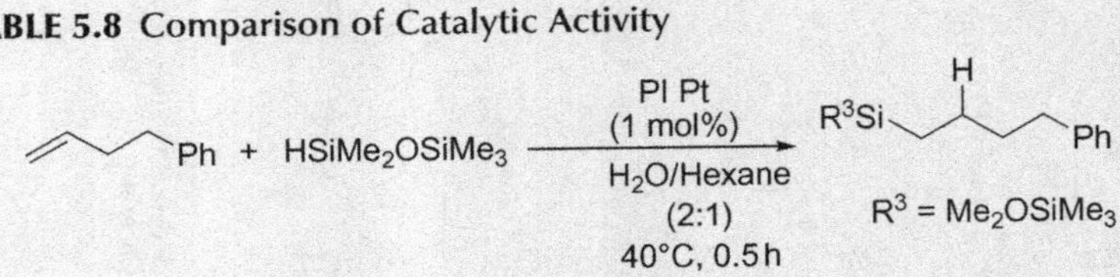

"Pt"	Additive (1 mol%)	Yield (%) [Leaching (%)]
PI Pt	–	99 (not detected)
Pt/C	–	99
$Pt(PPh_3)_4$	–	7
PtO_2	–	87
PI Pt	PPh_3	0
PI Pt	$PH_3P{=}O$	Quant.

Reprinted from H. Hagio, M. Sugiura, S. Kobayashi, Immobilization of a platinum catalyst using the polymer incarcerated (PI) method and application to catalytic reactions, Synlett (5) (2005) 813–816 with permission of Thieme.

TABLE 5.9 Hydrosilylation of Several Substrates Using PI Pt

Substrate	Product	Time (h)	Yield (%)
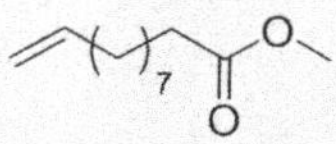	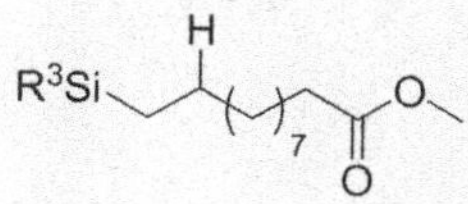	3	90
	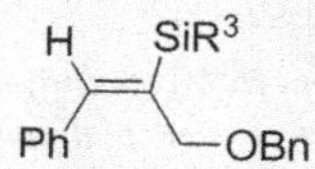	17	Quant.
	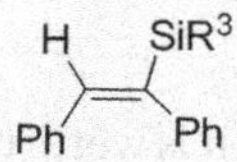	1	Quant.
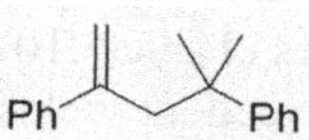	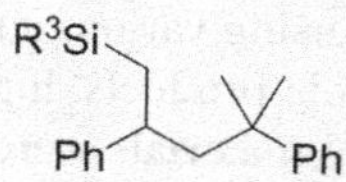	17	81
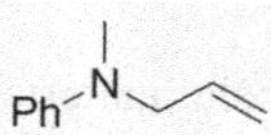	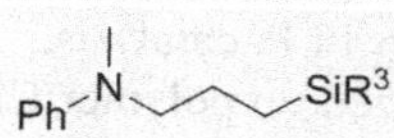	17	83
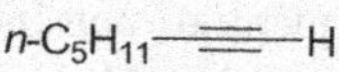	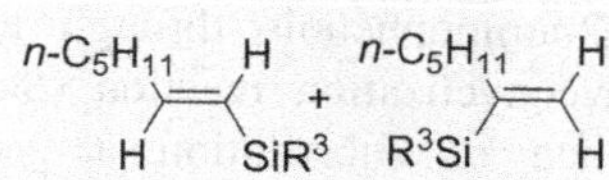	16	82

Reprinted from H. Hagio, M. Sugiura, S. Kobayashi, Immobilization of a platinum catalyst using the polymer incarcerated (PI) method and application to catalytic reactions, Synlett (5) (2005) 813–816 with permission of Thieme.

while $PH_3P{=}O$ had no effect. The procedure could be generalized for hydrosilylation of various substrates (Table 5.9) without any Pt leaching.

In another example, polymer-incarcerated Pt catalyst, PI Pt, was successfully used as a reusable catalyst for aerobic oxidation of hydroquinone and catechol derivatives including electron-deficient and electron-rich ones under ambient temperature [39]. The catalyst preparation procedure is depicted in Scheme 5.10. It was found that the Pt loading (~0.05 to ~0.1 mmol Pt g^{-1}) did not influence the particle size and the catalytic performance dramatically. A biphasic mixture containing chloroform and water (27:1) was used as solvent. To elucidate whether the molecular oxygen was necessary for reaction, a model reaction was carried out under N_2 atmosphere. Negligible yields under the later condition proved that the presence of molecular oxygen was essential and the

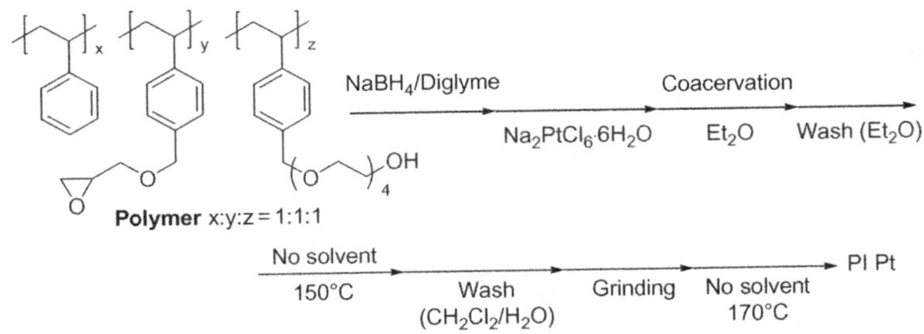

SCHEME 5.10 Catalyst preparation. *(Reprinted from H. Miyamura, et al., Aerobic oxidation of hydroquinone derivatives catalyzed by polymer-incarcerated platinum catalyst, Angew. Chem. Int. Ed. 47 (2008) 8093–8095 with permission of Wiley-VCH Verlag.)*

oxidation did not occur through H_2 elimination. It is worth noting that performing the reaction under atmospheric oxygen furnished the desired product. The generality of the process was investigated by using various substrates (Table 5.10). In all cases, high selectivity, negligible side-products, high to excellent yields, and complete consumption of the starting materials were reported. Moreover, PI Pt showed high chemoselectivity, and oxidation of *p*-methoxyanisole and *p*-methoxyphenol did not proceed under PI Pt catalysis.

Recently, Pt nanoclusters immobilized on polymer/CB composite material (PI/CB-Pt) have been used as a catalyst for the synthesis of benzoxazole derivatives from N-substituted 2-aminophenols through a sequential aerobic oxidation-enolization-oxidative cyclization reaction (Scheme 5.11) [40]. It was found that the presence of substitution at an aromatic ring of N-substituted 2-aminophenols was essential for achieving good catalytic activity through stabilizing *o*-quinines intermediate. The amount of water also affected the yield of reaction, and optimization of this factor was required for improving reaction yield. Although the presence of water was considered as a key factor for the catalytic activity of the catalyst, its high concentration reduced the yield of reaction (Table 5.11). The plausible reaction mechanism (Scheme 5.12) consisted of transformation of N-substituted 2-aminophenols to corresponding *o*-quinines followed by enolization to phenolic imine and subsequent oxidative cyclization. In this context, the probability of direct oxidation of N-substituted 2-aminophenols to phenolic imine was discounted due to relative difficulty of aerobic oxidation of amines to imines.

4.1 Polymer-Incarcerated Ni

Using novel polymer incarcerated Ni nanoparticles that contained *N*-heterocyclic carbenes (NHCs), the Corriu-Kumada-Tamao (CKT) reaction was catalyzed efficiently with no need for external addition of a ligand [41]. The design of the

TABLE 5.10 PI Pt-Catalyzed Aerobic Oxidation of Hydroquinones and Catechols

1,4-Hydroquinone or Catechol $\xrightarrow[\text{O}_2\text{ (1 atm), rt, CH(D)Cl}_3/\text{H}_2\text{O}]{\text{PI Pt}}$ *p*-Quinone or *o*-Quinone

Product	PI Pt Amount (mol% Pt)	T(h)	Yield (%)	Product	PI Pt Amount (mol% Pt)	T (h)	Yield (%)
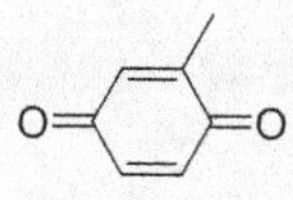	0.1 0.05	1 5 1	>99 90 >99	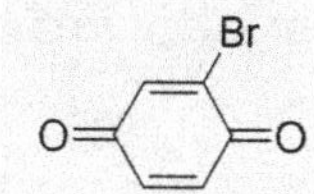	1	0.5	80
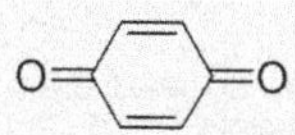	0.1	2	75	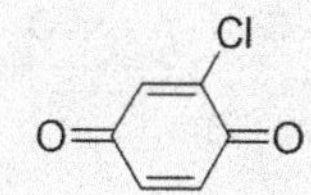	0.5	7	93
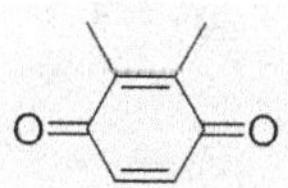	0.2	3.75	89	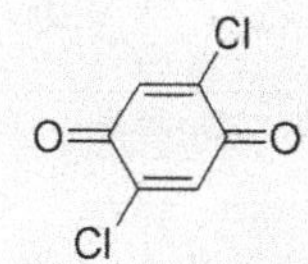	1	3	99
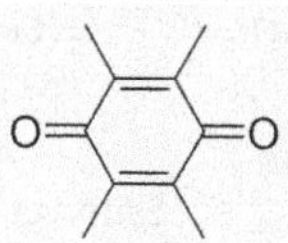	0.1	2	93	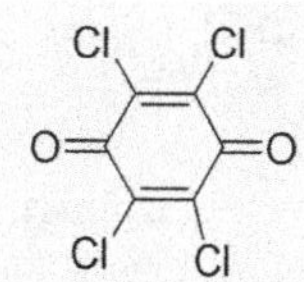	1 0.5	3 3	99 90

Continued

TABLE 5.10 PI Pt-Catalyzed Aerobic Oxidation of Hydroquinones and Catechols—cont'd

Product	PI Pt Amount (mol% Pt)	T(h)	Yield (%)	Product	PI Pt Amount (mol% Pt)	T (h)	Yield (%)
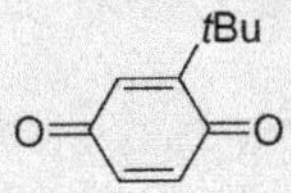	0.1	3.5	>99	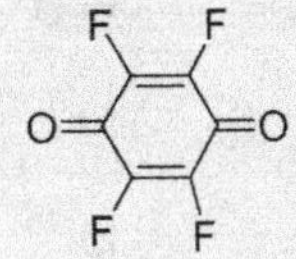	1	3	65
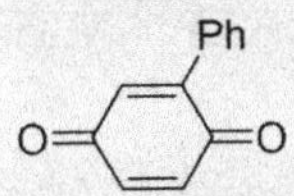	0.1	1	94	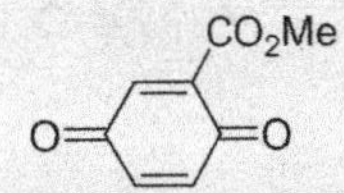	2	3	92
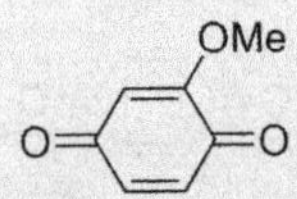	0.1	1	75	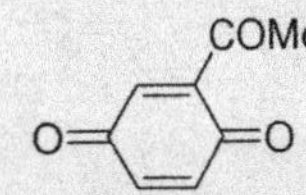	2	6	88
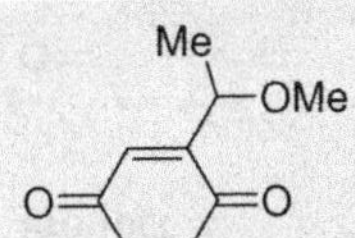	1	3	76	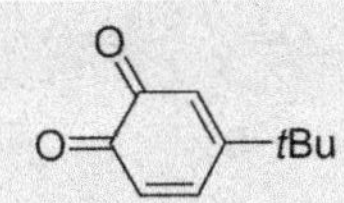	1	3	98
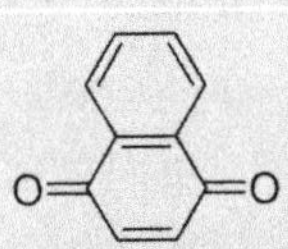	0.1	2	>99	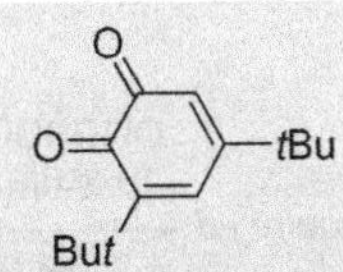	1	8	>99

Reprinted from H. Miyamura, et al., Aerobic oxidation of hydroquinone derivatives catalyzed by polymer-incarcerated platinum catalyst, Angew. Chem. Int. Ed. 47 (2008) 8093–8095 with permission of Wiley-VCH Verlag.

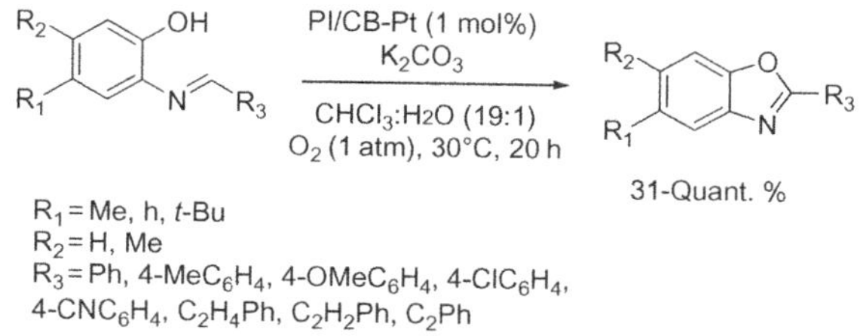

SCHEME 5.11 Sequential oxidative transformation of N-substituted 2-aminophenols. *(Adapted from W.-J. Yoo, H. Yuan, S. Kobayashi, Oxidative transformation of N-substituted 2-aminophenols to 2-substituted benzoxazoles catalyzed by polymer-incarcerated and carbon stabilized platinum nanoclusters, Can. J. Chem. 90 (2012) 306–313 with permission of NRC Research Press.)*

TABLE 5.11 Optimization of Reaction Conditions

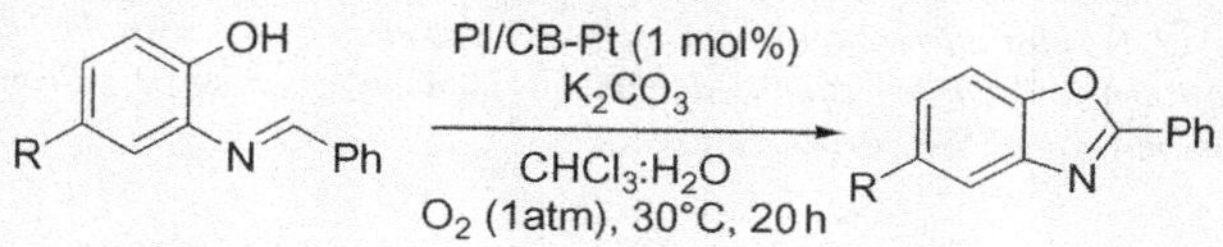

R	$CHCl_3$:H_2O	Conc. (mol/L)	Yield (%)
H	9:1	0.25	<5
Me	9:1	0.25	28
Me	9:1	0.05	35
Me	19:1	0.05	48
Me	1:0	0.05	28

Reprinted from W.-J. Yoo, H. Yuan, S. Kobayashi, Oxidative transformation of N-substituted 2-aminophenols to 2-substituted benzoxazoles catalyzed by polymer-incarcerated and carbon stabilized platinum nanoclusters, Can. J. Chem. 90 (2012) 306–313 with permission of NRC Research Press.

catalyst was based on using two novel copolymers with active cross-link moieties consisting of an imidazole part (2) and an alkyl bromide part (3) (Fig. 5.2). Upon cross-linking, embedded NHCs would be furnished. The process showed a broad scope; various substrates including vinyl, aryl, and alkyl halides with aryl or alkyl and Knochel-type Grignard reagent could be applied (Table 5.12). High yields, simplicity of work-up, reusability of catalyst, as well as no Ni leaching were other merits of this protocol. Notably, using Ni incarcerated polymer prepared from copolymer 1 and CB as a second support (Fig. 5.2) led to lower yields and Ni leaching. This observation was attributed to the positive effect of NHC ligands on the Ni-NPs.

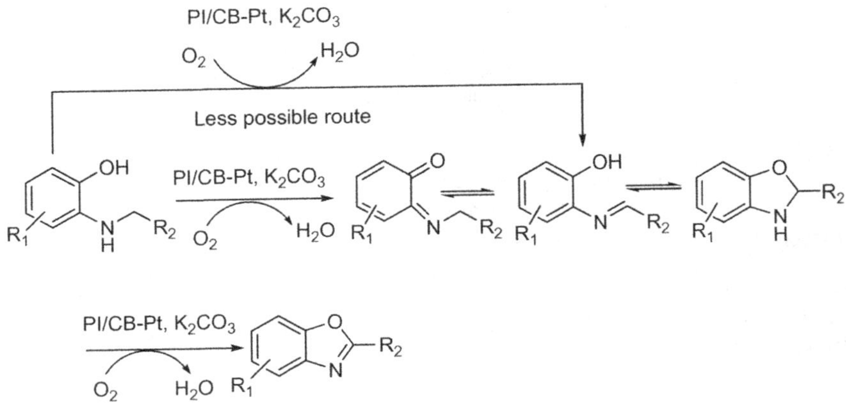

SCHEME 5.12 Tentative mechanism for PI/CB-Pt-catalyzed sequential oxidative transformation of N-substituted 2-aminophenols. *(Reprinted from W.-J. Yoo, H. Yuan, S. Kobayashi, Oxidative transformation of N-substituted 2-aminophenols to 2-substituted benzoxazoles catalyzed by polymer-incarcerated and carbon stabilized platinum nanoclusters, Can. J. Chem. 90 (2012) 306–313 with permission of NRC Research Press.)*

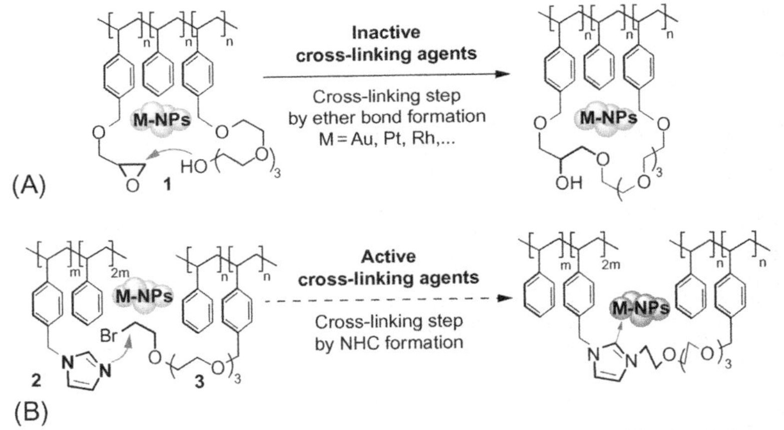

FIG. 5.2 Concept of PI M-NPs using active cross-linking agents. *(Reprinted from J.F. Soule, H. Miyamura, S. Kobayashi, Copolymer-incarcerated nickel nanoparticles with N-heterocyclic carbene precursors as active cross-linking agents for Corriu-Kumada-Tamao reaction, J. Am. Chem. Soc. 135 (2013) 10602–10605 with permission of the American Chemical Society.)*

4.2 Polymer-Incarcerated Bimetallic Catalyst

4.2.1 Imine Synthesis

The Kobayashi research group proved the utility of a carbon black-stabilized, polymer incarcerated Au/Pd catalyst for direct synthesis of imines from a

TABLE 5.12 The Scope of CKT Reaction

R-X (1)	R′-MgX′ (2)	Product (3)
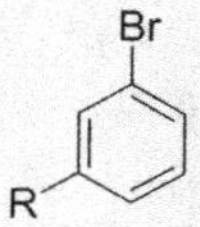 R= a)- Me b)- H c)- OMe d)- CF_3 e)- Cl	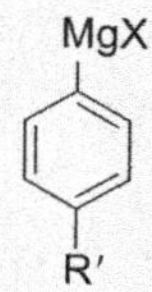 R′= a)- H b)- F c)- ^{t}Bu d)- OMe	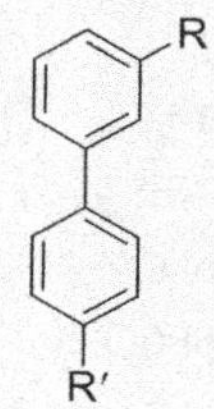 aa) 92% ba) 98% ca) 91% da) 74% ea) 89% cb) 86% cc) 79% ea) 89%
1,3-Dichlorobenzene	1a	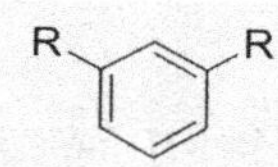 R=Ph, 88%
1-Chloro-4-fluorobenzene	1c	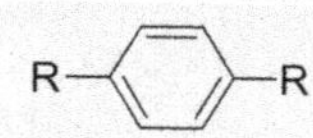 R=4-^{t}B-(C_6H_4)-, 85%
2-Naph-Br	1c	2—Naph(C_6H_4)-4-^{t}Bu 96%
1-Naph-Br	*c*-Hex-MgCl	1-Naph-^{t}Bu 84%
2-Naph-Br	^{t}Bu-MgBr	2-Naph-^{t}Bu 71%
3-bromoquinoline	1a	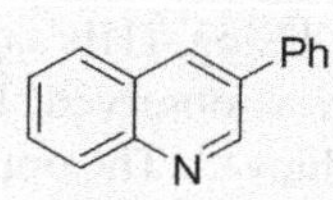 79%
8-chloroquinoline	1a	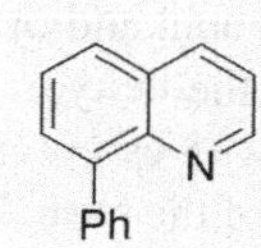 68%
2-Py-Cl	*n*-Hex-MgCl	2-Py-*n*-Hex 93%
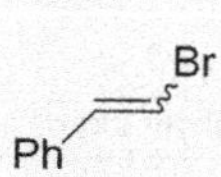 E/Z=6.1	*c*-Hex-MgCl	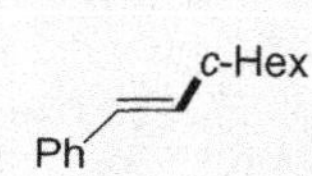 E/Z>99.1, 93%

Continued

TABLE 5.12 The Scope of CKT Reaction—cont'd

R-X (1)	R′-MgX′ (2)	Product (3)
Oct-I	1a	Oct-Ph 97%
Oct-I	$PhCH_2$-MgCl	Oct-CH_2Ph 88%
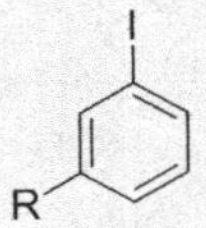 R= o)-CN p)-CO_2Et	Ph-MgCl.LiCl	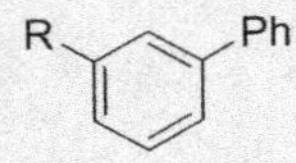 oh, 80% ph, 69%
Oct-I	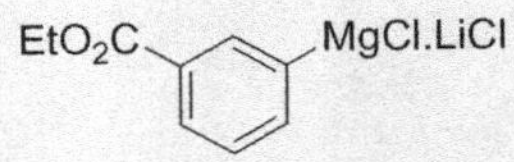	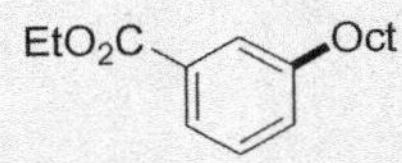 52%

Adapted from J.F. Soule, H. Miyamura, S. Kobayashi, Copolymer-incarcerated nickel nanoparticles with N-heterocyclic carbene precursors as active cross-linking agents for Corriu-Kumada-Tamao reaction, J. Am. Chem. Soc. 135 (2013) 10602–10605 with permission of the American Chemical Society.

tandem oxidative process, TOP, of alcohols and amines (Scheme 5.13). In this novel catalyst, PICB-Au/Pd, styrene-based copolymers were used for immobilization of the Au/Pd catalyst and CB served as a second support [42]. It was shown that PICB-Au/Pd had superior catalytic activity compared to PICB-Au/Co or PICB-Au/Pt. Additionally; the solvent affected the reaction significantly. Using THF as a solvent, high conversion and low selectivity toward imine was observed. In MeOH, however, high selectivity and low reactivity was achieved. The optimum results were obtained in the presence of a mixture of THF-trifluoroethanol (TFE). This was attributed to the role of TFE as a Brønsted acid and/or a hydrogen-bond donor to promote the most difficult step of TOP, the dehydration step. It is worth noting that the presence of a base, NaOH, was essential for the reaction and no product was observed in the absence of the base. In comparison to previously reported TOP, which required a high reaction temperature, this reusable catalyst could catalyze the reaction under a mild reaction condition. Furthermore, the ICP analysis ruled out the

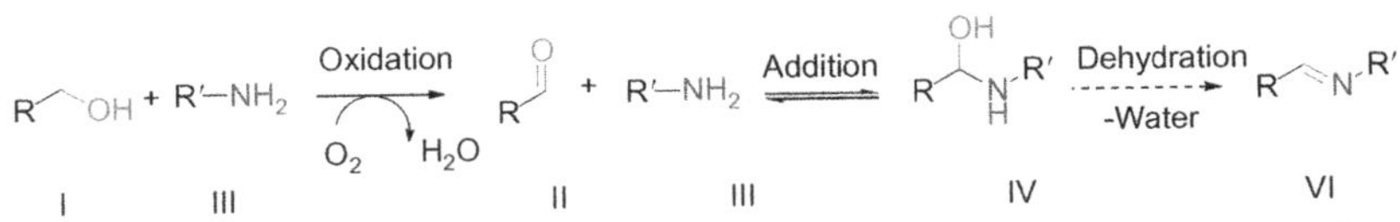

SCHEME 5.13 Tandem oxidative process (TOP) between alcohols and amines. *(Reprinted from J.-F. Soule, H. Miyamura, S. Kobayashi, Selective imine formation from alcohols and amines catalyzed by polymer incarcerated gold/palladium alloy nanoparticles with molecular oxygen as an oxidant. Chem. Commun. 49 (2013) 355–357 with permission of the Royal Society of Chemistry.)*

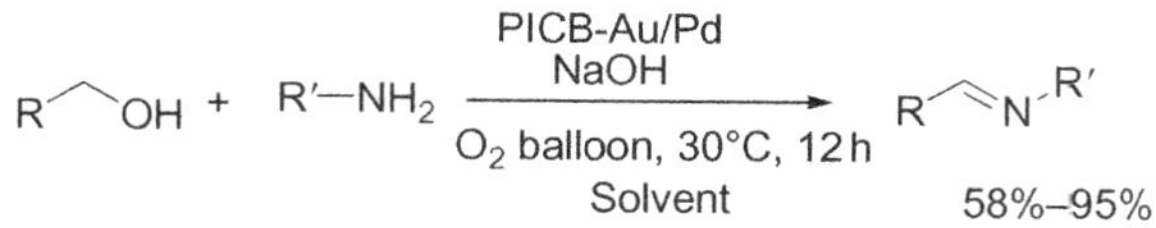

R = 4-MeC_6H_4, 4-$NO_2C_6H_4$, 4-OMeC_6H_4, 2-OHC_6H_4, Ph, 4-ClC_6H_4, 2-Pyridyl, 2-Furfuryl, C_7H_{15}, EthynylBn, VinylBn

R′ = Bn, Ph, 3-ClC_6H_4, 4-MeC_6H_4, C_8H_{17}, C_6H_{13}, C_2H_4(*i*-Pr), 4-OMeC_6H_4, 4-OMeBn

Solvent : THF–Trifluoroethanol (4:1)

SCHEME 5.14 Oxidative imine formation reaction: substrate scope. *(Reprinted from J.-F. Soule, H. Miyamura, S. Kobayashi, Selective imine formation from alcohols and amines catalyzed by polymer incarcerated gold/palladium alloy nanoparticles with molecular oxygen as an oxidant. Chem. Commun. 49 (2013) 355–357 with permission of the Royal Society of Chemistry, and S. Sadjadi, M.M. Heravi, Pd(0) encapsulated nanocatalysts as superior catalytic systems for Pd-catalyzed organic transformations, RSC. Adv. 6 (2016) 88588–88624 with permission of the Royal Society of Chemistry.)*

possibility of Pd or Au leaching. The generality of the procedure was confirmed by using various substrates (Scheme 5.14). It is obvious that a broad range of substrates such as heterocycle containing, saturated (as well as unsaturated) alcohols could tolerate this reaction with double or triple bonds intact.

A CB-stabilized polymer-incarcerated Au-Pd bimetallic nanocluster with boron, PI/CB-Au/Pd/B, was employed as an efficient and reusable catalyst for promoting a cascade oxidation-addition reaction of allylic alcohols with 1,3-dicarbonyl compounds under a mild reaction condition (Table 5.13) [43]. It was proposed that the boron (B) immobilization proceeds through the formation of a tetravalent boron complex through the esterification of $NaBH_4$ with the alcohol functional groups present on the polymer backbone (Scheme 5.15).

To confirm the merit of the bimetallic catalyst, the cascade reaction was carried out using of PI/CB-Au/B and PI/CB-Pd/B catalysts. The results indicated the superior catalytic activity of the bimetallic PI/CB-Au/Pd/B system. Because the tandem reaction could proceed in the relatively anhydrous condition, the metal leaching that could occur in aqueous media was controlled and rendered the catalyst reusable. The catalytic performance of the novel bimetallic catalyst was compared with the mixture of PI/CB-Au/Pd and PI/CB-B (Table 5.14). It is obvious that the reaction rate was higher in the case of mixed catalyst system, which is plausibly due to the decrease in the rate of the allylic oxidation that arose from reduced surface area of the polymer-CB composite material. It was also established that the boron-free catalyst resulted in faster oxidation reaction. Monitoring the cascade reaction by the ^{1}HNMR technique did not detect the α,β-unsaturated product, indicating that the latter would react with 1,3-dicarbonyl compounds to furnish the Michael adduct instead of undergoing side reactions. Notably, air could also be used as a terminal oxidant.

Combining heterogeneous polymer-immobilized bimetallic Pt/Ir nanoclusters, PI/CB-Pt/Ir, and 4-*tert*-butylcatechol that served as a simple redox-active

TABLE 5.13 Scope of the PI/CB-Au/Pd/B-Catalyzed One-Pot Sequential Reaction

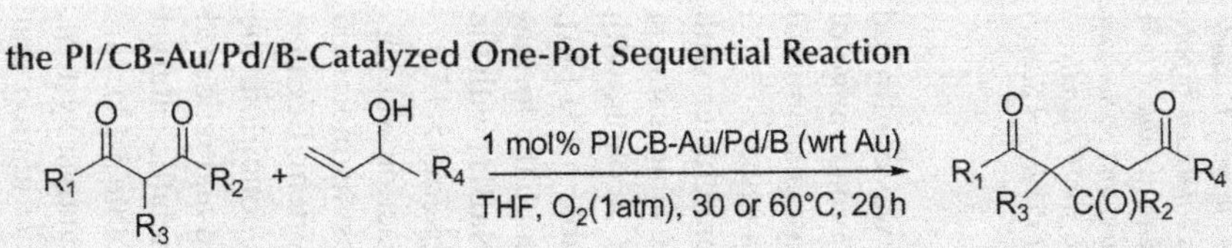

Donor	Alcohol	Product	Yield (%)
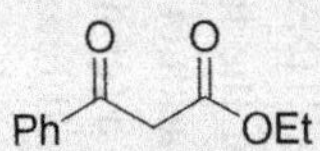	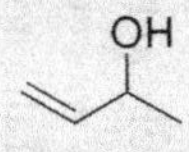	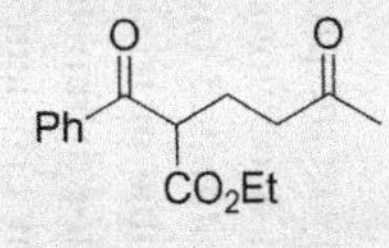	95
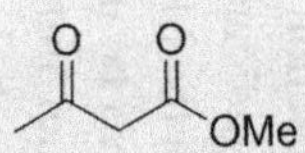	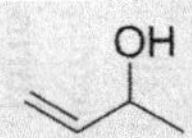	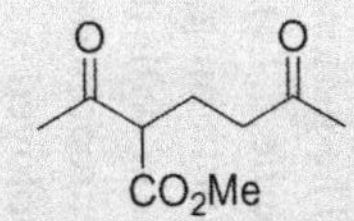	98
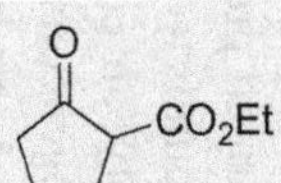	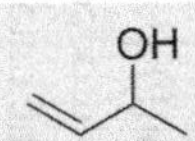	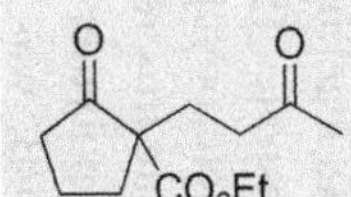	97
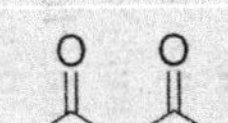	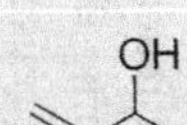	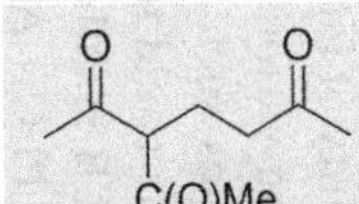	82

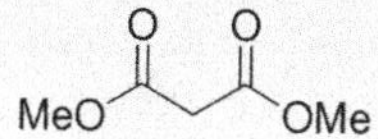	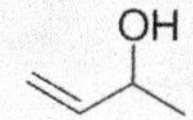	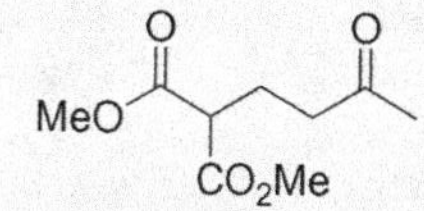	88
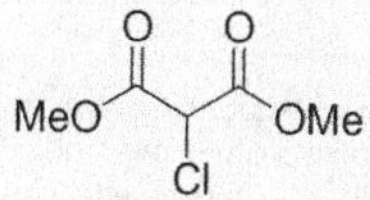	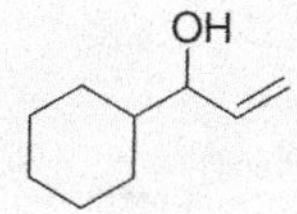	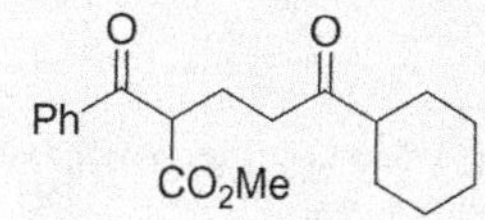	87
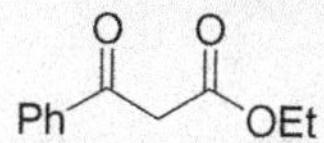	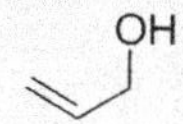	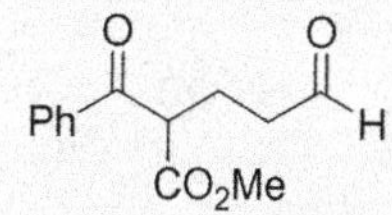	83
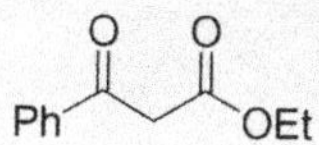	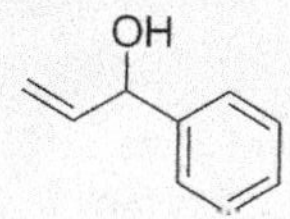	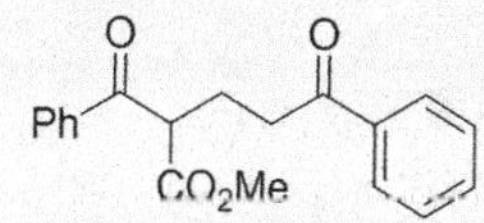	95
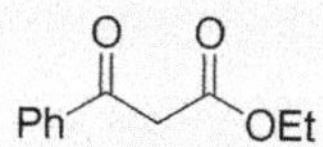	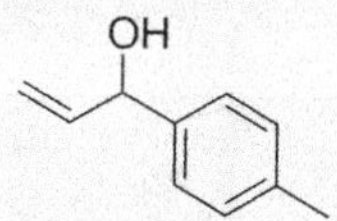	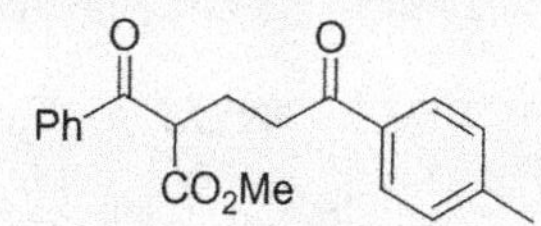	94
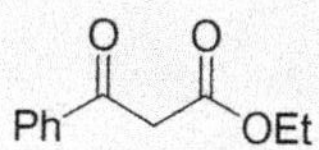	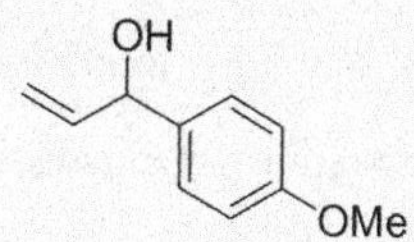	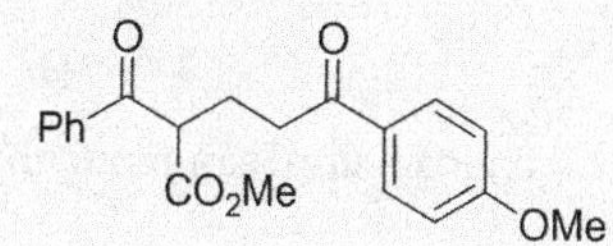	89

Continued

TABLE 5.13 Scope of the PI/CB-Au/Pd/B-Catalyzed One-Pot Sequential Reaction—cont'd

Donor	Alcohol	Product	Yield (%)
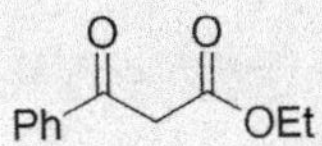	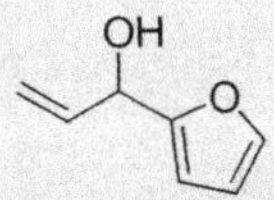	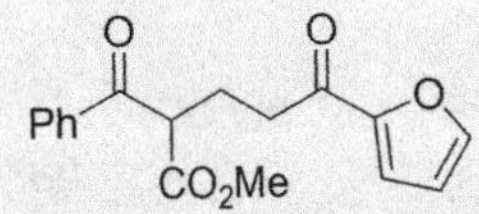	93
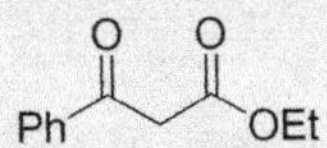	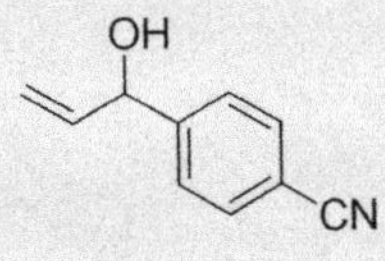	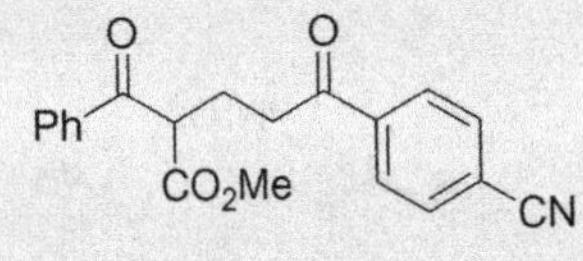	98
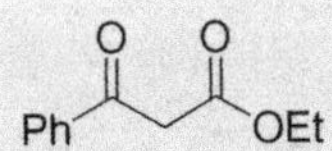	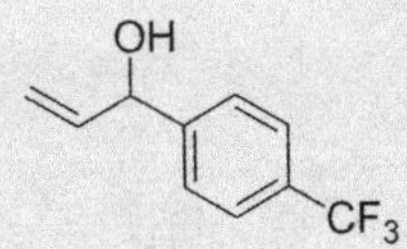	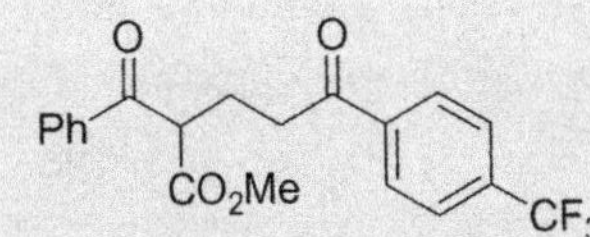	98

Reprinted from W.-J. Yoo, H. Miyamura, S. Kobayashi, Polymer-incarcerated gold-palladium nanoclusters with boron on carbon: a mild and efficient catalyst for the sequential aerobic oxidation-michael addition of 1,3-dicarbonyl compounds to allylic alcohols, J. Am. Chem. Soc. 133 (2011) 3095–3103 with permission of the American Chemical Society.

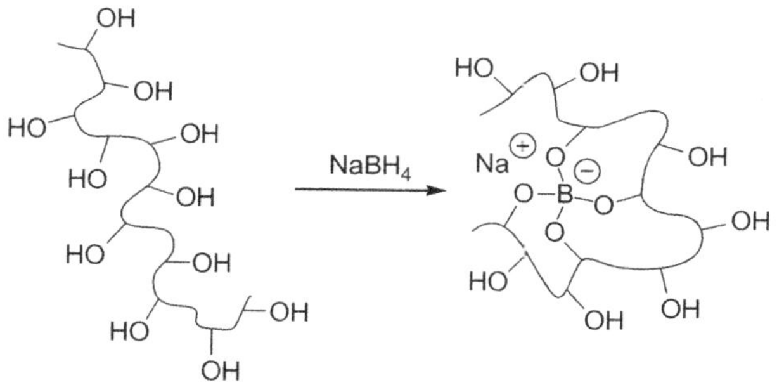

SCHEME 5.15 Immobilization of boron. *(Reprinted from W.-J. Yoo, H. Miyamura, S. Kobayashi, Polymer-incarcerated gold-palladium nanoclusters with boron on carbon: a mild and efficient catalyst for the sequential aerobic oxidation-michael addition of 1,3-dicarbonyl compounds to allylic alcohols, J. Am. Chem. Soc. 133 (2011) 3095–3103 with permission of the American Chemical Society.)*

TABLE 5.14 Comparison Between the Bifunctional Composite Catalyst and Mixed Catalyst System for the Sequential Oxidation–Michael Reaction

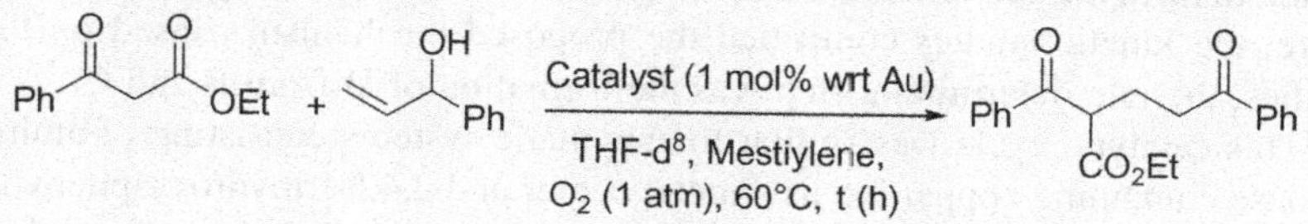

	Yield (%)	
T (h)	PI/CB-Au/Pd/B	PI/CB-Au/Pd + PI/CB-B
0.5	Not detected	Not detected
1.5	<5	6
4	13	18
8	35	49
22	>95	>95

Reprinted from W.-J. Yoo, H. Miyamura, S. Kobayashi, Polymer-incarcerated gold-palladium nanoclusters with boron on carbon: a mild and efficient catalyst for the sequential aerobic oxidation-michael addition of 1,3-dicarbonyl compounds to allylic alcohols, J. Am. Chem. Soc. 133 (2011) 3095–3103 with permission of the American Chemical Society, and S. Sadjadi, M.M. Heravi, Pd(0) encapsulated nanocatalysts as superior catalytic systems for Pd-catalyzed organic transformations, RSC. Adv. 6 (2016) 88588–88624 with permission of the Royal Society of Chemistry.

organic cofactor, Kobayashi et al. developed a novel metalloenzyme-like cooperative catalytic system for the aerobic oxidation of amines to imines under mild reaction conditions [44]. A broad range of secondary and primary amines containing electron-donating and electron-withdrawing functional groups could be oxidized under an optimized reaction condition (Table 5.15). However, more catechol was needed for less active substrates.

The authors studied the mechanism of oxidation precisely. Among various plausible reaction pathways, the rout contains initial formation of redox-active species B (Fig. 5.3) through oxidation of 4-*tert*-butylcatechol in the presence of PI/CB-Pt/Ir, base, and O_2 was suggested. It was found that the amine oxidation occurred at a faster rate during this initial period. It was believed that the misbehaved kinetics at the early stages of the reaction could be due to some chemical modification of the catechol by the starting material. HR-MS analysis confirmed the presence of an adduct generated from the 1,4- or 1,6-addition of amine to the oxidized form of 4-*tert*-butylcatechol followed by a $2e^-$ oxidation reaction (B1). Observing benzoxazoles as side products as well as various complexes derived from oxidized catechols and amines rationalized the covalent modification of B with amine C. After this initial process, the reaction system established a steady-state catalytic cycle in which the amine reactant C was expected to form D under equilibrium, and this complex underwent aerobic oxidation to form the desired imine E and regenerate the active catalyst B. Furthermore, the kinetic studies confirmed the proposed mechanism. Based on the studies, the rate determining step was the formation of D from B and C.

This catalytic cycle was similar to enzymatic systems consisting of amine oxidase containing copper ion as a metal center and 2,4,5-trihydroxy-phenylalanine-quinone (TPQ) cofactors in terms of using an oxidizing metal catalyst and a quinone-type co-catalyst. However, the biological system could not oxidize the secondary amines, indicating the difference of these two catalytic pathways. In enzymatic systems, dehydration of amines occurs through proton transfer, while in the PI/CB-Pt/Ir catalytic cycle presumably the quinone hemiaminal intermediate D was complexed to the metal nanocluster surface and underwent hydride transfer to afford the desired imine (Fig. 5.4B-(i)). It should be noted that the oxidation of amines in this reaction could occur through a single-electron-transfer mechanism (Fig. 5.4B-(ii)), although the use of radical inhibitors 2,2,6,6-tetramethylpiperidinoxyl and 2,6-di-tert-butyl-p-cresol only slightly diminished the yield of the desired imine.

Recently, this research group reported N-alkylation of primary amides with benzyl alcohols by using a synergistic tandem catalytic system consisting of polymer-incarcerated bimetallic Au/Pd nanoparticle catalysts with CB as a secondary support and $Ca(OTf)_2/Ba(OTf)_2$ Lewis acid (Fig. 5.5) [45]. Conducting the reaction in the presence of bimetallic alloy nanoparticles, PI/CB-Au/Pd, and the mixture of PI/CBAu and PI/CB-Pd catalysts indicated the superior catalytic activity of the former, implying that the catalytic activity of bimetallic nanoparticles is different from monometallic ones. It was postulated that in a bimetallic

TABLE 5.15 Amine Oxidation Catalyzed by PI/CB-Pt/Ir and 4-*tert*-Butylcatechol[a]

Substrate	Product	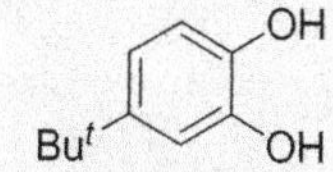 (mol%)	Isolated Yield (%)	GC Yield (%)[b]
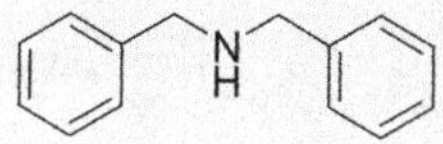	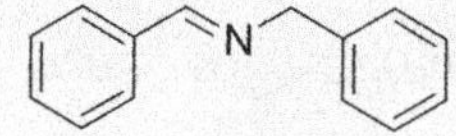	20	78	88 (3)
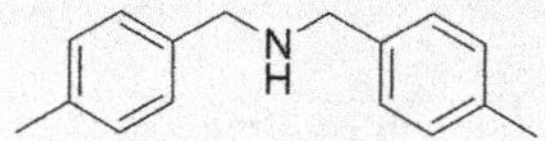	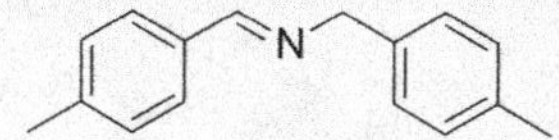	15	84	87
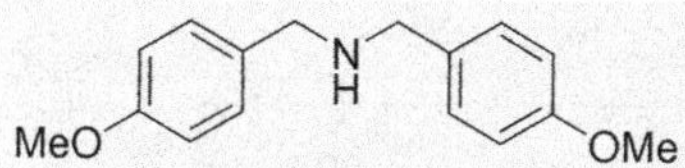	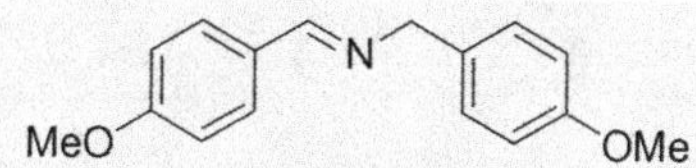	15	72	78
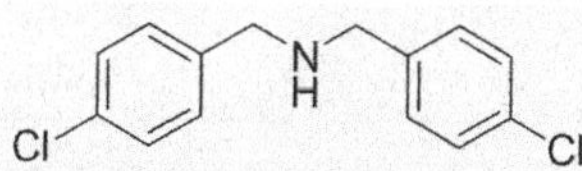	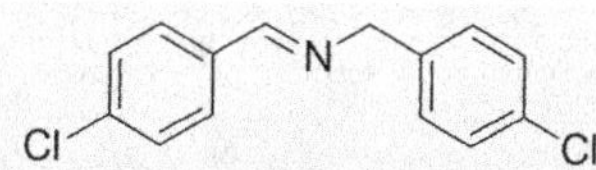	30	82	88[c]
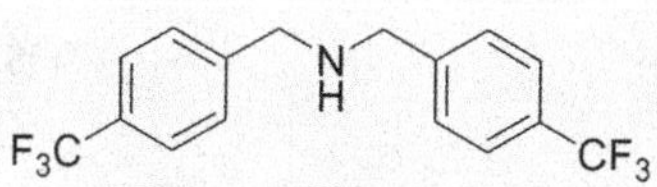	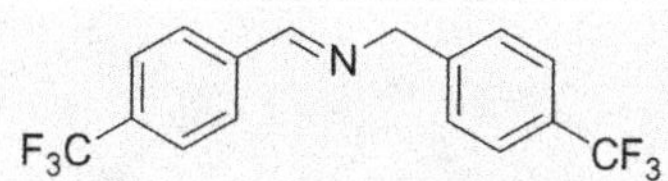	50	80	75
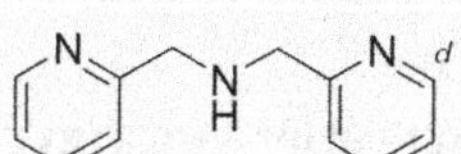 [d]	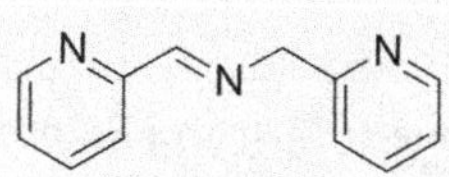	15	81	73[c]

Continued

TABLE 5.15 Amine Oxidation Catalyzed by PI/CB-Pt/Ir and 4-*tert*-Butylcatechol—cont'd

Substrate	Product	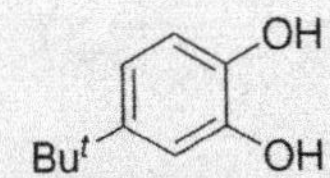 (mol%)	Isolated Yield (%)	GC Yield (%)
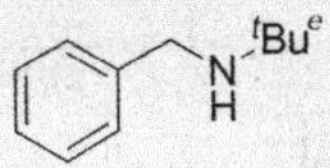 [e]	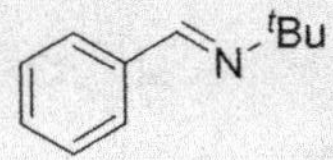	35	74	85
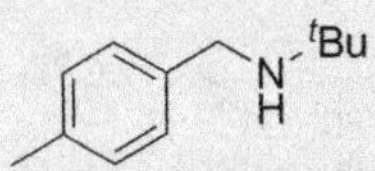	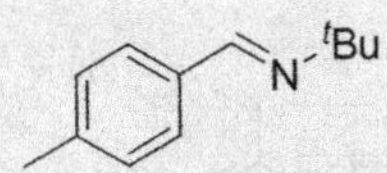	25	85	88
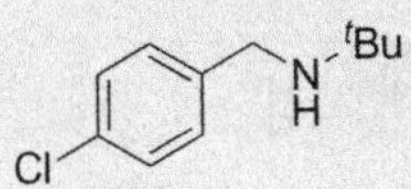	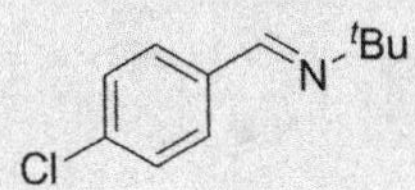	40	90	93
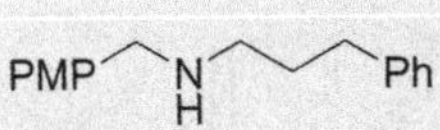	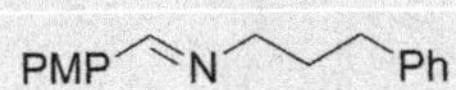	20	69	64[c]
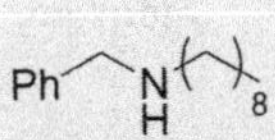		20	70	67[c]
Tetrahydroquinoline	Quinoline	30	82	81
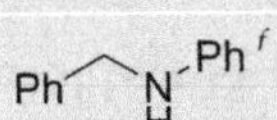 [f]	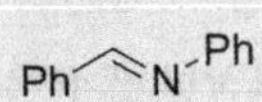	60	65	60[c]

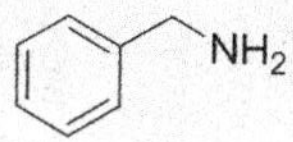	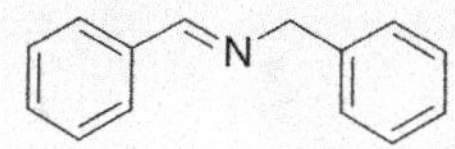	40	74	77(8)
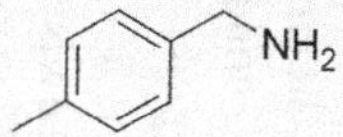	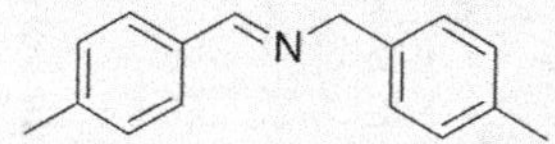	40	75	80
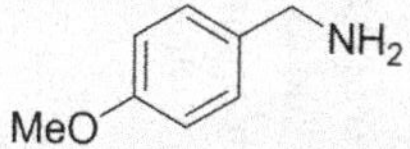	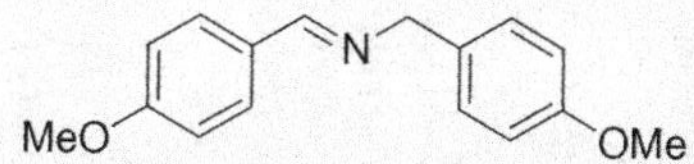	30	73	80
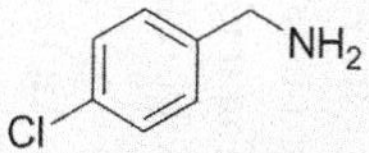	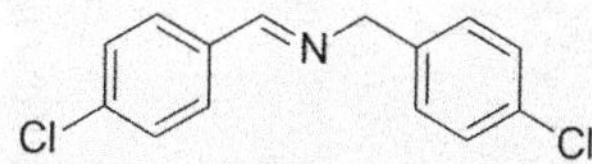	40	52	60
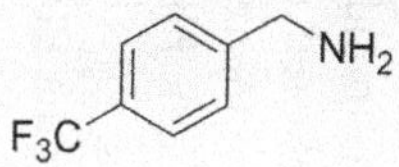	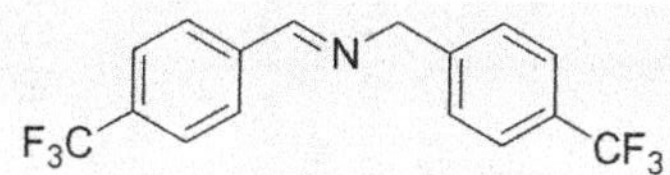	50	55	57

[a]*Reaction conditions: secondary amines (0.25 mmol) or primary amines (0.5 mmol), PI/CB-Pt/Ir (0.5 mol%), and K_2CO_3 (0.5 equiv.) in $CHCl_3$:H_2O (9:1), c = 0.5 M, under a balloon pressure of oxygen gas for 16 h at 35°C.*
[b]*Determined by GC analysis using anisole as the internal standard; the selectivity with respect to benzoxazole is shown in parentheses.*
[c]*Determined by ^1H NMR analysis using 1,1,2,2-tetrachloroethane as the internal standard.*
[d]*Using 0.75 equivalent of K_2CO_3.*
[e]*No leaching was observed for Pt (detection limit: 0.12%) and Ir (detection limit: 0.03%) by ICP analysis.*
[f]*Using 0.1 equivalent of K_2CO_3.*
Reprinted from H. Yuan, et al., Discovery of a metalloenzyme-like cooperative catalytic system of metal nanoclusters and catechol derivatives for the aerobic oxidation of amines, J. Am. Chem. Soc. 134 (2012) 13970–13973 with permission of the American Chemical Society.

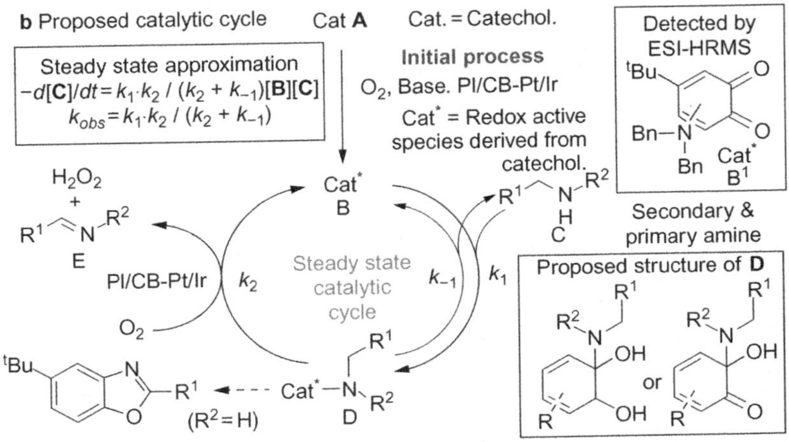

FIG. 5.3 Proposed and denied catalytic cycle. *(Reprinted from H. Yuan, et al., Discovery of a metalloenzyme-like cooperative catalytic system of metal nanoclusters and catechol derivatives for the aerobic oxidation of amines, J. Am. Chem. Soc. 134 (2012) 13970–13973 with permission of the American Chemical Society.)*

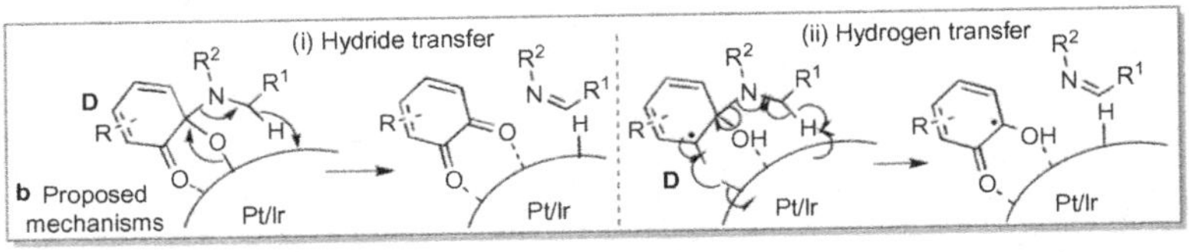

FIG. 5.4 Proposed catalytic mechanism. *(Reprinted from H. Yuan, et al., Discovery of a metalloenzyme-like cooperative catalytic system of metal nanoclusters and catechol derivatives for the aerobic oxidation of amines, J. Am. Chem. Soc. 134 (2012) 13970–13973 with permission of the American Chemical Society.)*

alloy system the proximity of Au to Pd resulted in a more profound effect of Au nanoparticles on Pd. The polarization of electric charge on the surface of the alloyed bimetallic nanoparticle, arising from the difference in electronegativity between Au and Pd, can also rationalize the observed superior catalytic activity. The authors investigated the effect of various additives such as $MgSO_4$, $Mg(OTf)_2$, $Sc(OTf)_3$, KOTf, $Ba(OTf)_2$, and $Yb(OTf)_3$. The results demonstrated the essential role of the acidity of additive in catalytic performance. In this regard, Lewis acidity was found to be more pivotal than Brønsted acidity. Additionally, $Ba(OTf)_2$ was selected as a co-catalyst due to its efficiency in low amounts. The substrate scope of the reaction under an optimized condition was established by using various substrates with electron-donating and electron-withdrawing functional groups.

It is worth noting that both the PI/CB-Au/Pd catalyst and Lewis acid were reusable and no leaching of Au and Pd was detected. To elucidate the

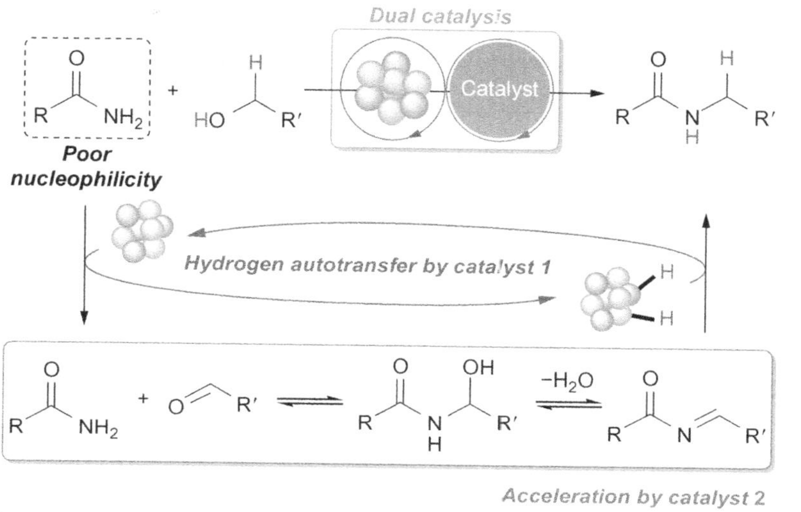

FIG. 5.5 Proposed dual catalysis for the challenging *N*-alkylation of primary amides via hydrogen autotransfer. *(Reprinted from G.C.Y. Choo, H. Miyamura, S. Kobayashi, Synergistic cascade catalysis by metal nanoparticles and Lewis acids in hydrogen autotransfer, Chem. Sci. 6 (2015) 1719–1727 with permission of the Royal Society of Chemistry)*

synergistic effect between Lewis acid and PI/CB-Au/Pd, the authors screened and compared the hydrogen autotransfer reactions of 4-methylbenzyl alcohol and benzamide under various reaction conditions in the presence and absence of $Ba(OTf)_2$ (Figs. 5.6 and 5.7).

Monitoring the formation/consumption of 4-methylbenzyl alcohol, the *N,N′*-diamide, tolualdehyde, the desired product as well as two side products (di(4-methylbenzyl) ether and xylene) implied that the formation of tolualdehyde and the two side products occurred only when PI/CB-Au/Pd was present. In this condition, the rate of formation of the desired product was slow while a large amount of tolualdehyde was formed after 3 h (Fig. 5.6). Furthermore, the regular formation of xylene was observed and almost no *N,N′*-diamide was obtained. Conversely, in the presence of $Ba(OTf)_2$ (Fig. 5.7); the yield of the desired product was higher, and the xylene, desired product, and the ether were formed faster. Also, consumption of 4-methylbenzyl alcohol was accelerated. Additionally, the formation of *N,N′*-diamide was observed at the first 3 h of the reaction and then returned to almost zero thereafter. The amount of tolualdehyde increased during this period and reached a steady-state concentration of 50%. This implied that an induction period existed and that a certain amount of the aldehyde had to be first accumulated before the desired product started to form. A similar phenomenon was observed even when a Lewis acid was not present (Fig. 5.6).

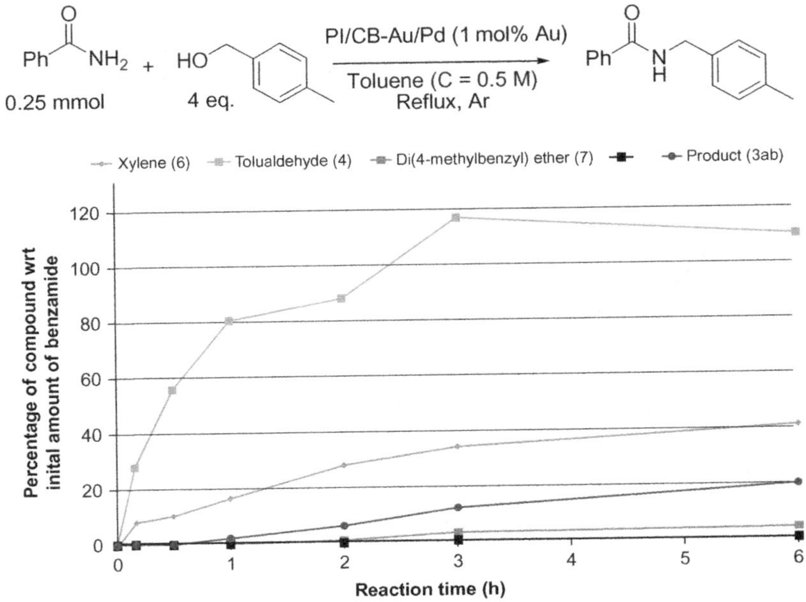

FIG. 5.6 Reaction profile with 4-methylbenzyl alcohol as substrate and no Lewis acid as co-catalyst. *(Reprinted from G.C.Y. Choo, H. Miyamura, S. Kobayashi, Synergistic cascade catalysis by metal nanoparticles and Lewis acids in hydrogen autotransfer, Chem. Sci. 6 (2015) 1719–1727 with permission of the Royal Society of Chemistry.)*

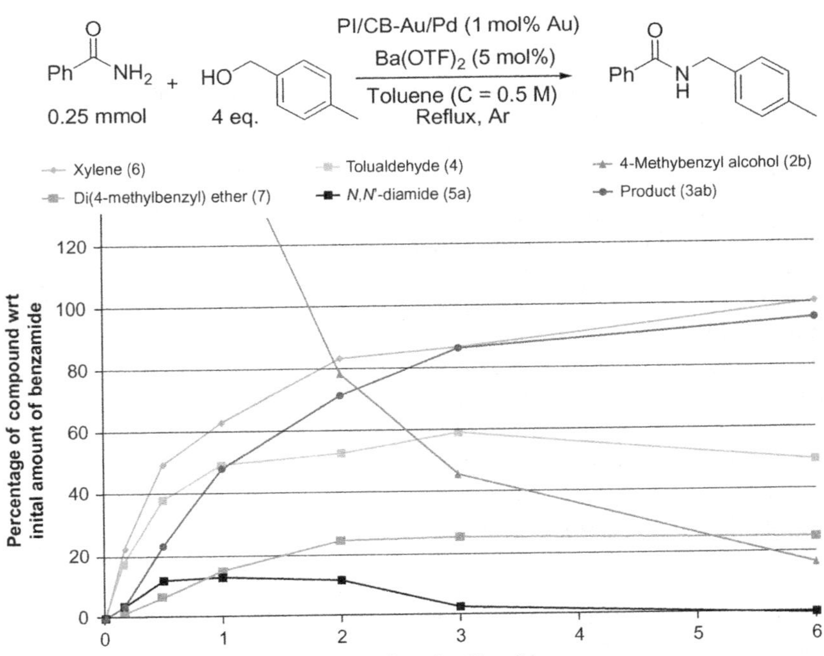

FIG. 5.7 Reaction profile with 4-methylbenzyl alcohol as substrate and $Ba(OTf)_2$ as the co-catalyst. *(Reprinted from G.C.Y. Choo, H. Miyamura, S. Kobayashi, Synergistic cascade catalysis by metal nanoparticles and Lewis acids in hydrogen autotransfer, Chem. Sci. 6 (2015) 1719–1727 with permission of the Royal Society of Chemistry.)*

Recent research by Kobyashi et al. explored a new catalytic system based on a layered heterogeneous bifunctional chiral catalyst containing metal nanoparticles and a chiral Jorgensen-Hayashi-type organocatalyst (OC) supported on different polymers for sequential aerobic oxidation-asymmetric Michael reactions between primary allylic alcohols and dibenzyl malonate [46]. The procedures for the preparation of catalysts are depicted in Scheme 5.16. To obtain polymer bead 5, which possessed activity for an asymmetric Michael reaction, Jorgensen-Hayashi-type OCs 1 were immobilized with acrylamides as co-monomers by suspension copolymerization with optimized amounts of *N-tert*-butylacrylamide, *N,N'* methylenebismethacrylamide, and dimethyl 2,2′-azobis(2-methylpropionate) (V-601 initiator). Incorporation of the Au-Pd bimetallic alloy nanoparticle was achieved via two pathways, Scheme 5.16B and C, in which the order of the layers was reversed. In the catalyst (PI(Au/Pd)-CO) 6, polymer 2 formed a coating around polymer beads 5, while in (IOC/PI/CB-Au/Pd) 7, PI/CB-Au/Pd catalyst was coated with polyacrylamide that contains OC 1. The optimized ratio of

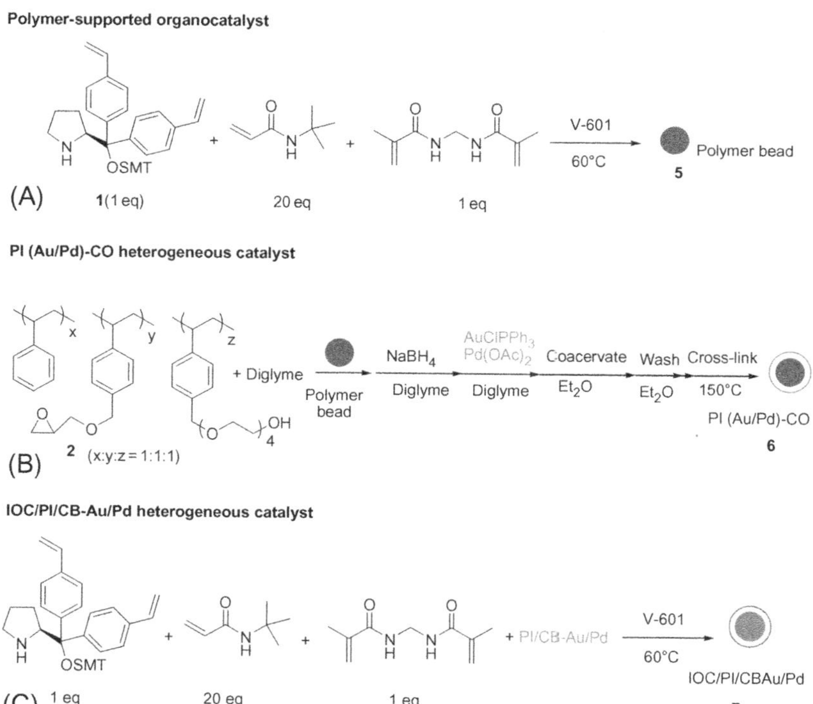

SCHEME 5.16 Fabrication procedure for (A) polymer-supported OC (**5**); (B) PI(Au/Pd)-CO catalyst (**6**); (C) IOC/PI/CB-u/Pd catalyst (**7**). The darker color represents the layer containing OC 1 while the lighter represents the layer containing Au-Pd alloy NPs. *(Reprinted from H. Miyamura, et al., A heterogeneous layered bifunctional catalyst for the integration of aerobic oxidation and asymmetric C-C bond formation, Chem. Commun. 49 (2013) 9917–9919 with permission of the Royal Society of Chemistry.)*

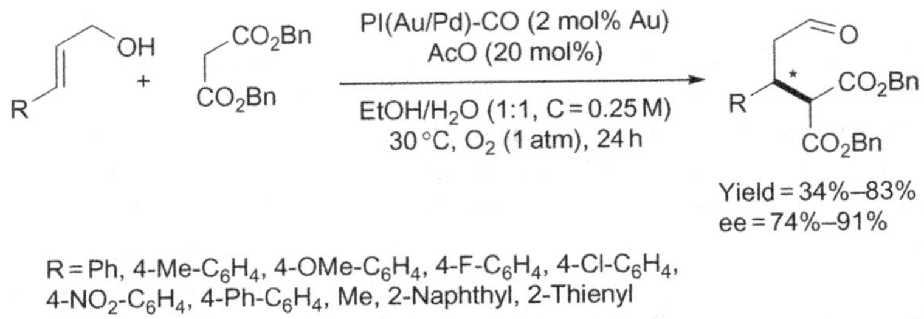

SCHEME 5.17 Substrate scope. *(Reprinted from H. Miyamura, et al., A heterogeneous layered bifunctional catalyst for the integration of aerobic oxidation and asymmetric C—C bond formation, Chem. Commun. 49 (2013) 9917–9919 with permission of the Royal Society of Chemistry.)*

Au:Pd was determined as 1:1. Additionally, it was found that each metallic nanoparticle as sole catalyst was not effective. The substrate scope was investigated by using a variety of allylic alcohols (Scheme 5.17). The authors showed that the catalytic activity strongly depended on the order of the layers of the catalyst (Table 5.16). No catalytic activity was observed in the case of IOC/PI/CB-Au/Pd catalyst 7 (Table 5.16, entry 2) in which the core bimetallic alloy nanoparticle was surrounded by an OC outer shell. To elucidate whether this observation was due to deactivation of a bimetallic alloy nanoparticle by OC in the course of catalyst preparation, the authors examined the effect of unsupported OC on the bimetallic alloy nanoparticle (Table 5.16). Although the deactivating effect of OC was confirmed, catalytic activity was obtained to some extent, implying that poisoning of the bimetallic alloy nanoparticle via some kind of coordination by acrylamides could result in zero catalytic activity (entry 2). Using the procedure depicted in Scheme 5.16B, all unfavorable interaction could be prevented.

4.2.2 Amide Synthesis

Kobayashi et al. reported the utility of CB-stabilized polymer-incarcerated gold/cobalt (PICB-Au/Co) and gold (PICB-Au) nanoparticles as reusable catalysts for green synthesis of amides from the tandem oxidation reaction of amine and alcohol derivatives in the presence of molecular oxygen as a terminal oxidant (Schemes 5.18 and 5.19) [47]. The catalyst was synthesized according to the procedure illustrated in Scheme 5.20. The CB role can be explained in terms of increasing catalyst surface area and preventing aggregation. In the case of activated alcohols (such as propargyl, and allylic alcohols) and primary amines, PICB-Au/Co exhibited high catalytic activity while PICB-Au was more efficient for secondary amines and aliphatic alcohols. It was proved that the catalysts could be recovered and reactivated simply by washing and heating under a solvent-free condition and reused for five reaction runs without significant loss of activity.

The scope of the abovementioned tandem oxidative process (TOP) was extended to amino alcohols for the synthesis of lactams by using molecular

TABLE 5.16 Optimization of Conditions for the Asymmetric TOP

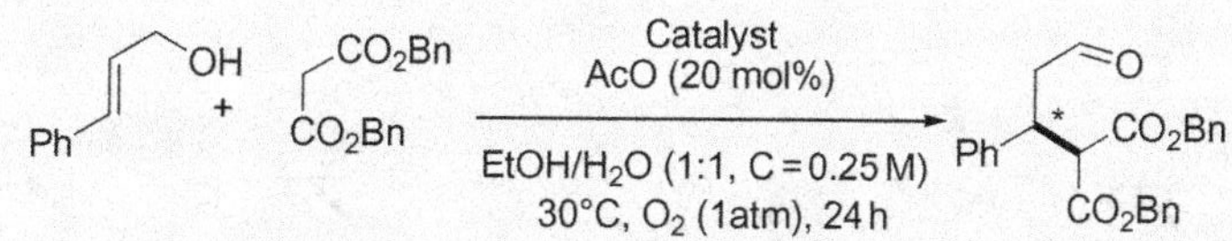

Entry	Catalyst	Conversion (%)	Yield (%)	ee (%)
1	6 (Au=1, Pd=1, OC=5)	76	Not determined	91
2	7 (Au=1, Pd=1, OC=5)	No reaction	–	–
3	PI/CB-Au/Pd (Au, Pd=1) & 1 (5)	39	Not determined	85
4	6(Au=2, Pd=2, OC=10)	97	75	90

Reprinted from H. Miyamura, et al., A heterogeneous layered bifunctional catalyst for the integration of aerobic oxidation and asymmetric C—C bond formation, Chem. Commun. 49 (2013) 9917–9919 with permission of the Royal Society of Chemistry.

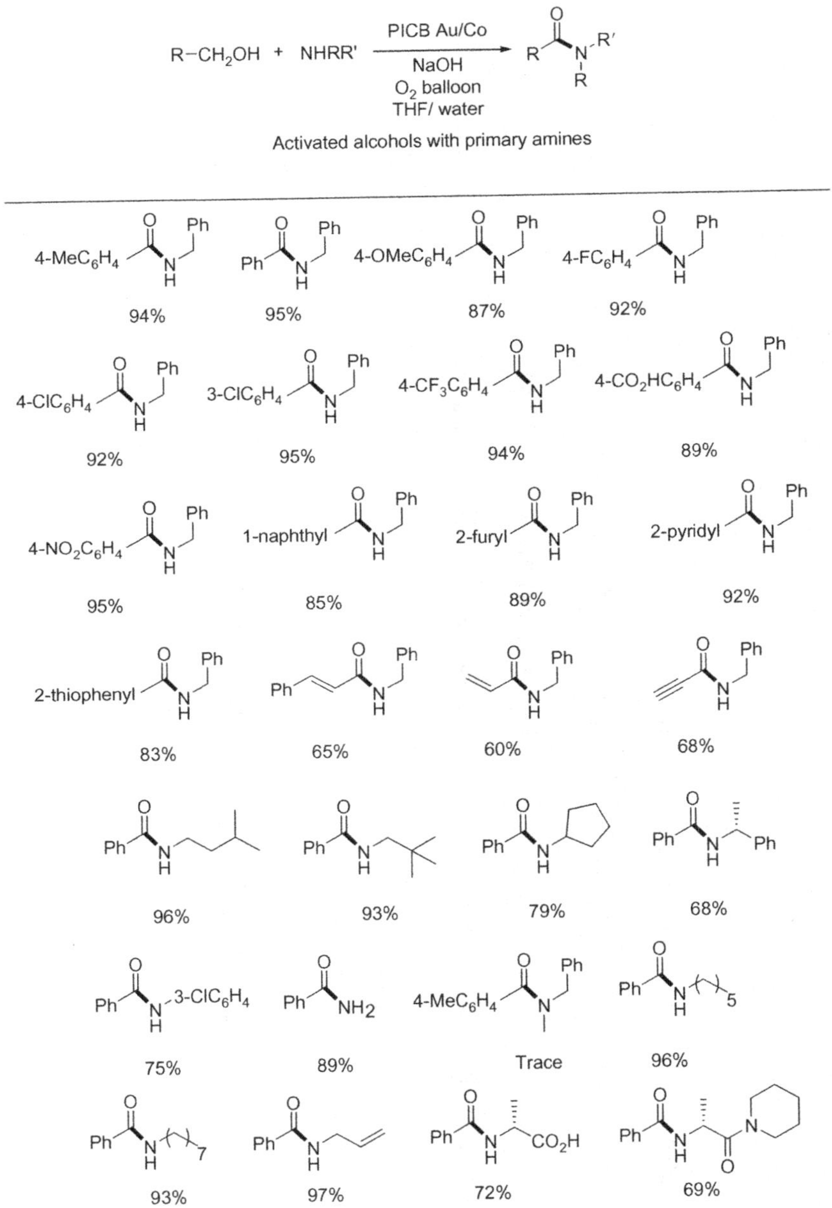

SCHEME 5.18 Scope of the reaction. *(Reprinted from J.-F. Soule, H. Miyamura, S. Kobayashi, Direct amidation from alcohols and amines through a tandem oxidation process catalyzed by heterogeneous-polymer-incarcerated gold nanoparticles under aerobic conditions, Chem. Asian J. 8 (2013) 2614–2626 with permission of Wiley.)*

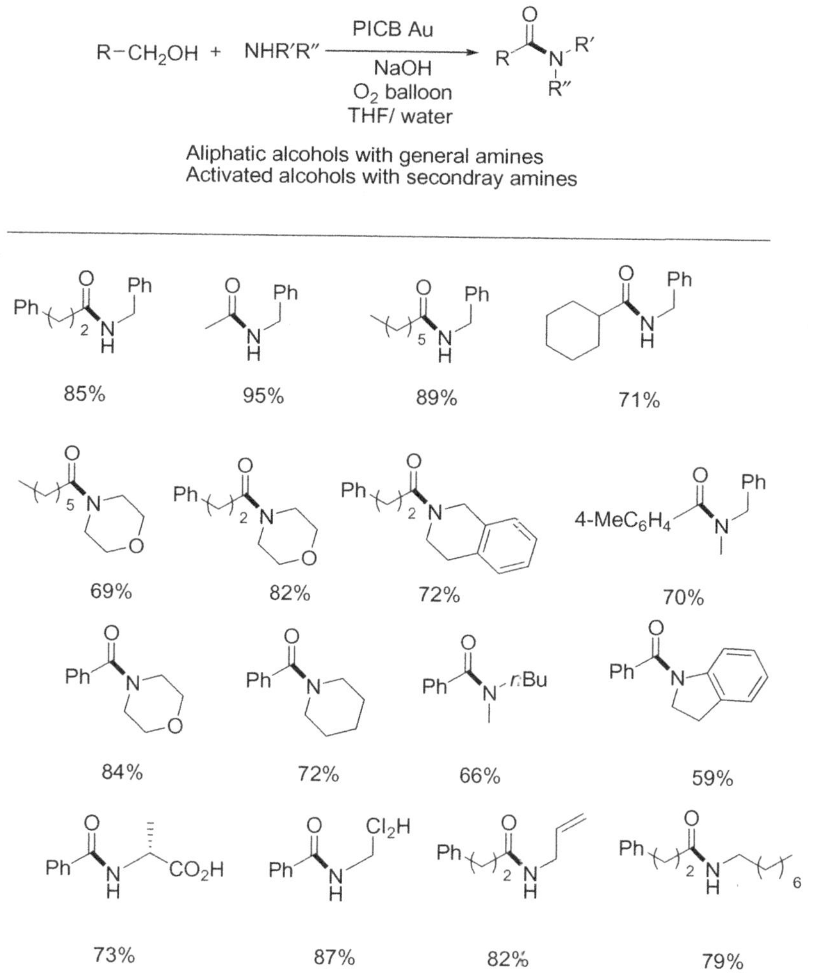

SCHEME 5.19 Scope of the reaction. *(Reprinted from J.-F. Soule, H. Miyamura, S. Kobayashi, Direct amidation from alcohols and amines through a tandem oxidation process catalyzed by heterogeneous-polymer-incarcerated gold nanoparticles under aerobic conditions, Chem. Asian J. 8 (2013) 2614–2626 with permission of Wiley.)*

oxygen as a terminal oxidant (Scheme 5.21) [48]. The use of amino alcohol as a starting reagent could result in the formation of various side products such as amino acids, amide or imine oligomers, and cyclic imines. However, this reaction under PICB-Au/Co and PICB-Au catalysis proceeded selectively to furnish the desired lactams. The reaction variables such as amino alcohol concentration, reaction temperature, and catalyst type were optimized. In the case of aliphatic amino alcohols, PICB-Au was more efficient while benzylic amino alcohols furnished the desired lactams under PICB-Au/Co catalysis.

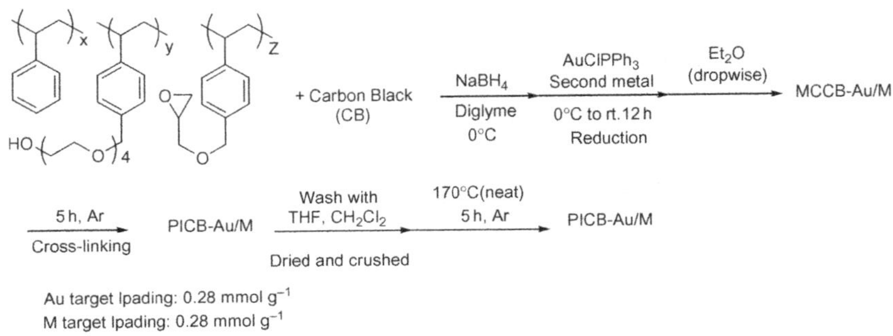

SCHEME 5.20 Catalyst preparation. *(Reprinted from J.-F. Soule, H. Miyamura, S. Kobayashi, Direct amidation from alcohols and amines through a tandem oxidation process catalyzed by heterogeneous-polymer-incarcerated gold nanoparticles under aerobic conditions, Chem. Asian J. 8 (2013) 2614–2626 with permission of Wiley.)*

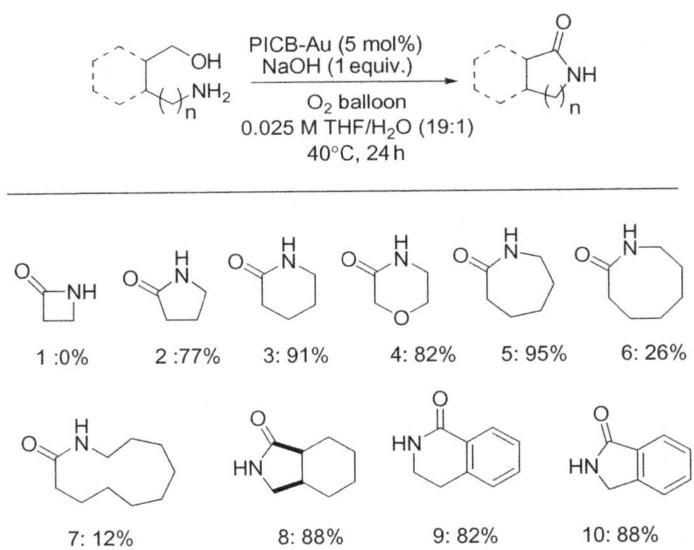

SCHEME 5.21 Scope of the reaction. For 9 and 10, PICB-Au/Co was used instead of PICB-Au, and the reaction was performed at room temperature. *(Reprinted from J.-F. Soule, H. Miyamura, S. Kobayashi, Selective lactam formation from amino alcohols using polymer-incarcerated gold and gold/cobalt nanoparticles as catalysts under aerobic oxidative conditions, Asian J. Org. Chem. 1 (2012) 319–321 with permission of Wiley-VCH Verlag.)*

The polymer incarceration method has been exploited for developing chiral Rh/Ag nanoparticles, PI/CB Rh/Ag, with utility for asymmetric 1,4-addition of arylboronic acids to enones (Table 5.17) [49]. CB was used as the second support to increase the surface area of the catalyst. The reaction was performed in the presence of a chiral ligand. Compared to a monometallic

TABLE 5.17 Scope of Reaction

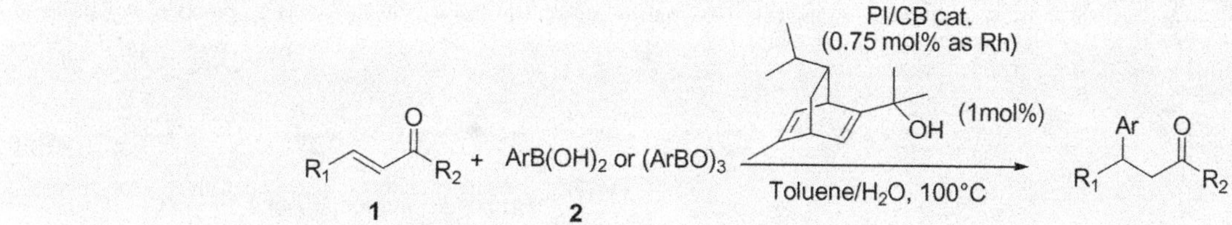

1	2 (eq.)	Catalyst	Time (h)	Yield (%)	ee (%)	Rh Leaching
2-Cyclohexenone	Phenylboronic acid (1.5)	Rh/Ag 9[a]	7	Quant.	98	ND
2-Cyclohexenone	2-Methylphenylboronic acid (2.0)	Rh/Ag 9	24	81	96	ND
2-Cyclohexenone	3-Methylphenylboronic acid (1.5)	Rh/Ag 9	9	99	96	ND
2-Cyclohexenone	3,5-Dimethylphenylboronic acid (1.5)	Rh/Ag 9	11	96	97	ND
2-Cyclohexenone	4-*tert*-Butylphenylboronic acid (1.5)	Rh/Ag 9	6	40	–	1.4%
		Rh/Ag 15[b]	24	89	97	ND
2-Cyclohexenone	4-Fluorophenylboronic acid (1.5)	Rh/Ag 9	24	87	98	ND
2-Cyclohexenone	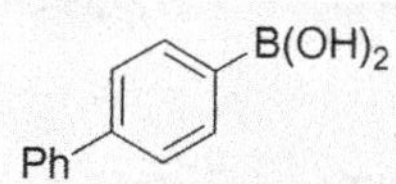 (1.5)	Rh/Ag 9	10	Quant.	98	ND
2-Cyclohexenone	Naphthalen-1-yl-1-boronic acid (2.0)	Rh/Ag 9	12	90	95	ND
2-Cyclohexenone	2-Methoxyphenylboronic acid (1.5)	Rh/Ag 9	7	95	93	ND

Continued

TABLE 5.17 Scope of Reaction—cont'd

1	2 (eq.)	Catalyst	Time (h)	Yield (%)	ee (%)	Rh Leaching
2-Cyclohexenone	3-Methoxyphenylboronic acid (1.5)	Rh/Ag 9	10	93	98	ND
2-Cyclohexenone	4-Methoxyphenylboronic acid (1.5)	Rh/Ag 9	7	82	97	ND
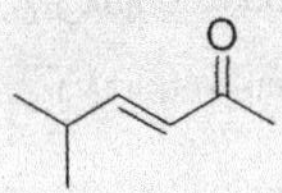	Phenylboronic acid (2.0)	Rh/Ag 15	12	91	92	ND
2-Cyclopentenone	Phenylboronic acid (1.5)	Rh/Ag 9	7	86	74	ND
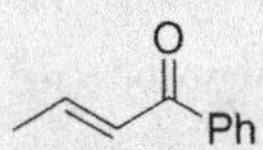	Phenylboronic acid (2.0)	Rh/Ag 15	12	72	92	ND
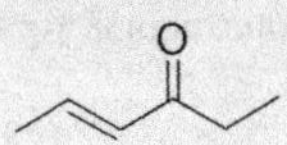	Phenylboronic acid (2.0)	Rh/Ag 15	12	82	96	ND
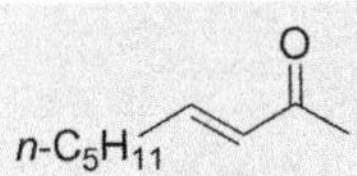	Phenylboronic acid (2.0)	Rh/Ag 15	12	70	95	ND
		Rh 8	12	83	95	ND

[a]*Rh/Ag = 1:1.*
[b]*Rh/Ag = 1:3.*
Reprinted from T. Yasukawa, H. Miyamura, S. Kobayashi, Polymer-incarcerated chiral Rh/Ag nanoparticles for asymmetric 1,4-addition reactions of arylboronic acids to enones: remarkable effects of bimetallic structure on activity and metal leaching, J. Am. Chem. Soc. 134 (2012) 16963–16966 with permission of the American Chemical Society.

system, a bimetallic system effectively reduced the Rh leaching. To elucidate the role of Ag in the catalytic activity and Rh leaching suppression, STEM and EDS analyses were carried out. The results demonstrated poor dispersion of small nanoparticles in a Rh mono metallic catalyst. Conversely, well-dispersed nanoparticles over the support were observed in the case of a Rh/Ag bimetallic catalyst. EDS analyses established a random ratio of Rh/Ag in nanoparticles indicating that the effects of the bimetallic structure might be explained by the size effect and the existence of metal-metal interactions. It was believed that the reaction proceeded on the surface of the bimetallic nanoparticles, because there were no significant morphological changes in the catalyst after use and after heating.

5 CONCLUSION

Recent advances in immobilization of monometallic and bimetallic nanocluster catalysts such as Au, Pd, Pt, Au/Co, Rh/Ag, and Au/Pd via the polymer incarceration method were presented in this chapter. As discussed, various polymers favorably containing a benzene ring, including polystyrene-based polymers and polysilane with benzene moiety, could be used for this purpose. However, the utility of some polymers without aryl groups was also established for polymer incarceration. In comparison to many heterogeneous catalysts, obtained through other immobilization methods, polymer-incarcerated catalysts benefit from several advantages including high catalytic activity, simple recovery and reuse, and avoided or negligible metal leaching. The catalytic activity of these catalysts can be even more improved by using a second support or functionalization of a polymer structure. It is clear from the many discussed examples that polymer-incarcerated catalysts can effectively catalyze various organic reactions. It is also predicted that research into the development and utilization of this class of catalysts will continue to grow in the future.

REFERENCES

[1] S.V. Ley, et al., Microencapsulation of osmium mTetroxide in polyurea, Org. Lett. 5 (2003) 185–187.

[2] M. Benaglia, Recoverable and Recyclable Catalysts, Wiley, Chichester, 2009.

[3] R. Akiyama, S. Kobayashi, "Microencapsulated" and related catalysts for organic chemistry and organic synthesis, Chem. Rev. 109 (2009) 594–642.

[4] S. Kobayashi, S. Nagayama, A microencapsulated Lewis acid. A new type of polymer-supported Lewis acid catalyst of wide utility in organic synthesis, J. Am. Chem. Soc. 120 (1998) 2985–2986.

[5] S. Nagayama, M. Endo, S. Kobayashi, Microencapsulated osmium tetraoxide. A new recoverable and reusable polymer-supported osmium catalyst for dihydroxylation of olefins, J. Org. Chem. 63 (1998) 6094–6095.

[6] S. Kobayashi, R. Akiyama, Renaissance of immobilized catalysts. New types of polymer-supported catalysts, "microencapsulated catalysts", which enable environmentally benign and powerful high-throughput organic synthesis, Chem. Commun. (2003) 449–460.

[7] Y. Miyazaki, S. Kobayashi, Novel gold(III) polymer-supported catalyst for indole library synthesis, J. Comb. Chem. 10 (2008) 355–357.

[8] R. Akiyama, S. Kobayashi, Microencapsulated palladium catalysts: allylic substitution and Suzuki coupling using a recoverable and reusable polymer-supported palladium catalyst, Angew. Chem. Int. Ed. 40 (2001) 3469–3471.

[9] S. Kobayashi, T. Ishida, R. Akiyama, Catalytic asymmetric dihydroxylation using phenoxyethoxymethyl-polystyrene (PEM)-based novel microencapsulated osmium tetroxide (PEM-MC OsO_4), Org. Lett. 3 (2001) 2649–2652.

[10] B.M. Choudary, et al., Microencapsulated bismuth(III) triflate catalyst for organic transformations, J. Mol. Catal. A 212 (2004) 237–243.

[11] A. Lattanzi, N.E. Leadbeater, Microencapsulated VO(acac)2: preparation and use in allylic alcohol epoxidation, Org. Lett. 4 (2002) 1519–1521.

[12] M.L. Kantam, et al., Heterogeneous catalytic sulfimidation using immobilized Cu(acac)2, Adv. Synth. Catal. 347 (2005) 641–645.

[13] R. Akiyama, S. Kobayashi, A novel polymer-supported arene–ruthenium complex for ring-closing olefin metathesis, Angew. Chem. Int. Ed. 42 (2002) 2602–2604.

[14] R. Saladino, et al., Preparation and structural characterization of polymer-supported methylrhenium trioxide systems as efficient and selective catalysts for the epoxidation of olefins, J. Org. Chem. 67 (2002) 1323–1332.

[15] C.K.Y. Lee, et al., Efficient batch and continuous flow Suzuki cross-coupling reactions under mild conditions, catalysed by polyurea-encapsulated palladium (II) acetate and tetra-n-butylammonium salts, Chem. Commun. (2005) 2175–2177.

[16] S.V. Ley, et al., Polyurea-encapsulated palladium(II) acetate: a robust and recyclable catalyst for use in conventional and supercritical media, Chem. Commun. (2002) 1134–1135.

[17] S.V. Ley, et al., Recyclable polyurea-microencapsulated Pd(0) nanoparticles: an efficient catalyst for hydrogenolysis of epoxides, Org. Lett. 5 (2003) 4665–4668.

[18] S. Kobayashi, H. Miyamura, Polymer-incarcerated metal(0) cluster catalysts, Chem. Rec. 10 (2010) 271–290.

[19] T. Inasaki, et al., Polymer-incarcerated palladium with active phosphine as recoverable and reusable Pd catalyst for the amination of aryl chlorides, Synlett (2007) 3209–3213.

[20] H. Miyamura, S. Kobayashi, Tandem oxidative processes catalyzed by polymer-incarcerated multimetallic nanoclusters with molecular oxygen, Acc. Chem. Res. 47 (2014) 1054–1066.

[21] H. Miyamura, T. Yasukawa, S. Kobayashi, Preparation of polymer incarcerated gold nanocluster catalysts (PI-Au) and their application to aerobic oxidation reactions of boronic acids, alcohols, and silyl enol ethers, Tetrahedron 70 (2014) 6039–6049.

[22] T. Yasukawa, H. Miyamura, S. Kobayash, Simple homopolymer-incarcerated gold nanoclusters prepared by self-assembled encapsulation with aluminum reagents as crosslinkers: catalysts for aerobic oxidation reactions, Chem. Lett. 44 (2015) 50–52.

[23] H. Miyamura, S. Kobayashi, Polymer-incarcerated metals: highly reactive, recoverable, and multifunctional nanocluster catalysts for organic synthesis, Aldrichim. Acta 46 (2013) 3–19.

[24] H. Miyamura, S. Kobayashi, α-Hydroxylation of 1,3-dicarbonyl compounds catalyzed by polymer-incarcerated gold nanoclusters with molecular oxygen, Chem. Lett. 41 (2012) 976–978.

[25] H. Miyamura, et al., Aerobic oxidation of amines catalyzed by polymer-incarcerated Au nanoclusters: effect of cluster size and cooperative functional groups in the polymer, Bull. Chem. Soc. Jpn. 84 (2011) 588–599.

[26] H. Miyamura, et al., Polymer incarcerated gold catalyzed aerobic oxidation of hydroquinones and their derivatives, Chem. Lett. 37 (2008) 360–361.

[27] W.-J. Yoo, et al., Facile preparation of 2-substituted benzoxazoles and benzothiazoles via aerobic oxidation of phenolic and thiophenolic imines catalyzed by polymer-incarcerated platinum nanoclusters, Adv. Synth. Catal. 353 (2011) 3085–3089.

[28] R. Akiyama, S. Kobayashi, The polymer incarcerated method for the preparation of highly active heterogeneous palladium catalysts, J. Am. Chem. Soc. 125 (2003) 3412–3413.

[29] R. Nishio, M. Sugiura, S. Kobayashi, Semi-hydrogenation of alkynes using phosphinated polymer incarcerated (PI) palladium catalysts, Org. Biomol. Chem. 4 (2006) 992–995.

[30] K. Okamoto, R. Akiyama, S. Kobayashi, Suzuki-miyaura coupling catalyzed by polymer-incarcerated palladium, a highly active, recoverable, and reusable Pd catalyst, Org. Lett. 6 (2004) 1987–1990.

[31] R. Nishio, M. Sugiura, S. Kobayashi, Novel polymer incarcerated palladium with phosphinated polymers: active catalyst for suzuki-miyaura coupling without external phosphines, Org. Lett. 7 (2005) 4831–4834.

[32] R. Nishio, et al., Synthesis of acridone derivatives using polymer-supported palladium and scandium catalysts, J. Comb. Chem. 8 (2006) 459–461.

[33] H. Oyamada, et al., Development of polysilane-supported palladium/alumina hybrid catalysts and their application to hydrogenation, Org. Biomol. Chem. 6 (2008) 61–65.

[34] H. Miyamura, et al., Aerobic oxidation of alcohols at room temperature and atmospheric conditions catalyzed by reusable gold nanoclusters stabilized by the benzene rings of polystyrene derivatives, Angew. Chem. 119 (2007) 4229–4232.

[35] H. Miyamura, T. Yasukawa, S. Kobayashi, Aerobic oxidative esterification of alcohols catalyzed by polymer-incarcerated gold nanoclusters under ambient conditions, Green Chem. 12 (2010) 776–778.

[36] C. Lucchesi, et al., Aerobic oxidation of alcohols under mild conditions catalyzed by novel polymer-incarcerated carbon-stabilized gold nanoclusters, Adv. Synth. Catal. 350 (2008) 1996–2000.

[37] Y. Miyazaki, H. Hagio, S. Kobayashi, Practical access to the polymer incarcerated platinum (PI Pt) catalyst and its application to hydrogenation, Org. Biomol. Chem. 4 (2006) 2529–2531.

[38] H. Hagio, M. Sugiura, S. Kobayashi, Immobilization of a platinum catalyst using the polymer incarcerated (PI) method and application to catalytic reactions, Synlett (5) (2005) 813–816.

[39] H. Miyamura, et al., Aerobic oxidation of hydroquinone derivatives catalyzed by polymer-incarcerated platinum catalyst, Angew. Chem. Int. Ed. 47 (2008) 8093–8095.

[40] W.-J. Yoo, H. Yuan, S. Kobayashi, Oxidative transformation of N-substituted 2-aminophenols to 2-substituted benzoxazoles catalyzed by polymer-incarcerated and carbon stabilized platinum nanoclusters, Can. J. Chem. 90 (2012) 306–313.

[41] J.F. Soule, H. Miyamura, S. Kobayashi, Copolymer-incarcerated nickel nanoparticles with N-heterocyclic carbene precursors as active cross-linking agents for Corriu-Kumada-Tamao reaction, J. Am. Chem. Soc. 135 (2013) 10602–10605.

[42] J.-F. Soule, H. Miyamura, S. Kobayashi, Selective imine formation from alcohols and amines catalyzed by polymer incarcerated gold/palladium alloy nanoparticles with molecular oxygen as an oxidant, Chem. Commun. 49 (2013) 355–357.

[43] W.-J. Yoo, H. Miyamura, S. Kobayashi, Polymer-incarcerated gold-palladium nanoclusters with boron on carbon: a mild and efficient catalyst for the sequential aerobic oxidation-Michael addition of 1,3-dicarbonyl compounds to allylic alcohols, J. Am. Chem. Soc. 133 (2011) 3095–3103.

[44] H. Yuan, et al., Discovery of a metalloenzyme-like cooperative catalytic system of metal nanoclusters and catechol derivatives for the aerobic oxidation of amines, J. Am. Chem. Soc. 134 (2012) 13970–13973.

[45] G.C.Y. Choo, H. Miyamura, S. Kobayashi, Synergistic cascade catalysis by metal nanoparticles and Lewis acids in hydrogen autotransfer, Chem. Sci. 6 (2015) 1719–1727.

[46] H. Miyamura, et al., A heterogeneous layered bifunctional catalyst for the integration of aerobic oxidation and asymmetric C—C bond formation, Chem. Commun. 49 (2013) 9917–9919.

[47] J.-F. Soule, H. Miyamura, S. Kobayashi, Direct amidation from alcohols and amines through a tandem oxidation process catalyzed by heterogeneous-polymer-incarcerated gold nanoparticles under aerobic conditions, Chem. Asian J. 8 (2013) 2614–2626.

[48] J.-F. Soule, H. Miyamura, S. Kobayashi, Selective lactam formation from amino alcohols using polymer-incarcerated gold and gold/cobalt nanoparticles as catalysts under aerobic oxidative conditions, Asian J. Org. Chem. 1 (2012) 319–321.

[49] T. Yasukawa, H. Miyamura, S. Kobayashi, Polymer-incarcerated chiral Rh/Ag nanoparticles for asymmetric 1,4-addition reactions of arylboronic acids to enones: remarkable effects of bimetallic structure on activity and metal leaching, J. Am. Chem. Soc. 134 (2012) 16963–16966.

FURTHER READING

S. Sadjadi, M.M. Heravi, Pd(0) encapsulated nanocatalysts as superior catalytic systems for Pd-catalyzed organic transformations, RSC Adv. 6 (2016) 88588–88624.

Chapter 6

Selenium-Containing Catalysts Encapsulated in Protein Systems

Tiezheng Pan and Junqiu Liu
Jilin University, Changchun, China

Contents

1 INTRODUCTION

The 21st century has witnessed considerable progress in the research area of selenium-containing catalysts. In historical terms, the first selenium-containing protein, glutathione peroxidase (GPx), was discovered in 1957 [1]. However,

Encapsulated Catalysts. http://dx.doi.org/10.1016/B978-0-12-803836-9.00006-7

selenium's biochemical role was not established until 1973, when the element was identified as an essential component of the active site of GPx [2,3]. In organisms, selenium exists in the form of selenocysteine (Sec), and the proteins containing Sec are known as selenoproteins. For instance, GPx is a well-known selenoenzyme that functions as an antioxidant and catalyzes the reduction of harmful peroxide to protect organisms from cellular damage. Besides, a lack of selenoproteins can result in a number of diseases such as Alzheimer's disease, cardiac shock, and atherosclerosis [4–6]. These selenoproteins exist in all major forms of organisms such as eukaryotes and bacteria, providing critical support of catalysis for life.

As the first extensively studied selenoprotein, GPx is an ideal medicine against oxidative damage due to its distinguished antioxidative capacity. However, the defects such as instability and poor availability have limited its application. Consequently, developing manmade GPx mimics with robust constitution and low expense has attracted more and more attention. From small selenocompounds to supramolecular selenoenzymes [7,8], scientists have made great efforts to construct a number of selenium-containing catalysts in consideration of the important biological roles and promising applications. Compared with small molecules, selenium-containing catalysts based on protein scaffolds showed obvious superiorities, such as much higher catalytic rates, stronger substrate affinity, and better adaptability. Also, the protein scaffolds allow minimal modification to access a new property [9]. In this chapter, we will provide a brief review of selenium-containing catalysts (taking GPx as example) based on protein systems, including antibodies, enzymes, allosteric proteins, and protein assemblies. These models have offered us a great platform for understanding the catalytic mechanism and the relationship between structure and function.

1.1 The Nature of Glutathione Peroxidase

Mimicking nature's creation requires sufficient studies of the native selenoprotein GPx. In 1973, the selenoprotein nature of GPx finally explained why selenium is essential for defense against oxidants in vertebrate organisms [2,3]. Since then, GPxs, defined as proteins with high sequence similarity, have been detected in almost every domain of life. A crystallography study indicated that GSH can bind to the reduced GPx through a noncovalent bond, while the oxidized GPx binds to GSH through an Se-S covalent bond [10]. In the catalytic cavity of bovine cGPx, the surrounding amino acid residues Arg167, Gln130, and Trp148 contribute to the stabilization of GSH through electrostatic force and salt bridges. The structure of the catalytic cavity is illustrated in Fig. 6.1.

The selenium-containing GPx catalyzes the reduction of H_2O_2 and organic hydroperoxides at the expense of GSH. After decades of research, the present mechanistic understanding of this selenoenzyme, which is widely relevant to other types of GPx, has been developed and reviewed as Eqs. (6.1), (6.2) [11].

$$ROOH + 2GSH \rightarrow ROH + H_2O + GSSG \tag{6.1}$$

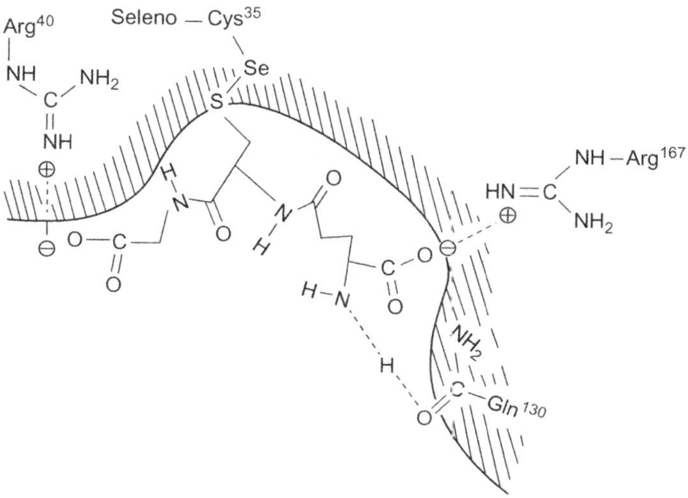

FIG. 6.1 The structure of cGPx catalytic cavity [10]. *(Reproduced from O. Epp, R. Ladenstein, A. Wendel, The refined structure of the selenoenzyme glutathione peroxidase at 0.2-nm resolution, Eur. J. Biochem. 133 (1983) 51-69 with permission from Wiley-VCH.)*

$$H_2O_2 + 2GSH \rightarrow 2H_2O + GSSH \tag{6.2}$$

As a multiple substrate enzyme catalytic reaction, its kinetics could be a ping-pong mechanism or a sequential mechanism. In a ping-pong mechanism [12,13], the catalytic cycle begins at the reduced state of the enzyme (E-SeH). Then E-SeH is oxidized by peroxide into the form of E-SeOH, consequently reduced by GSH into the Se-S complex (E-Se-SG), and it is finally reduced by another molecule of GSH to turn back to its origin state (E-SeH). The catalytic cycle has been illustrated as Eqs. (6.3)–(6.5) [11]. In the cycle, the rate-limiting reaction is the transition from E-Se-SG to E-SeH, while the fastest step is the oxidation of E-SeH by ROOH. Actually, the active site Sec locates on the exposed position of the GPx catalytic cavity, resulting in all kinds of peroxides being able to access and oxidize the Sec easily. That's also why GPx hardly exhibits substrate specificity for different peroxides [14].

$$\text{ROOH} + \text{ESeH} \xrightarrow{k_1} \text{ESeOH} + \text{ROH} \tag{6.3}$$

$$\text{ESeOH} + \text{GSH} \xrightarrow{k_2} \text{ESeSG} + H_2O \tag{6.4}$$

$$\text{ESeSG} + \text{GSH} \xrightarrow{k_3} \text{ESeH} + \text{GSSG} \tag{6.5}$$

The catalytic kinetics of natural GPx is in accordance with Eq. (6.6) (Dalziel's equation) [15,16].

$$\frac{E_o}{V} = \frac{\Phi}{[\text{RSH}] + \Phi_2/[\text{ROOH}]} \tag{6.6}$$

The kinetic equation could also be written in Eq. (6.7) (Segel's form) [17].

$$V = \frac{V_{mGX}[\mathrm{GSH}][\mathrm{ROOH}]}{K_{mGSH}[\mathrm{ROOH}] + K_{mH_2O_2}[\mathrm{GSH}] + [\mathrm{ROOH}] + [\mathrm{GSH}]} \tag{6.7}$$

By varying the concentration of one substrate while keeping the concentration of the other constant, double-reciprocal plots of the initial velocity versus substrate concentration yield a series of parallel linear plots.

On the other hand, Carsol's group proposed another sequential mechanism in 1997 [18]. In sequential mechanism, different substrates bind to the enzyme to form a multivariate complex before the release of products. The catalytic procedure is illustrated as Eq. (6.8) (Eq. (6.8) adapted from Ref. [18] with the permission of Wiley).

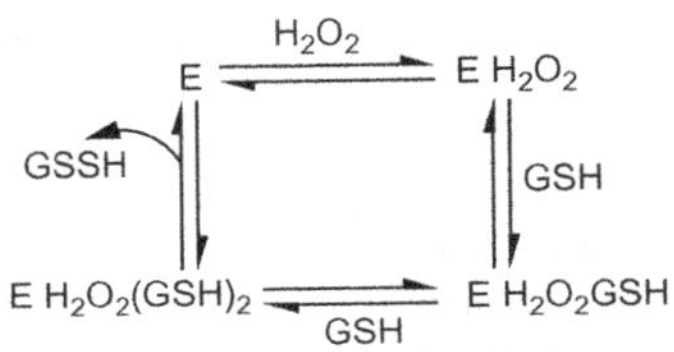

(6.8)

1.2 Selenium Encapsulation Into Protein Systems

The encapsulation of selenium (in the form of Sec) into scaffolds is essential to construct selenium-containing catalysts. In eukaryotes, Sec is encoded by a UGA stop codon in naturally occurring selenoproteins, which is recognized by a specific selenocysteyl-tRNA. But efficient suppression of the stop codon requires a special translation factor and specific recognition elements within the mRNA. Sec synthesis starts from serine (Ser) that is charged on the tRNASec by the conventional seryl-tRNA synthetase to yield the Ser-tRNASec. Before its conversion to Sec on the tRNASer, the serial moiety is phosphorylated to phosphoserine (Sep) by the O-phosphoseryl-tRNASer kinase (PSTK) that has been identified by comparative genomics. The Sep-tRNASec is further converted to Sec-tRNASec by the selenocysteine synthase SecS (SepSecS) that utilizes mono-selenophosphate as the selenium donor to catalyze the conversion [19]. Mono-selenophosphate is the product of selenophosphate synthetase (SPS2), which, in eukaryotes, very often harbors an active site Sec (Fig. 6.2).

Because it's difficult to express Sec by traditional recombinant DNA technology, the chemical modification method has been developed for Sec incorporation into proteins. The first example in the field of selenoprotein design is the chemical conversion of the active site Ser residue in subtilisin into a Sec [20]. The hydroxyl group of Ser221 can be selectively modified to introduce distinct functional groups into the active site of subtilisin, resulting in new activities. The reaction steps are illustrated as Fig. 6.3.

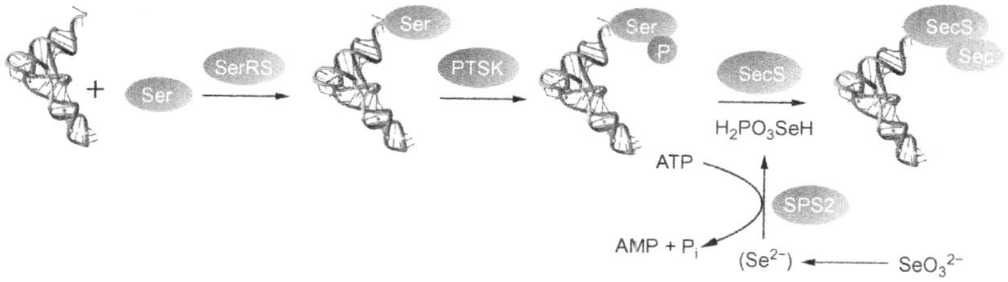

FIG. 6.2 Schematic representation of biosynthesis of Sec in eukaryotes. *(Adapted from C. Allmang, L. Wurth, A. Krol, The selenium to selenoprotein pathway in eukaryotes: more molecular partners than anticipated, Biochim. Biophys. Acta 1790 (2009) 1415–1423 with the permission of Elsevier.)*

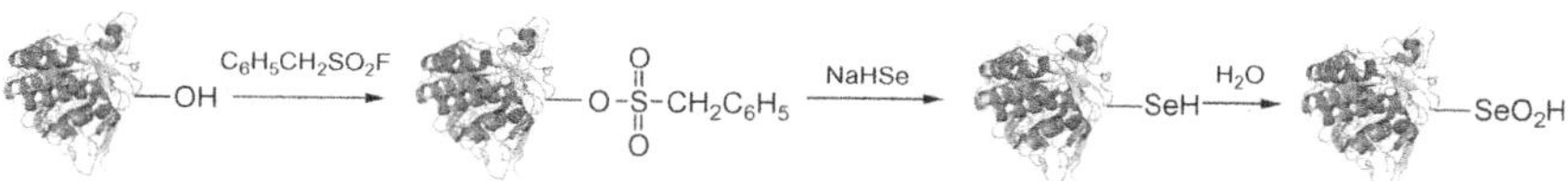

FIG. 6.3 Chemical incorporation of Sec on subtilisin [20].

This semisynthesized selenosubtilisin exhibits significant GPx activity, and the 3-D structure of the artificial selenoenzyme at 2.0 Å resolution has confirmed the overall structural integrity of the modified binding pocket, demonstrating that chemical modification can be a viable and practical method for Sec incorporation [21].

Compared with chemical modification, artificial bioincorporation of Sec has its own advantages such as site specificity and biocompatibility. Up to now, the most suitable approach for artificial bioincorporation of selenium is the auxotrophic expression technique. In 1994, Böck discovered the aminoacylation between tRNACys and Sec, indicating that Sec could bind to tRNACys in peptide biosynthesis instead of cysteine when cysteine is deficient in the culture medium. Inspired by this discovery, Böck developed an auxotrophic *Escherichia coli* strain that cannot produce cysteine itself [22]. Therefore, this strain introduces Sec into the biosynthesized protein when Sec is added into the medium instead of cysteine. In this case, Sec was successfully encapsulated into thioredoxin using the auxotrophic expression system. The elucidation of this specific Sec insertion method has opened up an exciting window of selenium-containing catalysts design.

The overall understanding of selenoproteins and selenium encapsulation allows us to imitate the structure of natural selenoenzymes based on protein scaffolds. Moreover, it is the foundation of construction of artificial selenoenzymes with distinguished features such as high catalytic efficiency, thermal stability, and adjustable activity. Based on nature's principles, different kinds of selenium-containing catalysts encapsulated in protein systems have been developed. Catalytic antibodies, enzymes, allosteric proteins, and protein assemblies are widely utilized as scaffolds, demonstrating a variety of remarkable properties and promising potential for antioxidative reagents.

2 SELENIUM-CONTAINING CATALYSTS ENCAPSULATED IN ANTIBODIES

Due to the absence of the binding site of the substrate GSH, many former studies showed that GPx mimics with small molecule weights generally exhibited limited activity. However, a substrate binding site can be easily generated using a standard monoclonal antibody preparation technique, inspiring biochemists to work on enzyme-like catalysts based on catalytic antibodies (also known as abzyme) [23]. First, the immune system provides a method to generate high affinity antibodies for antigens. When the substrate analog is used as hapten, tightly substrate-binding antibodies can be generated using monoclonal technology. Second, by chemical modification or genetic engineering, the catalytic active center Sec can be introduced into the substrate-binding cavity. Thus a highly efficient artificial enzyme with both an active site and substrate-binding site is obtained. This method is anticipated to be applied to construct any kind of tailor-made biocatalyst with amazing substrate selectivity and function diversity. Since the late 1980s, a series of selenium-containing abzymes have been constructed through standard monoclonal antibody preparation techniques. In addition, a phage antibody library was also used for the production of antibodies and antibody fragments such as scFv. So far, the activities of some selenium-containing catalytic antibodies have reached or even exceeded the level of native enzymes.

2.1 Development of Selenium-Containing Monoclonal Catalytic Antibodies

In early studies, the GPx active site Sec was incorporated into the protein scaffold by chemical modification. Most of the N-terminals in the antibodies' variable region contain several Ser residues. The alcoholic groups of Ser that are located in an antihuman IgM (Fcμ) (rabbit Ig fraction) specific site are selectively activated by phenylmethylsulfonyl fluoride (PMSF) and displaced with hydrogen selenide (H_2Se). Thus, Ser residues can be chemically converted into Sec [24].

The titer of the modified antibody is similar to that of the unmodified antibody. Also, the GPx activity of this modified antibody was seventy times higher than that of ebselen (PZ51), a well-studied small selenocompound used as an antioxidant drug (Table 6.1). However, the activity was still less efficient than that of the native GPx.

Lacking the substrate binding site, the antihuman IgM (Fcμ) cannot firmly bind to GSH. Thus this chemical modified catalytic antibody only displayed limited GPx activity. But it is easy to generate a substrate binding site using

TABLE 6.1 Activity of Artificial Mimics and Native Enzyme

Compound/Enzyme	Activity (U/μM)
Mutated antihuman IgM (Fcμ)	71.4
Ebselen (PZ51)	0.99
Native GPx	5780

Adapted from G. Luo, Z. Zhu, L. Ding, G. Gao, Q. Sun, Z. Liu, T. Yang, J. Shen, Generation of selenium-containing abzyme by using chemical mutation, Biochem. Biophys. Res. Commun. 198 (1994) 1240–1247 with the permission of Elsevier.

a standard monoclonal antibody preparation technique. First, a series of hydrophobic groups are grafted on the polar groups of GSH. Next, artificial antigens prepared from the modified substrates are used to produce the corresponding antibodies with the substrate binding ability. Finally, the catalytic group is incorporated into the hydrophobic cavity. In this way, an abzyme with a native enzyme-like substrate recognition cavity can be obtained.

Therefore, a series of substrate analogs were synthesized and used as haptens [25]. At first, S-substituted dinitrophenyl glutathione was prepared as hapten (GSH-S-DNP) through a reacting thiol group of GSH with 2,4-dinitrochlorobenzene (DNCB). Then, the hapten was cross-linked to bovine serum albumin (BSA) by glutaraldehyde to obtain the whole antigen [26]. Next, the whole antigen GSH-S-DPN-BSA immunized the BALB/c mice, and a hybridoma was produced by fusion of immunized mouse spleen cells with myeloma cells. GSH-S-DNP-specific monoclonal antibodies 4A4 was screened out by ELISA. Lastly, the Ser residues in V-domains of the 4A4 IgG antibody were activated by PMSF and treated with H_2Se, resulting in the conversion of Ser into Sec. It was demonstrated that about 33 GSH-S-DNP hapten molecules were linked to each BSA molecule, and the mutated 4A4 IgG displayed a very high GPx activity that rivals native GPx and is 1100-fold more than that of ebselen (PZ51). Further experiments also showed that the ranges of optimum pH and temperature of the selenium-containing abzyme were wider than those of the native GPx, and the stability of the abzyme in storage was higher than that of native GPx [27].

In this selenium-containing abzyme preparation method, the hapten molecule could be crucial to the catalytic efficiency of the abzyme, because the hapten determined the recognition between the substrate and abzyme. To further explore this method, three new haptens with different hydrophobic side chains were synthesized and the corresponding antibodies were also prepared [25]. The structures of these haptens were fully demonstrated using element analysis, IR, and NMR, and three related antibodies Se-3G5, Se-2F3, and Se-5C9, were also obtained. These abzymes all displayed remarkable GPx activity.

It is summarized that the affinity of abzymes for GSH tends to decrease as the modification level of GSH increases. However, the change trend of GPx activities is the opposite. In this viewpoint, a larger hydrophobic cavity was created by GSH modified with bigger hydrophobic groups, which allowed the second GSH molecule to access and resulted in higher activities. Using the same method, selenium-containing catalytic antibodies Se-4G3 and Se-1C8 were also constructed using GSH-S-DNPMe and GSH-S-DNPBu as haptens, respectively. And they both exhibited higher GPx activities than that of native GPx from rabbit liver.

2.2 Development of Selenium-Containing Single-Chain Variable Fragments of Abzyme

Although a series of selenium-containing catalytic antibodies with remarkable GPx activities have been successfully constructed, their major defects such as less availability and large mass must not be denied [28]. Single-chain variable fragments of antibodies (scFv) are recombinant polypeptides composed of an antibody heavy-chain variable sequence (V_H) and a light-chain variable sequence (V_L), joined together by a short peptide linker. This protein is one of the smallest antibody fragments that retain the ability to bind antigen. Some of the obvious advantages of scFv are as follows: (1) It can be expressed and precisely modified by genetic engineering; (2) Its low molecular weight makes it easy to be analyzed and suitable as drug candidates; (3) It has relatively low immunogenicity due to lacking the Fc portion. Also, GPx mimics based on scFv can be produced through the proteolysis of a monoclonal antibody and subsequent chemical modification from Ser to Sec.

Accordingly, to further research the high catalytic efficiency of the 2F3 monoclonal antibody, Liu and coworkers demonstrated that the selenium-containing Fv fragment (Se-2F3-scFv) displays a high GPx activity of 3394 U/μmol, the same order of magnitude as native GPx [29]. First, the variable region of the 2F3 antibody has been sequenced, cloned, and expressed as 2F3-scFv in different bacterial strains. The heavy and light variable region genes were cloned separately and incorporated into the same vector and connected with a linker $(Gly_4Ser)_3$. Then, the construction encodes a hexahistidine tag at the C-terminus of the V_L fragment for blotting. The 2F3-scFv was expressed as inclusion bodies. The experimental data showed that the Fv fragment containing a GSH binding site can be successfully generated by pepsin digestion from the antibody. The association constants between 2F3 and GSH/GSSG were about 10-fold that of the active 2F3-scFv by fluorescence quench assay, indicating that 2F3-scFv retains the conformation of the variable regions of 2F3. However, a bacterially produced 2F3-scFv fragment usually has a lower affinity than the native and intact antibody owing to incorrect folding. As a result, the activity of Se-2F3-scFv was lower than that of the selenium-containing abzyme 2F3, which was 24,300 U/μmol. In order to improve the expression quantity, Du et al. focused on increasing the expression level of soluble Se-2F3-scFv by using chemical chaperones. The results indicated that both glycerol and β-cyclodextrin greatly increase the expression level of soluble 2F3-scFv [30]. Later on, 2F3-scFv was also found to bind both GSH and CDNB, which are the substrates of native glutathione S-transferase. Furthermore, it was demonstrated that 2F3-scFv indeed displayed GST activity as well [31].

2.3 Development of Selenium-Containing Human Catalytic Antibodies

Abzymes based on selenium-containing mouse catalytic antibodies have been constructed and demonstrated to possess high activity. However, these GPx mimics cannot be applied as drug candidates due to immunogenicity. As long as these heterogeneous molecules are injected into the human circulatory system, they will cause intense immune response. The ultimate goal for biochemists is to develop enzyme-like catalysts with higher efficiency and less immunogenicity. Therefore, the human antibody was utilized to construct new selenium-containing abzymes.

The phage antibody library is an important method for the production of human antibodies, and it offers a new kind of monoclonal antibody or antibody fragments that bind specific substrates without antigenicity in vitro. Back in 2005, phage antibodies B8, H6, and C1 with a GSH binding site were obtained from the library using enzyme-linked immunosorbent assay (ELISA) analysis with four rounds of selection against three haptens: S-2,4-dinitrophenyl tbutyl ester [GSH-S-DNP-Bu (B)]; S-2,4-dinitrophenyl *t*-hexyl ester [GSH-S-DNP-He (H)]; and S-2,4-dinitrophenyl cycle-hexyl ester [GSH-S-DNP-cHe (C)] [32]. Their gene sequences were screened out from the phage display antibody library. To further optimize the sequences of the phage library, DNA shuffling was applied to construct a secondary library, and another positive phage antibody B9-scFv was screened out by another panning against GSH-S-DNP-Bu. After analysis and expression, the expressed phage clone B8, H6, C1, and B9 were purified and modified into selenium-containing abzymes. Their GPx activities are listed in Table 6.2.

The experimental results showed that human antibodies could also bear high GPx activities with less antigenicity to human beings. Moreover, owing to the low molecular weight of the scFv, it can be easily produced in bacterial cells and has lower retention time in nontarget tissues compared with monoclonal antibodies. Thus, all of these properties mentioned above endow the human catalytic antibody scFv with great potential for further development.

In conclusion, the standard monoclonal antibody preparation technique provides a convenient and highly efficient method to construct enzyme-like catalytic antibodies. In this strategy, the substrate binding site is created by immune reaction so that the abzyme can recognize the substrate to bear high activity. Many selenium-containing catalytic antibodies with excellent GPx activities

TABLE 6.2 GPx Activity of Seleno-Abzymes and Native Enzyme

Species	Phage (cfu/mL)	Protein (μg/mL) Nonbinding Se	Protein (μg/mL) Binding Se	GPx Activity (U/μM)	kD (μM/L)
Se-B8	10^{13}	260	220	3012±68	14±0.2
Se-H6	10^{13}	271	260	2102±68	12±0.2
Se-C1	10^{13}	253	241	694±68	12±0.2
Se-B9	10^{13}	260	216	3560±68	13±0.2
Native GPx	–	–	–	5780	–

Adapted from Y. Mu, D. Song, Y. Li, H. Zhang, W. Li, G. Luo, Q. Jin, Selection of phage antibodies with GPx activity by combination of phage displayed antibody library with chemical modification and their characterization using a surface plasmon resonance biosensor, Talanta 66 (2005) 181–187 with the permission of Elsevier.

have been developed in this method, and some of the artificial abzyme even displayed higher activities than that of native GPx, holding great promise for medicinal applications. Furthermore, scFv and human homologous antibodies with lower immunogenicity were utilized to produce a GPx mimic. With such various practical materials and possibilities, we believe more successful abzymes will be developed in the future.

3 SELENIUM-CONTAINING CATALYSTS ENCAPSULATED IN ENZYMES

Artificial design and simulation of selenium-containing catalysts has become an increasingly attractive topic, not only for understanding the intrinsic biological evolution process and the relationship between the structure and function, but also for preparing efficient drug candidates for related diseases. Up to now, a number of GPx mimics, ranging from simple organic compounds to complex engineered proteins, have been developed [7,23]. In contrast to synthetic molecules, enzyme as a natural source undoubtedly presents the most versatile scaffold to design and construct GPx mimics, for its unique properties provide a new possibility to develop more efficient and advanced biocatalysts [33,34]. Besides, the existing sophisticated catalytic system allows minimal modification to access another new function [9]. In the first decade of the 21st century, important progress has been made for the design of artificial GPx mimics, and some of these artificial selenoenzymes showed amazing GPx activity rivaling native ones. Using the gifts from nature, a novel approach to construct artificial selenium-containing catalysts has been developed.

3.1 Development of Selenium-Containing Subtilisin

The intrinsic quality of enzymatic catalysis is the combination of substrate recognition and intramolecular catalysis. The bacterial Ser protease subtilisin has proven to be a serviceable scaffold with a deep and narrow substrate binding cavity [20]. Also, in subtilisin, three amino acid residues, Asp32, His64, and Ser221, together form a "catalytic triad" that increases the nucleophilic characteristics of the catalytically critical hydroxyl group of Ser221, eventually enhancing the activity. Accordingly, the hydroxyl group of Ser221 can be selectively modified into Sec. This is also the first example in the field of semisynthesized selenoprotein design; the semisynthesized selenosubtilisin displayed significant GPx-like redox activity that is 70,000-fold faster compared to that of diselenide, a well-studied selenocompound [20,21].

To clarify the catalytic mechanism of this selenium-containing catalyst, a detailed kinetic study has been conducted by varying the concentration of one substrate while that of the other remains unchanged [35]. Double-reciprocal plots of the initial velocity versus substrate concentration yielded a series of paralleled linear plots, which revealed a ping-pong mechanism. Also, the enzymatic Sec was shown to be crucial for this activity. This mechanism indicated that in the proposed catalytic cycle it processes the substrates via selenic acid, selenenyl sulfide, and selenol forms of selenosubtilisin (Fig. 6.4).

Even though the semisynthesized selenosubtilisin has proven to be a successful example, its GPx activity is still relatively low compared to native GPx. In 2005, our group reengineered the active site of subtilisin by chemical conversion of the catalytically essential serine into a tellurocysteine and produced the semisynthetic telluroprotein, tellurosubtilisin [36]. The tellurosubtilisin was approximately two times more efficient than selenosubtilisin for GPx activity. However, the semisynthesis method cannot provide complete control of the enzyme structure. Therefore, we further constructed another selenosubtilisin by computer-aided design and an auxotrophic expression system [37]. In comparison with the chemically modified method, biosynthesis exhibits

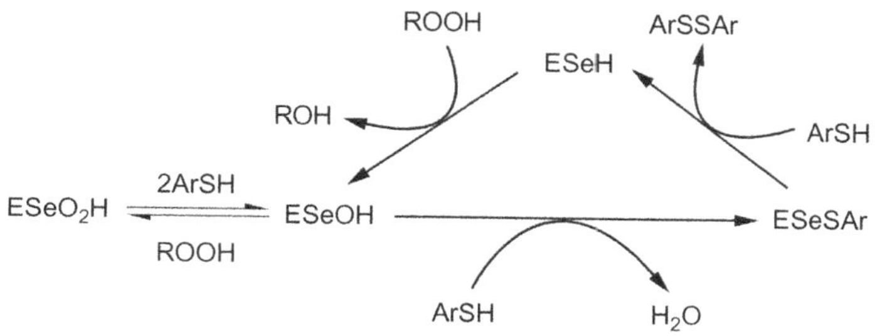

FIG. 6.4 The catalytic mechanism of selenosubtilisin. *(Adapted from I. Bell, M. Fisher, Z. Wu, D. Hilvert, Kinetic studies on the peroxidase activity of selenosubtilisin, Biochemistry 32 (1993) 3754–3762 with the permission of the American Chemical Society.)*

obvious advantages: Sec could be directly incorporated into active sites of enzymes to overcome the nonspecificity generated by chemical modification. This study provides an important strategy for specifically targeting and operating selenium in the active site of an enzyme.

3.2 Development of Selenium-Containing Glutaredoxin

The encapsulation technique of Sec into protein scaffolds by genetic engineering provides an opportunity for designing enzymes with improved and unusual catalytic properties. Besides subtilisin, other homologous GSH-binding proteins were also considered as the scaffolds of GPx mimics. Glutaredoxin (Grx) is a kind of oxidoreductase that catalyzes deglutathionylation and other redox reactions. It was first discovered in *E. coli* as GSH-dependent hydrogen donors for ribonucleotide reductase[38]. Accordingly, Grx may offer a model scaffold to endow GPx properties by engineering a catalytic Sec residue while preserving the binding specificity for GSH.

In this study [39], first the active residue Cys48 in the Grx domain of mouse thioredoxin-glutathione reductase was replaced by Sec through an auxotrophic expression system (Fig. 6.5). Then the engineered selenoprotein was tested in a coupled enzyme system, and the results showed that the selenium-containing Grx exhibited a remarkable level of GPx catalytic activity (2723 U μmol^{-1}) toward the reduction of H_2O_2 with GSH, which can rival naturally occurring GPx. For the first time, a small selenoenzyme with such remarkable GPx activity was achieved. In addition, the kinetic assay demonstrated that its catalytic behavior also fits a ping-pong mechanism, and the optimal temperature and pH of selenol-Grx were close to native GPx.

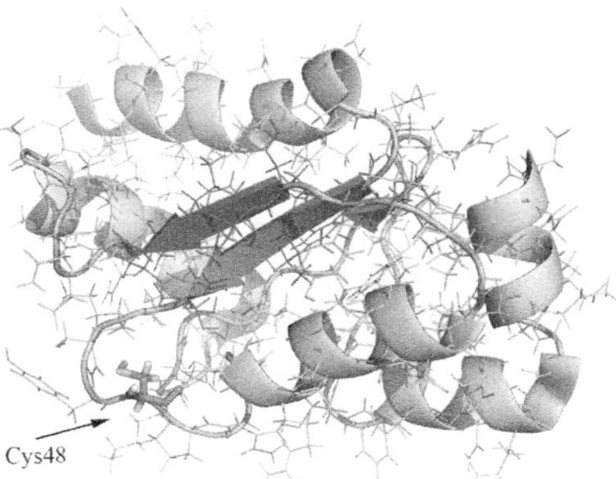

FIG. 6.5 The structure of Grx and the position of Cys48. *(Adapted from Y. Ge, Z. Qi, Y. Wang, X. Liu, J. Li, J. Xu, J. Liu, J. Shen, Engineered selenium-containing glutaredoxin displays strong glutathione peroxidase activity rivaling natural enzyme, Int. J. Biochem. Cell Biol. 41 (2009) 900–906 with the permission of Elsevier.)*

The experimental data showed that seleno-Grx is a novel GPx-like biocatalyst designed from the evolutionarily related Trx fold protein Grx. This study provides strong evidence for the hypothesis that Grx and GPx are derived from the same ancestor. The designed seleno-Grx is a small protein, and it has obvious advantages for further structure and function research and potential medical applications.

3.3 Development of Selenium-Containing Glutathione-S Transferase

Although protein-based GPx mimics like selenosubtilisin have been constructed by chemical modification and genetic engineering of natural enzymes, they did not display higher efficiency than native GPx. To do better than nature, glutathione-S transferase (GST) was used as scaffold to construct a highly efficient selenoenzyme [40,41]. GSTs are important phase II detoxification enzymes found mainly in the cytosol, and both GST and GPx belong to the thioredoxin superfamily due to the thioredoxin fold in their structure, which endows them with a similar GSH binding domain. The orientation of the catalytic residues of GST and GPx are the same, though their functions are quite different. In this regard, the rat theta class (rGST T2-2) and the human zeta1-1 class (hGST Z1-1) derived GSTs were chemically modified into Se-rGST T2-2 and Se-hGST Z1-1. The experimental data showed that the GPx activity of Se-rGST T2-2 was slightly lower than that of native GPx from rabbit liver, but the activity of Se-hGST T1-1 was about 1.5 times higher (Table 6.3). This work demonstrated that GST as a homologous protein could offer an excellent scaffold for designing efficient GPx mimics.

Furthermore, it's worth mentioning that tellurium-containing GST was also produced through genetic engineering [42]. Tellurocysteine (TeCys) has an intrinsically lower redox potential (−850 mV versus Ag/AgCl) than that of

TABLE 6.3 Catalytic Activity of GPx Mimics Based on GST

GPx Mimic	GPx Activity (U/μM)
Ebselen	0.99
Se-rGST T2-2	5406
Se-hGST Z1-1	8620
Native GPx	5780

Adapted from K. Zheng, P. Board, X. Fei, Y. Sun, S. Lv, G. Yan, J. Liu, J. Shen, G. Luo, A novel selenium-containing glutathione transferase zeta1-1, the activity of which surpasses the level of some native glutathione peroxidases, Int. J. Biochem. Cell. Biol. 40 (2008) 2090–2097 with the permission of Elsevier.

Sec (the catalytic center of GPx; −640 mV versus Ag/AgCl). As a result, proteins containing TeCys residues showed even higher activities than selenoproteins in many redox reactions. This first successful incorporation of TeCys into a protein to form an efficient GPx model can be viewed as a viable general route for the synthesis of telluroenzymes with GPx activity.

As described above, the structure of the natural enzyme not only shows us the essence of artificial enzyme design, but also offers the most versatile scaffolds to construct selenium-containing catalysts. Through minimal modification, the enzyme-based GPx mimics displayed significant activity (the same magnitude as native GPx). Also, the existing substrate binding cavity contributes to the improvement of activity. These studies have promoted the design of selenium-containing catalysts to a new level, and it is anticipated that selenium-containing catalysts will be endowed with more desirable catalytic performance.

4 SELENIUM-CONTAINING CATALYSTS ENCAPSULATED IN ALLOSTERIC PROTEINS

An ideal catalyst should not only show high catalytic efficiency, but it should also adjust its activity in a smart way. Since the beginning of the 21st century, enormous efforts have been made to develop selenium-containing catalysts based on protein scaffolds. In the first stage, for example in our group, we tried to develop several new methods to construct high-efficiency GPx mimics rivaling natural enzymes. Since a series of GPx mimics with high activities were achieved, we have started making efforts to build smart selenoenzymes for dynamic regulation of the ROS level to meet the physiological demands. The emerging smart artificial enzyme is used to meet the demand for dynamic and intelligent control of its biological function in various metabolic pathways. This has led to the development of enzyme mimics made of stimuli-responsive materials able to change their structures and properties in response to external stimuli such as temperature [43], pH [44], and light [45]. These stimuli or triggers applied to induce the reversible switching of enzyme activity are essential parts of the design strategy. Actually, this strategy has also been widely used by nature: an allosterically responsive site is directly associated with the catalytic domain of an enzyme. By employing the same principle, the reengineering of naturally allosteric proteins seems to be a more powerful approach to design artificial "switch" enzymes for allosteric regulation of their activity.

Designing a GPx active site on heterogeneous proteins requires precise control of the protein structures. De novo computer-aided design provides a method for orienting catalytic groups and setting a proper microenvironment (e.g., hydrophilicity/hydrophobicity, hydrogen-bonds, and electrostatic complementarity) to stabilize substrate recognition, and thereby it allows molecular design of tailor-made selenoenzymes with broader and more diverse properties for meeting high demand. In our work, a Ca^{2+}-responsive artificial selenoenzyme was constructed by computational design and engineering of recoverin with the active center of GPx [46]. At first, using molecular docking, we determined the low-energy interaction mode between this region and the GSH substrate. Then, Ala128 was selected and mutated to Cys (A128C) for the incorporation of Sec. To further enhance the recognition capacity of GSH when designing the binding site of this GPx mimic, glutamine at position 50 was replaced by an arginine (Q50R) to increase the binding specificity of the other carboxyl group of GSH.

Based on the above-mentioned computational design, a series of modification on the recoverin protein through genetic engineering has been made. By combining the recognition capacity for the substrate glutathione (GSH) and steric orientation of catalytic selenium moiety, the designed GPx mimic, selenium-containing recoverin, exhibited high GPx activity (the same level of some native GPx, e.g., pGPx in human plasma), which can be switched on/off by Ca^{2+}-induced allosterism of recoverin protein (Fig. 6.6). Also, the catalytic activity of this enzyme is completely reversible after multiple Ca^{2+}-bound and Ca^{2+}-free cycles.

Furthermore, to evaluate the switching behavior of this smart enzyme model in cells, mitochondria were used as a biological model to evaluate the protecting effects of seleno-recoverin against oxidative damage. The seleno-recoverin showed high protecting capacity for mitochondria against oxidative damage, and this process can be turned on and off by a calcium ion. We know that the mitochondrion is the main place of aerobic respiration and the storage of calcium ion in cells, with both ROS and calcium ion acting as oxidative stress-response elements and signal transducing messengers. The content of the calcium ion in cells is associated with the concentration change of ROS. Thus, it's anticipated that this calcium ion-switched artificial GPx model has the potential ability of spontaneous response to the concentration change of ROS in cells.

In nature, enzymes are regulated by conformational changes in their allosteric domain, a stimulus-responsive part of the protein that is distant from its active site. Studies of already developed de novo design principles will

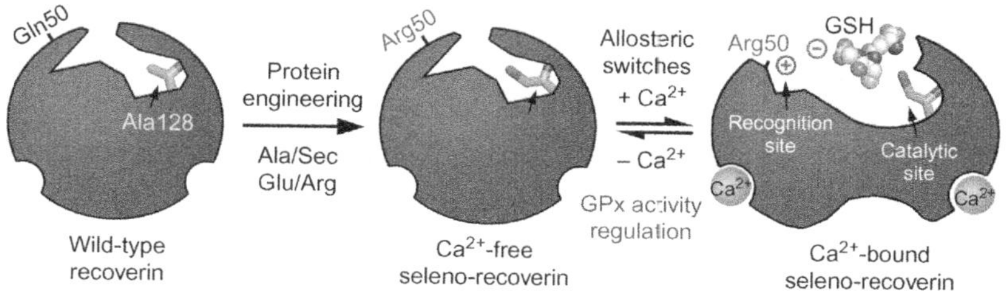

FIG. 6.6 Construction of Ca^{2+}-responsive artificial selenoenzyme based on allosteric protein recoverin [46]. *(Adapted from C. Zhang, T. Pan, C. Salesse, D. Zhang, L. Miao, L. Wang, Y. Gao, J. Xu, Z. Dong, Q. Luo, J. Liu, Reversible Ca^{2+} switch of an engineered allosteric antioxidant selenoenzyme, Angew. Chem. Int. Ed. 53 (2014) 13536–13539 with the permission from Wiley-VCH.)*

undoubtedly lead to refinement of allosteric protein design, and chemical modifications and genetic engineering offer the methods to realize the sophisticated design. Inspired by nature's wisdom, we believe that utilizing the allosteric conformational change to precisely control the enzyme mimics could provide new possibilities to develop smart and advanced biocatalysts.

5 SELENIUM-CONTAINING CATALYSTS ENCAPSULATED IN PROTEIN ASSEMBLIES

Protein self-assembly into exquisite, complex, yet highly ordered architectures represents the supreme intelligence of nature. Following nature's wisdom, to fabricate protein self-assemblies with dominant advantages such as biocompatibility, biodegradability, structural durability, and functional versatility represents an arising research hotspot. Various strategies have been devoted to construct high-ordered protein nanostructures. Some typical interactions, such as coiled-coil interactions [47], metal-mediated interactions [48], and host-guest interactions [49], have been adopted in the fabrication of protein self-assemblies. Also, the morphologies of these artificial protein assemblies range from simple nanowire/nanotube to sophisticated nanoring [50,51]. Because structure determines function, it is not surprising that protein assemblies always bear special characteristics such as stability and external stimulus-responsibility [52,53]. Therefore, the development of protein assembly might provide valuable assets to construct durable, smart, and multifunctional material and scaffolds.

Selenium-containing catalysts have been extensively developed based on both small organic molecules and macromolecules such as proteins, and now the development of protein assembly offers higher level scaffolds for enzyme design. De novo computer-aided design further extended the frameworks to nonhomogeneous proteins by precisely orienting catalytic groups and establishing a proper microenvironment (e.g., hydrophilicity/hydrophobicity, hydrogen-bonds, and electrostatic complementarity) so as to stabilize substrate recognition, thereby allowing the molecular design of selenoenzymes with broader and more diverse properties to meet the high demand. Taking advantage of the valuable properties of protein assemblies, a series of successful designs have been brought into practice.

5.1 Construction of a Highly Stable Selenium-Containing Nanoenzyme

Although natural enzymes have been extensively studied, their applications are usually limited due to intrinsic disadvantages such as instability, short half-life, and difficulty of preparing them. This problem could be fixed by constructing artificial enzymes with robust protein assemblies as scaffolds. Recently, a novel selenium-containing nanoenzyme was developed using a thermostable protein

SP1 as a scaffold [34]. By combining the artificial selenoenzyme design concepts of the synergy of substrate recognition and catalysis with computer simulations, Sec was incorporated into SP1 protein monomers to obtain a stable and efficient selenoenzyme (Fig. 6.7).

SP1 is a ring-like homo-oligomeric protein, consisting of 12 subunits that are tightly bound together via hydrophobic interactions forming two stacked hexameric rings, with a diameter of 11 nm and a central hole of 2–3 nm. This structural feature enables the SP1 protein to be distinguished from other proteins, giving an extremely high thermal and chemical stability to tolerate unnatural conditions. First, the optimal active site for Sec incorporation was selected from a protein-substrate binding mode according to the low-energy complex conformations. Then, the active Sec was implanted at Ala57 using a cysteine auxotrophic expression system. SDS-PAGE, CD, and TEM characterization all indicated that the newly developed selenoenzyme maintained the same

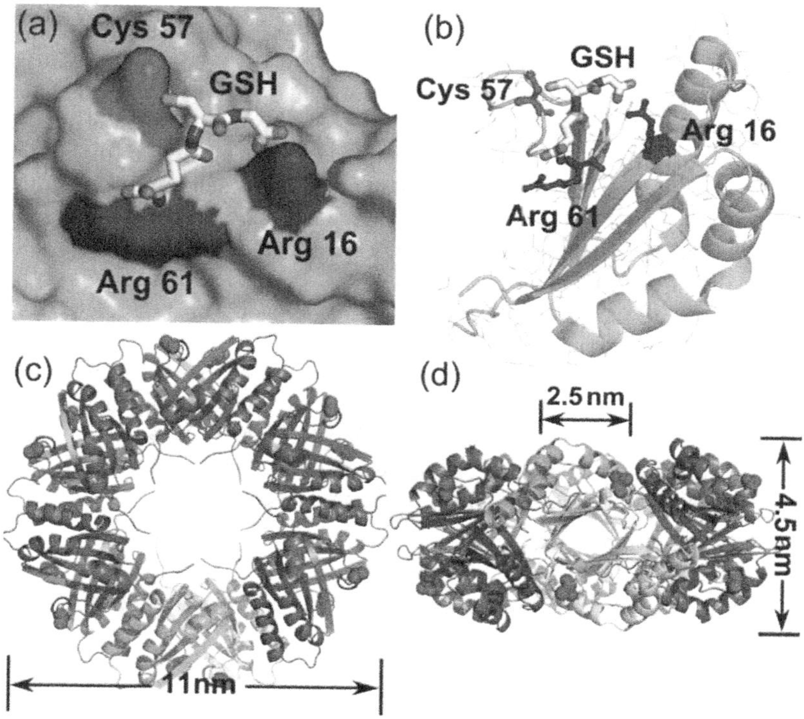

FIG. 6.7 Computer simulation of SP1 [34]. (A) The surface of catalytic cavity to bind substrate GSH. (B) A single subunit of SP1 with the reactive amino acids of the active center. The crystal structure of the SP1 nanoring with the Sec groups on the edge is shown from top view (C) and side view (D). *(Reproduced from L. Miao, X. Zhang, C. Si, Y. Gao, L. Zhao, C. Hou, O. Shoseyov, J. Liu, Construction of a highly stable artificial glutathione peroxidase on a protein nanoring, Org. Biomol. Chem. 12 (2014) 362–369 with permission from The Royal Society of Chemistry.)*

ring-like structure with wild-type SP1. Next, enzymatic experiments showed that the Seleno-ddSP1-57Cys exhibited a remarkable GPx activity of 390.6 μmol min^{-1} $μmol^{-1}$, which was higher than that of pGPx from human plasma. Furthermore, the thermostability of the SP1 protein makes it possible that Se-ddSP-57Cys was able to remain in a ring-like structure at a high temperature and maintain enzyme activity at 85°C for several days. Also, this nanoscale selenoenzyme had modest antioxidative capacity to protect mitochondria against oxidative damage.

5.2 Construction of a Protein-Dendrimer Hybrid Nanowire with GPx and SOD Activities

Besides natural protein assemblies, we also constructed sophisticated artificial protein assemblies as nanoenzyme scaffolds to bear new characteristics. The protein assembly skeletons pass their intrinsic properties to the artificial nanoenzymes, but creating multifunction cooperative nanostructures that understand the principles of biological systems still remains a challenge for scientists. Recently, a novel poly(amino amide) dendrimer mediated cricoid SP1 protein nanowire has been constructed as a scaffold for the design of a dual-enzyme cooperative antioxidative system [53]. Specifically, the catalytic center of GPx Sec was mutated to the outer surface of SP1, while the superoxide dismutase mimic (manganese porphyrin) was covalently reacted with the PAMAM dendrimer. The dual-enzyme nanowires showed significantly high GPx activity and SOD activity for their integrated catalytic surface, which was approaching the native rabbit liver GPx and Cu-Zn SOD values (Fig. 6.8).

For evaluating the cooperative antioxidative capacity of the protein nanowires, a mitochondria oxidative stress assay was performed. The antioxidative capacity of the dual-enzyme cooperative nanowires was significantly higher than that of any other single-enzyme system; the cooperation of GPx and SOD mimics could more effectively scavenge H_2O_2 and O^{2-}• in the system and protect organisms from being damaged. Also, the dual-enzyme cooperative system showed good biocompatibility, excellent cell uptake behavior, and ultra-low cell cytotoxicity.

5.3 Construction of a Selenium-Containing Nanoenzyme on a Virus Nanotube/Nanodisk

Among the known biological nanocomponents, the virus has become a versatile tool for nanomaterial and nanodevice design. In 2012, our group reported a novel artificial nanoenzyme with GPx-like function based on the tobacco mosaic virus (TMV) [50]. TMV comprises 2130 coat proteins that can assemble into a 300-nm long tubular virus with a diameter of 18 nm and a 4-nm center channel, presenting a perfect scaffold for the incoporation of Sec groups to achieve an antioxidant nanoenzyme.

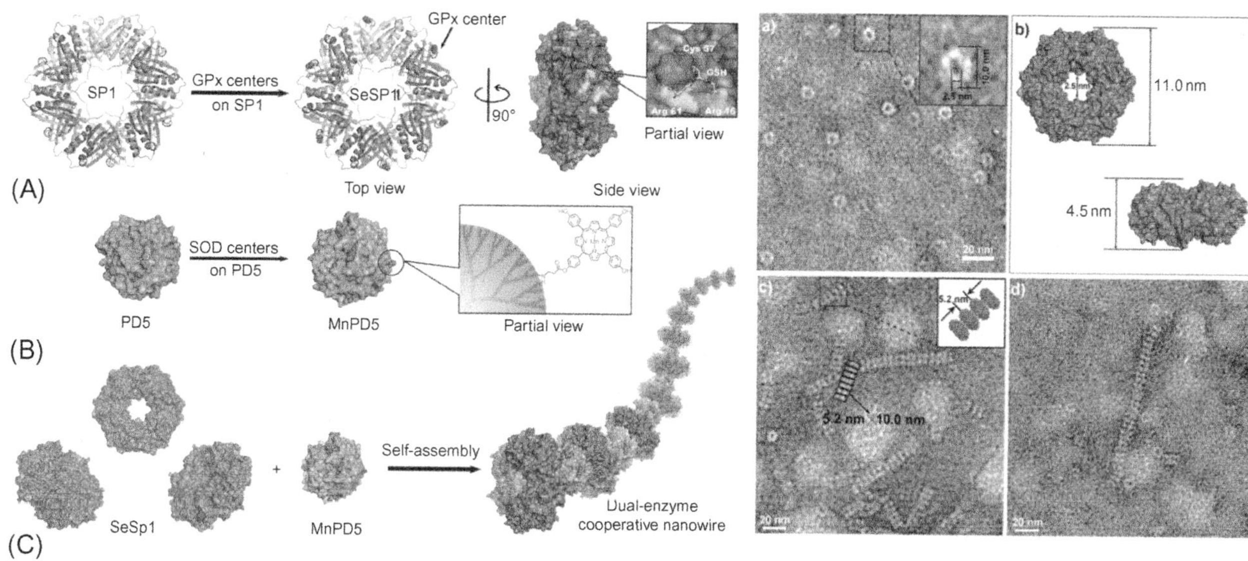

FIG. 6.8 Schematic representation of SP1-dentrimer hybrid nanowire with GPx and SOD activities [53]. (A) GPx catalytic center selenocysteine is designed on the outer surface of SP1. (B) SOD catalytic center manganese porphyrin is modified on the surface of PD5. (C) Assembly model of SeSP1 MnPD5 nanowire with SeSP1 and MnPD5. *(Reproduced with permission from H. Sun, L. Miao, J. Li, S. Fu, G. An, C. Si, Z. Dong, Q. Luo, S. Yu, J. Xu, J. Liu, Self-assembly of cricoid proteins induced by "soft nanoparticles": an approach to design multienzyme-cooperative antioxidative systems, ACS Nano 9 (2015) 5461–5469, Copyright 2015. American Chemical Society.)*

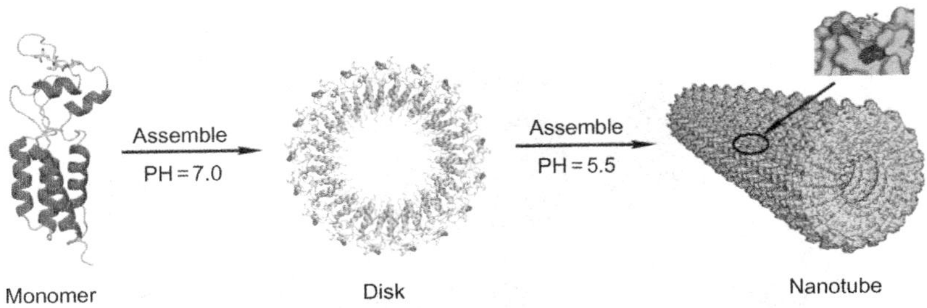

FIG. 6.9 Construction of the GPx catalytic center on a TMV nanodisk/nanotube [50]. *(Reproduced with permission from C. Hou, Q. Luo, J. Liu, L. Miao, C. Zhang, Y. Gao, X. Zhang, J. Xu, Z. Dong, J. Liu, Construction of GPx active centers on natural protein nanodisk/nanotube: a new way to develop artificial nanoenzyme. ACS Nano 6 (2012) 8692–8701, Copyright 2012. American Chemical Society.)*

By combining the artificial selenoenzyme design concept of the synergy of substrate recognition and catalysis with computer simulation, the main catalytic components of GPx were fabricated on TMV protein monomers. Through direct self-assembly of the functionalized viral coat proteins, the multi-GPx centers were installed on these well-defined nanodisks or nanotubes (Fig. 6.9). The resulting organized nanoaggregates exhibited remarkable GPx activities, even approaching the level of natural GPx. The antioxidation study on the subcellular mitochondrial level demonstrated that the virus-based nanoenzyme exerted excellent capacity for protecting cells from oxidative damage. The plant virus-derived nanoenzymes have low immune responses in mammals, providing potential pharmaceuticals for diseases caused by oxidative damage such as cardiovascular disease.

The combination of protein assembly and selenoenzyme design has yielded successful examples of robust, adjustable selenium-containing nanoenzymes. Benefiting from de nova computer-aided design, these nanoenzymes exhibited remarkable GPx activity and antioxidative capacity to protect mitochondria from ROS. Also, the large scale protein assemblies as scaffolds contributed to their activity enhancement, thermal stability, and multifunctionalization. As more and more protein assemblies with diverse morphologies and properties are developed, there will be limitless possibilities for selenium-containing nanoenzymes.

6 SUMMARY AND OUTLOOK

In this chapter, we first introduced the structure and catalytic behaviors of the natural selenium-containing catalyst GPx, and we then briefly discussed the methods to construct an artificial selenoenzyme encapsulated in protein systems. Catalytic antibodies, enzymes, allosteric proteins, and protein assemblies utilized as scaffolds and the related successful examples were separately highlighted. Nowadays, selenium-containing catalysts with remarkable substrate affinity, high catalytic efficiency, thermal stability, and adjustable activity have been extensively developed. Some of these features are even superior to natural selenoenzymes. However, the growing need for special antioxidative application will increase the demand for the properties of selenium-containing catalysts. Determining how to further advance this field is still a challenge. The recent emergence of various sophisticated materials may provide possibilities to construct antioxidant selenoenzymes with new properties, while more flexible and innovative design strategies are urgently needed to maximize their potential. It is anticipated that the designed selenium-containing catalysts could act as potent antioxidants in the treatment of human diseases and have practical applications in the biomedical field.

REFERENCES

[1] G. Mills, Hemoglobin catabolism: I. Glutathione peroxidase, an erythrocyte enzyme which protects hemoglobin from oxidative breakdown, J. Biochem. 229 (1957) 189–197.

[2] J. Rotruck, A. Pope, H. Ganther, W. Hoekstra, Selenium: biochemical role as a component of glutathione peroxidase, Science 179 (1973) 588–590.

[3] L. Flohé, W. Günzler, H. Schock, Glutathione peroxidase: a selenoenzyme, FEBS Lett. 32 (1973) 132–134.

[4] A. Spector, Oxidative stress-induced cataract: mechanism of action, FASEB J. 9 (1995) 1173–1182.

[5] D. Salvemini, Z. Wang, J. Zweier, A. Samouilov, H. Macarthur, T. Misko, M. Currie, S. Cuzzocrea, J. Sikorski, D. Riley, A nonpeptidyl mimic of superoxide dismutase with therapeutic activity in rats, Science 286 (1999) 304–306.

[6] S. Melov, J. Ravenscroft, S. Malik, M. Gill, D. Walker, P. Clayton, D. Wallace, B. Malfroy, S. Doctrow, G. Lithgow, Extension of life-span with superoxide dismutase/catalase mimetics, Science 289 (2000) 1567–1569.

[7] A. Muller, E. Cadenas, P. Graf, H. Sies, A novel biologically active seleno-organic compound-I glutathione peroxidase-like activity in vitro and antioxidant capacity of PZ51 (Ebselen), Biochem. Pharmacol. 33 (1984) 3235–3239.

[8] Z. Dong, Q. Luo, J. Liu, Artificial enzymes based on supramolecular scaffolds, Chem. Soc. Rev. 41 (2012) 7890–7908.

[9] O. Moroz, Y. Moroz, Y. Wu, A. Olsen, H. Cheng, K. Mack, J. McLaughlin, E. Raymond, K. Zhezherya, H. Roder, I. Korendovych, A single mutation in a regulatory protein produces evolvable allosterically regulated catalyst of nonnatural reaction, Angew. Chem. Int. Ed. 52 (2013) 6246–6249.

[10] O. Epp, R. Ladenstein, A. Wendel, The refined structure of the selenoenzyme glutathione peroxidase at 0.2-nm resolution, Eur. J. Biochem. 133 (1983) 51–69.

[11] H. Ganther, R. Kraus, Oxidation states of glutathione peroxidase, Methods Enzymol. 107 (1984) 593–602.

[12] L. Flohé, G. Loschen, W. Guenzler, E. Eichele, Glutathione peroxidase V. The kinetic mechanism. Hoppe-Seylers Z, Physiol. Chem. 353 (1972) 987–999.

[13] F. Ursini, M. Maiorino, M. Valente, L. Ferri, C. Gregolin, Purification from pig liver of a protein which protects liposomes and biomembranes from peroxidative degradation and exhibits

glutathione peroxidase activity on phosphatidylcholine hydroperoxides, Biochim. Biophys. Acta 710 (1982) 197–211.

[14] L. Flohé, Glutathione peroxidase, Basic Life Sci. 49 (1988) 663–668.

[15] K. Dalziel, The interpretation of kinetic data for enzyme-catalysed reactions, Biochem. J. 114 (1969) 547–556.

[16] K. Dalziel, Initial steady state velocities in the evaluation of enzyme-coenzyme-substrate reaction mechanisms, Acta Chem. Scand. 11 (1957) 1706–1723.

[17] I. Segel, Enzyme Kinetics: Behavior and Analysis of Rapid Equilibrium and Steady-State Enzyme Systems, Wiley-Interscience, New York, 1993.

[18] M. Carsol, I. Pouliquen-Sonaglia, G. Lesgards, G. Marchis-Mouren, A. Puigserver, M. Santimone, A new kinetic model for the mode of action of soluble and membrane-immobilized glutathione peroxidase from bovine erythrocytes, Eur. J. Biochem. 247 (1997) 248–255.

[19] C. Allmang, L. Wurth, A. Krol, The selenium to selenoprotein pathway in eukaryotes: more molecular partners than anticipated, Biochim. Biophys. Acta 2009 (1790) 1415–1423.

[20] Z. Wu, D. Hilvert, Selenosubtilisin as a glutathione peroxidase mimic, J. Am. Chem. Soc. 112 (1990) 5647–5648.

[21] K. House, R. Dunlap, J. Odom, Z. Wu, D. Hilvert, Structural characterization of selenosubtilisin by ^{77}Se NMR spectroscopy, J. Am. Chem. Soc. 114 (1992) 8573–8579.

[22] S. Muller, H. Senn, B. Gsell, W. Vetter, C. Baron, A. Böck, The formation of disenlenide bridges in proteins by incorporation of selenocysteine residues: biosynthesis and characterization of $(Se)_2$-thioredoxin, Biochemistry 33 (1994) 3404–3412.

[23] S. Pollack, J. Jacobs, P. Schultz, Selective chemical catalysis by an antibody, Science 234 (1986) 1570–1573.

[24] G. Luo, Z. Zhu, L. Ding, G. Gao, Q. Sun, Z. Liu, T. Yang, J. Shen, Generation of selenium-containing abzyme by using chemical mutation, Biochem. Biophys. Res. Commun. 198 (1994) 1240–1247.

[25] G. Luo, X. Ren, J. Liu, Y. Mu, J. Shen, Towards more efficient glutathione peroxidase mimics: substrate recognition and catalytic group assembly, Curr. Med. Chem. 13 (2003) 1151–1183.

[26] Z. Zhu, L. Ding, G. Luo, Z. Liu, Q. Sun, T. Yang, J. Shen, Some physicochemical and enzymatic properties of selenium-containing abzyme, Biochem. Biophys. Res. Commun. 202 (1994) 1645–1650.

[27] G. Luo, L. Ding, Z. Zhu, G. Gao, Q. Sun, Z. Liu, T. Yang, J. Shen, A new strategy for generating selenium-containing abzyme, Ann. N. Y. Acad. Sci. 750 (1995) 277–283.

[28] J. Huston, M. Tai, J. McCartney, P. Keck, H. Oppermann, Antigen recognition, and targeted delivery, by the single-chain Fv, Cell Biophys. 22 (1993) 189–224.

[29] J. Liu, C. Shi, G. Luo, Z. Liu, G. Zhang, W. Ma, J. Shen, An Fv catalytic antibody with high glutathione peroxidase activity, Mater. Sci. Eng. C 10 (1999) 131–134.

[30] X. Du, Y. Sun, F. Lin, K. Zheng, K. Wang, T. Lin, J. Liu, J. Shen, G. Luo, Chemical chaperones increasing expression level of soluble single-chain Fv antibody (scFv2F3), Chem. Res. Chin. Univ. 23 (2007) 69–75.

[31] K. Zhang, T. Zang, W. Yang, Y. Sun, Y. Mu, J. Liu, J. Shen, G. Luo, Single chain antibody displays glutathione S-transferase activity, J. Biol. Chem. 281 (2006) 12516–12520.

[32] Y. Mu, D. Song, Y. Li, H. Zhang, W. Li, G. Luo, Q. Jin, Selection of phage antibodies with GPx activity by combination of phage displayed antibody library with chemical modification and their characterization using a surface plasmon resonance biosensor, Talanta 66 (2005) 181–187.

[33] O. Makhlynets, E. Raymond, I. Korendovvych, Design of allosterically regulated protein catalysts, Biochemistry 54 (2015) 1444–1456.

[34] L. Miao, X. Zhang, C. Si, Y. Gao, L. Zhao, C. Hou, O. Shoseyov, J. Liu, Construction of a highly stable artificial glutathione peroxidase on a protein nanoring, Org. Biomol. Chem. 12 (2014) 362–369.

[35] I. Bell, M. Fisher, Z. Wu, D. Hilvert, Kinetic studies on the peroxidase activity of selenosubtilisin, Biochemistry 32 (1993) 3754–3762.

[36] S. Mao, Z. Dong, J. Liu, X. Li, X. Liu, G. Luo, J. Shen, Semisynthetic tellurosubtilisin with glutathione peroxidase activity, J. Am. Chem. Soc. 127 (2005) 11588–11589.

[37] J. Li, X. Liu, Y. Ji, Z. Qi, Y. Ge, J. Xu, J. Liu, G. Luo, J. Shen, Biosynthesis of selenosubtilisin: a novel way to target selenium into the active site of subtilisin, Chin. Sci. Bull. 53 (2008) 2454–2461.

[38] A. Holmgren, Hydrogen donor system for *Escherichia coli* ribonucleoside-diphosphate reductase dependent upon glutathione, PNAS 73 (1976) 2275–2279.

[39] Y. Ge, Z. Qi, Y. Wang, X. Liu, J. Li, J. Xu, J. Liu, J. Shen, Engineered selenium-containing glutaredoxin displays strong glutathione peroxidase activity rivaling natural enzyme, Int. J. Biochem. Cell Biol. 41 (2009) 900–906.

[40] X. Ren, P. Jemth, P. Board, G. Luo, B. Mannervik, J. Liu. K. Zhang, J. Shen, A semisynthetic glutathione peroxidase with high catalytic efficiency: selenoglutathione transferase, Chem. Biol. 9 (2002) 789–794.

[41] K. Zheng, P. Board, X. Fei, Y. Sun, S. Lv, G. Yan, J. Liu, J. Shen, G. Luo, A novel selenium-containing glutathione transferase zeta1-1, the activity of which surpasses the level of some native glutathione peroxidases, Int. J. Biochem. Cell Biol. 40 (2008) 2090–2097.

[42] X. Liu, L. Silks, C. Liu, M. Ollivault-Shiflett, X. Huang, J. Li, G. Luo, Y. Hou, J. Liu, J. Shen, Incorporation of tellurocysteine into glutathione transferase generates high glutathione peroxidase efficiency, Angew. Chem. Int. Ed. 48 (2009) 2020–2023.

[43] X. Huang, Y. Yin, X. Jiang, Y. Tang, J. Xu, J. Liu, J. Shen, Construction of smart glutathione peroxidase mimic based on hydrophilic block copolymer with temperature responsive activity, Macromol. Biosci. 9 (2009) 1202–1210.

[44] M. Sagermann, R. Chapleau, E. DeLorimier, M. Lei, Using affinity chromatography to engineer and characterize pH-dependent protein switches, Protein Sci. 18 (2009) 217–228.

[45] T. Shimoboji, E. Larenas, T. Fowler, S. Kulkarni, A. Hoffman, P. Stayton, Photoresponsive polymer-enzyme switches, PNAS 99 (2002) 16592–16596.

[46] C. Zhang, T. Pan, C. Salesse, D. Zhang, L. Miao, L. Wang, Y. Gao, J. Xu, Z. Dong, Q. Luo, J. Liu, Reversible Ca^{2+} switch of an engineered allosteric antioxidant selenoenzyme, Angew. Chem. Int. Ed. 53 (2014) 13536–13539.

[47] P. Harbury, J. Plecs, B. Tidor, T. Alber, P. Kim, High-resolution protein design with backbone freedom, Science 282 (1998) 1462–1467.

[48] J. Brodin, X. Ambroggio, C. Tang, K. Parent, T. Baker, F. Tezcan, Metal-directed, chemically tunable assembly of one-, two- and three-dimensional crystalline protein arrays, Nat. Chem. 4 (2012) 375–382.

[49] J. Carlson, S. Jena, M. Flenniken, T. Chou, R. Siegel, C. Wagner, Chemically controlled self-assembly of protein nanorings, J. Am. Chem. Soc. 128 (2006) 7630–7638.

[50] C. Hou, Q. Luo, J. Liu, L. Miao, C. Zhang, Y. Gao, X. Zhang, J. Xu, Z. Dong, J. Liu, Construction of GPx active centers on natural protein nanodisk/nanotube: a new way to develop artificial nanoenzyme, ACS Nano 6 (2012) 8692–8701.

[51] Y. Bai, Q. Luo, W. Zhang, L. Miao, J. Xu, H. Li, J. Liu, Highly ordered protein nanorings designed by accurate control of glutathione S-transferase self-assembly, J. Am. Chem. Soc. 135 (2013) 10966–10969.

[52] T. Luo, K. Kiick, Noncovalent modulation of the inverse temperature transition and self-assembly of elastin-b-collagen-like peptide bioconjugates, J. Am. Chem. Soc. 137 (2015) 15362–15365.

[53] H. Sun, L. Miao, J. Li, S. Fu, G. An, C. Si, Z. Dong, Q. Luo, S. Yu, J. Xu, J. Liu, Self-assembly of cricoid proteins induced by "soft nanoparticles": an approach to design multienzyme-cooperative antioxidative systems, ACS Nano 9 (2015) 5461–5469.

Chapter 7

Novel Core-Shell Nanoparticle@Metal-Organic Framework Composite Materials as Heterogeneous Catalysts

Fei Ke* and **Junfa Zhu**[†]
[*]Anhui Agricultural University, Hefei, China, [†]*University of Science and Technology of China, Hefei, China*

Contents

1 INTRODUCTION

Porous metal-organic frameworks (MOFs), also known as porous coordination polymers (PCPs), are a fascinating class of porous crystalline materials whose structure is defined by inorganic vertices and organic struts [1]. The most important advantages of MOFs with respect to other inorganic or organic porous materials are the large internal surface areas (up to 6000 $m^2\ g^{-1}$), uniformly structured cavities, ultralow densities, and large density of active sites [2]. MOFs have

Encapsulated Catalysts. http://dx.doi.org/10.1016/B978-0-12-803836-9.00007-9

attracted a great deal of attention in recent years, not only for their fundamental scientific interest but also for numerous attractive applications in the fields of selective adsorption and separation, chemical sensing, gas storage, drug delivery, and catalysis, which have been discussed in various reviews [3–6]. For the catalysis applications, the structure, composition, and pore size of MOFs can be judiciously tailored to facilitate the uptake of the specific guest molecules. However, there are still several issues that limit the utilization of MOFs, such as low chemical stability, poor thermal stability, and solubility under certain conditions [7]. Therefore, it remains an important challenge to improve the performance of MOFs-based catalysts.

Recently, combining MOFs with multifunctional nanoparticles (NPs) has been proposed to possess advanced properties originating from the two different components and to overcome their individual shortages [8]. This strategy has become an important tool for expanding MOF-based applications due to their synergism effect [9,10]. NPs/MOF composites can be synthesized by using MOFs as templates to prepare NPs within their cavities [11]. Although these materials have shown significant catalytic properties, there are several inherent limitations [12]. First, the size and morphology of the NPs, which are very important factors in the function of NPs for catalytic properties, are uncontrollable. Second, the porous structure of the MOFs is sometimes damaged during the harsh treatment conditions. Third, the nonselective deposition at the outer surface of the MOFs' support is found. To express more control over this process, a preferred strategy was reported where the MOFs were grown around preformed NPs [13]. Research on MOF-based core-shell heterostructures with MOF as a shell and multifunctional NPs as core are one of the most effective ways to obtain the synergistic effects of the MOFs and NPs for catalytic applications. To date, core-shell NP@MOF heterostructure materials have been successfully prepared with various active guests, including metal NPs, metal oxide, silica, polymers, carbon, and so on [14–19]. Compared to other porous shell materials, the main advantages of the MOFs as shell materials for catalysis can be summarized as follows: (i) The coordinatively unsaturated metal sites allow strong interactions with organic molecules; (ii) The polyfunctional organic ligands can interact with active NPs; and (iii) The porous structures provide a confinement effect and shape selectivity [20]. Moreover, the aggregation of the NP core with small size and high surface energy can be dramatically limited after the MOF shell coating, and then the stability and chemical activity of the NP core are preserved.

This chapter deals comprehensively with the NP@MOF core-shell heterostructure materials in the field of catalysis, with special focus on the components, structures, and the synergistic effects of the core-shell structure on their catalytic performance (Fig. 7.1). Considering there are many NPs that can be employed as the core materials, this chapter provides examples mostly from the different types of NP cores to illustrate the properties of these core-shell NP@MOF materials as

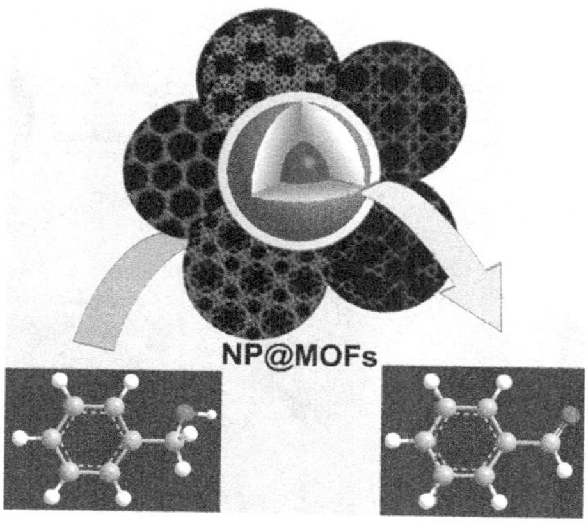

FIG. 7.1 Schematic illustration of NP@MOFs core-shell heterostructures as catalytic materials.

heterogeneous catalysts. In addition, the methods for synthesis of the encapsulated NPs cores with different sizes and compositions, as well as the thickness and structure of the MOF shells, will be addressed briefly.

2 OVERVIEW OF SYNTHESIS OF NP@MOFs CORE-SHELL HETEROSTRUCTURES

Up to now, there are four main approaches for the encapsulation of NPs into MOFs. The first approach is the versatile step-by-step assembly strategy. For example, Qiu et al. reported a series of NPs@MOF (e.g., Fe_3O_4@$Cu_3(btc)_2$, Fe_3O_4@MIL-100(Fe), Au@MIL-100(Fe)) core-shell microspheres with a tailored MOF shell by using the versatile step-by-step strategy (Fig. 7.2) [21,22]. In this method, functionalization of the NP core with the mercaptoaceticacid (MAA) molecules before the assembly process was found to play a crucial role in the initial stage of the step-by-step assembly, because no core-shell structures could be obtained using the unfunctionalized NPs. The growth of the MOF shell on the MAA-functionalized NPs should be initiated first by the metal ions binding to the carboxylate groups of the MAA on the NP surface, to which then the organic ligands from the solution bind. Finally, the MOF shell thickness can be easily controlled by tuning the step-by-step assembly processes. The structure, composition, and function of the MOF shell can also be judiciously tailored by choosing different metal vertices and organic struts. In the second approach the presynthesized NPs are used as seeds to induce the nucleation of the MOF shell. For example, Hupp and coworkers reported various surfactant-capped NPs with different sizes, shapes, and compositions to be enshrouded by ZIF-8 crystals (Fig. 7.3) [23]. The polyvinylpyrrolidone (PVP)

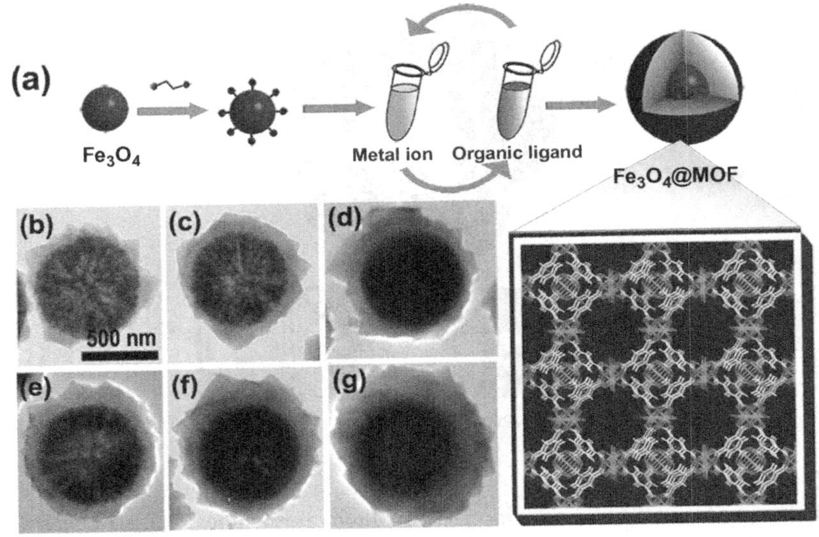

FIG. 7.2 Schematic representation of the step-by-step assembly strategy (A), and TEM images of Fe_3O_4@[$Cu_3(btc)_2$] core-shell magnetic microspheres after (B) 10, (C) 20, (D) 25, (E) 30, (F) 40, and (G) 50 assembly cycles at 25°C. *(Reprinted from F. Ke, L.-G. Qiu, Y.-P. Yuan, X. Jiang, J.-F. Zhu, Fe_3O_4@MOF core-shell magnetic microspheres with a designable metal-organic framework shell, J. Mater. Chem. 22 (2012) 9497–9500, Copyright 2012, with permission of the Royal Society of Chemistry.)*

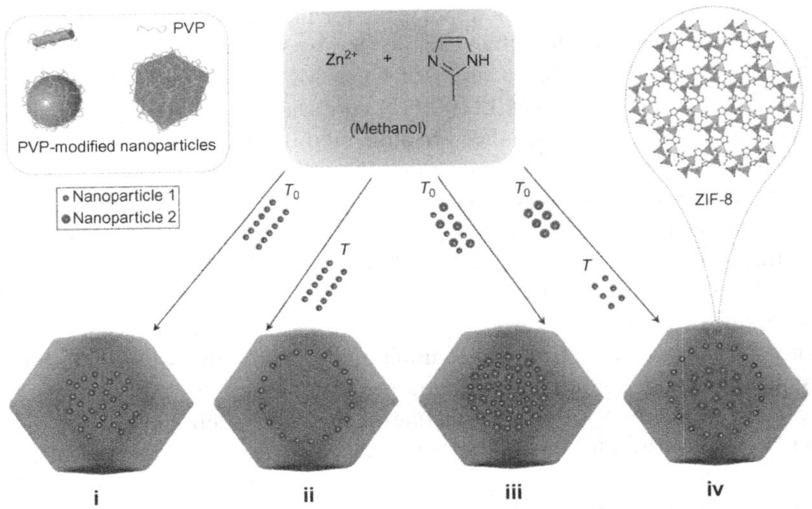

FIG. 7.3 Scheme of the controlled encapsulation of various PVP-capped NPs with different sizes, shapes, and compositions in ZIF-8 crystals. *(Reprinted from G. Lu, S.Z. Li, Z. Guo, O.K. Farha, B.G. Hauser, X.Y. Qi, et al., Imparting functionality to a metal-organic framework material by controlled nanoparticle encapsulation, Nature Chem. 4 (2012) 310–316, Copyright 2012, with permission of Nature Publishing Group.)*

modified NPs were mixed with a methanol solution of the ZIF-8 precursors, and then the NPs were well encapsulated into ZIF-8 crystals. In this method, PVP is an amphiphilic and nonionic polymer, which serves as a surfactant to stabilize NPs in polar solvents and as a capping agent to assist in controlling the size and shape of NPs during the formation of NP@ZIF-8. The spatial distribution of the embedded NPs within ZIF frameworks can be controlled by adjusting the sequence. By using this strategy, Oh et al. also reported a controlled encapsulation method that enabled carboxylate-terminated silica [16] or polystyrene [24] spheres to be embedded by the MOFs. After mixing carboxylate-terminated silica (polystyrene) spheres with an *N*,*N*-dimethylformamide (DMF) (methanol) solution of MOF precursors, the formed silica@MOF (polystyrene@ZIF-8) core-shell structures were collected by centrifugation. The carboxylate groups on the spheres' surface (silica, polystyrene) first interacted with metal ions and initiated the growth of MOFs on the surfaces of the spheres. These interactions induce a regular growth of MOF shells on the spheres' surface and the shell thickness in core-shell structures can be judiciously controlled by adjusting the amount of reactants (MOFs precursors and silica or polystyrene spheres). The third approach is the one-pot method. Tang and coworkers reported a facile one-pot method to prepare Au@MOF-5 core-shell NPs, in which the single Au core is coated with a uniform MOF-5 shell [25]. Different from the above two methods in which the presynthesized NPs used as seeds to induce the nucleation of MOF crystals, the Au@MOF-5 core-shell NPs were synthesized by directly mixing both the Au and MOF precursors in the reaction solution containing DMF, PVP, and ethanol. $HAuCl_4$ was first reduced to Au NPs by DMF. Subsequently, MOF-5 formed and spontaneously grew on the surface of the PVP-modified Au NPs, and then uniform core-shell Au@MOF-5 NPs were produced. In this reaction, DMF, PVP, and ethanol were indispensable to the preparation of the core-shell structures. DMF was the solvent for the organic ligands solution and determines whether MOF crystals can be produced. Moreover, it also served as a reducing agent for the synthesis of the Au NP core. PVP served as a stabilizer of Au NPs and ethanol can change the coordination environment of metal ions; MOF then grows preferentially around the Au NP core. The fourth approach is the self-template method. The metal or metal oxide templates can provide metal ions through sacrificing themselves and then initiate the growth of MOFs without any surface modification. For example, ZnO@ZIF-8 nanorods and nanotubes with core-shell heterostructures have been successfully obtained via a simple self-template strategy by using ZnO nanorods and nanotubes as sacrificial templates, respectively [26]. In this process, ZnO nanorods and nanotubes not only act as the template, but also provide Zn(II) ions for the formation of ZIF-8. Further, Huo and coworkers reported various NP@MOF (e.g., Au@HKUST-1, Pd@HKUST-1, PtAuDNP@HKUST-1) yolk-shell structures by using NP@Cu_2O core-shell nanostructures as the self-template [12]. Thanks to the protection of the Cu_2O shell and the gradual release of Cu ions, NPs could be encapsulated easily without being aggregated or dissolved in the reaction process.

3 NP@MOFs CORE-SHELL HETEROSTRUCTURES IN CATALYSIS

Depending on the types of the NP cores and MOF shells, NP@MOF core-shell heterostructures have been extensively applied for many applications [27]. Among these attractive applications, here, we are particularly interested in utilizing these core-shell heterostructures as active materials for application in heterogeneous catalysis. Encapsulation within porous MOF shells makes the NPs core catalytically active and at the same time keeps them stable in the harsh reaction conditions. In addition, a few methods have been reported for systematic control over the pore sizes of the MOF shells, making it easy to control the mass transfer of the reactants and products [28]. Here, we will discuss the state-of-art progress in NP@MOF core-shell heterogeneous catalysis categorized by the components of the inorganic NP cores.

3.1 Metal-NP@MOF Core-Shell Heterostructures in Catalysis

Recently, there have been an increasing number of reports that demonstrate the potential of single-metal-NP@MOF core-shell structures for applications in catalysis. Incorporating inorganic metal NPs within porous MOFs is an important means to enhance the catalytic performance. For example, Tsung and coworkers reported novel Pd@ZIF-8 nanostructures for gas phase heterogeneous catalysis with selectivity control [29]. The Pd NPs were first coated with a layer of Cu_2O as the sacrificial template, which could be etched simultaneously and spontaneously during the formation of the layer of ZIF-8. The as-synthesized Pd@ZIF-8 core-shell nanostructures were applied to the gas phase hydrogenation of ethylene, cyclohexene, and cyclooctene. The catalytic results showed high activity for ethylene and cyclohexene hydrogenations but not for cyclooctene hydrogenation, demonstrating that the ZIF-8 shell exhibits a highly effective molecule size-selective property. The size of the ethylene molecule is small (2.5 Å) enough to diffuse through the pore aperture of the ZIF-8 shell (3.4 Å), while the size of cyclooctene molecule (5.5 Å) is much bigger than the size of the shell pore. Further, Yaghi et al. also reported three Zr(IV)-based single crystals of nano-MOFs (nMOF-801, nUiO-66, nUiO-67) grown in monocrystalline form around metal NPs. The PVP-modified Pt NPs with a 2.5 nm size were subjected to reactions yielding Zr(IV)-based MOFs as a single crystal within which the Pt NPs are embedded [30]. The gas-phase hydrogenative conversion of methylcyclopentane (MCP) was used as a model reaction system to characterize the catalytic performance of the Pt$\subset$nMOF core-shell catalysis. The pore sizes progressively increased from nMOF-801 (5.4 and 7.0 Å) to nUiO-66 (6.8 and 7.2 Å) and nUiO-67 (9.6 and 12.6 Å). Significantly, the orders of conversion rate of MCP strongly varied between the different Pt$\subset$nMOF nanocatalysis. The Pt$\subset$nUiO-66 core-shell nanocatalysis showed selectivity to C_6-cyclic hydrocarbons such as cyclohexane (6.0 Å) and benzene

(5.9 Å). In the case of the Pt⊂nMOF-801, which possessed the smallest pore size, C_6-cyclic hydrocarbons were not furnished, while the largest pore of Pt⊂nUiO-67 core-shell nanocatalysis provided the same C_6-cyclic hydrocarbons products as observed over the Pt⊂nUiO-66, but with different selectivity (Fig. 7.4). These experiments clearly prove that the pore size of the MOF shell plays an important role in molecular shape selective catalysis.

Besides applications in gas-phase catalysis, metal-NP@MOF core-shell heterostructures are also candidates for liquid-phase catalysis. Metal NPs have been widely studied in liquid-phase catalysis. However, because of their high surface energies, metal NPs tend to aggregate in catalytic reactions, leading to gradual decrease of their initial activities. Encapsulation within porous MOF shells makes the metal NPs catalytically active and at the same time keeps them stable even under harsh reaction conditions. Zhu et al. reported a simple

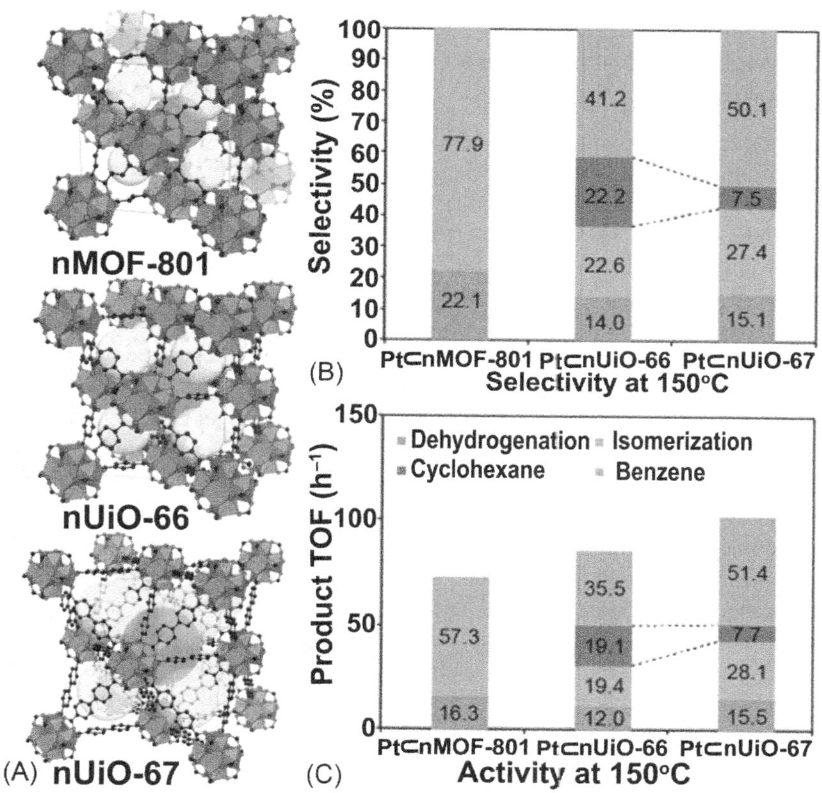

FIG. 7.4 Crystal structure of isoreticular Zr-based single crystals of MOFs (nMOF-801, nUiO-66, nUiO-67) (A). Product selectivity (B) and turnover frequency (C) obtained by Pt⊂nMOF core-shell catalysis. *(Reprinted from K. Na, K.M. Choi, O.M. Yaghi, G.A. Somorjai, Metal nanocrystals embedded in single nanocrystals of MOFs give unusual selectivity as heterogeneous catalysts, Nano Lett. 14 (2014) 5979–5983, Copyright 2014, with permission of the American Chemical Society.)*

encapsulation strategy for the fabrication of novel Au@MIL-100(Fe) core-shell nanocatalysts, which showed enhanced catalytic property in the conversion of 4-nitrophenol (4-NP) to 4-aminophenol (4-AP) [22]. Au@MIL-100(Fe) core-shell nanocatalysts with controllable MIL-100(Fe) shell thickness were fabricated by using a versatile step-by-step fashion. Catalytic results show that the Au@MIL-100(Fe) core-shell nanocatalysts exhibited much higher catalytic activity than those of pure Au, MIL-100(Fe), and even their mixture for the same reaction time. The adsorption of 4-NP over the MOF shells gives rise to a local high concentration of 4-NP, which increased the contact probability with the catalyst and thus enhanced the catalytic efficiency, suggesting that the MOF shell enhanced the catalytic activity via a synergistic effect. Moreover, this high catalytic activity remained almost unchanged after five catalytic cycles, indicating that the Au@MOF nanocatalysts possessed long-term stability with these reaction conditions.

3.2 Metal-Oxide-NP@MOFs Core-Shell Heterostructures in Catalysis

In principle, the preparation methods of metal-oxide-NP@MOF core-shell heterostructures are similar to that of metal-NP@MOF core-shell heterostructures. However, to date, there have been only a few studies on the magnetic metal-oxide-NPs@MOF core-shell heterostructures for heterogeneous catalysis. The combination of magnetic metal oxide NPs (e.g., Fe_3O_4 and Co_3O_4) and porous MOFs appears to be of tremendous interest owing to the versatility of the collective functionalities [31,32]. This type of core-shell heterostructure not only shows high catalytic efficiency, but also can be separated and recovered efficiently from the reaction solution by simply applying an external magnetic field. In the recent work of Zhu and coworkers, novel Fe_3O_4@MIL-100 (Fe) core-shell magnetic microspheres have been fabricated through a simple step-by-step assembly strategy [31]. Catalytic results demonstrated that the as-synthesized Fe_3O_4@MIL-100(Fe) core-shell magnetic microspheres can be used as a general catalyst with remarkable activities for a number of Claisen-Schmidt condensation reactions. Furthermore, the core-shell magnetic microspheres can be easily separated from the reaction mixture and can be reused without significant change in the catalytic activity after five reaction cycles. Therefore, compared to traditional catalysts, such Fe_3O_4@MIL-100 (Fe) core-shell magnetic catalysts are ideal recyclable catalysts for an organic reaction that derives from a combination of a superparamagnetic Fe_3O_4 core for efficient magnetic separation and a porous MIL-100(Fe) shell for high catalytic activity. Based on a similar process, two other magnetic core-shell heterostructures, HKUST-1@Fe_3O_4 [33] and Fe_3O_4@ZIF-8 [34], were successfully obtained based on the magnetic Fe_3O_4 NPs decorated with carboxyl groups and modified with an anionic polyelectrolyte, respectively. The modified magnetic Fe_3O_4 cores can adsorb metal ions to initiate nucleation and then grow a thin layer of MOFs to form the magnetic core-shell heterostructures. HKUST-1@Fe_3O_4 core-shell heterostructures were tested as highly efficient catalysts for the oxidation of benzylic C—H bonds up to 99% conversion and 99% selectivity (Fig. 7.5) [33], while Fe_3O_4@ZIF-8 core-shell heterostructures as catalysts could be easily filled into a capillary microreactor with the help of an external magnetic field [34]. Moreover, the microreactor showed excellent catalytic activity at a shorter residence time for the Knoevenagel condensation reaction of benzaldehyde and ethyl cyanoacetate.

In addition to these above metal-NP@MOFs and metal-oxide-NP@MOF core-shell heterostructures, some other types of NPs have also been embedded into the MOFs to construct NP@MOF core-shell heterostructures for catalysis. For example, Dong et al. reported a facile self-assembly strategy for fabrication of a copper hydroxysulfates@MOF (CHS@$Cu_3(BTC)_2$) core-shell structure [35]. The as-synthesized CHS@$Cu_3(BTC)_2$ core-shell structure exhibited excellent catalytic performance for an aldehyde acetalization reaction with high yields and selectivities. The core-shell CHS@$Cu_3(BTC)_2$ structure showed cooperative activity and reusability by integrating not only the CHS core with high catalytic activity, but also the $Cu_3(BTC)_2$ shell with large area and high porosity. Webley and coworkers reported a novel water dispersible

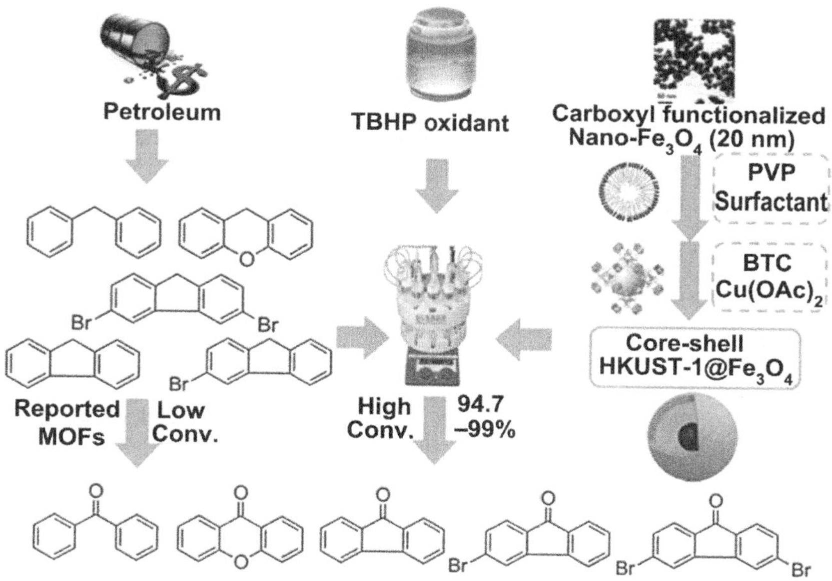

FIG. 7.5 Scheme of the oxidation of benzylic hydrocarbons over HKUST-1@Fe_3O_4 core-shell heterostructures. *(Reprinted from Y. Chen, X. Huang, X. Feng, J. Li, Y. Huang, J. Zhao, et al., Facile fabrication of magnetically recyclable metal-organic framework nanocomposites for highly efficient and selective catalytic oxidation of benzylic C-H bonds, Chem. Commun. 50 (2014) 8374–8377, Copyright 2014, with permission of the Royal Society of Chemistry.)*

polymer@MOF (P@MOF) core-shell composite NPs by a two-step approach at ambient conditions [36]. The Pd-P@MOF exhibited higher activities than that of Pd-MOF for the reduction reaction of 4-NP to 4-AP, which is because to its lower mass transfer resistance. Considering the excellent properties and limitless choice of both NPs and MOFs, it is expected that this new type of core-shell heterostructures will open a new doorway in the field of catalysis.

3.3 Two-or-More-Component-NP@MOF Core-Shell Heterostructures in Catalysis

Integration of two or more components into a greater assemblage or object for applications poses a major challenge to materials chemists. Encapsulation of different NPs with different functions has been found to be an attractive nanocomposite system, not only due to their enhancement of the intrinsic performance but also because of their novel properties and new functions that cannot be achieved for the single-component-NP-core systems [27]. Therefore, encapsulation of two or more components into MOFs has become an attractive area for the multifunctional nanomaterials.

Controllable synthesis of the two-or-more-component-NP@MOF core-shell structured catalysts remains at a nascent stage. To obtain better synthetic controllability, this type of core-shell heterostructure should be prepared in a stepwise manner [37]. Multifunctional Au-Fe_3O_4@MIL-100(Fe) core-shell catalysts were synthesized by a versatile stepwise approach [38,39]. The MOF shell thickness in the core-shell structures can be controlled by adjusting the assembly cycle number (Fig. 7.6A–J). This core-shell magnetic catalyst shows excellent dual functions and not only can undergo rapid catalytic reduction of nitrobenzene but also can be easily recycled by an external magnetic field. The values of kinetic rate constant k for Au-Fe_3O_4@MIL-100(Fe) catalysts with five assembly cycles was calculated to be 5.53 min^{-1}, which is 2 and 168 times higher than that of pure Au-Fe_3O_4 and MIL-100(Fe), respectively [38]. To deepen the understanding of the role of the core-shell structures in the enhancement of the reduction of nitrobenzene over Au-Fe_3O_4@MIL-100(Fe), the core-shell magnetic catalysts were investigated by x-ray absorption spectroscopy (XAS) (Fig. 7.6L and M). The XAS results showed that electron donation from MIL-100(Fe) to Au NPs made MIL-100(Fe) more acidic, which facilitated the activation of H_2 through the Lewis acid to base interaction, leading to improved catalytic activity of the core-shell catalysts. Very recently, perfect Pd/ZnO@ZIF-8 [40] and metal-mSiO_2@ZIF-8 core-shell nanostructures (metal = Pt, Pd, Ru, Ag, $Pt_{53}Ru_{47}$) [37] were reported by a versatile stepwise approach. The Pd/ZnO@ZIF-8 core-shell microsphere was fabricated by a novel ZnO support-induced encapsulation strategy [40]. ZnO not only served as the support for the trapping of Pd NPs, but also provided nucleation and growth sites for the growth of the ZIF-8 shell by producing Zn(II) sources.

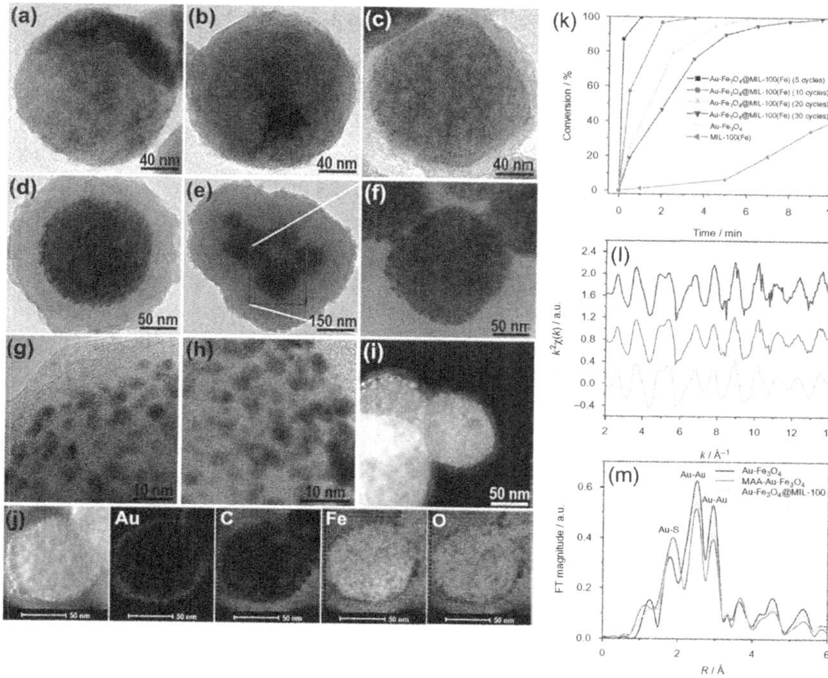

FIG. 7.6 TEM images of Au-Fe_3O_4 (A) and Au-Fe_3O_4@MIL-100(Fe) core-shell magnetic NPs after 5 (B), 10 (C), 20 (D), and 30 (E and F) assembly cycles. HRTEM and HAADF-STEM images of Au-Fe_3O_4@MIL-100(Fe) after 10 (G, I) and 20 (H, J) assembly cycles and the corresponding elemental mappings. Catalytic conversion of 4-NP to 4-AP over magnetic core-shell NPs (K). Au L_3-edge k^2-weighted extended X-ray absorption fine structure oscillations [$k^2\chi(k)$] (L) and their Fourier transforms (M) core-shell magnetic NPs. *(Reprinted from F. Ke, L. Wang, J. Zhu, Multifunctional Au-Fe_3O_4@MOF core-shell nanocomposite catalysts with controllable reactivity and magnetic recyclability, Nanoscale 7 (2015) 1201–1208, Copyright 2015, with permission of the Royal Society of Chemistry.)*

The as-synthesized Pd/ZnO@ZIF-8 core-shell nanostructures displayed excellent performance in terms of size-selectivity and antipoisoning in the liquid-phase hydrogenation of alkenes. The core-shell catalyst showed high activity for the hydrogenation of 1-hexene (1.7 Å) but no conversion for the hydrogenation of cyclooctene (5.4 Å). The prepared metal-mSiO_2@ZIF-8 core-shell nanocatalysts also showed selective-catalysis activity in liquid hydrogenation of alkenes. Before the coating of ZIF-8, the Pd-mSiO_2 catalysts exhibited good catalytic activity for both styrene (100% conversion) and cyclohexene (100% conversion) within 1 h at room temperature. After the coating of ZIF-8, the Pd-mSiO_2@ZIF-8 core-shell nanocatalysts still exhibited a good activity for styrene (71.1% conversion), but the conversion of cyclohexene was only 5.7% [37].

3.4 NP@MOFs Core-Shell Heterostructures in Photocatalysis

With the rapid development of industry, environmental pollution and energy shortages have led to an awareness of a potential global crisis [41]. Therefore, the development of pollution-free technologies has become an urgent task. Recently, photocatalysis has received much attention due to its low cost, ease of use, and environmental friendliness [42]. MOFs can act as photocatalysts for light absorption and excited-state applications in energy conversion processes. If the MOF is employed as a light-sensitive semiconductor, the metal-oxide cluster of the framework can be considered as discrete quantum dots (QDs), which are stabilized and interconnected by the organic linkers acting as photon antennae [43]. However, the photoactivity of MOFs is not as effective as that of traditional inorganic semiconductor NPs. If one combines MOFs and traditional semiconductor NPs together, the semiconductor@MOFs core-shell photocatalyst are expected to possess advanced performance originating from the two components. In this regard, the encapsulation of inorganic semiconductor NPs in MOFs has been reported previously. Qiu and coworkers reported a novel magnetic recyclable Fe_3O_4@MIL-100(Fe) core-shell photocatalyst for the photodegradation of methylene blue (MB) dye (Fig. 7.7) [44]. Moreover, the

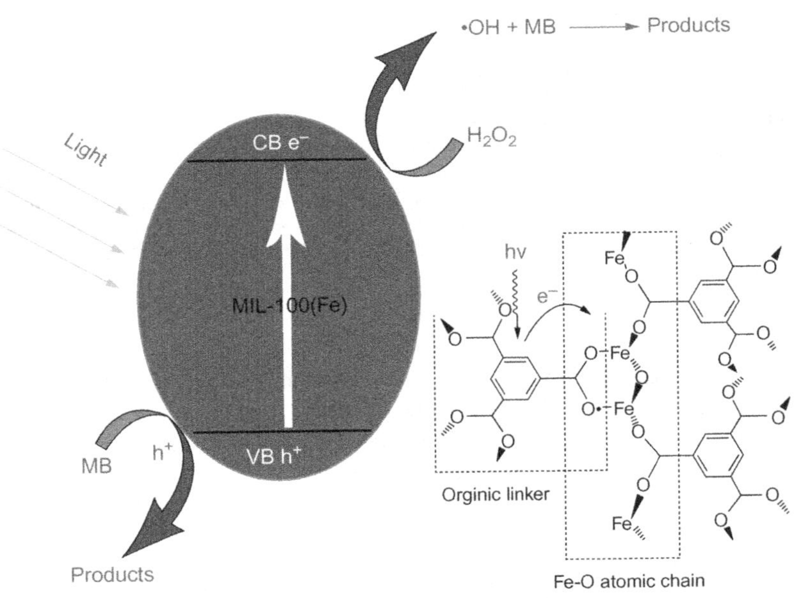

FIG. 7.7 Schematic illustration of MIL-100(Fe) photocatalyst for the photodegradation of MB dye under light irradiation. *(Reprinted from C.-F. Zhang, L.-G. Qiu, F. Ke, Y.-J. Zhu, Y.-P. Yuan, G.-S. Xu, et al., A novel magnetic recyclable photocatalyst based on a core-shell metal-organic framework Fe_3O_4@MIL-100(Fe) for the decolorization of methylene blue dye, J. Mater. Chem. A 1 (2013) 14329–14334, Copyright 2013, with permission of the Royal Society of Chemistry.)*

core-shell photocatalyst showed remarkable photocatalytic activity for the MB photodegradation over traditional TiO_2 and C_3N_4 photocatalysts under UV-vis light and visible light irradiation, respectively. Later, Gong and coworkers prepared ZnO@ZIF-8 core-shell nanospheres by the self-template method, and the photocatalysis of ZnO@ZIF-8 was also investigated by the photocatalytic degradation of MB under UV irradiation [45]. The activity of ZnO@ZIF-8 core-shell nanospheres was shown to be similar to ZIF-8 but higher than that of ZnO during the degradation of MB, which was due to the molecule size selectivity of the ZIF-8 shell. Because the molecule size of MB was larger than the pore size of ZIF-8, the photodegradation was mainly caused by the ligand-to-metal charge transfer (LMCT) mechanism of the ZIF-8 shell.

Besides these traditional metal oxide semiconductor NPs, semiconductor QDs have broad absorption spectra. They have attracted much attention in recent years due to their high luminescence quantum yields, large extinction coefficients, and excellent photostabilities [46]. Their optical properties especially can be systematically tunable via changes in their particle sizes. Compared with pure QDs, the QDs@MOF core-shell heterostructures have shown great advantages because of their synergistic effect [8]. In QDs@MOF core-shell heterostructures, QD particles can be stabilized against photochemical degradation, and the photocatalytic performance can also be enhanced due to the enhanced light absorption, more efficient separation of photogenerated electron-hole pairs, and increased surface area [42]. Up to now, there have been a few pioneering studies on QDs@MOF heterostructures for enhanced light harvesting or size-selective molecular sensing applications [23,46–48]. However, the utilization of QDs@MOF heterostructures as photocatalysts has rarely been reported. Recently, Zhu and Wang reported a novel system based on CdS-based QDs embedded on MOFs (e.g., CdS-MIL-100(Fe) [42], CdS/MIL-101(Cr) [49]). This system resulted in a significant increase in the photocatalytic efficiency of CdS. CdS-MIL-100(Fe) has been synthesized by a facile solvothermal method [42]. The resulting CdS-MIL-100(Fe) exhibited excellent photostability and enhanced photocatalytic activity with respect to the selective oxidation of benzyl alcohol to benzaldehyde under visible-light irradiation. The mechanism of synergistically enhanced photocatalytic activity of CdS after MOF hybridization may be explained as follows (Fig. 7.8): Under visible-light irradiation, CdS and MOFs are both photoexcited to generate charge carriers. The photogenerated electrons of CdS can be transferred to MOFs, then the electrons react with adsorbed oxygen to form superoxide radicals. At the same time, the adsorbed organic reactive substrates interact with the holes to form corresponding radical cations, which further react with superoxide radicals to give rise to the final products. CdS-MIL-101(Cr) has been also synthesized by the solvothermal method [49]. MIL-101(Cr) did not exhibit any photocatalytic activity. Significantly, the photocatalytic activity of CdS embedded on MIL-101(Cr) for hydrogen evolution under visible light was highly enhanced in comparison with the pure CdS.

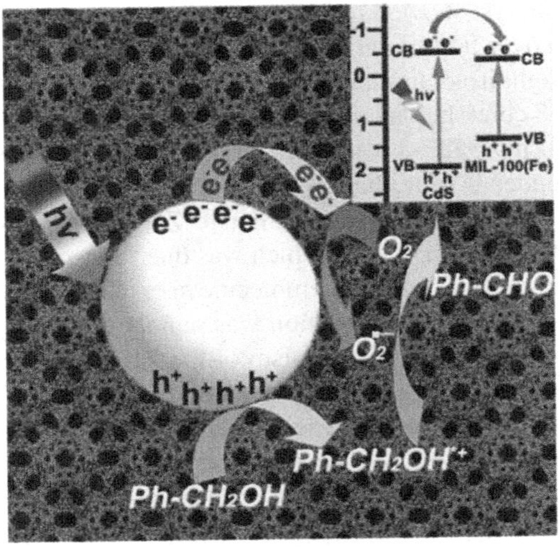

FIG. 7.8 Schematic illustration of CdS-MIL-100(Fe) nanocomposites for the selective oxidation of benzyl alcohol to benzaldehyde under visible-light irradiation. *(Reprinted from F. Ke, L.H. Wang, J.F. Zhu, Facile fabrication of CdS-metal-organic framework nanocomposites with enhanced visible-light photocatalytic activity for organic transformation, Nano Res. 8 (2015) 1834–1846, Copyright 2015, with permission of Springer.)*

3.5 Enzyme@MOF Heterostructures in Catalysis

Enzymes are naturally evolved biocatalysts, which possess great potential as environmentally friendly catalysts in various industries, such as pharmaceuticals, fuel production, and food processing [50]. However, the engineering of enzymes is severely hampered by their low operational stability and difficult recovery and reuse. Enzyme immobilization has been proven as an effective way to enhance enzyme stability as well as to facilitate their separation and recovery for reuse while maintaining their outstanding activity and selectivity [51]. In this respect, extensive attention has been paid to develop novel types of porous materials for enzyme encapsulation. As mentioned before, MOFs are a class of porous materials with a wide range of unique functions, which can be decorated with functional organic groups for specific interactions with biomolecules. This feature makes them appealing to accommodate enzymes for catalytic applications [52].

In 2011, Ma et al. reported the successful immobilization of an enzyme (et al. microperoxidase-11, MP-11) into mesoporous Tb-TATB frameworks. The resulting MP-11@Tb-mesoMOF exhibited superior enzymatic catalytic performances for the conversion of 3,5-di-*t*-butyl-catechol (DTBC) to the corresponding *o*-quinone product [53]. The pore sizes of the mesoporous Tb-TATB were about 3.9 and 4.7 nm, which were large enough for MP-11encappsulation ($3.3 \times 1.7 \times 1.1$ nm). The values of the kinetic rate for MP-11@Tb-mesoMOF were calculated to be 7.58×10^{-5}mM/s, which were 2 and 29 times higher than that of MP-11@MCM-41 and pure Tb-mesoMOF, respectively. In the first 2 min, the free MP-11 in HEPES buffer solution showed a fast initial rate of

8.93×10^{-4}mM/s. However, the activity dropped rapidly after only 3 min due to the aggregation in solution. Moreover, the high reaction rate of MP-11@Tb-mesoMOF remained almost unchanged after six cycles, while the activity of MP-11@MCM-41 decreased abruptly with more than 60% activity lost after only the first cycle. Further, Shieh and coworkers reported a new concept to impart new functions to biocatalysts by combining enzymes and MOFs [54]. They used a de novo approach to embed enzyme (catalase, CAT) molecules into the ZIF-90 crystals, in which the ZIF-90 crystals were synthesized in the presence of catalase molecules (Fig. 7.9A). This method allowed MOFs with pore

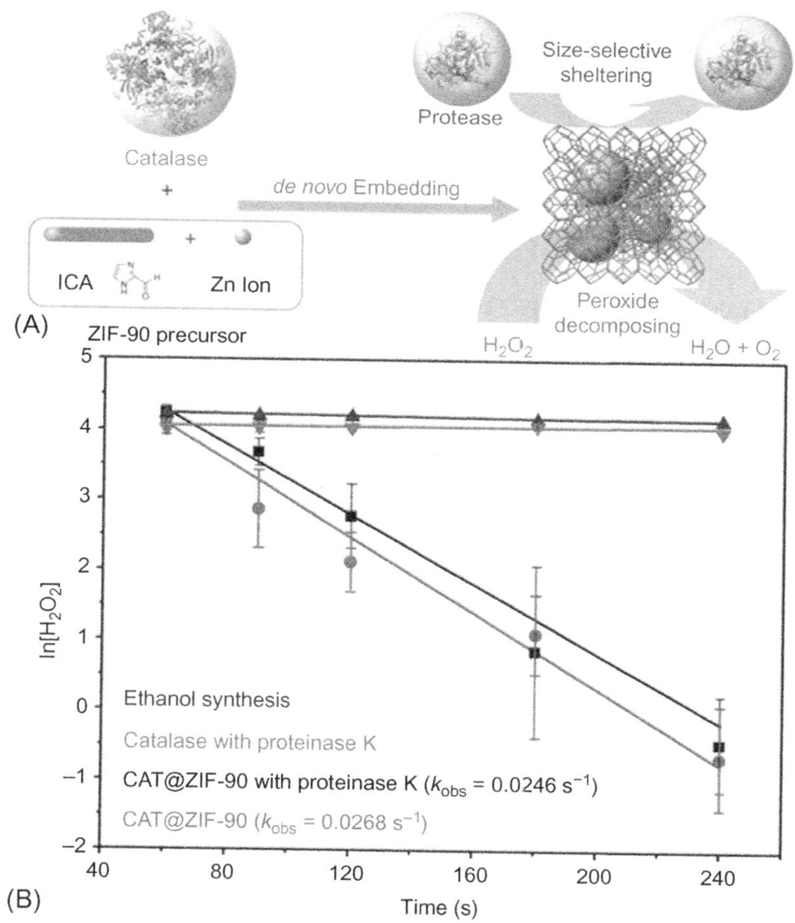

FIG. 7.9 (A) Water-based synthesized of ZIF-90 with encapsulated catalase enzyme and its functional activity. (B) Kinetics of degradation of H_2O_2. *(Reprinted from F.-K. Shieh, S.-C. Wang, C.-I. Yen, C.-C. Wu, S. Dutta, L.-Y. Chou, et al., Imparting functionality to biocatalysts via embedding enzymes into nanoporous materials by a de novo approach: size-selective sheltering of catalase in metal-organic framework microcrystals, J. Am. Chem. Soc. 137 (2015) 4276–4279, Copyright 2015, with permission of the American Chemical Society.)*

sizes smaller than the size of the enzymes to be used. The peroxide-decomposing activity of CAT@ZIF-90 was lower than for pure catalase, which might be caused by mass transport limitations or the nonoptimized interface between ZIF-90 and catalase. Significantly, the CAT@ZIF-90 showed a unique size-sheltering function to catalase and protected catalase from the inhibitor proteinase K. After incubated CAT@ZIF-90 and pure catalase with proteinase K, the pure catalase was inhibited immediately, while the activity of CAT@ZIF-90 was maintained (Fig. 7.9B). Very recently, Kim et al. reported the synthesis of functional biocompatible MIL-88A hollow spheres with versatility, high scalability, and fine size control by using a facile in-droplet microfluidic approach [55]. Particularly, to further ascertain the encapsulation capabilities, three different enzymes (glycerol dehydrogenase, horseradish peroxidase, and acetylcholinesterase) were encapsulated inside the single-shell hollow spheres under mild conditions. The catalytic activity and the recyclability of the enzymes@MIL-88A hollow composite spheres were superior to the nonencapsulated free enzymes. These examples introduced in this chapter have shown great promise of using such enzymes@MOF porous materials for catalysis in the near future, due to their increased enzyme stability and activity.

4 CONCLUSION AND OUTLOOK

In summary, the design and syntheses of MOFs-based core-shell heterostructures are promising and effective ways to achieve novel property synergies of inorganic NP cores and MOF shells. The most important advantages of NP@MOF heterostructures with respect to other core-shell materials, such as NP@SiO_2, NP@polymer, and NP@metal oxide are tunable pore sizes of the MOF shell, limitless selection of composition, and multifunctional properties of both NP cores and MOF shells. Because of these outstanding properties, NP@MOF core-shell heterostructures are promising materials for application in heterogeneous catalysis. The previous sections have illustrated various synthetic strategies for the NP@MOF core-shell heterostructures and many applications of these core-shell materials in heterogeneous catalysis. However, the design and construction of NP@MOF core-shell heterostructures for catalysis still remain at a nascent stage, and this type of core-shell material represents opportunities as well as some challenges in many areas of research. First, in order to further improve the catalytic performance of MOF-based core-shell catalysts, the development of general synthetic strategies for well-defined NP@MOF core-shell heterostructures is very important. Technically, it is easy to coat or graft polymers on the surface of MOF crystals. However, it is a great challenge to fabricate core-shell structures with an MOFs shell, which may be attributed to the higher tendency of homogeneous nucleation and crystal growth of MOFs in the solution phase rather than on the surface of NPs. Second, the ability of controlling the incorporated NPs' size, shape, and composition in the core-shell structures is highly desirable; these are often key parameters for application in catalysis with high activity and selectivity. The combination of different NP sizes, shapes, and compositions with MOFs for catalysis will be a primary research topic in the future. Third, although the catalytic activity of the MOF-based core-shell catalysts has been significantly improved, the mechanism of synergistically enhanced activity is still unclear. So far, the interaction between the NP cores and the MOF shells during the catalysis still is not well understood. Therefore, it remains an important challenge to understand the electronic structure and chemical bonding evolutions of the MOF-based core-shell catalysts. Though some challenges exist now, the rapid development of MOF-based core-shell heterostructures will open a new doorway for utilizing MOF-based materials in the field of heterogeneous catalysis.

ACKNOWLEDGMENTS

This work was supported by the National Natural Science Foundation of China (NSFC, 21501003, U1232102), Natural Science Foundation of Anhui Province (1608085QB27), China Postdoctoral Science Foundation funded project (2015M581973), the Scientific Research Foundation of Anhui Agricultural University (Grant No. yj2015-26), and the Youth Research Foundation of Anhui Agricultural University (2014zr001).

REFERENCES

[1] O.K. Farha, J.T. Hupp, Rational design, synthesis, purification, and activation of metal-organic framework materials, Acc. Chem. Res. 43 (2010) 1166–1175.

[2] H.C. Zhou, J.R. Long, O.M. Yaghi, Introduction to metal-organic frameworks, Chem. Rev. 112 (2012) 673–674.

[3] J.-R. Li, R.J. Kuppler, H.-C. Zhou, Selective gas adsorption and separation in metal-organic frameworks, Chem. Soc. Rev. 38 (2009) 1477–1504.

[4] P. Horcajada, R. Gref, T. Baati, P.K. Allan, G. Maurin, P. Couvreur, et al., Metal-organic frameworks in biomedicine, Chem. Rev. 112 (2012) 1232–1268.

[5] D. Farrusseng, S. Aguado, C. Pinel, Metal-organic frameworks: opportunities for catalysis, Angew. Chem. Int. Ed. 48 (2009) 7502–7513.

[6] S. Qiu, M. Xue, G. Zhu, Metal-organic framework membranes: from synthesis to separation application, Chem. Soc. Rev. 43 (2014) 6116–6140.

[7] A. Dhakshinamoorthy, M. Alvaro, H. Garcia, Commercial metal-organic frameworks as heterogeneous catalysts, Chem. Commun. 48 (2012) 11275–11288.

[8] Q.-L. Zhu, Q. Xu, Metal-organic framework composites, Chem. Soc. Rev. 43 (2014) 5468–5512.

[9] X. Gu, Z.-H. Lu, H.-L. Jiang, T. Akita, Q. Xu, Synergistic catalysis of metal-organic framework-immobilized Au-Pd nanoparticles in dehydrogenation of formic acid for chemical hydrogen storage, J. Am. Chem. Soc. 133 (2011) 11822–11825.

[10] Q.-L. Zhu, J. Li, Q. Xu, Immobilizing metal nanoparticles to metal-organic frameworks with size and location control for optimizing catalytic performance, J. Am. Chem. Soc. 135 (2013) 10210–10213.

[11] Y.-Z. Chen, Q. Xu, S.-H. Yu, H.-L. Jiang, Tiny Pd@Co core-shell nanoparticles confined inside a metal-organic framework for highly efficient catalysis, Small 11 (2015) 71–76.

[12] Y. Liu, W. Zhang, S. Li, C. Cui, J. Wu, H. Chen, et al., Designable yolk-shell nanoparticle@MOF petalous heterostructures, Chem. Mater. 26 (2014) 1119–1125.

[13] J. Zhou, P. Wang, C. Wang, Y.T. Goh, Z. Fang, P.B. Messersmith, et al., Versatile core-shell nanoparticle@metal-organic framework nanohybrids: exploiting mussel-inspired polydopamine for tailored structural integration, ACS Nano 9 (2015) 6951–6960.

[14] L. Chen, B. Huang, X. Qiu, X. Wang, R. Luque, Y. Li, Seed-mediated growth of MOF-encapsulated Pd@Ag core-shell nanoparticles: toward advanced room temperature nanocatalysts, Chem. Sci. 7 (2016) 228–233.

[15] Y. Chen, Z. Xiong, L. Peng, Y. Gan, Y. Zhao, J. Shen, et al., Facile preparation of core-shell magnetic metal organic framework nanoparticles for the selective capture of phosphopeptides, ACS Appl. Mater. Interfaces 7 (2015) 16338–16347.

[16] C. Jo, H.J. Lee, M. Oh, One-pot synthesis of silica@coordination polymer core-shell microspheres with controlled shell thickness, Adv. Mater. 23 (2011) 1716–1719.

[17] K.A. McDonald, J.I. Feldblyum, K. Koh, A.G. Wong-Foy, A.J. Matzger, Polymer@MOF@MOF: "grafting from" atom transfer radical polymerization for the synthesis of hybrid porous solids, Chem. Commun. 51 (2015) 11994–11996.

[18] R. Chen, J. Zhang, Y. Wang, X. Chen, J.A. Zapien, C.-S. Lee, Graphitic carbon nitride nanosheet@metal-organic framework core-shell nanoparticles for photo-chemo combination therapy, Nanoscale 7 (2015) 17299–17305.

[19] Y. Li, J. Tang, L. He, Y. Liu, Y. Liu, C. Chen, et al., Core-shell upconversion nanoparticle@metal-organic framework nanoprobes for luminescent/magnetic dual-mode targeted imaging, Adv. Mater. 27 (2015) 4075–4080.

[20] P. Hu, J.V. Morabito, C.-K. Tsung, Core-shell catalysts of metal nanoparticle core and metal-organic framework shell, ACS Catal. 4 (2014) 4409–4419.

[21] F. Ke, L.-G. Qiu, Y.-P. Yuan, X. Jiang, J.-F. Zhu, Fe_3O_4@MOF core-shell magnetic microspheres with a designable metal-organic framework shell, J. Mater. Chem. 22 (2012) 9497–9500.

[22] F. Ke, J. Zhu, L.-G. Qiu, X. Jiang, Controlled synthesis of novel Au@MIL-100(Fe) core-shell nanoparticles with enhanced catalytic performance, Chem. Commun. 49 (2013) 1267–1269.

[23] G. Lu, S.Z. Li, Z. Guo, O.K. Farha, B.G. Hauser, X.Y. Qi, et al., Imparting functionality to a metal-organic framework material by controlled nanoparticle encapsulation, Nat. Chem. 4 (2012) 310–316.

[24] H.J. Lee, W. Cho, M. Oh, Advanced fabrication of metal-organic frameworks: template-directed formation of polystyrene@ZIF-8 core-shell and hollow ZIF-8 microspheres, Chem. Commun. 48 (2012) 221–223.

[25] L. He, Y. Liu, J. Liu, Y. Xiong, J. Zheng, Y. Liu, et al., Core-shell noble-metal@metal-organic-framework nanoparticles with highly selective sensing property, Angew. Chem. Int. Ed. 52 (2013) 3741–3745.

[26] W.-w. Zhan, Q. Kuang, J.-z. Zhou, X.-j. Kong, Z.-x. Xie, L.-s. Zheng, Semiconductor@metal-organic framework core-shell heterostructures: acase of ZnO@ZIF-8 nanorods with selective photoelectrochemical response, J. Am. Chem. Soc. 135 (2013) 1926–1933.

[27] Y. Liu, Z. Tang, Multifunctional nanoparticle@MOF core-shell nanostructures, Adv. Mater. 25 (2013) 5819–5825.

[28] F. Ke, L. Wang, J. Zhu, An efficient room temperature core-shell AgPd@MOF catalyst for hydrogen production from formic acid, Nanoscale 7 (2015) 8321–8325.

[29] C.-H. Kuo, Y. Tang, L.-Y. Chou, B.T. Sneed, C.N. Brodsky, Z. Zhao, et al., Yolk-shell nanocrystal@ZIF-8 nanostructures for gas-phase heterogeneous catalysis with selectivity control, J. Am. Chem. Soc. 134 (2012) 14345–14348.

[30] K. Na, K.M. Choi, O.M. Yaghi, G.A. Somorjai, Metal nanocrystals embedded in single nanocrystals of MOFs give unusual selectivity as heterogeneous catalysts, Nano Lett. 14 (2014) 5979–5983.

[31] F. Ke, L.-G. Qiu, J. Zhu, Fe_3O_4@MOF core-shell magnetic microspheres as excellent catalysts for the Claisen-Schmidt condensation reaction, Nanoscale 6 (2014) 1596–1601.

[32] W. Wang, Y. Li, R. Zhang, D. He, H. Liu, S. Liao, Metal-organic framework as a host for synthesis of nanoscale Co_3O_4 as an active catalyst for CO oxidation, Catal. Commun. 12 (2011) 875–879.

[33] Y. Chen, X. Huang, X. Feng, J. Li, Y. Huang, J. Zhao, et al., Facile fabrication of magnetically recyclable metal-organic framework nanocomposites for highly efficient and selective catalytic oxidation of benzylic C-H bonds, Chem. Commun. 50 (2014) 8374–8377.

[34] T. Zhang, X. Zhang, X. Yan, L. Kong, G. Zhang, H. Liu, et al., Synthesis of Fe_3O_4@ZIF-8 magnetic core-shell microspheres and their potential application in a capillary microreactor, Chem. Eng. J. 228 (2013) 398–404.

[35] H. Gao, Y. Luan, K. Chaikittikul, W. Dong, J. Li, X. Zhang, et al., A facile in situ self-assembly strategy for large-scale fabrication of CHS@MOF yolk/shell structure and its catalytic application in a flow system, ACS Appl. Mater. Interfaces 7 (2015) 4667–4674.

[36] K. Xie, Q. Fu, Y. He, J. Kim, S.J. Goh, E. Nam, et al., Synthesis of well dispersed polymer grafted metal-organic framework nanoparticles, Chem. Commun. 51 (2015) 15566–15569.

[37] B. Xi, Y.C. Tan, H.C. Zeng, A general synthetic approach for integrated nanocatalysts of metal-silica@ZIFs, Chem. Mater. 28 (2016) 326–336.

[38] F. Ke, L. Wang, J. Zhu, Multifunctional Au-Fe_3O_4@MOF core-shell nanocomposite catalysts with controllable reactivity and magnetic recyclability, Nanoscale 7 (2015) 1201–1208.

[39] H.-j. Zhang, S.-d. Qi, X.-y. Niu, J. Hu, C.-l. Ren, H.-l. Chen, et al., Metallic nanoparticles immobilized in magnetic metal-organic frameworks: preparation and application as highly active, magnetically isolable and reusable catalysts, Catal. Sci. Technol. 4 (2014) 3013–3024.

[40] L. Lin, T. Zhang, H. Liu, J. Qiu, X. Zhang, In situ fabrication of a perfect Pd/ZnO@ZIF-8 core-shell microsphere as an efficient catalyst by a ZnO support-induced ZIF-8 growth strategy, Nanoscale 7 (2015) 7615–7623.

[41] H. Tong, S. Ouyang, Y. Bi, N. Umezawa, M. Oshikiri, J. Ye, Nano-photocatalytic materials: possibilities and challenges, Adv. Mater. 24 (2012) 229–251.

[42] F. Ke, L.H. Wang, J.F. Zhu, Facile fabrication of CdS-metal-organic framework nanocomposites with enhanced visible-light photocatalytic activity for organic transformation, Nano Res. 8 (2015) 1834–1846.

[43] C.-K. Lin, D. Zhao, W.-Y. Gao, Z. Yang, J. Ye, T. Xu, et al., Tunability of band gaps in metal-organic frameworks, Inorg. Chem. 51 (2012) 9039–9044.

[44] C.-F. Zhang, L.-G. Qiu, F. Ke, Y.-J. Zhu, Y.-P. Yuan, G.-S. Xu, et al., A novel magnetic recyclable photocatalyst based on a core-shell metal-organic framework Fe_3O_4@MIL-100(Fe) for the decolorization of methylene blue dye, J. Mater. Chem. A 1 (2013) 14329–14334.

[45] B. Yu, F. Wang, W. Dong, J. Hou, P. Lu, J. Gong, Self-template synthesis of core-shell ZnO@ZIF-8 nanospheres and the photocatalysis under UV irradiation, Mater. Lett. 156 (2015) 50–53.

[46] S. Jin, H.-J. Son, O.K. Farha, G.P. Wiederrecht, J.T. Hupp, Energy transfer from quantum dots to metal-organic frameworks for enhanced light harvesting, J. Am. Chem. Soc. 135 (2013) 955–958.

[47] D. Esken, S. Turner, C. Wiktor, S.B. Kalidindi, G. Van Tendeloo, R.A. Fischer, GaN@ZIF-8: selective formation of gallium nitride quantum dots inside a zinc methylimidazolate framework, J. Am. Chem. Soc. 133 (2011) 16370–16373.

[48] D. Buso, J. Jasieniak, M.D.H. Lay, P. Schiavuta, P. Scopece, J. Laird, et al., Highly luminescent metal-organic frameworks through quantum dot doping, Small 8 (2012) 80–88.

[49] J. He, Z. Yan, J. Wang, J. Xie, L. Jiang, Y. Shi, et al., Significantly enhanced photocatalytic hydrogen evolution under visible light over CdS embedded on metal-organic frameworks, Chem. Commun. 49 (2013) 6761–6763.

[50] D. Feng, T.-F. Liu, J. Su, M. Bosch, Z. Wei, W. Wan, et al., Stable metal-organic frameworks containing single-molecule traps for enzyme encapsulation, Nat. Commun. 6 (2015) 5979.

[51] X. Wu, M. Hou, J. Ge, Metal-organic frameworks and inorganic nanoflowers: a type of emerging inorganic crystal nanocarrier for enzyme immobilization, Catal. Sci. Technol. 5 (2015) 5077–5085.

[52] X. Wu, J. Ge, C. Yang, M. Hou, Z. Liu, Facile synthesis of multiple enzyme-containing metal-organic frameworks in a biomolecule-friendly environment, Chem. Commun. 51 (2015) 13408–13411.

[53] V. Lykourinou, Y. Chen, X.-S. Wang, L. Meng, H. Tran, L.-J. Ming, et al., Immobilization of MP-11 into a mesoporous metal-organic framework, MP-11@mesoMOF: a new platform for enzymatic catalysis, J. Am. Chem. Soc. 133 (2011) 10382–10385.

[54] F.-K. Shieh, S.-C. Wang, C.-I. Yen, C.-C. Wu, S. Dutta, L.-Y. Chou, et al., Imparting functionality to biocatalysts via embedding enzymes into nanoporous materials by a de novo approach: size-selective sheltering of catalase in metal-organic framework microcrystals, J. Am. Chem. Soc. 137 (2015) 4276–4279.

[55] G.-Y. Jeong, R. Ricco, K. Liang, J. Ludwig, J.-O. Kim, P. Falcaro, et al., Bioactive MIL-88A framework hollow spheres via interfacial reaction in-droplet microfluidics for enzyme and nanoparticle encapsulation, Chem. Mater. 27 (2015) 7903–7909.

Chapter 8

Encapsulated Porphyrins as Catalysts for Organic Synthesis

Pietro Tagliatesta and Marilena Carbone
University of Rome Tor Vergata, Rome, Italy

Contents

1 INTRODUCTION

Much attention has been recently devoted to the biomimetic catalysis by synthetic metalloporphyrins encapsulated in organic or inorganic cages [1,2]. The encapsulation has several advantages: by mimicking the properties of the natural enzymes, it enables the transformations of organic substrates at higher speed and larger stereoselectivity [3].

Synthetic metalloporphyrins in different assemblies [4] are well known for their catalytic properties in many important organic reactions, such as, for example, the oxidation of organic substrates [5–12], the cyclopropanation of olefins [13–19], the carbonyl ylide 1,3-dipolar cycloaddition reactions of α-diazoketones [20], the insertion of carbene into the S—H bond [21,22], the amination of hydrocarbons [23–25], and the olefination of aldehydes [26]. All the above cited reactions were performed in homogeneous organic solutions and in many cases the recycling of the catalyst, if possible, must undergo a tedious chromatographic recovering.

Furthermore many porphyrin catalysts are very sensitive to the reagents used in the reactions, such as hydrogen peroxide, diazocompounds, etc. and are inactivated giving a low TON (turnover number). This fact can be sometimes avoided by using a scaffold of organic or inorganic that protects the core

Encapsulated Catalysts. http://dx.doi.org/10.1016/B978-0-12-803836-9.00008-0

of the catalyst from the side reactions caused by the reagents or the reactive intermediates. The natural enzymes are very efficient in directing the reactions to give excellent stereochemical selectivities, and the goal of the research in this field is to reach such results using catalysts bearing active metals enclosed in well-established structures. The steric hindrance of the active site of the catalyst can afford good selectivities with high TON. Recently, the direction of the research in this field was to connect the properties of the soluble organic catalysts with those of inorganic materials showing well defined three-dimensional structures. For such a purpose active metalloporphyrins were encapsulated in different inorganic materials, providing good catalysts for several organic reactions. Such materials span from zeolites to molecular sieves and other well-shaped inorganic compounds. Furthermore, some other catalysts were built by the interactions of the metalloporphyrins with small molecules to provide adducts that are useful for giving good stereochemical results. The aim of this chapter is to present the contributions of different research groups in this field, looking at the literature of the last 20 years (up to 2015). We hope to stimulate new ideas in the new generation of researchers.

2 PORPHYRINS ENCAPSULATED IN MOLECULAR SIEVES

Mesoporous silicate sieves of the type MCM-41 are good candidates for obtaining encapsulated porphyrin catalysts. In 1998 Stein and coworkers [27] reported on the catalytic properties of *meso*-tetrakis(5-trimethylammoniopentyl)porphyrin (TMAP) that was encapsulated in MCM-41 by the hydrothermal synthesis or by a surfactant ⇔ porphyrin ion-exchange reaction with a previously synthesized molecular sieve. The metalloporphyrin was retained inside the sieves by both the two methods and further metalated with Cu^{2+}, Ni^{2+}, or Fe^{2+}. Finally a MCM-41 intercalated copper porphyrin was used as a catalyst for the peroxidation of α-naphthol violet by hydrogen peroxide. The segregation of each molecule of the metalloporphyrin prevents its self-oxidation, giving better results and stabilizing the catalyst by its incorporation in the channels of the sieves. Such approach has been used by Xia and coworkers [28] by encapsulating cationic Mn(III) porphyrins into MCM-41 and DMY, which is an NaY zeolite (Fig. 8.1).

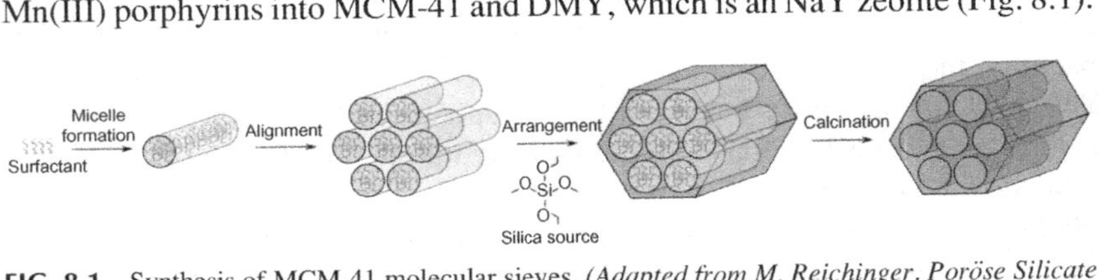

FIG. 8.1 Synthesis of MCM-41 molecular sieves. *(Adapted from M. Reichinger, Poröse Silicate mit hierarchischer Porenstruktur: Synthese von mikro-/mesoporösen MCM-41 und MCM-48 Materialien aus zeolithischen Baueinheiten des MFI-Gerüst strukturtyps, PhD Thesis, Institut für Geologie, Mineralogie und Geophysik, Bochum, Germany, 2007, p. 10; J.S. Beck, J.C. Vartuli, W.J. Roth, M.E. Leonowicz, C.T. Kresge, K.D. Schmitt, C.T.W. Chu, D.H. Olson, E.W. Sheppard, S.B. McCullen, J.B. Higgins, J.L. Schlenker, J. Am. Chem. Soc 114 (1992) 10834–10843 with permission of the American Chemical Society.)*

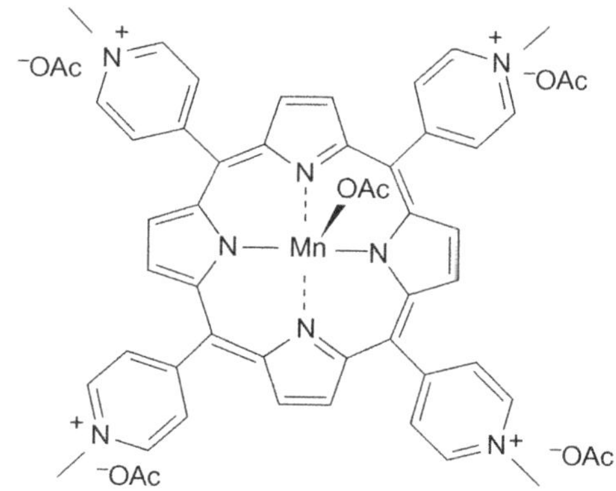

FIG. 8.2 [*meso*-Tetrakis(1-methyl-4-pyridinio)porphyrinato] manganese(III)pentaacetate, MnT-MPyP. *(Adapted from Z. Li, C.-G. Xia, X.-M. Zhang, Preparation and catalysis of DMY and MCM-41 encapsulated cationic Mn(III)-porphyrin complex. J. Mol. Catal. A: Chem. 185 (2002) 47–56 with permission of Elsevier.)*

The used metalloporphyrin was the 5,10,15,20-tetrakis(1-methyl-4-piridinio)porphyrinato manganese(III) pentaacetate (see Fig. 8.2) and was inserted in the inorganic materials by simple absorption in a heterogeneous solution. The obtained complexes have been used as oxidation catalysts in the epoxidation of styrene and cyclohexene by iodosilbenzene mimicking the function of the cytochrome P450 enzyme [29]. The encapsulated catalyst was more efficient and selective than the simple metalloporphyrins in a homogeneous solution. This fact can again explain the effect of the environment in which the metalloporphyrin works. Furthermore, the bleaching and leaching of the catalyst by the oxidant was low after few recycling, but prolonged reuse resulted in the loss of activity.

Nur and his group [30] reported the encapsulation of an iron porphyrin in mesoporous A1-MCM-41 with the ratio Si/Al = 20 or poly(methacrylic acid) and its use for the oxidation of benzene to phenol. The metalloporphyrin used in this investigation was the iron(III)-*tetra*-(4-pyridyl)-porphyrin whose structure is depicted in Fig. 8.3.

The encapsulation was obtained by standard methods and the obtained catalysts were tested in the oxidation of benzene by hydrogen peroxide as an oxygen source. The activity of the catalysts was much higher than that of the simple metalloporphyrin in a homogeneous solution.

The authors claim that this effect was due to the coordination of the metalloporphyrin to the sieve or polymer. Between the two reported catalytic systems, the polymeric material with an encapsulated catalyst was more active than the molecular sieve containing porphyrin in the first experiment, but the reuse experiments showed a loss of activity for the former system. This is probably due to the leaching of the porphyrin from the polymeric material.

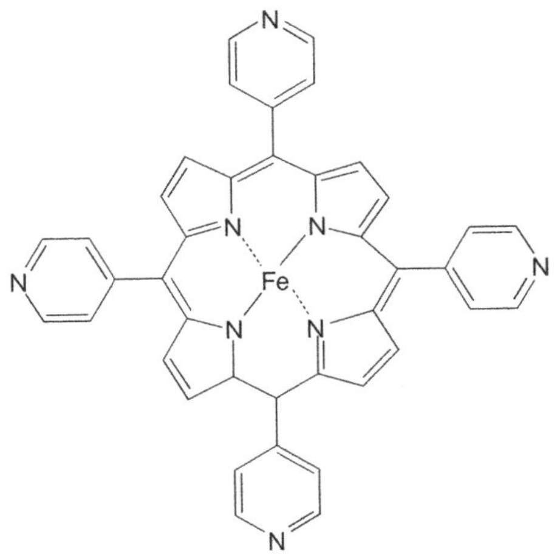

FIG. 8.3 Iron(III)-*tetra*-(4-pyridyl)-porphyrin. *(Adapted from H. Nur, H. Hamid, S. Endud, H. Hamdan, Z. Ramli, Iron-porphyrin encapsulated in poly(methacrylic acid) and mesoporous Al-MCM-41 as catalyst in the oxidation of benzene to phenol. Mater. Chem. Phys. 96 (2006) 337–342 with permission of Elsevier.)*

A similar approach has been used for immobilizing the Co, Fe, and Mn tetraphenylporphyrin into MCM-41 and using the catalyst for the epoxidation of cyclohexene by hydrogen peroxide [31]. The characterization of the obtained systems showed that the sylanol groups present in the inorganic material were able to anchor the metalloporphyrins.

The oxidation results showed a low conversion of the substrate that went from 11.8 in the case of cobalt to 25.8 for iron, with selectivities more pronounced toward cyclohexanol and cyclohexenone than to the epoxycyclohexane. These results were similar to those reported in the literature [32] and seemed to be governed by the diffusion of the reagent into the mesopores of the inorganic support.

Al-MCM-41 and V-MCM-41 molecular sieves containing different amounts of Al and V were used for the encapsulation of cationic manganese(III) *meso*-tetrakis(4-trimethylammoniophenyl)porphyrin (MnTAPP) [33]. Such materials were tested in the oxidation of cyclohexene and styrene by iodosylbenzene (PhIO). The reaction products were different and depended on the catalyst. With an Al-MCM-41 catalyst only allylic oxidation products were detected from styrene with no trace of the epoxide, while in the case of cyclohexene both epoxide and allylic oxidation products were found. In the case of a V-MCM-41 catalyst with styrene, the epoxide and a small quantity of allylic oxidation were obtained, while with cyclohexene both products were found. The obtained conversion was influenced by the charge of vanadium and aluminum in the support but the selectivities remain almost unchanged. The catalysts were recycled three times without a large loss of activity.

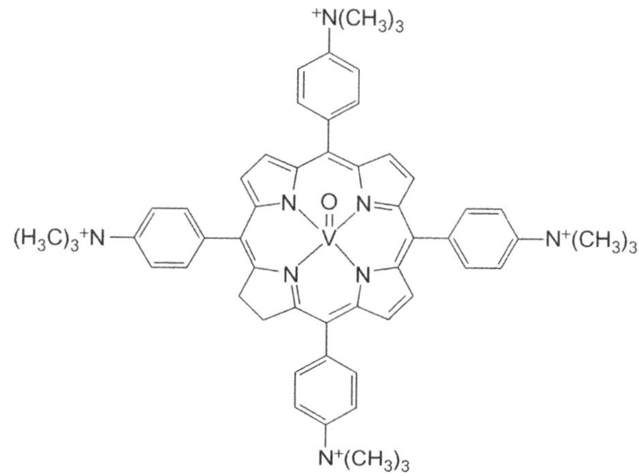

FIG. 8.4 *meso*-Tetrakis-(4-trimethylammoniophenyl)-porphyrinato oxovanadium(IV) tetrachloride (VOATPP). *(Adapted from A.K. Rahiman, K. Rajesh, K.S. Bharathi, S. Sreedaran, V. Narayanan, K. Rajesh, Cationic vanadyl porphyrin-encapsulated mesoporous Al/V-MCM-41 as heterogeneous catalyst for the oxidation of alkenes, Inorg. Chim. Acta 362 (2009) 1810–1818 with permission of Elsevier.)*

The same authors reported a similar investigation on the catalytic properties of a water soluble vanadyl porphyrin (VOATPP) encapsulated in Al or V MCM-41 with different amounts of both metals [34]. The inorganic support contains Brönsted/Lewis active sites that strongly interact with the absorbed metalloporphyrin due to the trimethylammonium groups on the macrocycle, whose structure is reported in Fig. 8.4.

The morphology of the supported Al-MCM-41 shows that no changes occurred to the initial support after the functionalization with the metalloporphyrin. Fig. 8.5 illustrates the SEM images of supported and unsupported Al-MCM-41.

The binding of the metalloporphyrins to the inorganic material is proposed by the authors using electrostatic or Lewis interactions and is reported in Fig. 8.6A and B.

The epoxidation reactions were performed using three different solvents: acetonitrile, dichloromethane, and methanol with PhIO as the oxidant.

The substrates used for the epoxidations were styrene and cyclohexene and the reactions were conducted at room temperature. In methanol the yields of the oxidation products for both the substrates were very low, and this was due to the formation of $PhIO_2$ as a side-product. This compound, which was not reactive, was formed by the competition between PhIO and the alkenes for the active intermediate. The oxidation results using the encapsulated metalloporphyrin were compared with those obtained from the catalyst in a homogeneous solution. The conversion of cyclohexene using V(50)-VOATPP or Al(50)-VOATPP

FIG. 8.5 SEM images of (A) Al-MCM-41 and (B) Al(50)-VOATPP. *(Adapted from A.K. Rahiman, K. Rajesh, K.S. Bharathi, S. Sreedaran, V. Narayanan, K. Rajesh, Cationic vanadyl porphyrin-encapsulated mesoporous Al/V-MCM-41 as heterogeneous catalyst for the oxidation of alkenes, Inorg. Chim. Acta 362 (2009) 1810–1818, with permission of Elsevier.)*

was almost comparable spanning from 40.2% to 45.6% while the allylic oxidation products were the main compounds obtained. The epoxide was obtained in less than 15% for both the catalysts. The situation in the case of styrene was almost different with 43.5% of conversion for V(50)-VOATPP and 18.5% for Al(50)-VOATPP where the formation of the epoxide was 63.0% and 31.4%, respectively. The authors claimed that the observed effect on the

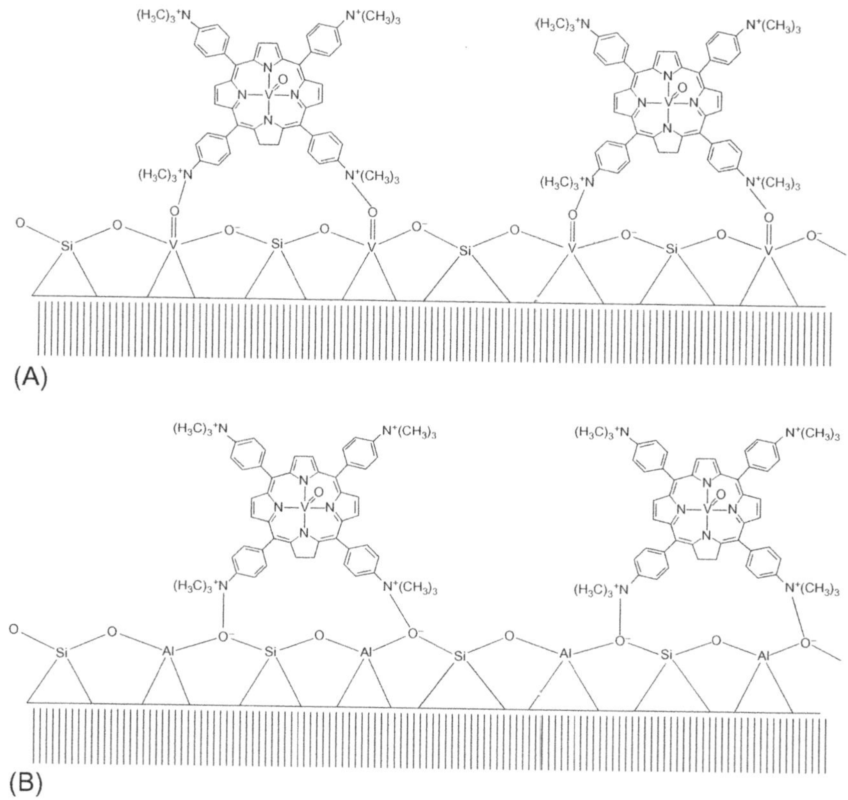

FIG. 8.6 (A) Schematic representation of anchoring of VOATPP on V-MCM-41. (B) Schematic representation of anchoring of VOATPP on Al-MCM-41. *(Adapted from A.K. Rahiman, K. Rajesh, K.S. Bharathi, S. Sreedaran, V. Narayanan, K. Rajesh, Cationic vanadyl porphyrin-encapsulated mesoporous Al/V-MCM-41 as heterogeneous catalyst for the oxidation of alkenes, Inorg. Chim. Acta 362 (2009) 1810–1818, with permission of Elsevier.)*

selectivity was due to the Al^{3+} acidic sites that resulted in the opening of the epoxide ring to give benzaldehyde as the final product. The catalysts were also tested in recycling experiments, giving good results with almost 90% of the initial activity maintained and almost the same selectivities.

The same authors reported a paper on the catalytic activity of the cationic manganese(III) porphyrin (MnTAPP) encapsulated in a titanium-substituted mesoporous molecular sieve with the macrocycle free base as reported in Fig. 8.4 for the oxovanadium derivative [35].

Cyclohexene and styrene were used as substrates for the oxidation reactions by PhIO catalyzed by the encapsulated porphyrin. Comparing with the results obtained in a homogeneous solution with the same metalloporphyrin or with the encapsulated catalyst in noncontaining titanium molecular sieves, the immobilized catalyst afforded higher conversion with a maximum yield of 52.4% for cyclohexene and 39.8% for styrene using Ti25-MnTAPP (25% of titanium). The selectivities were slightly different and the cyclohexene formed around 30% of the epoxide with large amounts of the corresponding ketone and alcohol. In the case of styrene the yield of the epoxide was 30%–35% but with less amounts of side products.

The authors suggested a reaction mechanism in line with what was proposed for the same processes catalyzed by the metalloporphyrin in a homogeneous solution [5,36] (Scheme 8.1).

The encapsulated MnTAPP exhibited good recycling performance with almost no loss of activity. The UV-Vis experiments also demonstrated that the catalyst had a high resistance to leaching from the sieve, indicating that the binding of the metalloporphyrin to the support was effective.

A paper from Narayanan and coworkers reported the catalytic activity of Fe, Mn, and V *meso*-tetrakis(4-trimethylammoniophenyl)porphyrin, TAPP encapsulated into ordered hexagonal mesoporous molecular sieves in the oxidation reactions of styrene by PhIO [37]. These heterogeneous catalysts were built

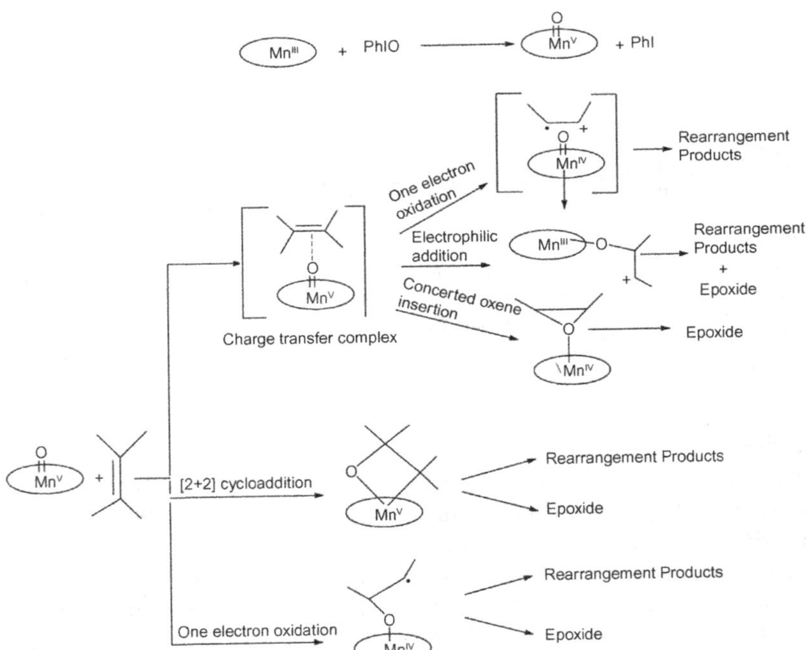

SCHEME 8.1 Reaction mechanism for the epoxidation catalyzed by manganese porphyrin complexes. *(Adapted from B. Meunier, Metalloporphyrins as versatile catalysts for oxidation reactions and oxidative DNA cleavage, Chem. Rev. 92 (1992) 1411–1456, with permission from American Chemical Society.)*

by using Si, Al, Ti, and V-MCM-41 that interacted strongly with the metal complexes by electrostatic interaction of the quaternary sites with the Brönsted/Lewis acidic sites.

The reaction results were similar for almost all the catalysts except Al(25)-MnTAPP, which gave phenylacetaldehyde as the major product while styrene epoxide was the dominant product in the other cases with high yields. The recycling experiments were used to check the chemical robustness of the heterogeneous systems. The results established that the catalyst maintained good activity even after three cycles.

Amino-functionalized MCM-41 was also used as a substrate to covalently link Mn(III)-*meso*-tetrakis(pentafluorophenyl)porphyrin. The hybrid system was used for the hydroxylation of naphthalene by 3-chloroperbenzoic acid (m-CPBA) [38]. The characterization of the heterogeneous catalyst showed that the manganese porphyrin was axially coordinated to the amino group. The oxidation catalyst showed a remarkable catalytic activity and could be reused several times without loss of its activity.

Recently an interesting application of the properties of the Fe_3O_4 has been reported by a Brazilian group [39]. The SiO_2 was used to form a layer on the Fe_3O_4 nanoparticles that could be separated from the reaction bulk using a magnet; after that, 3-aminopropyltriethoxysilane was anchored to the substrates and was used to covalently link the metalloporphyrin as previously reported by Che et al. for a ruthenium porphyrin [40]. The catalysts were used for the oxidation of (*Z*)-cyclooctene and cyclohexane with PhIO, giving high yields in epoxidation of the first substrate and hydroxylation of the second one. Recycling experiments for the epoxidation of cyclooctene showed a small loss of the activity even after eight cycles.

A paper from the group of Che [40] introduces a new concept in this field of research, the binding of a ruthenium chiral porphyrin to the ordered mesoporous molecular sieves (MCM-41 and MCM-48) for the enantioselective epoxidation and cyclopropanation of olefins.

The chiral metalloporphyrin was the [5,10,15,20-tetrakis-(1*S*,4*R*,5*R*,8*S*)-1,2,3,4,5,6,7,8-octahydro-1,2,5,8-dimethano-anthracen-9-yl]porphyrin Ru(II) complex, $H_2(D_4\text{-Por}^*)$ which was previously used in a homogeneous solution for the same reactions. This resulted in excellent results in terms of reactivity and enantioselectivity, but with high costs for its separation from the reaction products [41].

The binding of the catalyst to the sieves was obtained by using a modification of the surface of the sieves or silica by the use of 3-aminopropyltriethoxysilane. This group is able to make the complexation of the ruthenium in the porphyrin and retains the catalyst by a strong interaction. On Scheme 8.2 the general pathway to obtain the immobilization of the catalyst is depicted.

The two molecular sieves have very different structures because the MCM-41 silica phase is characterized by one-dimensional straight channels while

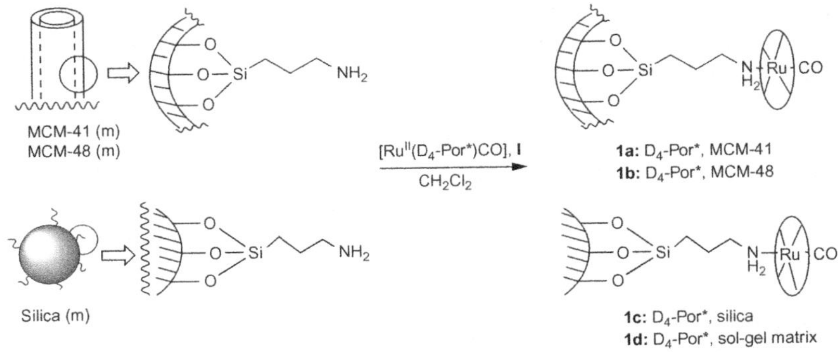

SCHEME 8.2 Reaction strategy for obtaining the immobilization of the catalyst on the surface of support. *(Adapted from J.L. Zhang, Y.L. Liu, C.M. Che, Chiral ruthenium porphyrin encapsulated in ordered mesoporous molecular sieves (MCM-41 and MCM-48) as catalyst for asymmetric alkene epoxidation and cyclopropanation, Chem. Commun. (2002) 2906–2907, with permission of the Royal Chemical Society.)*

MCM-48 contains a three-dimensional structure that has a larger surface area that can be easily functionalized by the reactants.

The epoxidation reactions were performed for several different olefins in benzene by using 2,6-dichloropyrine-*N*-oxide as an oxygen donor. The conversions were generally good or excellent (72%–99%) unless **1d** (10%) was used, which consists of catalyst bound to the sol-gel matrix, when using styrene as substrates. Furthermore, the most significant results were obtained measuring the ee, which in some cases reached almost 80%. More information was obtained from the epoxidation of cholesterol acetate using free metalloporphyrin as a catalyst in a homogeneous solution or with **1a** and **1b**. With the encapsulated catalyst there was no reaction while using the corresponding metalloporphyrin in a solution, there was good conversion (99%), and high ee (up to 91%) was obtained. This results suggested that the substrate could not reach the catalyst deeply encapsulated in the sieve. The encapsulated catalyst can be reused with almost the same conversions but with a small decrease in the ee.

The intramolecular cyclopropanation reaction of *trans*-cinnamyl diazoacetate by using **1b** gave high enantioselectivity (85%) and the catalyst was reused four times with only a slight decrease in ee.

An alternative mesoporous material for porphyrin encapsulation is the SBA-15, Santa Barbara Amorphous type material, in which the silica nanoparticles have pores of 4.6–30 nm arranged in hexagonal arrays. It owes its name to the University of California, Santa Barbara, the place where it was first produced.

Porphyrin encapsulation into SBA-15 for catalytic purposes was performed by Zhang et al. [42], who covalently anchored Pd-porphyrin functionalized ionic liquid into the channels of mesoporous material, through ion-pair electrostatic interaction between imidazolium-cationic and Pd porphyrin-anionic moieties. The modified material was prepared by post synthesis surface sol-gel polymerization, or a one-pot sol-gel procedure. The modified SBA-15 materials were efficient and recyclable catalysts for the cross-coupling of aryl iodides or activated aryl bromides with ethyl acrylate without activity loss and Pd leaching even after nine runs.

In a work by Pinto et al. [43] solid catalysts were prepared by the heterogenization of Mn porphyrins (MnP) on mesoporous silica SBA-15 by two strategies: immobilization of the three cationic isomers of Mn(III) *N*-methylpyridinium porphyrins (MnTM-X-PyPCl5, X=2,3,4) by electrostatic interaction, or anchoring of the three neutral isomers of Mn(III) *N*-pyridylporphyrins(MnT-X-PyPCl, X=2, 3, 4) on chloropropyl-functionalized mesoporous silica SBA-15Cl via covalent bonding. The resulting MnP-supported materials with MnP loadings of 0.3% w/w were explored as catalysts for hydroxylation of cyclohexane using PhIO as an oxygen donor. The heterogenized catalysts were more efficient and selective toward alcohol than the corresponding homogeneous systems. The covalently immobilized catalysts SBA-15Cl/MnT-X-PyPCl (X=2, 3, 4) showed total oxidation yields and selectivities slightly higher than those of SBA-15/MnTM-X-PyPCl5. In both solid systems, the efficiency and selectivity remained essentially unaffected by the immobilized MnP isomer. The solid catalysts were recovered and reused in recycling studies, showing high resistance against oxidative destruction with no significant changes in catalytic efficiency upon recycling.

Iron porphyrins were immobilized on SBA-type mesoporous silica matrixes by Zanatta et al. [44], for application in hydrocarbon oxidation with PhIO molecules, used as oxygen donors during the oxidation reactions. The catalytic assays demonstrated the high catalytic potential of iron porphyrins immobilized on SBA-15 and attested to their biomimetic behavior.

They correlated the surface changes with catalyst structure and activity and found that the Si-OH groups play an important part in the adsorption of PhIO molecules. The trimethylsilyl end-capped groups negatively affected the catalytic performance.

3 METAL PORPHYRINS IMMOBILIZED WITH LAYERED DOUBLE HYDROXIDES (LDH) AND MINERAL CLAYS

Layered double hydroxide (LDH) mineral clays are easily recyclable inorganic supports for the metalloporphyrins and were tested in oxidation reactions by Nakagaki and coworkers. They proposed anionic iron porphyrins immobilized on exfoliated layered double hydroxides for the oxidation of cyclooctene and cyclohexene [45]. The purposely synthesized base porphyrins [5,10,15,20-tetrakis(2,6-difluorphenyl-3-sulfonatophenyl) porphyrinate] tetrasodium ($Na_4[H_2(TDFSPP)]$ and [5,10,15,20-tetrakis (2-fluoro-6-chlorophenyl-3-sulfonatophenyl) porphyrinate] tetrasodium, $Na_4[H_2(TCFSPP)]$ underwent metalation by $[FeCl_2{\cdot}4H_2O]$ in DMF to afford the second

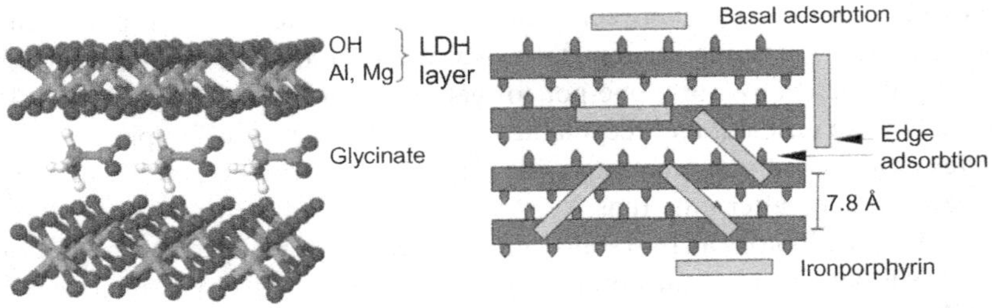

FIG. 8.7 Schematic representation of the anionic iron porphyrins immobilized on exfoliated LDH, intercalated with glycinate. *(Adapted from S. Nakagaki, M. Halma, A. Bail, G.G. Carbajal Arízaga, F. Wypych, First insight into catalytic activity of anionic iron porphyrins immobilized on exfoliated layered double hydroxides, J. Colloid Interface Sci. 281 (2005) 417–423 with permission from Elsevier.)*

generation porphyrins Fe(TDFSPP) and Fe(TCFSPP). The catalysts were then obtained by subsequent reaction with the Al-Mg layered hydroxides exfoliated with glycine in a basic environment (pH 10). According to the X-ray diffraction patterns, the basal distance of the LDH (7.8 Å) does not change upon interaction with the Fe porphyrins, thus indicating an immobilization of the macrocycles on the edges and on the external basal planes (Fig. 8.7), rather than an intercalation between the layers. The performance of the catalysts was tested using PhIO as an oxidizing agent. The yield for the epoxidation of cyclooctene was high for both of the immobilized catalysts, in other words, an average of 76% for the Fe(TDFSPP) and 87% for the Fe(TCFSPP), though the TONs are rather low [5,18–47]. The oxidation of cyclohexane by PhIO, in the presence of the immobilized iron porphyrins, showed a high alcohol vs. ketone selectivity. The yield was rather low for the immobilized Fe(TCFSPP), in other words, at best 15% for a catalyst/oxidant ratio 1:100 but higher using the immobilized Fe(TDFSPP) (20%–45%). It remarkably increased upon reuse, reaching 80%–85%.

Kaolinite was used to immobilize Fe(TDFSPP) as a catalyst for the oxidation of cyclooctene, cyclohexane, and heptane [47]. Kaolinite is a layer hydrated aluminosilicate of the 1:1 type with two distinct interlayer surfaces. On one side of the layer, aluminum atoms are octahedrally coordinated to corner oxygen atoms and hydroxyl groups (gibbsite-like structure). On the other side, silicon atoms are tetrahedrically coordinated to the oxygen atoms to form silica-like structures. The adjacent layers are connected via hydrogen bonds between aluminols (Al—OH) and siloxanes (Si—O), which may hinder the intercalation process especially of large molecules. In this case, it is expected that only the surface would be grafted. Therefore, the delamination of kaolinite was performed by mechanochemical intercalation of urea and ultrasound-assisted treatment with water, to promote a small number of stacked layers with a large basal surface area. This, however, was not yet sufficient to link the Fe(TDFSPP) to the inorganic support and a subsequent silanization of the delaminated kaolinite was performed with 3-aminopropyltriethoxysilane (3-APTS), leading to a successful Fe(TDFSPP) immobilization by electrostatic interaction. The resulting catalyst was firstly tested in the epoxidation of cyclooctene with PhIO, showing a high resistance, a high yield (97%), and a high TON (488). They were also proved to be interestingly selective, because the oxidation of cyclohexane yielded 90% of cyclohexanol vs. 3% cyclohexanone and a TON of 504, for high PhIO concentrations. The TON as well as the selectivity to alcohol products dramatically decreased in the oxidation of a linear alkane, such as *n*-heptane (TON 65, 48% alcohol, at high PhIO concentration). The reuse of the catalysts in the second reaction was quite successful, showing higher yields than those of the first run of the reactions for all the substrates. Attempts to use H_2O_2 as an oxidizing agent failed, probably due to the dismutation of Fe.

Yield, selectivity, and regioselectivity in the case of heptane oxidation by PhIO could be significantly improved by morphological control of the kaolinite substrate [48]. Kaolinite may be delaminated and intercalated with urea and subsequently with n-hexylamine to afford single layers and smaller stacks that can curl up, to some extent, and rearrange in tubular or scroll structures, thus providing a "halloysite-like" type of support. The curled tubes have a silica-like surface and an inner surface with the gibbsite-like structure. Two types of porphyrins were then immobilized on the modified kaolinite substrate: the anionic Fe(TDFSPP) and the neutral [5, 10, 15, 20-tetrakis (pentafluorphenyl) porphyrinate iron(III)] chloride (Fe(TPFPP)). The TEM characterization upon porphyrin linkage confirmed the formation of kaolinite nanotubes (or scroll-like structures), with the predominance of layered or nonexfoliated kaolinite. X-ray diffraction patterns indicated two distances of the basal planes in the curled up structures, 8.5 and 9.8 Å, which were compatible with diameter of the straight sections and bending of the nanotubes, respectively (Fig. 8.8). These dimensions ruled out an intercalation of the iron porphyrins inside the nanotubes, whereas they could be hosted either inside the curl or on the outer nanotube walls. The yield of oxidation products of cyclohexane and heptane by PhIO

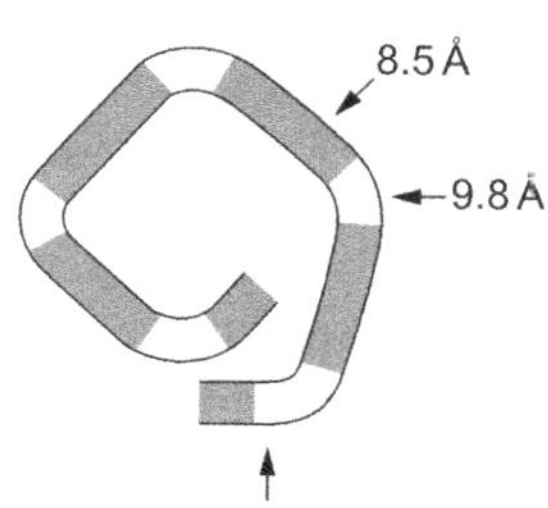

FIG. 8.8 Curled structure of the kaolinite nanotubes. *(Adapted from S. Nakagaki, G. Sippel Machado, M. Halma, A.A. dos Santos Marangon, K.A. Dias de Freitas Castro, N. Mattoso, F. Wypych, Immobilization of iron porphyrins in tubular kaolinite obtained by an intercalation/delamination procedure, J. Catal. 242 (2006) 110–117 with permission from Elsevier.)*

was strongly catalyst and solvent dependent. The immobilized neutral Fe(TPFPP) afforded very marginal yields of products, indicating either that the iron porphyrin was firmly attached inside the tubes and in layered kaolinite crystal agglomerates, or that both the fifth and sixth coordination positions of the iron were blocked by the bond of ligands from the support, preventing interaction with the oxidant. At variance with this, the immobilized anionic Fe(TDFSPP) had higher yields, in comparison with the nonimmobilized one, in both oxidation processes; there was a 71% alcohol yield in the oxidation of cyclohexane, 72% alcohol yield in the oxidation of heptane, and a remarkable 100% regioselectivity on the C2.

A successful intercalation of porphyrins into LDH structures was achieved with hydrotalcite-type of materials. Employed over more than three decades as catalysts and catalysts supports in secondary and fine chemistry, hydrotalcites are layered materials ideal for encapsulating porphyrins as well as creating bifunctional catalysts, in combination with them, because of the presence of basic sites.

First generation free-base anionic porphyrin $[H_2(TSPP)]^{4-}$, its iron(III) porphyrin $[Fe(TSPP)]^{3-}$, and the second generation anionic complexes $[Fe(TDFSPP)]^{3-}$ and $[Fe(TDCSPP)]^{3-}$ were intercalated into the layered double hydroxide Zn_nAl-LDH ($n=2$, 4 or 5) by coprecipitation at a constant pH [49]. The materials were used as catalysts in the oxidation of cyclooctene, cyclohexene, and cyclohexane by PhIO. The catalytic activity of $[Fe(TDCSPP)]$-Zn_2Al-LDH was higher than that of the homogeneous $[Fe(TDCSPP)]^{3-}$, but the opposite effect was observed in the case of $[Fe(TDFSPP)]$-Zn_2Al-LDH. Although $[Fe(TDFSPP)]^{3-}$ and $[Fe(TDCSPP)]^{3-}$ are structurally similar, their intercalation into Zn_2Al-LDH likely results in different chemical environments, leading to distinct catalytic activities.

In 2014 Zhou et al. [50] investigated the efficiency of tetra(4-sulfonatophenyl)porphinato-cobalt(II) anion (CoTSPP) intercalated into a Zn_2AlO_4 hydrotalcite in a Baeyer-Villiger (BV) reaction, in other words, an oxidation of ketones to esters or lactones. Usually carried out with strong oxidants such as peracids, *tert*-butylhydroperoxide and H_2O_2 systems, the BV reaction can be accomplished with a milder oxidant, such as O_2/aldehyde, in two steps, through using a catalyst. The generally accepted reaction scheme for the O_2/aldehyde oxidation system consists of peracid formation from the aldehyde and O_2 and the oxidation of the reactant by the peracid. In this framework, CoTSPP-Zn_2AlO_4 can catalyze both steps of the reaction, the metalloporphyrin intervening in the first step, and the hydrotalcite in the second one, through the basic sites. The hybrid bifunctional catalyst was probed with in the oxidation of cyclohexanone to ε-caprolactone under different reaction conditions in the presence of benzaldehyde, proving to be both active and selective.

Later on Zhou et al. [51] also probed other bifunctional hybrid catalysts originated from metalloporphyrins (MTSPP; M=Co, Fe, and Mn) intercalated with Zn_2AlO_4-LDH synthesized by coprecipitation, in alcohol oxidation reactions.

The reactions were carried out using O_2 as an oxidizing agent in the presence of isobutyraldehyde. The bifunctional catalysts were probed with respect to the oxidation of benzyl alcohol, and it was found that the CoTSPP has the best performance in terms of activity and selectivity. A variety of aromatic and aliphatic alcohols were used as substrates of the designed catalysts, exhibiting excellent activity, broad applicable scope, and good stability in the oxidation. Tests on the effects the hydrotalcite surface basicity were made by synthesizing catalysts with different cation ratios.

Curled structures are present also in chrysotile, an alternative source of highly hydroxylated silica, also classified in the kaolin/serpentine mineral 1:1 group. It has tri-octahedral site occupancy, where aluminum atoms, which form a compact brucite-like layer, are substituted. The mismatch of the brucite-like octahedral sheet with the silica tetrahedral sheet causes the curvature of the layers, which can roll into tight tubes. This material was used for immobilizing Fe-porphyrins upon a drastic leaching of concentrated acid that caused the removing of the brucite-like layer, resulting in an excellent source of hydrated disordered silica with physicochemical characteristics similar to silica gel. Fe(TDFSPP) and Fe(TCFSPP) were then linked to the resulting disordered silica via the 3-APTS and probed with respect to cyclohexane oxidation with PhIO [52]. The immobilization has opposite effects on the two Fe-porphyrins, with a higher alcohol yield for the TDFSPP/Si-3-APTS catalyst as compared to the free TDFSPP, and a lower yield for TCFSPP/Si-3-APTS as compared to free TCFSPP.

4 PORPHYRINS ENCAPSULATED INTO ZEOLITES

Zeolites are silicoaluminates with a microporous structure that can typically accommodate cations and are extensively used both as adsorbents and in catalysis [53,54], due to the molecular-size channels and pores in a three-dimensional network of well-defined crystalline structures that confers shape and size selectivity. Of the existing 229 different types of zeolites, the most used and best suited for the porphyrin accommodation are zeolites X (Z-X) and zeolite Y (Z-Y). The introduction of the porphyrins into Z-X and Z-Y cavities can be achieved by a type of synthesis proposed by Balkus et al. [55], where the zeolite is actually built around the selected porphyrin to afford a structure with an electrostatic interaction between the host anionic silicoaluminate and the guest cationic porphyrin. Compared to the template synthesis, it has the advantage of avoiding free ligand or impurities such as polypirrins in the zeolite cages.

Rosa et al. [56] used this method to fill the Z-X cavities with two types of iron porphyrins: the Fe(III) 5,10,15,20-tetrakis(4-*N*-methylpyridyl)porphyrin (FeP1) and the Fe (III) 5-mono(2,6-dichlorophenyl) 10,15,20-tris(4-*N*-methylpyridyl) porphyrin (FeP2), where one of the 4-*N*-methylpyridyl substituents is replaced by a 2,6-dichlorophenyl one. The resulting catalysts are very effective in the oxidation of (*Z*)-cyclooctane with PhIO, which was selectively transformed into *cis*-epoxycyclooctane. The conversion yields when using the free porphyrins were 50% (FeP1) or 60% (FeP2) and went up to 86% and 95% when using the encapsulated counterparts. The conversion of cyclohexane and adamantane was less efficient and selective. In the former case, both cyclohexanol and cyclohexanone were formed; in the latter case the products were alcohols with hydroxyl groups either in position 1 or 2. In all cases, the encapsulated porphyrins performed better than the free ones.

Moghadam et al. [57] used the method proposed by Balkus for the encapsulation of manganese (III) tetrakis[4-*N*-methylpyridinum]porphyrin both in Z-X and in Z-Y. The size of nanocages of zeolites is 1.3 nm, and the size of metalloporphyrin is 1.2 nm. It is expected that insertion of metalloporphyrin occurs without spatial strain. The XRD patterns indeed indicate that the encapsulation conditions have little impact on the zeolite hosts, such as the crystallinity of zeolite-encapsulated metallocomplexes. Such parameters are almost identical to that of the parent Z-X and Z-Y. The load of porphyrin, instead, is different for the two types of zeolites. This can be due to the Si/Al ratio in the two cases, in other words, when the Al content of zeolite becomes higher, the capacity of cations exchange is enhanced. Therefore, the number of sites that are occupied by cations increases. The obtained catalysts were tuned both with respect to solvent and the axial ligand. Oxidations with $NaIO_4$ were then performed on a number of cyclic and liner alkenes, either under mechanical stirring or under microwave irradiation. A high conversion rate and selectivity were observed with both catalysts, the microwave irradiation shortening the reaction time, when using the cyclic substrate. The yield decreased with the size in the case of linear alkenes.

Karimipour et al. [58] encapsulated a [*meso*-tetrakis(3-pyridyl) porphyrinato]iron(III) chloride complex into NaY zeolite (Fe(T-3-PyP)@NaY, Fig. 8.9), as a heterogeneous "ship-in-a-bottle" type catalyst.

The catalytic activity was examined for the epoxidation of cyclohexene by $PhI(OAc)_2$ in CH_3CN/H_2O (5:1) and compared to that of nonencapsulated Fe(T-3-PyP) as a homogenous catalyst. It was found that the heterogeneous catalyst Fe(T-3-PyP)@NaY was stable and reusable for several times, and it provided a mild condition and exhibited high activity and selectivity in the oxidation of alkenes to epoxides (16%–94%). Furthermore, high yields were obtained in the oxidation of 4-nitrobenzylalcohol to 4-nitrobenzaldehyde (97%), in the oxidative dehydrogenation of diethyl 4-(2,6-dichlorophenyl)-2,6-dimethyl-1,4-dihydro-3,5-pyridinedicarboxylate to the corresponding pyridine (100%), and moderate yield in the oxidation of diphenylacetic acid to benzophenone (64%).

5 3-D ARCHITECTURES BY SELF-ASSEMBLING

Several pathways were followed to create cages for hosting metalloporphyrins, simultaneously proving size selectivity of the reagents and/or products in

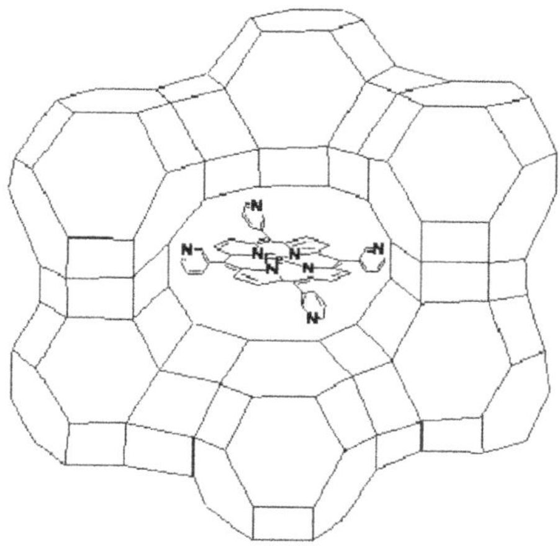

FIG. 8.9 [*meso*-Tetrakis(3-pyridyl)porphyrinato] iron(III) chloride encapsulated on NaY Zeolite [Fe(T-3-PyP)@NaY] was synthesized and characterized as a heterogeneous "ship-in-a-bottle" type catalyst. *(Adapted from G. Karimipour, M. Rezaei, D. Ashouri, Zeolite encapsulated Fe-poprhyrin for catalytic oxidation with iodobenzene diacetate (PhI(OAc)$_2$). J. Mex. Chem. Soc. 57 (2013) 276–282 with permission from the Mexican Chemical Society.)*

catalytic processes. Among them, self-assembling of macrostructures provides an efficient route to tune cages structural characteristics by varying the chemical building blocks.

Otte et al. self-assembled a Nitschke-type of M_8L_6 cubic cage or "molecular flask" to host metalloporphyrins for catalyzing size-selective radical-type cyclopropanation reactions [59]. The cube has a face centered structure, with Fe^{3+} at the edges and Zn^{2+} and with 16 iron(III) trifluoromethanesulfonate ($Fe(OTf\ )_2$) as counterions. Successively, the counterions were replaced by iron trifliminde($Fe(NTf_2)_2$) to improve the solubility in polar solvents, so that the encapsulation of tetra(4-pyridyl)metalloporphyrins (M(TPyP)) could be carried on [60].

The obtained compound is a hydrophilic moiety with a hydrophobic cage that can act as a phase-transfer catalyst and pull hydrophobic substrates into the aqueous phase. Its reactivity was probed with respect to the cyclopropanation of styrene derivatives, 2-vinylnaphtalene with benzyl-ethyl- and *tert*-butyl diazoacetates in different solvent mixtures and it was found that the water/acetone mixture increased the catalytic performance dramatically. Furthermore, the caged catalyst showed a preference for cyclopropanation of styrenes over other vinylic substrates, and a marked size-selectivity, with a preference for cyclopropanation of smaller over bulky substrates. Bulky substrates reacted slower than the smaller ones, thus allowing size-selective competition reactions.

Yang et al. [61] used manganese and nickel derivatives of 5,10,15,20-tetrakis (3,5-biscarboxylphenyl)porphyrin (MH_8OCPP, with $M = Mn^{III}Cl$ or Ni^{II}) as

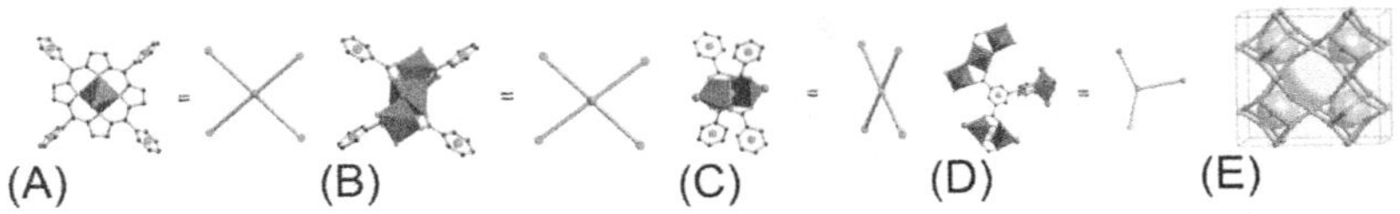

FIG. 8.10 A sketch of the Mn-containing SBUs, their linkage, and the overall structure: (A) $Mn^{III}N_4Cl$, (B) trinuclear $Mn_3(COO)_4(\mu\text{-}H_2O)_2(H_2O)_6$, (C) binuclear Mn^{II} $2(COO)_4Cl_2$, and (D) linkage of the SBUs to the 3-coordinated node to form (E) the net from four nodes. (A)–(D) Small cavities are represented, whereas in (E) the central one is sketch as a central ball. *(Adapted from X.-L. Yang, M.-H. Xie, C. Zou, Y. He, B. Chen, M. O'Keeffe, C.-D. Wu, Porous metalloporphyrinic frameworks constructed from metal 5,10,15,20-tetrakis(3,5-biscarboxylphenyl)porphyrin for highly efficient and selective catalytic oxidation of alkylbenzenes, J. Am. Chem. Soc. 134 (2012) 10638–10645 with permission from American Chemical Society.)*

building blocks of porous metal organic frameworks (MOF). Three isomorphous compounds were investigated, which were obtained by different combination of (MH_8OCPP) and $MnCl_2$ of $CdCl_2$ in the synthesis process, in other words, $[Mn_5Cl_2(MnClOCPP)(DMF)_4(H_2O)_4]\cdot 2$ DMF·8 $CH_3COOH\cdot 14H_2O$ (ZJU-18), $[Mn_5Cl_2(NiOCPP)(H_2O)_8]\cdot 7DMF\cdot 6CH_3COOH\cdot 11H_2O$ (ZJU-19), and $[Cd_5Cl_2(MnClOCPP)(H_2O)_6]\cdot 13DMF\cdot 2CH_3COOH\cdot 9H_2O$ (ZJU-20). They crystallize in an orthorhombic structure made up of an octatopic Mn^{III}Cl-OCPP metalloligand joined to two metal-containing secondary building units (SBUs): binuclear $Mn_2(COO)_4$ and trinuclear $Mn_3(COO)_4(\mu\text{-}H_2O)_2$ (for ZJU-18 and 19). The equivalent Cd-based SUBs build the frame of ZJU-20. The overall spatial arrangement of the structure is depicted in Fig. 8.10 along with the constituent SBUs.

This structure has cavities about 21.3 Å in diameter and corresponding pore windows approximately 11.5 Å in diameter, which contribute to the size selectivity of the substrate.

The catalytic activity of the three compounds was probed, first with respect to the oxidation of ethylbenzene to acetophenone, using *tert*-butyl hydroperoxide (TBHP) as an oxidizing agent and a mixture of acetonitrile, acetic acid, and water as a solvent. ZJU-18 had a conversion rate of 99% and also displayed a very high selectivity, in other words, 99%. At variance with this ZJU-19 (with NiOCPP) had a marginal conversion rate (9%) and lower selectivity (80%), and ZJU-20 had a 69% conversion rate and 95% selectivity. In all the cases, the MOF compounds exhibited both higher activity and selectivity than the corresponding primary and secondary building units. Once it was established that ZJU-18 had better catalytic performance as compared to ZJU-19 and 20, it was chosen for the oxidation of larger substrates in the same experimental conditions. Therefore, the substrates propylbenzene, 1,2,3,4, tetrahydronaftalene, fluorene, diphenylmethane, 1,4-phyenlethylbenzene were probed for the oxidation into the corresponding ketone. The conversion rate decreased gradually with the increase of the size of the substrate, to the level that ZJU-18 exhibited even lower catalytic activity than the molecular Mn-Cl-Me_8OCPP catalyst for larger substrates such as diphenylmethane and 1,4 ethylphenylbenzene. This can be related to the difficult access of substrate to the porous cavities, thus showing size and site selectivity, besides high catalytic activity for small substrates.

Zeolite-like metal organic frameworks (ZMOFs) can be successfully synthesized by assembling rigid and directional single-metal-ion-based molecular building blocks, the key of the structure being tetrahedral building units (TBUs). ZMOFs are topologically analogous to pure inorganic zeolites, because they are anionic, chemically stable in aqueous media, and possess extra-large cavities suited for hosting metalloporphyrins for applications in catalysis. Alkordi et al. [62] obtained a ZMOF topologically analogous to zeolite RHO based on InN_4 tetrahedral units, and the large voids inside its periodic R-cages were used for encapsulation of cationic porphyrins. In particular, the reaction of $In(NO_3)_3{\cdot}xH_2O$ and 4,5-imidazoledicarboxylic acid (H_3ImDC) was carried out in an *N,N*-dimethylformamide (DMF)/acetonitrile (CH_3CN) mixture in the presence of 5,10,15,20-tetrakis(1-methyl-4-pyridinio)porphyrin tetra(*p*-toluenesulfonate) ($[H_2TMPyP]$ $[p\text{-tosyl}]_4$). The overall structure is sketched in Fig. 8.11. Mn, Co, Cu, and Zn were introduced in the structure by PSM, and the Mn-ZMOFs then was tested as a catalyst for cyclohexane oxidation by *tert*-butyl hyperoxide (TBHP). The total yield of cyclohexanol and cyclohexanone after 24 h was 90%, with a TON of 25.

The nature of the building blocks and the porphyrin to encapsulate in the MOFs determines the properties of the resulting catalysts as well as the type of catalyzed reactions. The epoxidations of *trans* and *cis*-stilbene, for instance, are catalyzed by two different types of catalysts. Zhang et al. [63] investigated the reaction of *trans*-stilbene with TBHP by employing a porph@MOM-10, made of a Cd(II) carboxylate framework that contains CdTMPyP cations (TMPyP = *meso-tetra*(*N*-methyl-4-pyridyl)porphyrin tetratosylate). Both the framework and the caged porphyrin can be subjected to the PSM of the metal moieties, with retention of the parent structure. The porph@MOM-10 is sketched in Fig. 8.12.

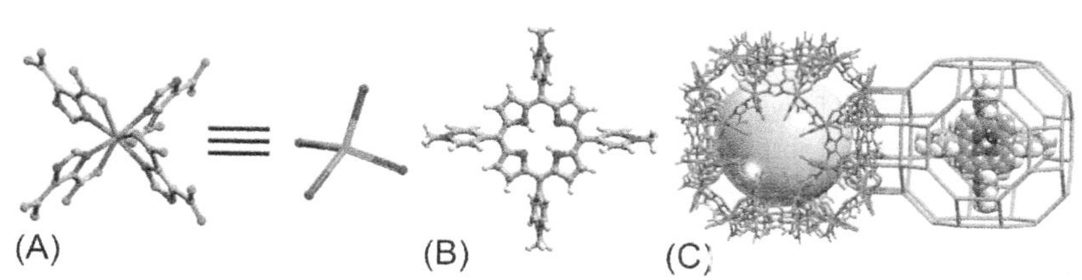

FIG. 8.11 (A) Eight-coordinate MBB that could be represented as a TBU, (B) $[H_2TMPyP]^{4+}$ porphyrin, (C) crystal structure of rho-ZMOF (left), and schematic presentation of $[H_2TMPyP]^{4+}$ porphyrin ring enclosed in rho-ZMOF R-cage (right). *(Adapted from M.H. Alkordi, Y. Liu, R.W. Larsen, J.F. Eubank, M. Eddaoudi, Zeolite-like metal-organic frameworks as platforms for applications: on metalloporphyrin-based catalysts, J. Am. Chem. Soc. 130 (2008) 12639–12641 with permission of the American Chemical Society.)*

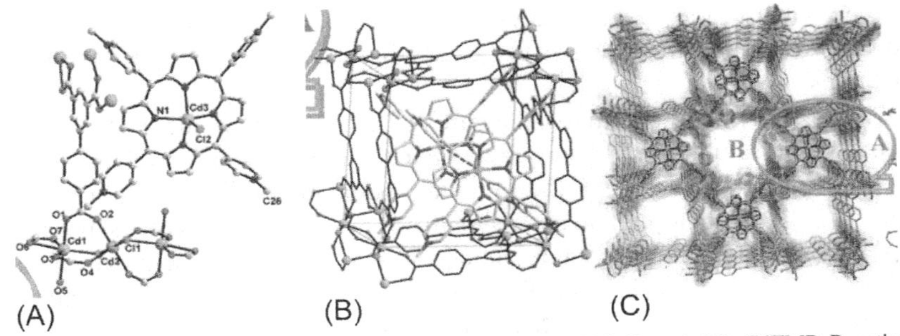

FIG. 8.12 (A) Components of the framework porph@MOM-10 and (B) CdTMPyP cations trapped in cuboid boxes in porph@MOM-10. (C) Projection of the structure of porph@MOM-10 along the *c*-axis, with the two types of channels A and B. *(Adapted from Z. Zhang, L. Zhang, L. Wojtas, P. Nugent, M. Eddaoudi, M.J. Zaworotko, Templated synthesis, postsynthetic metal exchange, and properties of a porphyrin-encapsulating metal-organic material, J. Am. Chem. Soc. 134 (2012) 924–927 with the permission of American Chemical Society.)*

The frame contains two types of Cd(II) (Cd1 and Cd2), with different reactivity. Cd1 possesses distorted octahedral geometry through four carboxylate oxygen atoms and two μ_2-chloride anions, whereas Cd2 adopts a distorted octahedral geometry via coordination to four carboxylate oxygen atoms, an aqua ligand, and a μ_2-chloride anion (Fig. 8.12A). The CdTMPyP cations are trapped in cuboid boxes (Fig. 8.12B). The projection of the structure along the *c* axis (Fig. 8.12C) shows that there is a 1:1 ratio of two types of square channels: (A) 12.6 Å × 12.6 Å; (B) 11.9 Å × 11.9 Å. CdTMPyP cations are stuck in the channels A separated from each other by 10.3 Å, whereas channels B are occupied by solvent molecules. PSM was performed by immersing porph@MOM-10 either in $MnCl_2$/methanol or in $CuCl_2$/methanol. The Cd of the porphyrin could be completely substituted by both metals, whereas the Cd of the frame was only partially substituted when using Cu, then yielding Mnporph@MOM-10-Mn and Cuporph@MOM-10-CdCu.

The properties as catalysts are strongly metal dependent, with 79% conversion of *trans*- stilbene after 12 h cycles by copper porph@MOM-10-CdCu, TON = 182, 75% by manganese porph@MOM-10-Mn, TON = 178, and <10% by porph@MOM-10.

Lee et al. [64] proposed a self-sorting strategy for the assembling of rigid porphyrin boxes possessing cavities to encapsulate catalytically active achiral manganese porphyrin dimers for the epoxidation of *cis*-stilbene. The building blocks of the catalytic box are reported in Fig. 8.13 and are labeled A, T, T′, M, and C, which is a chiral variant of A. A is an Sn(IV) dimer porphyrin, M is an Mn(II) dimer porphyrin, T is a diacetylene-linked (porphyrin)Zn trimer, and T′ is analogous of the Zn trimer porphyrin, with the diacetyl groups replaced by a phenylene substituent. When 2 equiv. of A are combined with 4 equiv. of the diacetylene-linked (porphyrin)Zn trimer T, the rigid 16-porphyrin AAT_4 "box" is resulted. The analogous combination of A with the phenylene-linked

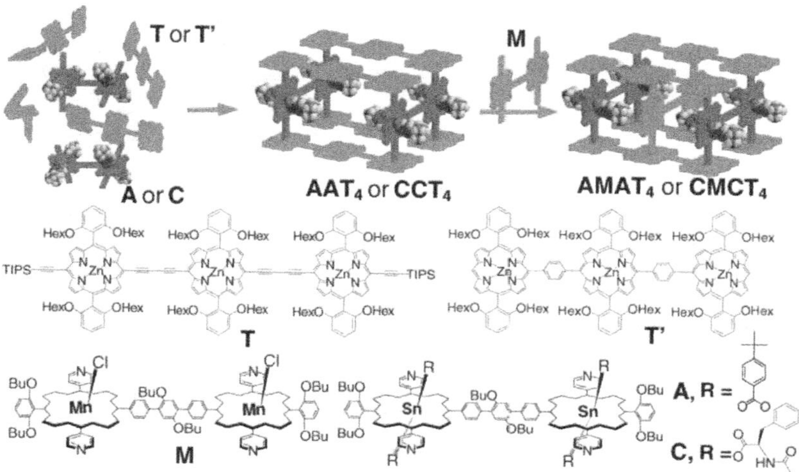

FIG. 8.13 Schematic representation of the steric self-sorting assembly of supramolecular catalytic boxes, $AMAT_4$ and $CMCT_4$. *(Adapted from S.J. Lee, S.-H. Cho, K.L. Mulfort, D.M. Tiede, J.T. Hupp, S.B.T. Nguyen, Cavity-tailored, self-sorting supramolecular catalytic boxes for selective oxidation, J. Am. Chem. Soc. 130 (2008) 16828–16829 with the permission of American Chemical Society.)*

(porphyrin)Zn trimer T′ yielded the corresponding AAT'_4. Ligand A has a large axial steric hindrance due to the carboxylate ligand; therefore, the middle zinc porphyrins in the T components of AAT_4 remain unbound, leaving an unoccupied cavity inside the box. This space could be uniquely occupied by the manganese dimer M to generate a supramolecular catalyst box $AMAT_4$. The use of the C ligand instead of A provides chiral moieties to the box.

The supramolecular boxes $AMAT_4$ and $AMAT'_4$ were tested in size-selective catalysis, by employing them in the epoxidation of *cis*-stilbene vs. *cis*-3,3′,5,5′-*tetra*(*tert*-butyl)stilbene. Due to access inhibition of the larger olefin to the metal center by the cavity modifier, *cis*-stilbene is 5.5-fold more reactive when using the catalytic boxes than the sterically bulky *cis*-3,3′,5,5′-tetra(*tert*-butyl)stilbene. Furthermore, the enantioselectivity of $CMCT_4$ (the chiral version of $AMAT_4$) was tested in the catalytic sulfoxidation of methyl *p*-tolyl sulfide. Methyl *p*-tolyl sulfoxide was obtained with 12% ee. The enantiomeric excess increased to 14% when the smaller catalytic box CMCT′4 was used.

MOF may be employed also in other types of reactions, such as acyl-transfer. Schultz et al. [65] employed a mixed-ligand strategy for incorporating metalloporphyrins into frameworks featuring large pores using the tetratopic, "octa"-oxygen ligand 1,2,4,5-tetrakis(4-carboxyphenyl), which has the advantage of high chelating ability, providing large stability and thus permanent microporosity to the resulting MOF. In particular, they obtained a zinc-based, pillared paddlewheel MOF with the formula [Zn2(**1**)(Zn-**2**)] (Fig. 8.14C), by combining $Zn(NO_3)_2{\cdot}6H_2O$, 1,2,4,5-tetrakis(4-carboxyphenyl)benzene

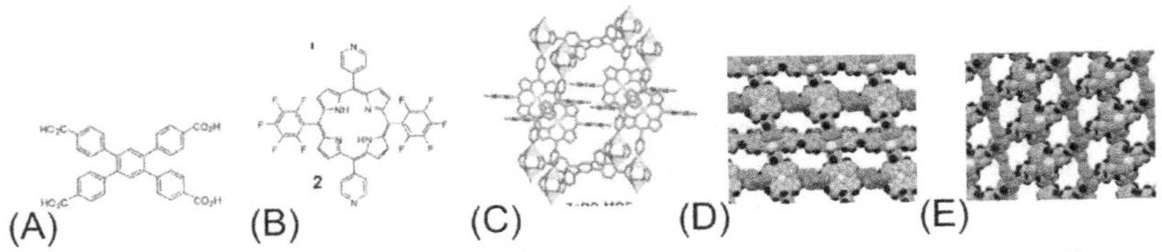

FIG. 8.14 (A) 1,2,4,5-Tetrakis(4-carboxyphenyl)benzene, (B) (5,15-dipyridyl-10,20-bis(pentafluorophenyl))porphyrin, (C) stick representation of the unit cell (polyhedra=Zn connected through oxygen atoms; fluorine atoms lie in the horizontal planes). (d) Space-filling model of the crystal structure of ZnPO-MOF showing channels down the a crystallographic axes and (E) along the *b* axis. *(Adapted from A.M. Shultz, O.K. Farha, J.T. Hupp, S.B.T. Nguyen, A catalytically active, permanently microporous MOF with metalloporphyrin struts, J. Am. Chem. Soc. 131 (2009) 4204–4205 with permission from American Chemical Society.)*

(**1**) (Fig. 8.14A) and (5,15-dipyridyl-10,20-bis(pentafluorophenyl))porphyrin (**2**) (Fig. 8.14B) under solvothermal conditions, while the free-base porphyrin **2** was simultaneously metalated. The crystal structure reveals large channels in three directions (Fig. 8.14D and E) that are occupied by a substantial amount of disordered solvent. ZnPO-MOF comprises a single independent framework, in contrast with other pillared paddlewheel MOFs where doubly or triply catenated structures are observed.

The catalytic activity was evaluated in terms of acyl-transfer reaction transfer from *N*-acetylimidazole (NAI) to 3-pyridylcarbinol (3-PC). A 2420-fold rate enhancement relative to the un-catalyzed reaction was observed. A control experiment with a simpler MOF, Zn_2(*p*-terephthalate)$_2$(4,4′-bipyridine), showed little catalytic effect.

The MOF boxes can be shaped to mimic nature. This is, for instance, the case of MOMs that mimic heme enzymes in terms of both structure and reactivity [66]. The MOMzyme-1 class (metal organic material enzyme) is based on HKUST-1, into which the catalytically active metalloporphyrins are selectively encapsulated within one of the three existing polyhedral cages. HKUST-1 is formed through the assembly of benzene-1,3,5-tricarboxylate anions and Cu(II) or Zn(II) cations, and it is well-suited to serve as a platform for heme biomimetic chemistry since its topology affords three distinctly different polyhedral cages capable of entrapping guest molecules (Fig. 8.15). Either iron(III) tetrakis(4-sulphonatophenyl)porphyrin, Fe4SP, or manganese(III) tetrakis(4-sulphonatophenyl)porphyrin, Mn4SP, was encapsulated within the octahedral cage which is most suited based on cage size and symmetry, whereas the remaining cavities allow small molecules to reach the active site for catalysis similarly to channels in heme proteins. The materials are labeled Fe4SP@HKUST-1(Cu or Zn) or Mn4SP@HKUST-1(Cu or Zn).

Subsequently the H_2O_2 degradation was probed, using ABTS (2,2′ azido-di(3-ethylbenzthiazoline)-6-sulfonate) as redox indicator.

In spite of the structural similarities, the catalytic performances are not yet at the level of natural enzymes. The initial rate $ABTS^{+\bullet}$ formation by the

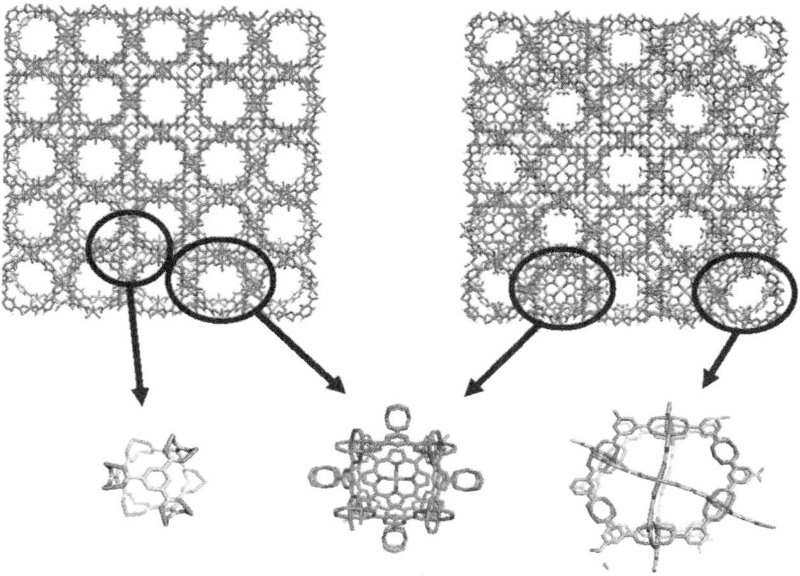

FIG. 8.15 Diagram of the encapsulation of metalloporphyrins within the octahemicatohedral cages of HKUST-1. The other cavities associated with the framework are also illustrated. *(Adapted from R.W. Larsen, L. Wojtas, J. Perman, R.L. Musselman, M.J. Zaworotko, C.M. Vetromile, Mimicking heme enzymes in the solid state: metalorganic materials with selectively encapsulated heme, J. Am. Chem. Soc. 133 (2011) 10356–10359 with the permission of American Chemical Society.)*

Fe4SP@HKUST-1(Cu) is lower than that observed for microperoxidase-11, horse heart myoglobin, or Fe4SP (all in solution), while the maximum yield of $ABTS^{+\bullet}$, relative to horse heart myoglobin, is comparable to that of microperoxidase-11 and Fe4SP in a solution.

5.1 Other Encapsulation and Intercalation Methods

More methods for porphyrin encapsulation useful in catalysis include the mimicking the functions of the natural heme group like those of a cytochrome enzyme, by assembling molecular squares around the selected porphyrin, by constructing cavity-containing porphyrins, embedding the macrocycle in polymer matrices, or by dendritic encapsulation

Cavity-containing manganese porphyrins were built by using molecular clips. Elemans et al. [67] used the molecular clip sketched in Fig. 8.16, which is equipped with a porphyrin roof situated symmetrically above the receptor that creates a rigid and relatively closed cavity with a diameter of approxmately 9 Å.

The outside of the cavity in this kind of construction also requires a stabilizing axial ligand, prior to the use in catalysis. In cytochrome P_{450}, the axial coordination is featured by a cysteine thiolate. The strength of axial ligand

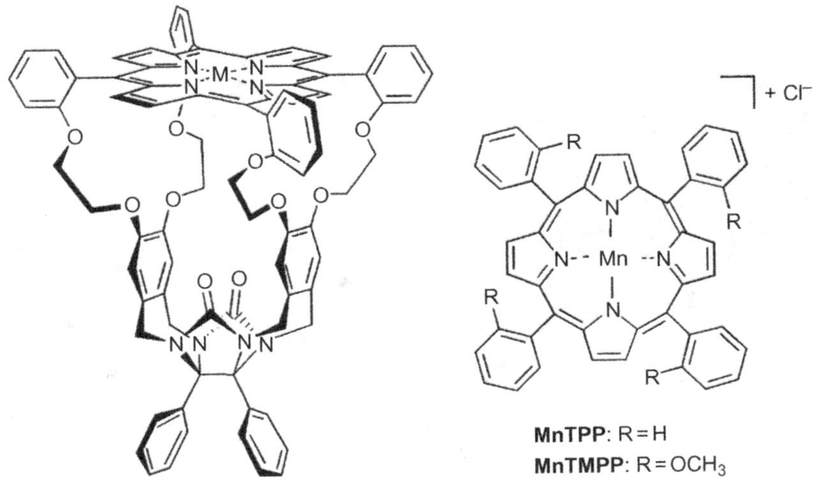

FIG. 8.16 Left panel: A sketch of the manganese substituted molecular clipped porphyrin (M = Mn(III)Cl). Right panel: the roof porphyrins used to cap the catalyst's cavity. *(Adapted from J.A.A.W. Elemans, E.J.A. Bijsterveld, A.E. Rowan, R.J.M. Nolte, A host-guest epoxidation catalyst with enhanced activity and stability, Chem. Commun. (2000) 2443–2444 with the permission of the Royal Chemical Society.)*

binding is related to the enhancement of the activity of the catalyst, since the electron donating properties of the ligand facilitate the formation of the Mn (V)-oxo reaction intermediate. Therefore, in most of the synthetic applications of the catalysts mimicking cytochrome P450, pyridine, imidazole, or phenolate derivatives were used to enhance the activity and stereoselectivity of the catalytic center. The epoxidation of α-pinene, *cis*-stilbene, and *trans*-stilbene was carried out using either pyridine or butyl-pyridine as an axial ligand. Regio- and stereoselective oxidation was achieved when, for example, *cis*-stilbene is oxidized, with a major preference for the production of the *cis*-epoxide.

A method of obtaining metalloporphyrin catalysts easier to handle and exhibiting an improved catalytic activity is anchoring them on polymer support. Microencapsulation is a classic method for immobilizing catalysts onto polymers on the basis of physical envelopment by the polymers and the catalysts, which renders them highly dispersible in common organic solvents. Naik et al. [68] investigated the microencapsulation of Mn(III) complexes of *meso*-tetraphenylporphyrin (MnTPP)Cl and *meso*-tetrakis(2,6-dichlorophenyl)porphyrin Mn(TDCPP)Cl in polystyrene, via electronic interactions between the aromatic electrons of the benzene rings of the polymer matrix and the empty orbital of porphyrin catalysts.

These encapsulated catalysts (MCMPs) exhibit enhanced activity for the epoxidation of alkenes such as styrene and α-methyl styrene in the presence of $NaIO_4$, $KHSO_5$, and NaOCl as oxidants, with the highest conversion rate and selectivity being achieved with $NaIO_4$. These catalysts give high turnover frequencies and can be quantitatively recovered by simple filtration and reused without loss of activity.

Enhanced stability in the reaction medium under oxidizing conditions can be achieved by the intercalation of metalated tetraphenylporphyrins in α-zirconium hydrogen phosphate (metal = Fe, Co, Mn, Na) [69]. The obtained catalysts were tested in a cyclohexene and *cis*-stilbene reaction with dioxygen as oxidizing agents, in the presence of isobutyraldehyde. It was observed that the matrix protects the metal complexes from degradation, saving the activity of the catalytic system. For comparison purposes, the degradation of the starting metal complexes in a solution was verified in the reaction media. Epoxides were predominantly observed as products of the oxidation of cyclohexene with intercalated complexes, whereas allylic oxidation products were observed in lower yields. The product distributions depended on the metal present in the complexes. The ratios of *cis*- vs. *trans*-stilbene oxide and benzaldehyde also depended on the intercalated metal complex. Furthermore, it was inferred that reaction proceeds via a free radical mechanism, since the addition of a free radical inhibitor stops the reaction.

Dendrimers are repetitively highly branched molecules that represent unique artificial building blocks through a fine control over both their size and molecular architecture, and therefore their three-dimensional nanoscale structure. This allows the creation of an encapsulating scaffold around a catalytic center through one synthesis strategies [2], in other words, convergent, divergent, and click chemistry methods. The channels through the typical branched structure are selective pathways for the reactants to get in and the products get out. Epoxidation reactions of a variety of alkenes by dendritic-porphyrin catalysts were reported by Zhang et al. [70]. They performed the binding of the Fréchet-type poly(benzyl ether) dendrons [G-n] to carbonylruthenium(II) *meso*-tetraphenylporphyrin [Ru(tpp)(CO)] using covalent ethereal bonds to form a series of dendritic ruthenium(II) porphyrins [Ru(tpp)(CO)]-[G-*n*]*m* ($m=4$, $n=1, 2$; $m=8$, $n=0$–2) (Fig. 8.17). The subsequent catalysis tests were performed on styrene, *trans*-/*cis*-stilbene, 2,2-dimethylchromene, cyclooctene, and unsaturated steroids such as cholesteryl esters and estratetraene derivative, using with 2,6-dichloropyridine *N*-oxide (Cl_2pyNO) as an oxidizing agent. It was found that all the reported dendritic ruthenium porphyrins work as highly selective catalysts for the epoxidation of alkenes. The chemo- or diastereoselectivity increases with the number of dendrons attached to [Ru(tpp)(CO)], and complex $[Ru(tpp)(CO)]$-$[G\text{-}2]_8$ exhibits remarkable selectivity and a high turnover number.

The cyclopropanation of styrene and its parasubstituted derivatives with ethyl diazoacetate catalyzed by $[Ru(tpp)(CO)]$-$[G\text{-}2]_8$ were found to be highly selective for the *trans* product.

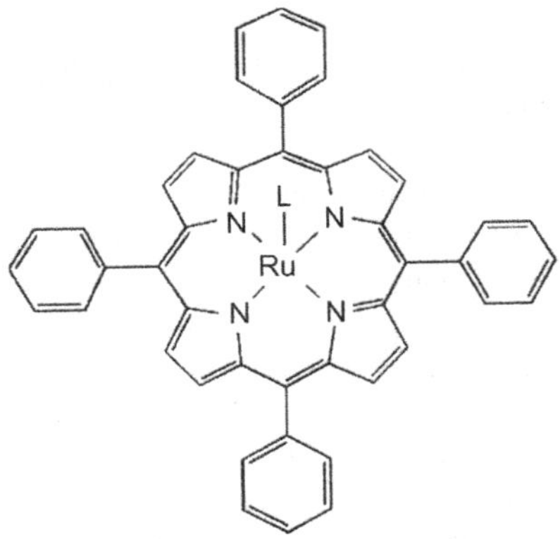

FIG. 8.17 Schematic structures of the dendritic ruthenium porphayrins: (a) [Ru(tpp)(CO)] building block. *(Adapted from J.-L. Zhang, H.-B. Zhou, J.-S. Huang, C.-M. Che, Dendritic ruthenium porphyrins: a new class of highly selective catalysts for alkene epoxidation and cyclopropanation, Chem. Eur. J. 8 (2002) 1554–1562 with permission of Wiley.)*

REFERENCES

[1] J.Y. Lee, O.K. Farha, J. Roberts, K.A. Scheidt, S.T. Nguyen, J.T. Hupp, Metal-organic framework materials as catalysts, Chem. Soc. Rev. 38 (2009) 1450–1459.

[2] S. Hecht, J.M.J. Frechet, Dendritic encapsulation of function: applying nature's site isolation principle from biomimetics to materials science, Angew. Chem. Int. Ed. 40 (2001) 74–91.

[3] B.F. Abrahams, B.F. Hoskins, D.M. Michail, R. Robson, Assembly of porphyrin building blocks into network structures with large channels, Nature 369 (1994) 727–729.

[4] R. Matassa, M. Carbone, R. Lauceri, R. Purrello, R. Caminiti, Supramolecular structure of extrinsically chiral porphyrin hetero-assemblies and achiral analogues, Adv. Mater. 19 (22) (2007) 3961–3967.

[5] B. Meunier, Metalloporphyrins as versatile catalysts for oxidation reactions and oxidative DNA cleavage, Chem. Rev. 92 (1992) 1411–1456.

[6] D. Mansuy, Activation of alkanes: the biomimetic approach, Coord. Chem. Rev. 125 (1993) 129–141.

[7] M.W. Grinstaff, M.G. Hill, J.A. Labinger, H.B. Gray, Mechanism of catalytic oxygenation of alkanes by halogenated iron porphyrins, Science 264 (1994) 1311–1313.

[8] P.E. Ellis Jr., J.E. Lyons, Selective air oxidation of light alkanes catalyzed by activated metalloporphyrins—the search for a suprabiotic system, Coord. Chem. Rev. 105 (1990) 181–193.

[9] J.F. Bartoli, O. Brigaud, P. Battioni, D. Mansuy, Hydroxylation of linear alkanes catalysed by iron porphyrins: particular efficacy and regioselectivity of perhalogenated porphyrins, J. Chem. Soc. Chem. Commun. (1991) 440–442.

[10] T. Wijesekara, A. Matsumoto, D. Dolphin, D. Lexa, Perchlorinated and highly chlorinated meso-tetraphenylporphyrins, Angew. Chem. Int. Ed. Engl. 29 (1990) 1028–1030.

[11] M.N. Carrier, C. Scheer, P. Gouvine, J.F. Bartoli, P. Battioni, D. Mansuy, Biomimetic hydroxylation of aromatic compounds: hydrogen peroxide and manganese-polyhalogenated porphyrins as a particularly good system, Tetrahedron Lett. 31 (1990) 6645–6648.

[12] B. Meunier, A. Robert, G. Pratviel, J. Bernadou, in: K.M. Kadish, K.M. Smith, R. Guilard (Eds.), The Porphyrin Handbook, Academic Press, Amsterdam, 2000.

[13] J. Robbins Wolf, C.G. Hamaker, J.-P. Djukic, T. Kodadek, L.K. Woo, Shape and stereoselective cyclopropanation of alkenes catalyzed by iron porphyrins, J. Am. Chem. Soc. 117 (1995) 9194–9199.

[14] J.L. Maxwell, K.C. Brown, D.W. Bartley, T. Kodadek, Mechanism of the rhodium porphyrin-catalyzed cyclopropanation of alkenes, Science 256 (1992) 1544–1547.

[15] H.J. Callot, C. Piechocki, Cyclopropanation using rhodium (III) porphyrins: large cis vs trans selectivity, Tetrahedron Lett. 21 (1980) 3489–3492.

[16] P. Tagliatesta, A. Pastorini, Remarkable selectivity in the cyclopropanation reactions catalysed by an halogenated iron meso-tetraphenylporphyrin, J. Mol. Catal. A Chem. 198 (2003) 57–61.

[17] C.G. Hamaker, J.-P. Djukic, D.A. Smith, L.K. Woo, Mechanism of cyclopropanation reactions mediated by (5,10,15,20-tetra-p-tolylporphyrinato)osmium(II) complexes, Organometallics 20 (2001) 5189–5199.

[18] P. Tagliatesta, A. Pastorini, Electronic and steric effects on the stereoselectivity of cyclopropanation reactions catalysed by rhodium meso-tetraphenylporphyrins, J. Mol. Catal. A Chem. 185 (2002) 127–133.

[19] E. Galardon, P. Le Maux, G. Simonneaux, Insertion of ethyl diazoacetate into N—H and S—H bonds catalyzed by ruthenium porphyrin complexes, Chem. Commun. (1997) 2455–2456.

[20] C.-Y. Zhou, W.-Y. Yu, C.-M. Che, Ruthenium(II) porphyrin catalyzed tandem carbonyl ylide formation and 1,3-dipolar cycloaddition reactions of α-diazo ketones, Org. Lett. 4 (2002) 3235–3238.

[21] E. Galardon, P. Le Maux, G. Simonneaux, Cyclopropanation of alkenes with ethyl diazoacetate catalysed by ruthenium porphyrin complexes, J. Chem. Soc. Chem. Commun. (1997) 927–928.

[22] E. Galardon, S. Roue', P. Le, G. Maux, Simonneaux, Asymmetric cyclopropanation alkenes and diazocarbonyl insertion into SH bonds catalyzed by a chiral porphyrin Ru(II) complex, Tetrahedron Lett. 39 (1998) 2333–2334.

[23] F. Ragaini, A. Penoni, E. Gallo, S. Tollari, C. Li Gotti, M. Lapadula, E. Mangioni, S. Cenini, Amination of benzylic CH bonds by arylazides catalyzed by CoII–porphyrin complexes: a synthetic and mechanistic study, Chem. Eur. J. 9 (2003) 249–259.

[24] S.K.-Y. Leung, W.-M. Tsui, J.-S. Huang, C.-M. Che, J.-L. Liang, N. Zhu, Imido transfer from Bis(imido)ruthenium(VI) porphyrins to hydrocarbons: effect of imido substituents, C—H bond dissociation energies, and RuVI/V reduction potentials, J. Am. Chem. Soc. 127 (2005) 16629–16640.

[25] J.E. Jones, J.V. Ruppel, G. Gao, T.M. Moore, X. Peter Zhang, Cobalt-catalyzed asymmetric olefin aziridination with diphenylphosphoryl azide, J. Org. Chem. 73 (2008) 7260–7265.

[26] G.A. Mirafzal, G.L. Cheng, L.K. Woo, A new and efficient method for the selective olefination of aldehydes with ethyl diazoacetate catalyzed by an iron(II) porphyrin complex, J. Am. Chem. Soc. 124 (2002) 176–177.

[27] B.T. Holland, C. Walkup, A. Stein, Encapsulation, stabilization, and catalytic properties of flexible metal porphyrin complexes in MCM-41 with minimal electronic perturbation by the environment, J. Phys. Chem. 102 (1998) 4301–4309.

[28] Z. Li, C.-G. Xia, X.-M. Zhang, Preparation and catalysis of DMY and MCM-41 encapsulated cationic Mn(III)-porphyrin complex, J. Mol. Catal. A: Chem. 185 (2002) 47–56.

[29] P.R. Ortiz de Montellano (Ed.), Cytochrome P450: Structure, Mechanism, and Biochemistry, second ed., Plenum, New York, 1995.

[30] H. Nur, H. Hamid, S. Endud, H. Hamdan, Z. Ramli, Iron-porphyrin encapsulated in poly(methacrylic acid) and mesoporous A1-MCM-41 as catalyst in the oxidation of benzene to phenol, Mater. Chem. Phys. 96 (2006) 337–342.

[31] A.A. Costa, G.F. Ghesti, J.L. de Macedo, V.S. Braga, M.M. santos, J.A. Dias, S.C.L. Dias, Immobilization of Fe, Mn, and Co tetraphenylporphyrins complexes in MCM-41 and their catalytic activity in cyclohexene oxidation reaction by hydrogen peroxide, J. Mol. Catal. A: Chem. 282 (2008) 149–157.

[32] J. Poltowicz, K. Pamin, L. Matachowski, E.M. Serwicka, R. Mokaya, Y. Xia, Z, Oxidation of cyclooctane over Mn(TMPyP) porphyrin-exchanged Al, Si-mesoporous molecular sieves of MCM-41 and SBA-15, Catal. Today 114 (2006) 287–292.

[33] A.K. Rahiman, K. Rajesh, K.S. Bharathi, S. Sreedaran, V. Narayanan, Catalytic oxidation of alkenes by manganese(III) porphyrin-encapsulated Al, V, Si-mesoporous molecular sieves, Inorg. Chim. Acta 362 (2009) 1491–1500.

[34] A.K. Rahiman, K. Rajesh, K.S. Bharathi, S. Sreedaran, V. Narayanan, K. Rajesh, Cationic vanadyl porphyrin-encapsulated mesoporous Al/V-MCM-41 as heterogeneous catalyst for the oxidation of alkenes, Inorg. Chim. Acta 362 (2009) 1810–1818.

[35] A.K. Rahiman, K. Rajesh, K.S. Bharathi, S. Sreedaran, V. Narayanan, Manganese(III) porphyrin-encapsulated Ti, Si-mesoporous molecular sieves as heterogeneous catalysts for the epoxidation of alkenes, Appl. Catal. A Gen. 314 (2006) 216–225.

[36] D. Mansuy, Cytochromes P-450 and model systems: great diversity of catalyzed reactions, Pure Appl. Chem. 66 (1994) 737–744.

[37] A.K. Rahiman, K.S. Bharathi, S. Sreedaran, V. Narayanan, Epoxidation of styrene by Fe, Mn and V metalloporphyrins encapsulated Si, Al, Ti and V-Mcm-41, Catal. Lett. 127 (2009) 175–182.

[38] F. Yang, S. Gao, C. xiong, H. Wang, J. Chen, Y. Kong, Coordination of manganese porphyrins on amino-functionalized MCM-41 for heterogeneous catalysis of naphthalene hydroxylation, Chin. J. Catal. 36 (2015) 1035–1041.

[39] F.B. Zanardi, I.A. Barbosa, P.C. de Sousa Filho, L.D. Zanatta, D.L. da Silva, O.A. Serra, Y. Iamamoto, Manganese porphyrin functionalized on Fe_3O_4@nSiO_2@MCM-41 magnetic composite: structural characterization and catalytic activity as cytochrome P450 model, Microporous Mesoporous Mater. 219 (2016) 161–171.

[40] J.L. Zhang, Y.L. Liu, C.M. Che, Chiral ruthenium porphyrin encapsulated in ordered mesoporous molecular sieves (MCM-41 and MCM-48) as catalyst for asymmetric alkene epoxidation and cyclopropanation, Chem. Commun. (2002) 2906–2907.

[41] C.M. Che, J.S. Huang, F.W. Lee, Y. Li, T.S. Lai, H.L. Kwong, P.F. Teng, W.S. Lee, W.C. Lo, S.M. Peng, Z.Y. Zhou, Asymmetric inter- and intramolecular cyclopropanation of alkenes catalyzed by chiral ruthenium porphyrins. Synthesis and crystal structure of a chiral metalloporphyrin carbene complex, J. Am. Chem. Soc. 123 (2001) 4119–4129.

[42] J. Zhang, G.-F. Zhao, Z,. Popovic, Y. Lu, Y. Liu, Pd-porphyrin functionalized ionic liquid-modified mesoporous SBA-15: an efficient and recyclable catalyst for solvent-free Heck reaction, Mater. Res. Bull. 45 (2010) 1648–1653.

[43] V.H. Araújo Pinto, S. Rebouças, G.M. Ucoskia, E.H. de Fariac, B. Freitas Ferreira, R.A. Silva San Gild, S. Nakagaki, Mn porphyrins immobilized on non-modified andchloropropyl-functionalized mesoporous silica SBA-15 as catalysts forcyclohexane oxidation, Appl. Catal. A Gen. 526 (2016) 9–20.

[44] L.D. Zanatta, I.A. Barbosa, F.B. Zanardi, P.C. de Sousa Filho, L.B. Bolzon, A.P. Ramos, O.A. Serra, Y. Iamamoto, Hydrocarbon oxidation by iron-porphyrin immobilized on SBA-15 as biomimetic catalyst: role of silica surface, RSC Adv. (2016), http://dx.doi.org/10.1039/x0xx00000x. in press.

[45] S. Nakagaki, M. Halma, A. Bail, G.G. Carbajal Arízaga, F. Wypych, First insight into catalytic activity of anionic iron porphyrins immobilized on exfoliated layered double hydroxides, J. Colloid Interface Sci. 281 (2005) 417–423.

[46] (a) M. Reichinger, Poröse Silicate mit hierarchischer Porenstruktur: Synthese von mikro-/mesoporösen MCM-41 und MCM-48 Materialien aus zeolithischen Baueinheiten des MFI-Gerüst strukturtyps, PhD Thesis, The Institut für Geologie, Mineralogie und Geophysik, Bochum, Germany, 2007, p.10. (b) J.S. Beck, J.C. Vartuli, W.J. Roth, M.E. Leonowicz, C.T. Kresge, K.D. Schmitt, C.T.W. Chu, D.H. Olson, E.W. Sheppard, S.B. McCullen, J.B. Higgins, J.L. Schlenker, J. Am. Chem. Soc. 114 (1992) 10834–10843.

[47] S. Nakagaki, F.L. Benedito, F. Wypych, Anionic iron(III) porphyrin immobilized on silanized kaolinite as catalyst for oxidation reactions, J. Mol. Catal. 217 (2004) 121–131.

[48] S. Nakagaki, G. Sippel Machado, M. Halma, A.A. dos Santos Marangon, K.A. Dias de Freitas Castro, N. Mattoso, F. Wypych, Immobilization of iron porphyrins in tubular kaolinite obtained by an intercalation/delamination procedure, J. Catal. 242 (2006) 110–117.

[49] M. Halma, K. Aparecida Dias de Freitas Castro, C. Taviot-Gueho, V. Prévot, C. Forano, F. Wypych, S. Nakagaki, J. Catal. 257 (2008) 233–243.

[50] W. Zhou, P. Tian, F. Sun, M. He, Z. Chen, Efficient catalysis of the aerobic Baeyer–Villiger oxidation over a bifunctional catalyst based on cobalt tetraphenylporphyrin intercalated into Zn_2Al hydrotalcite Asian, Asian J. Org. Chem. 4 (2014) 33–37.

[51] W.-Y. Zhou, P. Tian, F. Sun, M.-Y. He, Q. Chen, Highly efficient transformation of alcohol to carbonyl compounds under a hybrid bifunctional catalyst originated from metalloporphyrins and hydrotalcite, J. Catal. 335 (2016) 105–116.

[52] M. Halma, A. Bail, F. Wypych, S. Nakagaki, Catalytic activity of anionic iron(III) porphyrins immobilized on grafted disordered silica obtained from acidic leached chrysotile, J. Mol. Catal. A 243 (2006) 44–51.

[53] M. Turan, Application of nanoporous zeolites for the removal of ammonium from wastewaters: a review, in: Low-Dimensional and Nanostructured Materials and Devices, Springer, 2016, pp. 477–504.

[54] M. Hartman, W. Schwieger, Hierarchically-structured porous materials: from basic understanding to applications, Chem. Soc. Rev. 45 (2016) 3311–3312.

[55] K.J. Balkus Jr., A.G. Gabrielov, S.L. Bell, F. Bedioui, L. Roué, J. Devynck, Zeolite encapsulated cobalt(II) and copper(II) perfluorophthalocyanines. Synthesis and characterization, Inorg. Chem. 34 (1994) 67–72.

[56] I.L. Viana Rosa, C.M.C.P. Manso, O.A. Serra, Y. Iamamoto, Biomimetical catalytic activity of iron(III) porphyrins encapsulated in zeolite Y, J. Mol. Catal. A 160 (2000) 199–208.

[57] M. Moghadam, S. Tangestaninejad, V. Mirkhani, I. Mohammadpoor-Baltork, M. Moosavifar, Host (nanocavity of zeolite-Y or X)-guest (manganese (III) tetrakis[4-N-methylpyridinum] porphyrin) nanocomposite materials as efficient catalysts for biomimetic alkene epoxidation with sodium periodate: shape-selective epoxidation of linear alkenes, J. Mol. Catal. A 302 (2009) 68–75.

[58] G. Karimipour, M. Rezaei, D. Ashouri, Zeolite encapsulated Fe-poprhyrin for catalytic oxidation with iodobenzene diacetate ($PhI(OAc)_2$), J. Mex. Chem. Soc. 57 (2013) 276–282.

[59] M. Otte, P.F. Kuijpers, O. Troeppner, I. Ivanović-Burmazović, J.N.H. Reek, B. de Bruin, Encapsulation of metalloporphyrins in a self-assembled cubic M_8L_6 cage: a new molecular flask for cobalt porphyrin-catalysed radical-type reactions, Chem. Eur. J. 19 (2013) 10170–10178.

[60] M. Otte, P.F. Kuijpers, O. Troeppner, I. Ivanović-Burmazović, J.N.H. Reek, B. de Bruin, Encapsulated cobaltporphyrin as a catalyst for size-selective radical-type cyclopropanation reactions, Chem. Eur. J. 20 (2014) 1–6.

[61] X.-L. Yang, M.-H. Xie, C. Zou, Y. He, B. Chen, M. O'Keeffe, C.-D. Wu, Porous metalloporphyrinic frameworks constructed from metal 5,10,15,20-tetrakis(3,5-biscarboxylphenyl)porphyrin for highly efficient and selective catalytic oxidation of alkylbenzenes, J. Am. Chem. Soc. 134 (2012) 10638–10645.

[62] M.H. Alkordi, Y. Liu, R.W. Larsen, J.F. Eubank, M. Eddaoudi, Zeolite-like metal-organic frameworks as platforms for applications: on metalloporphyrin-based catalysts, J. Am. Chem. Soc. 130 (2008) 12639–12641.

[63] Z. Zhang, L. Zhang, L. Wojtas, P. Nugent, M. Eddaoudi, M.J. Zaworotko, Templated synthesis, postsynthetic metal exchange, and properties of a porphyrin-encapsulating metal-organic material, J. Am. Chem. Soc. 134 (2012) 924–927.

[64] S.J. Lee, S.-H. Cho, K.L. Mulfort, D.M. Tiede, J.T. Hupp, S.B.T. Nguyen, Cavity-tailored, self-sorting supramolecular catalytic boxes for selective oxidation, J. Am. Chem. Soc. 130 (2008) 16828–16829.

[65] A.M. Shultz, O.K. Farha, J.T. Hupp, S.B.T. Nguyen, A catalytically active, permanently microporous MOF with metalloporphyrin struts, J. Am. Chem. Soc. 130 (2008) 16828–16829.

[66] R.W. Larsen, L. Wojtas, J. Perman, R.L. Musselman, M.J. Zaworotko, C.M. Vetromile, Mimicking heme enzymes in the solid state: metal organic materials with selectively encapsulated heme, J. Am. Chem. Soc. 133 (2011) 10356–10359.

[67] J.A.A.W. Elemans, E.J.A. Bijsterveld, A.E. Rowan, R.J.M. Nolte, A host-guest epoxidation catalyst with enhanced activity and stability, Chem. Commun. (2000) 2443–2444.

[68] R. Naik, P. Joshi, S. Umbarkar, R.K. Deshpande, Polystyrene encapsulation of manganese porphyrins: highly efficient catalysts for oxidation of olefins, Catal. Commun. 6 (2005) 125–129.

[69] M.E. Niño, S.A. Giraldo, E.A. Páez-Mozo, Olefin oxidation with dioxygen catalyzed by porphyrins and phthalocyanines intercalated in α-zirconium phosphate, J. Mol. Catal. A 175 (2001) 139–151.

[70] J.-L. Zhang, H.-B. Zhou, J.-S. Huang, C.-M. Che, Dendritic ruthenium porphyrins: a new class of highly selective catalysts for alkene epoxidation and cyclopropanation, Chem. Eur. J. 8 (2002) 1554–1562.

Chapter 9

Palladium Complexes and Nanoparticles Encapsulated by Functionalized Mesoporous Silica Materials: A Promising Hybrid Catalyst in Organic Transformations

Esmail Doustkhah and Sadegh Rostamnia
University of Maragheh, Maragheh, Iran

Contents

1 INTRODUCTION

Palladium encapsulation through mesoporous materials is an important topic in catalysis. This is due to the rapid aggregation and reduction of Pd ions. To circumvent these issues, palladium species are need to be conjugated and stabilized on a support. In addition, when Pd species are supported on a surface, their recovery makes the catalytic system more precious rather than homogeneous catalysts. Furthermore, the possibility of reusing of these catalysts for several consecutive

Encapsulated Catalysts. http://dx.doi.org/10.1016/B978-0-12-803836-9.00009-2

cycles makes the process cost-effective and favorable. These advantages are also effective factors in the development of green chemistry [1–3].

There are various methods for the encapsulation of Pd species. When using Pd nanoparticles (NPs), because of their generation inside the pore channels, aggregation and growing of the particle size have been suppressed thanks to encapsulation of Pd NPs inside the pore channels. When using the Pd ions, the type of ligand used for stabilization of Pd species and preventing the Pd leaching is crucial. Silica itself has the ability to support Pd species without use of any ligand. However, in this method, the catalyst undergoes more Pd leaching and as an alternative, Pd ions are reduced to Pd NPs. Pd^{2+} and Pd NPs are catalytically active species in which their catalytic domain can be identical or not identical. In Fig. 9.1, some total approaches for the encapsulation of palladium species inside the mesoporous silica (MS) materials have been illustrated (Scheme 9.1) [4–6].

2 UNFUNCTIONALIZED MS FOR PD ENCAPSULATION

In this method, there is no postmodification or co-condensation with functional precursors. The silanol groups on the surface of the silica are responsible for supporting the Pd species. In this approach, Pd is usually applied as NPs to be anchored within the pores and will be encapsulated by pore walls. Therefore, leaching of Pd will significantly decrease. Encapsulation of the generated Pd NPs within the pores of silica upon particle formation is a promising approach to preserve Pd NPs. However, as silanol groups are the only agents playing the role of ligand, the control of particle size and stabilization of Pd NPs are somehow difficult compared to that of functionalized mesopores.

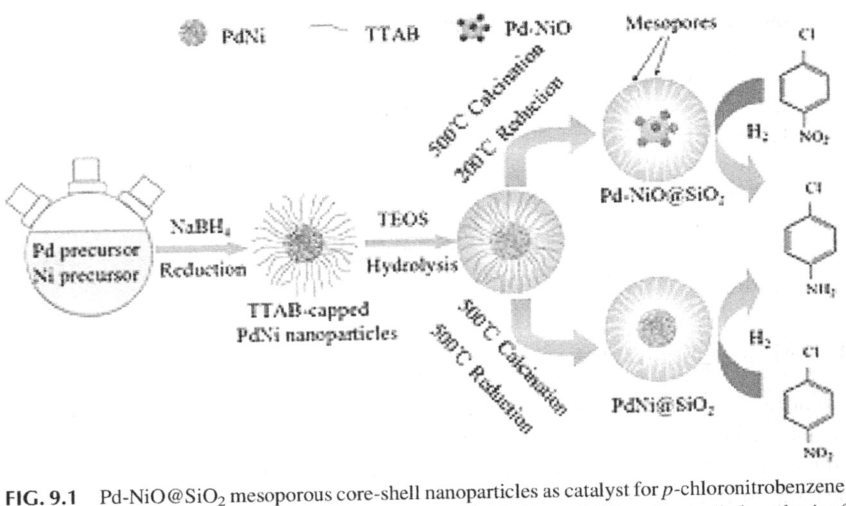

FIG. 9.1 Pd-NiO@SiO_2 mesoporous core-shell nanoparticles as catalyst for *p*-chloronitrobenzene hydrogenation with H_2. *(Reproduced from H. Liu, K. Tao, C. Xiong, S. Zhou, Controlled synthesis of Pd-NiO@SiO$_2$ mesoporous core-shell nanoparticles and their enhanced catalytic performance for p-chloronitrobenzene hydrogenation with H$_2$, Catal. Sci. Technol., 5 (2015) 405–414, with permission of Royal Society of Chemistry.)*

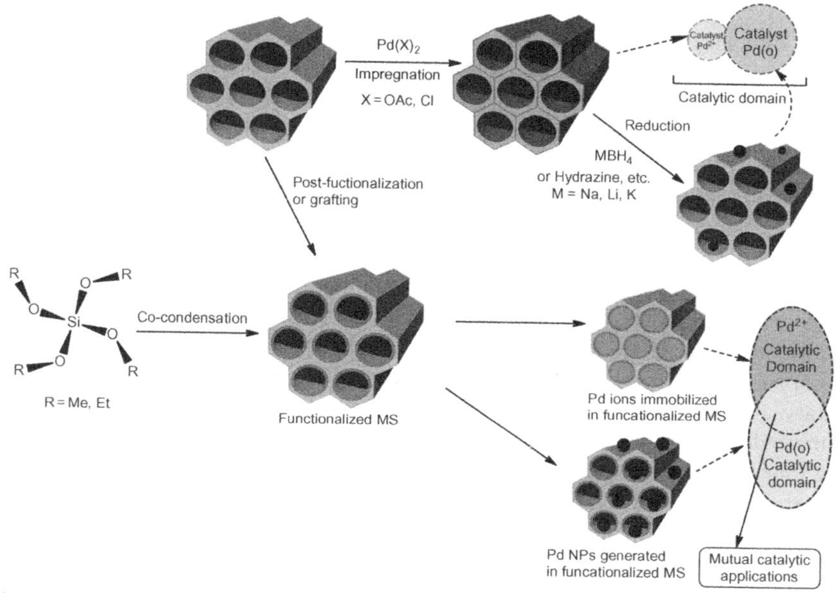

SCHEME 9.1 A general outline for the synthesis of Pd-encapsulated mesoporous silica catalysts.

It is proved that in some cases, bimetallic catalysts exhibit superior catalytic activity compared to their individual counterparts [7]. As an example, supporting Pd and Au in MS prepared by an organic impregnation-hydrogen reduction approach showed unique catalytic activity in the hydrogenation of cinnamaldehyde. The effects of MS support on the dispersion of Pd/Au, the catalyst performance, as well as the effect of the addition of gold to Pd on the catalytic activity were all evaluated by Liao [8]. The authors also investigated the effects of reaction parameters such as solvent, pressure, and temperature on the catalyst activity. It was established that the molar ratio of gold to palladium on the support had a significant effect on the catalytic performance. Using a molar ratio of 0.2:1, the catalytic activity increased four times higher than that of Pd/MS (without Au as a promoter) and eight times higher than that of the commercial Pd/C catalyst. This observation can be attributed to the synergistic effect of Pd and Au as well as the high dispersion of active components [8].

Au supported (Au/SBA-15) and Au-Pd supported (Au-Pd/SBA-15) materials, synthesized by impregnation and grafting methods, were also studied as the catalysts. An Au-Pd/SBA-15 sample, prepared by the grafting method, had high dispersed NPs inside the channels of SBA-15 and exhibited excellent catalytic activity for the selective oxidation of benzyl alcohol to benzaldehyde. The channels of SBA-15 were excellent barriers for preventing Au and Au-Pd NPs from agglomeration and leaching. Moreover, the incorporation of Pd to Au/SBA-15 decreased the size of Au NPs in both methods [9].

SBA-15 and MSU-2, both as MS materials, were incorporated for encapsulation of a specific palladium complex [$PdCl_2$(cod)] (cod = 1,5-cyclooctadiene). The obtained catalyst was applied for the catalysis of the Suzuki-Miyaura coupling reaction. Pd complex encapsulation was achieved in the absence of any grafted ligand to the surface of MS and the Pd complex was inserted into the pore channels by the impregnation method. Characterization of synthesized materials by different techniques indicated that the palladium NPs remained impregnated in the silica. Interestingly, both catalysts exhibited high recyclability and could be used for five consecutive catalytic runs without a significant amount of Pd leaching from the heterogeneous surface of MS and the loss of the catalytic activity. This observation confirmed the true heterogeneity of the catalysts [10].

Sayari reported the utility of pore-expanded silica MCM-41 as a support for encapsulation of monodispersed, highly active Pd NPs. For preparation of the catalyst, pore-expanded MCM-41 was synthesized by introducing *N,N*-dimethyldecylamine within the pores during the postsynthesis. This drove the metallic cations inside the pores before their reduction. Owing to the fact that the amine was weakly supported, its elimination was achieved by washing the catalyst with ethanol. The obtained Pd NPs thus were found to be highly active, recyclable, and stable catalysts for the Suzuki coupling reaction in both organic and aqueous media.

In another attempt, Zhou and coworkers applied an interesting strategy for fabrication of core-shell mesoporous nanocatalysts with Pd-NiO nanocore and MS shells. The hybrid system was successfully synthesized by a sol-gel method [11]. The process included introduction of surfactant-capped PdNi alloy NPs during condensation of tetraethylorthosilicate (TEOS) followed by the removal of surfactants over calcination at 500°C. Calcination at a higher temperature converted metallic Ni to NiO. In this step, H_2 reduction at 200°C led to the formation of Pd-NiO core inside the MS shell while reduction at 500°C resulted in the formation of Pd-Ni alloy nanocore inside the MS shell. The obtained Pd-NiO@SiO_2 nanocatalysts exhibited excellent catalytic performance for *p*-chloronitrobenzene hydrogenation with H_2. The enhancement in the catalytic activity of Pd-NiO@SiO_2 was related to the strong interaction between Pd and NiO in the core (Fig. 9.1).

Pd species can be embedded inside the structure of MS by a one-step sol-gel method. In this method, Pd ions are added during the synthesis of MS from the condensation of the silica precursor. For example, $PdCl_2$ can be added in the course of SBA-15 synthesis, in other words, hydrolysis and condensation of TEOS in the presence of P123. Calcination at 500°C and subsequent reduction of Pd oxide to Pd NPs under H_2 at 200°C can furnish Pd/SBA-15 in which the Pd NPs are supported onto SBA-15.

Conversion of glycerol to the key intermediate, 1,2,3-tribromopropane and the introduction of the alkyl group using a Suzuki coupling reaction was achieved by one-step sol-gel synthesized Pd/SBA-15. Different wt% of Pd (10%, 15%, and 20%) led to different catalytic behaviors. 20 wt% of Pd in Pd/SBA-15 resulted in good catalytic activity (complete conversion and 64% selectivity of 1,2,3-tribromopropane) at 90°C with methylboronic acid. The recyclability of the catalyst up to 5th cycle was investigated. In 20 wt% of Pd/SBA-15, a significant leaching of palladium was observed in the first run, which decreased the catalytic activity [12]. However, no significant loss of the catalytic activity was observed in later runs.

Composites of silica-carbon mesopores can exhibit superior behavior for the encapsulation of Pd rather than each of the individual component, MSs, and carbonaceous materials. This can be due to simultaneous effects of each incorporated component. An example of these composites in which Pd was supported by the impregnation method exhibited superior catalytic activity rather than single mesoporous carbon or silica in the Heck coupling reaction of chlorobenzene and styrene at 100°C and in the Ullmann coupling reaction of chlorobenzene at 30°C under aqueous conditions. This heterogeneous catalyst was stable and exhibited high NP dispersion and insignificant Pd leaching over reusing for more than 20 times [13].

Incorporation of Fe_3O_4 NPs is an interesting method for production of magnetically recoverable nanaocatalysts [14]. The composite of carbon nanotube with MS containing magnetic NPs is another efficient example that was applied for the encapsulation of the Pd species. This catalyst was examined in the Suzuki-coupling reaction. The induced magnetism facilitated the separation of the catalyst from the reaction media. In addition, the catalyst could be reused for several times with preserving significant activity. Moreover, the existence of a MS layer made the Pd catalyst more stable and prevented Pd leaching from the CNT support (Fig. 9.2) [15].

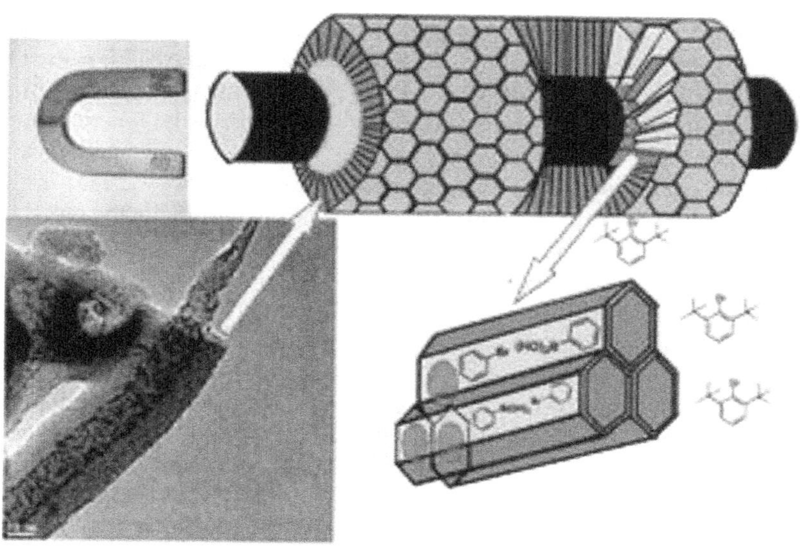

FIG. 9.2 The composite of carbon nanotube with MS containing magnetic NPs for encapsulation of the Pd species. *(Reproduced from W. Zhang, X. Chen, T. Tang, E. Mijowska, Superstable magnetic nanoreactors with high efficiency for Suzuki-coupling reactions, Nanoscale, 6 (2014) 12884–12889, with the permission of the Royal Society of Chemistry.)*

3 THE ROLE OF FUNCTIONAL GROUPS OF THE MS SURFACE

3.1 Amino-Functionalized MSs

Multifunctionalized MSs are important supports for the efficient encapsulation of Pd ions and NPs. In other words, the multiple functional groups are complementary to each other and can minimize the leaching of Pd species from the support. Some ligands such as phosphorous and carbine-based ligands can denote catalytic potential to the palladium to some extent.

Co-condensation of TEOS, (3-aminopropyl)triethoxysilane (APTES) and phenyltriethoxysilane (PTES) in the presence of P123 furnishes bifunctionalized amino-phenyl functionalized SBA-15 (NH_2&Ph-SBA-15), which is a promising support for encapsulation of Pd. The generated Pd NPs inside the NH_2&Ph-SBA-15 mesoporous (Pd/NH_2&Ph-SBA-15) didn't lead to the collapse of the hexagonal mesoporous array and it preserved two organic functional groups on its surface. In comparison with unfunctionalized SBA-15, this catalyst showed higher catalytic activity and selectivity in an Ullman coupling reaction under similar aqueous conditions. The addition of a phenyl group could lead to the uniform dispersion of Pd NPs inside the pore channels. On the other hand, phenyl groups could increase the mass transfer inside the pores of the catalyst. Recoverability and recyclability of the catalyst was at least six times without significant loss of activity and selectivity (Fig. 9.3) [16].

Palladium oxide is also a catalytically active species especially in oxidation reactions. Calcination of Pd complexes at higher temperatures can furnish PdO. To prevent this catalytic species from agglomeration, it is favored to generate it on the surface of a support. In this regard, MCM-41 can be considered as a suitable candidate for embedding Pd oxide NPs inside its pores. APTES is used to functionalize the surface of MCM-41 and support Pd ions. To prevent Pd from direct attachment to silanol groups of the surface, trimethylsilyl functionalities were incorporated for capping the surface. Finally its calcination at a higher temperature in air led to the formation of PdO NPs ($\sim$ 2 nm) inside MCM-41 pores. This catalyst was used in methane combustion in the presence of O_2 (Fig. 9.4) [17].

Aminofunctionalized SBA-15 with a new organosiloxane precursor was found to be promising ligands for supporting Pd ions and developing a new catalyst for promoting the Suzuki coupling reaction under aqueous conditions with no need to have an inert atmosphere. The amine precursor, the analogous of the ethylenediamine ligand, was more active and selective than mono amine functionality (APTES). Also, it led to an insignificant amount of Pd leaching from the surface of SBA-15 and could be used for several times with excellent recyclability (Fig. 9.5) [18].

For some complexes such as Grubbs-Hoveyda metathesis catalyst, the chemical structure of the ligand is so important in some especial reactions that the parent form the complex should be preserved [19]. Thus, their encapsulation into a heterogeneous surface should be accompanied by preservation of their

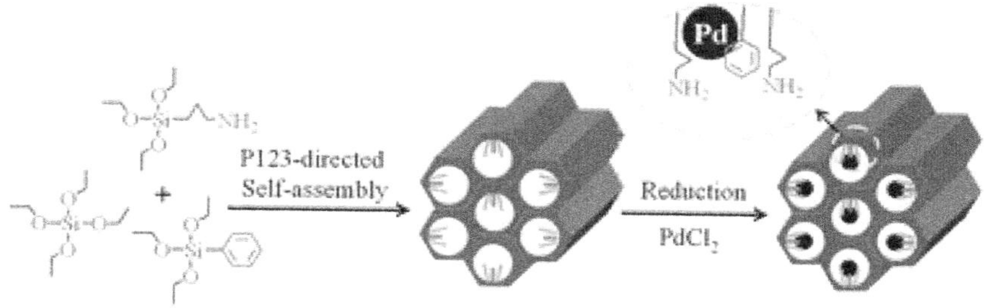

FIG. 9.3 Bifunctionalized phenylamine-based mesoporous silica for Pd nanoparticles generations. *(Reproduced from J. Huang, J. Yin, W. Chai, C. Liang, J. Shen, F. Zhang, Multifunctional mesoporous silica supported palladium nanoparticles as efficient and reusable catalyst for water-medium Ullmann reaction, New J. Chem. 36 (2012) 1378–1384, with the permission of the Royal Society of Chemistry.)*

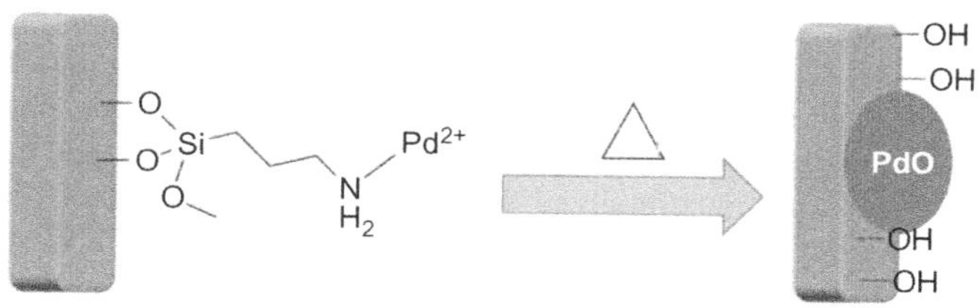

FIG. 9.4 PdO inside MS for CH_4 oxidation. *(Reproduced from S. Zribi, B. Albela, L. Bonneviot, M.S. Zina, Surface engineering and palladium dispersion in MCM-41 for methane oxidation, Appl. Catal., A 502 (2015) 195–203, with the permission of Elsevier.)*

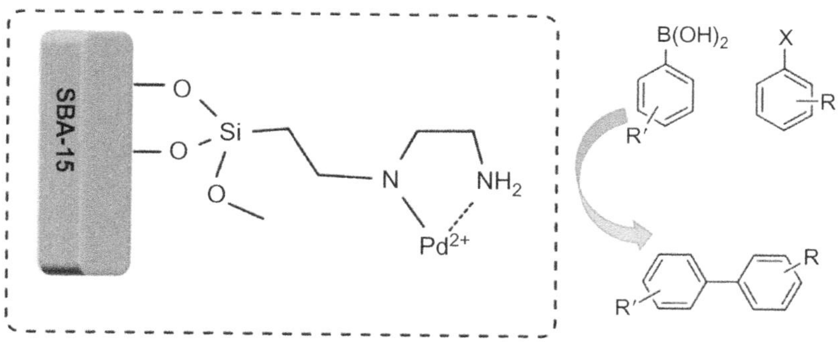

FIG. 9.5 Ethylenediamine for Pd encapsulation inside SBA-15. *(Reproduced from S. Rostamnia, H. Xin, Pd(OAc)$_2$@SBA-15/PrEn nanoreactor: a highly active, reusable and selective phosphine-free catalyst for Suzuki–Miyauracross-coupling reaction in aqueous media, Appl. Organomet. Chem. 27 (2013) 348–352, with the permission of John Wiley and Sons.)*

main chemical structure. This can be achieved by tethering or electrostatic immobilization of the Pd complex.

Pd(pyca)(PPh_3)(OTs) [pyca = 2-picolinate] is a complex that is anchored inside aminoporpylated mesoporous MCM-41, MCM-48, and SBA-15 by tethering to amine functionality. These materials were successfully designed as active and selective catalysts for the carbonylation of different aryl olefins and alcohols. The main concerns in this regard, the stability and true heterogeneous nature of all the anchored catalysts, were studied by investigating the recyclability of the catalyst, which was confirmed for several consecutive runs (Fig. 9.6) [20].

Aldoxime can act as an excellent ligand for generation of the Pd complex. Cyanofunctionalized SBA-15 was reacted with hydroxylamine to produce covalently anchored amino-aldoxime SBA-15. The latter was used to encapsulate Pd NPs inside the pores. The catalyst was a novel phosphine-free recyclable heterogeneous catalyst applicable in the Suzuki coupling reaction of aryl halides (I, Br, Cl) and phenylboronic acid. The products were obtained in excellent yields under mild conditions at extremely low palladium loading (~0.2 mol%). The catalyst showed good heterogeneous behavior and could be recovered by simple filtration and reused for at least six times without significant loss of its activity (Fig. 9.7) [21].

As mentioned before, mesoporous composites are good candidates for the catalytic cases. The mesoporous carbon nanosphere (MCN) was encapsulated by aminofunctionalized MS for supporting Pd species. Hydrophobic mesoporous carbon with aminofunctionalized MS was coordinated to a palladium precursor. Subsequently, the reduction of palladium precursor to Pd NPs led to the formation of a hybrid nanocatalyst with a hydrophobic carbon core and

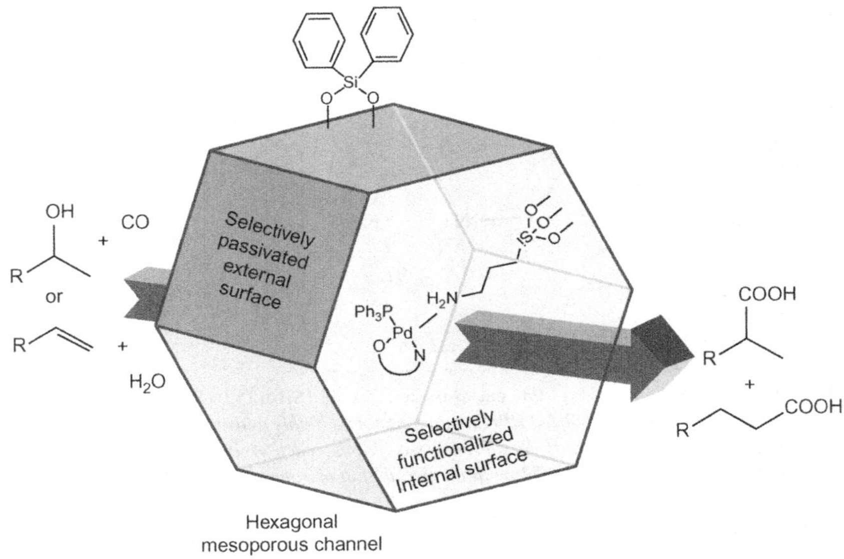

FIG. 9.6 Encapsulation of Pd complex inside MS. *(Reproduced from B.R. Sarkar, R.V. Chaudhari, Anchored Pd-complexes in mesoporous supports: synthesis, characterization and catalysis studies for carbonylation reactions, Catal. Today, 198 (2012) 154–173, with the permission of Elsevier.)*

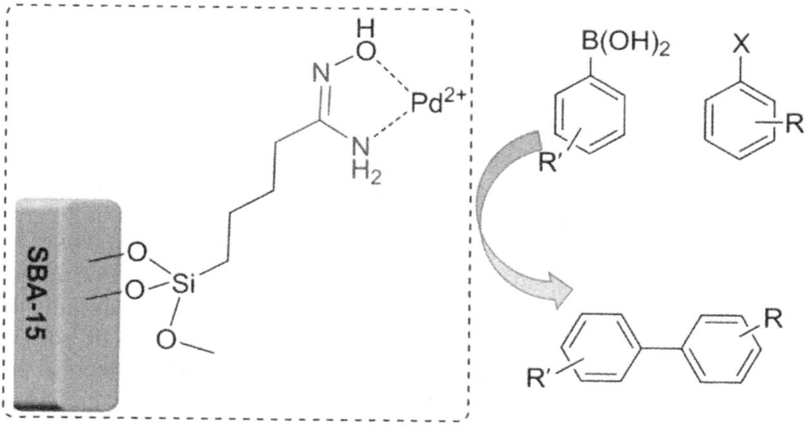

FIG. 9.7 Amino-aldoxime SBA-15 for encapsulation of Pd. *(Reproduced from R. Ghorbani-Vaghei, S. Hemmati, H. Veisi, Pd immobilized on amidoxime-functionalized mesoporous SBA-15: a novel and highly active heterogeneous catalyst for Suzuki–Miyaura coupling reactions, J. Mol. Catal. A: Chem. 393 (2014) 240–247, with the permission of Elsevier.)*

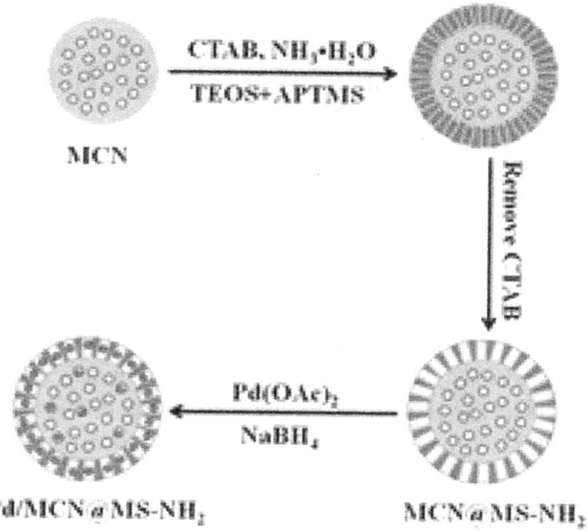

FIG. 9.8 Pd NPS within the mesoporous shell of MCN. *(Reproduced from F. Zhang, S. Chen, H. Li, X.-M. Zhang, H. Yang, Pd nanoparticles embedded in the outershell of a mesoporous core-shell catalyst for phenol hydrogenation in pure water, RSC Adv. 5 (2015) 102811–102817, with the permission of the Royal Society of Chemistry.)*

aminofunctionalized silica shell. This catalyst had high catalytic activity and selectivity towards hydrogenation of phenol to cyclohexanone under 1 atm of H_2 at 80°C for 3 h in pure water (conversion > 99% and selectivity 99%). Furthermore, the catalyst was highly recoverable and could be reused for at least six times without significant loss of the catalytic activity (Fig. 9.8) [22].

3.2 Sulfur-Based Functional Groups

Thiol functional groups are important ligands for supporting and stabilizing Pd. 3-Mercaptopropyltrimethoxysilane (MPTMS) is a suitable precursor for functionalization of the MS surface by thiol functionality. This precursor is widely applied for the functionalization of various types of silica mesostructures with trimethoxy- moiety. On the other hand, some other thiol precursors have been utilized for the functionalization of the mesopores.

Shimizu [23] incorporated MPTMS for thio-functionalization of FSM-16 MS, and then for the encapsulation of Pd(II), to afford Pd@FSM-SH. This catalyst exhibited active, stable, and recyclable heterogeneous features for the catalysis of Heck and Suzuki cross-coupling reactions. The catalytic activity of the Pd species in Pd@FSM-SH was compared with MPTMS-functionalized nonporous silica-supported Pd(II) (Pd-SH-SiO_2), unfunctionalized FSM-16 supported $Pd(OAc)_2$ (Pd-FSM), and previously reported heterogeneous catalysts (Pd zeolite and Pd/C). It was proved that the aggregation behavior of Pd species in the reaction drastically depended upon the support. Recovery of Pd-SH-FSM after Heck and Suzuki reactions led to lower aggregation rather than other catalysts. Moreover, a large fraction of the Pd(II) species were converted to Pd clusters after the Suzuki reaction in nonporous silica. In ligand-free Pd-FSM, Pd species tended to aggregate to bulkier species. The activity order of the recycled catalysts were Pd-SH-FSM > Pd-SH-SiO_2 > Pd-FSM, which indicated that the encapsulation of Pd(II) within the thiol functionalized porous framework avoided Pd aggregation. Therefore, the thiol ligand played a significant role in controlling the size distribution of generated Pd NPs and preventing aggregation of the coordinated Pd complexes. This phenomenon resulted in high durability and recyclability of Pd@FSM-SH.

Functionalization of SBA-15 with MPTMS also leads to generation of a suitable support for the encapsulation of Pd ions [24]. Likewise, this catalyst exhibited an excellent catalytic behavior in Suzuki-Miyaura and Mizoroki-Heck coupling reactions. Studies proved Pd leaching less than 3 ppb after recovery. Conversely, use of aminopropylated silica in the catalysis of the Mizoroki-Heck reaction had a significant amount of leaching (35 ppm), uncovering the importance of the thiol ligand to maintain the active state of Pd on the surface. Furthermore, hot filtration and three-phase tests confirmed that the reaction was occurring predominantly via surface-bound Pd species. In comparison, palladium in thiofunctionalized SBA-15 (through MPTMS) was less active than its corresponding palladium supported on KIT-6 with thiofunctionality in Suzuki coupling. Also, the higher recyclability of KIT-6 compared to SBA-15 was attributed to a greater redistribution of Pd to the external surface of the material, which resulted in Pd NPs not being constrained by the size of the pores. These Pd NPs were then available for catalysis after the material collapsed, whereas in the case of SBA-15-based materials, the Pd was captured inside the pores and, therefore, catalytic activity ceases when the mesostructure collapses. In these catalysts, the Ostwald ripening phenomenon leads to larger particles and reduced catalytic activity [25].

The method of MPTMS incorporation onto the mesoporous surface is also crucial in the stability of the MS-based catalysts. Moreover, the reaction conditions such as solvent and reaction atmosphere can influence the stability of the catalyst. In comparison, a grafting approach produces a significantly more stable catalyst, while incorporation of thiol by the co-condensation method gives a less stable material under the reaction conditions [26].

In fact, among sulfur based ligands, thiofunctionalized MS materials are the most common types for immobilizing Pd species inside the pore channels. Also, Pd ions and NPs can be immobilized and generated on the sulfonated surface. These can also be used as catalysts in coupling reactions with high catalytic activity. SBA-15/PrSO_3K is an example that was applied for the immobilization of Pd NPs inside the pore channels. The catalytic activity of this catalyst was investigated for the Suzuki coupling reaction. This heterogeneous palladium precatalyst could be separated from the reaction mixture and reused at least 11 times [27]. The two types of the catalysts including SBA-15/PrSO_3Pd and SBA-15/PrSO_3PdNPs were also applied for the Heck arylation reaction of conjugate alkenes with aryl halides under aerobic conditions with good to excellent yields. These supported palladium catalysts were recoverable and reusable for several times, showing the true heterogeneity of the catalysts in the reactions (Fig. 9.9) [28].

3.3 Phosphorus-Based Functional Groups

Palladium complexes can be encapsulated by coordinating to phosphinite ligands on silica SBA-15 ($PdCl_2(PPh_2)_2$@SBA-15), and can be used as the catalysts for the double carbonylation of aryl iodides with secondary and primary amines to produce α-ketoamides. High conversions (up to 80%) and excellent selectivities (up to 96%) towards products were obtained by using K_2CO_3 as a base, MEK or DMF as a solvent, and a 1 mol% [Pd] catalyst. By changing the ligand -PPh_2 to -PCy_2, a similar selectivity and reactivity was obtained without loss of the activity. Also, the catalyst, $PdCl_2(PPh_2)_2$@SBA-15, showed excellent recyclability up to 3 cycles without loss of the activity in the catalysis [29].

In phosphorus-based catalysts, it is difficult to determine the amounts of the two coordination modes of the Pd nucleus, whether the Pd is coordinated to one or two phosphorus atoms. The ^{31}P double-quantum filtered (DQ-filtered) method in solid-state NMR can be used for studying palladium coordination to phosphorous ligands. Applying the DQ-filtered method, the amounts of the two different kinds of palladium coordination modes can be determined. Also, the interatomic distance of two ^{31}P nuclei bonded to a palladium nucleus can be estimated (Scheme 9.2).

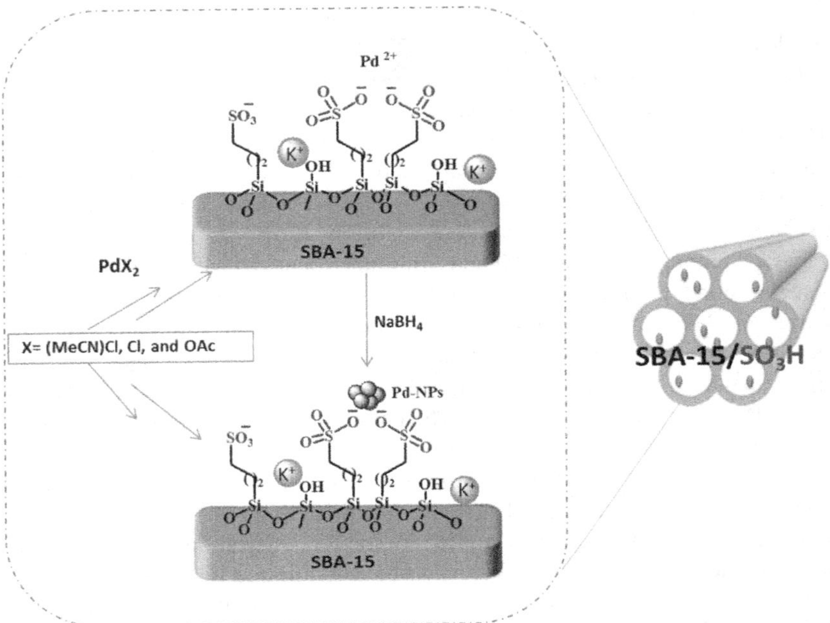

FIG. 9.9 SBA-15/PrSO$_3$Pd$_{NPs}$ chemical structure and production pathway. *(Reproduced from S. Rostamnia, T. Rahmani, Ordered mesoporous SBA-15/PrSO$_3$Pd and SBA-15/PrSO$_3$PdNP as active, reusable and selective phosphine-free catalysts in C-X activation Heck coupling process, Appl. Organomet. Chem. 29 (2015) 471–474, with the permission of John Wiley and Sons.)*

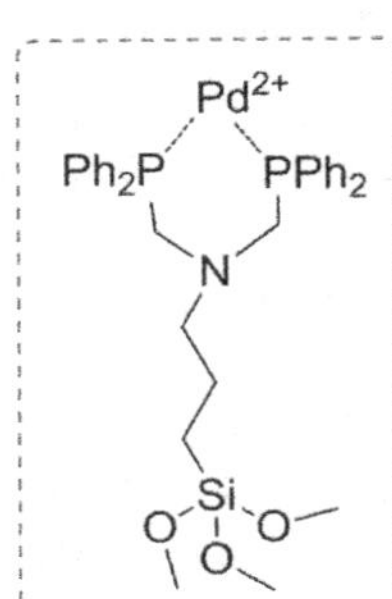

SCHEME 9.2 *(Reproduced from M. Genelot, N. Villandier, A. Bendjeriou, P. Jaithong, L. Djakovitch, V. Dufaud, Palladium complexes grafted onto mesoporous silica catalysed the double carbonylation of aryl iodides with amines to give α-ketoamides, Catal. Sci. Technol. 2 (2012) 1886–1893, with the permission of the Royal Society of Chemistry.)*

A series of mono-, di- and tri-phosphinite ligands functionalized on modified MSs were synthesized, characterized, and subsequently incorporated to immobilize Pd^{2+} (Table 9.1). Well Pd NP dispersion with narrow size distribution was achieved within these heterogeneous catalytic systems. The catalytic activities of these Pd-contained catalysts were investigated in a typical Heck

TABLE 9.1 Pd Encapsulation Inside MS Through Various Types of Phosphorous-Based Ligands

Entry	Catalyst	Phosphorous Loading (mmol/g)	Pd Loading (mmol/g)
1	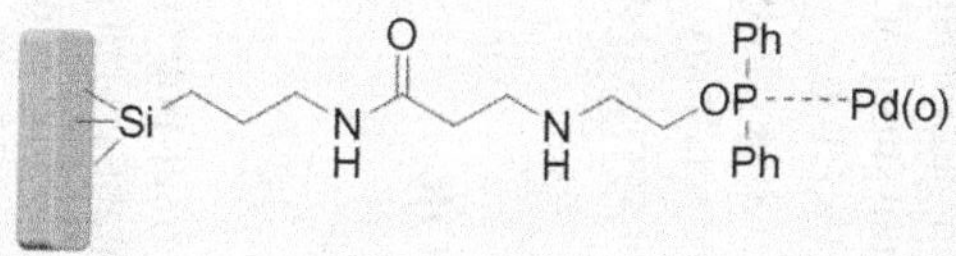	0.5	0.06
2	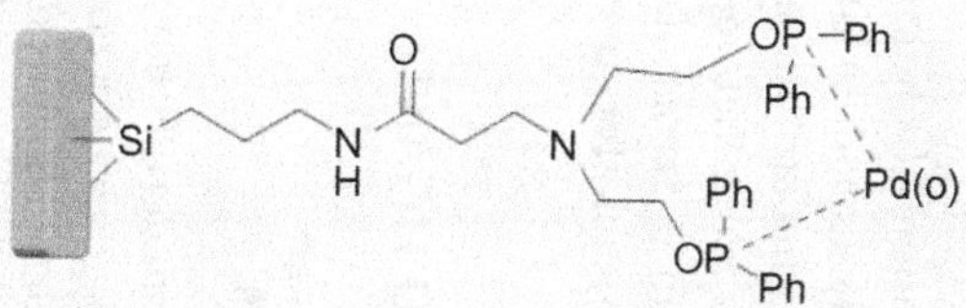	3.5	0.34
3	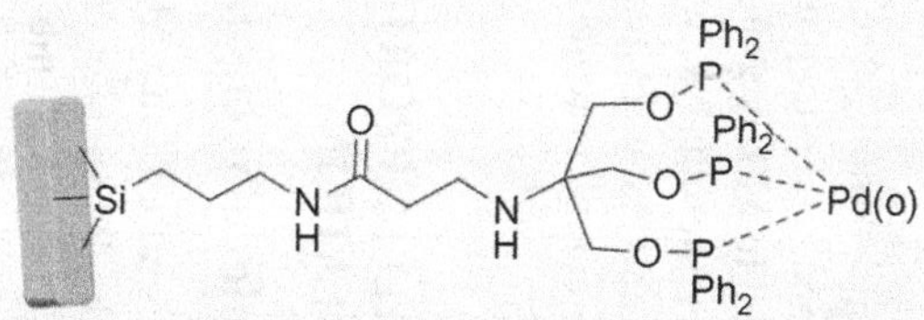	2.3	0.26

(Reproduced from F. Farjadian, M. Hosseini, S. Ghasemi, B. Tamami, Phosphinite-functionalized silica and hexagonal mesoporous silica containing palladium nanoparticles in Heck coupling reaction: synthesis, characterization, and catalytic activity, RSC Adv. 5 (2015) 79976–79987, with the permission of the Royal Society of Chemistry.)

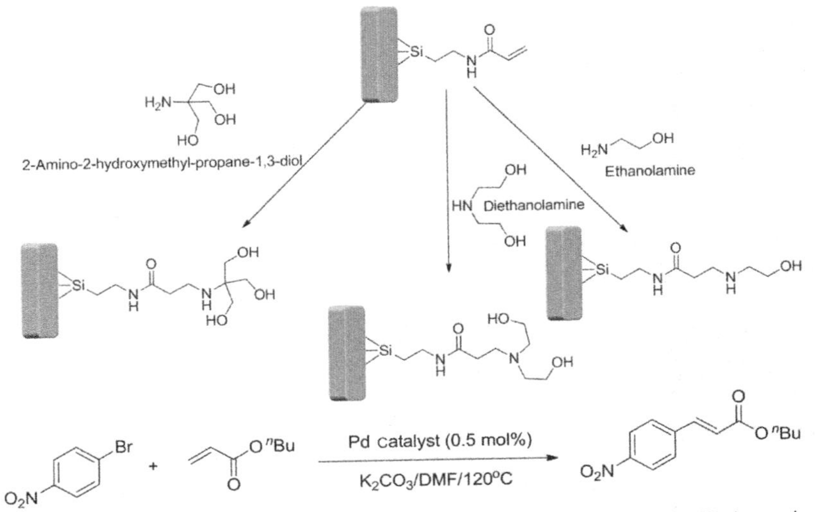

SCHEME 9.3 Amidoalcohol ligands in Pd encapsulation and its catalysis in the Heck reaction. *(Reproduced from F. Farjadian, M. Hosseini, S. Ghasemi, B. Tamami, Phosphinite-functionalized silica and hexagonal mesoporous silica containing palladium nanoparticles in Heck coupling reaction: synthesis, characterization, and catalytic activity, RSC Adv. 5 (2015) 79976, with the permission of the Royal Society of Chemistry.)*

coupling reaction. Among the aminoalcohol and aminophosphinite-based ligands, difunctionalized phosphinite ligands exhibited superior results for being postfunctionalized to MS by generating uniform and well-dispersed Pd NPs. To elucidate whether the catalyst was a true heterogeneous catalyst, the reaction with identical catalysts were tested in terms of recyclability and reusability. Accordingly, the catalysts were reusable for several cycles (Scheme 9.3) [30].

4 ADVANCED SUPPORTED LIGANDS FOR PD ENCAPSULATION

4.1 Supramolecular Ligands

Supramolecules grafted onto MS materials have great potential for grabbing and stabilizing the Pd species. To generate a covalently anchored supramolecule on the surface of MS, the surface synthesis method can be useful. This method can be used to introduce various generations (G1, G2, and G3) of dendrons on MS. However, the relatively bulkier size of dendrons as well as pore size restrictions limit the growth of the dendron to some extent. This system can lead to a very insignificant amount of leaching of Pd because of the functional-rich array.

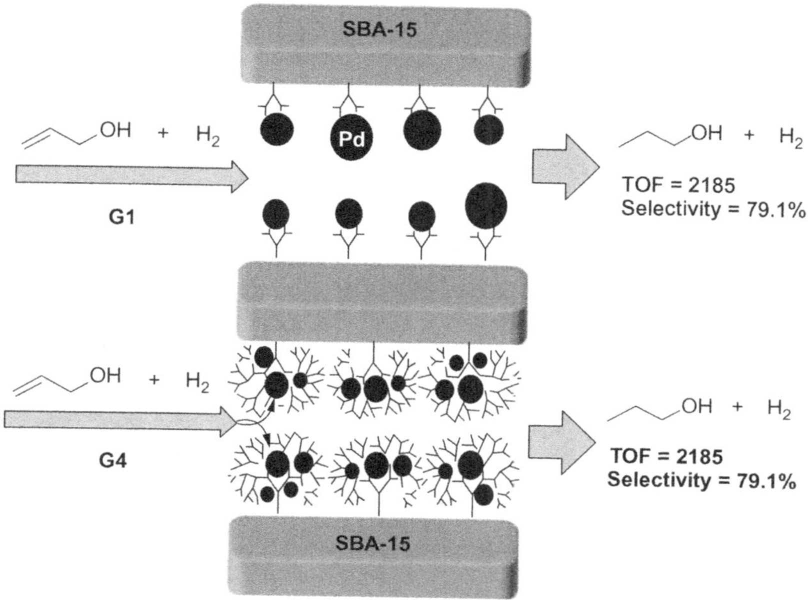

FIG. 9.10 Pd NPs inside dendron-supported SBA-15 and hydrogenation of allyl alcohol. *(Reproduced from Y. Jiang, Q. Gao, Heterogeneous hydrogenation catalyses over recyclable Pd(0) nanoparticle catalysts stabilized by PAMAM-SBA-15 organic-inorganic hybrid composites, J. Am. Chem. Soc. 128 (2006) 716–717, with the permission of the American Chemical Society.)*

MS SBA-15 was subjected to aminofunctionalization and subsequent Michael addition of amine to methyl acrylate followed by amidation of ethylenediamine with methyl esters to generate dendron G*n*-PAMAM-SBA-15 ($n = 1$–4). In the next step, Pd(0) NPs were produced and stabilized among the branches of the anchored dendrimer for further use as a catalyst. The activity of the Pd(0)-G4-PAMAM-SBA-15 catalyst was 1.5 times higher than the fourth-generation PAMAM encapsulated Pd(0) homogeneous catalyst. These catalysts were also easily recoverable, reusable for several times, and stable for one month in air, confirming the high catalytic efficiency (Fig. 9.10) [31].

A new supramolecular ligand based on the 2,4,6-trialkoxy-1,3,5-triazine was designed and synthesized for modification of SBA-15 and subsequent Pd ion encapsulation. This catalyst was tested in the cross-coupling of copper-free Sonogashira and Mizoroki-Heck reactions. The reason that the use of copper for Sonogashira was eliminated in this catalyst was the basic nature of the ligand that played the role of copper. This catalyst showed superior performance from the viewpoint of reaction time, isolation, Pd loading (0.62 mmol%), and yields of the products as compared to the earlier versions of heterogeneous Pd/SBA-15. The catalyst could be recycled and reused for five times without any remarkable loss of the catalytic activity (Fig. 9.11) [32].

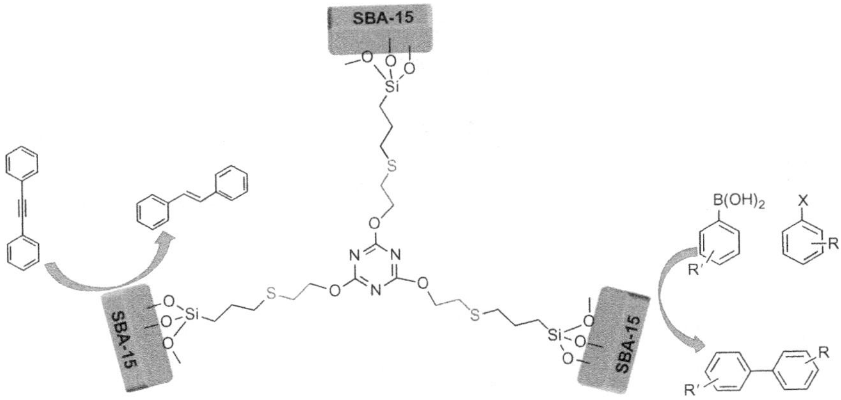

FIG. 9.11 Triazine-based supramolecule for Pd encapsulation. *(Reproduced from C. Singh, K. Jawade, P. Sharma, A.P. Singh, P. Kumar, Carboncarbon bond forming reactions: application of covalently anchored 2,4,6-triallyloxy-1,3,5-triazine (TAT) Pd(II) complex over modified surface of SBA-15 to Heck, Suzuki, Sonogashira and Hiyama cross coupling reactions, Catal. Commun. 69 (2015) 11–15, with the permission of Elsevier.)*

In another example, the reaction of biguanides with trichlorotriazine and ATPES has produced a supramolecule that was subsequently grafted to the surface of SBA-15 to encapsulate the Pd species. This dendron-like ligand conjugated with Pd ions led to the formation of an efficient catalyst for Suzuki-Miyaura coupling reactions and was found to exhibit excellent heterogeneous catalytic activity in green media; it could also be easily separated and reused for several times (Fig. 9.12) [33].

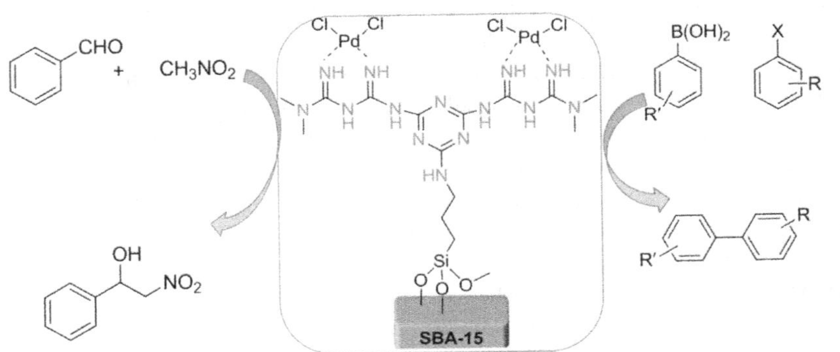

FIG. 9.12 A superbase dendron for Pd encapsulation in SBA-15. *(Reproduced from H. Veisi, D. Kordestani, S. Hemmati, A.R. Faraji, H. Veisi, Catalytic applications of an organosuperbase dendron grafted on mesoporous SBA-15 and a related palladium complex in Henry and Suzuki–Miyaura coupling reactions, Tetrahedron Lett. 55 (2014) 5311–5314, with the permission of Elsevier.)*

4.2 Heterocyclic Ligands

Pyridine-based heterocyle, as an example, has been used as ligand by reacting with azido-functionalized SBA-15 through the "*click*" reaction. In this regard, the produced triazole itself served as a ligand for Pd supporting. Therefore, the click-triazole acted both as a stable linker and an efficient chelator. The catalyst exhibited excellent catalytic activity for the aerobic oxidation of alcohols, and the product was obtained in up to 98% yield (Fig. 9.13) [34].

SBA-15 can also be functionalized by DABCO for efficient supporting of Pd species inside the pores. This catalyst has been utilized for homocoupling of terminal alkynes at 50°C in acetonitrile for 5 h. Likewise, it showed catalytic activity after several recoveries and reuses. Because DABCO possess two tertiary amines, one of them was used for anchoring to the surface of SBA-15 and another one was used as a ligand for Pd complex formation and, therefore, encapsulation of Pd (Scheme 9.4) [35].

Heteroleptic and homoleptic *N*-heterocyclic carbene complexes of Pd are important classes of complexes due to their unique catalytic activities rather than typical ligands. If one uses one type of *N*-heterocyclic carbene for complexation with Pd(II), the final complex will be a homoleptic carbene complex. Synthesis and characterization of homoleptic complex on a heterogeneous surface would be easier and more favorable than heteroleptic ones. In an example, SBA-15 was incorporated as a heterogeneous surface for anchoring *N*-methylimidazolium through the reaction of functionalized propylchloride and *N*-methylimidazole. Finally, this hybrid material was used for the production of homoleptic *N*-heterocyclic carbene ionic liquids of pore channels. For the first time, this system was used as a catalyst for the Hiyama coupling reaction with lower mol% Pd rather than ones reported in the literature. The main

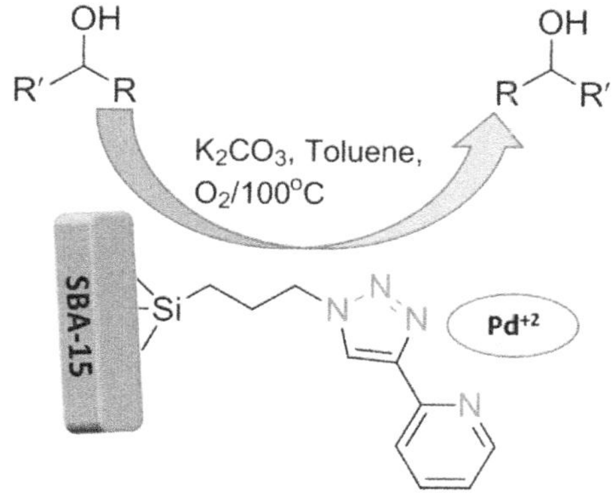

FIG. 9.13 Azido-bridged pyridine in Pd encapsulation inside SBA-15. *(Reproduced from G. Zhang, Y. Wang, X. Wen, C. Ding, Y. Li, Dual-functional click-triazole: a metal chelator and immobilization linker for the construction of a heterogeneous palladium catalyst and its application for the aerobic oxidation of alcohols, Chem. Commun. 48 (2012) 2979–2981, with the permission of the Royal Society of Chemistry.)*

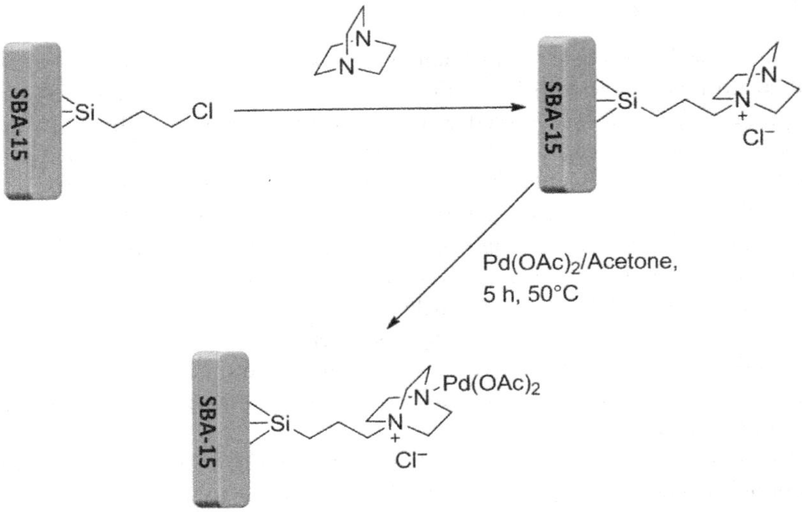

SCHEME 9.4 DABCO modified SBA-15 for Pd encapsulation. *(Reproduced from H. Li, M. Yang, Q. Pu, Palladium with spindle-like nitrogen ligand supported on mesoporous silica SBA-15: a tailored catalyst for homocoupling of alkynes and Suzuki coupling, Microporous Mesoporous Mater. 148 (2012) 166–173, with the permission of Elsevier.)*

reason for requiring a lower amount of Pd mol% in the reaction was the unique nature of the catalyst and simultaneous incorporation of tetrabutylammonium fluoride (TBAF) and Cs_2CO_3, which were supposed to activate the C-Si bond. Recyclability tests indicated that this catalyst had reusability for 5 runs without decrease of the activity. In addition, simultaneous use of TBAF and NaOH did not collapse the silica mesostructure of SBA-15 in the first run (Fig. 9.14) [36].

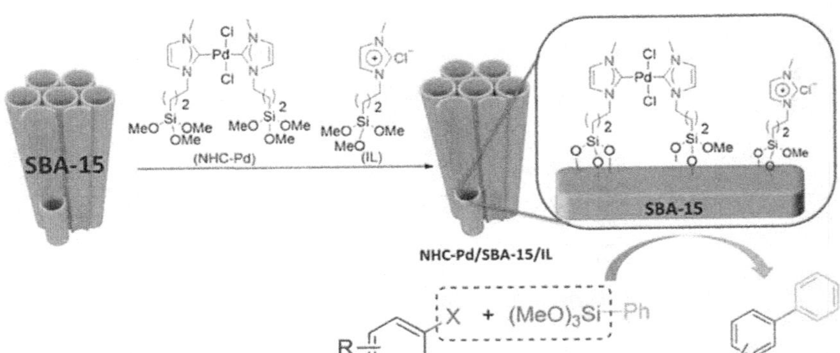

FIG. 9.14 NHC-Pd/SBA-15/IL in the catalysis of the Hiyama coupling reaction. *(Reproduced from S. Rostamnia, H. Golchin Hossieni, E. Doustkhah, Homoleptic chelating N-heterocyclic carbene complexes of palladium immobilized within the pores of SBA-15/IL (NHC–Pd@SBA-15/IL) as heterogeneous catalyst for Hiyama reaction, J. Organomet. Chem. 791 (2015) 18–23, with the permission of Elsevier.)*

4.3 Miscellaneous Ligands

Guanidine grafted ligands are also another type of excellent agent for the encapsulation of Pd species inside the MS materials. These ligands can ligate to palladium ion through amine to generate a stable complex [37,38]. Metformin is a biguanide and bidendate ligand that is more efficient and interesting for grafting to the heterogeneous surface of MSs and, consequently, for encapsulation of Pd species. This material was employed as an efficient catalyst for promoting the Suzuki cross-coupling reaction and aerobic oxidation of benzyl alcohols. The proposed catalyst by the Alizadeh group (Fig. 9.15) was excellent in the viewpoint of catalytic activity and reusability for various cycles in air in the studied reactions. By investigating the structure of the recovered catalyst, it was found that the mesostructure of SBA-15 dealt with insignificant collapse and good dispersion of in situ generated Pd NPs within the SBA-15 structure, which indicates that Pd NPs were preserved (Fig. 9.15) [39].

N-methyliminodiacetic acid (MIDA) is a type of ligand that was first reported by Guo for catalytic use, as Pd^{2+}-MIDA complex, in a Hiyama coupling reaction [40]. Inspired by this work, we supported its analogous complex onto the mesoporous SBA-15 (Pd@SBA-PIDA) surface. A new catalyst based on the Pd(II) supported SBA-15 was prepared by incorporation of APTES on SBA-15 followed by post modification with chloroacetate and finally supporting Pd ions. Lobster-like *N*-propyliminodiacetate (PIDA) herein acted as a ligand onto the surface of SBA-15 to form a stable complex with Pd ions to catalyze the Suzuki-Miyaura coupling reaction. The insignificant leaching of Pd, simple recovery, and recyclability of the catalyst over the several consecutive runs were some advantages of this catalyst (Fig. 9.16) [41].

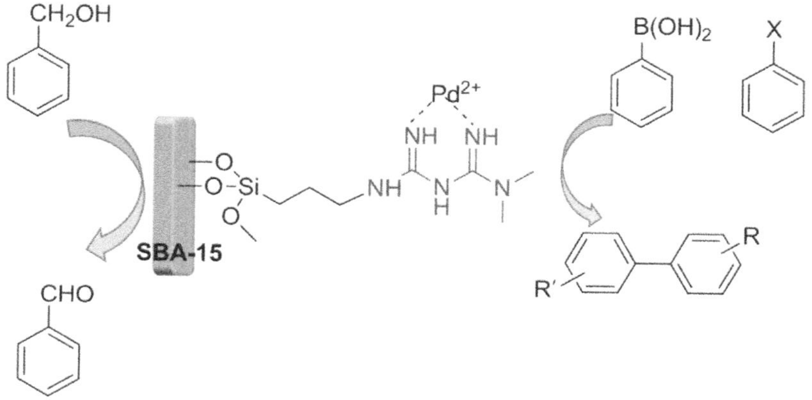

FIG. 9.15 Pd with biguanide-supported SBA-15. *(Reproduced from A. Alizadeh, M.M. Khodaei, D. Kordestania, M. Beygzadeh, A biguanide/Pd-decorated SBA-15 hybrid nanocomposite: synthesis, characterization and catalytic application, J. Mol. Catal. A: Chem. 372 (2013) 167–174, with the permission of Elsevier.)*

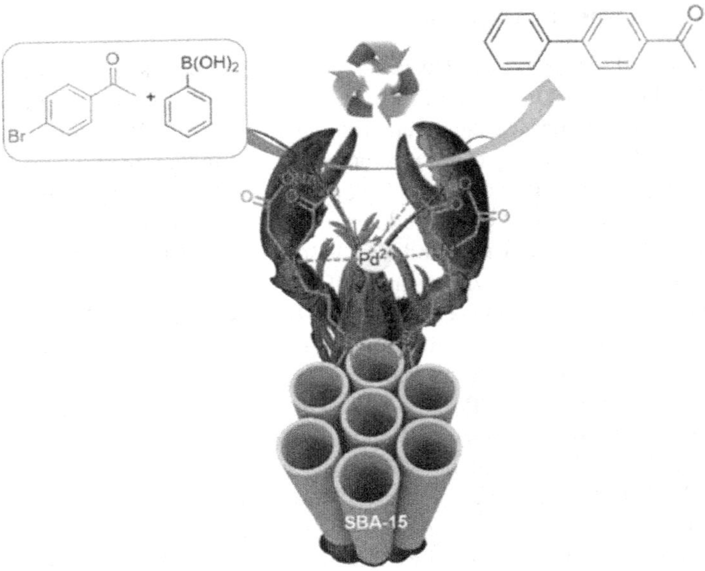

FIG. 9.16 Catalysis of Suzuki coupling by Pd@SBA-PIDA. *(Reproduced from S. Rostamnia, E. Doustkhah, R. Luque, Covalently bonded PIDA on SBA-15 as robust Pd support: water-tolerant designed catalysts for aqueous Suzuki couplings, Chem. Select. 2 (2017) 329–334.)*

Among the ligands, Schiff bases are favorable ligands for the immobilization and supporting Pd species onto the heterogeneous surfaces whether in nonporous silica or MS [42,43]. These ligands are usually prepared by initial functionalization of MS with amine through using ATPES. In the second step, the aminofunctionalized surface can furnish the Schiff base by reacting with a ketone or aldehyde that preferably possesses another heteroatom such as hydroxyl, sulfur, or nitrogen in the structure. These complexes are usually bidendate ligands for Pd complexes. Ketones or aldehyde can include 2-pyridine carbaldehyde, thiophene-2-carbaldehyde, salicylaldehyde, etc. (Fig. 9.17).

5 ELECTROSTATIC IMMOBILIZATION INSIDE THE MS CAVITY

For the first time, we functionalized SBA-15 with double-charged DABCO for immobilization of Pd ions through electrostatic immobilization. Synthesis of double-charged DABCO included the reaction of CPTMS and DABCO under an inert atmosphere at a higher temperature. This new hybrid nanocatalyst was then used for Suzuki-Miyaura coupling reaction, and Pd@SBA-15/IL_{DABCO} resulted in excellent yields in a short reaction time at 80°C and under aqueous

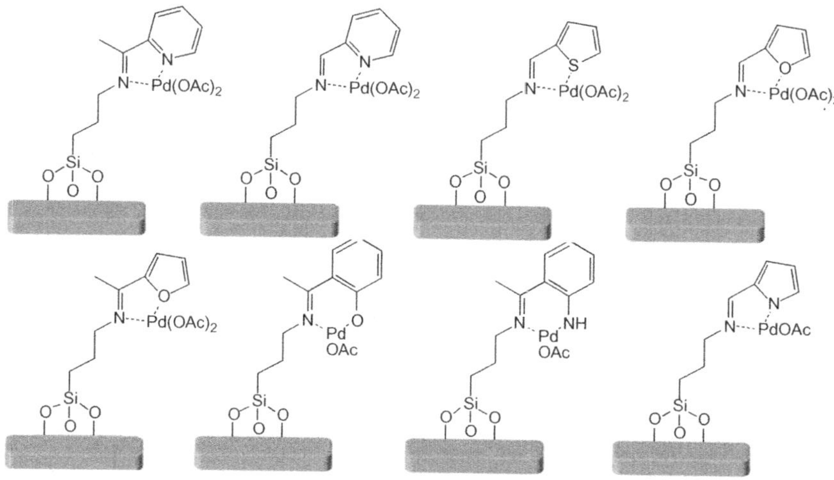

FIG. 9.17 Some examples of Pd-based catalysts. *(Reproduced from S. Paul, J.H. Clark, Structure-activity relationship between some novel silica supported palladium catalysts: a study of the Suzuki reaction, J. Mol. Catal. A: Chem. 215 (2004) 107–111; S. Bhunia, R. Sen, S. Koner, Anchoring of palladium(II) in chemically modified mesoporous silica: An efficient heterogeneous catalyst for Suzuki cross-coupling reaction, Inorg. Chim. Acta, 363 (2010) 3993–3999, with the permission of Elsevier.)*

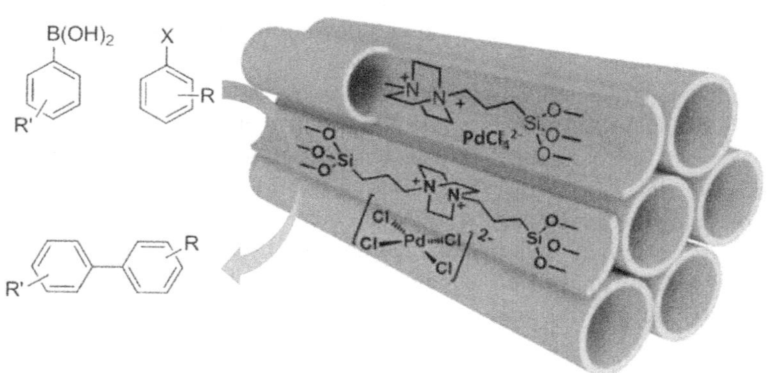

FIG. 9.18 Double-charged DABCO for electrostatic Pd encapsulation. *(Reproduced from S. Rostamnia, E. Doustkhah, B. Zeynizadeh, Cationic modification of SBA-15 pore walls for Pd supporting: Pd@SBA-15/IL$_{DABCO}$ as a catalyst for Suzuki coupling in water medium, Microporous Mesoporous Mater. 222 (2016) 87–93, with the permission of Elsevier.)*

conditions. To elucidate its true heterogeneity, the catalyst was tested at least for 9 reaction runs. The results established no significant loss of the activity (Fig. 9.18) [44].

One strategy for the electrostatic immobilization of Pd complexes is adjusting the pH of MS dispersion and then adsorption of the Pd-ammonium

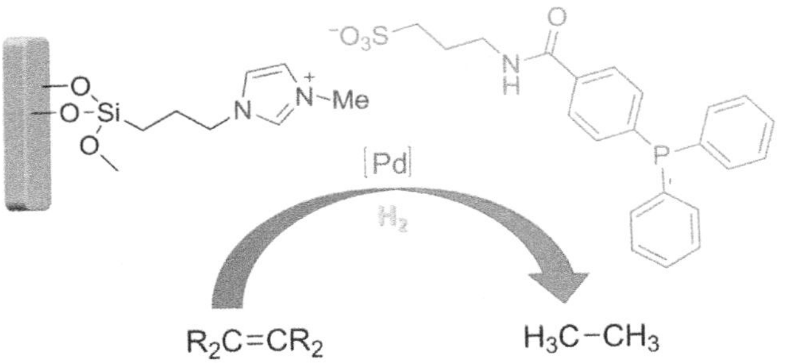

SCHEME 9.5 Hydrogenation of olefins with the catalysis of a phosphorous-based electrostatically immobilized catalyst. *(Reproduced from L. Wang, D. Dehe, T. Philippi, A. Seifert, S. Ernst, Z. Zhou, M. Hartmann, R.N.K. Taylor, A.P. Singh, M. Jia, W.R. Thiel, Electrostatic grafting of a triphenylphosphine sulfonate on SBA-15: application in palladium catalyzed hydrogenation, Catal. Sci. Technol. 2 (2012) 1188–1195, with the permission of the Royal Society of Chemistry.)*

complex into the MS [45]. By altering the pH, the point of zero change of silica can be found and then the metal amines can be adsorbed onto the surface of silica. In this method, a larger amount of Pd can be adsorbed into the pore channels compared to the impregnation method. This electrostatic immobilization of Pd complex can be implemented on SBA-15 by adjusting the pH of the media. The Pd complex can be deposited uniformly on the internal surface area (pore channels) of SBA-15, forming very well-dispersed (1.3 to 2.0 nm) NPs after reduction, with small standard deviation and with estimated loadings of 10 wt% [46].

A sulfonated triphenylphosphine ligand was designed to generate an electrostatically immobilized Pd-complex. Covalently grafted *N*-imidazoliumon SBA-15 was the counterpart of that anionic sulfonated triphenylphosphine Pd-complex ($PdCl_2(CNPh)_2$). The resulting hybrid material efficiently catalyzed olefin hydrogenation under mild conditions. The catalyst exhibited excellent activity, selectivity, stability, and reusability (10 cycles) without any loss of the activity. Compared to an analogous but covalently grafted palladium, a higher catalytic activity was observed for $PdCl_2(CNPh)_2$ in hydrogenation under the similar conditions. Atomic absorption spectroscopy (AAS) exhibited insignificant palladium leaching from support after several reaction cycles. This showed that the catalyst was truly heterogeneous. In electrostatic immobilization, The Pd complex had high mobility; hence, the catalytic activity could be as high as the homogenous counterpart (Scheme 9.5) [47].

6 ENCAPSULATION INSIDE PERIODIC MESOPOROUS ORGANOSILICA

Dialkylimidaozolium-based periodic mesoporous organosilica (PMO) (Im-PMO) was first synthesized by Karimi's group. They used it for supporting various catalytically active species. In this regard, they encapsulated Pd species by supporting it to this PMO. Imidazolium herein can act as a ligand that is covalently embedded within the framework of pore walls. In addition, $PdCl_2$ can be converted to a $PdCl_4^{2-}$ anion with the help of free Cl^- anions of the imidazolium counter ion. Pd supported Im-PMO afterwards was used for various reactions including Suzuki and Heck coupling reactions, and also oxidation of alcohols to aldehyde [48]. We also developed a different morphology for the synthesis of Im-PMO to immobilize Pd ions. Then, it was used for a C-S bond coupling reaction under aqueous conditions [49].

Some ligands such as phosphorous and carbene-based building blocks are very precious and can render Pd unique catalytic nature. Covalent grafting of carbene-based complexes of Pd can deal with some difficulties and sometimes can decrease the catalytic activity. However, electrostatic immobilization of these complexes is favorable and preserves the catalytic activity as it is homogeneous.

Rajabi and Thiel used sulfonated phenylene-based PMO (Ph-PMO) for electrostatic immobilization of Pd *N*-heterocyclic carbene complex in which the complex of Pd was cationic [50]. This hybrid material was applied as a catalyst for Suzuki-Miyaura coupling between low active aryl chlorides and phenylboronic acid under heterogeneous and aerobic conditions. This catalyst was advantageous since there were few catalysts that could catalyze the coupling of aryl chlorides and phenylboronic acid. Stability of the catalyst was also investigated by the recyclability test. It was indicated that the catalyst could be reused at least six times without any loss of the activity. Analysis showed no leaching, implying real heterogeneous catalysis (Scheme 9.6) [50].

Co-condensation of the $(MeO)_3Si$-H precursor with phenyl-bridged PMO [Ph-PMO-H] via a soft-template approach led to the formation of Si-H functionalities on the surface of Ph-PMO-H. This functionality caused in situ formation of Pd NPs inside the pores. The Ullmann reaction, in water as medium, was used to investigate the catalytic performance of Pd/Ph-PMO-H. This method simplified the process of the catalyst synthesis and prevented the detrimental effect of reductants on the mesostructure. Pd/Ph-PMO-H as a catalyst had excellent catalytic activity and selectivity towards the Ullmann coupling reaction. The hydrophobic nature of pore wall structures owing to the presence of phenylene bridges was more effective on the catalytic activity. This hydrophobic effect could accelerate the mass transfer inside the pores and consequently, an improvement in the catalytic activity could be observed. Furthermore, the catalyst could be recovered and recycled for five times without significant loss of the activity and selectivity (Fig. 9.19) [51].

Bhaumik designed and synthesized a new organosilane bridge for the synthesis of a new PMO and subsequent supporting of Pd inside the pore structure. The bridge could be obtained by Vilsmeier-Haack formylation of

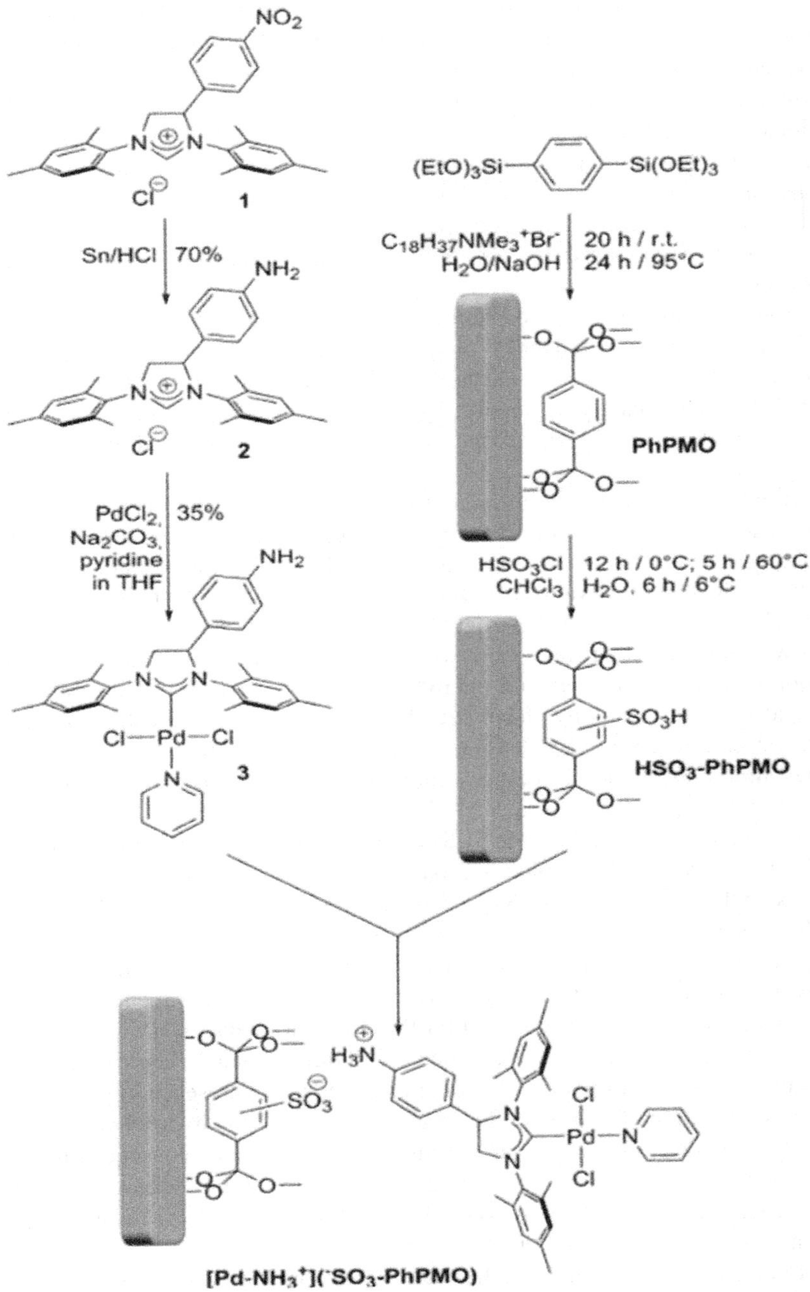

SCHEME 9.6 Synthesis procedure of electrostatically immobilized Pd-NHC carbene within the Ph-PMO-SO_3^-. *(Reproduced from F. Rajabi, D. Schaffner, S. Follmann, C. Wilhelm, S. Ernst, W.-R. Thiel, Electrostatic grafting of a palladium n-heterocyclic carbene catalyst on a periodic mesoporous organosilica and its application in the suzuki–miyaura reaction, ChemCatChem, 7 (2015) 3513–3518, with the permission of Wiley-VCH Verlag G.)*

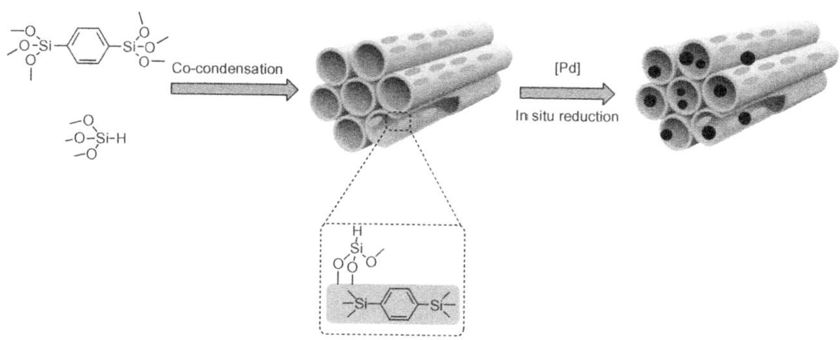

FIG. 9.19 In situ production of Pd NPs inside Ph-PMO-H. *(Reproduced from F. Zhang, J. Yin, W. Chai, H. Li, Self-assembly of palladium nanoparticles on periodic mesoporous organosilica using an in situ reduction approach: catalysts for ullmann reactions in water, ChemSusChem 3 (2010) 724–727, with the permission of Wiley-VCH Verlag G.)*

phloroglucinol in the first step, which was followed by its Schiff base condensation with APTES. This bridge with two salicylimine functionalities could be inserted into PMO walls after co-condensation of the silica source in the presence of a surfactant. Two bidendate ligands in each molecule of the bridge were utilized to support Pd(II) inside the PMO. ^{13}C and ^{29}Si solid state MAS NMR confirmed the successful synthesis of mesoporous PMO. This Pd-PMO showed excellent catalytic activity and selectivity in Mizoroki-Heck reactions for the synthesis of a series of aromatic and aliphatic olefins [52].

One of the interesting methods of Pd encapsulation by mesoporous materials is the co-condensation of an organosilane-based Pd complex with a silica precursor (e.g., TEOS) in the presence of a surfactant. However, for the synthesis of such materials, co-condensation conditions should be very mild and the removal of surfactant should be achieved very smoothly through ethanol. Also, the temperature of the synthesis reaction should be set at room temperature. In this regard, Pd^{2+} was used to synthesize an organosilane complex by the complexation of Pd ions, pyridine and diphenylphosphine, and free amine, which were linked to propyltriethoxysilane from one side and to Pd from another side. These organosilane-based Pd complexes were co-condensed with TEOS under neutral conditions at room temperature in the presence of Triton X-100 as a surfactant. The obtained hybrid mesoporous materials, MSU-X, by this approach may be analogous to PMO materials. These materials were studied as catalysts for Suzuki coupling reaction at room temperature under solvent-free conditions. In comparison, pyridine- and diphenylphosphine-based complexes indicated higher catalytic activities (Fig. 9.20) [53].

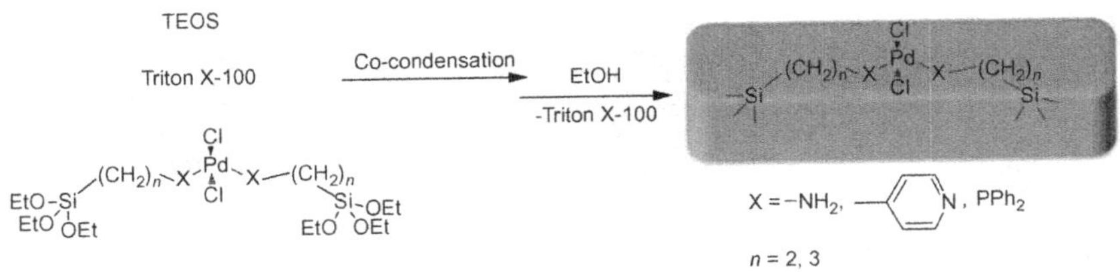

FIG. 9.20 Pd encapsulation within MSU-X. *(Reproduced from N. Linares, A.E. Sepulveda, M.C. Pacheco, J.R. Berenguer, E. Lalinde, C. Najera, J. Garcia-Martinez, Synthesis of mesoporous metal complex-silica materials and their use as solvent-free catalysts, New J. Chem. 35 (2011) 225–234, with the permission of the Royal Society of Chemistry.)*

7 CONCLUSION

Pd encapsulation inside MS materials can be achieved by various methods depending upon the application of Pd in the catalysis. On the other hand, they provide true heterogeneous catalysts for many types of reactions such as cross-coupling reactions and oxidation and reduction reactions by showing their reusability and minimum amount of Pd leaching from the heterogeneous surface. Generation of Pd NPs inside the pore channels provide a real Pd encapsulated catalyst that intensively avoids aggregation. Pore size tuning and diversity of the modified ligands can keep the gateway of research in this area open.

REFERENCES

[1] M. Opanasenko, P. Stepnicka, J. Cejka, Heterogeneous Pd catalysts supported on silica matrices, RSC Adv. 4 (2014) 65137–65162.

[2] S. Rostamnia, E. Doustkhah, Nanoporous silica-supported organocatalyst: a heterogeneous and green hybrid catalyst for organic transformations, RSC Adv. 4 (2014) 28238–28248.

[3] S. Rostamnia, E. Doustkhah, H. Golchin-Hosseini, B. Zeynizadeh, H. Xin, R. Luque, Efficient tandem aqueous room temperature oxidative amidations catalysed by supported Pd nanoparticles on graphene oxide, Catal. Sci. Technol. 6 (2016) 4124–4133.

[4] S. Rostamnia, E. Doustkhah, B. Zeynizadeh, Exfoliation effect of PEG-type surfactant on Pd supported GO (SE-Pd(nanoparticle)/GO) in cascade synthesis of amides: a comparison with Pd(nanoparticle)/rGO, J. Mol. Catal. A Chem. 416 (2016) 88–95.

[5] S. Rostamnia, B. Zeynizadeh, E. Doustkhah, H.G. Hosseini, Exfoliated Pd decorated graphene oxide nanosheets (PdNP–GO/P123): non-toxic, ligandless and recyclable in greener Hiyama cross-coupling reaction, J. Colloid Interface Sci. 451 (2015) 46–52.

[6] S. Rostamnia, E. Doustkhah, Z. Karimi, S. Amini, R. Luque, Surfactant-exfoliated highly dispersive pd-supported graphene oxide nanocomposite as a catalyst for aerobic aqueous oxidations of alcohols, ChemCatChem 7 (2015) 1678–1683.

[7] J. Zhang, H. Wang, A.K. Dalai, Development of stable bimetallic catalysts for carbon dioxide reforming of methane, J. Catal. 249 (2007) 300–310.

[8] X. Yang, D. Chen, S. Liao, H. Song, Y. Li, Z. Fu, Y. Su, High-performance Pd–Au bimetallic catalyst with mesoporous silica nanoparticles as support and its catalysis of cinnamaldehyde hydrogenation, J. Catal. 291 (2012) 36–43.

[9] C.Y. Ma, B.J. Dou, J.J. Li, J. Cheng, Q. Hu, Z.P. Hao, S.Z. Qiao, Catalytic oxidation of benzyl alcohol on Au or Au–Pd nanoparticles confined in mesoporous silica, Appl. Catal., B 92 (2009) 202–208.

[10] A. Balbin, F. Gaballo, J. Ceballos-Torres, S. Prashar, M. Fajardo, G.N. Kaluderovic, S. Gomez-Ruiz, Dual application of Pd nanoparticles supported on mesoporous silica SBA-15 and MSU-2: supported catalysts for C-C coupling reactions and cytotoxic agents against human cancer cell lines, RSC Adv. 4 (2014) 54775–54787.

[11] H. Liu, K. Tao, C. Xiong, S. Zhou, Controlled synthesis of Pd-NiO@SiO_2 mesoporous core-shell nanoparticles and their enhanced catalytic performance for p-chloronitrobenzene hydrogenation with H_2, Catal. Sci. Technol. 5 (2015) 405–414.

[12] V. Udayakumar, A. Pandurangan, A novel route for the synthesis of alkanes from glycerol in a two step process using a Pd/SBA-15 catalyst, RSC Adv. 5 (2015) 78719–78727.

[13] Y. Wan, H. Wang, Q. Zhao, M. Klingstedt, O. Terasaki, D. Zhao, Ordered mesoporous Pd/silica–carbon as a highly active heterogeneous catalyst for coupling reaction of chlorobenzene in aqueous media, J. Am. Chem. Soc. 131 (2009) 4541–4550.

[14] V. Polshettiwar, R. Luque, A. Fihri, H. Zhu, M. Bouhrara, J.-M. Basset, Magnetically recoverable nanocatalysts, Chem. Rev. 111 (2011) 3036–3075.

[15] W. Zhang, X. Chen, T. Tang, E. Mijowska, Superstable magnetic nanoreactors with high efficiency for Suzuki-coupling reactions, Nanoscale 6 (2014) 12884–12889.

[16] J. Huang, J. Yin, W. Chai, C. Liang, J. Shen, F. Zhang, Multifunctional mesoporous silica supported palladium nanoparticles as efficient and reusable catalyst for water-medium Ullmann reaction, New J. Chem. 36 (2012) 1378–1384.

[17] S. Zribi, B. Albela, L. Bonneviot, M.S. Zina, Surface engineering and palladium dispersion in MCM-41 for methane oxidation, Appl. Catal., A 502 (2015) 195–203.

[18] S. Rostamnia, H. Xin, $Pd(OAc)_2$@SBA-15/PrEn nanoreactor: a highly active, reusable and selective phosphine-free catalyst for Suzuki–Miyauracross-coupling reaction in aqueous media, Appl. Organomet. Chem. 27 (2013) 348–352.

[19] P. Schwab, R.H. Grubbs, J.W. Ziller, Synthesis and applications of $RuCl_2(CHR')(PR_3)_2$: the influence of the alkylidene moiety on metathesis activity, J. Am. Chem. Soc. 118 (1996) 100–110.

[20] B.R. Sarkar, R.V. Chaudhari, Anchored Pd-complexes in mesoporous supports: synthesis, characterization and catalysis studies for carbonylation reactions, Catal. Today 198 (2012) 154–173.

[21] R. Ghorbani-Vaghei, S. Hemmati, H. Veisi, Pd immobilized on amidoxime-functionalized mesoporous SBA-15: a novel and highly active heterogeneous catalyst for Suzuki–Miyaura coupling reactions, J. Mol. Catal. A Chem. 393 (2014) 240–247.

[22] F. Zhang, S. Chen, H. Li, X.-M. Zhang, H. Yang, Pd nanoparticles embedded in the outershell of a mesoporous core-shell catalyst for phenol hydrogenation in pure water, RSC Adv. 5 (2015) 102811–102817.

[23] K.-i. Shimizu, S. Koizumi, T. Hatamachi, H. Yoshida, S. Komai, T. Kodama, Y. Kitayama, Structural investigations of functionalized mesoporous silica-supported palladium catalyst for Heck and Suzuki coupling reactions, J. Catal. 228 (2004) 141–151.

[24] C.M. Crudden, M. Sateesh, R. Lewis, Mercaptopropyl-modified mesoporous silica: a remarkable support for the preparation of a reusable, heterogeneous palladium catalyst for coupling reactions, J. Am. Chem. Soc. 127 (2005) 10045–10050.

[25] S. MacQuarrie, B. Nohair, J.H. Horton, S. Kaliaguine, C.M. Crudden, Functionalized mesostructured silicas as supports for palladium catalysts: effect of pore structure and collapse on catalytic activity in the Suzuki-Miyaura reaction, J. Phys. Chem. C 114 (2009) 57–64.

[26] J.D. Webb, S. MacQuarrie, K. McEleney, C.M. Crudden, Mesoporous silica-supported Pd catalysts: an investigation into structure, activity, leaching and heterogeneity, J. Catal. 252 (2007) 97–109.

[27] S. Rostamnia, T. Rahmani, H. Xin, Pd($PrSO_3$)$_2$@SBA-15 and Pd-NPs($PrSO_3$)@SBA-15 hybrid materials: a highly active, reusable, and selective interface catalyst for C–X activations in air and water, J. Ind. Eng. Chem. 32 (2015) 218–224.

[28] S. Rostamnia, T. Rahmani, Ordered mesoporous SBA-15/$PrSO_3$Pd and SBA-15/$PrSO_3$PdNP as active, reusable and selective phosphine-free catalysts in C-X activation Heck coupling process, Appl. Organomet. Chem. 29 (2015) 471–474.

[29] M. Genelot, N. Villandier, A. Bendjeriou, P. Jaithong, L. Djakovitch, V. Dufaud, Palladium complexes grafted onto mesoporous silica catalysed the double carbonylation of aryl iodides with amines to give α-ketoamides, Catal. Sci. Technol. 2 (2012) 1886–1893.

[30] F. Farjadian, M. Hosseini, S. Ghasemi, B. Tamami, Phosphinite-functionalized silica and hexagonal mesoporous silica containing palladium nanoparticles in Heck coupling reaction: synthesis, characterization, and catalytic activity, RSC Adv. 5 (2015) 79976–79987.

[31] Y. Jiang, Q. Gao, Heterogeneous hydrogenation catalyses over recyclable Pd(0) nanoparticle catalysts stabilized by PAMAM-SBA-15 organic-inorganic hybrid composites, J. Am. Chem. Soc. 128 (2006) 716–717.

[32] C. Singh, K. Jawade, P. Sharma, A.P. Singh, P. Kumar, Carboncarbon bond forming reactions: application of covalently anchored 2,4,6-triallyloxy-1,3,5-triazine (TAT) Pd(II) complex over modified surface of SBA-15 to Heck, Suzuki, Sonogashira and Hiyama cross coupling reactions, Catal. Commun. 69 (2015) 11–15.

[33] H. Veisi, D. Kordestani, S. Hemmati, A.R. Faraji, H. Veisi, Catalytic applications of an organosuperbase dendron grafted on mesoporous SBA-15 and a related palladium complex in Henry and Suzuki–Miyaura coupling reactions, Tetrahedron Lett. 55 (2014) 5311–5314.

[34] G. Zhang, Y. Wang, X. Wen, C. Ding, Y. Li, Dual-functional click-triazole: a metal chelator and immobilization linker for the construction of a heterogeneous palladium catalyst and its application for the aerobic oxidation of alcohols, Chem. Commun. 48 (2012) 2979–2981.

[35] H. Li, M. Yang, Q. Pu, Palladium with spindle-like nitrogen ligand supported on mesoporous silica SBA-15: a tailored catalyst for homocoupling of alkynes and Suzuki coupling, Microporous Mesoporous Mater. 148 (2012) 166–173.

[36] S. Rostamnia, H. Golchin Hossieni, E. Doustkhah, Homoleptic chelating N-heterocyclic carbene complexes of palladium immobilized within the pores of SBA-15/IL (NHC–Pd@SBA-15/IL) as heterogeneous catalyst for Hiyama reaction, J. Organomet. Chem. 791 (2015) 18–23.

[37] H. Yang, X. Han, Z. Ma, R. Wang, J. Liu, X. Ji, Palladium-guanidine complex immobilized on SBA-16: a highly active and recyclable catalyst for Suzuki coupling and alcohol oxidation, Green Chem. 12 (2010) 441–451.

[38] H. Yang, Z. Ma, Y. Qing, G. Xie, J. Gao, L. Zhang, J. Gao, L. Du, A periodic mesoporous hybrid material with a built-in palladium complex: an efficient catalyst for the Suzuki coupling and alcohol oxidation, Appl. Catal., A 382 (2010) 312–321.

[39] A. Alizadeh, M.M. Khodaei, D. Kordestania, M. Beygzadeh, A biguanide/Pd-decorated SBA-15 hybrid nanocomposite: synthesis, characterization and catalytic application, J. Mol. Catal. A Chem. 372 (2013) 167–174.

[40] M. Guo, L. Qi, Q. Zhang, Z. Zhu, W. Li, X. Li, MIDA as a simple and highly efficient ligand for palladium-catalyzed Hiyama cross-coupling of aryl halides, Org. Biomol. Chem. 12 (2014) 7136–7139.

[41] S. Rostamnia, E. Doustkhah, R. Luque, Covalently bonded PIDA on SBA-15 as robust Pd support: water-tolerant designed catalysts for aqueous Suzuki couplings, Chem. Select 2 (2017) 329–334.

[42] S. Paul, J.H. Clark, Structure-activity relationship between some novel silica supported palladium catalysts: a study of the Suzuki reaction, J. Mol. Catal. A Chem. 215 (2004) 107–111.

[43] S. Bhunia, R. Sen, S. Koner, Anchoring of palladium(II) in chemically modified mesoporous silica: an efficient heterogeneous catalyst for Suzuki cross-coupling reaction, Inorg. Chim. Acta 363 (2010) 3993–3999.

[44] S. Rostamnia, E. Doustkhah, B. Zeynizadeh, Cationic modification of SBA-15 pore walls for Pd supporting: Pd@SBA-15/IL_{DABCO} as a catalyst for Suzuki coupling in water medium, Microporous Mesoporous Mater. 222 (2016) 87–93.

[45] L. Jiao, J.R. Regalbuto, he synthesis of highly dispersed noble and base metals on silica via strong electrostatic adsorption: I. Amorphous silica, J. Catal. 260 (2008) 329–341.

[46] L. Jiao, J.R. Regalbuto, The synthesis of highly dispersed noble and base metals on silica via strong electrostatic adsorption: II. Mesoporous silica SBA-15, J. Catal. 260 (2008) 342–350.

[47] L. Wang, D. Dehe, T. Philippi, A. Seifert, S. Ernst, Z. Zhou, M. Hartmann, R.N.K. Taylor, A.P. Singh, M. Jia, W.R. Thiel, Electrostatic grafting of a triphenylphosphine sulfonate on SBA-15: application in palladium catalyzed hydrogenation, Catal. Sci. Technol. 2 (2012) 1188–1195.

[48] B. Karimi, D. Elhamifar, J.H. Clark, A.J. Hunt, Palladium containing periodic mesoporous organosilica with imidazolium framework (Pd@PMO-IL): an efficient and recyclable catalyst for the aerobic oxidation of alcohols, Org. Biomol. Chem. 9 (2011) 7420–7426.

[49] S. Rostamnia, E. Doustkhah, R. Bulgar, B. Zeynizadeh, Supported palladium ions inside periodic mesoporous organosilica with ionic liquid framework (Pd@IL-PMO) as an efficient green catalyst for S-arylation coupling, Microporous Mesoporous Mater. 225 (2016) 272–279.

[50] F. Rajabi, D. Schaffner, S. Follmann, C. Wilhelm, S. Ernst, W.R. Thiel, Electrostatic grafting of a palladium n-heterocyclic carbene catalyst on a periodic mesoporous organosilica and its application in the Suzuki–Miyaura reaction, ChemCatChem 7 (2015) 3513–3518.

[51] F. Zhang, J. Yin, W. Chai, H. Li, Self-assembly of palladium nanoparticles on periodic mesoporous organosilica using an in situ reduction approach: catalysts for ullmann reactions in water, ChemSusChem 3 (2010) 724–727.

[52] A. Modak, J. Mondal, V.K. Aswal, A. Bhaumik, A new periodic mesoporous organosilica containing diimine-phloroglucinol, Pd(II)-grafting and its excellent catalytic activity and trans-selectivity in C-C coupling reactions, J. Mater. Chem. 20 (2010) 8099–8106.

[53] N. Linares, A.E. Sepulveda, M.C. Pacheco, J.R. Berenguer, E. Lalinde, C. Najera, J. Garcia-Martinez, Synthesis of mesoporous metal complex-silica materials and their use as solvent-free catalysts, New J. Chem. 35 (2011) 225–234.

Chapter 10

Design and Characterization of Hierarchical Porous Inorganic Solids for Enzyme Immobilization: Potential Application in Industrial Processes

Claudia Bernal* and Monica Mesa†

*Universidad de La Serena, La Serena, Chile, †Universidad de Antioquia, Medellín, Colombia

Contents

Encapsulated Catalysts. http://dx.doi.org/10.1016/B978-0-12-803836-9.00010-9

1 INTRODUCTION

Enzymes are catalysts of increasing importance for the industrial production of chemicals and drugs, and they are also used as biosensors. However, it is necessary to design new strategies to stabilize and reuse them in order to have efficient, cheap, and robust biocatalysts that can be implemented on a large scale and at industrial level. The enzyme immobilization onto solid supports is one of the most applied methodologies to accomplish these objectives. Hierarchical porous silica is an interesting enzyme support because it exhibits a high surface area, thermal and mechanical stability, ease of handling, ability to withstand high flow rates in continuous reactors, nontoxicity, and high resistance against microbial attack and organic solvents. Furthermore, the control of pore size, morphology, and chemical characteristics of its surface offers new opportunities for preparing biocatalysts; this is an active and current field of research.

This chapter is devoted to the synthesis, characterization, and application of hierarchical inorganic porous materials as enzyme support (Scheme 10.1). The following topics will be developed:

- Historical background of hierarchical silica application as immobilization support. The emphasis is placed on the evaluation of known materials for enzyme immobilization by an adsorption pathway.
- Design of hierarchical porous silica for the stabilization of model molecules such as β-galactosidases, which have high molecular weights and therefore need materials with large pores for their encapsulation. These silicas can be chemically modified for the immobilization of β-galactosidases by covalent bonding. The improvement of thermal stability and nonlixiviation is highlighted. These β-galactosidase/silica biocatalysts are used for the hydrolysis of whey.

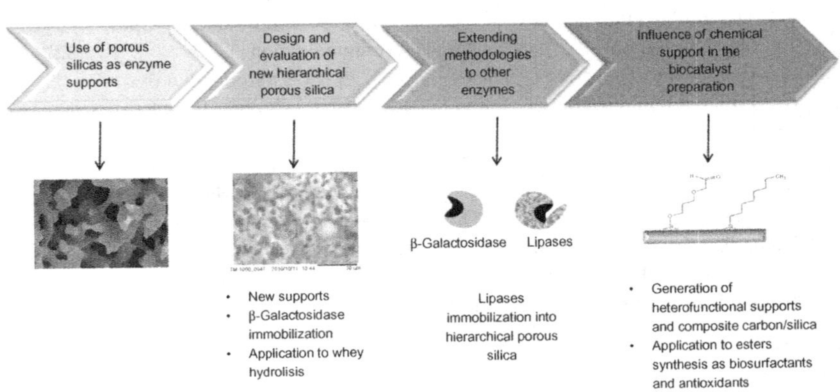

SCHEME 10.1 Main topic developed in the book chapter

- Application of these silicas and immobilization pathways for immobilization of other enzymes like lipases, expanding the knowledge about stabilization pathways and mechanisms to other systems.
- Chemical modification of the silica surface chemistry for obtaining heterofunctional supports and synthesis of carbon/silica composite materials for immobilization, stabilization, and hyperactivation of lipases. These biocatalysts can be applied in the synthesis of sugar-based surfactants.

2 HISTORICAL BACKGROUND ON THE APPLICATION OF POROUS SILICA AS IMMOBILIZATION SUPPORT

After the discovery of organized mesoporous silica in1992 [1], the synthesis of these porous materials has been widely studied with the aim to control the porous structure and morphology. Porous silica is an interesting material as support for enzyme immobilization due to its high surface area, thermal and mechanical stability, nontoxicity, and high resistance against microbial attacks and organic solvents. Furthermore, the possibility of adjusting its pore sizes and modifying its chemical surface shows that it is an excellent option for enzyme immobilization.

Currently, it is possible to find several kinds of silica materials in respect to pore size. Some examples of these materials are the MCM-X and SBA-X mesoporous silica families, which exhibit pores between 2 and 10 nm. The mesoporous silica templated by a surfactant has been used as support for enzyme immobilization [2,3]. However, the small pore size is the main disadvantage of typical mesoporous silica such as SBA-15, SBA-16, or MCM-41 materials for encapsulating the enzymes inside of the pores [4]. This disadvantage is due to the fact that the enzyme not only needs the space for being lodged, but it also needs enough space to undergo catalysis efficiently, which involves tridimensional movement, and free ingress and egress of substrates and products from the solution toward the biocatalyst. Designing silicas that accomplish the desired porosity and surface chemistry for the immobilization and stabilization of enzymes into their pores is currently an active research topic for producing immobilized biocatalysts that exhibit high catalytic performance and low diffusional restrictions. In this sense, hierarchical porous silica, which combines large mesopores and macropores in an interconnected structure, has gained ground in the last year (Fig. 10.1) [5].

There are many synthesis pathways for obtaining porous silica and every methodology searches to optimize some characteristics of materials. For example, in order to use a material as support for enzyme immobilization, it is desired to have a high surface area and large pore size. Specifically, silicas with large mesopores have been achieved by hydrothermal pathways by adding swelling agents [6] and a cosolvent [7] to the reaction medium, or by the spinodal method [8]. Every method has its intrinsic advantages and disadvantages according to reaction conditions. Bernal et al. reported the modified synthesis of MCM-41 silica by addition of ethyl acetate at low temperatures in comparison with

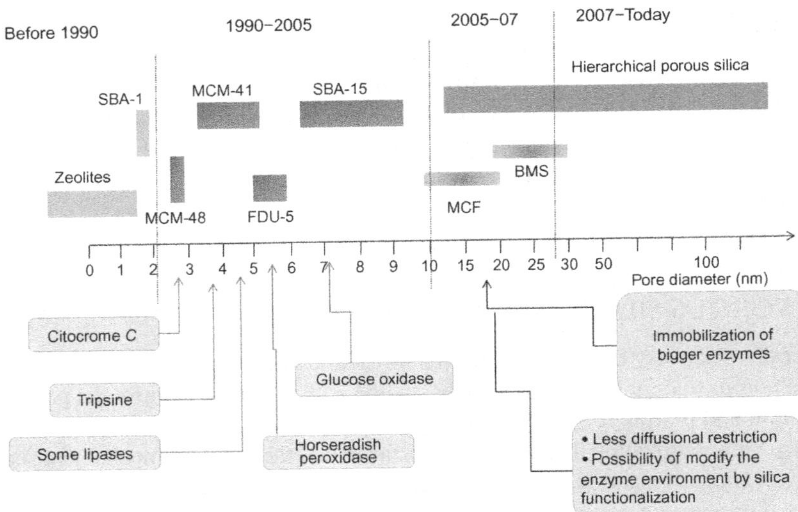

FIG. 10.1 Timeline of enzyme immobilization in materials with different pore sizes.

typical reaction conditions; the authors obtained a hierarchical macro/mesoporous silica suitable as enzyme support, through a cheap and fast methodology [9]. This new silica exhibits a surface area similar to bimodal mesopore silica (BMS), showing large mesopores (10–40 nm) and macropores in the same material. This combination of porosity solves some complications of typical materials due to the presence of small mesopores such as diffusional restrictions and low protein loading.

3 HIERARCHICAL POROUS SILICA FOR β-GALACTOSIDASE IMMOBILIZATION

3.1 Design of Hierarchical Porous Silica for the Stabilization of Model Molecules Such as β-Galactosidases

β-Galactosidases (E.C. 3.2.1.23) are hydrolases, which can be extracted from plants (peach, apricot, almond, etc.), animal organs (intestine, skin, and brain tissue), yeasts, fungi, and bacteria such as *Kluyveromyces lactis* [10,11], *Aspergillus oryzae* [12], and *Bacillus circulans* [13]. The source, extraction, and purification processes determine their physicochemical characteristics (molecular weight, isoforms, and dimensions) and pH-temperature intrinsic stability. The temperature and pH in which they exhibit the optimum activity vary in function of the commercial sources, but the specificity remains the same [14]. The mentioned β-galactosidases from microorganisms have molecular weights higher than 150 kDa and hydrodynamic diameters higher than 10 nm; therefore, they can be classified as large enzymes. There are some reviews about the immobilization of β-galactosidases [15,16].

The dimensions and/or molecular weight of the enzyme are key parameters for determining the desired pore size of silica that will be used as immobilization support, allowing the following:

- The diffusion of the enzyme inside of the pores, where they will be encapsulated.
- The geometrical congruence that leads to an active form of the encapsulated enzyme, in order to enable the conformational changes during biocatalysis.

Silica with hierarchical macro/mesoporosity is an example of material that accomplishes these two objectives for the immobilization of globular enzymes with a molecular weight higher than 100 kDa, such as β-galactosidases [9]. In this report, the synthesis route is designed in an aqueous solution from sodium silicate (SiO_2) in the presence of a cetyltrimethylammonium bromide (CTAB) surfactant and ethyl acetate (EtAc). The latter can form droplets that are the template for macropores. The CTAB/SiO_2 and EtAc/SiO_2 molar ratios modulate the porosity and morphology of the silica by controlling the polycondensation pH and silica aggregation (Fig. 10.2).

The immobilization of β-galactosidase derived from *K. lactis* was carried out in hierarchical macro/mesoporous silica, reaching efficiency until 80% depending on the pore size [17]. In this case, the presence of macropores facilitates the diffusion into the pores (Table 10.1).

Another key parameter for designing porous silica and selecting the immobilization pH for the encapsulation of β-galactosidases by adsorption is the surface charge of the enzyme. The adsorption kinetic and loading capacity at pH 6.0 in silica was affected by the zeta potential of the enzymes, being higher for β-galactosidase derived from *B. circulans* than for *K. lactis:* The greater difference between the zeta potential between *B. circulans* (−25.0 mV) and silica (−11.0 mV) favors the attraction of the enzyme onto the support. For *K. lactis* β-galactosidase (−12.5 mV), this zeta value is similar to that of silica, explaining the low loading of this enzyme [18].

The silanol groups of the silica are responsible of the weak electrostatic and Van der Waals interactions with amino, hydroxyl, and amido groups of the enzymes (Fig. 10.3). These kinds of interactions help to preserve the tridimensional structure of the enzyme when it is immobilized by adsorption, as it is seen by the preservation of the same pH and temperature for the maximal activity (7.0 ± 0.5 and 38 ± 1°C, respectively), compared with the soluble β-galactosidases [17,18]. However, the lixiviation is not avoided for these biocatalysts under the reaction conditions applied for the hydrolysis of lactose.

3.2 Chemical Modification of Silicas for the Immobilization of β-Galactosidases by Covalent Bonding

The enzyme stability against lixiviation and thermal denaturation is gained by covalent immobilization. The formation of Schiff bases and posterior reduction is one of the pathways that can be used for enzyme immobilization when the

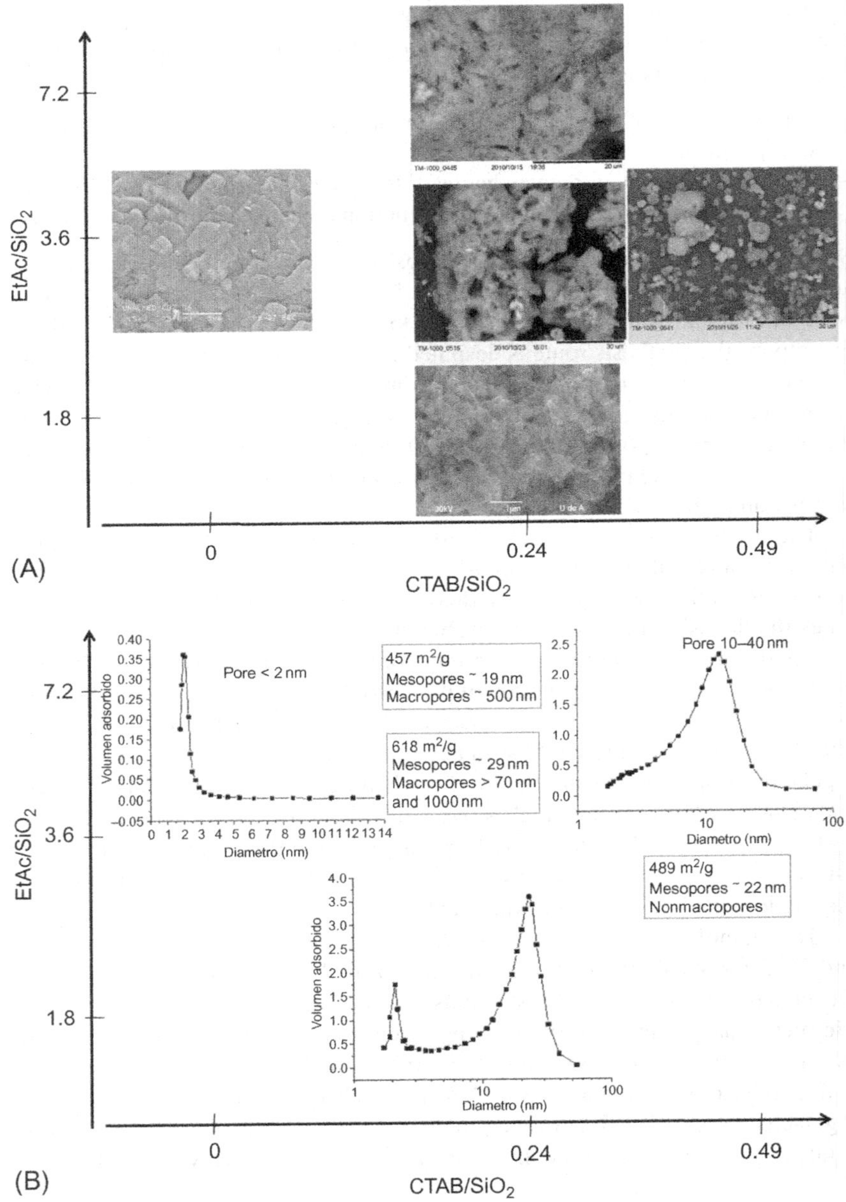

FIG. 10.2 Variation of the morphology (A) and porosity (B) of hierarchical macro/mesoporous silica in a function of the reaction mixture composition. *(Adapted from C. Bernal, M. Mesa, L. Sierra, Synthesis of new silicas with high stable and large mesopores and macropores for biocatalysis applications, Mater. Sci. Eng. C, 32 (6) (2012) 1380–1385, with permission of Elsevier.)*

TABLE 10.1 Immobilization Results of β-Galactosidases From a *B. circulans* as a Function of the Porous Characteristics of Hierarchical Macro/Mesoporous Silica

Material	Pore Diameter (nm)	Immobilization Yield (%)	Immobilization Time (h)	Lixiviation (48 h) %
A	29	80	6	20
C	21	38	5	12
B	19	30	7	6

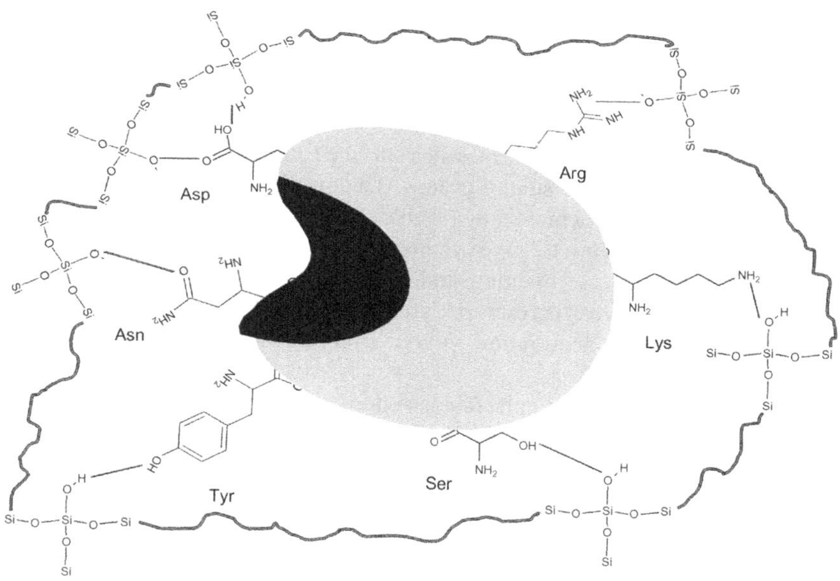

FIG. 10.3 Schematic representation of the interactions of the amino acids of the enzymes with the silanol groups of the silica supports during the immobilization by adsorption at pH 7. The silanol groups can be protonated or negatively charged.

enzyme is rich in lysine groups, as it is the case for β-galactosidases and with a support that bears glyoxyl groups (Fig. 10.4) [19]. Specifically, glyoxyl groups exhibit low steric hindrances for interacting with the enzyme molecules due to their alkyl chain and their high stability, even at alkaline pHs, allowing the enzyme immobilization. This interaction leads to stiffening of the enzyme, which enhances the thermal stability.

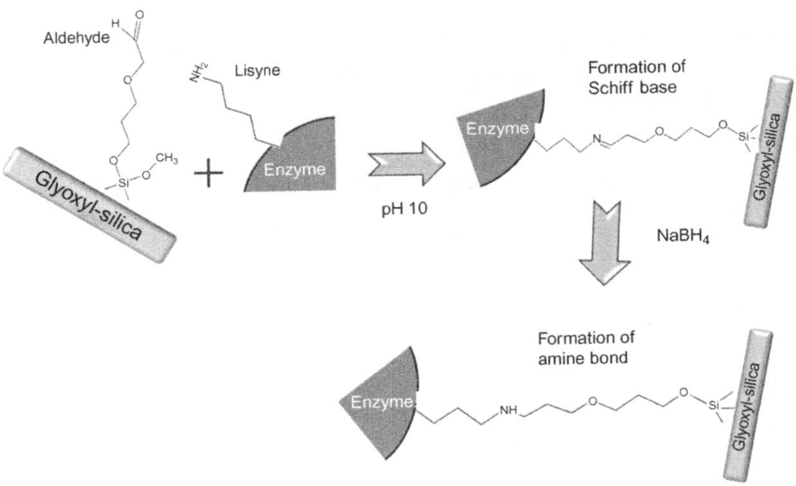

FIG. 10.4 Covalent attachment of enzymes to silica modified with glyoxyl groups.

The chemistry and charge of the silica surface can be modified by attaching organic groups through the silanol groups. The grafting of the epoxide groups onto the silica surface is achieved by (3-glycidoxypropyl)methyldiethoxysilane [20], and the conversion to glyoxyl groups for enzyme immobilization is obtained by ring epoxide opening and diol oxidation, as in other supports [21]. The density of glyoxyl groups per gram of support can be varied by changing the oxidant agent concentration without affecting the porous characteristics of the silica [18].

The immobilization of β-galactosidase derived from *B. circulans* by covalent attachment has a cost in terms of immobilization yield and expressed activity; however, the thermal stability is outstanding, compared with the immobilization by adsorption in hierarchical porous silica. This corroborates that porosity of the support is the first factor for allowing the diffusion of enzyme molecules during the immobilization process, and the second factor is the chemistry of the surface that defines the stiffening of the enzyme. However, the effect of being encapsulated inside the pores leads to a different result in the stabilization factor compared with the results in glyoxyl-agarose support, which is attributed to difference in geometrical congruence in both materials. This is a consequence of differences in pore morphologies and sizes for the macroporous agarose and macro/mesoporous silica (Fig. 10.5).

The advantages of having a suitable porosity and surface chemistry in silica are well seen after the immobilization of β-galactosidase derived from *B. circulans* by multipoint covalent attachment in hierarchical meso/macroporous silica, with the controlled density of glyoxyl groups. Besides high loading capacity (around 60%–80%) and expressed activity (around 50%–80%)

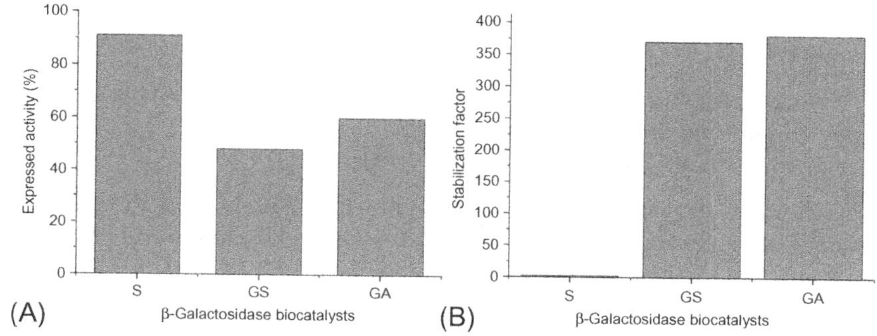

FIG. 10.5 Expressed activity percentage (A) and stabilization factor (B) for β-galactosidase from *B. circulans* immobilized in silica (S), glyoxyl-silica (GS), and glyoxyl-agarose (GA). *(Adapted from C. Bernal, L. Sierra, M. Mesa, Improvement of thermal stability of β-galactosidase from Bacillus circulans by multipoint covalent immobilization in hierarchical macro-mesoporous silica, J. Molec. Catal.B Enzym. 84 (2012) 166–172, with permission of Elsevier.)*

promoted by the presence of macropores and large mesopores, the thermal stability at 60°C improves up to 370-fold in comparison with the free enzyme when the silica bears 1.0 mmol of glyoxyl groups per gram [22]. On the other hand, the glyoxyl group density on the silica surface affects the thermal stabilization degree: an optimum functionalization degree confers rigidity to the enzyme without distortions of its active site (Fig. 10.6) [18,22]. The lixiviation is completely avoided with the covalent immobilization, independent of the glyoxyl group density, which is an advantage for the biocatalyst reuse.

3.3 Behavior of β-Galactosidase/Silica Biocatalyst

The biocatalysis with encapsulated enzymes exhibits differences in the kinetic behavior, compared with the free forms. This is a consequence of multiple

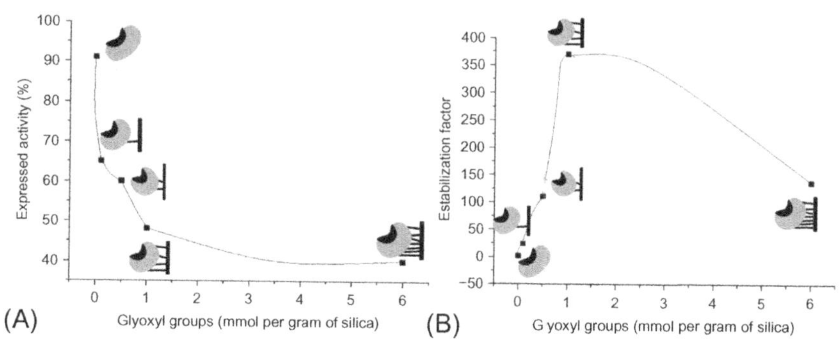

FIG. 10.6 Schematic representation of the enzyme expressed activity (A) and thermal stability (B) in a function of the density of glyoxyl groups in glyoxyl-silica support.

effects related to the immobilization process in solid supports [23,24], such as the following:

- Hindering of conformational changes due to the interaction with the support.
- Restrictions on the substrate transport.
- Steric constraints for biomolecular reactions.

As an example, the lactose hydrolysis slows down, as it is shown with the lower apparent maximum velocity for β-galactosidase derived from *B. circulans* covalently attached to glyoxyl-silica (GS) in comparison with the free enzyme. The substrate affinity of the immobilized enzyme decreases, as it is shown by the increase in Michaelis-Menten constant, Km (Fig. 10.7). The enzyme stiffening and the distortion of its active site affect the Km [25].

The β-galactosidase/silica biocatalyst converts the 95% of lactose into glucose and galactose in 15 min when the reaction is carried out at 25°C in a batch process. This biocatalyst behaves similarly to the free enzyme, with the advantage that it can be used under these conditions for more than 10 cycles since it maintains 90% of its initial activity. The lactose diffusion restrictions in the first hour at 4°C, which slow down the lactose conversion, must be taken into account for designing the reactors for the industrial applications.

4 EXPANDING THE KNOWLEDGE ABOUT STABILIZATION PATHWAYS AND MECHANISMS TO OTHER ENZYMES

Hydrolases are an important group of enzymes. Among them, the lipases have several applications in the food and pharmaceutical industry [26]. However, the scale of the reaction at an industrial level is limited by the enzyme cost, its reuse impossibility, and low stability of these enzymes against some reaction conditions (organic solvent, temperature, pH).

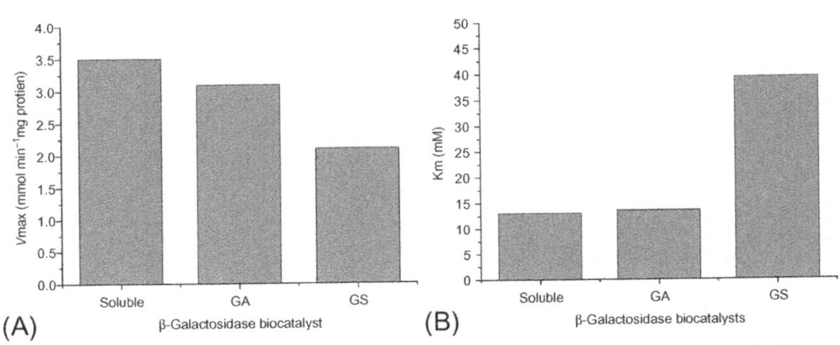

FIG. 10.7 Comparison of (A) the apparent maximum velocity (V_{max}) and (B) Michaleis-Menten constant (Km) for the soluble β-galactosidase from *B. circulans* and immobilized in glyoxyl-silica (GS) or glyoxyl-agarose. *(Adapted from C. Bernal, P. Urrutia, A. Illanes, L. Wilson, Hierarchical meso-macroporous silica grafted with glyoxyl groups: opportunities for covalent immobilization of enzymes, New Biotechnol. 30 (5) (2013) 500–506, with permission of Elsevier.)*

Lipases are triacylglycerol acylhydrolase enzymes (EC3.1.1.3) that catalyze the hydrolysis of acylglycerols at the water-lipid interface, playing an important role in fat digestion. They can be originated from a variety of sources: microbial [27], plant [28], and animal kingdoms [29]. These enzymes are applied mainly in biodiesel production, but they are used in the synthesis of products for the pharmaceutical, food, and chemical industries. The most interesting applications, on chemical and pharmaceutical industry, require the organic solvents different from water and higher temperatures than those commonly used on enzymatic reactions, which can affect the enzyme stability. Immobilization is an interesting strategy to stabilize these enzymes [26].

Lipases are generally immobilized in a hydrophobic support because these enzymes exhibit an interfacial mechanism by which their active center is opened when the enzyme comes in contact with a hydrophobic surface. This occurs through reorganization of their tridimensional structure, increasing the lipase activity [30], where the main advantage is the enzyme hyperactivation (Fig. 10.8). Covalent immobilization using glyoxyl groups is possible if the lipases exhibit their lysine in an external surface. This methodology gives extra thermal stabilization with respect to hydrophobic immobilization due to covalent bond formation.

Lipases derived from *Candida antarctica B* (CalB) and from *Alcaligenes* sp. (QL) were immobilized in hierarchical macro/mesoporous silica. The immobilization process for both enzymes in GS was different in comparison with other hydrolases such as β-galactosidases in the same support [31]. Since these lipases have lower lysine groups in their external surface, the CalB immobilization was slower (the process was 48 h) and the QL lipase needed an extra step before the immobilization process: amination of their amino acid was done with the aim to increase the amino groups on an external surface, favoring the formation of

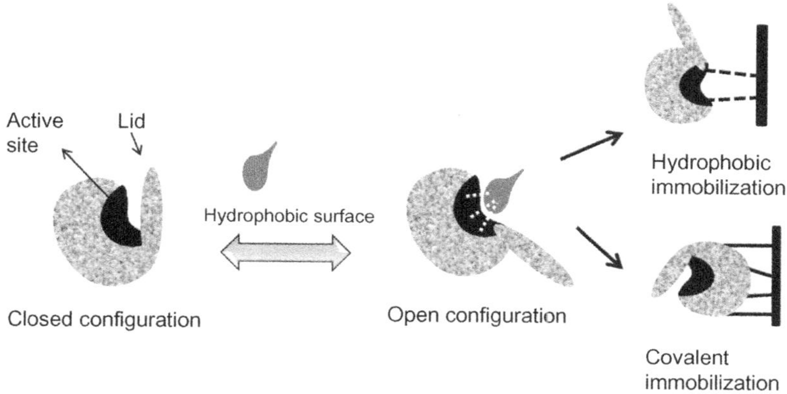

FIG. 10.8 Interfacial activation of lipases by contact with hydrophobic surfaces and their probable orientation in support when the enzyme is immobilized by hydrophobic or covalent immobilization pathway.

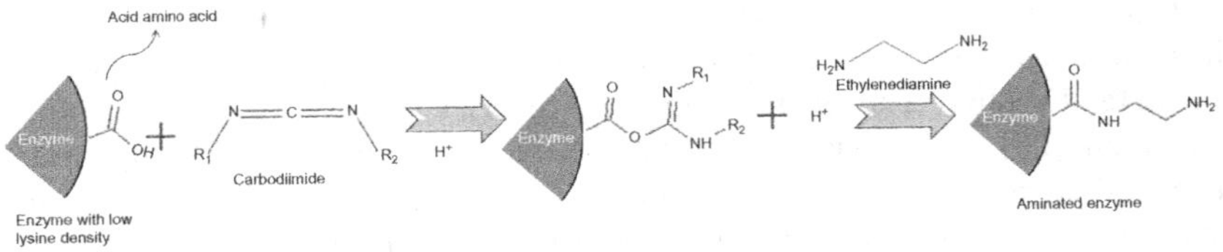

FIG. 10.9 Chemical reactions involved in the amination of acid amino acids of the enzymes. *(Adapted from R.C. Rodrigues, O. Barbosa, C. Ortiz, A. Berenguer-Murcia, R. Torres, R. Fernandez-Lafuente. Amination of enzymes to improve biocatalyst performance: coupling genetic modification and physicochemical tools. RSC Adv. 4 (2014) 38350–38374, with permission of the Royal Society of Chemistry.)*

Schiff bases between enzyme and support (Fig. 10.9) [31,32]. Even this chemical modification of enzyme is simple, and it is only carried out when the enzyme cannot be immobilized under standard methodologies since it generally decreases the enzyme activity.

The comparison of the behavior of enzymes immobilized in silica and agarose modified with the same chemical functionality shows interesting results. The CalB lipase exhibited a higher immobilization yield when it was immobilized in GS (45.5%) than in glyoxyl-agarose (21.4%). These results are due to the fact that the inorganic material exhibits a higher glyoxyl group concentration than agarose [22], allowing the efficient lipase immobilization. The trend was similar for QL lipase despite the fact that this biocatalyst exhibits a lower immobilization yield, probably because the amination affects its intrinsic activity (Fig. 10.10). Thermal stability at 55°C under nonreactive conditions in an aqueous medium is improved when both lipases are immobilized in silica support, in comparison with their soluble counterparts and agarose biocatalysts. This can be explained in terms of the higher enzyme rigidification as a consequence of the stronger interaction between lipase and GS.

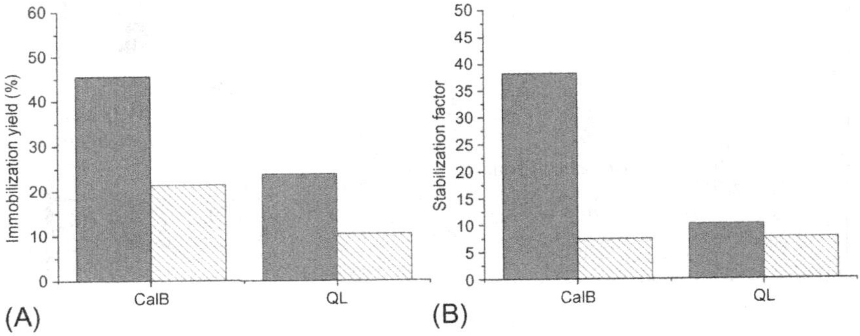

FIG. 10.10 Immobilization yield (A) and stabilization factor at 55°C (B) for lipases from *Candida antarctica* B (CalB) and *Alcaligenes* sp. (QL) immobilized in glyoxyl-silica and glyoxyl-agarose. *(Adapted from C. Bernal, P. Urrutia, A. Illanes, L. Wilson, Hierarchical meso-macroporous silica grafted with glyoxyl groups: opportunities for covalent immobilization of enzymes, New Biotechnol. 30 (5), (2013) 500–506, with permission of Elsevier.)*

TABLE 10.2 Stability of CalB Lipase Biocatalysts in Different Pure Organic Solvents at 40°C

Solvent	Catalyst	Half-Life (h)	Stabilization Factor
Acetone	Glyxoxyl-agarose	4.0	1.8
	Glyoxyl-silica	62.1	31
tert-Butanol	Glyxoxyl-agarose	5.6	1.1
	Glyoxyl-silica	24.0	4.9
Dioxane	Glyxoxyl-agarose	4.7	0.7
	Glyoxyl-silica	20.7	3.0

Lipases are frequently used as biocatalysts for synthetic reaction, for which it is necessary to expose the enzyme to aggressive conditions such as organic solvent, in many cases resulting in lipase inactivation in a short time [26]. The main advantage of covalent immobilization into siliceous support is the extra stabilization against organic solvent in comparison with those obtained from other methodologies of immobilization. For example, lipase derived from CalB immobilized in GS exhibits a higher stability when it is incubated at 40°C in a different solvent in comparison with the glyoxyl-agarose counterpart (Table 10.2). This behavior can be explained based on the fact that the silica support is stable in the presence of an organic solvent, while agarose is a hydrogel that can suffer from damage when exposed to similar conditions due to the dehydration of the agarose that is surrounding the enzyme.

In the same way, the use of this biocatalyst for interesting synthesis shows that inorganic materials exhibit a big advantage due to their inert characteristics. For example, CalB lipase was used as a biocatalyst in lactose oleate synthesis, and the conversion of oleic acid to ester was better when lipase was immobilized in GS since this material can conserve its porous characteristic in comparison with agarose support, under the evaluated reaction conditions (Fig. 10.11).

In general, immobilization of lipases in GS shows that this support helps obtain excellent biocatalysts due to the presence of a higher surface area and adequate pore size for the immobilization of enzymes, improving thermal stability in both organic and aqueous media; this generates better biocatalysts than when glyoxyl-agarose is used as support.

5 HETEROFUNCTIONAL SILICEOUS SUPPORTS

During the immobilization process, the enzyme is oriented into the pores depending on the chemical surface of support, whereby some variation in this parameter can influence the catalytic parameters of the obtained biocatalyst including: expressed activity, stability, and selectivity [33]. When

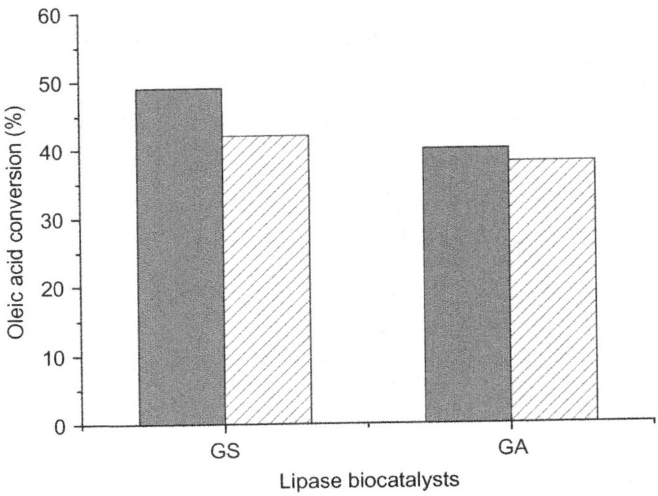

FIG. 10.11 Oleic acid conversion catalyzed by CalB lipase immobilized in glyoxyl-silica and glyoxyl-agarose using *tert*-butanol (*pattern*) or acetone (*gray*) at 40°C for 48 h.

monofunctional supports are used, the enzyme is immobilized through the face of protein where there is a high density of suitable groups for the interaction between enzyme and support. For example, with GS the enzymes are immobilized through the ε-amino groups of lysines and the aldehydes of the support; or when octyl silica is used, the hydrophobic amino acids interact with octyl groups of the material (Fig. 10.4).

The immobilization in this kind of monofunctional material has many advantages:

- The functionalization of the supports is a simple procedure.
- The immobilization is carried out in one step.

However, if the enzyme does not have suitable groups, it cannot be immobilized in those supports or it is only possible to improve just one catalytic parameter. This is the case of QL lipase, which must be aminated to be immobilized in GS [31], which improves its thermal stability but reduces its the expressed activity [22].

Heterofunctional supports are materials that have two or more kinds of functional groups on their surface, allowing the enzyme to interact with the same support in different ways. There are different supports with these characteristics; most of them use one functional group with the aim to adsorb the enzyme with one specific orientation (electrostatic, hydrophobic, complex formation) and the other groups generate a covalent bond between enzyme and support (glyoxyl or epoxide groups) (Fig. 10.12) [33]. The use of this kind of support brings advantages because of the homogeneity of orientation of

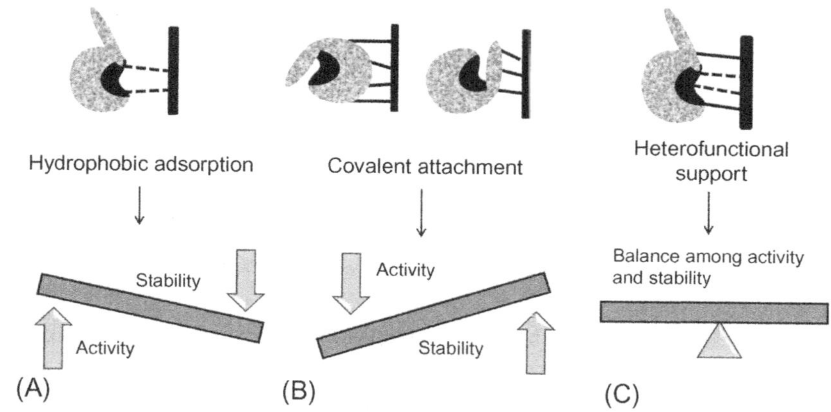

FIG. 10.12 Enzyme orientation into monofunctional support by hydrophobic adsorption (A) and covalent attachment (B), and enzyme orientation into heteterofunctional support (C) and its effects on the enzyme activity/stability.

the enzyme into the support, that is, high expressed activity, and high thermal stability. However, the process should be carried out in two steps: the first is enzyme adsorption (generally at pH 7.0), and the second is the formation of a covalent bond (at basic pH, near to 10.0), which could be affect the enzyme and increase the immobilization time and cost.

5.1 Heterofunctional Silica Synthesis and Its Use for Lipase Immobilization. Application in the Synthesis of Sugar-Based Surfactants

Many enzymes have been immobilized in heterofunctional supports, most of them hydrolases, with excellent results [33]. In particular, lipases are excellent study models for this methodology for the following reasons:

- Despite exhibiting an interfacial mechanism of activation [30], whereby their immobilization in hydrophobic surfaces is really easy, this methodology does not give an increase in their stability for industrial applications.
- Their immobilization by covalent attachment using glyoxyl support is difficult due to the low content of lysines in their external surface, causing a biocatalyst with low expressed activity [31].

Considering these issues, silica has been functionalized with octyl groups that favor the hydrophobic adsorption and aldehyde or epoxide groups for a covalent bond [34,35] (Fig. 10.13) to be used as support for lipase immobilization. This modification can be carried out in one step, and it is similar to those carried out for producing monofunctional supports, since both functionalizing reactants (3-glycidoxypropyl)methyldiethoxysilane and trimethoxy(octyl)silane are

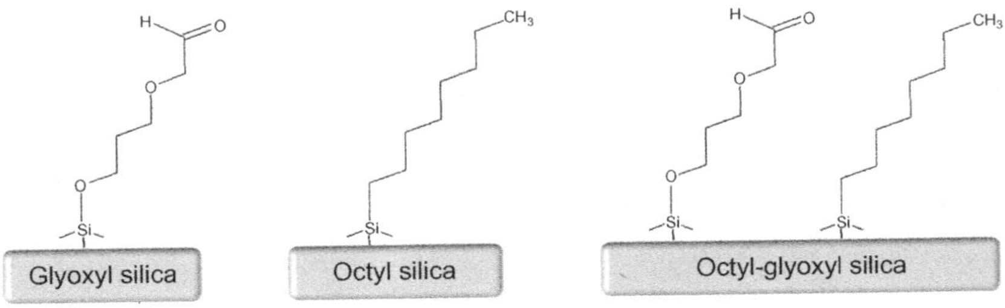

FIG. 10.13 Organic groups in mono and heterofunctional silica supports.

mixed at the same time with the silica material [34]. Bernal et al. functionalized silica with glyoxyl (100%), octyl-glyoxyl (50%–50%), and octyl (100%), and verified the covalent bonding of these groups to the siliceous structure [22] without affecting the porous characteristics, leading to a suitable pore size for lipase immobilization [34].

5.1.1 *Lipase Immobilization and Catalytic Behavior*

Lipases derived from *Pseudomonas stutzeri* (Ps) and *Alcaligenes* sp. (QL) were selected as study models due to their substrates' promiscuity and stability against some organic solvent in esterification reactions [34]. The immobilization of these lipases onto GS, octyl-glyoxyl silica (OGS), and octyl-silica (OS) produced biocatalysts that exhibited different behaviors depending on the chemical surface of support (Fig. 10.14). The values of immobilization yield with respect to protein loading indicate that a biocatalyst obtained with OS support is more selective for lipases than one obtained with GS support for both commercial extracts. It is possible that protein molecules that can interact with the OS support by hydrophobic interaction are fewer than those that interact by lysine residues, whereby a pseudopurification is carried out. The same behavior is observed when OGS support is used, because in this case the first interaction between lipase and silica is the same as in the OS support. This is correlated with expressed activity of the obtained biocatalyst: those prepared by covalent attachment exhibit low activity with respect to biocatalysts from OS and OGS supports (Fig. 10.12), indicating that the hydrophobic interaction is favorable for this kind of enzyme. This trend is more important for QL lipase, since the immobilization of this enzyme in GS support is not feasible without an additional step of chemical amination of its external surface [31].

The heterofunctional support offers a high stabilization of lipases in comparison with hydrophobic monofunctional supports. Half-life is improved significantly when biocatalysts are incubated in organic solvents at high temperatures, simulating reaction conditions (Fig. 10.15). For example, the stability of Ps lipase immobilized in OGS, OS, and GS exhibit different trends, which confirms the importance of enzyme orientation and extra stiffens of enzyme into the support [34]. Stability of Ps lipase immobilized in OS and GS supports

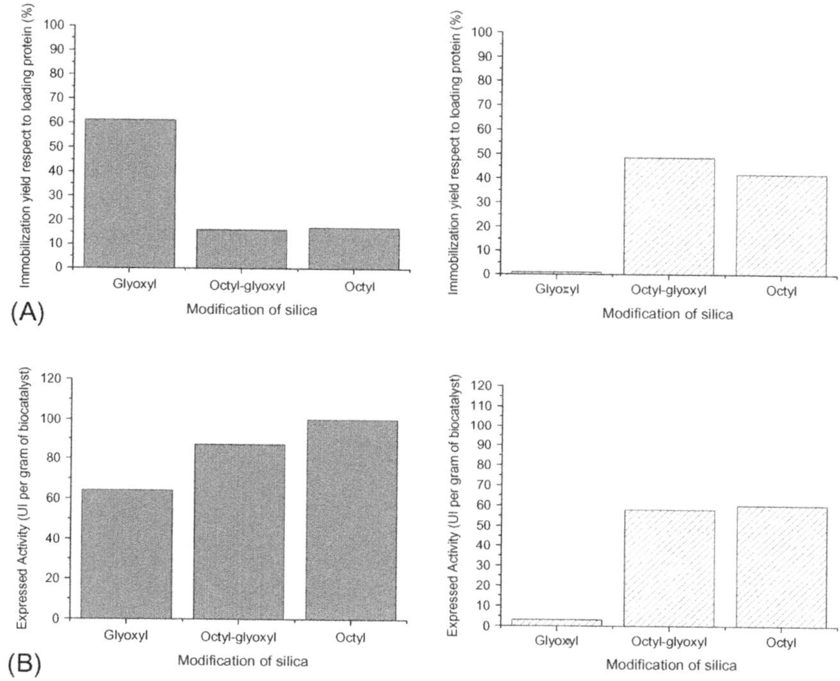

FIG. 10.14 Catalytic behavior of lipase from *Pseudomonas stutzeri* (*gray*) and *Alcaligenes* sp. (*pattern*) immobilized in silica with different chemical surfaces: (A) immobilization yield with respect to protein loading and (B) expressed activity.

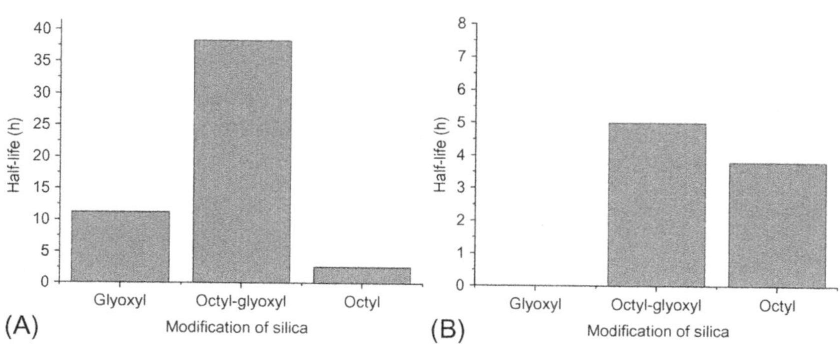

FIG. 10.15 Half-life of lipase from (A) *Pseudomonas stutzeri* and (B) *Alcaligenes* sp. immobilized in GS, OGS, and OS support in 100% acetone at 40°C. *(Adapted from C. Bernal, A. Illanes, L. Wilson, Improvement of efficiency in the enzymatic synthesis of lactulose palmitate, J. Agr. Food Chem. 63 (14) (2015) 3716–3724, with permission of the American Chemical Society.)*

against 100% acetone at 40°C is the expected behavior when it is immobilized in a monofunctional support. The half-life of GS biocatalyst is five times higher than those of the OS biocatalyst (Fig. 10.15) probably because covalent interactions among lysine and aldehyde groups in GS support favor a better stabilization than just hydrophobic interactions [34,36]. When Ps lipase is immobilized in OGS support, the stability is higher than for OS and GS biocatalysts (20-fold and 4-fold, respectively, Fig. 10.15). The enzyme structure stabilization in this heterofunctional support is probably carried out by the site active face, since the first step during the immobilization favors that all enzyme molecules are adsorbed by the hydrophobic pocket and the covalent bonding is formed with the lysine surrounding this site (Fig. 10.12). QL lipase behaves in a similar way. However, in this case the stabilization achieved with OGS support is more important because this enzyme can be immobilized by covalent attachment, whereby the immobilization in the heterofunctional support offers excellent protection that cannot be achieved using OS or GS supports, due to it low density of lysines (Fig. 10.15).

5.1.2 *Operating Performance in Successive Batches for Enzymatic Synthesis of a Lactulose Palmitate Monoester Surfactant*

There are many interesting reactions for lipase application. The synthesis of sugar fatty acid ester is one of them. These compounds are obtained from sugar and fatty acid, generating a molecule with hydrophilic-hydrophobic balance suitable to be used as a surfactant in the food and pharmaceutical industry. The chemical surface of the support where the lipase is immobilized influences the synthesis yield and the operating performance of the biocatalyst [34,37]. For example, QL and Ps lipases immobilized in siliceous supports with different chemical modifications (Table 10.3) were tested as biocatalysts for lactulose palmitate synthesis. The conversion of palmitic acid to monoester in acetone

TABLE 10.3 Palmitic Acid Conversion Into Lactulose Palmitate With Different Lipases Immobilized in Monofunctional and Heterofunctional Porous Silica Supports

Support	*Pseudomonas stutzeri* Lipase	*Alcaligens* sp. Lipase
Soluble	5,2	3,1
GS	20,4	na
OGS	33,5	21,8
OS	29,7	24,3

na this biocatalyst was not used for lactulose palmitate synthesis due to its low activity.

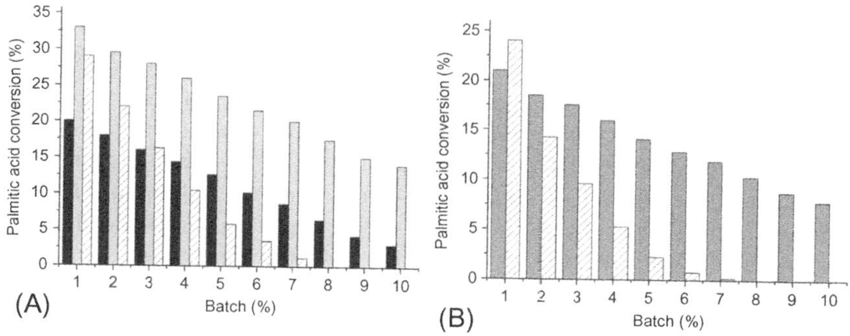

FIG. 10.16 Operating performance in successive batches in 100% acetone at 40°C during 48 h for the enzymatic synthesis of lactulose palmitate monoester surfactant using lipases from (A) *Pseudomonas stutzeri* and (B) *Alcaligenes* sp. immobilized in GS (*black*), OGS (*gray*), and OS (*pattern*) supports. *(Adapted from C. Bernal, A. Illanes, L. Wilson, Improvement of efficiency in the enzymatic synthesis of lactulose palmitate, J. Agr. Food Chem. 63 (14) (2015) 3716–3724, with permission of the American Chemical Society.)*

at 40°C was very low when a soluble enzyme was used. The lipases are probably inactivated under the synthetic conditions. However, the conversion was highly increased when these lipases were immobilized; even there were differences in the palmitic acid conversion among biocatalysts (Table 10.3). The advantages of the immobilization in heterofunctional supports are highlighted when the reaction is carried out in successive batches (Fig. 10.16): the conversion decrease is slow and lower from batch to batch when both lipases are immobilized in OGS supports than that immobilized in the OS or GS monofunctional supports. It confirms that the covalent bond formed between the enzyme and this support offers a higher stability under these reaction conditions.

Immobilization of lipases in heterofunctional support favors the expression of high activity, as seen in the OS support, combined with the high thermal stability of GS supported biocatalyst, which result in better conversions from batch to batch that when other biocatalysts are used (soluble enzyme, commercial enzyme, or enzyme immobilized in monofunctional supports).

6 CARBON/SILICA COMPOSITE SYNTHESIS AND ITS USE FOR LIPASE IMMOBILIZATION

The tendency of lipases being adsorbed and hyperactivated in hydrophobic surfaces expands the research for developing immobilization supports by using carbon and related materials. Nowadays, there is a growing interest in carbon/silica composite materials for enzyme immobilization because the balance between their hydrophilic-lipophilic characteristics allows their dispersion in the aqueous solution during the immobilization process and the stabilization and hyperactivation of enzymes such as lipases.

6.1 A Short Introduction of Synthesis and Characteristics of Porous Carbon/Silica Composites

The siliceous or carbonaceous materials are synthetized by different pathways influencing the porous, morphological and chemical characteristics of final material. Why is the combination of siliceous and carbonaceous materials—in other words, carbon/silica composites—interesting as support for enzyme immobilization? The answer is so simple, considering that a composite material combines the advantages of each counterpart in a unique material.

The route of synthesis of mesoporous silica templated by surfactants [38] and a soft or hard nanocasting process for the synthesis of porous carbons and nonsiliceous materials [39,40] has motivated the preparation pathways of carbon/silica composite materials: If the silica or the carbon is not eliminated from the intermediate product during this pathway, then a composite material is obtained. With the aim of obtaining a porous composite material, hydrothermal synthesis followed by calcination can be adopted. In an example with sucrose and Triton X-100 as a carbon source and tetraethylortosilicate for silica formation, the concomitant polycondensation at a controlled temperature of 100°C modulates the porous properties of the final material. The formed mesoporosity results in nanoparticle aggregation. Then, the solvent elimination and carbonization define the chemical and thermal nature of the composite. The gases formed during the polymerization of carbonaceous species in the systems contribute to the formation of micropores (Fig. 10.17) [41].

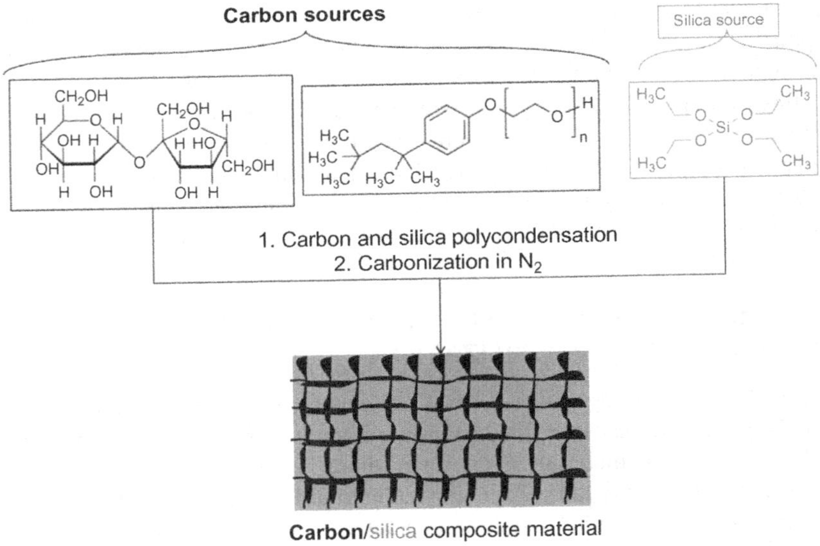

FIG. 10.17 Schematic representation of the pathway for obtaining carbon/silica composite materials.

6.1.1 Lipase Immobilization

There are some reports about the use of carbon/silica composite materials as immobilization supports. Some examples include the immobilization of whole cells or cells lysates of a recombinant strain *Escherichia coli* producing lipase from *Thermomyces lanuginosus* in nanocarbon-in-silica composites for its use in triglyceride hydrolysis and interesterification [42]. Similar nanocarbon-in-silica adsorbents, prepared by the synthesis of carbon nanofibers in silica xerogel during the pyrolysis of alkanes, confer stability for lipase from *T. lanuginosus* in the tributyrine hydrolysis [42]. The immobilization of lipases in silica using multiwalled carbon nanotubes as additives during the sol-gel process prevents the inactivation and enhances the stability of *Candida rugosa, Candida antarctica type B, and T. lanuginosus* lipases [43].

The use of porous carbon/silica composite gives additional advantages. For example two kinds of composites were used as immobilization support for lipase derived from *Alcaligenes* sp. One of them had small mesopores material named small mesoporous composite (SMC) between 2 and 7 nm, and the other had large mesopores material named large mesoporous composite (LMC) between 4 and 20 nm [44]. This difference in porosity allows obtaining biocatalysts with different characteristics. The SMC support favors the selective adsorption of the protein and the high expressed activity, achieving a biocatalyst with 110 IU per gram with respect to 81 IU per gram for the LMC biocatalyst. This is due to the lipase immobilization mainly being in the external surface of SMC support, which avoids the diffusional restriction imposed by the porosity in the LMC one. Both kinds of supports offer an extra thermal stability for the enzyme under aqueous and organic media; however, in the LMC case the large mesopores are more beneficial than the smaller ones, because a stronger confinement effect into the pores favors the preservation of the enzyme activity for more time (Fig. 10.18). On the other hand, the use of these two kinds of porous composite materials for the synthesis of palmitoyl ascorbate highlights the importance of the porosity of the support in the biocatalytic application, only the biocatalyst prepared in the composite support with pores of larger diameter are active. This is due to the pore size allowing free diffusion of the substrate and products during biocatalysis [44].

7 CONCLUDING REMARKS

The porous siliceous solid materials have been attractive for their use as enzyme immobilization supports; therefore, the control of their characteristics has motivated research in the last few years. Taking into account the enzyme properties and the final application for the biocatalysts prepared by immobilization in the solid support, the design of the support (porosity and surface chemistry) and the immobilization pathway can be accomplished. In this chapter large molecules such as β-galactosidases have been discussed as models for studying the effect

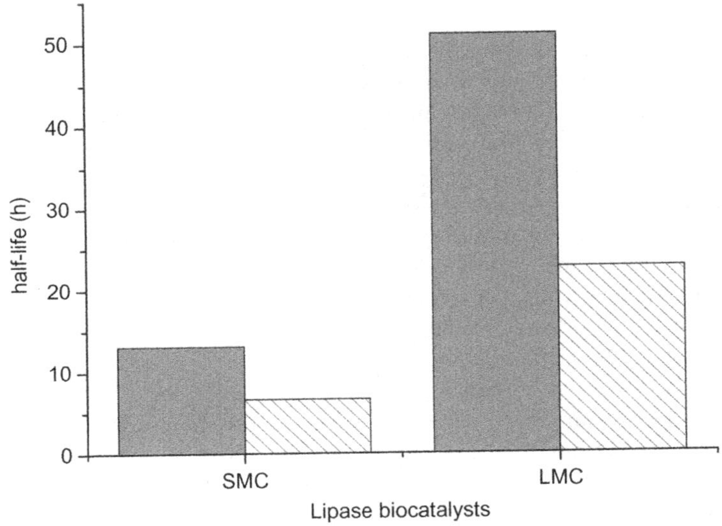

FIG. 10.18 Half-life comparison of lipase from *Alcaligenes* sp. immobilized in a silica/carbon composite with different porosity, under the following conditions: (*gray*) at 55°C and pH 7.0 in a 25 mM sodium phosphate buffer; (*pattern*) at 40°C in 100% *tert*-butanol (with 4% of equilibrium moisture). *(Adapted from M. Higuita, C. Bernal, M. Mesa, Interplay of carbon–silica sources on the formation of hierarchical porous composite materials for biological applications such as lipase immobilization, Mater. Sci. Eng. C, 43 (2014) 199–206, with permission of Elsevier.)*

of the porosity and confinement on hierarchical macro/mesoporous silica, after immobilization by adsorption or covalent attachment. The prepared biocatalysts have reached thermal and mechanical stability, allowing their reuse in processes such as whey hydrolysis. This knowledge has been expanded to other enzymes such as lipases and other kinds of materials, such as porous carbon/silica composites or heterofunctional supports. The applications of the prepared biocatalysts with β-galactosidades and lipases for performing industrially-attractive reactions are promissory in cost-efficiency terms. Nowadays, every immobilization system (enzyme-support-immobilization conditions) is treated as a special case, which makes the production of enzyme-solid biocatalysts with the desired properties in terms of stability/activity so lengthy. Nowadays, the development of general protocols based on the properties of both support and enzyme is highly desired. This is only the beginning and continuous research in this field will allow this objective to be achieved.

ACKNOWLEDGMENTS

This work was financed by Conicyt Chilean Agency through Fondecyt Program (Proyecto de Iniciacion 11150050).

REFERENCES

[1] J.S. Beck, J.C. Vartuli, W.J. Roth, M.E. Leonowicz, C.T. Kresge, K.D. Schmitt, C.T.W. Chu, D.H. Olson, E.W. Sheppard, A new family of mesoporous molecular sieves prepared with liquid crystal templates, J. Am. Chem. Soc. 114 (1992) 10834–10843.

[2] M. Hartmann, X. Kostrov, Immobilization of enzymes on porous silicas-benefits and challenges, Chem. Soc. Rev. 42 (15) (2013) 6277–6289.

[3] N. Carlsson, H. Gustafsson, C. Thörn, L. Olsson, K. Holmberg, B. Åkerman, Enzymes immobilized in mesoporous silica: a physical-chemical perspective, Adv. Colloid Interf. Sci. 205 (2014) 339–360.

[4] C.-H. Lee, T.-S. Lin, C.-Y. Mou, Mesoporous materials for encapsulating enzymes, Nano Today 4 (2) (2009) 165–179.

[5] D. Xin, H. Junhui, Spherical silica micro/nanomaterials with hierarchical structures: synthesis and applications, Nanoscale 2 (2011) 3984–4002.

[6] D.Ş. Karaman, T. Gulin-Sarfraz, J. Zhang, J.M. Rosenholm, One-pot synthesis of pore-expanded hollow mesoporous silica particles, Mater. Lett. 143 (2015) 140–143.

[7] X. Zhou, A. Duan, Z. Zhao, Y. Gong, H. Wu, Y. Wei, G. Jiang, J. Liu, The influence of cosurfactants on the synthesis of hierarchically porous silica materials in a P123/cosurfactant/1,3,5-trimethylbenzene/water four-component microemulsion system, Mater. Lett. 133 (2014) 228–231.

[8] R. Takahashi, S. Sato, T. Sodesawa, A. Yachi, Bimodal porous silica prepared from water glass by inducing phase separation, J. Ceram. Soc. Jpn. 109 (7) (2001) 577–579.

[9] C. Bernal, M. Mesa, L. Sierra, Synthesis of new silicas with high stable and large mesopores and macropores for biocatalysis applications, Mater. Sci. Eng. C 32 (6) (2012) 1380–1385.

[10] D. Cavaille, D. Combes, Characterization of β-galactosidase from *Kluyveromyces lactis*, Biotechnol. Appl. Biochem. 22 (1995) 55–64.

[11] Q.Z.K. Zhou, X.D. Chen, Effects of temperature and pH on the catalytic activity of the immobilized β-galactosidase from *Kluyveromyces lactis*, Biochem. Eng. J. 9 (2001) 33–40.

[12] Y. Tanaka, A. Kagamiishi, A. Kiuchi, T. Horiuchi, Purification and properties of β-galactosidase from Aspergillus oryzae, J. Biochem. 77 (1975) 241–247.

[13] A. Vetere, S. Paoletti, Separation and characterization of three β-galactosidases from Bacillus circulans, Biochim. Biophys. Acta 1380 (1998) 223–231.

[14] M.L. Richmond, J.I. Gray, C.M. Stine, Beta-galactosidase: review of recent research related to technological application, nutritional concerns, and immobilization, J. Dairy Sci. 64 (9) (1981) 1759–1771.

[15] P.S. Panesar, S. Kumari, R. Panesar, Potential applications of immobilized β-galactosidase in food processing industries, Enzyme Res. 2010 (2010) 1–16.

[16] Q. Husain, β Galactosidases and their potential applications: a review, Crit. Rev. Biotechnol. 30 (1) (2010) 41–62.

[17] C. Bernal, L. Sierra, M. Mesa, Application of hierarchical porous silica with a stable large porosity for b-galactosidase immobilization, ChemCatChem 3 (2011) 1948–1954.

[18] C. Bernal, L. Sierra, M. Mesa, Design of β-galactosidase/silica biocatalysts: impact of the enzyme properties and immobilization pathways on their catalytic performance, Eng. Life Sci. 14 (1) (2014) 85–94.

[19] R.M. Blanco, J.J. Calvete, J. Guisán, Immobilization-stabilization of enzymes; variables that control the intensity of the trypsin (amine)-agarose (aldehyde) multipoint attachment, Enzyme Microb. Technol. 11 (1989) 353–359.

[20] P. Vejayakumaran, I.A. Rahman, C.S. Sipaut, J. Ismail, C.K. CheeJ, Structural and thermal characterizations of silica nanoparticles grafted with pendant maleimide and epoxide groups, J. Colloid Interface Sci. 328 (2008) 81–91.

[21] C. Mateo, J.M. Palomo, M. Fuentes, L. Betancor, V. Grazu, F. López-Gallego, B.C.C. Pessela, A. Hidalgo, G. Fernández-Lorente, R. Fernández-Lafuente, J.M. Guisán, Glyoxyl agarose: a fully inert and hydrophilic support for immobilization and high stabilization of proteins, Enzyme Microb. Technol. 39 (2006) 274–280.

[22] C. Bernal, L. Sierra, M. Mesa, Improvement of thermal stability of β-galactosidase from Bacillus circulans by multipoint covalent immobilization in hierarchical macro-mesoporous silica, J. Mol. Catal. B Enzym. 84 (2012) 166–172.

[23] J. Szczodrak, Hydrolysis of lactose in whey permeate by immobilized β-galactosidase from Kluyveromycesfragilis, J. Mol. Catal. B Enzym. 10 (2000) 631–637.

[24] T. Haider, Q. Husain, Immobilization of β-galactosidase from Aspergillus oryzae via immunoaffinity support, Biochem. Eng. J. 43 (2009) 307–314.

[25] A. Illanes, Enzyme Biocatalysis, Principles and Applications, Springer, Dordrecht, 2008. 108–112.

[26] S. Sharma, S.S. Kanwar, Organic solvent tolerant lipases and applications. Sci. World J. 2014 (2014), No. 625258. http://dx.doi.org/10.1155/2014/625258.

[27] K.-E. Jaeger, S. Ransac, B.W. Dijkstra, C. Colson, M. Vanheuvel, O. Misset, Bacterial lipases, FEMS Microbiol. Rev. 15 (1) (1994) 29–63.

[28] K.D. Mukherjee, M.J. Hills, Lipases from plants, in: P. Woolley, S.B. Petersen (Eds.), Lipases: Their Structure, Biochemistry and Application, Cambridge University Press, Cambridge, 1994, pp. 49–75.

[29] F. Carrière, Y. Gargouri, H. Moreau, S. Ransac, E. Rogalska, R. Verger, Gastric lipases: cellular, biochemical and kinetic aspects, in: P. Wooley, S.B. Petersen (Eds.), Lipases: Their Structure, Biochemistry and Application, Cambridge University Press, Cambridge, 1994, pp. 181–205.

[30] J.M. Palomo, J.M. Guisan, Different strategies for hyperactivation of lipase biocatalysts, in: G. Sandoval (Ed.), Lipases and Phospholipases Methods in Molecular Biology, vol. 861, Springer, Mexico, 2012, pp. 329–341.

[31] C. Bernal, P. Urrutia, A. Illanes, L. Wilson, Hierarchical meso-macroporous silica grafted with glyoxyl groups: opportunities for covalent immobilization of enzymes, New Biotechnol. 30 (5) (2013) 500–506.

[32] R.C. Rodrigues, O. Barbosa, C. Ortiz, A. Berenguer-Murcia, R. Torres, R. Fernandez-Lafuente, Amination of enzymes to improve biocatalyst performance: coupling genetic modification and physicochemical tools, RSC Adv. 4 (2014) 38350–38374.

[33] J.M. Bolivar, C. Mateo, C. Godoy, B.C.C. Pessela, D.S. Rodrigues, R.L.C. Giordano, R. Fernandez-Lafuente, J.M. Guisan, The co-operative effect of physical and covalent protein adsorption on heterofunctional supports, Process Biochem. 44 (7) (2009) 757–763.

[34] C. Bernal, A. Illanes, L. Wilson, Heterofunctional hydrophilic-hydrophobic porous silica as support for multipoint covalent immobilization of lipases: application to lactulose palmitate synthesis, Langmuir 30 (12) (2014) 3557–3566.

[35] N. Guajardo, C. Bernal, L. Wilson, Z. Cabrera, Asymmetric hydrolysis of dimethyl-3-phenylglutarate in sequential batch reactor operation catalyzed by immobilized Geobacillusthermocatenulatus lipase, Catal, Today 255, (2015) 21–26.

[36] A. Suescun, N. Rueda, J.C.S. dos Santos, J.J. Castillo, C. Ortiz, R. Torres, O. Barbosa, R. Fernandez-Lafuente, Immobilization of lipases on glyoxyl–octyl supports: improved stability and reactivation strategies, Process Biochem. 50 (8) (2015) 1211–1217.

[37] C. Bernal, A. Illanes, L. Wilson, Improvement of efficiency in the enzymatic synthesis of lactulose palmitate, J. Agric. Food Chem. 63 (14) (2015) 3716–3724.

[38] C. Gérardin, J. Reboul, M. Bonne, B. Lebeau, Ecodesign of ordered mesoporous silica materials, Chem. Soc. Rev. 42 (2013) 4217–4255.

[39] D. Gu, F. Schüth, Synthesis of non-siliceous mesoporous oxides, Chem. Soc. Rev. 43 (2014) 313–344.

[40] Y. Xia, Z. Yang, R. Mokaya, Templated nanoscale porous carbons, Nanoscale 2 (2010) 639–659.

[41] M. Higuita, C. Bernal, M. Mesa, Interplay of carbon–silica sources on the formation of hierarchical porous composite materials for biological applications such as lipase immobilization, Mater. Sci. Eng. C 43 (2014) 199–206.

[42] G.A. Kovalenko, A.B. Beklemishev, L.V. Perminova, A.L. Mamaev, N.A. Rudina, S.I. Moseenkov, V.L. Kuznetsov, Immobilization of recombinant E. coli thermostable lipase by entrapment inside silica xerogel and nanocarbon-in-silica composites, J. Mol. Catal. B Enzym. 98 (2013) 78–86.

[43] S.H. Lee, T.T.N. Doan, K. Won, S.H. Ha, Y.-M. Koo, Immobilization of lipase within carbon nanotube–silica composites for non-aqueous reaction systems, J. Mol. Catal. B Enzym. 62 (2010) 169–172.

[44] C. Bernal, S. Escobar, L. Wilson, A. Illanes, M. Mesa, Carbonaceous–siliceous composite materials as immobilization support for lipase from Alcaligenes sp.: application to the synthesis of antioxidants, Carbon 74 (2014) 96–103.

Chapter 11

Zeolite-Encapsulated Catalysts: Challenges and Prospects

David Farrusseng and Alain Tuel
Univ Lyon, Université Claude Bernard Lyon 1, Villeurbanne, France

Contents

1 INTRODUCTION

Zeolites are crystalline microporous materials with open frameworks built of corner-sharing TO_4 tetrahedra, where T is generally Si or Al. The flexibility of the T-O-T linkage allows a large variety of zeolite structures to be prepared by simply changing synthesis conditions, for example, the composition of the gel, the nature of the structure directing agent (SDA), or the temperature [1]. Zeolite structures can be regarded as an assembly of a finite number of

Encapsulated Catalysts. http://dx.doi.org/10.1016/B978-0-12-803836-9.00011-0

structural units such as single and double rings as well as small cages. These structures all have in common the presence of well-defined pores and/or cavities or molecular dimensions, typically in the 0.4–1.2 nm range. Such cavities have been used for several decades as nanoreactors to entrap and isolate catalysts such as transition metal complexes, organometallic compounds, or oxide and metal nanoparticles in a highly ordered and uniform local environment.

The encapsulation of catalysts into the framework cavities of molecular sieves has received considerable attention during the last decades [2–12]. One of the most important motivations was to prepare hybrid materials that combined the advantages of the catalyst itself with those of stable heterogeneous supports such as recyclability and handling. Depending on the nature and size of the catalyst, different strategies have been developed for the immobilization in zeolite voids; they include the following:

- Crystallization of the zeolite in the presence of a metal complex, precursor, or a metal colloid. The zeolite is formed around the catalyst but the method necessitates that precursors are stable under synthesis conditions, in other words, generally high pH values and temperature. This method has been particularly employed for encapsulating of metal complexes in FAU-type zeolites [13–17] and metal nanoparticles in small pore zeolites [18–25].
- More generally, encapsulation is performed by postsynthesis modifications using standard techniques such as impregnation, grafting, or ion exchange.

Besides immobilization, the encapsulation in zeolite cavities stabilizes fragile catalysts, which are susceptible to deactivate rapidly by sintering, poisoning, polymerization, or degradation during the reaction. In the case of metal complexes such as porphyrins or phthalocyanines, which are deactivated by dimerization under reaction conditions, heterogenization not only allows a much easier recovery of the catalysts from reaction mixtures, but also prevents fast and irreversible deactivation by isolating catalytic sites from each other [26–28]. It is also well-known that zeolite frameworks can stabilize small metal nanoparticles, particularly against sintering at high temperature under harsh reaction conditions [29–35]. If the catalyst is immobilized without attachment to the zeolite framework, it retains its chemical and structural environment. Assuming that reaction rates are not diffusion-limited, zeolite-encapsulated catalysts should thus surpass the activity exhibited in their nonencapsulated form.

In many cases, the role of zeolite framework is not limited to host molecules or nanoparticles, but it also significantly influences the catalytic activity. In the particular case of metal complexes, even in the absence of a chemical bond between the catalyst and the zeolite framework, the latter influences the reactivity by modifying the local geometry around active sites as compared to free catalysts. Such modifications can emerge from steric constraints imposed by the size of zeolite cavities as previously reported for copper or cobalt phthalocyanines encapsulated in the supercages of Y zeolites [15,16,36,37]. Encapsulation forces the complex to adopt a nonplanar geometry, which leads to a higher reactivity towards molecular oxygen in oxidation reactions. The presence of charges in the zeolite framework as well as extra-framework cations can also influence the electronic and redox properties of the encapsulated complex.

The zeolite framework also influences the rate and selectivity of catalytic reactions by allowing selective access of molecules to intracrystalline active sites based on size or shape [38–42]. Shape selectivity depends of course on the structure of the zeolite used since only molecules capable of diffusing through the pores can react with the catalytic sites and escape from the solid. Similar properties exist in natural systems used for the biomimetic oxidation of hydrocarbons. It was demonstrated that the selectivity of the reaction arose from the steric effects imposed by the environment of the enzyme active site upon substrate approach [43–46]. Numerous papers and reviews have discussed the effect of encapsulation of a variety of metal complexes and clusters in zeolites on their stabilities and catalytic properties in comparison with their nonencapsulated analogs.

This chapter aims to focus on the various approaches that have been developed during the last decades for the encapsulation of homogeneous and heterogeneous catalysts in zeolite cavities in order to improve or modify their catalytic properties. From the heterogenization of transition metal complexes to the stabilization of uniform oxide or metal nanoparticles, the benefit of encapsulation on structural and catalytic properties of the catalyst will be illustrated by selected relevant examples of the literature. Moreover, emerging developments in the field of yolk-shell and core-shell catalysts with zeolite shells as well as encapsulation in zeolites with hierarchical porosities will be addressed.

2 METAL COMPLEXES IN ZEOLITES

Metal complexes have been used for many decades in catalysis and biomimetic chemistry [47–63]. Although these complexes exhibit generally good catalytic activities and selectivities, their encapsulation in the cavities of inorganic matrices offers many advantages. First, encapsulation immobilizes the complex, which means that separation from the reaction mixture and recovery of the catalyst are greatly facilitated. Second, complexes are isolated from each other, and deactivation processes due to the formation of dimers with μ-oxo or -peroxo bridges are considerably reduced. Finally, the presence of micropores of nanometer size in zeolites prevents large molecules to react with encapsulated catalytic sites and the structure imposes a restriction on the size of reactants, products, or transition states. In the particular case of small mononuclear metal complexes such as metal carbonyls, encapsulation can be performed directly by incorporating precursors in the pores and cavities of the zeolite [64–75]. In contrast, most of metal complexes are larger than the zeolite cavities in which they are encapsulated, which explains why they are commonly denoted as "ship-in-a-bottle" complexes. This ensures a certain stability to the catalyst by minimizing leaching through the restricted pore openings of the framework but necessitates the development of specific synthesis methods.

2.1 Methods for the Encapsulation of Metal Complexes in Zeolite Frameworks

Because of their size, preformed metal complexes such as porphyrins or phthalocyanines cannot be introduced in the small pores of the majority of zeolite frameworks. Alternative encapsulation methods have been developed in which one of the components—the complex or the zeolite—is formed in the presence of the other component.

2.1.1 The Zeolite Synthesis Route

In the "zeolite synthesis route," a metal complex is introduced in the synthesis gel and entrapped in the zeolite framework during crystallization. Depending on the natures of the zeolite structure and the metal complex, the latter can act either as a true templating molecule or it can be simply incorporated in the voids of the structure. This approach, in which the "bottle" is built around the "ship," provides many advantages over other methods of encapsulation, in particular a better dispersion of complexes throughout the zeolite framework. Moreover, if the complex is thermally and chemically stable under synthesis conditions, its environment in the zeolite framework is predictable, unique and well-defined. However, some critical points strongly limit the application of this method to a small number of zeolite structures and complexes. First, the topology of the zeolite framework is essential: encapsulation is only possible in large cavities in which the complex is intact and easily accessible to reactants. Some of the best candidates are X and Y zeolites with the FAU topology, which consists of supercages of c.1.2 nm diameter accessible by 12 MR circular windows of 0.74 nm (Fig. 11.1) [76]. Second, complexes have to be soluble and stable under the harsh conditions of pH and temperature that are generally required to crystallize zeolites. Finally, the zeolite must be "ready to use" without a further calcination step that could damage the complex. In the case where the synthesis requires additional organic molecules to direct the structure, these molecules have to be removed by soft treatments such as solvent extraction under mild conditions.

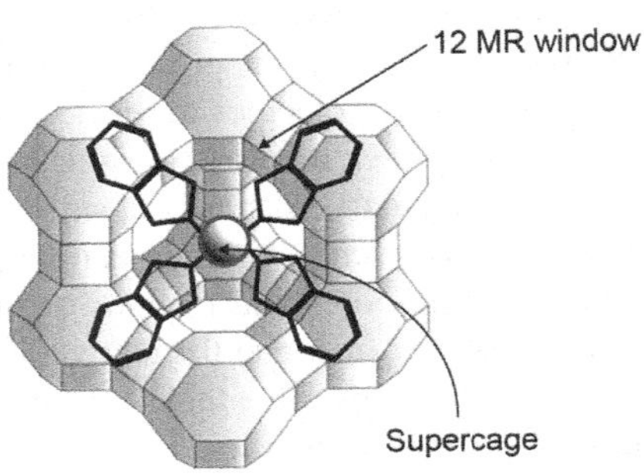

FIG. 11.1 Framework cavity of Y zeolite with encapsulated phthalocyanine.

Derivatives of the bis(cyclopentadienyl)cobalt(III) cation are famous examples of complexes that act as true structure directing molecules in the synthesis of zeolites. The nonsubstituted complex makes clathrasil phases with NON, AST, or DOH framework types or UTD-12, a one-dimensional zeolite with a structure related to ZSM-48 [77–79]. The addition of one methyl group on each cyclopentadienyl ring leads to UTD-18, a one-dimensional 12 MR zeolite related to SSZ-31. More substituted templates such as bis(pentamethylcyclopentadienyl)cobalt(III) are particularly interesting since they allowed the crystallization of the first extra-large pore high-silica zeolite UTD-1, with a one-dimensional system of pores delimited by 14 rings [80–82]. However, the low dimensionality of these zeolites makes the metal complexes inaccessible through the pore system; therefore they cannot be used as catalysts for organic reactions. More recently, a copper-amine complex was used as a structure directing agent for the direct synthesis of Cu-SSZ-13 zeolite [83]. Cu-SSZ-13, a zeolite with the CHA framework type, is an active catalyst for selective catalytic reduction (SCR) of NOx with NH_3 [84–87]. As compared to Cu introduction by ion exchange in a standard SSZ-13 zeolite, both the Cu content and dispersion were enhanced, yielding higher catalytic performance in the NH_3-SCR reaction.

More generally, the complex does not participate in the mechanism of zeolite formation and it is simply occluded in the cavities during crystallization of the framework. Most of the examples concern complexes such as phthalocyanines in X and Y zeolites, whose synthesis does not require the presence of organic templating molecules. Ru and Fe perfluorinated or Co, Cu, and Fe perchlorinated phthalocyanines were incorporated in the cavities of zeolites NaX and NaY using this method [15–17,88–91]. Catalyst loading is relatively low and corresponds to a maximum of approximately 1–2 complexes per 100 cavities of the zeolite framework. Nevertheless, low metal loading is an advantage to maintain a good accessibility to reactants during catalytic reactions. In the particular case of Cu, chlorinated $CuCl_{16}Pc$ complexes could be successfully incorporated in the channels of zeolite L as well [37]. The zeolite synthesis method was also applied for encapsulating metal porphyrins (Fe and Mn) in FAU-type zeolites [13,14]. High encapsulation levels were achieved with porphyrins having cationic substituents, which were explained by strong electrostatic interactions with anionic (alumino) silicate species in solution.

For the first time a slightly modified method was used to encapsulate Jacobsen's catalysts in MCM-22 [92]. The as-made form of MCM-22 consists of two-dimensional layers separated by hexamethyleneimine molecules and maintained together by hydrogen-bonded surface silanol groups. The final structure with MWW topology is obtained by heating the precursor at a moderate temperature, which leads to the condensation of Si-OH groups and the formation of large ellipsoidal supercages of c.0.71 × 1.82 nm. The Jacobsen's catalyst was intercalated between the layers of the zeolite precursor and the obtained material was carefully heated at 280°C. Upon condensation,

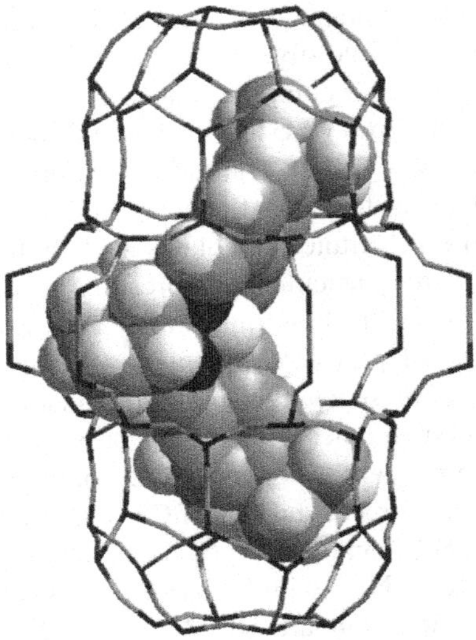

FIG. 11.2 Jacobsen's catalyst encapsulated in the ellipsoidal cage of MCM-22 zeolite. *(Adapted from G. Gbery, A. Zsigmond, K.J. Balkus, Jr., Catal. Lett. 74 (1–2) (2001) 77–80, with permission of Springer.)*

complexes were encapsulated in the supercages with no possibility to escape through the 0.4 × 0.54 nm 10-ring windows (Fig. 11.2).

The same synthesis route has been used more recently to encapsulate an (*R*, *R*)-(−)-*N*,*N'*-bis(3,5-di-*tert*-butylsalicylidene)-1,2-cyclohexanediaminocobalt(II) catalyst in MCM-22 [93].

2.1.2 The Flexible Ligand Route

Following this route, the complex is synthesized inside the zeolite cavity by reaction of ligands with a metal center (Fig. 11.3) [8,12,94–105]. The metal is generally introduced in the zeolite by ion exchange and the solid is eventually dehydrated prior to complexation. Although the flexible ligand route is one of the most versatile and easy methods to immobilize a metal complex in a zeolite, it is limited to small ligands that can diffuse in the zeolite pores. However, once formed, metal complexes become generally too large to escape from the framework, particularly in the case of X and Y zeolites in which large cavities are accessible by windows of a c.0.74 nm diameter.

The family of organic ligands that have been used to complex metal species in X and Y zeolites is huge and diverse: it comprises small monodentate ligands such as NO, CO, amines, or phosphines [6,17,106–116] as well as larger ones like bipyridine [117–122], polyamines [123], and Schiff bases [124–134]. The preparation, characterization, and applications of the corresponding materials have been widely reported in the literature. For example, encapsulation of transition

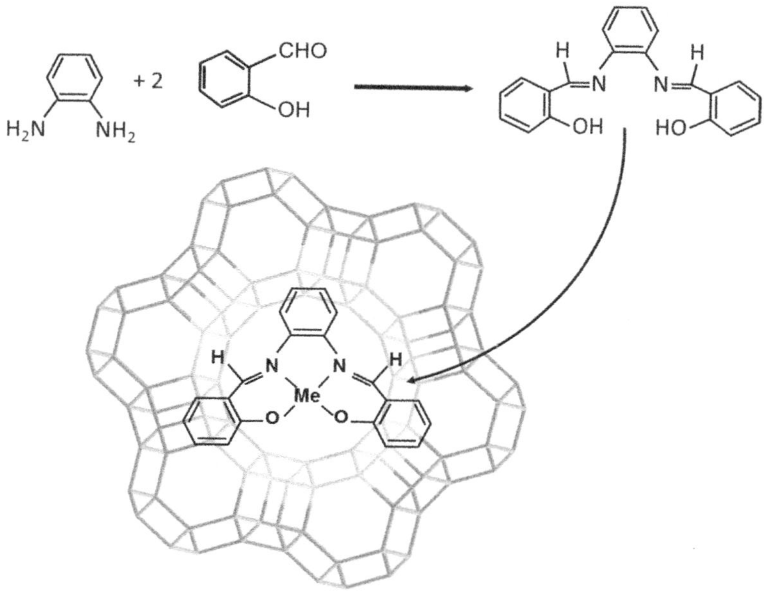

FIG. 11.3 Encapsulation of Me(II) salophen complexes in Y zeolite using the "flexible ligand method."

metal salen complexes (where salen is *N,N′*-bis(salicylidene)ehtylenediamine) has been particularly studied because salen ligands offer the desired flexibility [135–141]. Moreover, many of these complexes possess interesting catalytic properties in the oxidation of olefins and aromatic compounds (see Section 2.3).

2.1.3 The Template Synthesis Route

In some cases, ligands are too large to pass through the zeolite pores but some of their constituents can. Ligands are thus formed in situ and they can complex metal ions preliminarily introduced by ion exchange. The method has been essentially used to prepare metallophthalocyanines in zeolite Y; four dicyano-benzene molecules are introduced into zeolite cavities where they can cyclize around a metal ion to form the complex [142–149]. The reaction is generally performed at a relatively high temperature in the presence of an excess of dicyanobenzene. In contrast to direct encapsulation of the phthalocyanine by the "zeolite synthesis route," the catalysts are often contaminated by metal complexes formed on the outer surface of the crystals as well as unreacted metal species and organic molecules. Moreover, the method is limited to nonsubstituted phthalocyanines because substituted dicyanobenzenes can hardly enter the zeolite pores. Nevertheless, hexadecafluoro iron phthalocyanines $FePcF_{16}$ could be prepared following this method by reaction of tetrafluorophthalonitrile molecules on iron carbonyl species introduced in NaY [15]. The main disadvantage of using metal carbonyls or ferrocenes as metal sources is the possible presence of clusters in the zeolite formed by thermal decomposition of the precursors.

2.2 Effects of the Zeolite Framework on the Encapsulated Complex

As already mentioned above, encapsulation of metal complexes in zeolite frameworks offers many advantages compared to the corresponding homogeneous compounds.

2.2.1 *Heterogenization*

One of the most important advantages of encapsulation is that the catalyst becomes heterogeneous and can be easily recovered and separated from the reaction mixture. Evidence for a pure heterogeneous catalysis was demonstrated during the oxybromination of toluene with H_2O_2 in the presence of copper phthalocyanines $CuCl_{16}P_c$ encapsulated in NaX zeolite [37]. When the zeolite was removed by filtration from the reaction mixture after 4 h, the toluene conversion immediately stopped, whereas it continued in the presence of the catalyst (Fig. 11.4).

The experiment also clearly showed that copper phthalocyanines did not leach out from the zeolite cavities under reaction conditions because no Cu was detected in the liquid phase after filtration.

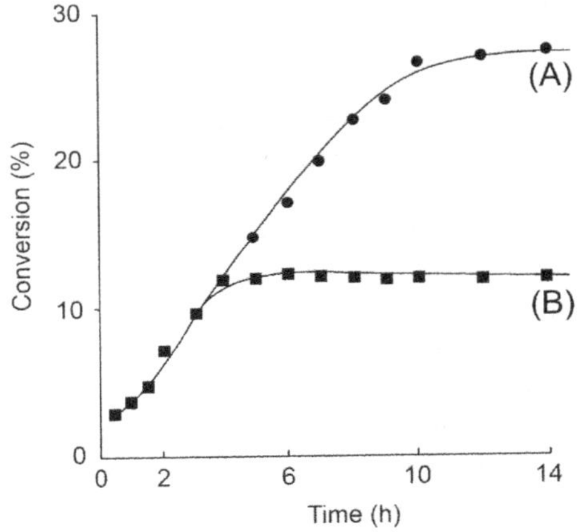

FIG. 11.4 Kinetics of toluene oxybromination in the presence of a $CuCl_{16}Pc$–Na-X catalyst (A) and when the catalyst is removed from the reaction mixture at a 4 h reaction time (B). *(Adapted from R. Raja, P. Ratnasamy, J. Catal. 170 (1997) 244–253, with permission of Elsevier.)*

2.2.2 Site Isolation

When metal complexes are entrapped in the cavities of the zeolite, they are isolated and cannot react with each other under reaction conditions. As a consequence, site isolation enhances the catalytic activity by preventing the formation of dimers, which are known to be catalytically inactive in oxidation reactions [26–28]. This was clearly illustrated in the oxidation of cyclohexane in the presence of perfluorinated ruthenium phthalocyanines [16]. The free complex is moderately active because of the rapid formation of dimers, evidenced by a change in color from blue to green and a shift of characteristic UV-Vis bands. In contrast, when complexes are encapsulated in the cavities of NaX zeolite, they become significantly more active and do not show any sign of deactivation after 7 days. Moreover, encapsulation of the complex increases the cyclohexanol selectivity (98.4% vs. 67% for the homogeneous catalyst) and the peroxide efficiency as well (from 55% to 97%). Site isolation can also be performed by bonding a mononuclear complex of Re, Rh, Os, or Ir to oxygen atoms of the zeolite framework [7,71,75,150–154].

2.2.3 Distortion of the Complexes in Zeolite Cavities

Complexes such as phthalocyanines are generally larger than the supercage of X and Y zeolites in which they can be occluded. Therefore, the encapsulated catalyst generally adopts a nonplanar geometry due to space constraints imposed by zeolite walls [15,16,36,37]. Deviation of the geometry from planarity is generally evidenced experimentally by shifts in the UV-Vis bands and hyperfine structures in EPR spectra as compared to the free complexes. It has been confirmed by computer modeling in the case of Cu, Co, and Fe-based phthalocyanines; in the case of $FeF_{16}P_c$ in NaX the complex adopts a saddle-type distortion with angles of 34.5° out of the plane (Fig. 11.5) [15].

Distortion leads to a quasitetrahedral symmetry, which admix d_{xy} and d_z^2 orbitals and decreases the electron density on the metal ion, thereby enhancing the reactivity of the complex toward oxygen in oxidation reactions [155]. Similar conclusions were obtained on $CuCl_{16}Pc$ complexes in NaX by computer modeling and molecular strain energy minimization calculations [37].

2.2.4 Effect of Charge Compensating Cations in Zeolite Cavities

The nature of alkali cations in zeolites can modify the electronic density on metal active sites as well as their redox potential, and it then influences the overall catalytic activity. In the case of zeolite Y-encapsulated Fe(III)-Schiff-base complexes, the activity (TON) in the oxidative coupling of 2-naphthol increases with the size of the cation (Fig. 11.6) [138]. Data were explained by the energy of the interaction between the cation and the molecule, small cations interacting more strongly and lowering diffusivity of 2-naphthol in the zeolite pores. The variation of the extra framework alkali cation in catalysts containing

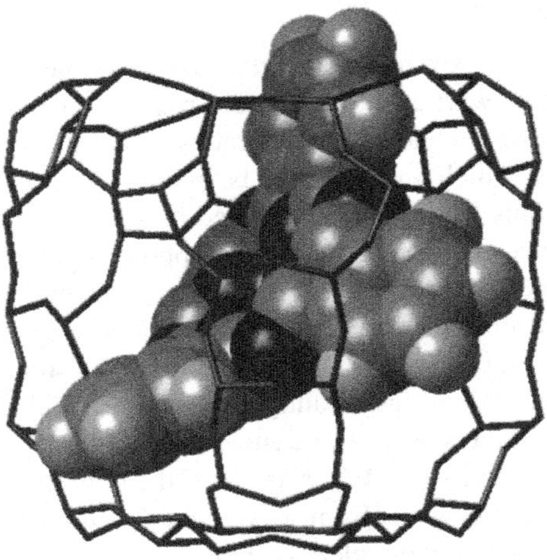

FIG. 11.5 Molecular graphics representation of CoPc within the supercage of zeolite Y having the minimum energy geometry with saddle-type distortion. *(Reproduced from S. Ray, S. Vasudevan, Inorg. Chem. 42 (2003) 1711–1719, with permission of American Chemical Society.)*

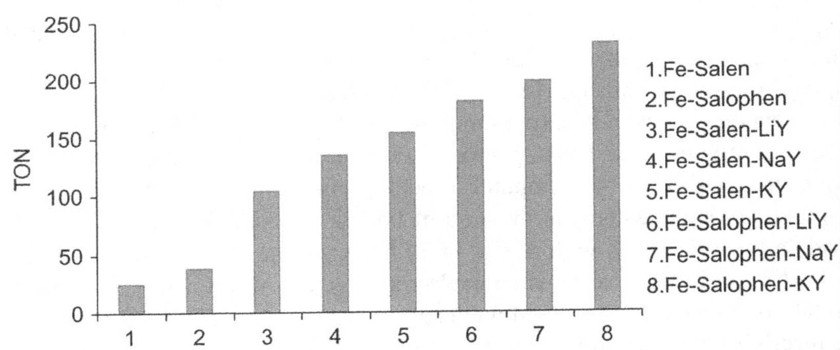

FIG. 11.6 TON for the catalytic oxidative coupling of 2-napthol using Fe(III) Schiff-base complexes in Y zeolites with various cations. *(Reproduced from K.K. Bania, D. Bharali, B. Viswanathan, R.C. Deka, Inorg. Chem. 51 (3) (2012) 1657–1674, with permission of American Chemical Society.)*

tris(2,2′-bipyridine)iron(II) complexes in the cages of zeolite Y had a similar effect during the photocatalytic oxidation of styrene [122].

Alkali cations can also affect the selectivity of a reaction by simply modifying the pore opening of the zeolite. Cobalt cyanide complexes $Co(CN)_5^{3-}$ are known to reversibly bind oxygen, and they can be used to selectively adsorb oxygen from air. When these complexes are formed in NaY, Co species rapidly oxidize to Co(III) through an intermolecular mechanism involving two Co centers in neighboring cavities.

Exchanging CoNaY zeolite with the large Cs^+ cations prior to the introduction of cyanide ligands decreases the mean diameter of the 12 MR windows between adjacent cavities and thus stabilizes $Co(CN)_5^{3-}$ species by preventing bimolecular reactions [156]. Large cations such as Rb^+ or NH_4^+ can also affect the selectivity during the competitive oxidation of alkanes with different sizes (see Section 2.2.6).

2.2.5 *Influence of the Acidity of the Zeolite*

The acidity of the host zeolite may greatly affect the catalytic performance of the encapsulated catalyst by favoring consecutive reactions. *Cis*-manganese(II) bis-2,2′-bipyridyl complexes $[cMn(bpy)_2]^{2+}$ in FAU-type zeolites catalyze the epoxidation of linear and cycloalkenes with hydrogen peroxide. In the particular case of 1-hexene, the oxide is the main product but diol is also obtained by a secondary acid-catalyzed ring opening. At 20% conversion, the diol selectivity increases from 14% on $Mn(bpy)_2X$ to 40% on $Mn(bpy)_2Y$, probably because of the higher acidity of MnY zeolite compared to MnX.

2.2.6 *Selectivities*

Size Selectivity by Exclusion of Reactants:

The accessibility to the confined space in which complexes are encapsulated controls the nature of molecules that can effectively diffuse to the catalyst and react with active sites. In contrast to molecules in solution, the pore opening of the zeolite can discriminate between molecules of different sizes, and the largest molecules are not allowed to react with the complex [16]. An example of size exclusion was reported by the oxidation of the large cyclooctene over Jacobsen's catalyst encapsulated in MCM-22 [92]. Using the homogeneous nonencapsulated complex, the cycloalkene is oxidized mainly into the corresponding alcohol with a conversion of 57% after 20 h. When the complex is inside the ellipsoidal supercages of MCM-22, less than 1% of the substrate is converted at the same reaction time, indicating that the molecule cannot pass through the 10-MR windows of the framework (4 × 5.4 Å in size). The preferential oxidation of small substrates in a mixture of cycloalkanes was also reported over NaY zeolite encapsulated iron phthalocyanine catalysts. During the competitive oxidation of cyclohexane and cyclododecane, the selectivity for the smaller substrate (60:40) was attributed to the molecular sieving properties of the zeolite. This selectivity could even be increased to 90:10 when the pore size was narrowed by exchanging Na^+ for larger cations such as NH_4^+ or Rb^+ [142].

Product Selectivity:

Even when the substrate is small enough to reach the metal complex, the product selectivity may be different from that observed in solution if some of the transition states are too bulky to form in the confined space of the cavity.

Such a transition state shape selectivity was observed during the hydrogenation of butadiene over the EDA complex of iron phthalocyanine ($Na_4^+FePc^{4-}$) in NaY [157]. At 6% conversion, the *trans/cis* ratio of 2-butene was 2.3 over the intrazeolite complex, whereas it was only 0.5 when the complex was dispersed on the outer surface of the crystals. This was explained by the specific distorted configuration of the complex in the cavity, which favored the formation of the *syn*-type π-allyl intermediate leading to *trans*-2-butene.

Regioselectivity:

Regioselectivity results from the constraints imposed on the molecule by the zeolite framework. Indeed, the molecule is generally extended in the direction of the channels and oriented in such a way that it can easily diffuse in the pores. Therefore, when the molecule passes through the 12 MR window of the FAU-type framework, positions near the ends preferentially react with the encapsulated catalyst. This was illustrated during the oxidation of 2,5-dimehtylhexane over vanadium (IV) picolinate peroxo complexes in the supercages of zeolite Y [158]. The ratio between products oxidized at positions 1 and 3 increased by a factor of 2 as compared to the homogeneous catalyst. Regioselectivity was also observed during the oxidation of *n*-alkanes with long chains over Fe phthalocyanines in zeolite Y. Compared to the homogeneous complex, the C2 position was preferentially oxidized and both the C2/C3 and C2/(C4+C5) ratios increased with the alkane chain length [159].

2.3 Catalytic Activities of Zeolite Encapsulated Metal Complexes

2.3.1 Oxidation Reactions

Zeolite encapsulated metalloporphyrins, metallophthalocyanines, and metal Schiff-base complexes have been extensively used as oxidation catalysts because they mimic enzymatic systems, in particular monoxygenase enzymes of the cytochrome P-450 family [11,43,160,161]. The selectivity of natural enzymes in biomimetic oxidations being attributed to steric effects imposed by the structure of the proteins, the concept of encapsulation originally was in reference to the replacement of the protein skeleton of the enzyme by a zeolite cavity [44–46]. In addition to providing a confined space for the catalytic centers, zeolite frameworks also influence the reaction selectivities by controlling the diffusion of molecules in the micropores. One of the very first applications of encapsulated complexes in catalysis concerned oxidation reactions with iodobenzene over iron phthalocyanine immobilized in X and Y zeolites [142]. Though catalytic results were not remarkable, they clearly demonstrated the role of the zeolite host on the stability and selectivity of the catalyst.

Metallophthalocyanines and metalloporphyrins encapsulated in the supercages of FAU-type zeolites have been widely investigated in catalytic oxidation of hydrocarbons. Encapsulated Cu, Co, and Ru phthalocyanines catalyze the oxidation of cyclic alkanes such as cyclohexane with aqueous H_2O_2 or *tert*-butylhydroperoxide (TBHP) as oxidants [8,13,16,17,101,104,143,162–164]. Conversions are generally higher with TBHP but both oxidants give the corresponding alcohol and ketone in similar amounts. Nevertheless, ruthenium perfluorophthalocyanine $RuF_{16}Pc$ in zeolite X showed a high activity with a selectivity for cyclohexanone above 98% at a 70% conversion [16]. In contrast, $Fe(proline)_2$ complexes in NaX yielded 100% cyclohexanol at a 70% conversion in the presence of $TBHP/O_2$ as an oxidant (Fig. 11.7) [165].

FePcY also catalyzes the direct transformation of cyclohexane to adipic acid with TBHP [166]. Copper phthalocyanines containing electron-withdrawing groups (Cl, NO_2) immobilized in the cavities of NaX also catalyze the aerobic oxidation of cyclohexane in a mixture of cyclohexanol, cyclohexanone, and acids [17]. Iron porphyrins have also been reported to catalyze the oxidation of cycloalkanes with various oxygen donors [167–169]. A few examples of cycloalkane oxidations with dioxygen have been reported over various Co complexes in zeolite NaX [167,170]. Compared to cycloalkanes, examples of oxidations of linear alkanes using encapsulated phthalocyanines are scarce. Various metallophthalocyanines for which most of ring hydrogens have been substituted by electron-withdrawing groups are able to catalyze the oxidation of methane with O_2/TBHP at 273 K when encapsulated in X or Y zeolites [171]. In the best cases methane conversion reaches 5% with a selectivity in methanol + formaldehyde above 98%. The direct oxidation of methane to methanol has also been reported on a series of transition metal substituted salen complexes encapsulated in zeolites X and Y [172]. As in the case of phthalocyanines, methanol and formaldehyde were predominantly formed in the presence of a mixture of TBHP and oxygen as an oxidant. Propane could be

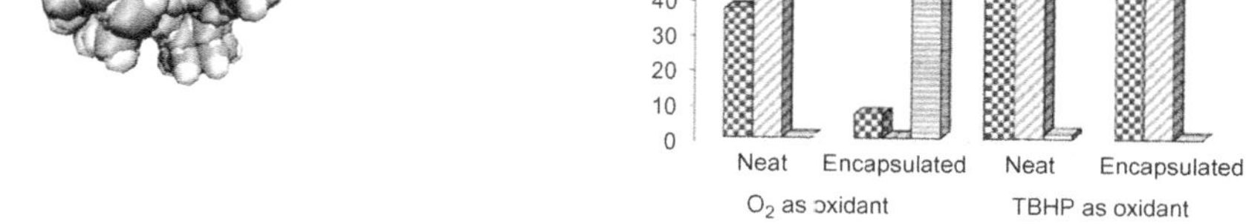

FIG. 11.7 Graphical representation of a Fe-Proline complex within the supercage of zeolite X and catalytic oxidation of cyclohexane using both neat and zeolite encapsulated $Fe(proline)_2$ complexes in the presence of air and TBHP as oxidants. *(Reproduced from J. Dzierzak, E. Bottinelli, G. Berlier, E. Gianotti, E. Stulz, R.M. Kowalczyk, et al., Chem. Commun. 46 (2010) 2805–2807; J. Dzierzak, M. Lefenfeld, R. Raja, Top. Catal. 52 (2009) 1669–1676, with permission of Springer and Royal Society of Chemistry.)*

selectively converted to a mixture of isopropanol and acetone over copper, cobalt, or iron substituted phthalocyanines in NaX or NaY using O_2/TBHP as an oxidant [173]. The oxidation of cyclic and linear alkanes has also been reported over various vanadium (IV) encapsulated catalysts [158,174]. In the case of picolinate peroxo complexes, cyclohexylhydroperoxide was initially formed and decomposed to a mixture of alcohol and ketone. Unfortunately, those catalysts were not stable and the complex progressively leached from the support under reaction conditions (Fig. 11.8).

In the presence of hydrogen peroxide, the initial bis-picolinate complex was dissociated and the mono-picolinate complex formed in the zeolite cavity was not large enough to be retained by spatial confinement. Encapsulated catalysts also catalyze the oxidation of substituted aromatics such as xylenes or alkylbenzenes [174–178]. For example, *p*-xylene was selectively oxidized to terephthalic acid over μ_3-oxo-bridged Co/Mn clusters in Zeolite–Y [176]. The high activity of the bimetallic complex as compared to the homonuclear complex was explained by a more facile redox behavior of Co between +2 and +3 oxidation states. Various copper complexes in zeolite Y have been used to selectively oxidize ethylbenzene to acetophenone with H_2O_2 or TBHP. Compared to the nonencapsulated catalyst, ring hydroxylation was negligible, thus featuring the role of the zeolite in regioselectivity [105,178].

Zeolite encapsulated metal complexes are also excellent catalysts for epoxidation reactions with iodosylbenzene, hydrogen peroxide, or alkyl hydroperoxides as oxygen donors. Among all reactions reported in the literature, epoxidations of cyclohexene and styrene have been particularly studied [13,14,117,179–183]. For both substrates, the product selectivity depends on the nature of the catalyst and oxidant. With anhydrous TBHP, cyclohexene is generally converted selectively to the corresponding oxide at a low

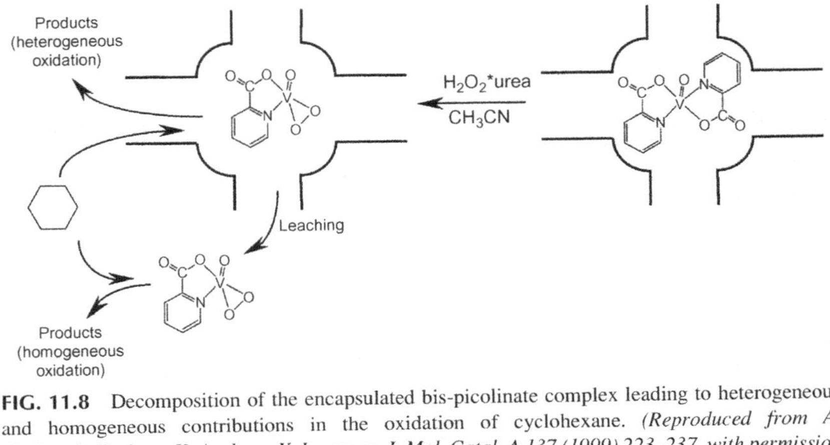

FIG. 11.8 Decomposition of the encapsulated bis-picolinate complex leading to heterogeneous and homogeneous contributions in the oxidation of cyclohexane. *(Reproduced from A. Kozlov, A. Kozlova, K. Asakura, Y. Iwasawa, J. Mol. Catal. A 137 (1999) 223–237, with permission of Elsevier.)*

conversion. Increasing the reaction time favors the formation of the diol by ring opening, and finally adipic acid. When hydrogen peroxide is used as an oxidant, the diol selectivity increases and it depends on the intrinsic acidity of the zeolite framework. This is clearly illustrated by the epoxidation of different linear and cyclic alkenes over *Cis*-$[Mn(bpy)_2]^{2+}$ in X and Y zeolites; the diol selectivity is systematically higher when the complex encapsulated in the more acidic MnY zeolite [117].

Products obtained from the oxidation of styrene also depend on reaction conditions (solvent, oxidant), the nature of the complex, and the acidity of the zeolite framework. Mild oxidants such as TBHP generally favor the formation of styrene oxide. For example, 60.2% of styrene was converted into 78% styrene oxide and only 18% benzaldehyde within 7 h over NaY encapsulated Cu complexes of the Schiff base derived from 2-acetylpyridine and *o*-aminophenol [184]. The use of Mn(III) complexes in FAU-type zeolites led to higher selectivities in benzaldehyde, with significant amounts of phenyl acetaldehyde [117]. When aqueous hydrogen peroxide is used as an oxidizing agent, styrene oxide selectivities are usually quite low and products such as benzaldehyde or 1-phenylethane-1,2-diol become predominant. For example, a Zn(II) complex with thio-Schiff base ligands in zeolite Y converts 98% of styrene after 12 h with a selectivity in diol of 65% [185].

Zeolite immobilized complexes are also efficient catalysts for hydroxylation reactions, particularly phenol hydroxylation with hydrogen peroxide. Liquid-phase hydroxylation of phenol with H_2O_2 to catechol and hydroquinone (*ortho* and *para* dihydroxybenzenes, respectively) has been reported over a wide range of complexes encapsulated in zeolites X and Y [174,186–194] (Scheme 11.1).

Catechol is usually the major product formed with a ratio Cat/Hyd between the two isomers in the range of 2–3. However, very high selectivities in catechol have been reported, even at relatively high conversion levels. For example, various complexes of amidate ligand 1,2-bis(2-hydroxybenzamido)ethane (H_2hybe) show selectivities in catechol higher than 90% for phenol conversions around 40% [192] (Table 11.1).

Similar catechol selectivities have been observed on Cu thio-Schiff base complexes in NaY for conversions below 20% [194].

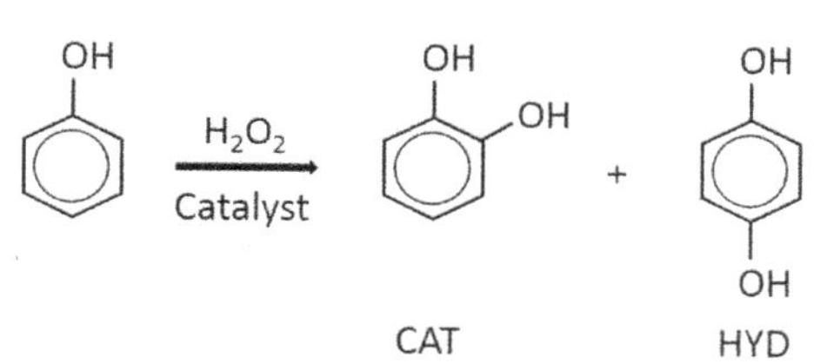

SCHEME 11.1 Phenol conversion into catechol and hydroquinone.

TABLE 11.1 Conversions and Selectivities in Hydroquinone and Catechol in the Hydroxylation of Phenol Over Cu, Ni, and Zn Complexes in Y Zeolite

		Selectivity (%)	
Catalyst	Conversion (%)	Catechol	Hydroquinone
[Cu(hybe)]-Y	40.14	90.9	9.1
[Ni(hybe)]-Y	36.5	90.4	9.6
[Zn(hybe)]-Y	32.66	89.0	11

Adapted from M.R. Maurya, S.J.J. Titinchi, S. Chand, Catal. Lett. 89 (3–4) (2003) 219–227 of Springer.

Many encapsulated complexes catalyze the oxidation of organic sulfides in sulfoxides [195,196]. Vanadium complexes are particularly active in the oxidation of methyl phenyl sulfide with H_2O_2 to the corresponding sulfoxide, with selectivities close to 90% at almost complete substrate conversion [196]. Quite similar selectivities and conversions were obtained over Cu and Mn chiral C_2-multidentate complexes immobilized in USY using iodosylbenzene as an oxidant [195].

There are also many reports on alcohol oxidations catalyzed by zeolite encapsulated metal complexes, in particular the oxidation of benzyl alcohol [105,165,197–199]. Though many complexes have shown significant catalytic activities, the case of Fe^{3+} proline in NaX is particularly interesting since the alcohol is almost totally converted in the presence of TBHP with benzaldehyde as the sole product (99+ %) [165]. More recently, Cu-salen complexes in NaY have been used for the catalytic oxidation and conversion of Kraft Lignin with CH_3COOOH into phenolic compounds, essentially 2-methoxy phenol [135].

2.3.2 Hydrogenations

Besides oxidation reactions, there are relatively few examples of hydrogenation reactions catalyzed by encapsulated metal complexes. As previously mentioned, EDA complexes of iron phthalocyanine ($Na_4^+FePc^{4-}$) in NaY catalyze the hydrogenation of butadiene to 2-butene with a *trans*/*cis* ratio of 2-butene of 2.3, different from that observed when the complex is dispersed on the outer surface of the crystals (*trans*/*cis* = 0.5) [157]. Pd(II) salen complexes in X or Y zeolites catalyze the hydrogenation of 1-hexene or cyclohexene to the corresponding alkanes [98]. In the case of 1-hexene, the reaction is complicated by the fact that the olefin is also isomerized to *trans*-2-hexene. Linear and cyclic alkenes have also been hydrogenated over rhodium complexes immobilized in

NaY [200]. In particular, Rh Salen encapsulated in zeolite FAU possesses high selectivities for the hydrogenation of only one double bond of 1,5-cyclooctadiene to cyclooctene [201]. Ir, Ru, and Os carbonyls in zeolite Y have also been reported to catalyze the hydrogenation of CO or CO_2 [73,74,202,203]. Such complexes are generally not very stable in zeolite cavities and some of the carbonyl ligands can be easily removed, leading to supported metal complexes with oxygen atoms of the zeolite framework bonded to the metal. Those heterogeneous ligands rigidify the structure of the complex, making reactivity different from that of free complexes in solution. Upon heating, decarbonylation can be complete and nude atoms migrate and agglomerate into metallic clusters inside the zeolite cavities or on the outer surface of the crystals. Decarbonylation is actually an excellent method to prepare metal nanoparticles in zeolite frameworks, as will be discussed below.

3 METAL OXIDE NANOPARTICLES IN ZEOLITES

Examples of zeolite-encapsulated metal oxide nanoparticles as heterogeneous catalysts are less common than those of the corresponding metal species. Metal oxide nanoparticles are commonly an intermediate formed during the preparation of encapsulated metal nanoparticles. Encapsulation is generally performed by postsynthesis methods, namely ion exchange, incipient wetness impregnation (IWI) with the corresponding metal salt, chemical vapor deposition (CVD), or in situ thermal decomposition of metal carbonyl compounds. However, in the case of zeolites with small or medium pore sizes, the direct synthesis of the zeolite in the presence of metal precursors in the gel has been reported; it will be discussed in the next section devoted to metal nanoparticles. Iron/molybdenum oxide was introduced by CVD from $FeCl_3$ and MoO_2Cl_2 in various zeolites with an MFI framework type and used for the oxidation of *p*-xylene and 3-fluorotoluene in the corresponding aldehydes (Scheme 11.2) [204].

Incorporation of TiO_2 in the pores of Beta zeolites by a cold $TiCl_4$ plasma process led to photocatalysts for the oxidation of methanol in the gas phase [205]. The catalytic rate was one order of magnitude higher than that observed on a conventional P25 titanium oxide catalyst. The addition of ruthenium chloride in the synthesis gels of LTA or MFI zeolites produced materials with subnanometric RuO_2 nanoparticles in the pores of the structure [21]. EXAFS and XANES spectroscopies indicated that $RuCl_3$ precursors were fully converted to RuO_2 during hydrothermal synthesis. LTA-encapsulated RuO_2clusters that were c.1 nm in diameter have been used as a catalyst for the oxidation of methanol to formaldehyde and methylformate. Methanol oxidative dehydrogenation turnover rates were much higher than those observed on RuO_2 clusters deposited on silica.

Molybdenum oxide nanoparticles in acidic zeolites are also famous for catalyzing the aromatization of methane to benzene. In this reaction, molybdenum

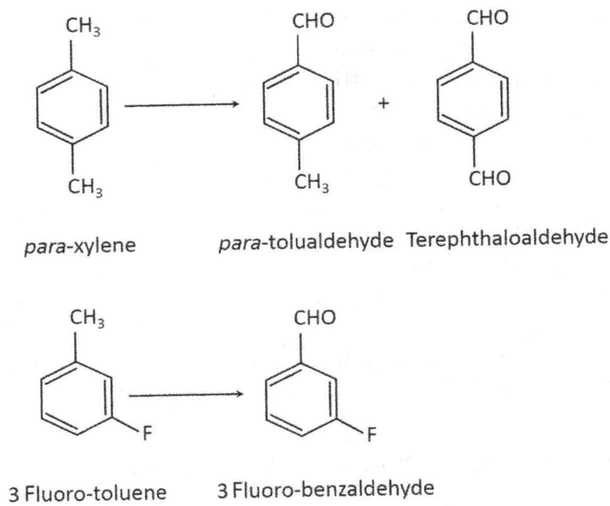

SCHEME 11.2 Main products in the oxidation of *para*-xylene and 3-fluoro-toluene.

oxide nanoparticles are converted to molybdenum carbide Mo_2C, which are the active species for methane activation. The Bronsted acidity of the zeolite promotes not only the methane activation on the molybdenum carbide sites, but also the oligomerization of the surface species towards primary products such as ethene, which are then converted into aromatics [206–209].

Another class of zeolite-encapsulated oxide nanoparticles concerns basic zeolites containing alkali metal oxides, particularly cesium oxide [210]. The presence of additional cesium oxide clusters in the cavities of Cs-exchanged zeolites such as CsX considerably increased the activity for the isomerization of 1-butene to 2-butene, with *cis*/*trans* ratios above 10 [211]. The lack of activity of Cs-exchanged zeolites as well as physicochemical properties of cesium oxide/CsX zeolites suggests that the strong basicity of the catalyst emerges from the presence of intrazeolitic particles. On the basis of different characterizations, it has been proposed that occluded particles could be a mixture of oxides, including Cs_2O, cesium peroxide (Cs_2O_2), and cesium superoxide (CsO_2) [212]. Those zeolites have been used to prepare cyclic carbonates from the addition of CO_2 to epoxides [213]. The reactivity of the zeolite strongly depends on the strength of basic sites: they must be strong enough to adsorb CO_2 but not too strong to allow desorption and subsequent reaction. Using CO_2 adsorption, it has been reported that basic sites of cesium oxides CsO_x occluded in CsX zeolites were not very strong and significantly weaker than those of bulk Cs_2O [214]. Their moderate strength, which likely results from the presence of several oxide phases inside the cavities, could explain their good activity in cycloaddition reactions.

4 METAL NANOPARTICLES IN ZEOLITES

4.1 Processes of Encapsulation of Metal Nanoparticles in Zeolites

4.1.1 Postsynthesis Introduction of Metal Precursors

Several decades ago, zeolites already appeared to be among the best candidates for the encapsulation of metal particles in isolated cavities or channels with nanometric dimensions. Metal particles are generally introduced in zeolites by standard postsynthesis methods such as ion exchange or impregnation with an ionic precursor followed by reduction. Ion exchange is restricted to cationic zeolite frameworks, in other words, those containing trivalent elements such as Al, B, Ga, etc. The amount of metal that can be introduced is limited by the exchange capacity, in other words, by the number of trivalent atoms per unit cell. Higher metal loading can be obtained by impregnating zeolite pores with a concentrated solution of metal precursor followed by evaporation of the solvent. Under such conditions, the amount of metal is not limited by the composition of the framework but by the pore volume, and it is theoretically possible to fill completely the zeolite pores with metal precursors. Impregnation is also used to incorporate metal precursors in neutral zeolites, typically all-silica zeolites with frameworks made of exclusively SiO_4 tetrahedra. Methods to obtain nanoparticles from exchanged or impregnated zeolites have been widely illustrated in the literature, and all aspects related to dispersion, size, and location of the nanoparticles as well as their catalytic activity have been discussed [6,215–219]. In particular, one of the most critical points of the preparation is the calcination/reduction process at a high temperature during which metal atoms can migrate on the outer surface of the crystals and agglomerate to larger particles [220–222]. Nevertheless, agglomeration can also occur inside the zeolite framework, and examples of metal crystals exceeding the size of the supercage size in faujasites have been reported [223]. Their formation requires local fragmentation of the zeolite framework around the particle.

Metals such as rhenium, iridium, osmium, molybdenum, or tungsten, which do not form cationic complexes in solution, can be incorporated in acidic zeolites in the form of neutral complexes such as metal carbonyls with zero-valent metal atoms or clusters [7,12,71,109,113,116,224]. These complexes are introduced into zeolites either via sorption from the vapor phase or by impregnation, depending on their volatility. Nanoparticles are then formed by simple decarbonylation of metal complexes in the presence of reducing agents such as H_2 or heating. Metal nanoparticles can also be obtained from isolated species, as is the case for Re clusters obtained by reduction of $SiO\text{-}ReO_3\text{-}Al$ species in HZSM-5 at 723 K [225]. Gold nanoclusters of less than 1 nm in the cavities of Y zeolites were also obtained by in situ treatment of $Au(CH_3)_2(C_5H_7O_2)$ precursors in flowing helium [226].

4.1.2 Direct Synthesis of Zeolites in the Presence of Metal Precursors

Among other things, the homogeneity of the population of metal nanoparticles in a zeolite crystal depends on how metal precursors can diffuse through the porosity. In a solution, the diffusion of hydrated cations is facilitated by three-dimensional arrays of large pores, which explains the success of FAU-type zeolites X and Y as hosts for encapsulation. In contrast, it is very difficult to introduce metals in medium or small-pore zeolites because hydrated precursors are too large to enter the channels, and most of metal species remain on the surface of the crystals and coalesce to form large particles upon activation and reduction at a high temperature. An alternative consists in crystallizing the zeolite in the presence of metal precursors or nanoparticles, a strategy similar to that used to encapsulate phthalocyanine complexes in X and Y zeolites [15–17,88–91]. Pt, Pd, Ru, and Rh clusters encapsulated in small pore zeolites with SOD (channels of 0.26×0.26 nm) and GIS (channels of 0.31×0.45 and 0.28×0.48 nm) framework types have been obtained following this method [18]. The same procedure was used to encapsulate a series of metal clusters (Pt, Ir, Pd, Rh, Re) within the small pores of LTA zeolite (channel of 0.41×0.41 nm) [19] (Fig. 11.9).

In all cases, metal precursors (introduced in the form of nitrates, chlorides, or sulfates) were stabilized by ammonia or amine ligands to avoid their precipitation as colloidal hydroxides prior to crystallization. As the addition of metal precursors influences the crystallization of zeolites, metal contents in the final zeolites were generally quite low, typically below 1%. For LTA, particle sizes were in the 1–2 nm range after reduction at 400°C, which is larger than the zeolite cages (max. 1.1 nm for the α-cage), and this suggested that that particles could occupy more than one cage in the framework. As already reported for the Y zeolite, these nanocavities in which particles are occluded could be

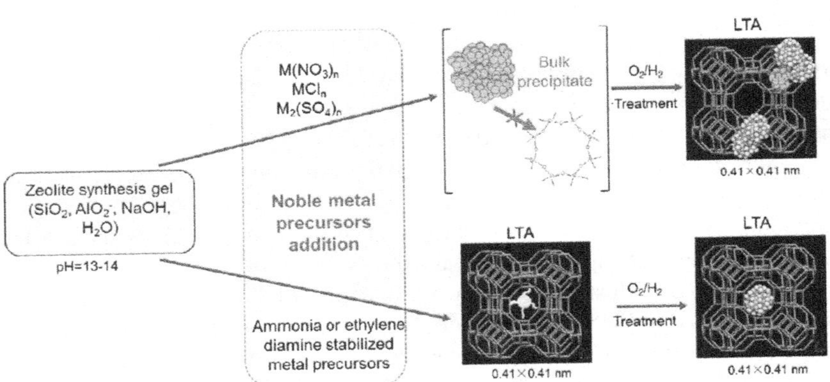

FIG. 11.9 Process for the encapsulation of metal clusters within LTA cavities. *(Adapted from Z. Wu, S. Goel, M. Choi, E. Iglesia, J. Catal. 311 (2014) 458–468, with permission of Elsevier.)*

defective regions formed during crystallization of the zeolite around the clusters [223]. In the case of zeolite NaA, the addition of $Pt(NH_3)_4Cl_2$ to the synthesis gel allowed the preparation of zeolites with metal contents up to 4 wt.% [24]. The size of platinum particles was approximately 1.4 nm, suggesting that they were mainly located in the cages of the framework. Increasing the Pt content in the synthesis gel had a strong influence on the size and morphology of the zeolite crystals: from 5 μm in the absence of Pt, the size decreased to 0.5 μm for the highest Pt loading. When metal precursors are not stable under synthesis conditions, metal can be introduced in synthesis gels directly in the form of colloidal suspensions of nanoparticles. The method was used to introduce nanometric gold clusters in silicalite-1 [227]. A gold colloid with a particle size of 1–2 nm was first immobilized in an amorphous silica matrix, which was then converted to silicalite-1 in the presence of a tetrapropylammonium hydroxide solution (TPAOH) at 100°C. After crystallization, most of the gold nanoparticles were encapsulated in zeolite crystals and possessed a size in the range of 1–2 nm. Similarly, cobalt oxide particles were embedded in ZSM-5 crystals by dissolution/recrystallization of a bulk SiO_2-supported Co catalyst in the presence of TPAOH [228].

4.2 Catalytic Reactions Over Zeolite Encapsulated Metal Nanoparticles

Zeolite encapsulated metal nanoparticles have been used as catalysts for a plethora of reactions including oxidations, hydrogenations, hydro-isomerizations, Fischer-Tropsch syntheses, etc. When the zeolite acts as a neutral host (pure silica or nonacidic zeolites), the catalytic activity depends only on the nature and size of metal nanoparticles. Nevertheless, the confinement of particles in micropores may induce shape and transition-state selectivities and modify the product selectivity as compared to catalysts dispersed on nonporous supports. In many cases, metals are introduced in acidic zeolites and the resulting materials act as bifunctional catalysts, combining the properties of the metal with those of the zeolite framework.

4.2.1 Oxidation Reactions

In the presence of oxidizing agents, most of metal nanoparticles undergo rapid oxidation, which explains why they are not considered to be real catalysts for oxidation reactions. However, the case of gold is particular since Au nanoparticles are stable in the presence of oxygen and they have shown to be particularly active and selective in oxidation reactions at low temperatures [229–237]. Au nanoparticles in silicalite-1 have been tested in the aerobic oxidation of benzaldehyde and 3,5-di-*tert*-butylbenzaldehyde in methanol, and the activity was compared with that of a standard Au/TiO_2 catalyst [227]. If the TiO_2-supported nanoparticles oxidize both substrates into their corresponding methyl esters, the zeolite-based catalyst is not able to oxidize the bulk substrate, which is too large to diffuse in the zeolite pores. A similar material with Au NPs of 2–3 nm in diameter has been used for the aerobic gas-phase oxidation of ethanol to benzaldehyde [238]. Au NPs in HY zeolite obtained by impregnation with $HAuCl_4$ followed by reduction have shown to be remarkably active and selective in the oxidation of 5-hydroxymethyl-2-furfural (HMF) into 2,5-furandicarboxylic acid (FDCA) [219]. HMF is first transformed into hydroxymethyl-2-furancarboxylic acid (HMFCA), which is further oxidized into FDCA. The yield of FDCA increased with the concentration of acidic OH groups in the supercage of the zeolite, but the Au particle size varied in the opposite direction. The best results were obtained for Au/HY, showing that the zeolite not only maintained the size of nanoclusters, but also promoted the catalytic activity by strong interaction between gold and acidic groups of the framework (Fig. 11.10).

4.2.2 Hydrogenation Reactions

In contrast to oxidations, many examples of hydrogenation reactions over Pt, Pd, Ir, Rh, or Ru nanoparticles in zeolites have been reported [239]. Previous works have largely demonstrated the influence of the nature of the metal, the particle size, and the acidity of the zeolite on the hydrogenation of aromatics. Recently, exceptional activities in the hydrogenation of benzene and its methyl-substituted aromatic compounds have been observed over Ru clusters stabilized

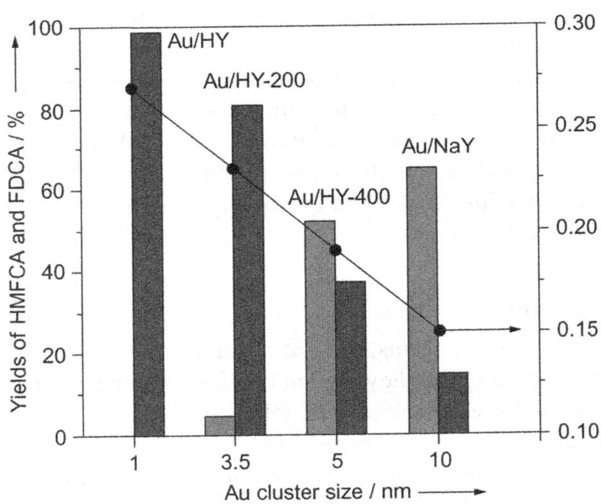

FIG. 11.10 Catalytic performance *(bars)* and the PyH^+ concentration (●) for Au nanoparticles in various Y zeolites with decreasing acidity. Yield of HMFCA = *dark gray bars*, yield of FDCA = *black bars*. *(Reproduced from J. Cai, H. Ma, J. Zhang, Q. Song, Z. Du, Y. Huang, J. Xu, Chem. Eur. J. 19 (2013) 14215–14223, with permission of Wiley-VCH Verlag.)*

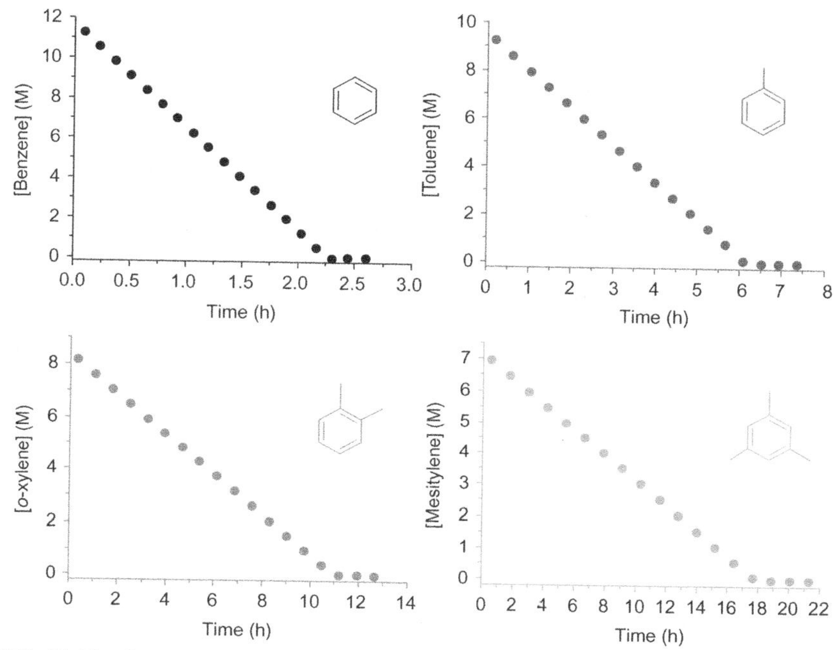

FIG. 11.11 Concentration of benzene, toluene, *o*-xylene, and mesitylene versus time during hydrogenation reactions catalyzed by Ru nanoparticles in colloidal FAU-type nanozeolite crystals. *(Adapted from M. Zahmakıran, Y. Tonbul, S. Ozkar, J. Am. Chem. Soc. 132 (2010) 6541–6549, with permission of American Chemical Society.)*

in the framework of nanosized Y zeolite (Fig. 11.11) [22,240]. The catalyst was prepared by ion-exchange of Ru^{3+} with Na^+ or TMA^+ cations of the zeolite followed by reduction with sodium borohydride in an aqueous solution at room temperature. Compared to nanoparticles dispersed on nonporous supports, the presence of the zeolite framework strongly influences the selectivity of reactions as well as the catalyst lifetime. Substrate selectivity is influenced by the zeolite pores, which act as molecular sieves and prevent large molecules to access the catalytic sites. A representative example concerns the hydrogenation of cycloalkenes over Rh nanoparticles in NaY, obtained by decarbonylation of carbonyl clusters [241]. During the competitive hydrogenation of cyclohexene and cyclododecene, the smallest substrate was fully hydrogenated after 6 h while cyclododecene conversion remained negligible, even after 16 h of reaction.

The molecular sieving properties of the zeolite can also be used to prevent poisoning of the encapsulated nanoparticles by sulfur-containing molecules. Pt particles in small pore zeolites with GIS and ANA framework types can catalyze the hydrogenation of ethylene in the presence of S-containing molecules. Compared to a standard Pt/SiO_2 catalyst for which the activity completely falls down in the presence of thiophene, Pt/ANA and Pt/GIS catalysts remain active with a decrease of hydrogenation rates by only 15%–30% [18]. The same observation was made on Pt and Rh-containing LTA zeolites during the hydrogenation of ethylene [19].

4.2.3 Dehydrogenation, Aromatization

Zeolite supported Pt particles have also been used in the aromatization of *n*-alkanes, principally in the conversion of *n*-hexane into benzene. One of the most selective zeolites in this reaction was KL, a LTL-type zeolite in which the acid sites had been neutralized by potassium ions [218,242,243]. The high benzene selectivity was attributed not only to the specific shape of the channels, but also to the basicity of KL zeolite.

4.2.4 Bifunctional Catalysis

Classical bifunctional catalysts are composed of metal nanoparticles dispersed in the cavities of acidic zeolites. The association of a dehydrogenation/hydrogenation function with the acidic properties of the zeolite framework makes them particularly interesting in petroleum chemistry, in particular in hydroisomerization and hydrocracking of alkanes. Selective hydroisomerization is a well-known process for improving the fuel quality of *n*-alkanes by converting them into branched isomers. Metal nanoparticles (Pt, Pd) catalyze the formation of intermediate olefins via dehydrogenation or the formation of products via hydrogenation. The acidic sites of the zeolite generate carbenium ions from olefins for the isomerization step. There have been many reports on the isomerization of *n*-alkanes from C_4 to C_{10} as well as other molecules such as ethylbenzene [244–248]. The balance between hydrogenating and acid functions and the zeolite pore structure are pertinent parameters that affect the activity, the selectivity, and the deactivation of the catalyst. An interesting reaction is the synthesis of methyl isobutyl ketone (MIBK) from acetone because it can be performed on both metal/acid (Pt/PdHZSM-5, PdHMCM-22) and metal/base (PdNaX or PdCsX) zeolite catalysts [249–256]. Though less studied than catalysts associating metal nanoparticles with the acid sites of a zeolite, metal/base catalysts have been used for the ring opening of *cis*-decalin over Ir- or Pt/LaX zeolites [203]. Carbon monoxide hydrogenation was also successfully performed over Ru particles encapsulated in NaA [254]. Nevertheless, the production of hydrocarbons from the reaction between CO and H_2, known as Fischer-Tropsch (FT) process, is generally performed over Co particles with an optimal size of c.6–8 nm. Such particles are too large to be encapsulated in zeolite cavities, but they can be embedded in the framework during hydrothermal crystallization. Examples of Co-embedded HZSM-5 zeolites have been reported in the literature, and the effect of the zeolite on the catalytic activity has been discussed [228]. In this case, the zeolite acts as a cracking catalyst to modify the product distribution of the FT reaction towards higher selectivities in the gasoline range (C_5–C_{11}). Advantages of Co-containing bulk and hierarchical zeolites in FT synthesis will be discussed in the following paragraph.

5 CATALYST ENCAPSULATION IN HIERARCHICAL ZEOLITES

When catalysts are encapsulated in zeolite cavities, they are supposed to be randomly distributed all throughout the crystals and accessible only via the micropores of the structure. However, even if the reactants are small enough to enter the zeolite micropores, their slow diffusion makes that active sites, far below the surface, do not participate in the catalytic processes. Therefore, if such encapsulated catalysts are more selective than supported catalysts to discriminate substrates according to their size, they are nevertheless generally less active to convert small molecules. Mass transport in zeolite crystals can be significantly improved by creating additional diffusion pathways such as a secondary array of mesopores, while keeping all properties and advantages of the microporous solid. Such zeolites are generally named "hierarchical" or "mesoporous" zeolites because they contain at least two populations of pores: the micropores of the zeolite and an additional system of pores with a size in the mesopore range. They are generally obtained by postsynthesis modifications, the most commonly used being dealumination under acidic conditions or desilication in highly alkaline solutions [255–257]. Besides facilitating diffusion of molecules to catalytic sites, the presence of mesopores significantly increases the external surface of the zeolite, which may have a direct impact on the size and dispersion of supported nanoparticles. When supported on mesoporous zeolites, metal nanoparticles are not only encapsulated in the structure, but also dispersed on the external surface of the mesopores. However, the synergy between the nanoparticle and the zeolite emerges from a close proximity between the metal and the zeolite framework, the latter acting as a solid acid or as a membrane by selecting the reactants and products on the basis of size and/or diffusivity.

5.1 Metal Nanoparticles in Hierarchical Zeolites for Bifunctional Catalysis

Since the beginning of the 21st century, there have been many reports on the benefits of using hierarchical zeolites instead of bulk crystals for bifunctional catalytic reactions [258–266]. Additional porosity mostly improves the catalyst performance, particularly in the case of medium pore zeolites with a monodimensional array of micropores. A typical example concerns zeolite ZSM-22 (TON topology), a 1-D zeolite with pore opening of 0.45×0.57 nm [258]. This zeolite is used for the isomerization of *n*-alkanes to branched paraffins, with a very high monobranching selectivity. The zeolite was treated with NaOH/HCl solutions to create intracrystalline mesopores and intercrystalline voids, with an increase of the mesopore surface area by a factor of 5 as compared to the non-treated ZSM-22. The treatment had no significant influence on the platinum dispersion and both the nontreated and treated solids contained Pt nanoparticles that were c.1 nm in size. The presence of mesopores increased the activity of the zeolite in *n*-octane hydroisomerization and significantly improved the mono-branching selectivity. The same observations were made over Pt-containing hierarchical MOR and BEA zeolites obtained by recrystallization of bulk crystals in the presence of surfactant molecules [259,260]. The advantage of using a hierarchical zeolite was also clearly evidenced during the alkylation of benzene with methanol over ZSM-5 [261]. This reaction produces mainly toluene, xylene, and trimethylbenzene but also ethylbenzene, whose separation from C_8 aromatics remains a challenge. Addition of platinum to the zeolite significantly reduces the formation of ethylbenzene but both the benzene conversion and the selectivity to toluene remain low due to diffusion limitations in the pores. Moreover, the catalyst deactivates rapidly by formation of coke on the external surface of the crystals. A hierarchical ZSM-5 zeolite was synthesized by a solvent evaporation assisted dry-gel route that led to aggregated nanocrystals of about 10–20 nm with interparticle mesoporosity. The zeolite was then impregnated with a solution of $H_2PtCl_6 \cdot 6H_2O$ (0.02 wt.% Pt) and directly reduced under hydrogen at 400°C for 2 h, resulting in Pt nanoparticles highly dispersed in the zeolite mesopores as evidenced by TEM. By combining the advantages of a mesoporous zeolite (slower deactivation by coking) with the specific properties of Pt nanoparticles towards suppression of ethylbenzene formation (hydrogenation of ethylene to ethane), it was possible to design a catalyst with improved activity and selectivity in xylenes as compared to a conventional ZSM-5. The benzene conversion was maintained at c.50% for more than 50 h with xylene selectivity at 30%–35% and a selectivity to ethylbenzene below 0.1% (Fig. 11.12).

The combination of gold nanoparticles with an oxidation catalyst such as TS-1 offers the possibility to generate in situ H_2O_2 from H_2 and O_2 on Au particles

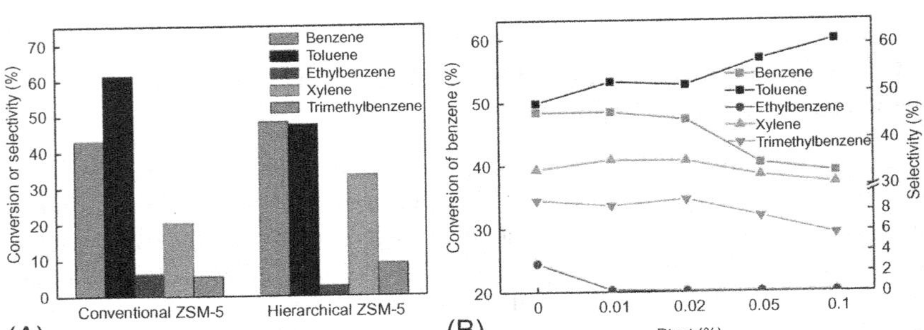

FIG. 11.12 Catalytic performances of conventional and hierarchical ZSM-5 in benzene alkylation with methanol (A) and Pt-modified hierarchical ZSM-5 prepared by the direct reduction method (B). *(Reproduced from H. Hu, Q. Zhang, J. Cen, X. Li, Catal. Lett. 145 (2015) 715–722, with permission of Springer.)*

and to use it directly to oxidize various substrates under mild conditions. TS-1, a titanium-containing silicalite-1, is indeed one of the most efficient catalysts for oxidation, epoxidation, and hydroxylation reactions in the presence of H_2O_2 [267,268]. In situ generation of H_2O_2 is a low-cost, environmentally friendly, and safe alternative to aqueous concentrated solutions. In Au/TS-1, H_2O_2 molecules formed on Au NPs react with framework Ti species to form peroxo species, and the efficiency of the catalyst is conditioned by the proximity between Au and Ti sites. Deposition of small and uniform Au particles on bulk TS-1 crystals is difficult as most of Au species remain on the outer surface of the crystals in the form of large aggregates [269]. Both the dispersion of gold nanoparticles and their proximity with framework titanium sites can be significantly improved by using mesoporous TS-1 zeolites. The advantage of using a mesoporous support was clearly illustrated in the oxidation of benzyl alcohol over TS-1-supported Au-Pd nanoparticles [270]. On bulk TS-1, impregnation with Au and Pd precursor solutions led to very large particles up to 300 nm exclusively located on the outer surface of the crystals. In contrast, mesoporous TS-1 with additional porosity in the range of 1–7 nm contained very small Au-Pd nanoparticles centered around 2.15 nm and located in the mesopores. The substrate conversion in the solvent-free oxidation of benzaldehyde increased by c.70% over the mesoporous support as compared to a nonporous zeolite.

5.2 Hierarchical Zeolites for the Transformation of Bulky Molecules

Catalytic applications of zeolites are normally restricted to molecules capable of entering the micropores, in other words, with a kinetic diameter smaller than 1 nm. Even when reactions occur on the surface or at the pore mouths of the zeolite, they are limited by the low external surface of the crystals. The creation of additional mesoporosity not only increases the external surface of the zeolite significantly, but it also greatly facilitates the accessibility of bulky molecules to pore mouths. The benefit of mesoporosity was clearly illustrated in the hydrogenation of naphthalene over mesoporous Pt-containing ZSM-5 [271]. A mesoporous ZSM-5 was prepared by adding a secondary porogen molecule to the synthesis gel and the dry solid was impregnated with an aqueous solution of $H_2PtCl_6{\cdot}6H_2O$. The mean mesopore diameter was 3.6 nm with a Pt dispersion of 33%, corresponding to particles of c.4 nm. With sizes comprised between 0.5 and 0.7 nm, naphthalene and its corresponding hydrogenation products hardly diffuse in the 10-ring channels of the zeolite. Therefore, quite low conversions were observed on bulk crystals with low decalin selectivity and high selectivity to cracking products at 300°C. Over Pt-loaded mesoporous ZSM-5, the naphthalene conversion is complete at 230°C and the decalin selectivity exceeds 97% between 250°C and 300°C (Fig. 11.13).

Moreover, it was shown that the mesoporous zeolite was much more sulfur-resistant than a standard Pt/Al_2O_3 catalyst when the reaction was performed in

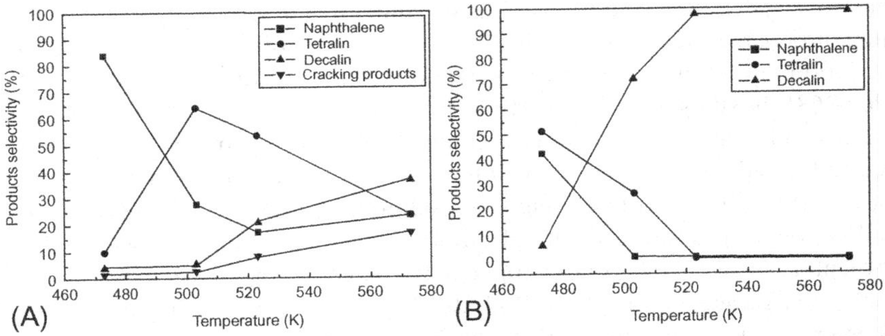

FIG. 11.13 Product selectivities of naphthalene hydrogenation over Pt-containing conventional (A) and mesoporous (B) ZSM-5. *(Reproduced from T. He, Y. Wang, P. Miao, J. Li, J. Wua, Y. Fang, Fuel 106 (2013) 365–371, with permission of Elsevier.)*

the presence of sulfur. When the reaction was performed with 3000 ppm dibenzothiophene (DBT) in the feed, the decalin selectivity decreased from 98.2% to 89.4% on the hierarchical zeolite, while it dropped from 83.6% to 51.4% on alumina. This was attributed to the strong acidity of the zeolite along with the presence of the microporous/mesoporous bimodal structure in the crystals. By contrast to Al_2O_3, some active Pt species in the micropores of the zeolite cannot be poisoned by large DBT molecules, and some of Pt sites located in the mesopores may be partially recovered by spillover hydrogen coming from the micropores.

The properties of metal-containing hierarchical zeolites strongly depend on the way particles are distributed in the porosity of zeolite crystals. Standard methods for introducing metal species such as ion-exchange or impregnation generally lead to broad particle size distributions, with nanoparticles both in zeolite micropores and mesopores. In the case of the small pore zeolite A, a strategy was developed to control the distribution of Pt particles in zeolite cavities and mesopores [272]. By incorporating platinum either during synthesis or after crystallization, it was possible to obtain three types of catalysts with the same Pt content, but with different sizes and locations of Pt particles. In the first type, all Pt were introduced during synthesis and almost all particles were located in the zeolite cages with a size around 1 nm. The second type of catalyst was obtained by postimpregnation of a zeolite synthesized in the absence of platinum, and Pt was introduced exclusively in the mesopores in the form of larger particles. A third type was prepared by incorporating half of Pt during synthesis and half in the mesopores. In the absence of sulfur, the best activity in naphthalene hydrogenation was obtained on the catalyst with small particles in zeolite cages, which was attributed to a better dispersion of Pt as compared to samples with Pt in the mesopores. In contrast, in the presence of H_2S or DBT, only the third type of catalyst remained highly active, which was attributed to the synergetic effect of Pt particles confined both in zeolite cages and larger mesoporous channels.

5.3 Hierarchical Zeolites for Fischer-Tropsch Synthesis

Fischer-Tropsch (FT) synthesis is a catalytic process that produces hydrocarbons from "syngas," a mixture of H_2 and CO obtained by steam reforming of natural gas, coal, or biomass [273]. The reaction is usually performed on nanoparticles of transition metals such as ruthenium, iron, or cobalt, but Co is by far the most widely used due to its low cost, high activity, and high selectivity to liquid fuels. FT synthesis is a structure-sensitive reaction with an optimum activity obtained for Co particle sizes between 4 and 6 nm [274]. The reaction leads to broad product distributions that nearly follow the Anderson-Schulz-Flory (ASF) distribution [275]. This distribution is not selective towards middle-distillate products such as gasoline or diesel fuel and the maximum selectivity for the gasoline range (C_5–C_{11}) does not exceed 45%. Attempts to modify the ASF chain growth probability and enhance the selectivity in middle isoparaffins using bifunctional catalysts have been more or less successful [276,277]. For example, physical mixtures of FT catalysts with zeolites do not always ensure sufficient proximity between Co particles and acidic sites of the zeolite to significantly limit the hydrocarbon chain length. Better results are obtained when Co particles are totally encapsulated inside zeolite crystals: after reaction, products have to pass through the zeolite membrane, which greatly enhances the hydrocracking and isomerization of long-chain paraffins.

As already mentioned in Section 4.2 for concerning bifunctional catalysts on zeolite encapsulated metal NPs, many papers report the use of Co nanoparticles in ZSM-5 to modify the product distribution in FT synthesis. Encapsulation in bulk zeolite crystals provides interesting results with an increase of selectivity in the gasoline range and a significant decrease in the C_{13}–C_{20} fraction. Moreover, the presence of the zeolite also increases the fraction of isoparaffins and reduces the formation of CO_2, methane and short-chain paraffins [228]. However, when small Co particles are located deep inside the crystals of zeolite, the residence time of long-chain hydrocarbons in the pores is increased and overcracking takes place, leading to high undesired hydrocarbons selectivity. Moreover, water formed during the reaction is less easily expelled, which accelerates the reoxidation and agglomeration of Co particles and leads to faster deactivation. Besides overcracking, diffusion may also restrict the access of primary hydrocarbons to the acid sites of the zeolite. The above reasons make that most of the recent examples of bifunctional catalysts for FT synthesis are based on hierarchical zeolites [278–280]. Moreover, the presence of mesopores helps the stabilization of Co particles with optimal dimensions between 4 and 8 nm, which is not possible in the cavities of microporous frameworks.

Mesoporous HZSM-5 zeolites can be prepared either by direct synthesis in the presence of porogen agents or by postmodification of bulk crystals. Co particles in mesoporous HZSM-5 have been obtained by incipient wetness impregnation of a mesoporous support, prepared in the presence of triblock copolymers as surfactants [281]. The catalyst was evaluated in an FT reaction and its activity was compared to that of a conventional microporous zeolite. As usually reported, the presence of mesopores increased the CO conversion as well as the C_{5+} selectivity and the fraction of isoparaffins. Under similar conditions, the gasoline selectivity increased by almost 20%, from 56.6% to 65.4%. Complete encapsulation of Co particles in mesoporous ZSM-5 has also been achieved using a replication technique from mesoporous carbons. A CMK-5 mesostructured carbon was prepared by replication of an Al-doped SBA-15 silica using furfuryl alcohol, and it was impregnated with a cobalt precursor solution and reduced to form Co nanoparticles [282] (Fig. 11.14). The Co-supported mesoporous carbon was then used as a hard template to crystallize mesoporous ZSM-5 and later eliminated upon calcination in air. A highly mesoporous ZSM-5 containing 10 wt.% Co in the form of particles of c.4–5 nm was then obtained and tested in FT synthesis. Compared to Co-supported mesoporous silicas, the encapsulated catalyst was much more selective in the C_5–C_{11} fraction as well as methane formation. The presence of the zeolite broke the ASF law product distribution with a sharp distribution between C_1 and C_{12} and an isoparaffin selectivity of 34.6%.

The influence of the acidity of the zeolite on the product distribution in FTS has been studied using hierarchical zeolites with different degrees of H^+ exchange [283]. Co-supported mesoporous H/Na-ZSM-5 zeolites were

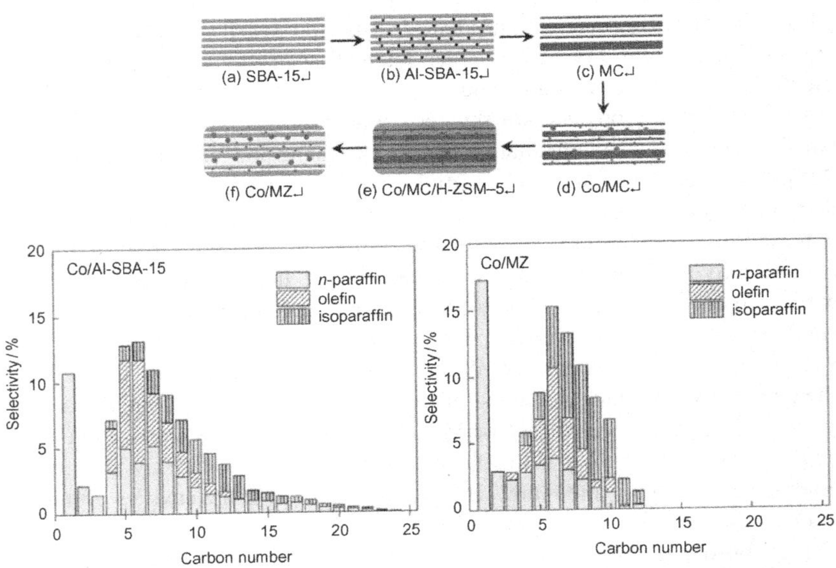

FIG. 11.14 Synthesis procedure for Co-containing mesoporous zeolite Co/MZ and FTS product distribution for mesoporous supports (Al/SBA-15 and mesoporous zeolite) loaded by 10 wt.% cobalt. *(Reproduced from C. Xing, W. Shen, G. Yang, R. Yang, P. Lu, J. Sun, et al., Catal. Commun. 55 (2014) 53–56, with permission of Elsevier.)*

obtained by desilication of a parent zeolite with a 0.5 M NaOH solution followed by partial ion exchange and impregnation with a cobalt precursor. As expected, increasing the acidity of the zeolite did not strongly change the CO conversion but significantly increased the selectivity in the C_{5-11} fraction as well as the ratio of isoparaffins to *n*-paraffins in the C_{5-11} fraction. Interestingly, the selectivities in methane and C_{2-4} fraction did not vary with the acidity, and the change in C_{5-11} selectivity occurred at the expense of C_{12+} selectivity. Moreover, catalysts were very stable with CO conversion and C_{5-11} selectivity of 36% and 65%, respectively, after more than 100 h of reaction.

The zeolite associated to metal NPs in the FT process is generally ZSM-5 in the protonic form because of its particular properties for cracking and isomerizing *n*-paraffins. Nevertheless, other zeolites have also been used, for example, Beta or even Y associated with metals such as Co, Ru, or Fe [284–288]. Ru nanoparticles in mesoporous H-Beta obtained by desilication with NaOH solutions gives 67% selectivity in the C_{5-11} fraction with a ratio between iso and normal paraffins > 3, with some values as high as those obtained over Co/H-ZSM-5 catalysts [284]. The selectivity to C_{5-11} hydrocarbons could even reach 77% by using the Na-form of mesoporous Beta zeolite, which was explained by the combination of a unique Lewis acidity and mesopores. Other examples of metal nanoparticles in nonacidic zeolites have clearly shown the interest of such catalysts for the production of diesel fuel (C_{10-20} fraction). A mesoporous Co/NaY zeolite with mesopores of 15 nm and cobalt particles of 8.4 nm gives a C_{10-20} selectivity of 60%, which is 50% higher than the maximum expected from the ASF distribution [286] (Fig. 11.15).

Such a high selectivity was explained by the hydrogenolysis of the C_{21+} fraction on Co particles and by the absence of Bronsted acid-catalyzed cracking reactions. Moreover, the catalyst was very stable; CO conversion remained constant around 35% after more than 500 hours of reaction with C_{10-20} selectivity higher than 60% and CH_4 and C_{2-4} selectivities below 5%.

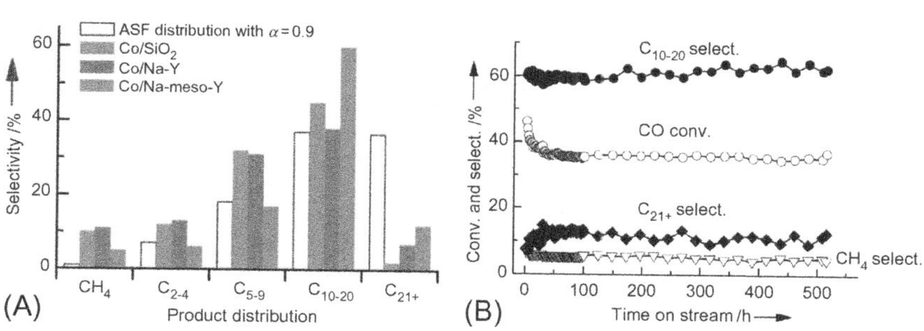

FIG. 11.15 (A) product selectivities for FT synthesis over cobalt catalysts loaded on SiO_2, Na-Y, and Na-meso-Y. The selectivity from ASF distribution ($\alpha = 0.9$) is also displayed. (B) Stability of the Co/Na-meso-Y catalyst in FT synthesis. *(Reproduced from X. Peng, K. Cheng, J. Kang, B. Gu, X. Yu, Q. Zhang, et al., Angew. Chem. Int. Ed. 54 (2015) 4553–4556, with permission of Wiley-VCH Verlag.)*

All the above examples concern zeolites in which the mesoporosity is intracrystalline and obtained by postsynthesis treatments of bulk crystals such as desilication. Another way to encapsulate metal nanoparticles is to take advantage of interparticle space between zeolite nanocrystals, in particular in zeolite nanosponges made of stacking nanosheets. Zeolite nanosheets can be obtained using templating molecules containing long-chain alkanes that prevent crystallization of the framework along one direction [289–291]. By the addition of standard MFI crystal seeds to the synthesis, a nanosponge morphology is obtained with a very narrow pore size distribution centered around 4 nm. After calcination and ion exchange, impregnation of zeolite nanosheets with a solution of cobalt nitrate leads to a catalyst with 10 wt.% Co and an average particle size of 4 nm, similar to the pore dimension [292]. Co-supported zeolite nanosponges were tested in FT synthesis and their performance was compared with that of solids obtained by impregnating a bulk ZSM-5 zeolite and a commercial γ-Al_2O_3. As for hierarchical zeolites, the nanosponge was the most active catalyst, with a high selectivity to branched paraffins in the gasoline range.

6 CORE-SHELL ZEOLITE-BASED CATALYSTS

For many applications, encapsulation of catalysts in zeolites is not necessarily restricted to their immobilization or stabilization in a confined space. The zeolite may also act as a molecular sieve by selecting molecules according to their size and shape or as a second component for transforming products formed on the catalyst. In both cases, the success of those hybrid catalysts implies that molecules pass through the zeolite barrier before or after reaction, which necessitates a close proximity between active sites and the zeolite framework. However, the case of FT synthesis has clearly demonstrated that internal zeolite cavities could be inappropriate to accommodate active catalysts, whose size may largely surpass a few nanometers. An alternative consists in using core-shell materials in which the catalyst is totally surrounded by a zeolite capsule (Scheme 11.3). As compared to postimpregnation of mesoporous zeolites, this

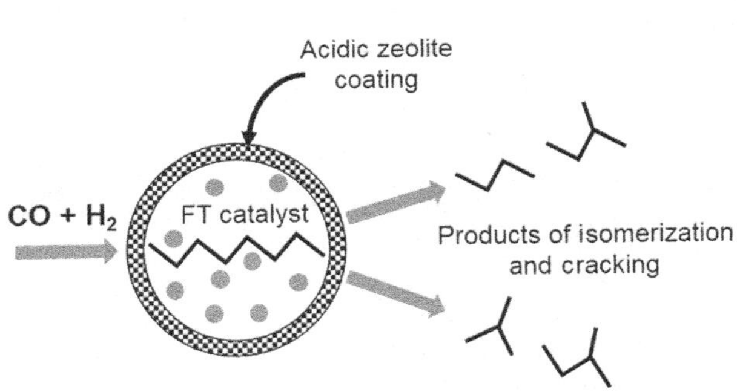

SCHEME 11.3 Isomerization and cracking of products obtained with an FT catalyst surrounded by an acidic zeolite shell.

method allows a strict control of the thickness of the zeolite coating as well as of the metal particle size prior to encapsulation. Capsule catalysts have been widely used to modify the product selectivity in FT synthesis [293–295].

The preparation of such materials is facilitated by the fact that the zeolite can be directly coated on the surface of a standard SiO_2-supported metal nanoparticles FT catalyst. Encapsulation of conventional FT synthesis catalysts such as Fe/SiO_2 or Co/SiO_2 by zeolites effectively improves the selectivity in light isoparaffins while reducing the formation of methane and heavy hydrocarbons. A typical example is given by the crystallization of H-Beta zeolite on the surface of a Co/Al_2O_3 catalyst containing 7 wt.% Co in the form nanoparticles with an average diameter of 15.7 nm [296]. The catalyst was coated with a uniform layer of zeolite of a c.16 μm thickness with SiO_2/Al_2O_3 molar ratio of 84.2 (Fig. 11.16). Compared to the noncoated catalyst, the presence of the zeolite decreased the CO conversion by less than 5% in the direct synthesis of isoparaffins from syngas. In contrast, it increased the isoparaffins' selectivity C_{iso}/C_n with $n > 3$ from 0.01 to 3.13 and totally suppressed the formation of products with $n > 10$. Similar results were obtained by crystallizing layers of H-ZSM-5 zeolite directly on the surface of fused-iron catalyst pellets or Co-containing activated carbons [297,298].

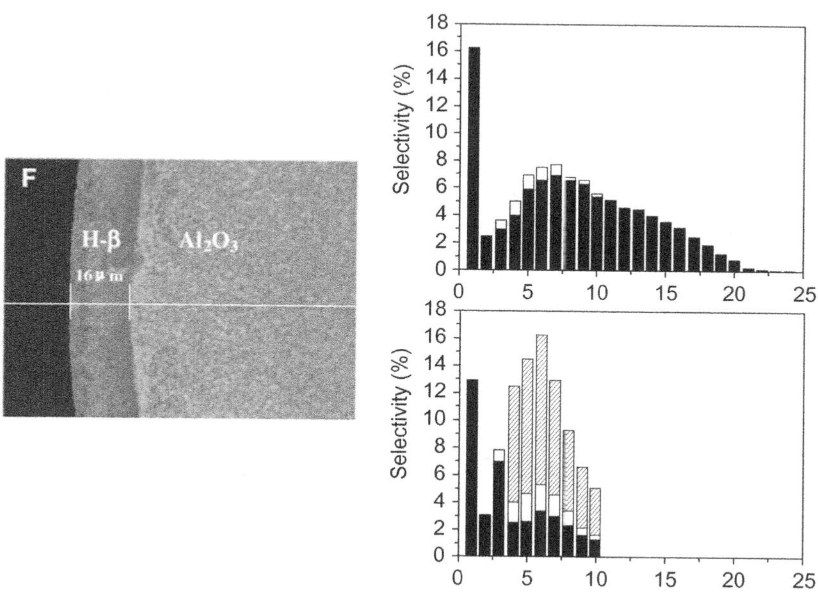

FIG. 11.16 SEM image of a cross-section of H-Beta zeolite on the surface of a Co/Al_2O_3 catalyst and product distribution with uncoated (top) and zeolite-coated (bottom) Co/Al_2O_3 catalyst. *White*: olefins; *black*: normal paraffins; *shadow*: isoparaffins. *(Reproduced from X. Li, J. He, M. Meng, Y. Yoneyama, N. Tsubaki, J. Catal. 265 (2009) 26–34, with permission of Elsevier.)*

Zeolite coatings have also been used as shape-selective molecular sieves to modify the selectivity of reactions that occur on the encapsulated catalyst. For such applications, the zeolite is generally purely siliceous to avoid chemical reactions during diffusion of the products through the membrane. For example, deposition of a layer of silicalite-1 with a thickness of 4 μm onto the surface of Pt/Al_2O_3 catalyst by a hydrothermal synthesis method allowed the selective hydrogenation of the para isomer in a mixture of xylenes [299] (Fig. 11.17).

The noncoated catalyst hydrogenates all three isomers without any selectivity and yields in *ortho*, *meta*, and *para* dimethylcyclohexane are similar. When equimolar *ortho*/*para* or *meta*/*para* mixtures are hydrogenated over the core-shell catalyst at 473 K, a high selectivity in *para* dimethylcyclohexane is observed in both cases, with *para*/*ortho* and *para*/*meta* selectivities of 17 and 13.6, respectively.

Coating an activated carbon supported catalyst with silicalite-1 significantly improved the selectivity to normal aldehyde in the hydroformylation of 1-hexene [300]. Increasing the thickness of the zeolite layer from 15 to 30 μm decreased the conversion of the olefin due to diffusion limitations but significantly increased the ratio of normal to iso aldehyde. High conversion levels along with high selectivities to *n*-heptanal could be obtained by further treating thin zeolite coats with tetraethyl orthosilicate, which blocked part of the gaps between zeolite crystals.

Highly selective hydrogenation reactions were also reported over Ni/SiO_2-silicalite-1 core-shell catalysts. In the competitive gas phase hydrogenation of an equimolar mixture of 1-hexene and 3,3-dimethylbutene, selectivities up to 300 were observed for 1-hexene conversion of 90% [301]. Similar results were obtained over a silicalite-1-coated Pt/TiO_2 catalyst [302]. The high selectivity to 1-hexene hydrogenation was attributed to the permeation resistance of the branched alkene through the silicalite-1 layer. In a similar way, a $Pt\text{-}Fe/SiO_2$ catalyst coated with a layer of zeolite A was used in the selective oxidation

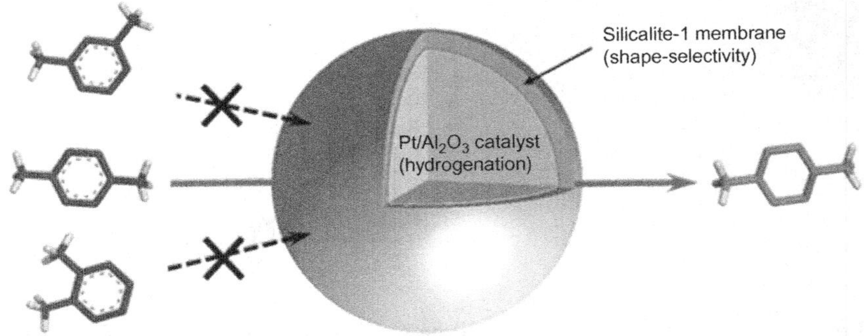

FIG. 11.17 Shape selective hydrogenation of xylene isomers over silicalite-1 coated Pt/Al_2O_3 catalyst. *(Reprinted from Y. Wu, Y. Chai, J. Li, H. Guo, L. Wen, C. Liu, Catal. Commun. 64 (2015) 110–113, with permission of Elsevier.)*

of CO in the presence of butane [303]. This reaction is particularly important to remove the undesired CO species from CO/CO_2/butane mixtures in the industrial production of maleic anhydride. When Pt-Fe particles of 4–10 nm in size were totally covered by a thick zeolite A membrane, butane was not oxidized below 400°C while CO was totally converted at 200°C.

Zeolite coatings can also act as a protection against molecules that can significantly affect the catalytic activity. Methane steam reforming can be performed in direct internal reforming-molten carbonate fuel cells (DIR-MCFC). Nevertheless, the nickel catalyst for methane steam reforming deactivates because of the high alkali content of the electrolyte (Li_2CO_3, K_2CO_3). Coating the Ni/Al_2O_3 catalyst with an 8-μm thick uniform layer of silicatite-1 significantly improved its resistance against alkali poisoning from molten carbonate electrolyte [304]. For the noncoated catalyst, poisoning resulted in a loss of 85% of the activity, while in the presence of the zeolite membrane, the initial activity was preserved, even after 6 h of reaction (Fig. 11.18).

Silicalite-1 coatings have also been used as hydrophobic membranes to improve the hydrothermal stability of silica-based catalysts when reactions are performed in water. A magnetically recoverable $Fe_3O_4@SiO_2$-supported Au nanoparticles catalyst, used in the reduction of 4-nitrophenol in $NaBH_4$ aqueous solution, was coated with silicalite-1 crystals, and its hydrothermal stability was compared with that of a similar compound coated with amorphous mesoporous silica [305]. As expected after a heat treatment at 140°C in water for 24 h, the mesoporous silica coating was totally destroyed and gold nanoparticles agglomerated on the surface of primary silica spheres. In contrast, in the presence of the hydrophobic and highly crystalline silicalite-1 membrane, gold nanoparticles were totally protected and both the particle size and the catalytic activity were maintained.

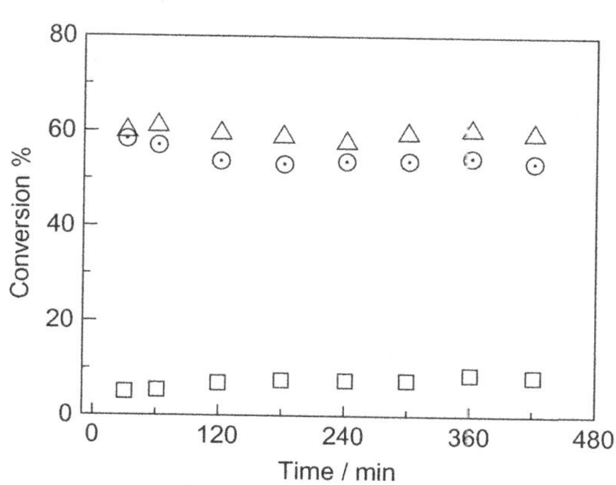

FIG. 11.18 Methane conversion during steam reforming over (△) nonpoisoned uncoated $Ni-Al_2O_3$, (○) poisoned Silicalite-1 coated $Ni-Al_2O_3$, and (□) poisoned uncoated $Ni-Al_2O_3$ catalysts. *(Reproduced from J. Zhou, X. Zhang, J. Zhang, H. Liu, L. Zhou, K. Yeung, Catal. Commun. 10 (2009) 1804–1807, with permission of Elsevier.)*

7 YOLK-SHELL ZEOLITE-BASED CATALYSTS

7.1 Polycrystalline Zeolite Shells

Silica-zeolite core-shell materials are generally obtained by a local recrystallization of the surface of silica spheres in the presence of structure-directing molecules. Silica particles are partially dissolved under high pH conditions and the corresponding silica species serve as nutriment for the crystallization of the zeolite. By controlling the rates of dissolution and recrystallization, it is possible to completely dissolve the original core and get hollow capsules, made of newly formed agglomerated zeolite nanocrystals. If cores (silica spheres or zeolite crystals) are loaded with metal nanoparticles prior to recrystallization, these nanoparticles will remain intact and encapsulated in hollow zeolite capsules. This strategy has been used to prepare Pt and Ag NPs encapsulated in zeolite microcapsules with an MFI structure [306]. A catalyst containing 3 wt.% Pt was used in the oxidation of cyclohexanol to cyclohexanone in the presence of benzothiazole, which is a well-known poison for Pt nanoparticles. The intact Pt@silicalite-1 catalyst was not affected by the presence of poisoning molecules and the reaction proceeded with 80% conversion and 90% selectivity. In contrast, if capsules were broken, Pt particles became accessible to poisoning molecules and both the conversion and selectivity dropped below 20%. A similar catalyst containing 9.5 wt.% Pt in the form of particles of c.22 nm was used in the liquid-phase oxidation of aromatic alcohols with different sizes. A small molecule (1-phenylethanol) was oxidized with quite similar rates over Pt-supported SiO_2 and Al_2O_3, Pt@silicalite-1 capsules as well as the corresponding crushed material. In contrast, the large reactant 1-(2-naphthyl) ethanol was oxidized only on catalysts where Pt particles were exposed, namely Pt/SiO_2, Pt/Al_2O_3 and crushed Pt@silicalite-1. The intact Pt@silicalite-1 catalyst was not only totally inactive, but it also prevented Pt nanoparticles from leaching during the reaction. The same type of catalyst containing Ag nanoparticles of 11 nm was used for the gas-phase oxidation of aliphatic alcohols [306]. The silicalite-1 capsule showed molecular sieve properties and was very efficient against poisoning when the oxidation of ethanol was performed in the presence of benzothiazole.

7.2 Single Crystal Hollow Zeolites

Encapsulation of metal nanoparticles in hollow zeolite spheres (or capsules) constitutes a very elegant way to protect them against leaching, poisoning, or sintering while maintaining a high catalytic activity. In contrast to nanometric cavities in zeolite frameworks, a large internal void is available for chemical reactions and such catalysts can be considered as nanoreactors in which the reaction is essentially governed by the permeability of the shell. Nevertheless, efficient polycrystalline zeolite shells are generally rather thick (>200 nm), which limits the overall reaction rate by diffusion transport limitations.

Attempts to prepare catalysts encapsulated in ultrathin zeolite shells were not successful because the thickness of polycrystalline zeolite shells is imposed by the individual crystal size and because thin polycrystalline zeolite shells contain many defects and are usually very fragile.

Recently, new strategies have been developed to prepare microcapsules made of zeolite single crystals. They are prepared either from core-shell zeolites or from bulk zeolite crystals displaying spatial heterogeneities in composition or structure [307–311]. For example, hollow ZSM-5 crystals have been obtained from bulk crystals with aluminum-zoning by taking advantage of the differences in solubility between Al-rich and Al-poor regions. Under alkaline conditions, the Al-poor core of the crystals is preferentially dissolved, leading to hollow structures with an Al-rich surface [310,311]. We have also reported the formation of hollow silicalite-1 crystals by dissolution and recrystallization of defective nanocrystals in the presence of TPAOH solutions at high temperatures [307,308]. Those techniques enable a strict control of the wall composition and thickness down to less than 10 nm. Compared to polycrystalline capsules, hollow single crystals are generally defect-free and possess excellent shape selective properties. Moreover, their very small wall thickness minimizes the effect of the zeolite on the reaction rate. Due to the small size of the pores (0.55 nm for MFI zeolites), direct incorporation of metal precursors into those hollow crystals by conventional impregnation or ion-exchange leads to large particles mainly located on the external surface. We have shown that the best way to encapsulate metal nanoparticles into these structures was to form hollow crystals in the presence of the corresponding metal precursors [312–315]. Silicalite-1 crystals impregnated with Au, Pt, Pd, etc. precursors could be transformed into hollow structures by recrystallization at high temperatures in the presence of TPAOH. Under reaction conditions (high temperature and pH value), an inner cavity is formed while metal species are reduced and precipitate in the form of a unique encapsulated nanoparticle (Fig. 11.19).

The mean particle size can be easily controlled between c.2 and 10 nm, while keeping a relatively narrow particle size distribution. Due to the synthesis method, metal particles are totally encapsulated into the hollow single crystals and their accessibility is restricted to molecules capable of diffusing through micropores of the zeolite (0.55 nm for MFI). As a consequence, they show excellent stability against sintering at high temperature. Compared to a supported catalyst for which the particle size increased from 7.7 to 17.5 nm when heated from 300°C to 750°C under hydrogen, the mean size of the encapsulated particles remained unchanged at c.10.7 nm. Pt nanoparticles encapsulated in silicalite-1 hollow crystals have been used as catalysts for the hydrogenation of methyl-substituted aromatic compounds with different sizes [313]. The absence of defects in the zeolite shell makes the catalyst very size-selective towards toluene hydrogenation. The bulky mesitylene (1,3,5-trimethylbenzene) is not converted at all while a conventional Pt/SiO_2 catalyst is active for both substrates with quite similar reaction rates (Fig. 11.20).

FIG. 11.19 SEM pictures of Au (left) and Pt (right) nanoparticles in hollow silicalite-1 single crystals.

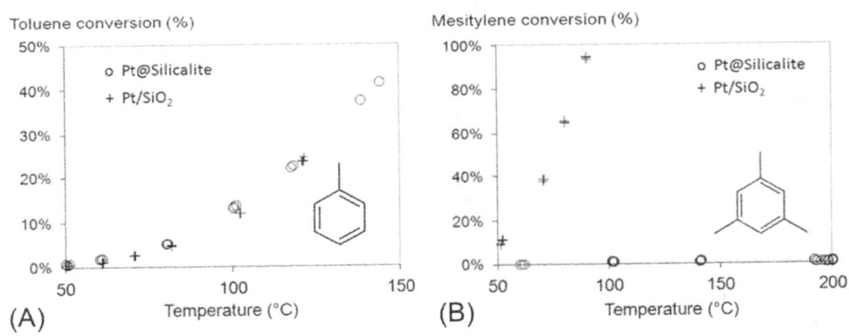

FIG. 11.20 Substrate conversion in the hydrogenation of toluene (A) and mesitylene (B) over Pt nanoparticles encapsulated in hollow silicalite-1 (Pt@silicalite) and supported on amorphous SiO_2 (Pt/SiO_2). *(Reproduced from S. Li, T. Boucheron, A. Tuel, D. Farrusseng, F. Meunier, Chem. Commun. 50 (2014) 1824–1826, with permission of Royal Society of Chemistry.)*

The same Pt/hollow silicalite-1 materials have shown a remarkable resistance to poisoning by propylene during the oxidation of carbon monoxide. In the absence of propylene, CO and O_2 conversions are similar to those observed on a Pt-supported catalyst in which all Pt NPs are directly accessible to molecules, without any size discrimination. When propylene is introduced in the feed, the activity of the supported catalyst drastically decreases while that of the encapsulated Pt particles remains high, with a decrease in CO conversion of only 5% at 360°C. Since both CO and propylene are smaller than the pore size of the zeolite, the poison resistance of the yolk-shell catalyst has been attributed to the large difference in the diffusion rates between CO and propylene. Bimetallic or metal-oxide nanoparticles encapsulated in hollow silicalite-1 have been obtained by co-impregnation of bulk crystals using two distinct metal precursors. In the case of Ni-Pt bimetallic nanoparticles, EDX confirmed that all nanoparticles encapsulated in hollow crystals contained the two metals with almost the same Ni/Pt ratio [316]. A catalyst containing bimetallic nanoparticles with a mean size of 4.57 nm and with an average composition of 1.5 wt.% Ni and 0.5 wt.% Pt was used in the dry reforming of methane. Compared to pure Ni nanoparticles, the addition of Pt significantly improved the stability of the catalyst and maintained high activities for CH_4 and CO_2 conversions. Moreover, the better dispersion of bimetallic particles and the absence of large aggregates on the surface of hollow crystals considerably reduced the amount of coke formed.

Besides metal nanoparticles, silicalite-1 hollow crystals have also been used to encapsulate heteropolyacids (HPA) [317] (Fig. 11.21). With HPA being too large to enter the zeolite pores, they were formed inside the cavities from

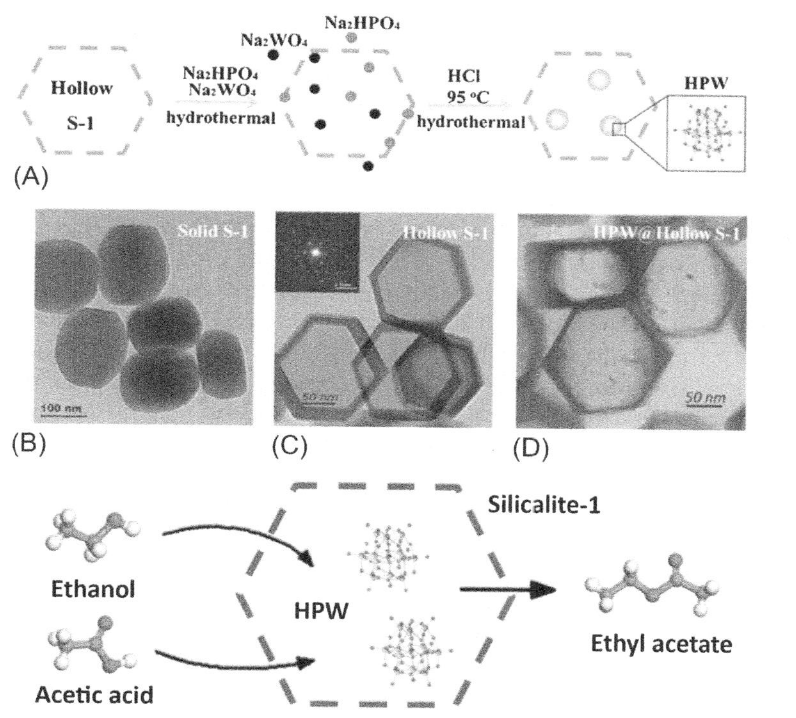

FIG. 11.21 Top: schematic preparation of HPW@Hollow S-1 catalysts (A) with a TEM image of silicalite-1 (B), hollow silicalite-1 (C), and HPW@Hollow S-1 (D). Bottom: schematic illustration of the esterification reaction inside HPW@Hollow S-1 nanoreactor. *(Reproduced from C. Dai, A. Zhang, J. Li, K. Hou, M. Liu, C. Song, et al., Chem. Commun. 50 (2014) 4848–4850, with permission of Royal Society of Chemistry.)*

reaction between small building blocks. For example, phosphotungstic acid (HPW) has been directly synthesized in silicalite-1 hollow crystals from Na_2WO_4 and Na_2HPO_4, which were preliminary introduced in the cavity by impregnation methods. The yolk-shell catalyst was used in the esterification between ethanol and acetic acid. In contrast to a catalyst obtained by a standard impregnation method, the encapsulated HPW was very stable and retained its initial activity after six runs. In both catalysts, HPW nanoparticles were likely dissolved in the acidic reactant solution and acted as a homogeneous catalyst. Nevertheless, dissolved species could not escape from hollow structures while they were leached out in the case of the impregnated solid.

Similarly, the addition of copper oxide to Pd nanoparticles in hollow silicalite-1 greatly improved the dispersion and the catalytic activity in the Suzuki-Miyaura reaction [314].

Metal or metal oxide nanoparticles have also been encapsulated in ZSM-5 hollow zeolites. Compared to silicalite-1 for which hollow structures are formed by preferential dissolution of the defect-rich core of the crystals, dissolution of ZSM-5 is based on differences in aluminum concentration [310,316].

Hollow structures can be obtained either by dissolution/recrystallization in the presence of TPAOH or under milder conditions using sodium carbonate solutions as desilicating agents. Desilication with Na_2CO_3 necessitates the starting ZSM-5 to be inhomogeneous, with a strong Al zoning in the crystals. Si-rich cores are preferentially dissolved and the obtained hollow structures usually possess more aluminum than the original bulk zeolite. In addition, desilication temperatures as well as pH values differ significantly between TPAOH and Na_2CO_3, which greatly influences the size and dispersion of encapsulated nanoparticles. In the particular case of Pt, TPAOH leads to a unique particle in the cavity while milder conditions form many small particles of c.2 nm [318]. Moreover, very thin shells obtained with sodium carbonate are generally so defective that Pt particles are accessible to bulk molecules such as mesitylene, which is not the case when desilication is performed with TPAOH.

Yolk-shell materials made of hollow zeolite single crystals can also encapsulate transition-metal nanoparticles and their corresponding oxides [319,320]. As for noble metals, the number and size of encapsulated nanoparticles depends on the desilication method. The main difference with noble metals is that particles are absent from inner cavities immediately after treatment with TPAOH or Na_2CO_3 and that their formation necessitates reduction at very high temperature. Similarly to Pt, Ni particles in hollow silicalite-1 showed outstanding size selectivity in the hydrogenation of substituted benzenes. Moreover, desilication with Na_2CO_3 enables the formation of yolk-shell structures with very thin zeolite shells and very high metal contents (more than 20 wt.% in the case of Co with regular particles of 10–15 nm), which opens the route to many catalytic applications requiring the use of high metal contents, for example, Fischer-Tropsch synthesis [319].

8 PERSPECTIVES

Well-defined cavities and pores of molecular dimensions present in the framework of zeolites constitute exceptional environments for the immobilization of catalysts. The huge number of zeolite structures as well as the possibility of modifying their composition result in the local environment around the encapsulated catalyst being physically or chemically adapted in order to optimize its properties. There has been very abundant literature showing the advantage of immobilizing catalysts such as metal complexes or nanoparticles on their catalytic activity. Nevertheless, if the presence of micropores is a real advantage for protecting catalysts against poisoning or sintering or for imposing selectivity, catalytic reactions are very often limited by diffusion, which constitutes a severe drawback to their utilization in industry. As a consequence, only the outer regions of the crystals are useful, and most of the encapsulated species deeply buried under the surface do not participate in the reaction. The recent discovery of new synthetic methods to develop additional mesoporosity in zeolite crystals has considerably contributed to improve the accessibility to internal active sites. These "hierarchical" zeolites, which associate mesopores to the original array of micropores, have essentially been used to support metal nanoparticles. Only a few examples concern the immobilization of metal complexes, but it is obvious that such applications will rapidly develop in the near future. One of the main obstacles to the encapsulation of complexes in hierarchical zeolites could be the lack of efficient and simple methods to prepare mesoporous FAU-type zeolites as compared to MFI. However, recent progress in the design of specific zeolite morphologies by direct synthesis or postsynthesis modifications suggests that new methods will probably be discovered soon. Besides facilitating the access to internal sites, mesopores can also be used as an additional surface to support nanoparticles, particularly when those particles are too large to enter the nanometric framework cavities. Mesoporous zeolites can be a real alternative to conventional mesoporous materials such as MCM-41 or SBA-15, in particular when reactions are performed under harsh hydrothermal conditions. Moreover, the close proximity between nanoparticles in mesopores and zeolite active sites is very favorable to the preparation of bifunctional catalysts. Pertinent examples of synergism between nanoparticles and a hierarchical acidic zeolite have been reported for the production of gasoline-range hydrocarbons in Fischer-Tropsch synthesis.

The recent development of new synthesis approaches to design zeolite with specific morphologies has emerged as a very promising field of research, particularly for catalytic applications. Famous examples concern the formation of ultrathin zeolite nanosheets using specific templates or mesoporosity induced by rotational intergrowth of very thin zeolite plates. Although a couple of papers have reported the use of Pt nanoparticles dispersed in MFI nanosheets for hydrogenation reactions, there is no doubt that these particular zeolite morphologies will find many applications as hosts for heterogeneous catalysts in the near future.

The concept of core/yolk-shell catalysts in which nanoparticles are totally encapsulated in zeolite shells is also particularly innovative. The zeolite layer deposited on the surface of a standard catalyst can be used either as a membrane to select reactants and/or products on the basis of size or as a second catalyst to change the product selectivity. Nevertheless, the development of such core/shell catalysts suffers from the difficulty to form a regular thin polycrystalline coating and to control the interface. Very often, effective shells need to be quite thick, which limits the overall reaction rate by diffusion transport limitations. The thickness of the zeolite shell can be considerably reduced when catalysts are encapsulated in hollow single crystals. Compared to polycrystalline shells, hollow single crystals are perfectly crystallized and their molecular sieving efficiency is already achieved with very thin walls. To date, such materials are limited to MFI-type zeolites, but work is in progress to try to extend the concept to other structures, particularly large-pore zeolites. One of the main drawbacks of the concept is the very low density of active sites per volume unit. In the case of metal nanoparticles, the catalyst contains approx. 1 particle per 10^6 nm^3, assuming very small zeolite crystals of c.100 nm in size. For comparison, zeolites in which only 1% of the cavities are occupied by particles possess a density of the order of 1 particle per 10^2 nm^3. Increasing the density can be achieved by decreasing the size of individual zeolite crystals. Nevertheless, very small zeolite crystals are not appropriate for desilication and the size of hollow crystals can hardly fall below 100 nm. Another way is to prepare multihollow crystals with many small cavities inside a crystal, each of them containing one particle. Under such conditions, controlling the size and number of cavities is of prime importance to maintain a narrow particle size distribution.

The use of hollow zeolite crystals as chemical nanoreactors is still an emerging field of research. As an example, hollow crystals could be used to include large soluble catalysts capable of transforming molecules inside the zeolite cavity. Such materials would provide all advantages of homogeneous catalysts (enantioselectivity) while being shape selective and easily recovered from the reaction mixture.

A real breakthrough in the use of yolk/shell catalysts could also be attained by combining the properties of hierarchical zeolites with hollow structures. Application of yolk-shell structures are indeed limited to very small molecules, and the combination of cavities with a hierarchical porosity inside the walls could open many opportunities in the selective transformation of bulky molecules.

Lastly, there is a real need for encapsulated metal complexes as solid catalysts for asymmetric synthesis. Indeed, in the absence of covalent attachment between complexes and the zeolite framework, it is expected that encapsulated asymmetric catalysts retain the properties they have in the solution, particularly enantioselectivity. Unfortunately, attempts to encapsulate chiral complexes for enantioselective reactions were not as successful as expected, and the resulting enantioselectivity is generally very low. A very promising way would be to encapsulate catalysts in chiral zeolite frameworks and to use the chirality of the host zeolite to increase the enantioselctivity.

REFERENCES

[1] Database of zeolite structures (http://www.iza-structure.crg/databases).

[2] F. Bedioui, Coord. Chem. Rev. 144 (1995) 39–68.

[3] S. Nakagaki, C.R. Xavier, A.J. Wosniak, A.S. Mangrich, F. Wypych, M.P. Cantao, I. Denicolo, L.T. Kubota, Colloids Surf. A 168 (3) (2000) 261–276.

[4] F. Van der Velde, I.W.C.E. Arends, R.A. Sheldon, J. Inorg. Biochem. 80 (1–2) (2000) 81–89.

[5] W.M.H. Sachtler, Catal. Today 15 (1992) 419–429.

[6] W.M.H. Sachtler, Acc. Chem. Res. 26 (1993) 383–387.

[7] M. Flytzani-Stephanopoulos, B.C. Gates, Annu. Rev. Chem. Biomol. Eng. 3 (2012) 545–574.

[8] K.J. Balkus Jr., A.K. Khanmamedova, K.M. Dixon, F. Bedioui, Appl. Catal. A: General 143 (1996) 159–173.

[9] A.K. Chandrakar, G.P. Dewangan, N. Chandraker, R.S. Thakur, Int. J. Adv. Res. Chem. Sci. 2 (7) (2015) 1–6.

[10] D.E. de Vos, F. Thibault-Starzyk, P.P. Knops-Gerrits, R.F. Parton, P.A. Jacobs, Macromol. Symp. 80 (1994) 157–184.

[11] B.V. Romanovski, Macromol. Symp. 80 (1994) 185–192.

[12] K.J. Balkus Jr., A.G. Gabrielov, J. Inclusion Phenom. Mol. Recognit. Chem. 21 (1995) 159–184.

[13] J.L. Viana Rosa, C.M.C.P. Manso, O.A. Serra, Y. Yamamoto, J. Mol. Catal. A: Chem. 160 (2000) 199–208.

[14] B.-Z. Zhan, X.-Y. Li, Chem. Commun. (1998) 349–350.

[15] A.G. Gabrielov, K.J. Balkus Jr., S.L. Bell, F. Bedioui, J. Devynck, Microporous Mater. 2 (1994) 119–126.

[16] K.J. Balkus Jr., M. Eissa, R. Levado, J. Am. Chem. Soc. 117 (1995) 10753–10754.

[17] R. Raja, P. Ratnasamy, Catal. Lett. 48 (1997) 1–10.

[18] S. Goel, Z. Wu, S.I. Zones, E. Iglesia, J. Am. Chem. Soc. 134 (2012) 17688–17695.

[19] Z. Wu, S. Goel, M. Choi, E. Iglesia, J. Catal. 311 (2014) 458–468.

[20] S. Altwasser, R. Gläser, A.S. Lo, P.-H. Liu, K.-J. Chao, J. Weitkamp, Microporous Mesoporous Mater. 89 (2006) 109–122.

[21] B.Z. Zhan, E. Iglesia, Angew. Chem. Int. Ed. 46 (2007) 3697–3700.

[22] M. Zahmakıran, Y. Tonbul, S. Ozkar, J. Am. Chem. Soc. 132 (2010) 6541–6549.

[23] S. Altwasser, R. Gläser, J. Weitkamp, Microporous Mesoporous Mater. 104 (2007) 281–288.

[24] H. Yang, H. Chen, H. Du, R. Hawkins, F. Craig, Z. Ring, O. Omotoso, V. Munoz, R. Mikula, Microporous Mesoporous Mater. 117 (2009) 33–40.

[25] J. Kecht, S. Mintova, T. Bein, Langmuir 24 (2008) 4310–4315.

[26] A. Zsigmond, F. Notheisz, J.E. Backvall, Catal. Lett. 65 (2000) 135–139.

[27] N. Kobayashi, in: C.C. Leznoff, A.B.P. Lever (Eds.), Phthalocyanines – Properties and Applications, Vol. 2, VCH Publishers, New York, 1993, p. 97.

[28] Y.C. Yang, J.R. Ward, R.P. Seiders, Inorg. Chem. 24 (1985) 1765–1769.

[29] R.T.K. Baker, J. Catal. 78 (1982) 473–476.

[30] P. Harris, Int. Mat. Rev. 40 (1995) 97–115.

[31] T.W. Hansen, A.T. DeLaRiva, S.R. Challa, A.K. Datye, Acc. Chem. Res. 46 (2013) 1720–1730.

[32] C.H. Bartholomew, Appl. Catal. A 107 (1993) 1–57.
[33] S.R. Challa, A.T. Delariva, T.W. Hansen, S. Helveg, J. Sehested, P.L. Hansen, F. Garzon, A.K. Datye, J. Am. Chem. Soc. 133 (2011) 20672–20675.
[34] T.W. Hansen, J.B. Wagner, P.L. Hansen, S. Dahl, H. Topsøe, C.J. Jacobsen, Science 294 (2001) 1508–1510.
[35] J. McCarty, G. Malukhin, D. Poojary, A. Datye, Q. Xu, J. Phys. Chem. B 109 (2005) 2387–2391.
[36] S. Ray, S. Vasudevan, Inorg. Chem. 42 (2003) 1711–1719.
[37] R. Raja, P. Ratnasamy, J. Catal. 170 (1997) 244–253.
[38] P.B. Venuto, Microporous Mater. 2 (5) (1994) 297–411.
[39] B. Smit, T.L.M. Maesen, Chem. Rev. 108 (10) (2008) 4125–4184.
[40] B. Smit, T.L.M. Maesen, Nature 451 (2008) 671–678.
[41] E.G. Derouane, J. Catal. 100 (2) (1986) 541–544.
[42] L.D. Rollmann, J. Catal. 47 (1) (1977) 113–121.
[43] D. Mansuy, P. Battioni, in: J. Reedijk (Ed.), Bioinorganic Catalysis, Marcel Dekker, New York, 1993, pp. 395–398.
[44] A.E. Shilov, J. Mol. Catal. 47 (1988) 351–362.
[45] B.R. Cook, T.J. Reinert, K.S. Suslick, J. Am. Chem. Soc. 108 (1986) 7281–7286.
[46] K. Suslik, B. Cook, M. Fox, J. Chem. Soc., Chem. Commun. (1980) 580–582.
[47] A. Maldotti, R. Amadelli, C. Bartocci, V. Carassiti, E. Polo, G. Varani, Coord. Chem. Rev. 125 (1–2) (1993) 143–154.
[48] A.M. Rocha Gonzalves, M.M. Pereira, J. Mol. Catal. A: Chemical 113 (1996) 209–221.
[49] A. Fuerstner, Chem. Soc. Rev. 38 (11) (2009) 3208–3221.
[50] M. Heitbaum, F. Glorius, I. Escher, Angew. Chem. Int. Ed. 45 (29) (2006) 4732–4762.
[51] A.J. Canty, Acc. Chem. Res. 25 (2) (1992) 83–90.
[52] P. Sehnal, R.J.K. Taylor, I.J.S. Fairlamb, Chem. Rev, 110 (2) (2010) 824–889.
[53] I.P. Beletskaya, A.V. Cheprakov, Chem. Rev, 100 (8) (2000) 3009–3066.
[54] A.F. Littke, G.C. Fu, Angew. Chem. Int. Ed, 41 (22) (2002) 4176–4211.
[55] E.A.B. Kantchev, C.J. O'Brien, M.G. Organ, Angew. Chem. Int. Ed, 46 (16) (2007) 2768–2813.
[56] V. Farina, Adv. Synth. Catal, 346 (13–15) (2004) 1553–1582.
[57] B.M. Trost, D.L. VanVranken, Chem. Rev, 96 (1) (1996) 395–422.
[58] D.J. Gorin, B.D. Sherry, F.D. Toste, Chem. Rev, 108 (8) (2008) 3351–3378.
[59] D.J. Gorin, F.D. Toste, Nature, 446 (7134) (2007) 395–403.
[60] W.A. Herrmann, C.W. Kohlpaintner, Angew, Chem. Int. Ed 32 (11) (1993) 1524–1544.
[61] B.M. Trost, Angew, Chem. Int. Ed 34 (3) (1995) 259–281.
[62] A. Hashmi, K. Stephen, Angew, Chem. Int. Ed 49 (31) (2010) 5232–5241.
[63] A.S.K. Hashmi, Gold Bull, 37 (1–2) (2004) 51–65.
[64] H. Forster, J. Seebode, J. Chem. Soc., Faraday Trans. I 79 (1983) 373–382.
[65] I. Kiricsi, I. Hannus, K. Varga, P. Fejes, J. Catal. 63 (1980) 501–504.
[66] C. Dossi, J. Schaefer, W.M.H. Sachtler, J. Mol. Catal. 52 (1989) 193–209.
[67] T. Bein, S.J. McLain, D.R. Corbin, K. Moller, G.D. Stucky, G. Woolery, D. Sayers, J. Am. Chem. Soc. 110 (1988) 1801–1810.
[68] K. Moller, T. Bein, J. Phys. Chem. 94 (1990) 845–853.
[69] C. Dossi, C.M. Tsang, W.M.H. Sachtler, R. Psaro, R. Ugo, Energy and Fuels 2 (1989) 468–474.
[70] B.C. Gates, Chem. Rev. 95 (1995) 511–522.
[71] J. Guzman, B.C. Gates, Dalton Trans. 17 (2003) 3303–3318.

[72] J.F. Goellner, B.C. Gates, G.N. Vayssilov, N. Rösch, J. Am. Chem. Soc. 122 (2000) 8056–8066.
[73] P.-L. Zhou, B.C. Gates, J. Chem. Soc., Chem. Commun (1989) 347–349.
[74] P.-L. Zhou, S.D. Maloney, B.C. Gates, J. Catal. 129 (1991) 315–329.
[75] P. Serna, B.C. Gates, Acc. Chem. Res. 47 (2014) 2612–2620.
[76] J.A. Hriljac, M.M. Eddy, A.K. Cheetham, J.A. Donohue, G.J. Ray, J. Solid State Chem. 106 (1) (1993) 66–72.
[77] K.J. Balkus, S. Shepelev, Microporous Mater. 1 (1993) 383–391.
[78] G. Van de Goor, C.C. Freyhardt, P.Z. Behrens, Anorg. Allg. Chem. 621 (1995) 311–322.
[79] K.J. Balkus Jr., A. Ramsaran, R. Szostak, M. Mitchell, in: Proceedings of the 12th International Zeolite Conference, Materials Research Society, Warrendale, PA, 1999, pp. 1931–1935.
[80] K.J. Balkus, M. Biscotto, A.G. Gabrielov, Stud. Surf. Sci. Catal. 105 (1997) 415–421.
[81] R.F. Lobo, M. Tsapatsis, C.C. Freyhardt, S. Khodabandeh, P. Wagner, C.-Y. Chen, K.J. Balkus, S.I. Zones, M.E. Davis, J. Am. Chem. Soc. 119 (36) (1997) 8474–8484.
[82] T. Wessels, C. Baerlocher, L.B. McCusker, E.J. Creyghton, J. Am. Chem. Soc. 121 (26) (1999) 6242–6247.
[83] L. Ren, L. Zhu, C. Yang, Y. Chen, Q. Sun, H. Zhang, C. Li, F. Nawaz, X. Meng, F.-S. Xiao, Chem. Commun. 47 (2011) 9789–9791.
[84] J.H. Kwak, R.G. Tonkyn, D.H. Kim, J. Szanyi, C.H.F. Peden, J. Catal. 275 (2010) 187–190.
[85] S.T. Korhonen, D.W. Fickel, R.F. Lobo, B.M. Weckhuysen, A.M. Beale, Chem. Commun. 47 (2011) 800–802.
[86] D.W. Fickel, E.D. Addio, J.A. Lauterbach, R.F. Lobo, Appl. Catal. B 102 (2011) 441–448.
[87] D.W. Fickel, R.F. Lobo, J. Phys. Chem. C 114 (2010) 1633–1640.
[88] K.J. Balkus Jr., S. Kowalak, K.T. Ly, ACS Symp. Ser. 499 (1992) 347–354.
[89] K.J. Balkus Jr., A.G. Gabrielov, S.L. Bell, F. Bedioui, L. Roue, J. Devynck, Inorg. Chem. 33 (1994) 67–72.
[90] K.J. Balkus, Jr., S. Kowalak, U. S. Patent 5,167,942, 1992.
[91] K.J. Balkus Jr., S. Kowalak, K.T. Ly, C.D. Hargis, Stud. Surf Sci. Catal. 69 (1991) 93–99.
[92] G. Gbery, A. Zsigmond, K.J. Balkus Jr., Catal. Lett. 74 (1–2) (2001) 77–80.
[93] K. Bogar, Cent. Eur. J. Chem. 3 (1) (2005) 63–71.
[94] N. Herron, Inorg. Chem. 25 (1986) 4714–4717.
[95] C. Bowers, P.K. Dutta, J. Catal. 122 (1990) 271–279.
[96] L. Gaillon, N. Sajot, F. Bedioui, J. Devynck, K.J. Balkus Jr., J. Electroanal. Chem. 345 (1993) 157–167.
[97] K.J. Balkus Jr., A.A. Welch, B.E. Gnade, Zeolites 10 (1990) 722–729.
[98] S. Kowalak, R.C. Weiss, K.J. Balkus Jr., J. Chem. Soc., Chem. Commun (1991) 57–58.
[99] M. Salavati-Niasari, A. Sobhani, J. Mol. Catal. A 285 (2008) 58–67.
[100] M. Salavati-Niasari, Z. Salimi, M. Bazarganipour, F. Davar, Inorg. Chim. Acta 362 (2009) 3715–3724.
[101] M. Salavati-Niasari, M. Shakouri-Arani, F. Davar, Microporous Mesoporous Mater. 116 (1–3) (2008) 77–85.
[102] C.R. Jacob, S.P. Varkey, P. Ratnasamy, Appl. Catal. A 168 (2) (1998) 353–364.
[103] M.R. Maurya, A.K. Chandrakar, S. Chand, J. Mol. Catal. A 274 (1–2) (2007) 192–201.
[104] C. Jin, W.B. Fan, Y.L. Jia, B.B. Fan, J.H. Ma, R.F. Li, J. Mol. Catal. A 249 (1–2) (2006) 23–30.
[105] K.O. Xavier, J. Chacko, K.K.M. Yusuff, Appl. Catal. A 258 (2) (2004) 251–259.
[106] Z. Sobalik, J. Dedecek, I. Ikonnikov, B. Wichterlova, Microporous Mesoporous Mater. 21 (4–6) (1998) 525–532.

[107] R. Ramprasad, K.C. Hass, W.F. Schneider, J.B. Adams, J. Phys. Chem. B 101 (35) (1997) 6903–6913.

[108] C.E. Ruggiero, S.M. Carrier, W.E. Antholine, J.W. Whittaker, C.J. Cramer, W.B. Tolman, J. Am. Chem. Soc. 115 (24) (1993) 11285–11298.

[109] S. Ozkar, G.A. Ozin, K. Moller, T. Bein, J. Am. Chem. Soc. 112 (26) (1990) 9575–9586.

[110] L. Crowfoot, G.A. Ozin, S. Ozkar, J. Am. Chem. Soc. 113 (6) (1991) 2033–2040.

[111] X.P. Li, G.A. Ozin, S. Ozkar, J. Phys. Chem. 95 (11) (1991) 4463–4476.

[112] G.A. Ozin, S. Ozkar, Adv. Mater. 4 (1) (1992) 11–22.

[113] G.A. Ozin, S. Ozkar, Chem. Mater. 4 (3) (1992) 511–521.

[114] H.O. Pastore, G.A. Ozin, A.J. Poe, J. Am. Chem. Soc. 115 (4) (1993) 1215–1230.

[115] C. Brémard, Coord. Chem. Rev. 178–180 (1998) 1647–1677.

[116] B.C. Gates, J. Mol. Catal. A 163 (1–2) (2000) 55–65.

[117] P.P. Knops-Gerrits, D. De Vos, F. Thibault-Starzyk, P.A. Jacobs, Nature 369 (6481) (1994) 543–546.

[118] W. Turbeville, D.S. Robins, P.K. Dutta, J. Phys. Chem. 96 (12) (1992) 5024–5029.

[119] P.P. Knops-Gerrits, D.E. DeVos, P.A. Jacobs, J. Mol. Catal. A 117 (1–3) (1997) 57–70.

[120] W.H. Quayle, G. Peeters, G.L. Deroy, E.F. Vansant, J.H. Lunsford, Inorg. Chem. 21 (6) (1982) 2226–2231.

[121] K. Maruszewski, J.R. Kincaid, Inorg. Chem. 34 (8) (1995) 2002–2006.

[122] K. Mori, K. Kagohara, H. Yamashita, J. Phys. Chem. C 112 (7) (2008) 2593–2600.

[123] D.E. De Vos, D.L. Vanoppen, X.Y. Li, S. Libbrecht, Y. Bruynseraede, P.P. Knops-Gerrits, P.A. Jacobs, Chem. Eur. J. 1 (2) (1995) 144–149.

[124] C. Baleizao, H. Garcia, Chem. Rev. 106 (9) (2006) 3987–4043.

[125] T. Joseph, S.B. Halligudi, C. Satyanarayan, D.P. Sawant, S. Gopinathan, J. Mol. Catal. A 168 (1–2) (2001) 87–97.

[126] T. Joseph, D.P. Sawant, C.S. Gopinath, S.B. Halligudi, J. Mol. Catal. A 184 (1–2) (2002) 289–299.

[127] D.E. De Vos, F. Thibault-Starzyk, P.A. Jacobs, Angew. Chem. Int. Ed. 33 (4) (1994) 431–433.

[128] P.K. Saha, B. Dutta, S. Jana, R. Bera, S. Saha, K. Okamoto, S. Koner, Polyhedron 26 (3) (2007) 563–571.

[129] R. Ferreira, M. Silva, C. Freire, B. de Castro, J.L. Figueiredo, Microporous Mesoporous Mater. 38 (2–3) (2000) 391–401.

[130] A. Domenech, P. Formentin, H. Garcia, M.J. Sabater, Eur. J. Inorg. Chem. 6 (2000) 1339–1344.

[131] M. Salavati-Niasari, S. Baskoutas, Synth. React. Inorg. Met. Org. Chem. 45 (12) (2015) 1826–1833.

[132] S.M.B. Hosseini-Ghazvini, P. Safari, A. Mobinikhaledi, M. Zendehdel, React. Kinet. Mech. Catal. 115 (2) (2015) 703–718.

[133] C.K. Modi, B.C. Gade, J.A. Chudasama, D.K. Parmar, H.D. Nakum, A.L. Patel, Spectrochim. Acta. Part A 140 (2015) 174–184.

[134] A.H. Ahmed, Rev. Inorg. Chem. 34 (3) (2014) 153–175.

[135] X.-F. Zhou, Environ. Prog. Sustain. Energy 34 (4) (2015) 1120–1128.

[136] I. Kuzniarska-Biernacka, P. Parpot, C. Oliveira, A.R. Silva, M.J. Alves, A.M. Fonseca, I.C. Neves, J. Phys. Chem. C 118 (33) (2014) 19042–19050.

[137] X.-F. Zhou, Rcs. Adv. 4 (53) (2014) 28029–28035.

[138] K.K. Bania, D. Bharali, B. Viswanathan, R.C. Deka, Inorg. Chem. 51 (3) (2012) 1657–1674.

[139] M. Bagherzadeh, M. Zare, J. Coord. Chem. 65 (22) (2012) 4054–4066.

[140] B.B. Fan, H.Y. Li, W.B. Fan, C. Jin, R.F. Li, Appl. Catal. A 340 (1) (2008) 67–75.
[141] P. Chen, B.B. Fan, M.G. Song, C. Jin, J.H. Ma, R.F. Li, Catal. Commun. 7 (12) (2006) 969–973.
[142] N. Herron, G.D. Stucky, C.A. Tolman, J. Chem. Soc., Chem. Commun. (1986) 1521–1522.
[143] N. Herron, J. Coord. Chem. 19 (1988) 25–38.
[144] B.V. Romanovsky, A.G. Gabrielov, Stud. Surf. Sci. Catal. 72 (1992) 443–455.
[145] B.V. Romanovsky, A.G. Gabrielov, J. Mol. Catal. 74 (1992) 293–303.
[146] G. Meyer, D. Wohrle, D. Mohl, G. Schultz-Ekloff, Zeolites 4 (1984) 30–34.
[147] H. Diegruber, P.J. Plath, G. Schultz-Ekloff, J. Mol. Catal. 24 (1984) 115–126.
[148] K.J. Balkus, C.D. Hargis, S. Kowalak, ACS Symp. Ser. 499 (1992) 347–354.
[149] M.A. Zanjanchi, K. Tabatabaeian, M. Moosavifar, J. Inclusion Phenom, Macro. Chem 40 (3) (2001) 193–198.
[150] C. Martinez-Macias, P. Serna, B.C. Gates, ACS Catal. 5 (10) (2015) 5647–5656.
[151] C. Martinez-Macias, M.Y. Chen, D.A. Dixon, B.C. Gates, Chem. Eur. J. 21 (33) (2015) 11825–11835.
[152] B.C. Gates, J. Catal. 328 (2015) 72–74.
[153] D. Yang, M.Y. Chen, C. Martinez-Macias, D.A. Dixon, B.C. Gates, Chem. Eur. J. 21 (2) (2015) 631–640.
[154] J.D. Kistler, N. Chotigkrai, P.H. Xu, B. Enderle, P. Praserthdam, C.Y. Chen, N.D. Browning, B.C. Gates, Angew. Chem. Int. Ed. 53 (34) (2014) 8904–8907.
[155] P. Ratnasamy, D. Srinivas, Catal. Today 141 (2009) 3–11.
[156] R.J. Taylor, R.S. Drago, J.P. Hage, Inorg. Chem. 31 (1992) 253–258.
[157] T. Kimura, A. Fukuoka, M. Ichikawa, Catal. Lett. 4 (1990) 279–286.
[158] A. Kozlov, A. Kozlova, K. Asakura, Y. Iwasawa, J. Mol. Catal. A 137 (1999) 223–237.
[159] R.F. Parton, L. Uytterhoeven, P. Jacobs, Stud. Surf. Sci. Catal. 59 (1991) 395–403.
[160] R.E. White, M.J. Coon, Ann. Rev. Biochem. 49 (1980) 315–356.
[161] I.C. Gunsalus, S.C. Sligar, Adv. Enzymol. 47 (1978) 1–44.
[162] R. Raja, J.M. Thomas, V. Dreyer, Catal. Lett. 110 (3–4) (2006) 179–183.
[163] C.K. Modi, P.M. Trivedi, J. Coord. Chem. 65 (3) (2012) 525–538.
[164] J. Dzierzak, E. Bottinelli, G. Berlier, E. Gianotti, E. Stulz, R.M. Kowalczyk, R. Raja, Chem. Commun. 46 (2010) 2805–2807.
[165] J. Dzierzak, M. Lefenfeld, R. Raja, Top. Catal. 52 (2009) 1669–1676.
[166] R.F. Patton, I.F.J. Vankelecom, M.J.A. Casselman, C.P. Bezouhanova, J.B. Uytterhoeven, P.A. Jacobs, Nature 370 (1994) 541–544.
[167] J. Haber, K. Pamin, J. Poltowicz, J. Mol. Catal. A 224 (2004) 153–159.
[168] R.F. Parton, G.J. Peere, P.E. Neys, P.A. Jacobs, R. Claessens, G.V. Baron, J. Mol. Catal. A 113 (1996) 445–454.
[169] M.N. Cele, H.B. Friedrich, M.D. Bala, ReacKinetMech Cat 111 (2014) 737–750.
[170] J. Połtowicz, K. Pamin, E. Tabor, J. Haber, A. Adamski, Z. Sojka, Applied Catalysis A: General 299 (2006) 235–242.
[171] S.P. Varkey, C.R. Jacob, Indian J. Chem. 38A (1999) 320–324.
[172] R. Raja, P. Ratnasamy, Appl. Catal. A 158 (1997) L7–L15.
[173] R. Raja, C.R. Jacob, P. Ratnasamy, Catal. Today 49 (1999) 171–175.
[174] G. von Willingh, H.S. Abbo, S.J.J. Titinchi, Catal. Today 227 (2014) 96–104.
[175] S. Deshpande, D. Srinivas, P. Ratnasamy, J. Catal. 188 (1999) 261–269.
[176] S.A. Chavan, D. Srinivas, P. Ratnasamy, J. Catal. 204 (2001) 409–419.
[177] D. Srinivas, S. Sivasanker, Catal. Surv. Asia 7 (2–3) (2003) 121–132.
[178] T.H. Bennur, D. Srinivas, S. Sivasanker, J. Mol. Catal. A 207 (2004) 163–171.

[179] Z. Li, S. Wu, Y. Ma, H. Liu, J. Hu, L. Liu, Q. Huo, J. Guan, Q. Kan, Transition Met Chem 38 (2013) 243–251.

[180] M.R. Maurya, B. Singh, P. Adão, F. Avecilla, J. Costa Pessoa, Eur. J. Inorg. Chem. 36 (2007) 5720–5734.

[181] I. Kuzniarska-Biernacka, A.M. Fonseca, I.C. Neves, Inorg. Chim. Acta 394 (2013) 591–597.

[182] Y. Yang, H. Ding, S. Hao, Y. Zhang, Q. Kana, Appl. Organometal. Chem. 25 (2011) 262–269.

[183] M. Silva, C. Freire, B. de Castro, J.L. Figueiredo, J. Mol. Catal. A 258 (2006) 327–333.

[184] M.R. Maurya, P. Saini, C. Haldar, A.K. Chandrakar, S. Chand, J. Coord. Chem. 65 (16) (2012) 2903–2918.

[185] H.S. Abbo, S.J.J. Titinchi, Top Catal. 53 (2010) 1401–1410.

[186] K.K. Bania, R.C. Deka, J. Phys. Chem. C 117 (2013) 11663–11678.

[187] S. Chavan, D. Srinivas, P. Ratnasamy, Top. Catal. 11/12 (2000) 359–367.

[188] S. Chavan, D. Srinivas, P. Ratnasamy, J. Catal. 192 (2000) 286–295.

[189] S. Seelan, A.K. Sinha, Appl. Catal. A 238 (2003) 201–209.

[190] M.R. Maurya, S.J.J. Titinchi, S. Chand, J. Mol. Catal. A 193 (2003) 165–176.

[191] M.R. Maurya, S.J.J. Titinchi, S. Chand, J. Mol. Catal. A 201 (2003) 119–130.

[192] M.R. Maurya, S.J.J. Titinchi, S. Chand, Catal. Lett. 89 (3–4) (2003) 219–227.

[193] M.R. Maurya, S.J.J. Titinchi, S. Chand, J. Mol. Catal. A 214 (2004) 257–264.

[194] H.S. Abbo, S.J.J. Titinchi, Top Catal 53 (2010) 254–264.

[195] M.J. Alcón, A. Corma, M. Iglesias, F. Sánchez, J. Mol. Catal. A 194 (2003) 137–152.

[196] M.R. Maurya, A.K. Chandrakar, S. Chand, J. Mol. Catal. A 278 (2007) 12–21.

[197] V.K. Bansal, P.P. Thankachan, R. Prasad, Appl. Catal. A 381 (2010) 8–17.

[198] K.N. Bhagya, V. Gayathri, J Porous Mater 20 (2013) 257–266.

[199] A. Mobinikhaledi, M. Zendehdel, P. Safari, ReacKinetMech Cat 110 (2013) 497–514.

[200] Á. Zsigmond, K. Bogár, F. Notheisz, J. Catal. 213 (2003) 103–108.

[201] S. Ernst, H. Disteldorf, X. Yang, Microporous Mesoporous Mater. 22 (1998) 457–464.

[202] S. Kawi, J.R. Chang, B.C. Gates, J. Catal. 142 (1993) 585–601.

[203] W.R. Hastings, C.J. Cameron, M.J. Thomas, M.C. Baird, Inorg. Chem. 27 (1988) 3024–3028.

[204] G. Centi, S. Perathoner, Top. Catal. 15 (2–4) (2001) 145–152.

[205] M. El-Roz, L. Lakiss, J. El Fallah, O.I. Lebedev, F. Thibault-Starzyk, V. Valtchev, Phys. Chem. Chem. Phys. 15 (2013) 16198–16207.

[206] F. Solymosi, J. Cserenyi, A. Szoke, T. Bansagi, A. Oszko, J. Catal. 165 (2) (1997) 150–161.

[207] D.J. Wang, J.H. Lunsford, M.P. Rosynek, Top. Catal. 3 (3–4) (1996) 289–297.

[208] J. Gao, Y.T. Zheng, J.M. Jehng, Y.D. Tang, I.E. Wachs, S.G. Podkolzin, Science 348 (6235) (2015) 686–690.

[209] J.J. Spivey, G. Hutchings, Chem. Soc. Rev. 43 (3) (2014) 792–803.

[210] R.J. Davis, J. Catal. 216 (2003) 396–405.

[211] J.C. Kim, H.-X. Li, C.-Y. Chen, M.E. Davis, Microporous Mater. 2 (1994) 413–423.

[212] M. Laspéras, H. Cambon, D. Brunel, I. Rodriguez, P. Geneste, Microporous Mater. 7 (1996) 61–72.

[213] M. Tu, R.J. Davis, J. Catal. 199 (1) (2001) 85–91.

[214] S.V. Bordawekar, R.J. Davis, J. Catal. 189 (2000) 79–90.

[215] L. Guczi, I. Kiricsi, Appl. Catal. A 186 (1999) 375–394.

[216] P. Gallezot, Catal. Sci. Rev. Eng. 20 (1979) 121–154.

[217] J. de Graaf, A.J. van Dillen, K.P. de Jong, D.C. Koningsberger, J. Catal. 203 (2001) 307–321.

[218] R.E. Jentoft, M. Tsapatsis, M.E. Davis, B.C. Gates, J. Catal. 179 (1998) 565–580.

[219] J. Cai, H. Ma, J. Zhang, Q. Song, Z. Du, Y. Huang, J. Xu, Chem. Eur. J. 19 (2013) 14215–14223.
[220] E. Bayram, J. Lu, C. Aydin, N.D. Browning, S. Ozkar, E. Finney, B.C. Gates, R.G. Finke, ACS Catal. 5 (6) (2015) 3514–3527.
[221] B. Coughlan, M.A. Keane, Zeolites 11 (1) (1991) 2–11.
[222] S. Ciccariello, A. Benedetti, F. Pinna, G. Strukul, W. Juszczyk, H. Brumberger, Phys. Chem. Chem. Phys. 1 (2) (1999) 367–372.
[223] A. Tonscheidt, P.L. Ryder, N.I. Jaeger, G. Schulz-Ekloff, Zeolites 16 (1996) 271–274.
[224] S. Kawi, J.R. Chang, B.C. Gates, J. Am. Chem. Soc. 115 (1993) 4830–4843.
[225] H.S. Lacheen, P.J. Cordeiro, E. Iglesia, Chem. Eur. J. 13 (2007) 3048–3057.
[226] J.C. Fierro-Gonzalez, Y. Hao, B.C. Gates, J. Phys Chem. C 111 (2007) 6645–6651.
[227] A.B. Laursen, K.T. Højholt, L.F. Lundegaard, S.B. Simonsen, S. Helveg, F. Schüth, M. Paul, J.-D. Grunwaldt, S. Kegnæs, C.H. Christensen, K. Egeblad, Angew. Chem. Int. Ed. 49 (2010) 3504–3507.
[228] J.-Y. Liu, J.-F. Chen, Y. Zhang, Catal. Sci. Technol. 3 (2013) 2559–2564.
[229] M. Haruta, Cat. Tech. 6 (2002) 102–115.
[230] A. Corma, H. Garcia, Chem. Soc. Rev. 37 (2008) 2096–2126.
[231] B.K. Min, C.M. Friend, Chem. Rev. 107 (2007) 2709–2724.
[232] G.J. Hutchings, M. Brust, H. Schmidbaur, Chem. Soc. Rev. 37 (2008) 1759–1765.
[233] Z. Ma, S. Dai, Nano Res. 4 (2011) 3–32.
[234] Y. Chenevière, V. Caps, A. Tuel, Appl. Catal. A: General 387 (2010) 129–134.
[235] V. Mendez, K. Guillois, S. Daniele, A. Tuel, V. Caps, J. Chem. Soc., Dalton Trans. 39 (2010) 8457–8463.
[236] M. Haruta, Chem. Rec. 3 (2003) 75–87.
[237] K. Guillois, L. Burel, A. Tuel, V. Caps, Appl. Catal.A: General 415–416 (2012) 1–9.
[238] J. Mielby, J.O. Abildstrøm, F. Wang, T. Kasama, C. Weidenthaler, S. Kegnæs, Angew. Chem. Int. Ed. 53 (2014) 12513–12516.
[239] E.J. Creyghton, R.S. Downing, J. Mol. Catal. A 134 (1998) 47–61.
[240] M. Zahmakiran, S. Ozkar, Langmuir 24 (2008) 7065–7067.
[241] I. Yamaguchi, T. Joh, S. Takahashi, J. Chem. Soc. Chem. Commun. (1986) 1412–1413.
[242] K.G. Azzam, G. Jacobs, W.D. Shafer, B.H. Davis, J. Catal. 270 (2) (2010) 242–248.
[243] R.J. Davis, Heterogen. Chem. Rev. 1 (1) (1994) 41–53.
[244] H. Liu, G.D. Lei, W.M.H. Sachtler, Appl. Catal. A 137 (1996) 167–177.
[245] J.A. Martens, R. Parton, L. Uytterhoeven, P.A. Jacobs, G.F. Froment, Appl. Catal. 76 (1991) 95–116.
[246] H.D. Setiabudi, A.A. Jalil, S. Triwahyono, J. Catal. 294 (2012) 128–135.
[247] M. Guisnet, Catal. Today 218–219 (2013) 123–134.
[248] M.J. Girgis, Y.P. Tsao, Ind. Eng. Chem. Res. 35 (1996) 386–396.
[249] M.A. Alotaibi, E.F. Kozhevnikova, I.V. Kozhevnikov, J. Catal. 293 (2012) 141–144.
[250] P.Y. Chen, S.J. Chu, W.C. Lin, K.C. Wu, C.L. Yang, Stud. Surf. Sci. Catal. 83 (1994) 481–488.
[251] P.P. Yang, Y.C. Shang, J.F. Yu, J. Wang, M.L. Zhang, T.G. Wu, J. Mol. Catal. A 272 (1–2) (2007) 75–83.
[252] A.C.C. Rodrigues, J.L.F. Monteiro, Appl. Catal. A 362 (1–2) (2009) 185–192.
[253] A.C.C. Rodrigues, J.L.F. Monteiro, Stud. Surf. Sci. Catal. 154 (2004) 2395–2402.
[254] S. Rabl, A. Haas, D. Santi, C. Flego, M. Ferrari, V. Calemma, J. Weitkamp, Appl. Catal. A 400 (1–2) (2011) 131–141.
[255] S. van Donk, A.H. Janssen, J.H. Bitter, K.P. de Jong, Catal. Rev. 45 (2003) 297–319.

[256] J.C. Groen, J.A. Moulijn, J. Pérez-Ramırez, J. Mater. Chem. 16 (2006) 2121–2131.
[257] J.C. Groen, W. Zhu, S. Brouwer, S.J. Huynink, F. Kapteijn, J.A. Moulijn, J. Perez-Ramırez, J. Am. Chem. Soc. 129 (2007) 355–360.
[258] D. Verboekend, K. Thomas, M. Milina, S. Mitchell, J. Perez-Ramırez, J.-P. Gilson, Catal. Sci. Technol. 1 (2011) 1331–1335.
[259] S.V. Konnov, Yu.V. Monakhova, E.E. Knyazeva, V.V. Yushchenko, O.A. Ponomareva, I.I. Ivanova, Petrol. Chem. 49 (1) (2009) 79–85.
[260] S.V. Konnov, I.I. Ivanova, O.A. Ponomareva, V.I. Zaikovskii, Microporous Mesoporous Mater. 164 (2012) 222–231.
[261] H. Hu, Q. Zhang, J. Cen, X. Li, Catal Lett 145 (2015) 715–722.
[262] S.-T. Tsai, C.-H. Chen, T.-C. Tsai, Green Chem. 11 (2009) 1349–1356.
[263] P.H. Chao, S.-T. Tsai, S.-L. Chang, I. Wang, T.-C. Tsai, Top Catal 53 (2010) 231–237.
[264] B.K. Modhera, M. Chakraborty, H.C. Bajaj, P.A. Parikh, Catal Lett 141 (2011) 1182–1190.
[265] M.N. Akhtara, N. Al-Yassira, S. Al-Khattafa, J. Cejka, Catal. Today 179 (2012) 61–72.
[266] R. Monteiro, C.O. Ania, J. Rocha, A.P. Carvalho, A. Martins, Appl. Catalysis A 476 (2014) 148–157.
[267] B. Notari, Adv. Catal. 41 (1996) 253–334.
[268] T.A. Nijhuis, B.J. Huizinga, M. Makkee, J.A. Moulijn, Ind. Eng. Chem. Res. 38 (1999) 884–891.
[269] B. Taylor, J. Lauterbach, W.N. Delgass, Catal. Today 23 (2007) 50–58.
[270] I. Moreno, N.F. Dummer, J.K. Edwards, M. Alhumaimess, M. Sankar, R. Sanz, P. Pizarro, D.P. Serrano, G.J. Hutchings, Catal. Sci. Technol. 3 (2013) 2425–2434.
[271] T. He, Y. Wang, P. Miao, J. Li, J. Wua, Y. Fang, Fuel 106 (2013) 365–371.
[272] C. Ju, Y. Wang, Y. Huang, Y. Fang, Fuel 154 (2015) 80–87.
[273] M.E. Dry, Catal. Today 71 (2002) 227–241.
[274] G.L. Bezemer, J.H. Bitter, H.P.C.E. Kuipers, H. Oosterbeek, J.E. Holewijn, X. Xu, F. Kapteijn, A.J. van Dillen, K.P. de Jong, J. Am. Chem. Soc. 128 (2006) 3956–3964.
[275] R.A. Friedel, R.B. Anderson, J. Am. Chem. Soc. 72 (1950) 1212–1215.
[276] N. Tsubaki, Y. Yoneyama, K. Michiki, K. Fujimoto, Catal. Commun. 4 (2003) 108–111.
[277] A. Martınez, J. Rollan, M.A. Arribas, H.S. Cerqueira, A.F. Costa, E.F.S. Aguiar, J. Catal. 249 (2007) 162–173.
[278] S. Sartipi, K. Parashar, M.J. Valero-Romero, V.P. Santos, B. van der Linden, M. Makkee, F. Kapteijn, J. Gascon, J. Catal. 305 (2013) 179–190.
[279] S. Sartipi, K. Parashar, M. Makkee, J. Gascon, F. Kapteijn, Catal. Sci. Technol. 3 (2013) 572–575.
[280] S. Sartipi, M. Alberts, M.J. Meijerink, T.C. Keller, J. Pérez-Ramirez, J. Gascon, F. Kapteijn, ChemSusChem 6 (2013) 1646–1650.
[281] C. Xing, J. Sun, G. Yang, W. Shen, L. Tan, P. Zhu, Q. Wei, J. Li, M. Kyodo, R. Yang, Y. Yoneyama, N. Tsubaki, Fuel Process Technol 136 (2015) 68–72.
[282] C. Xing, W. Shen, G. Yang, R. Yang, P. Lu, J. Sun, Y. Yoneyama, N. Tsubaki, Catal. Commun. 55 (2014) 53–56.
[283] K. Cheng, L. Zhang, J. Kang, X. Peng, Q. Zhang, Y. Wang, Chem. Eur. J. 21 (2015) 1928–1937.
[284] K. Cheng, J. Kang, S. Huang, Z. You, Q. Zhang, J. Ding, W. Hua, Y. Lou, W. Deng, Y. Wang, ACS Catal. 2 (2012) 441–449.
[285] J. Sun, X. Li, A. Taguchi, T. Abe, W. Niu, P. Lu, Y. Yoneyama, N. Tsubaki, ACS Catal. 4 (2014) 1–8.

[286] X. Peng, K. Cheng, J. Kang, B. Gu, X. Yu, Q. Zhang, Y. Wang, Angew. Chem. Int. Ed. 54 (2015) 4553–4556.

[287] C. Xing, G. Yang, M. Wu, R. Yang, L. Tan, P. Zhu, Q. Wei, J. Li, J. Mao, Y. Yoneyama, N. Tsubaki, Fuel 148 (2015) 48–57.

[288] K.A. Chalupka, S. Casale, E. Zurawicz, J. Rynkowski, S. Dzwigaj, Microporous Mesoporous Mater. 211 (2015) 9–18.

[289] M. Choi, K. Na, J. Kim, Y. Sakamoto, O. Terasaki, R. Ryoo, Nature 461 (2009) 246–260.

[290] W. Chaikittisilp, Y. Suzuki, R.R. Mukti, T. Suzuki, K. Sugita, K. Itabashi, A. Shimojima, T. Okubo, Angew. Chem. Int. Ed. 52 (2013) 3355–3359.

[291] X. Zhang, D. Liu, D. Xu, S. Asahina, K.A. Cychosz, K.V. Agrawal, Y.A. Wahedi, A. Bhan, S.A. Hashimi, O. Terasaki, M. Thommes, M. Tsapatsis, Science 336 (2012) 1684–1687.

[292] J.-C. Kim, S. Lee, K. Cho, K. Na, C. Lee, R. Ryoo, ACS Catal. 4 (2014) 3919–3927.

[293] N. Jiang, G. Yang, X. Zhang, L. Wang, C. Shi, N. Tsubaki, Catal. Commun. 12 (2011) 951–954.

[294] J. He, Z. Liu, Y. Yoneyama, N. Nishiyama, N. Tsubaki, Chem. Eur. J. 12 (2006) 8296–8304.

[295] G. Yang, H. Kawata, Q. Lin, J. Wang, Y. Jin, C. Zeng, Y. Yoneyama, N. Tsubaki, Chem. Sci. 4 (2013) 3958–3964.

[296] X. Li, J. He, M. Meng, Y. Yoneyama, N. Tsubaki, J. Catal. 265 (2009) 26–34.

[297] B. Sun, G. Yu, J. Lin, K. Xu, Y. Pei, S. Yan, M. Qiao, K. Fan, X. Zhang, B. Zong, Catal. Sci. Technol. 2 (2012) 1625–1629.

[298] F. Chen, W. Jin, D.-G. Cheng, X. Zhan, Y.S. Lin, J ChemTechnol Biotechnol 88 (2013) 2133–2140.

[299] Y. Wu, Y. Chai, J. Li, H. Guo, L. Wen, C. Liu, Catal. Commun. 64 (2015) 110–113.

[300] X. Li, Y. Zhang, M. Meng, G. Yang, X. San, M. Takahashi, N. Tsubaki, J. Membr. Sci. 347 (2010) 220–227.

[301] E.A. Khan, A. Rajendran, Z. Lai, Ind. Eng. Chem. Res. 49 (2010) 12423–12428.

[302] N. Nishiyama, K. Ichioka, D.-H. Park, Y. Egashira, K. Ueyama, L. Gora, W. Zhu, F. Kapteijn, J.A. Moulijn, Ind. Eng. Chem. Res. 43 (2004) 1211–1215.

[303] P. Collier, S. Golunski, C. Malde, J. Breen, R. Burch, J. Am. Chem. Soc. 125 (2003) 12414–12415.

[304] J. Zhou, X. Zhang, J. Zhang, H. Liu, L. Zhou, K. Yeung, Catal. Commun. 10 (2009) 1804–1807.

[305] X. Wang, Y. Cui, Y. Wang, X. Song, J. Yu, Inorg. Chem. 52 (2013) 10708–10710.

[306] N. Ren, Y.-H. Yang, J. Shen, Y.-H. Zhang, H.-L. Xu, Z. Gao, Y. Tang, J. Catal. 251 (2007) 182–188.

[307] Y. Wang, A. Tuel, Microporous Mesoporous Mater. 102 (2007) 80–85.

[308] Y. Wang, A. Tuel, Microporous Mesoporous Mater. 113 (2008) 286–295.

[309] D. Fodor, L. Pacosova, F. Krumeich, J.A. van Bokhoven, Chem. Commun. 50 (2014) 76–78.

[310] J.C. Groen, T. Bach, U. Ziese, A. Donk, K.P. de Jong, J.A. Moulijn, J. Perez-Ramirez, J. Am. Chem. Soc. 127 (2005) 10792–10793.

[311] C. Dai, A. Zhang, L. Li, K. Hou, F. Ding, J. Li, D. Mu, C. Song, M. Liu, X. Guo, Chem. Mater. 25 (2013) 4197–4205.

[312] S. Li, L. Burel, C. Aquino, A. Tuel, F. Morfin, J.-L. Rousset, D. Farrusseng, Chem. Commun. 49 (2013) 8507–8509.

[313] S. Li, T. Boucheron, A. Tuel, D. Farrusseng, F. Meunier, Chem. Commun. 50 (2014) 1824–1826.

[314] C. Dai, X. Li, A. Zhang, C. Liu, C. Song, X. Guo, RSC Adv. 5 (2015) 40297–40302.

[315] S. Li, C. Aquino, L. Gueudré, A. Tuel, Y. Schuurman, D. Farrusseng, ACS Catal. 4 (2014) 4299–4303.

[316] C. Mei, Z. Liu, P. Wen, Z. Xie, W. Hua, Z. Gao, J. Mater. Chem. 18 (2008) 3496–3500.

[317] C. Dai, A. Zhang, J. Li, K. Hou, M. Liu, C. Song, X. Guo, Chem. Commun. 50 (2014) 4848–4850.

[318] S. Li, A. Tuel, F. Meunier, D. Farrusseng, J. Catal. 332 (2015) 25–30.

[319] S. Li, A. Tuel, D. Laprune, F. Meunier, D. Farrusseng, Chem. Mater. 27 (2015) 276–282.

[320] D. Fodor, T. Ishikawa, F. Krumeich, J.A. van Bokhoven, Adv. Mater. 27 (2015) 1919–1923.

Chapter 12

Metal Catalysts Intercalated in Smectite Clays

Bing Sun*, Farooq-Ahmad Khan†, Georg Süss-Fink* and Bruno Therrien*

**University of Neuchâtel, Neuchâtel, Switzerland, †Lakehead University, Thunder Bay, ON, Canada*

Contents

1 AIM AND SCOPE

This chapter deals with the design of metal catalysts intercalated in smectite clays and their applications in the field of heterogeneous catalysis. Smectite clay characteristics such as molecular structure, chemical composition, and their properties including ion exchange, acidity, and swelling are discussed to give the necessary background for this type of catalyst support. Recent advances in the synthesis of smectite intercalated metal catalysts are reviewed in detail. The last section of this chapter focuses on the catalytic reactions by this type of catalyst in the preparation of fine chemicals. The goal of this chapter is to systematically review the recent progress of our knowledge in this particular field.

Encapsulated Catalysts. http://dx.doi.org/10.1016/B978-0-12-803836-9.00012-2

2 GENERAL INTRODUCTION

In 1936, a smectite clay (montmorillonite) was successfully prepared as the first hydrotreating catalyst (Houdry process). For decades, these montmorillonite-based catalysts were used in various petrochemical processes such as fluid catalytic cracking (FCC). At that time, they were essentially acid-treated clays [1]. In the early sixties, large pore Y-zeolites replaced clay-based catalysts in the petrochemical industry due to their efficiency and selectivity [2]. However, the preparation of an FCC catalyst still required large quantities of clay minerals along with zeolites for better physical properties of the catalyst [1]. By the end of the 20th century, clays once again gained the attention of researchers due to their remarkable range of functionality arising from the newly discovered pillaring process [3]. Large-scale production of fine chemicals is usually done via stirred tank reactors employing homogeneous catalysts and producing huge volumes of hazardous waste. New clean technologies such as heterogenization have a proven track record of minimizing waste production [4]. Generally, heterogeneous catalysts are solid compounds, and their separation from the reaction mixture is done in a straightforward manner via filtration instead of extraction—the source of much of the hazardous waste. Moreover, the water quench phase becomes unnecessary also [5]. Sometimes a heterogeneous catalyst can also promote a solventless process by acting as a solid solvent itself [6]. The selectivity of a heterogeneous catalyst can be controlled by tuning the nanoparticle dimensions and the nature (e.g., pore constraints) of the catalyst support [7]. This highly efficient and green technology requires versatile catalysts to suit a great number of chemical processing technologies [5]. Zeolites offer numerous advantages, but they are limited in terms of pore sizes (<10 Å), resulting in poor diffusion of compounds across the channels [8]. This is good for several petrochemical processes but problematic because the molecular dimension of products and reactants increases [4,5]. Unlike zeolites, mesoporous silica provides appropriate pore sizes, but it lacks chemical activities [8]. Similarly, the metal-organic frameworks (MOF) may not be chemically and thermally stable enough to be useful. Clay-based catalysts have a different but interesting set of properties. These smart materials have "soft" dimensional constraints and display remarkable regio- or shape selectivity in numerous organic reactions [9]. They have a considerable potential in the fine chemical process industry. However, clays were previously ignored as heterogeneous support due to the wide variability of natural clays. This has hampered the generation of a predictive and reliable set of data from earlier catalytic studies [10]. Little attention towards the engineering aspect of catalyst design to help them survive through the rigorous reactor conditions is another reason for limited development in this area [9]. Early work has focused instead on acid treatment of clays to exploit the Lewis-type acidity or performing cation exchange to boost the Brønsted acidity in clays [11]. However, the new "nano" age marked by the manipulation of catalytically active species at an ultra-small scale within clays has widened the reaction portfolio to a staggering degree [12,13].

Higher selectivity and improved efficiency from nanoparticle catalysts with the advantages of heterogeneous clay supports having "green" credentials are the main driving forces behind this rapidly expanding field [12]. Clay-based catalysts are now recognized as successful in producing the intended results for certain organic transformations [13,14] such as redox reactions, reactions involving reactive intermediates, and cycloadditions (such as Diels-Alder reactions).

2.1 What Is a Smectite Clay?

Clays are phyllosilicates ("phyllo" meaning sheet), and they are generally classified into three main groups: kaolin, micas, and smectites [15]. Smectites ("smectos" meaning soap) are expandable (swelling) clays. They are made up of parallel aluminosilicate sheets, where the particle size is in the micron range. Smectites have two basic building blocks [15]:

1. Silicon-oxygen tetrahedra
2. Magnesium-oxygen octahedral (brucite) or aluminum-oxygen octahedra (gibbsite)

The basic structure of smectite clays contains interconnected six-membered rings of SiO_4^{4-} tetrahedra that extend outward in infinite sheets. Three out of the four oxygens from each tetrahedra are shared with other tetrahedra [16,17]. This leads to a structural unit of $Si_2O_5^{2-}$. The corresponding tetrahedron and the tetrahedral layers are depicted in Fig. 12.1 [16].

Interconnected octahedra also form layers. Smectite clays have two types of octahedral sheets (Fig. 12.1) [16–18]:

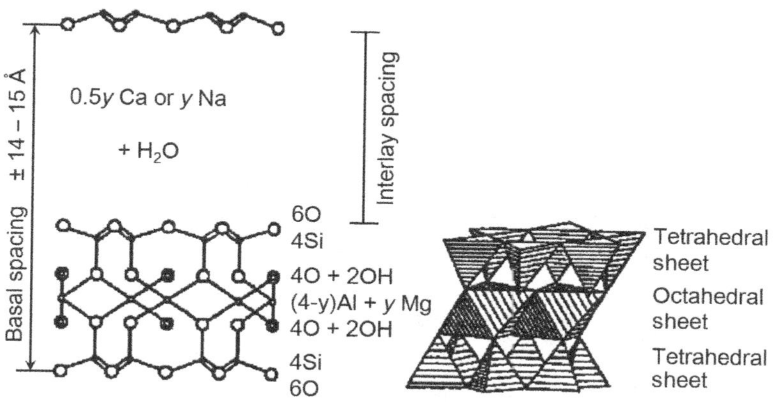

FIG. 12.1 Silicon-oxygen tetrahedra and the interconnected tetrahedral sheet found in smectite clays. *(Reprinted with permission from J.T. Kloprogge, J. Porous Mater. 5 (1998) 5–41, Copyright Springer.)*

- dioctahedral sheets, where trivalent Al^{3+} or Fe^{3+} is the dominant central atom in each octahedron
- trioctahedral sheets, where divalent Mg^{2+} or Fe^{2+} occupy the central site of the octahedron

Overall, the smectite group of clays has one octahedral layer sandwiched between two layers of tetrahedral layers. The way in which the layers come together results in many unsatisfied valences (charges) that can attract ions and dipolar molecules. Ion exchange can occur within the lattice layers and interlayer [17,18]. The most common members of the smectite clay family are montmorillonite, hectorite, saponite, nontronite, and beidellite [18].

Hectorite is a trioctahedral smectite with a brucite octahedral layer [19]. It has a three-layer sheet-like morphology that results from the two-dimensional condensation of silicic acid, two layers of SiO_4 tetrahedra being connected by a layer of MgO_6 octahedra. Partial replacement of the Mg^{2+} cations in the octahedral layers by Li^+ cations leads to an excess of anionic charges of the layers, which are compensated by Na^+ cations in the interlaminar space (Fig. 12.2). Hydratization of the interlaminar sodium cations to give $[Na(H_2O)_n]^+$ is responsible for the swelling of hectorite in water, since the interlaminar space is widened [18,19]. Contrary to hectorite, montmorillonite and beidellite are dioctahedral smectites with a gibbsite octahedral layer, and in nontronite the octahedral sheet is predominantly iron hydroxide [18]. The shapes of smectite clay particles vary from plate-like (montmorillonite, beidellite) to lath-like (hectorite, nontronite), and they are all highly polydisperse [20]. Smectite clay particles range in size from 0.1–2 mm (except for the synthetic laponites, which are much smaller, with diameters of approximately 25–40 nm) [20,21].

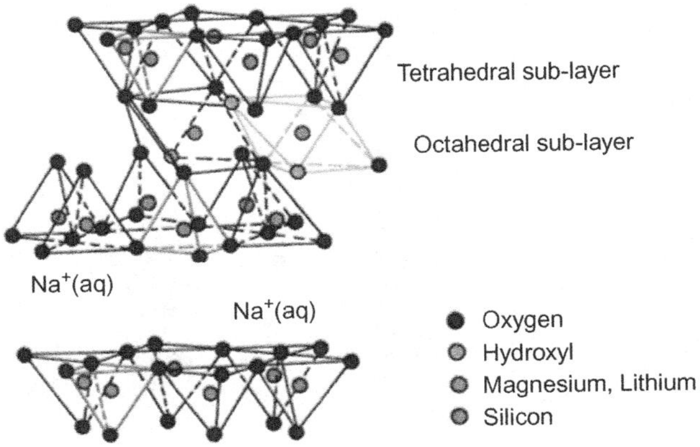

FIG. 12.2 Structural model according to Grim [18] of sodium hectorite $Mg_{5.5}Li_{0.5}Si_8O_{20}(OH)_4$ $Na \cdot nH_2O$, showing the anionic three-layer sheets and the interlaminar space containing sodium cations and water molecules.

2.2 Classification and Properties of Smectite Clays

Smectites are 2:1 type clays with 10–15 Å interlayer spacing [17]. Their general formula can be expressed as follows:

$$(Ca, Na, H)(Al, Mg, Fe, Zn)_2(Si, Al)_4O_{10}(OH)_2 \cdot xH_2O$$

Calcium, sodium and hydrogen hold an optional position in the final equation above. These elements are positive ions and the negatively charged particles of the clay attract ions [22].

Smectite clays can be divided into four subclasses [17,22] depending on the following factors:

(a) The type of octahedral layer that can be either dioctahedral or trioctahedral
(b) The predominant location of layer charge either on octahedral or tetrahedral sites

Examples of dioctahedral smectites include montmorillonite and beidellite. They have the layer charge predominantly in octahedral and tetrahedral sites, respectively [23]. Hectorite and saponite are considered trioctahedral smectites with the layer charge present in octahedral and tetrahedral sites, respectively [22,23]. Hectorite, montmorillonite, and saponite are the best known smectites, being widely studied for catalysis and used in various chemical processes.

In the general formulas in Scheme 12.1, x and y refer to layer charges due to substitution in tetrahedral and octahedral sites, respectively; R^{2+} and R^{3+} stand

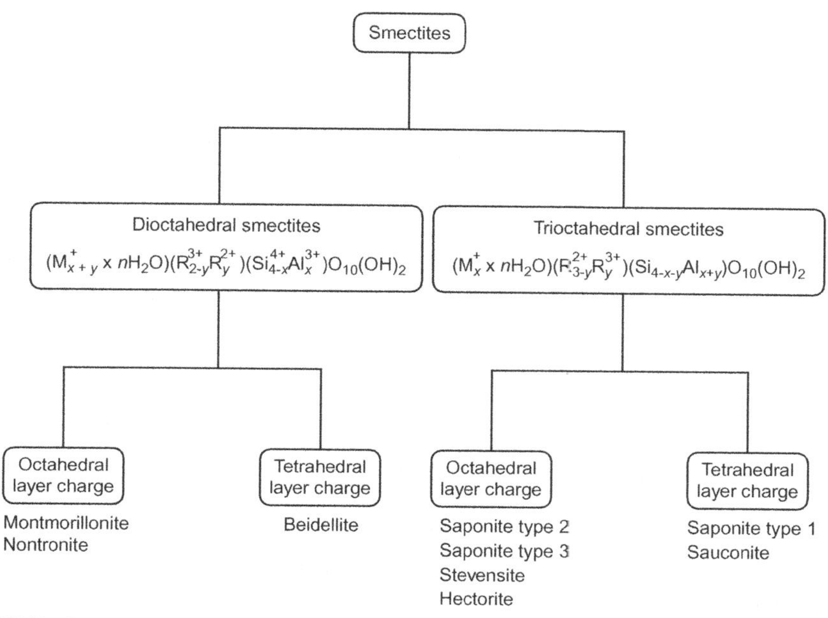

SCHEME 12.1 Classification of smectite clays [22–24].

for divalent and trivalent octahedral cations, respectively; and M^+ refers to a monovalent interlayer cation [22–24].

These clays possess a combination of exchangeable cations, intercalation, and swelling properties that makes them unique as compared to other catalyst supporting materials [25].

2.2.1 Hydration and Swelling

Smectite clays are hydrophilic in nature and possess a remarkable tendency to absorb water that pulls the clay layers apart beyond the intercalant size [26]. This interlayer separation can be measured using powder x-ray diffraction (XRD). The extent of this interlayer separation in the clays depends on the following factors [27]:

1. Layer charge
2. Exchange cations
3. Swelling agent

For example, Na^+ and Li^+ exchange cations in smectite layers are particularly susceptible to hydration in an aqueous medium, thereby becoming detached from the clay surface. In contrast, K^+ ions show reluctance to fully hydrate, which results in reduced tendency of K^+-saturated clays to expand [27]. Lamellar distance in clay supported catalysts can also be adjusted by the careful selection of reaction medium, which can ensure an increased stereoselectivity from the catalyst [28,29].

2.2.2 Ion-Exchange Capacity

The lattice of smectite clays has a residual negative charge balanced by interlayer cations. These loosely bound ions can be exchanged against external cations present in an aqueous medium [27]. The total capacity of clay to hold exchangeable cations is called the cation exchange capacity (CEC); it is measured in milliequivalents per 100 g of clay. Smectites usually have a high CEC due to their highest concentration of interlayer cations [17,30].

2.2.3 Acidity

Untreated smectites have the ability to donate protons and to catalyze organic reactions. The acidic properties of these clays arise from the dissociation of interlayer water molecules coordinated to the exchangeable cations [31]. It has been found that the H^+ ion concentration in the interlayer regions of these clays may be $\sim$10 mol dm^{-3} [32]. The extent of Brønsted acidity in smectite clays depends on many factors such as the dissociation of interlayer exchangeable cations and the degree of hydration of the clays. A greater polarizing ability of the interlayer cations increases both the strength and number of acid sites per

H_0 value (the proton-donating ability) [33]. Brønsted acid strength depends on the polarizing power of the interlayer exchangeable cations in the following order [34]:

$$H^+ > Al^{3+}, Fe^{3+} > Mg^{2+} > Ca^{2+} > Na^+ > K^+$$

Lewis acid sites are usually derived from iron or aluminum species located at the periphery of the crystal [17]. Halide- or sulfide-containing inorganic or organometallic complexes intercalated in smectite clays can also act as Lewis acids, if the clay layers are devoid of absorbed water (usually done at 600°C during the pillaring process). For this type of smectite catalysts, the Lewis acidity predominates if the reaction is performed in anhydrous solvent [17,35].

2.2.4 Surface Area and Porosity

The specific surface area (SSA) is an essential microstructural property. It is the quantitative assessment of the total area at disposal for surface reactions. Smectites have high effective surface area (650–800 m^2/g) [17]. Porosity in smectite clays can be modulated (1) by varying the size, shape, and concentration of the intercalated species; (2) by modifying the interlayer charge density; and (3) by partial exchange of inorganic cations [36].

2.3 Catalytic Activity of Smectite Clays

Smectite clays are capable of catalyzing many reactions. Their catalytic properties are governed by surface properties, type of exchangeable cations, and Lewis or Brønsted acidity [37]. Lewis acidity in smectite clays can be introduced via $ZnCl_2$, $FeCl_3$, $AlCl_3$, $TiCl_4$, or $SnCl_4$ treatment. These clays also can act as heterogeneous Brønsted acids after treatment with mineral acids such as HCl, HF, H_2SO_4, or the introduction of high charge density ions such as Cr^{3+} [33]. Such modifications enable them to behave as heterogeneous acid catalysts for reactions that are normally catalyzed by homogeneous Lewis or Brønsted acids [11]. They proved to be useful for many organic reactions such as Friedel-Crafts reactions. Their high surface area also renders them an efficient support for catalytically active species such as organometallic complexes or metallic nanoparticles. The catalytic activity of clay supports can be attributed to two decisive factors: the local concentration effect due to the adsorption of reactants on a clay surface and the low dimensionality responsible for high probability of reactant molecules meeting each other [33,37].

In conclusion, catalytic activity and reactivity pattern within smectite clay layers can be influenced and in part controlled through the choice of the following variables [33]:

1. Brønsted acidity,
2. Lewis acidity,
3. Presence of redox active species,

4. Choice of clay layer charge,
5. Introduction of catalytically active species (mainly transition metals),
6. Swelling solvents.

2.4 Smectite Clay Modifications for Catalysis

Naturally occurring smectite clays may not be efficient catalysts. Shaping the properties such as acidity, porosity, thermal stability, and mechanical strength of the catalyst can be done by clay modifications such as acid activation, layer charge manipulation, intercalation, pillaring, organic modification, and chemical grafting [17–20].

2.4.1 Acid Treatment

Smectite clay properties can be modified by their treatment with mineral acid solutions. The acidic attack improves the overall surface area, porosity, and the number of acidic sites in the clay [38b]. When smectite clays, e.g., montmorillonite are treated with hot mineral acid solutions such as HCl or H_2SO_4, the following changes occur in the clay [38]:

1. Interlayer exchangeable cations such as Na^+ or Ca^{2+} are replaced with H^+,
2. Cations from octahedral sheets (Mg^{2+}, Fe^{3+}, Al^{3+}) of the clay are leached out,
3. Staking of clay platelets is lost (delamination),
4. Noncrystalline silica with a three-dimensional cross-linked porous structure is formed.

The extent of dissolution reaction depends on many factors such as acid/clay ratio, reaction conditions, and the nature of octahedral cations [38d]. Aluminum-rich dioctahedral smectites dissolve slowly when compared to the magnesium-rich trioctahedral smectites [39]. In montmorillonite, the substitution of octahedral aluminum cations by magnesium cations improves their dissolution rate in the acids [39d]. A 1-h 50% sulfuric acid treatment of Wyoming bentonite usually results in little structural decompositions but relatively high acidity and catalytic activities [39e]. This bentonite clay is essentially composed of montmorillonite with high octahedral aluminum content. Acid-treated montmorillonites are solid sources of protons and act as heterogeneous acid catalysts in many organic reactions such as Friedel-Crafts alkylation, acylation, and isomerization [39]. Montmorillonite K-10 and montmorillonite KSF are commercially available acid-modified clays [39f]. Both are highly efficient catalysts being extensively used in industry. K-10 possesses a high surface area ($\sim$250 $m^2\ g^{-1}$) as compared to KSF (10 $m^2\ g^{-1}$) [40]. Acid-treated clays have also been chosen to immobilize other catalytically active species. For example, clayfen is an $Fe(NO_3)_3$-containing K-10 montmorillonite clay that is used as a mild, inexpensive NO^+ source and oxidizes aromatic rings to radical cations [41]. Claycop is another type of acid catalyst obtained by $Cu(NO_3)_2$ deposition on K-10 bentonite clay [42].

2.4.2 Layer Charge Manipulations

In dioctahedral-type smectite clays, Fe^{3+} is frequently found in the octahedral sheets. Reduction of Fe^{3+} to Fe^{2+} has a profound effect on the layer charge, which may subsequently improve many clay properties such as cation exchange capacity, specific surface area, and expandability [43a]. However, Fe^{2+} in smectites readily reoxidizes in air, but this can be annihilated by Li^+ fixation. This is another method of modifying the layer charge in clays. When dioctahedral smectites are heated in the presence of Li^+ cations, small quantities of these ions enter into the layer structure and decrease the layer charge accordingly [43], thus modifying the clay properties such as swelling ability and cation exchange capacity.

2.4.3 Intercalation

Intercalation is the insertion of ionic complexes in the silicate lamellae of clays. Intercalation can be verified by XRD patterns, showing an increase in the basal spacing of the layers [44a]. This is one of the most common clay alteration techniques in view of catalytic applications. Smectite-type clays are capable of accommodating different kinds of guest molecules via adsorption in their interlayer space (intercalation). The intercalation is a simple process usually achieved by dispersing the clay in a solution with an appropriate amount of guest ions [44]. This ion exchange against interlayer alkali ions depends on the CEC of the smectite clay. The cationic nature of the guest facilitates its electrostatic binding with polyanionic clay layers [44b]. Intercalated smectite catalysts are generally prepared either by indirect assembly of a catalyst complex in situ or a catalyst complex is directly introduced between the clay layers by ion exchange. For example, μ-peroxo Rh(II) complexes have been indirectly prepared by oxygenation of cationic $[Rh(PPh_3)_3]^+$ species trapped in the lamellae of montmorillonite [45]. Similarly, the monohydrido complex $[RhH(PPh_3)_3]^{2+}$ can be indirectly made by the interaction of molecular hydrogen with $[Rh(PPh_3)_3]^+$ immobilized in the interlamellar space of montmorillonite [46]. Pinnavaia et al. reported an indirect assembly of $RhH_2(PPh_3)_n{}^+$-montmorillonite in situ by reacting triphenyl phosphine with $Rh_2(OAc)^{x+}{}_{4-x}$ pre-intercalated montmorillonite. The method has several disadvantages, notably the pre-intercalated $Rh_2(OAc)^{x+}{}_{4-x}$ ions are partially accessible for reaction with the free ligand (PPh_3) due to the shielding of reactive sites inside the lamellar silicate layers [47]. Therefore, direct ion-exchange is generally preferable for preparing clay-intercalated metal complexes.

Shimazu et al. used lithium hectorite as a host for several palladium (II) complexes, such as $[Pd(NH_3)_4]^{2+}$ and $[Pd(PDT)_3]^{2+}$ (where PDT is 3-(2-pyridyl)-5,6-diphenyl-1,2,4-triazine) [48]. A direct exchange between lithium cations and the cationic complexes was observed. These intercalation complexes were used as catalysts for the hydrogenation of alkynes [48]. Giannelis and Pinnavia reported the first known example of an intercalated metal cluster catalyst $H_2Os_3(CO)_{10}(P—P^+)$-hectorite, which was formed by thermal decarbonylation of $H_2Os_3(CO)_{10}(P—P^+)$-hectorite. They also demonstrated the ion-exchange reaction of Na^+-hectorite with different cationic cluster complexes such as $Ru_3(CO)_9(P—P^+)_3$ $H_4Ru_4(CO)_8(P—P^+)_4$, $Ir_4(CO)_9(P—P^+)_3$, $OS_3(CO)_{11}(P—P^+)$ and $H_2Os_3(CO)_{10}(P—P^+)$, where $P—P^+$ is the phosphonium-phosphine ligand $Ph_2P(CH_2)_2P^+Ph_2(CH_2Ph)$, resulting in the formation of cluster-clay intercalation compounds with d_{001} spacings in the range of 22–27 Å and gallery heights of 12.5–17.5 Å [49]. Later on, Uematsu and coworkers demonstrated the first successful intercalation of ruthenium in hectorite using $[Ru(NH_3)_6]Cl_3$ as a catalyst precursor. The synthesized precatalyst was reduced under an H_2 atmosphere to obtain ruthenium nanoparticles intercalated in hectorite, which shows high selectivity for the hydrogenation of olefins [50].

Organic compounds have also been intercalated in smectite clays. For example, a chiral organocatalyst, (5*S*)-2,2,3-trimethyl-5-phenylmethyl-4-imidazolidinone monohydrochloride, has been successfully trapped in montmorillonite via cation exchange reaction; see Fig. 12.3 [51]:

The preparation of highly dispersed metal nanoparticles on a support has attracted much attention in the last decade [52]. The intercalation of metal ions as a catalyst precursor to a support often relates to a synthetic methodology, which requires a proper transformation of the metal ions to an active form [53].

2.4.4 Pillaring

Smectite clays such as hectorite and montmorillonite suffer physical collapse in harsh reaction conditions. At a high temperature, c.470 K, interlayer species are forced out and the clay lamellae coalesce with consequent loss of catalytic activity [54]. One way of preventing this collapse is by inserting thermally stable and robust inorganic moieties (pillars) that serve to keep the individual clay layers apart [54]. These pillars are usually generated via metalpolyoxycations such as a Keggin ion, $[Al_{13}O_4(OH)_{24}(H_2O)_{12}]^{7+}$ [16,55]. However, organic, organometallic, or inorganic complexes, preferably of a high positive charge, can also be employed as pillaring agents [16]. These positively charged cations

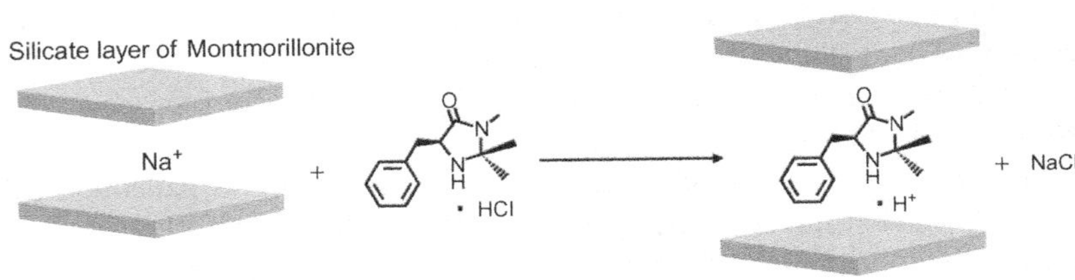

FIG. 12.3 Intercalation of an organic compound in montmorillonite lamella. *(Reproduced with permission from T. Mitsudome, K. Nose, T. Mizugaki, K. Jitsukawa, K. Kaneda, Tetrahedron Lett. 49 (2008) 5464–5466, Copyright 2008, Elsevier.)*

are intercalated between layers and then converted into stable metal oxide pillars through a calcination step. Pillars hold the aluminosilicate clay layers permanently apart. The generation of pillared clay results in creating extra porosity. The pores can be tailored by varying the size, shape, and concentration of the intercalated species [55].

Metal oxide pillared clays derived from the hydrolysis of Cr (III) hydroxides were found useful as cracking catalysts [56]. Pillared clays can also be employed as support for other transition metals catalysts. They are usually deposited on pillared clays by means of ionic exchange, incipient wetness, or dry impregnation of calcined samples [56b]. For example, Dékány and coworkers generated Pd^0 subcolloids in the size range of 2–10 nm by reduction of Pd^{2+} ions that were adsorbed on pillared montmorillonite (Figs. 12.4 and 12.5) [56c]. Alternatively, palladium nanoparticles can also be prepared in the interlamellar space of montmorillonite and saponite containing a pillaring agent [56d]. This alternative approach is summarized in Fig. 12.4.

2.4.5 Organic Modifications

Organic modifications of smectite clays facilitate the intercalation of metal nanoparticles by altering their swelling ability in organic solvents [57]. A suitable dispersion medium increases the lamellar basal spacing for guest species to an appreciable extent [58]. In the literature, there are usually two approaches for the preparation of metallic nanoparticles (M^0) containing organoclays. One is to prepare M^0/organoclay by adsorption of an organic unit such as quaternary ammonium surfactants on the M^0/clay, which is firstly prepared by means of cation exchange of M^{n+} for Na^+ followed by chemical reduction [59]. Alternatively, M^0/organoclay can also be prepared by the adsorption of M^{n+} on the organically modified clay, followed by a chemical reduction [59a,60].

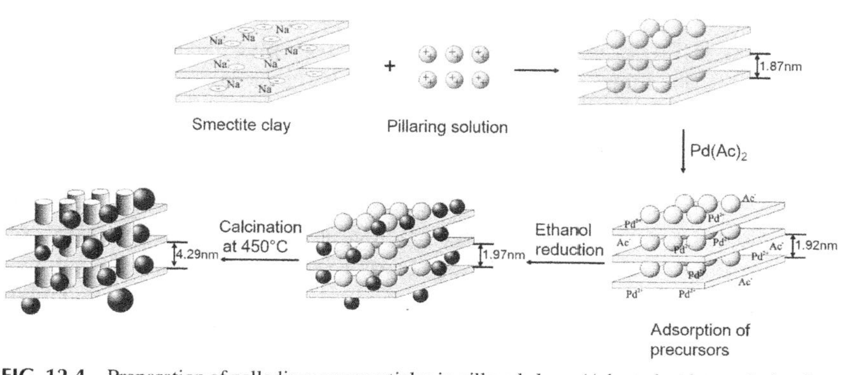

FIG. 12.4 Preparation of palladium nanoparticles in pillared clays. *(Adapted with permission from A. Szűcs, F. Berger, I. Dékány. Colloids Surf. A Physicochem. Eng. Asp. 174 (2000) 387–402, Copyright 2000, Elsevier.)*

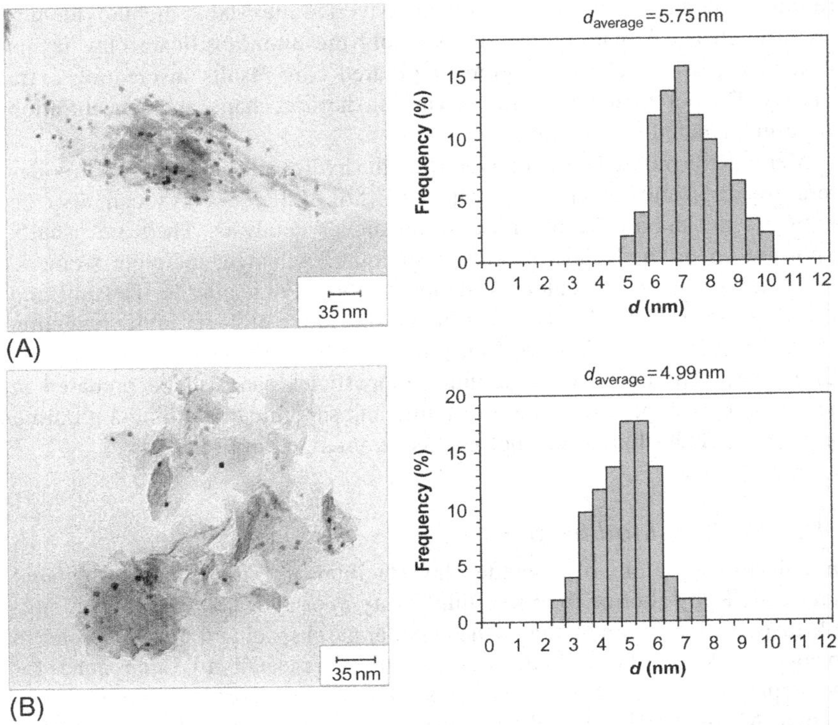

FIG. 12.5 TEM micrograph and particle size distribution of palladium nanoparticles in pillared montmorillonite (A) and pillared saponite (B). *(Adapted with permission from A. Szűcs, F. Berger, I. Dékány. Colloids Surf. A Physicochem. Eng. Asp. 174 (2000) 387–402, Copyright 2000, Elsevier.)*

For example, hectorite and montmorillonite were initially modified with chiral dihydrocinchonidine hydrochloride (DHCD·HCl) followed by their impregnation with H_2PtCl_6 and subsequent reduction to give enantioselective Pt nanoparticle containing catalysts [58b]. Similarly, cinchonidine has also been employed as a chiral modifier to prepare clay-supported Pt and Pd catalysts [61]. Bartók and coworkers have reported Pd^0 in the lamellae of an organoclay. The surfactant-modified clay was prepared by a cation exchange reaction between Na^+ montmorillonite and hexadecylammonium bromide. Palladium nanoparticles were then generated by ethanolic reduction of palladium(II) acetate in situ inside the interlamellar space of hexadecylammonium montmorillonite [29]. L. Zhou et al. used cetyltrimethylammonium bromide (CTAB) intercalated montmorillonite as a supporting matrix to prepare ruthenium catalyst containing worm-like ruthenium nanoparticles (Fig. 12.6) [59a].

Partially organically modified clay containing metallic nanoparticles have also been reported. For example, a hexamine rhodium complex precursor has been employed to synthesize partially organically modified montmorillonite containing metallic rhodium nanoparticles (Rh-POMM) as illustrated in

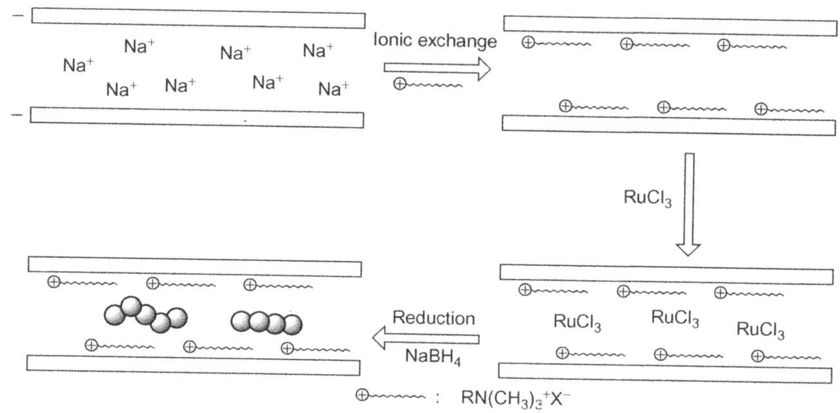

FIG. 12.6 Preparation of worm-like ruthenium nanoparticles in organically modified montmorillonite. *(Reproduced with permission from L. Zhou, X. Qi, X. Jiang, Y. Zhou, H. Fu, H. Chen, J. Colloid. Interface. Sci. 392 (2013) 201–205, Copyright 2012, Elsevier.)*

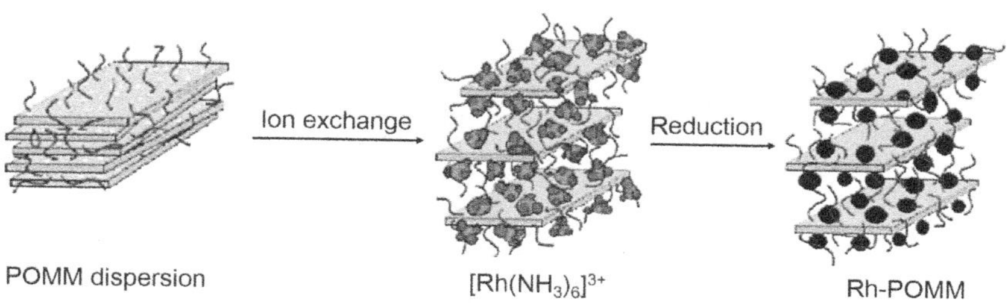

FIG. 12.7 Illustration showing generation of Rh-POMM. *(Adapted with permission from H.A. Patel, H.C. Bajaj, R.V. Jasra, J. Nanopart. Res. 10 (2008) 625–632, Copyright 2007, Springer.)*

Fig. 12.7 [62]. The frequently used quaternary ammonium cations to prepare organoclays are tetramethylammonium, trimethyl phenylammonium, benzyl trimethylammonium, and some other analogues [63].

Recently, organoclays containing dendrimers followed by in situ dispersion of metallic nanoparticles have been reported. Bolton polyol dendrimers were incorporated in montmorillonite to obtain dendrimers containing organoclays followed by in situ dispersion of Cu^0 and Pd^0 nanoparticles to produce metal organoclays (MOCs) (Fig. 12.8) [64].

3 NEW TRENDS IN SMECTITE-INTERCALATED METAL CATALYSTS

Nanoparticles are finely dispersed metal colloids (particle size 1–100 nm), which are highly reactive and thus interesting for catalytic purposes [65]. Their preparation, manipulation, and storage require protection and stabilization by suitable supports to prevent aggregation to bulk metal. Smectite clay minerals,

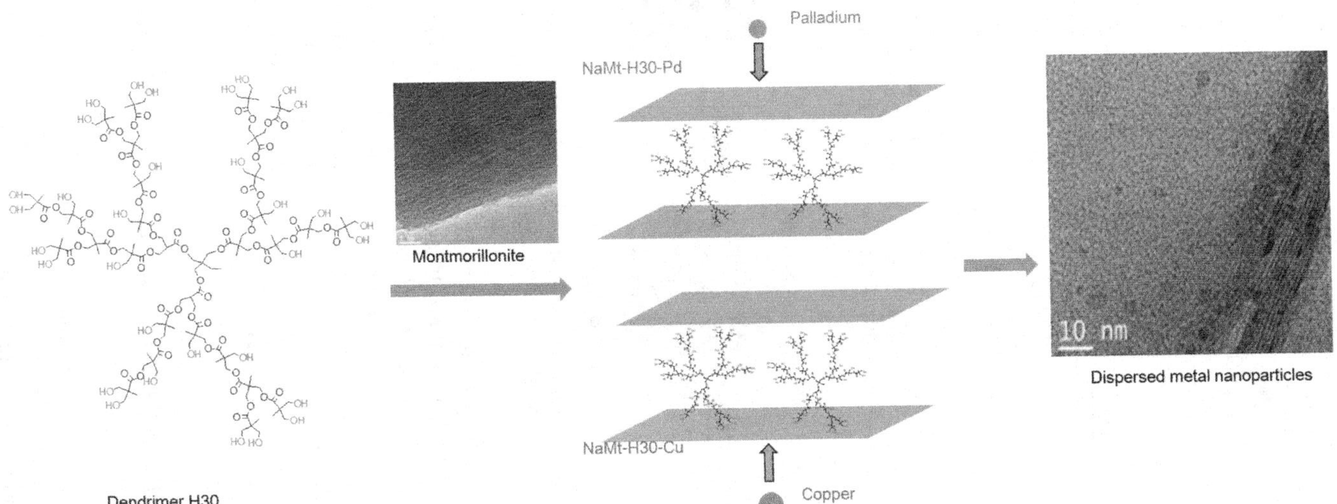

FIG. 12.8 Intercalation of metal nanoparticles in dendrimer containing montmorillonite. *(Reproduced with permission from N. Bouazizi, D. Barrimo, S. Nousir, R. Ben Slama, T.C. Shiao, R. Roy, A. Azzouz, J. Energy Inst. (2016), Copyright 2016, Elsevier.)*

featured by their unique swelling properties and flexible intercalation capacity, provide an ideal platform for immobilizing metal nanoparticles. The most studied smectite clays are montmorillonite and hectorite. In recent years, they have been widely used, as a matrix or as a host, for obtaining intercalated complexes and for stabilizing metal nanoparticles [65]. In clay minerals, the confined space between silicate lamellae has nanosized dimensions that limit the particles' growth during the Ostwald ripening process [66]. Steric constraints due to metal nanoparticles in smectite clay lamellae orient the reactant adsorption, thereby directing the reaction selectivity [67]. This section deals with recent advances in the preparation of smectite-intercalated metal catalysts mainly involving palladium, rhodium, ruthenium, platinum, gold, and copper. The effects of the synthetic method, the pretreatment of the support, and modification of the preparation as well as the choice of the metal precursors are discussed.

3.1 Palladium Catalysts

Although metallic nanoparticles are known for almost all metals, the vast majority of publications in the field is concerned with gold nanoparticles (the "New Gold Rush"), palladium being the second metal to be extensively studied in the form of nanoparticle containing catalysts [68]. The formation and properties of smectite supported palladium nanoparticles have been reported in detail by S. Papp, Dékány, Király, and Barthos and coworkers and by others [29,56c,69]. Palladium nanoparticles in clay minerals are often prepared by the reduction of an intercalated metal precursor, a method preferred due to its simplicity. For example, Király et al. developed a method of generating Pd^0 nanopaticles in the lamellae of a clay organocomplex. According to their method, the surfactant-modified clay was prepared by a cation exchange reaction between Na^+ montmorillonite and hexadecylammonium bromide. Palladium nanoparticles (2–12 nm) were then generated by ethanolic reduction of palladium(II) acetate in situ in the interlamellar space of hexadecylammonium montmorillonite [29]. This Pd-organoclay catalyst was found to be catalytically active for the olefin hydrogenation.

Well defined, stable and subnanoordered Pd clusters entrapped in clay lamellae have also been reported (Fig. 12.9). The Pd clusters were prepared by KBH_4 reduction of Pd^{2+} containing montmorillonite. The clusters were estimated to contain approximately ten Pd atoms, and their size remained unchanged after multiple catalytic runs [70].

Several organometallic complexes such as dichloro(1,5-cyclooctadiene)-palladium(II) or dichlorobis(benzonitrile)palladium(II) were also intercalated in montmorillonite to prepare Pd^0-clay catalysts for the carbonylation of 2-allylphenols [71]. Recently, Zeng and coworkers were capable of encaging palladium nanoparticles in chitosan modified montmorillonite to obtain recyclable catalysts (Fig. 12.10) [72].

Preformed palladium nanoparticles can also be intercalated in the silicate lamellae of smectite clay minerals. Immobilization of palladium nanoparticles

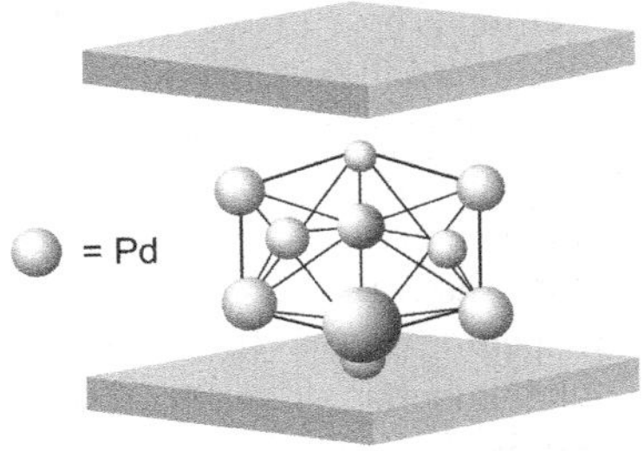

FIG. 12.9 Pd nanocluster containing only 10 atoms entrapped in montmorillonite lamellae. *(Reproduced with permission from T. Mitsudome, K. Nose, K. Mori, T. Mizugaki, K. Ebitani, K. Jitsukawa, K. Kaneda, Angew. Chem. Int. Ed. 46 (2007) 3288–3290, Copyright 2007, John Wiley and Sons.)*

in the organoclay host was achieved by the cation-exchange reaction between sodium montmorillonite and tetradecyltrimethylammonium bromide serving as a stabilizing agent in a palladium hydrosol (Fig. 12.11); the Pd-organoclay proved to be catalytically active in olefin hydrogenation in the liquid phase [73].

Laponite is a synthetic smectite clay. Laponite-supported palladium nanoparticles were obtained via impregnation of polyvinylpyrrolidone (PVP)-stabilized Pd nanoparticles in the clay (Fig. 12.12) [74].

Heterogenization of homogeneous palladium catalysts using smectite clay minerals as support has also been explored. For example, Shimazu et al. extensively studied intercalation of palladium complexes in hectorie and their catalytic properties. Lithium hectorite was employed as the host and several palladium (II) complexes such as $[Pd(NH_3)_4]^{2+}$ and $[Pd(PDT)_3]^{2+}$ where PDT is 3-(2-pyridyl)-5,6-diphenyl-1,2,4-triazine were used as guest compounds to obtain catalysts for the hydrogenation of alkynes [48]. Kitayama et al. prepared a sepiolite-containing $[Pd(NH_3)_4]^{2+}$ catalyst for the Suzuki coupling reaction in benign aqueous medium [75]. Bergman et al. successfully intercalated a well-defined cationic palladium complex into an organically modified synthetic silicate fluorohectorite. In situ polymerization of ethylene by the palladium catalyst was able to promote exfoliation to produce a polyethylene-silicate nanocomposite (Fig. 12.13). Delamination was confirmed by monitoring the reaction progress via powder XRD [76].

Park et al. intercalated the Wacker catalyst (palladium(II) chloride/copper(II) chloride) in montmorillonite and studied the catalytic activity of this material for many reactions such as low-temperature CO oxidation, carbonylation for the synthesis of ibuprofen or naproxen, oxidative carbonylation for the synthesis of diphenyl carbonate, and functionalization of methane [5]. Although the catalyst was transferred from the solution to the solid surface, the heterogenization had no significant effect on the reaction mechanisms of the Wacker catalyst. Recently, vanadia-pillared montmorillonite and titania-pillared laponite were also employed to immobilize the Wacker catalyst to

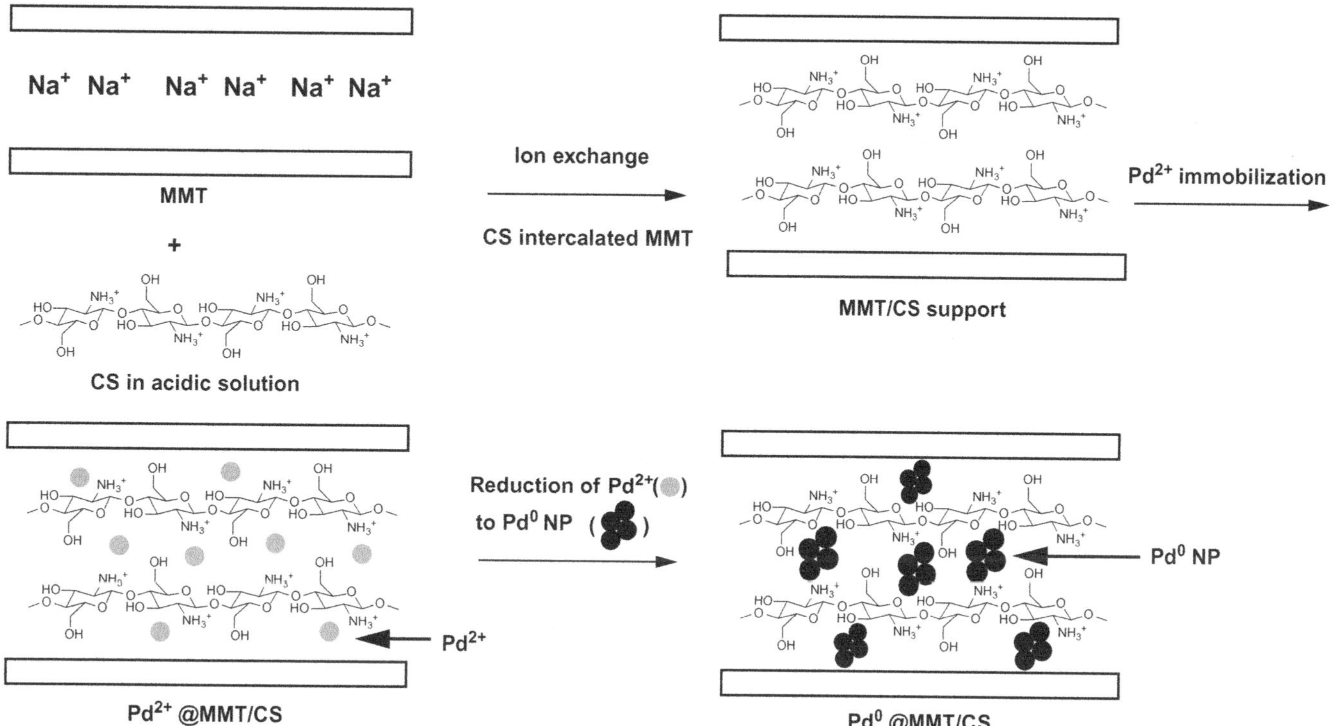

FIG. 12.10 Preparation of Pd^0 nanoparticles encaged in lamellae of chitosan-modified MMT (Pd@MMT/CS). *(Reprinted with permission from M. Zeng, Y. Wang, Q. Liu, X. Yuan, S. Zuo, R. Feng, J. Yang, B. Wang, C. Qi, Y. Lin, ACS Appl. Mater. Interfaces 8 (2016) 33157–33164, Copyright 2016, American Chemical Society.)*

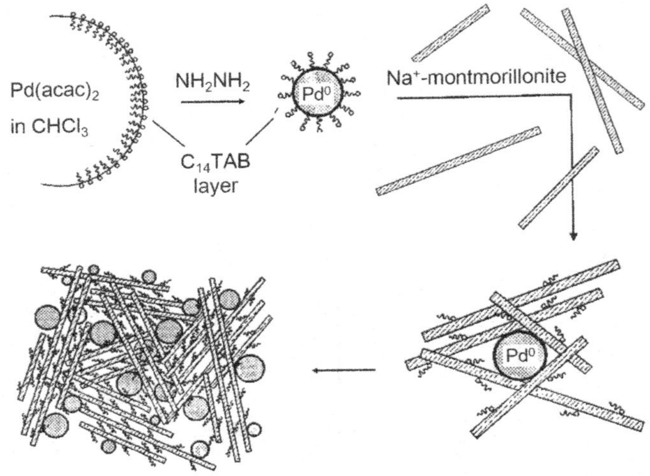

FIG. 12.11 Proposed mechanism for the successive formation of the Pd^0 hydrosol and the Pd-organoclay. *(Reproduced from Z. Király, B. Veisz, I. Dékány, Á. Mastalir, Z. Rázga, Chem. Commun. 19 (1999) 1925–1926, with permission from The Royal Society of Chemistry.)*

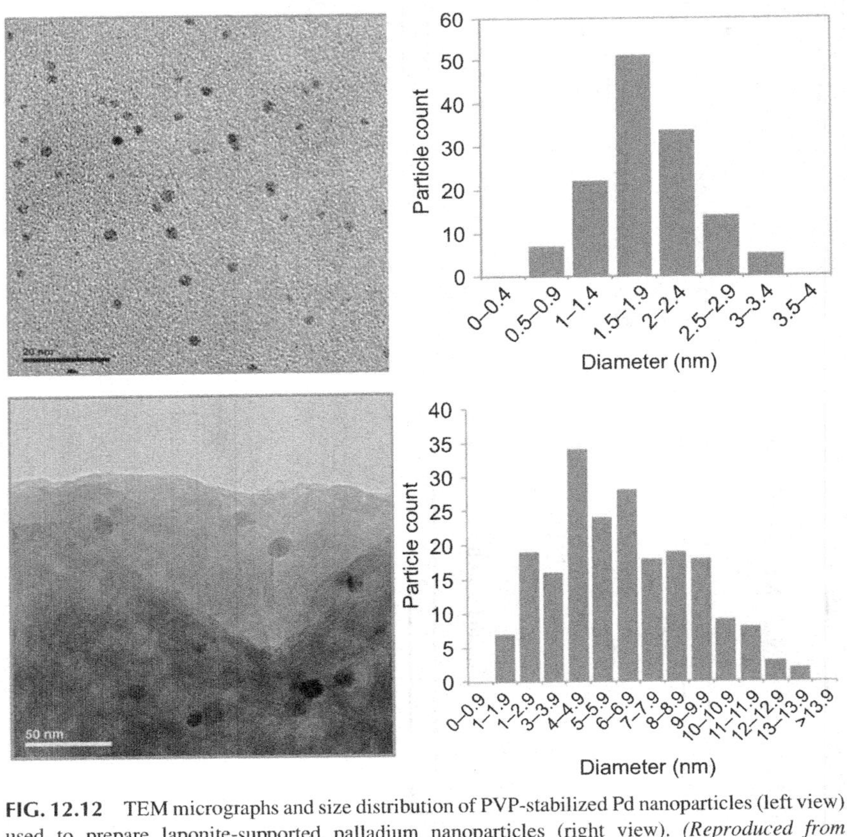

FIG. 12.12 TEM micrographs and size distribution of PVP-stabilized Pd nanoparticles (left view) used to prepare laponite-supported palladium nanoparticles (right view). *(Reproduced from A.V. Martinez, A. Leal-Duaso, J.I. Garcia, J.A. Mayoral, RSC Adv. 5 (2015) 59983–59990, with permission from The Royal Society of Chemistry.)*

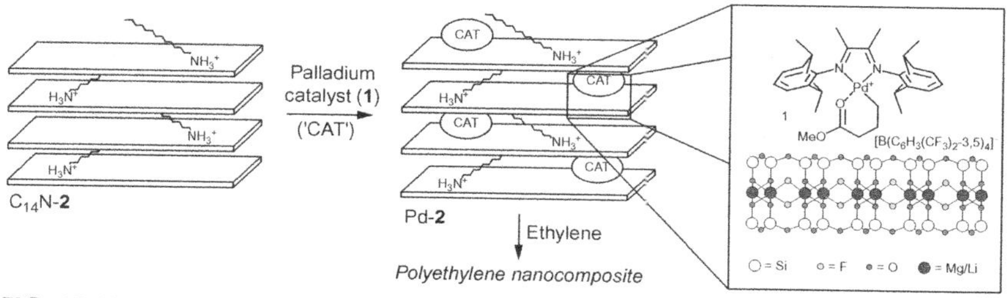

FIG. 12.13 Polyolefin-silicate nanocomposite containing palladium catalyst in fluorohectorite lamellae. *(Reproduced from J.S. Bergman, G.W. Coates, H. Chen, E.P. Giannelis, M.G. Thomas, Chem. Commun. 21 (1999) 2179–2180, with permission from The Royal Society of Chemistry.)*

study the ethylene oxidation [69b]. The performance of heterogenized palladium catalysts largely depends on a synergistic effect between both the clay support and the Pd complex [5].

3.2 Platinum Catalysts

In catalysis, platinum has been used for the hydrogenation of various functional groups such as nitro- or carbonyl-containing compounds and dehydrogenation reactions in organic synthesis, *syn*-addition on alkynes, and oxidation reactions [77]. Platinum is preferred over palladium to minimize hydrogenolysis when reducing nitro compounds to amines [78]. Moreover, it has good resistance to poisons such as sulfur, lead, or phosphorous during oxidation reactions [78d].

In the literature many methods are mentioned for the incorporation of platinum nanoparticles into smectite lamellae. They have been intercalated into silicate lamellae of clay minerals through cation exchange of a platinum precursor ion, surfactant-assisted reduction method, miscellar technique, supercritical fluid (SCF)-aided methods, and many more [79]. Among them, the SCF-aided method is a newly developed and effective route for the preparation of clay-supported Pt nanoparticles [79,80]. For example, Yang and coworkers incorporated uniform platinum nanoparticles into montmorillonite using $H_2PtCl_6 \cdot 6\,H_2O$ as the precursor in supercritical methanol, and the resulting Pt^0-containing montmorillonite catalyst was employed in hydrogenation reactions [79].

Harrison et al. demonstrated the reduction of $Pt(NH_3)_4{}^{2+}$ complexes exchanged on Wyoming montmorillonites. Metallic platinum nanoparticles (50–100 nm) were generated according to the following reaction [81]:

$$Pt(NH_3)_4{}^{2+} + H_2 \rightarrow Pt^0 + 2NH_4{}^+ + 2NH_3$$

Ionic-liquid-stabilized platinum nanoparticles have also been incorporated into silicate lamellae of clay minerals. Zap et al. demonstrated the preparation of platinum nanoparticles stabilized by an ionic liquid *viz.*

1-butyl-3-methylimidazolium tetrafluoroborate (Pt—BMI·BF_4). Intercalation of the nanoparticles in montmorillonite was accomplished through sonication [82]. Chirally modified clays containing platinum nanoparticles have also been reported. Mastalir et al. prepared enantioselective Pt nanoparticles immobilized in dihydrocinchonidine-modified montmorillonite, the mean crystallite size being 0.8 nm. The catalyst was mildly enantioselective for the asymmetric hydrogenation of ethyl pyruvate to give (*R*)-ethyl lactate [58b]. Synthetic hectorite was also used to intercalate a chiral modifier, e.g., dihydrocinchonidine hydrochloride (DHCD·HCl). Such a modification enabled clay swelling in organic solvents. The surface-modified hectorite was then impregnated with chloroplatinic acid (H_2PtCl_6) and the precursor was subsequently reduced via $NaBH_4$ to obtain monodispersed platinum nanoparticles with a mean crystallite size of 1.2 nm. The enantioselective catalyst was then exploited for the asymmetric hydrogenation of ethyl pyruvate to obtain predominantly (*R*)-ethyl lactate. External active sites due to the presence of Pt nanoparticles on the edges of the hectorite lamellae were found to be mainly responsible for the catalyst performance. The role of Pt particle size was regarded as less significant [58b].

3.3 Rhodium Catalysts

The Wilkinson complex, $[Rh(PPh_3)_3]^+$, is normally used for the selective hydrogenation of olefins. This complex was intercalated into montmorillonite to prepare heterogeneous rhodium catalysts. For example, the rhodium(II) monohydrido cation derived from the Wilkinson complex $[RhCl(PPh_3)_3]$ in silicate lamellae of montmorillonite was successfully used to hydrogenate cyclohexene [46]. Bentonite, the main constituent of the volcanic ash weathering product, is typically regarded as impure montmorillonite. Heterogenized Rh(I) triphenylphosphine complex (Rh-bentonite) has been prepared via ion exchange of a Hungarian Na^+-bentonite [83] with Wilkinson complex $[RhCl(PPh_3)_3]$. Acidic clays such as montmorillonite K-10 have also been employed to immobilize rhodium complexes [84] such as $[Rh(COD)(PPh_3)_2]^+$.

The covalent grafting of organic molecules with silanol groups on the clay surface can facilitate the subsequent anchoring of rhodium complexes, ultimately leading to the heterogenization of homogeneous catalysts. Saikia and coworkers reported montmorillonite-anchored Rh complexes to study the catalytic carbonylation of methanol (Monsanto's process). The corresponding clay was first activated by treatment with formic acid under controlled conditions to increase the surface area and pore volume prior to functionalization by APTMS (3-aminopropyl)-trimethoxysilane) and subsequent anchoring of $[Rh(CO)_2Cl]_2$ with APTMS-linked quinoline-type ligands (Fig. 12.14) [85].

Shimazu et al. prepared hectorite-supported complex catalysts by intercalating [Rh(P*—P*) (COD)]$^+$ (P*—P* = (*S*)-BINAP and (*S*)-(*R*)-BPPFA, COD = cyclooctadiene) into sodium hectorites for the asymmetric hydrogenation of α,β-unsaturated carboxylic acid esters [86]. Metallic rhodium nanoparticles have also been generated by reduction of organometallic rhodium complexes

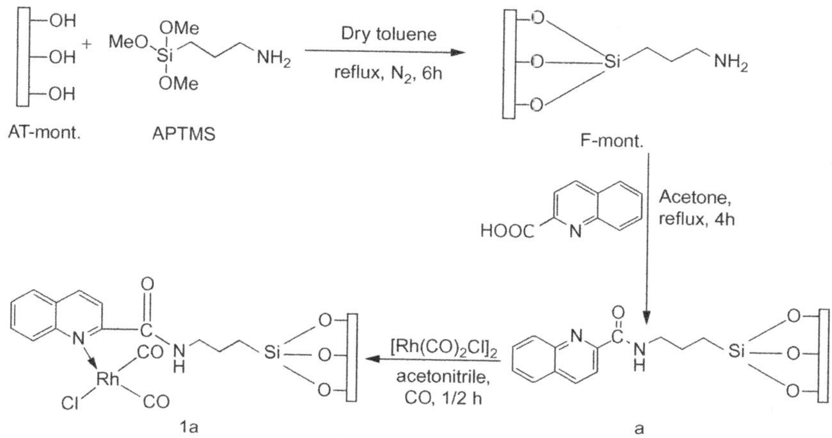

FIG. 12.14 Covalent grafting of organic molecules with silanol groups on a clay surface and subsequent anchoring of $[Rh(CO)_2Cl]_2$. *(Reproduced with permission from P.K. Saikia, P.P. Sarmah, B.J. Borah, L. Saikia, D.K. Dutta, J. Mol. Catal. A 412 (2016) 27–33, Copyright 2015, Elsevier.)*

in clay minerals. For example, Blanco and coworkers used molecular hydrogen to reduce intercalated $[Rh(ACE)_x(NBD)]^+$ (NBD = 2,5-norbornadiene, ACE = acetone) to prepare a metallic rhodium catalyst in montmorillonite [87]. Alper and coworkers intercalated a dimeric chloro(1,5-cyclooctadiene)-rhodium(I) complex in montmorillonite and then reduced it to prepare a metallic rhodium catalyst intercalated between clay layers (Fig. 12.15) [88].

Pillared smectite clays have been exploited to incorporate rhodium complexes such as $[Rh(NH_3)_5Cl]^{2+}$ and their subsequent reduction to generate metallic rhodium catalysts for hydroformylation reactions [89]. Rhodium-impregnated Al-pillared montmorillonite and Al-pillared saponite have been used to prepare Rh/Al pillared clay catalysts [90]. Reduction of the impregnated rhodium in pillared montmorillonite with molecular hydrogen was found to occur in two steps: one at low temperature and the other at high temperature. In saponite samples, the second step was not observed, suggesting that Rh did not interact strongly with this type of clay [56b].

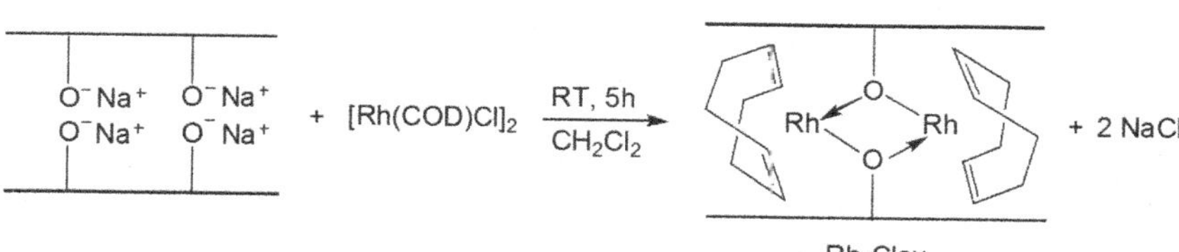

FIG. 12.15 Intercalation of dimeric chloro(1,5-cyclooctadiene)rhodium(I) complex in montmorillonite. *(Reprinted with permission from V.L.K. Valli, H. Alper, Chem. Mater. 7 (1995) 359–362, Copyright 1995, American Chemical Society.)*

3.4 Ruthenium Catalysts

Ruthenium, one of the cheapest noble metals, was only scarcely used in catalytic reactions until the recognition of two major discoveries (rewarded by Grubbs' Nobel Prize in 2001 and Noyori's Nobel Prize in 2005), which stimulated the catalytic chemistry of ruthenium [91]. Clay-supported ruthenium nanoparticles are important catalysts for hydrogenation reactions. Their catalytic performance depends on various factors such as nanoparticle size, nature of support, and reaction conditions. The development of simple and highly efficient methods to prepare monodispersed ruthenium nanoparticles in clay lamellae remains challenging. Various strategies have been adopted to prepare clay supported ruthenium nanoparticles. They are usually obtained by the reduction of Ru^{2+} or Ru^{3+} ions. Since the physical and chemical properties of metallic nanoparticles are often related to their size and shape [92], controlling their growth has been an area of active research for many years [93]. In the field of catalytic research, an effective size control of metal nanoparticles helps the rational design of catalyst for practical use [94]. The size and shape of nanoparticles are usually controlled by varying different reaction parameters. Süss-Fink and coworkers developed practical methods to efficiently control the growth of hectorite-supported ruthenium nanoparticles [95]. Water-soluble arene ruthenium complexes were used for generating nanoparticles of various size and shape. Thus, the dinuclear complex benzene ruthenium dichloride dimer dissolves in water with hydrolysis to provide, with successive substitution of chloro ligands by aqua ligands, a mixture of mononuclear benzene ruthenium complexes being in equilibrium (Scheme 12.2). The dication $[(C_6H_6)Ru(H_2O)_3]^{2+}$, which has been isolated as the sulfate and structurally characterized

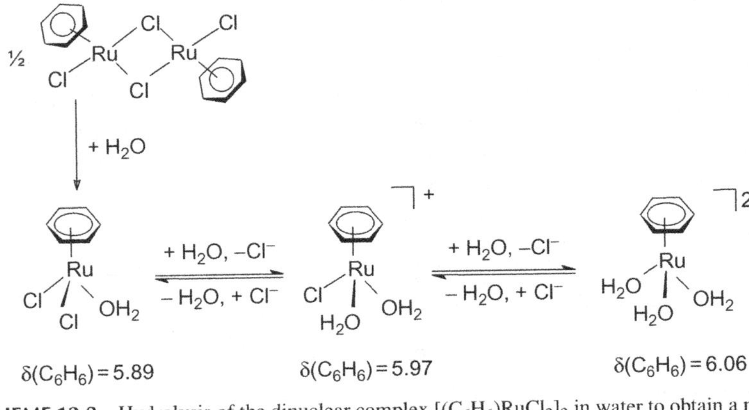

SCHEME 12.2 Hydrolysis of the dinuclear complex $[(C_6H_6)RuCl_2]_2$ in water to obtain a mixture of mononuclear benzene ruthenium complexes, the dicationic triaqua complex $[(C_6H_6)Ru(H_2O)_3]^{2+}$ being the major product. *(Adapted with permission from F.-A. Khan, A. Vallat, G. Süss-Fink, J. Mol. Catal. A 355 (2012) 168–173, Copyright 2011, Elsevier.)*

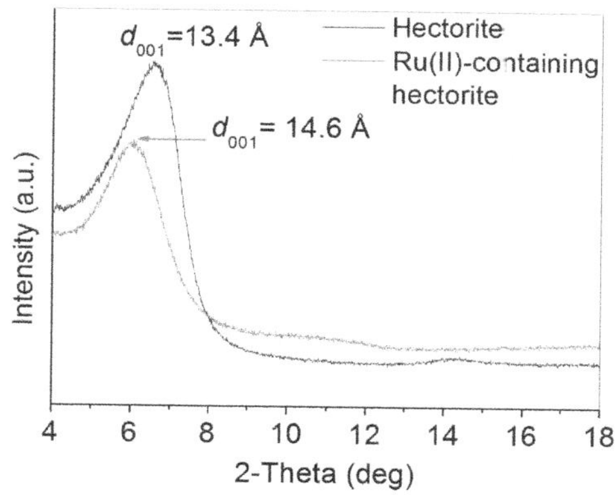

FIG. 12.16 XRD patterns of the sodium-containing hectorite and ruthenium(II)-containing hectorite in a scan range of 2θ angles from 4 degrees to 18 degrees. The calculated d-spacing (d_{001}) of Na-hectorite and Ru(II)-containing hectorite are 13.4 Å and 14.6 Å, respectively. *(Reproduced with permission from B. Sun, G. Süss-Fink, J. Organomet. Chem. 812 (2016) 81–86, Copyright 2015, Elsevier.)*

[96], is the major species present in the hydrolytic mixture over the pH range from 5 to 8 according to the NMR spectrum.

The intercalation of ruthenium in hectorite was accomplished via ion exchange by mixing the white sodium hectorite with a solution derived from the hydrolysis of the dinuclear ruthenium complex $[(C_6H_6)RuCl_2]_2$ in H_2O. Intercalation was confirmed by comparing x-ray diffraction (XRD) patterns of the parent sodium hectorite with that of ruthenium(II)-modified hectorite (Ru(II)@hectorite); showing a difference in the d-spacing (d_{001}) of the interlayers, which were 13.4 Å and 14.6 Å, respectively (Fig. 12.16). The swollen interlayer space was attributed to the interaction between the adsorbed ruthenium complexes and the intercalant in hectorite [58a].

Ru(II)@hectorite, which can be dried and stored in air, was reacted either with hydrogen under pressure (50 bar) at 100°C or refluxing alcohols by reduction of $[(C_6H_6)Ru(H_2O)_3]^{2+}$ to obtain the black ruthenium(0)-modified hectorite [95]. When yellow ruthenium(II)-containing hectorite was reduced with molecular hydrogen in an aqueous medium, a stable dispersion of hexagonally shaped ruthenium nanoparticles intercalated in hectorite was obtained, and the nanoparticles had a mean size of 38 nm with a wide range of size distribution ($\sigma = 11.6$) [95a]. However, the reduction with H_2 in alcohols alone produced more or less spherically shaped ruthenium particles (Fig. 12.17), and the nanoparticles obtained in aqueous methanol were also hexagonally shaped, the size of which depend on the ratio of methanol/water: For example, the reduction of ruthenium(II)-containing hectorite in an aqueous solution with 50% methanol resulted in relatively small hexagons with an average size up to 20.7 nm having a standard deviation $\sigma < 25\%$ of the mean particle size, which is fairly narrow as

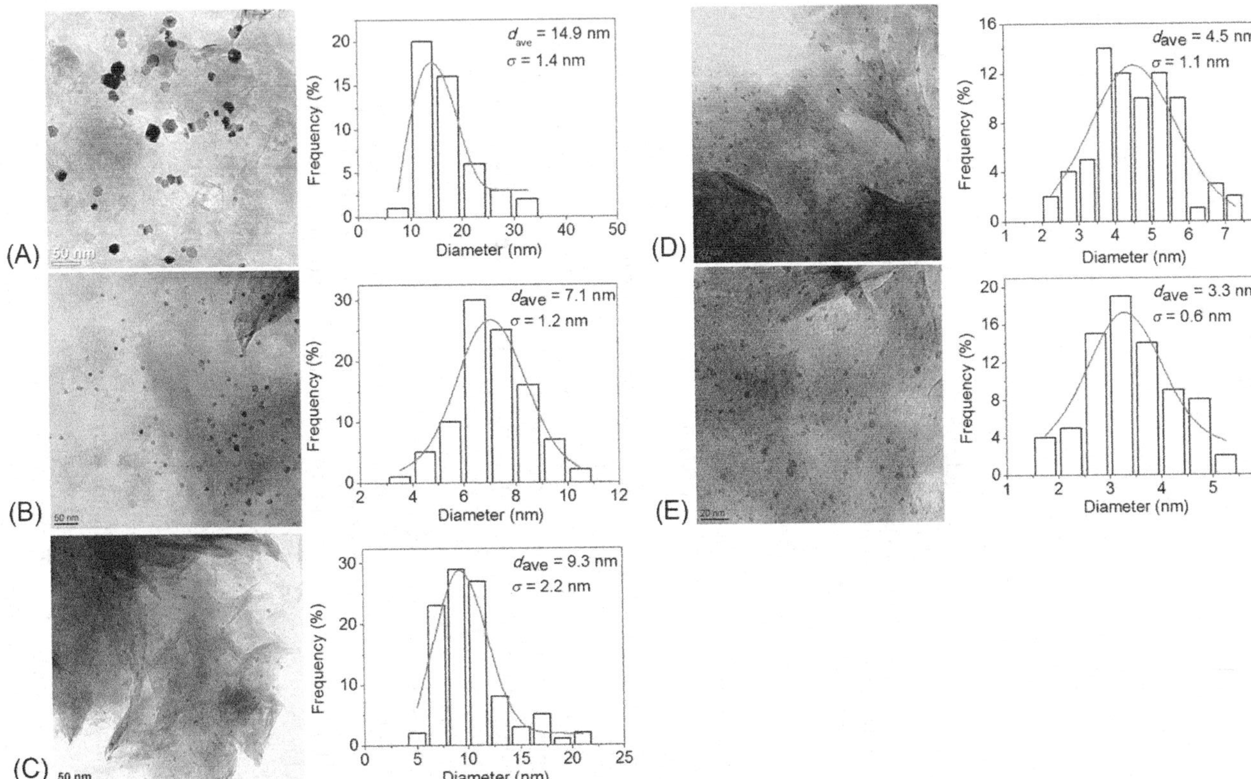

FIG. 12.17 TEM images of the hectorite-intercalated ruthenium nanoparticles reduced from a ruthenium(II)-containing hectorite precatalyst under a H_2 pressure of 50 bar for 14 h in the solvents (A) water/methanol (1:1), (B) methanol, (C) ethanol, (D) isopropanol, and (E) 2-butanol. The calculated mean particle sizes (d_{ave}) are 14.9 ± 1.4 nm, 7.1 ± 1.2 nm, 9.3 ± 2.2 nm, 4.5 ± 1.1 nm, and 3.3 ± 0.6 nm, respectively. *(Adapted with permission from G. Süss-Fink, F.-A. Khan, J. Boudon, V. Spassov, J. Clust. Sci. 20 (2009) 341–353, Copyright 2009, Springer.)*

compared to those observed in pure water. If the water content were inferior to 20%, hexagonally shaped nanoparticles were no longer observed. Reduction of ruthenium(II)-containing hectorite in refluxing alcohols afforded clusters of ruthenium nanoparticles in the clay [95a].

The growth of ruthenium nanoparticles in hectorite can be controlled by changing the nature of the arene substituent in ruthenium(II)-containing hectorite. For example, the reduction of $[(C_6H_6)_2Ru_2(OH)_3]^+$ containing hectorite obtains large nanoparticles with a wide range of size distribution as compared to that of $[(C_6Me_6)_2Ru_2(OH)_3]^+$ containing hectorite [97].

Instead of using organometallic ruthenium complexes as a precursor, a more general method to prepare supported ruthenium nanoparticles is the immobilization of Werner-type ruthenium(III) ions from $RuCl_3{\cdot}xH_2O$ on the support and the subsequent reduction by $NaBH_4$ or molecular H_2. According to the practical requirements for certain reactions, a hectorite-supported ruthenium(III) precatalyst (Ru(III)@hectorite) was prepared by the impregnation method with $RuCl_3{\cdot}xH_2O$ in water [95d]. As illustrated in Fig. 12.18, the interlayer spacing (d_{001}) of the sodium hectorite and Ru(III)@hectorite were calculated from x-ray diffraction patterns to be 13.4 Å and 15.2 Å, respectively. The swollen interlayer distance was attributed to the presence of $[Ru(H_2O)_6]^{3+}$ cations [95e].

The reduction of the ruthenium(III)-containing hectorite was carried out in water under an H_2 atmosphere at an elevated temperature or by a reducing agent such as $NaBH_4$. As shown in Fig. 12.19, the reaction of Ru(III)@hectorite with $NaBH_4$ in water at room temperature yielded the ruthenium nanoparticles that were clustered and thoroughly embedded in hectorite with a worm-like

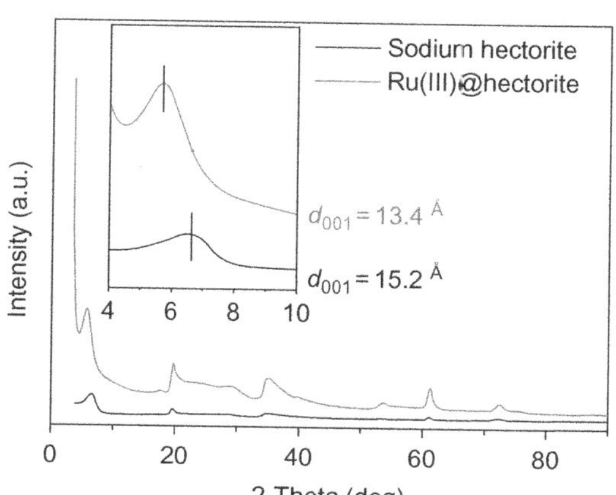

FIG. 12.18 XRD patterns of the sodium hectorite and the ruthenium(III)-containing hectorite precatalyst (Ru(III)@hectorite), of which the interlayer spacings (d_{001}) are 13.4 Å and 15.2 Å, respectively. The inset is the zoomed patterns in a scan range of 2θ angles from 4 degrees to 10 degrees. *(Adapted with permission from B. Sun, D. Carnevale, G. Süss-Fink, J. Organomet. Chem. 821 (2016) 197–205, Copyright 2016, Elsevier.)*

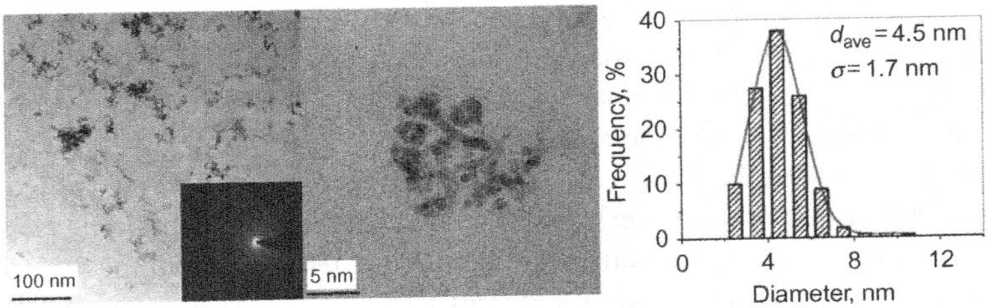

FIG. 12.19 TEM analysis of the hectorite supported ruthenium nanoparticles reduced by $NaBH_4$ in water at room temperature for 4 h. The mean particles size is calculated to be 4.5 ± 1.7 nm. *(Adapted with permission from B. Sun, D. Carnevale, G. Süss-Fink, J. Organomet. Chem. 821 (2016) 197–205, Copyright 2016, Elsevier.)*

morphology, similar to that of the ruthenium nanoparticles intercalated in montmorillonite modified by cetyltrimethylammonium bromide (CTAB) [59a]. The mean particle size is calculated to be 4.5 nm ($\sigma = 1.7$ nm), which accounts for the observed peak of Ru(011) in the x-ray diffraction pattern.

Mio and coworkers used an ionic liquid (IL) exchanged montmorillonite to prepare a Ru nanocatalyst. In their method, the IL *viz.* 1,1,3,3-tetramethylguanidinium trifluoroacetate ([TMG][TFA]) was first exchanged with Na^+ in the clay, and then Ru^{3+} ions were loaded onto the IL-exchanged MMT (TGM-MMT). Finally, Ru^0/TGM-MMT nanocatalyst was obtained by treatment with molecular hydrogen. The nanoparticles had an average particle size <3 nm and were believed to be stabilized by a combination of electrostatic and coordination forces (Fig. 12.20). This catalyst was found to be highly active for the hydrogenation of benzene probably due to synergistic effects of TMG, MMT, and Ru nanoparticles [98].

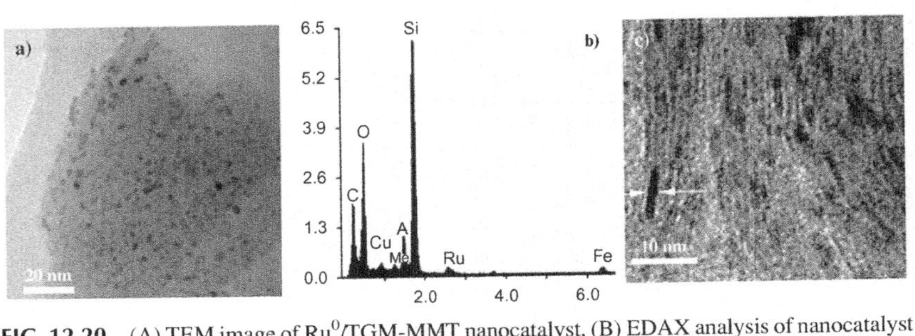

FIG. 12.20 (A) TEM image of Ru^0/TGM-MMT nanocatalyst, (B) EDAX analysis of nanocatalyst (C) HR-TEM image of a cross-section of nanocatalyst. The *white arrows* indicate the interlaminar space of montmorillonite intercalated with the Ru^0 nanoparticle. *(Adapted with permission from S. Miao, Z. Liu. B. Han, J. Huang, Z. Sun, J. Zhang, T. Jiang, Angew. Chem. Int. Ed. 45 (2005) 266–269, Copyright 2006, John Wiley and Sons.)*

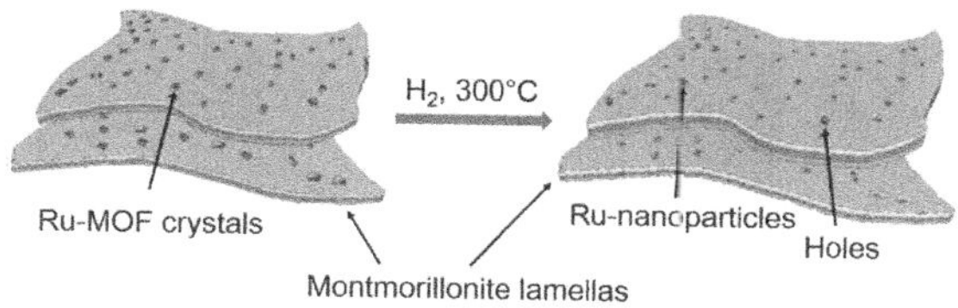

FIG. 12.21 In situ growth of Ru-MOF crystals in montmorillonite lamellae and the Ru nanoparticles produced after H_2-annealing. *(Reproduced with permission from D. Chen, M. Huang, S. He, S. He, L. Ding, Q. Wang, S. Yu, S. Miao, App. Clay Sci. 119 (2016) 109–115, Copyright 2015, Elsevier.)*

A metal-organic framework of ruthenium in montmorillonite was demonstrated by D. Chen et al. The solvothermal technique was employed to grow a ruthenium framework of the type $[Ru_2^{II,III}(BTC)_2X^xY_{1.5-x}]$ (BTC = 1,3,5-benzenetri-carboxylate, X, Y = Cl^- or OH^-) on montmorillonite (Fig. 12.21). The Ru-MOF-montmorillonite hybrids were annealed at 300°C with molecular hydrogen. The hybrid-containing ruthenium nanoparticles were found to be highly efficient catalysts for benzene hydrogenation reactions [99].

Mesoporous montmorillonite, postmodified by hydrochloric acid, was suggested to be an improved support for ruthenium nanoparticles, showing a high reactivity in the transfer hydrogenation of substituted nitrobenzenes to the corresponding anilines [100]. Remarkably, ruthenium nanoparticle-intercalated montmorillonite was able to catalyze the solvent-free alkene hydrogenation reaction with high reactivity and recyclability [101]. Another supported Ru catalyst by using cetyltrimethylammonium bromide (CTAB)-modified montmorillonite as support was reported to have high activity and selectivity in the hydrogenation of quinoline, the modification being responsible for the robust stability of ruthenium nanoparticles [59a]. Hectorite has also shown its potential to be a promising support for metal catalysts. Ruthenium nanoparticles intercalated in hectorite have a superior activity and selectivity for the selective hydrogenation of benzene [95a], toluene [95h], furfuryl alcohol [95b], and α,β-unsaturated ketones to their corresponding saturated products [95g].

3.5 Gold Catalysts

Gold has been the most fascinating metal throughout the history of mankind. However, it was considered catalytically inactive until the discovery of supported gold nanoparticles with high catalytic activity in CO oxidation at low temperature [68,102]. Later, gold nanoparticles were found to catalyze the selective oxidation of alcohol in water and the epoxidation of propylene gas [103]. These new discoveries spurred an era of nano gold rush. Today, the art of nano gold synthesis has become mature. In the literature, well-defined procedures for the preparation of gold nanoparticles in any size and morphology

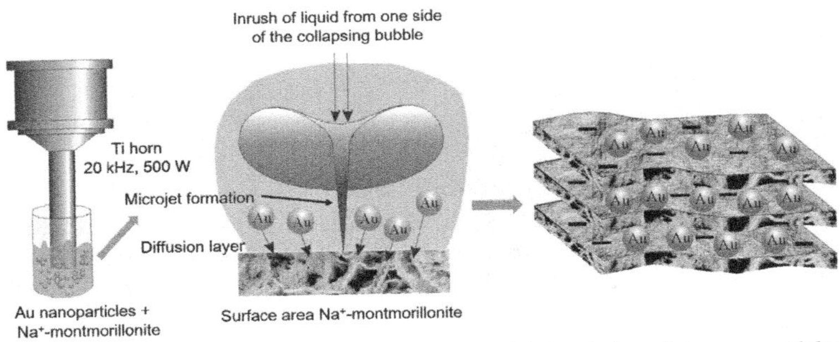

FIG. 12.22 Schematic illustration of sonication-induced intercalation of Au-nanoparticles into Na^+-montmorillonite. *(Reproduced with permission from V. Belova, H. Möhwald, D.G. Shchukin, Langmuir 24 (2008) 9747–9753, Copyright 2008, American Chemical Society.)*

are found. It is imperative to impregnate the preformed nanoparticles into a clay matrix, since the size and shape of the nanoparticles can be better controlled by separate standard preparation steps [104]. Sonication is one of the possible ways to intercalate gold nanoparticles into the interlamellar space of smectite clays [105]. Belova and coworkers demonstrated the intercalation of preformed gold nanoparticles into the clay matrix (montmorillonite) with ultrasonic irradiation (Fig. 12.22) [104].

The resulting clay/nano-Au composite was calcined at 800°C and characterized. Nearly spherical-shaped gold nanoparticles, with a size of 6 ± 0.5 nm were obtained and no aggregation of the gold nanoparticles was observed during calcination (Fig. 12.23) [104].

A combination of polyethylene glycol and the ultrasonic irradiation technique was also used for the preparation of Au/clay nanocomposites. The combined application of ultrasonic irradiation and polymer application accelerated the intercalation process and resulted in homogeneously dispersed gold nanoparticles of 6–8 nm diameter in montmorillonite (Fig. 12.24) [105].

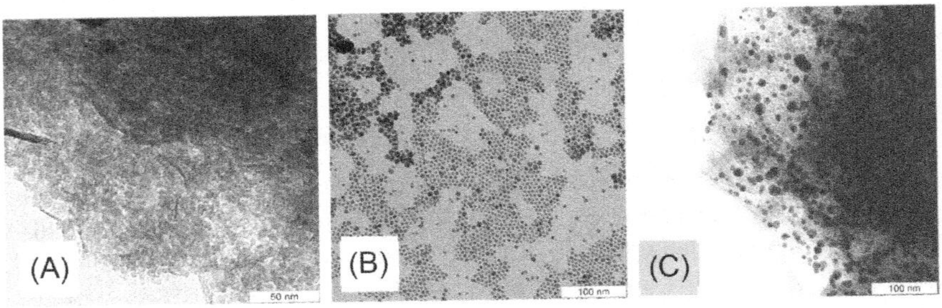

FIG. 12.23 TEM images of (A) pure Na^+-montmorillonite without Au nanoparticles; (B) preformed Au nanoparticles; (C) Na^+-montmorillonite loaded from 4.2 wt% Au colloid solution. *(Reproduced with permission from V. Belova, H. Möhwald, D.G. Shchukin, Langmuir 24 (2008) 9747–9753, Copyright 2008, American Chemical Society.)*

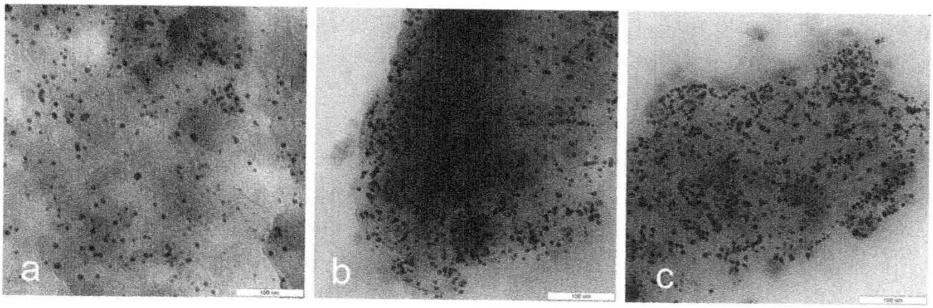

FIG. 12.24 TEM images of (A) Na^+-montmorillonite loaded from 4.5 wt% Au colloid solution; (B) Na^+-montmorillonite loaded from 8.5 wt% Au colloid solution; and (C) Na^+-montmorillonite loaded from 12 wt% Au colloid solution. *(Reproduced with permission from V. Belova, D.V. Andreeva, H. Möhwald, D.G. Shchukin, J. Phys. Chem. C 113 (2009) 5381–5389, Copyright 2009, American Chemical Society.)*

Gold nanoparticles have also been reported on polyethylenimine-modified montmorillonite [65], where the montmorillonite acts as solid support and the amino group of polyethylenimine serves simultaneously to both bind and reduce the gold ions from $HAuCl_4$. An overall literature survey points to the fact that the nature of the support significantly influences the gold nanoparticle dispersion and subsequently its catalytic activity. The size of gold nanoparticles is another factor capable of enhancing the catalytic activity, especially when the particle size is less than 6 nm. Further investigation of these factors may well reveal a far wider scope for catalysis by gold nanoparticles intercalated in clays [68].

3.6 Copper Catalysts

Copper nanoparticles are known to be capable of catalyzing various reactions such as azide-alkyne click reaction, oxidation, reduction, cross-coupling, C—H functionalization, and other miscellaneous reactions [106]. Smectite clays, especially pillared smectite clays, have been extensively used for supporting copper catalysts. Copper nanoparticles stabilized in silicate layers of montmorillonite were first reported by Malla et al. in the 1990s. Hydroxyl-copper-acetate-containing montmorillonite was reduced in refluxing ethylene glycol to give Cu^0 clusters entrapped in montmorillonite [107]. These clusters had a diameter of about 0.4 nm and the hybrid catalyst was found to be highly stable at elevated temperatures (500°C). Borah et al. demonstrated the stabilization of Cu^0-nanoparticles into the nanopores of acid-modified montmorillonite [108]. In a typical procedure, the clay was pretreated with sulfuric acid to generate the nanopores followed by successful loading of the metal precursor such as $Cu(CH_3COO)_2$ through an incipient wetness impregnation technique. Cu^{2+}-containing montmorillonite was then reduced by $NaBH_4$ to obtain copper nanoparticles that were evenly distributed on the support. The presence of nanopores on montmorillonite was attributed to limited nanoparticle growth, thus restricting their diameter to a size of less than 10 nm. This material was shown to efficiently catalyze the 1,3-dipolar cycloaddition of aliphatic azides with terminal alkynes in water at room temperature.

Copper-impregnated pillared clays have also been widely studied [56b]. Bahranowski et al. reported the preparation of Cu/Al-pillared montmorillonite. After calcination at 400°C, the Cu content was found to be 2.31 wt% and the copper species to be located in the interlayer region of montmorillonite, either as isolated Cu^{2+} cations anchored to the alumina pillars or as patches of amorphous CuO [109]. Bergaoui et al. introduced Cu^{2+} ions into Al-pillared saponite and studied the nature of the resulting Cu species. An EPR study revealed the presence of grafted species with the formula $[Cu(AlO)_n(H_2O)_{6-n}]^{2+}$ in wet samples and $[Cu(AlO)_n(H_2O)_{4-n}]^{2+}$ in dried samples of the pillared clay [110]. Bergaya and coworkers also demonstrated the synthesis of Cu—Al-pillared bentonite [111]. A mixture of aqueous Cu and Al nitrate was employed as a pillaring solution. Cu/(Al+Cu) ratios of up to 10% were used. In a typical procedure, the intercalation of the pillaring solution was carried out either by adding the solution to a suspension of the clay or by adding the powdered clay to the pillaring solution. After calcination, the pillared clays showed basal spacings of 17.0–18.0 Å containing only 0.39 wt% Cu, relatively lower than the amount of copper used in the pillaring solutions. The specific surface areas of the solids were about 109–159 m^2/g. The solids were used to study catalytic wet peroxide oxidation (CWPO) of phenol. The catalytic activity was found to be strongly dependent on the Cu content and the preparation method of Cu—Al-pillared clays. Almost negligible Cu leaching was observed during the reaction (30–40 ppm), suggesting that Cu was strongly associated with surface alumina in the clay (Fig. 12.25) [56b,111]. On the other hand, Cu-doped Al-pillared laponite (synthetic hectorite) was reported containing octahedral Cu species chemisorbed to an Al_{13} pillar through two coordination positions [56b,112].

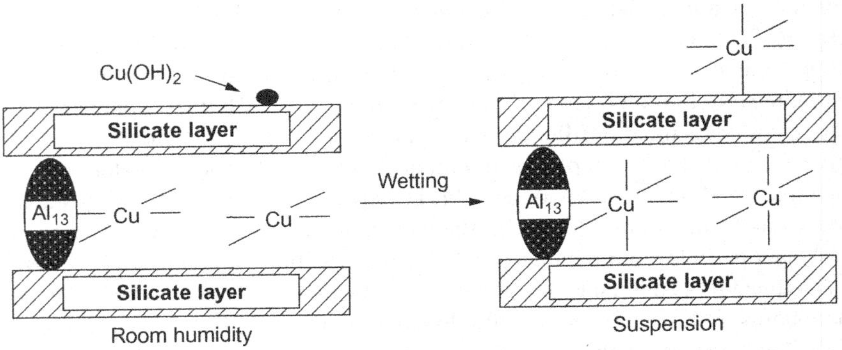

FIG. 12.25 Different possible forms of copper interaction with pillared clays. *(Reproduced with permission from A. Gil, L.M. Gandia, M.A. Vicente, Catal. Rev. Sci. Eng. 42 (2000) 145–212, Copyright 2000, Taylor & Francis.)*

4 CATALYTIC APPLICATIONS OF SMECTITE-INTERCALATED METAL CATALYSTS

This section outlines the catalytic characteristics of smectite-intercalated metal catalysts in some reactions such as hydrogenation, oxidation, isomerization, and Friedel-Crafts and Diels-Alder reactions. The effects of the synthetic method, the pretreatment of the support, and modification of the metal precursor on the reactivity and selectivity of the supported metal catalysts are summarized. A structure-performance relationship as well as the role of the support is discussed for the different reactions.

4.1 Hydrogenation Reactions

Catalytic hydrogenation plays an important role in the synthesis of fine chemicals. In particular, supported hydrogenation catalysts have found wide applications in industrial chemical production. The active species of the catalyst include metals, alloys, oxides, sulfides, nitrides, carbides, or mixtures of these [113]. To fulfill the multiple requirements for the hydrogenation of each functional group, supported catalysts exhibit distinct advantages such as high dispersion of metal species over supports, tenacious resistance against sintering under harsh conditions and poisoning, and ease of separation and recovery [114]. To date, there is evidence for interactions between the active phase and the support [114–121], and the research on this issue is continuously growing. The common materials used as support can be SiO_2, TiO_2, MgO, γ-Al_2O_3, carbon, clays, and zeolites [114,122,123]. By tuning the morphology of the support, the catalytic properties can possibly be tailored toward specific applications [124].

There have been a few reviews summarizing the applications of clay-supported catalysts in fine chemical syntheses [3,125–127]. Smectite is well-known for its swelling interlayers, the changeable interlaminar distance from which makes it a unique structure for the intercalation of metal ions and metal nanoparticles [58a].

Pinnavaia et al. found that the intercalation of homogeneous $Rh(PPh_3)_n^+$ precursors in hectorite significantly improves the catalytic selectivity in the hydrogenation of 1-hexene, whereas extensive isomerization to 2-hexene was observed with the analogous homogeneous system [128]. The difference in specificity between the intercalated and homogeneous catalyst systems may be related to the difference in the equilibrium between the active dihydride species and the monohydride rhodium complex as shown by the equation below [129]:

$$RhH_2(PPh_3)_n^+ \rightleftharpoons RhH(PPh_3)_n + H^+$$

(n = 2 or 3)

The dihydride is very active for terminal olefin hydrogenation but much less for isomerization, whereas the monohydride is active for both the hydrogenation and isomerization. Compared with the homogeneous system, the dihydride is favored in the intercalated state because of the surface Brønsted acidity, which is believed to arise from the dissociation of hydrated Na^+ ions present on the interlamellar surfaces. The presence of Brønsted acid sites favors the formation of cationic dihydrido complexes, which catalyze the hydrogenation of terminal olefins with little or no isomerization.

The immobilization of organometallic complexes on the support is governed by many factors, and at the same time it greatly affects the catalytic properties of the catalyst [130]. A systematic study on immobilizing iridium and rhodium complexes in smectite clays was carried out by Salagre and coworkers [84,131]. $[Ir(\mu\text{-}Cl)(COD)]_2+4PPh_3$, $[Ir(COD)(PPh_3)_2]BF_4$, $[Ir(COD)(PPh_3)_2]PF_6$ and $[Rh(COD)(PPh_3)_2]BF_4$ as organometallic precursors were introduced into montmorillonite K10, Na^+-montmorillonite and Li^+-hectorite. The results show that the amount of the cationic metal complexes immobilized in the montmorillonite K10 and Li^+-hectorite was higher than that in the Na^+-montmorillonite. However, the neutral metal complex was less immobilized in all the clays. The detailed characterization of the clay structure indicates that cationic metal complexes mainly reside on the external surface of the clay montmorillonite K10 and Li^+-hectorite by adsorption, whereas the complexes were partially intercalated into the interlaminar space of Na^+-montmorillonite as well as on the external surface, as interpreted in Scheme 12.3. In the hydrogenation of *N*-benzylidene aniline catalyzed by metal complexes adsorbed or intercalated in clays (Scheme 12.3), the activity of a fresh run is comparable to that of their homogeneous counterparts. However, the activity was obviously reduced for other runs. This reduction is believed to be due to the presence of intermediate species, which may be oxidized in air during the filtration step and reusing process. On the whole, immobilizing the organometallic complexes in the discussed clay supports afforded good catalytic activity and recyclability. Either the clay structures or the organometallic complexes have a considerable effect on the catalytic process.

The Schiff base Pt(II) complex, $PtL(H_2O)$ (L=isonicotinic acid (3-hydroxy-1,2-dimethyl-but-2-enylidene)-hydrazide), intercalated montmorillonite was found to be a robust catalyst for hydrogenation of aromatic nitro compounds at room temperature [132]. Nitrobenzene and its derivatives can be exclusively reduced to the corresponding product under a H_2 pressure of 1 atm within the microenvironment of the interlayer space (Scheme 12.4). Compared with its homogeneous counterpart, the intercalated catalyst exhibits superior activity and selectivity. The enhanced catalytic properties in a heterogeneous medium may be attributed to the unique interlayer space that localizes the substrate and catalyst in close proximity; however, it would not be the case in homogeneous reaction. Furthermore, by means of direct interaction of the complex and the clay, the intercalated catalyst can be recycled and reused three times without significant leaching and loss of its catalytic activity.

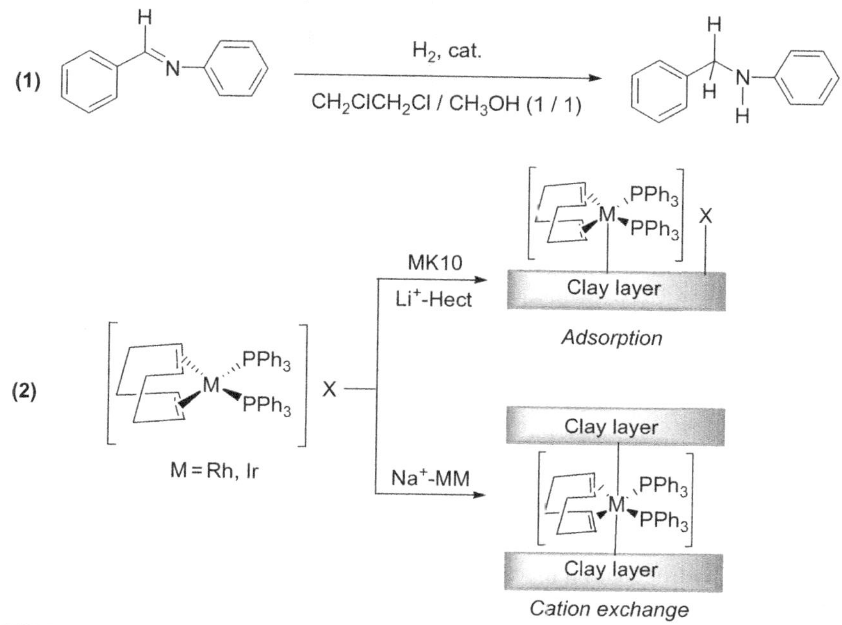

SCHEME 12.3 Hydrogenation of *N*-benzylidene aniline catalyzed by metal complexes adsorbed or intercalated in clays. *(Reproduced with permission from C. Claver, E. Fernandez, R. Margalef-Catala, F. Medina, P. Salagre, J.E. Sueiras, J. Catal. 201 (2001) 70–79, Copyright 2001, Elsevier.)*

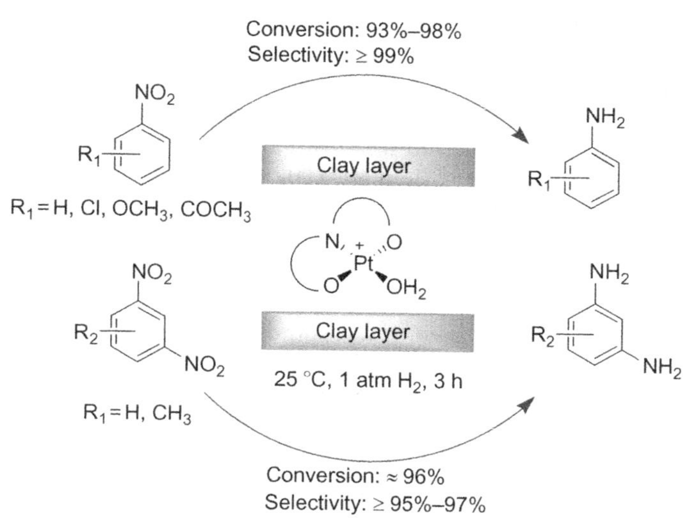

SCHEME 12.4 Hydrogenation of aromatic nitro compounds catalyzed by Schiff base Pt(II) complex montmorillonite. *(Adapted with permission from K. Parida, G.B.B. Varadwaj, S. Sahu, P.C. Sahoo, Ind. Eng. Chem. Res. 50 (2011) 7849–7856, Copyright 2011, American Chemical Society.)*

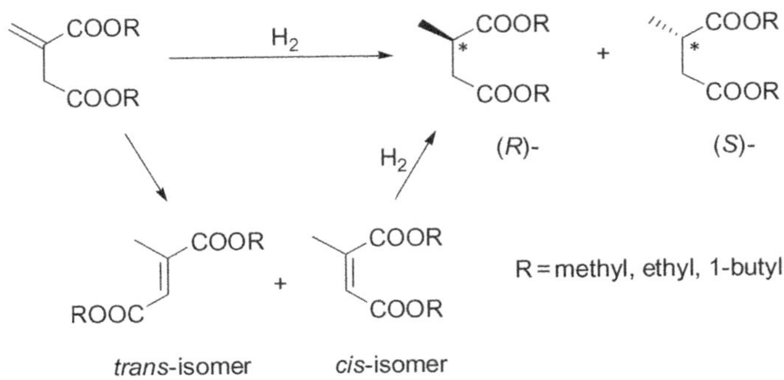

SCHEME 12.5 Rh complexes intercalated in hectorite for the catalytic asymmetric hydrogenation of itaconates. *(Adapted with permission from S. Shimazu, K. Ro, T. Sento, N. Ichikuni, T. Uematsu, J. Mol. Catal. A 107 (1996) 297–303, Copyright 1996, Elsevier.)*

Asymmetric hydrogenation in organic synthesis, particularly in the discovery and production of pharmaceuticals, has gained enormous and continuous attention [133,134]. Traditionally, homogeneous catalysts play a key role in asymmetric hydrogenation reactions. However, the intercalation of metal complexes into clay layers offers opportunities of enhancing the catalytic selectivity and of recycling the hydrogenating catalyst [135,136]. Shimazu et al. described the intercalation of Rh complexes in hectorite for the catalytic asymmetric hydrogenation of itaconates (Scheme 12.5) [86,137]. Via ion-exchange using different Rh complexes as precursors, a series of hectorite-intercalated catalysts were obtained, including $[Rh((S)\text{-BINAP})(COD)]^+$/HT, $[Rh((S)\text{-}(R)\text{-BPPFA})(COD)]^+$/HT and $[Rh((S,S)\text{-DIOP})(COD)]^+$/HT (HT = hectorite). By screening of the reaction parameters, the solvent and the size of the ester groups were regarded as important factors in affecting the enantiomeric excess (% ee). In mixed solvents, EtOH/cyclohexane (2/1) at 30°C for 144 h, dibutyl itaconate was catalytically hydrogenated to the corresponding (*R*)-methylsuccinate by $[Rh((S)\text{-BINAP})(COD)]^+$/HT, affording a 100% conversion and 57.7% ee. For the asymmetric hydrogenation of itaconate diesters by $[Rh((S)\text{-}(R)\text{-BPPFA})(COD)]^+$/HT in different solvents, the selectivity was ranging from 64.4% ee in MeOH to 88.3% ee in EtOH. In the case of $[Rh((S,S)\text{-DIOP})(COD)]^+$/HT as the catalyst, the highest selectivity (20.3% ee) was obtained when diethyl ester was employed. It was surprising that the less bulky methyl ester exhibited the lowest selectivity for the hectorite catalyst as compared to the homogeneous system. It was suggested that the chiral conformation of the Rh-complex in the hectorite layer might be more rigidly constrained than in solution, which may be better fitted with the diethyl ester derivatives.

Homogeneous catalysts are often used in industry in biphasic systems or by fixation on supports. The bridge between heterogeneous and homogeneous catalysis is now being built through the use of nanoparticles [138]. Indeed, selectivity is still the major problem in heterogeneous catalysis [139,140].

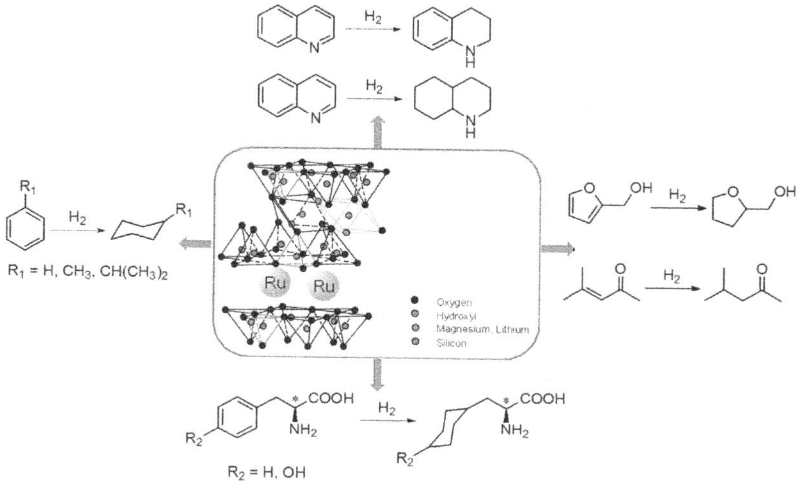

SCHEME 12.6 Collection of substrates used in hydrogenation catalyzed by hectorite-intercalated ruthenium nanoparticles [95].

Tailoring the nanoparticle morphology and adjusting the metal-support interaction may be precisely controlled to improve the selectivity.

Intercalation of organometallic complexes in solid supports to produce supported nanoparticles has been thoroughly discussed [141]. Süss-Fink et al. demonstrated a useful approach to intercalate ruthenium complexes in hectorite, which involved a reduction process to generate the active hectorite-intercalated ruthenium catalyst (Scheme 12.6) [95]. In the hydrogenation of benzene and alkyl-substituted benzenes, this catalyst was highly active, thus exclusively converting the benzene ring to cyclohexane. It was found that both the reduction process and the solvent influence the catalytic activity and selectivity. Interestingly, in the catalytic hydrogenation of quinoline, an aqueous medium was favorable for the production of 1,2,3,4-tetrahydroquinoline (1,2,3,4-THQ); however, a cyclohexane medium favors the formation of decahydroquinoline. This can be explained by the possible formation of H-bonds between water molecules and the π-electrons in the benzene ring of quinoline, which hampers the adsorption of an aromatic ring on the catalyst surface. The nitrogen atom would be likely to coordinate to the ruthenium surface and would receive the hydrogen atoms already present at the surface of the nanoparticles, thus yielding 1,2,3,4-THQ almost exclusively. Nevertheless, the absence of the H-bond with cyclohexane gives rise to the competitive adsorption of an aromatic ring and heteroaromatic ring on the metal surface, which leads to the fully saturated decahydroquinoline under harsh conditions [95c]. The hectorite-intercalated ruthenium can also catalyze the hydrogenation of aromatic amino acids to give the corresponding cyclohexyl-substituted amino acids without racemization.

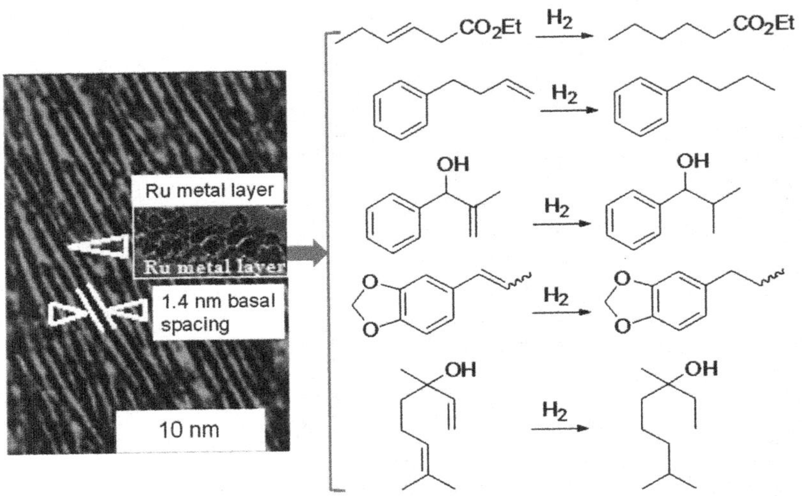

SCHEME 12.7 Ru nanoparticles intercalated into montmorillonite for solvent-free alkene hydrogenation reactions. *(Reproduced from P. Upadhyay, V. Srivastava, RSC Adv. 5 (2015) 740–745, with permission from The Royal Society of Chemistry.)*

Recycling experiments show that the hectorite-intercalated ruthenium catalyst is very stable and can be recovered and reused without loss of activity and selectivity.

Srivastava et al. reported a synthetic protocol for Ru nanoparticles intercalated into montmorillonite for solvent-free alkene hydrogenation reactions (Scheme 12.7) [101]. In this work, $[Ru(NH_3)_6]Cl_3$ was used to replace the interlayer Na^+, K^+, or Ca^{2+} ions of montmorillonite, and the intercalated Ru^{3+} cations were then reduced by $NaBH_4$. TEM photos show that the Ru nanoparticles were highly dispersed in the clay with a size distribution of 20.5 ± 2 nm. During the hydrogenation of alkenes, the reaction temperature, reaction time, and H_2 pressure were regarded as key factors governing the product yield. Under optimized conditions, both the functionalized and non-functionalized alkenes were exclusively converted to the corresponding product with good yields. The ability to recycle the catalyst nine times, together with low catalyst loading and high catalytic selectivity, are advantages of this protocol [101].

In order to insert and stabilize nanoparticles in clays, acid treatment of the clay to generate nanopores is a commonly used approach. Dutta et al. found that H_2SO_4 activation of montmorillonite generates nanopores up to about 10 nm in size, which can act as host and stabilize nanoparticles into the pores [142]. By reduction with ethylene glycol under reflux, Rh nanoparticles are homogeneously distributed into the nanopores in acid activated montmorillonite. Compared with other heterogeneous catalytic systems for the transfer hydrogenation of aromatic carbonyl compounds to the corresponding alcohols [143,144], this

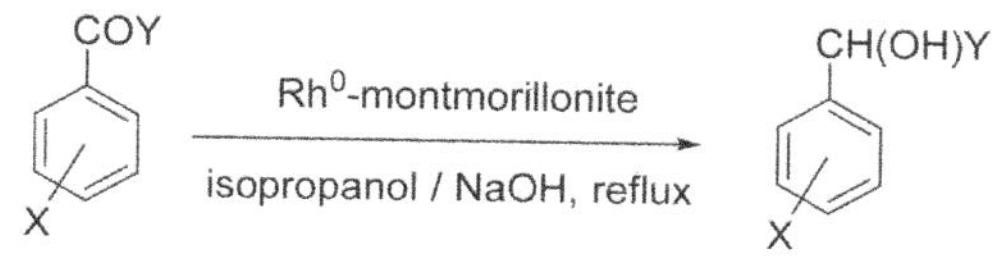

SCHEME 12.8 Transfer hydrogenation of aromatic carbonyl compounds to the corresponding alcohols catalyzed by Rh^0-nanoparticles supported on activated montmorillonite. *(Adapted with permission from P.P. Sarmah, D.K. Dutta, Appl. Catal. A: Gen. 470 (2014) 355–360, Copyright 2013, Elsevier.)*

catalyst exhibits high efficiency and stability (Scheme 12.8). It can also be recycled and reused for three times without loss of activity.

As good catalyst supports, clays possess various general advantages, such as high surface area and porosity. However, it would be more desirable that electron transfer between the support and nanoparticles occurs through a direct or mediated mechanism, which has been proven to sufficiently improve the catalytic activity of nanoparticles [145]. Xia et al. proposed a synthetic route for the preparation of citrate-stabilized Au nanoparticle self-assembled montmorillonite/polyaniline core/shell nanoflakes [146]. Hydrogen bonding between Au NPs and the PANI shell of MMT/PANI nanoflakes is believed to play a vital role in driving the self-assembly process. This catalyst shows excellent activity and selectivity for the hydrogenation of *p*-nitrophenol with $NaBH_4$. By a simple centrifugation and washing process, the catalyst can be recovered and reused after each run.

Recently, Sharma et al. reported a palladium-nanoparticle-intercalated montmorillonite catalyzing solvent-free hydrogenation of squalene (Scheme 12.9) [147]. A metal loading of approximately 6 wt% in the clay proved to be sufficient to catalyze the hydrogenation of squalene to squalane under a H_2 pressure of 6 bar at 200°C for 10 h, obtaining a conversion of 100% and selectivity of 100%. The efficiency of this catalyst was believed to arise from the formation of Pd nanoparticles with a dominating (111) plane, which promoted the reaction

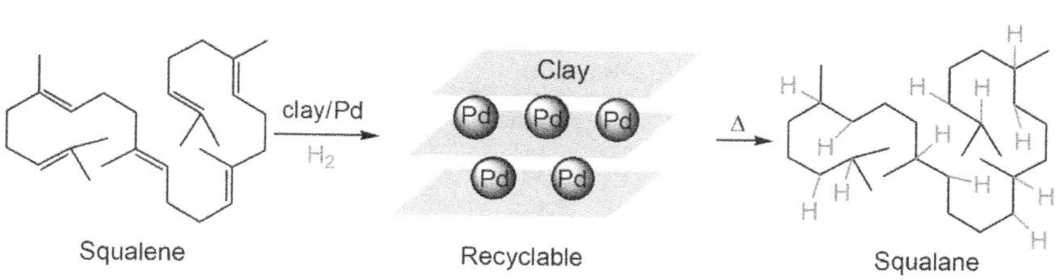

SCHEME 12.9 Palladium-nanoparticles-intercalated montmorillonite catalyzing the hydrogenation of squalene. *(Reproduced with permission from V.K. Soni, R.K. Sharma, ChemCatChem 8 (2016) 1763–1768, Copyright 2016, John Wiley and Sons.)*

at a low Pd/squalene ratio. Recycling experiments show that the palladium-intercalated montmorillonite catalyst can be reused up to five times without significant loss of activity and selectivity.

4.2 Oxidation Reactions

Partial oxidation reactions are widely applied in fine chemical production [148]. The use of stoichiometric inorganic reagents is often involved. The present stringent environmental standards urge the scientific community to develop new, environmentally benign methods. Although homogeneous catalysis provides powerful solutions, the problems relevant to corrosion, recovery, and reuse of the catalyst, on an industrial scale, reflect the limitations of these processes [149]. Heterogeneous catalysts are therefore in high demand regarding their value in easy separation and recyclability.

A catalytic liquid-phase oxidation of a side chain of antioxidizing cresols provides a clean route for several industrial and pharmaceutical chemicals [150,151]. There have been several heterogeneous catalysts containing copper or manganese supported on zeolite, carbon, or resins reported so far [152,153]. Rode et al. described the liquid-phase oxidation of *p*-cresol catalyzed by Co-clay prepared by different approaches (Scheme 12.10) [151,154,155]. Synthesized by an in situ precipitation method, Co-saponite with a Co weight percentage of 13% afforded the highest conversion of *p*-cresol with 92% selectivity to *p*-hydroxybenzaldehyde without any nonoxidized products. Interestingly, the oxidation of *p*-cresol catalyzed by Co-saponite under an inert (N_2) atmosphere proceeded, though the conversion was only 1%, which was possible only by utilizing the lattice oxygen of the Co-saponite. This strongly suggests that the

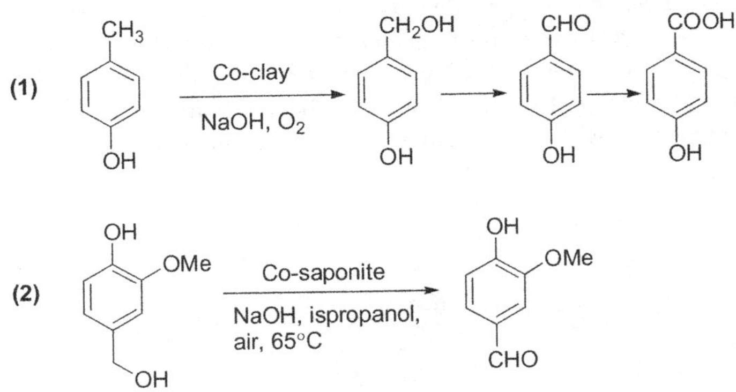

SCHEME 12.10 Co-clay catalyzed liquid phase oxidation of *p*-cresol and *p*-vanillyl alcohol. *(Reproduced from V.S. Kshirsagar, A.C. Garade, R.B. Mane, K.R. Patil, A. Yamaguchi, M. Shirai, C.V. Rode, Appl. Catal. A: Gen. 370 (2009) 16–23; A.C. Garade, N.S. Biradar, S.M. Joshi, V.S. Kshirsagar, R.K. Jha, C.V. Rode, Appl. Clay Sci. 53 (2011) 157–163, with permission from Elsevier.)*

Mars-van Krevelen pathway is operating for the liquid-phase oxidation of *p*-cresol over Co-saponite [156,157]. Another approach to the preparation of the Co catalyst involved the intercalation of cobalt-salen complexes into the interlamellar spaces of montmorillonite. XANES and EXAFS results indicated that cobalt atoms in Co(salen)-montmorillonite form two additional Co—O bonds upon intercalation while retaining the Co-salen structure. This catalyst showed excellent activity for the air oxidation of *p*-cresol, providing 88% selectivity to the oxidation products [155]. Furthermore, the liquid oxidation of *p*-vanillyl alcohol to *p*-vanillin with a conversion of 55% and selectivity of 99% was also achieved with Co-saponite (Scheme 12.10) [158]. The good catalytic properties are attributed to the increasing distribution ratios of Co_3O_4/CoO phases from 0.84 to 1.67, which is associated with an increase in Co loading from 5% to 13%.

The Baeyer-Villiger oxidation is of considerable interest in organic chemistry, since the products, lactones, or esters are important intermediates in the agrochemical and pharmaceutical industries [159,160]. Dutta et al. reported the synthesis of Fe_3O_4 magnetic nanoparticle-supported montmorillonite and its catalytic behaviors in Baeyer-Villiger oxidation of cyclic and aromatic ketones using H_2O_2 as an oxidant (Scheme 12.11) [161]. Prior to the introduction of Fe_3O_4 nanoparticles, montmorillonite was treated with acid to generate pores for hosting the nanoparticles. All the examined cyclic and aromatic ketones yielded the desired ester with good conversion. However, the cyclic

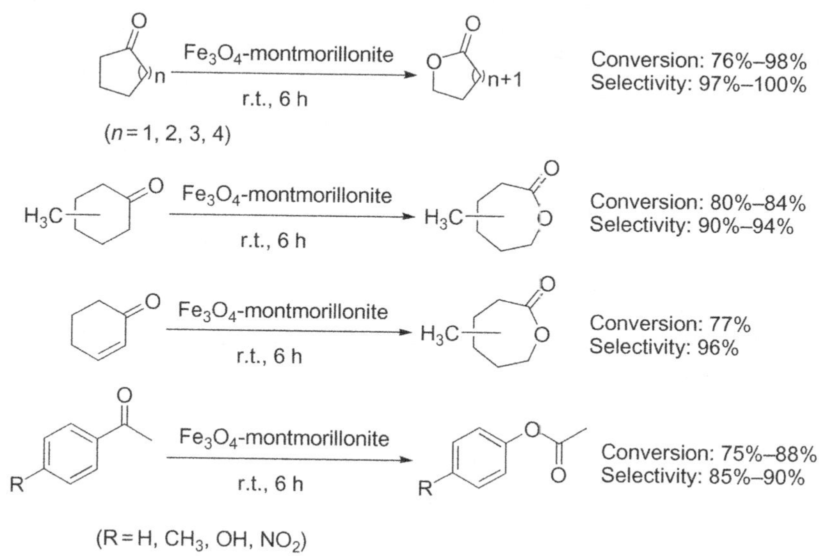

SCHEME 12.11 Fe_3O_4 magnetic nanoparticle-supported montmorillonite catalyzed Baeyer-Villiger oxidation of cyclic and aromatic ketones using H_2O_2 as the oxidant. *(Reproduced from P.K. Saikia, P.P. Sarmah, B.J. Borah, L. Saikia, K. Saikia, D.K. Dutta, Green Chem. 18 (2016) 2843–2850, with permission from The Royal Society of Chemistry.)*

ketones showed a higher conversion than the aromatic ones, which may be due to the ring strain present in the cyclic ketones making them less stable than the aromatics [161]. Moreover, the resonance effect in the aromatic ketones may also stabilize them, which is absent in the cyclic ones. The catalyst can be magnetically recovered and recycled for three runs without obvious loss of efficiency.

Kiuchi et al. described an easy and versatile method for the preparation of a gold nanoparticle catalyst from an aqueous alkaline solution of chloride-free gold acetate [162]. Au nanoparticles can be immobilized on the supports by impregnating them with the stocked solution and followed by calcination. For glucose oxidation in an aqueous solution at pH 9.5 and at 60°C, Au-saponite showed remarkable activity, better than Au nanoparticles supported on other supports such as Al_2O_3, CeO_2, TiO_2, and SiO_2. Besides the small size of Au nanoparticles, the interlaminar space of the clay was swollen by water. Large external surface area based on its structure and the excellent dispersion in water should have contributed to the high activity [162].

4.3 Diels-Alder Reactions

The Diels-Alder reaction is one of the most important methods in organic synthesis. Through a cycloaddition process between a conjugated diene (4e π-system) and a dienophile (2e π-system), a 6-membered ring system can be formed with specific regio- and stereochemical control [163]. It has long been known that Lewis acids catalyze the Diels-Alder reaction by activating the dienophile and increasing both regio and endo/exo selectivities [164]. In addition, the control over the structure of the enoate moiety of the dienophile by complexation with the Lewis acid allows good diastereofacial selectivities in asymmetric Diels-Alder reactions with chiral dienophiles [164–166]. Smectite clays of various compositions, such as montmorillonite, have often been used as solid acid catalysts. By varying the internal structure and the mole ratio of Brønsted or Lewis acid sites, smectite clays have found applications in catalyzing the Diels-Alder reactions. For example, in the hetero-Diels-Alder reaction of 2,3-dimethyl-1,3-butadiene with benzaldehyde and its derivatives catalyzed by the montmorillonite K10, a preheating treatment (250°C) of the catalyst significantly improved the catalytic activity (Scheme 12.12) [167]. The authors

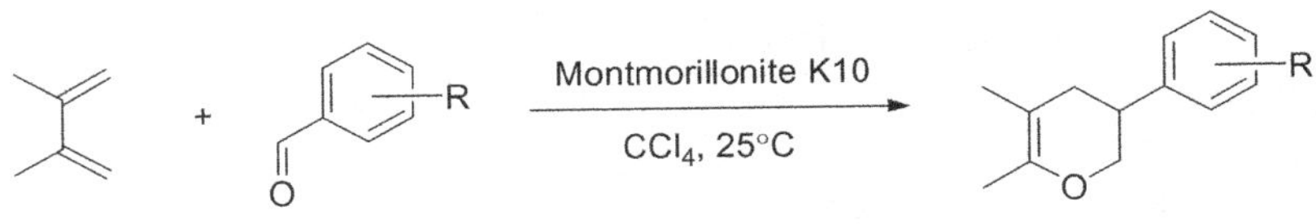

R = H, *o*-OMe, *o*-Me, *o*-OEt, *o*-F, *o*-Cl, *o*-Br, *o*-CN, *o*-NO_2, *p*-Cl, *p*-NO_2

SCHEME 12.12 The hetero-Diels-Alder reaction of dimethylbutadiene with benzaldehyde and its derivatives. *(Adapted with permission from M.R. Dintzner, A.J. Little, M. Pacilli, D.J. Pileggi, Z.R. Osner, T.W. Lyons, Tetrahedron Lett. 48 (2007) 1577–1579, Copyright 2007, Elsevier.)*

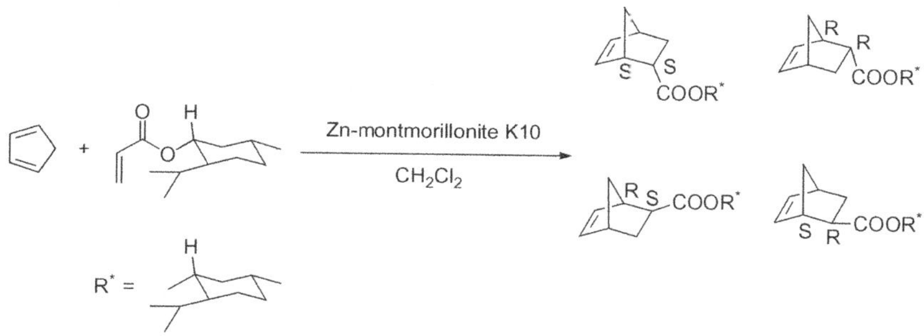

SCHEME 12.13 Zn-montmorillonite-catalyzed Diels-Alder reactions of cyclopentadiene. *(Adapted with permission from C. Cativiela, J.M. Fraile, J.I. García, J.A. Mayoral, E. Pires, F. Figueras, L.C. de Mènorval, Tetrahedron 48 (1992) 6467–6476, Copyright 1992, Elsevier.)*

suggested that the preheating process induces the collapse of the interior structure of montmorillonite by expelling water molecules, leading to a decrease of the Brønsted acidity and an increase of the Lewis acidity [167]. Consequently, the catalytic activity closely relevant to Lewis acidity is enhanced.

Besides serving as catalysts, smectite clays are widely used as support for organocatalysts. Metal species are commonly intercalated into the interlaminar space between the clay layers via ion-exchange. The homogeneous organocatalyst can therefore be heterogenized. Compared with other inorganic solid support such as silica, zeolite, and hydroxyapatite, smectite clays show superior catalytic activity and stability.

Mayoral et al. found that calcined Zn(II)-exchanged K10 montmorillonite is an efficient catalyst in the asymmetric Diels-Alder reaction between (−)-menthyl acrylate and cyclopentadiene (Scheme 12.13) [168,169]. Better conversions and diastereofacial selectivities were obtained with the calcined Zn(II)-montmorillonite (550°C, 1 h) compared with those without calcination. High temperature calcination not only eliminates the Brønsted acid sites but removes the organic radicals, which may promote diene dimerization via cation-radical intermediates [170]. Both endo/exo and diastereofacial selectivities increase at a lower reaction temperature (−20°C). Meanwhile, a dienophile model was proposed that displays an anti-enoate conformation where the isopropyl group of the (−)-menthol covers the re-face of the dienophile, leading to a preferential attack on the si-face by the diene [169].

While Lewis acids are often regarded to be beneficial for promoting the Diels-Alder reaction, there are cases where the presence of Lewis acids played a negative role. Mayoral et al. reported that in the Diels-Alder reaction of cyclopentadiene with *trans*-anethole catalyzed by Fe(III)-K10 montmorillonite (Fe(III)-K10), the polymerization of *trans*-anethole was simultaneously promoted by Brønsted acids and Lewis acids (Scheme 12.14) [171]. To shield the influence of the acidic sites, a silylation treatment of Fe(III)-K10 was performed to reduce the acidity. Surprisingly, the silylated clay catalyst (Fe(III)-K10-sily) selectively catalyzed the reaction of cyclopentadiene with *trans*-anethole

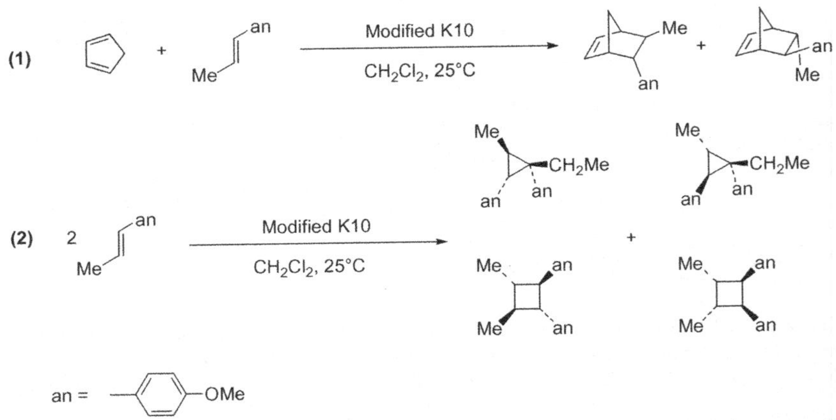

SCHEME 12.14 Pericyclic reaction of *trans*-anethole catalyzed by modified K10 montmorillonite. *(Reproduced from J.M. Fraile, J.I. García, J.A. Mayoral, T. Tarnai, P.J. Alonso, Chem. Commun. 32 (1996) 1981–1982, with permission from The Royal Society of Chemistry.)*

without the undesired polymerization, which indicated that the structural Fe(III) of the K10 can be reduced by the *trans*-anethole.

4.4 Friedel-Crafts Reactions

The Friedel-Crafts reaction is an important carbon-carbon bond forming approach in constituting fundamental intermediates in the fine chemical, pharmaceutical, fragrance, dye and agrochemical industries [172,173]. By electrophilic aromatic substitution, the aromatic and heteroaromatic compounds can be attached by alkyl or acyl groups catalyzed by heterogenous or homogeneous acid catalysts [172–174]. Alkyl halides, alcohols, sulfonates, and olefins are the usual alkylating agents, and acyl halides or acid anhydrides are the common acylation agents [127]. The conventional Lewis acids such as $ZnCl_2$, $AlCl_3$, $FeCl_3$, $SnCl_4$, and $TiCl_4$ are often homogeneously involved in the catalytic Friedel-Crafts reactions and require more than stoichiometric amounts. From a green chemistry point of view, supported Lewis acid catalysts, particular clays, and modified clays are showing great promise in Friedel-Crafts reactions.

Laszlo et al. described the use of Fe(III)- and Zn(II)-montmorillonites in Friedel-Crafts acylation [175–178]. The clay catalysts were prepared through ion-exchange of the interlamellar cations and through impregnation of the metal chlorides into montmorillonite K10. The acid Fe(III)-K10, obtained from the ion-exchange approach, showed excellent activity in the acylation of mesitylene and *p*-xylene with benzoyl chloride or benzoic anhydride (arene/acylating agent mole ratio = 7, ≥98 yield). The comparison of the catalytic performance of exchanged versus impregnated clay catalysts suggested that the preparation method is not

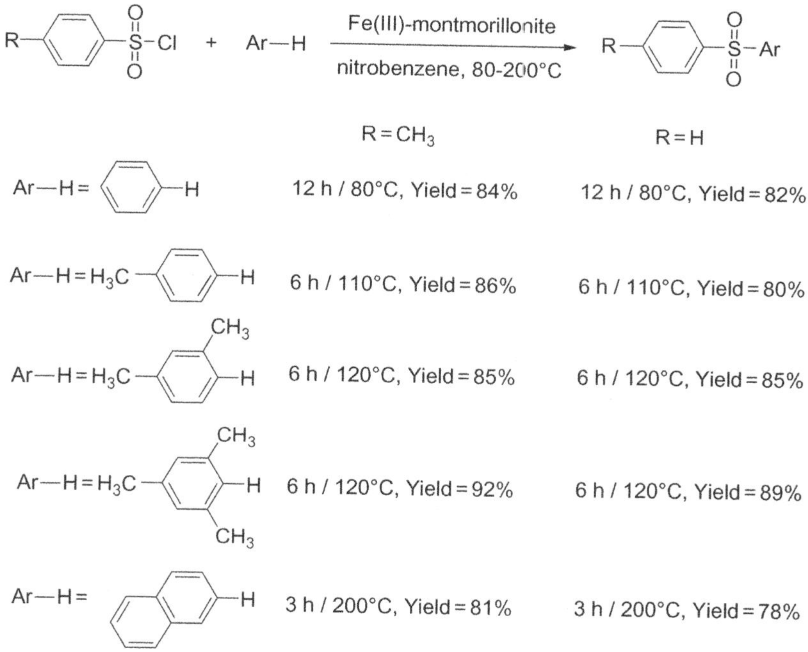

SCHEME 12.15 Friedel-Crafts sulfonylation of arenes with arylsulfonyl chlorides catalyzed by Fe(III)-montmorillonite. *(Adapted with permission from B.M. Choudary, N.S. Chowdari, M. Lakshmi Kantam, R. Kannan, Tetrahedron Lett. 40 (1999) 2859–2862, Copyright 1999, Elsevier.)*

important. After a fresh run, the ion-exchanged Fe(III)-K10 catalyst could be recycled and reused for up to four times without any special treatment.

Choudary et al. reported that the Fe(III)-exchanged montmorillonite catalyst was efficient to catalyze the Friedel-Crafts sulfonylation of arenes with arylsulfonyl chlorides to obtain the corresponding sulfones (Scheme 12.15) [179]. Notably, the *para*-selectivity for the *para*-isomers of ditolyl sulfone and phenyl tolyl sulfone could be superior to 95%, while the Lewis acid $AlCl_3$ only yields a mixture of isomers with the composition of *ortho*: *meta*: *para* = 28: 7: 65 [180]. Furthermore, the regioselectivity in sulfonylation of naphthalene with benzene sulfonyl chloride and toluene-*p*-sulfonyl chloride was also improved. Compared with the Zn-, Cu-, Al-exchanged montmorillonites in catalytic sulfonylation, Fe(III)-montmorillonite was found to be superior. The higher activity may be referred to the generated Lewis acid sites upon the introduction of Fe^{3+} in the montmorillonite and the Brønsted acid sites from the edges of broken layers in the clay [179]. The same catalyst was also successfully applied for catalyzing one pot Friedel-Crafts alkylation of arenes using alcohols (Scheme 12.16) [181]. Interestingly, in the case of toluene, the alkylation of toluene with cyclopentanol afforded the cyclopentyltoluene with a 62% *para* selectivity (*ortho*:

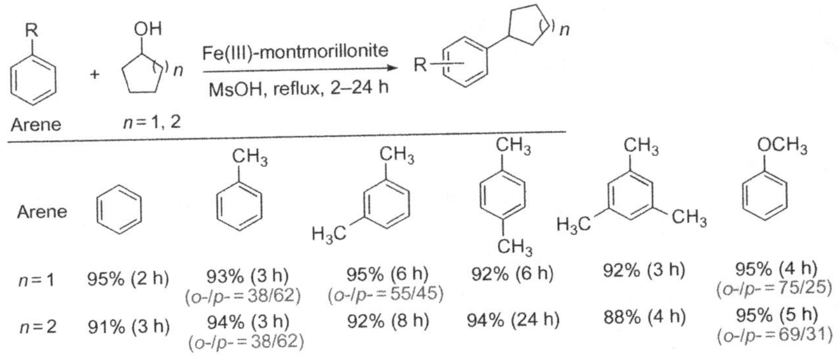

SCHEME 12.16 Fe(III)-montmorillonite catalyzed one pot Friedel-Crafts alkylation of arenes using alcohols. *(Adapted with permission from B.M. Choudary, B.P.C. Rao, N.S. Chowdari, M.L. Kantam, Catal Commun. 3 (2002) 363–367, Copyright 2002, Elsevier.)*

para = 38: 62); however, under similar conditions Sc(OTf $)_3$ catalyzes the alkylation with cyclopentyl methanesulfonate with only 50% paraselectivity [182]. The activity may be attributed to the shape selectivity of the clay. Considering the mechanism of metal-exchanged clay catalyzing alkylation of aromatic compounds by in situ generation of alkyl sulfonates, it is expected that the metal aqua complexes are formed and protonate the alcohol. Upon the interaction of protonated alcohol with methanesulfonic acid (MsOH), alkyl sulfonate and water are produced and the metal aqua complex can also be regenerated [183]. Meanwhile, the metal aqua complexes again protonate the alkyl sulfonate to generate the alkyl electrophile, which in turn attacks on aromatic nucleus to give alkyl substituted aromatic compounds and regenerates the MsOH to complete the catalytic cycle. Moreover, this catalyst can be recovered and reused for five cycles without activity loss.

Since the Fe^{3+}-exchanged montmorillonite K10 catalyst has gained much attention regarding its excellent catalytic activity, an insight into its real active

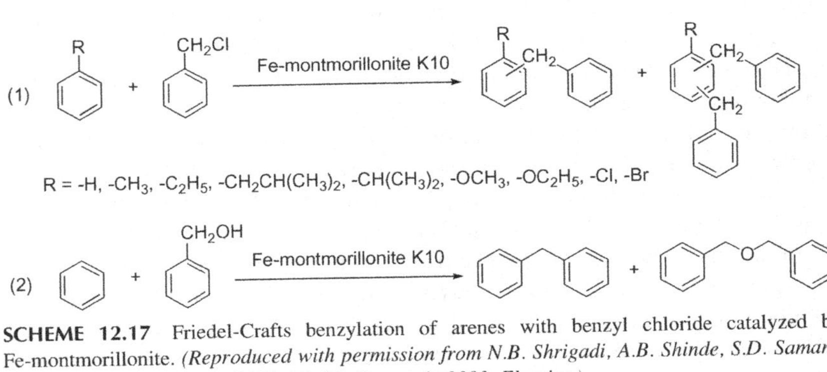

SCHEME 12.17 Friedel-Crafts benzylation of arenes with benzyl chloride catalyzed by Fe-montmorillonite. *(Reproduced with permission from N.B. Shrigadi, A.B. Shinde, S.D. Samant, Appl. Catal. A: Gen. 252 (2003) 23–35, Copyright 2003, Elsevier.)*

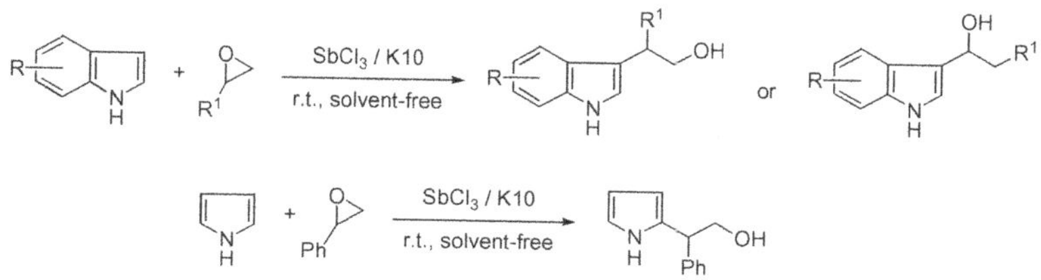

SCHEME 12.18 $SbCl_3$/K10 catalyzed Friedel-Crafts alkylation of nitrogen heterocycles with epoxides. *(Reproduced with permission from Y.H. Liu, Q.S. Liu, Z.H. Zhang, Tetrahedron Lett. 50 (2009) 916–921, Copyright 2008, Elsevier.)*

species requires rational understanding. Samant et al. carried out a study by comparing the catalytic performance of Fe_xO_y-montmorillonite and $FeCl_3$-montmorillonite in the Friedel-Crafts benzylation of different arenes using benzyl chloride (Scheme 12.17) [184]. Among the different iron oxides, γ-Fe_2O_3 showed the best catalytic activity for Friedel-Crafts benzylation of substituted aromatic compound using benzyl chloride. The difference in the activity of various Fe-oxides could be related to their lattice structure. At different calcination temperature, α-FeOOH is transformed into α-Fe_2O_3 [185] and γ-FeOOH into γ-Fe_2O_3 [186]. The phase transformation could lead to a decrease of the acidity. Consequently, a decrease on the activity was observed in the following order: γ-$Fe_2O_3 > Fe_3O_4 > \alpha$-$Fe_2O_3$ [184].

Zhang et al. reported that $SbCl_3$ supported on montmorillonite K10 is an efficient catalyst for Friedel-Crafts alkylation of nitrogen heterocycles such as indoles and pyrroles with epoxides (Scheme 12.18) [187]. Surprisingly, the Friedel-Crafts alkylation of indole with glycidyl phenyl ether catalyzed by $SbCl_3$/montmorillonite K10 could proceed at room temperature under solvent-free conditions, presenting a 58% yield of 1-(1*H*-indol-3-yl)-3-phenoxypropan-2-ol. However, the same reaction catalyzed by 10 mol % of $SbCl_3$ alone gave only a 20% yield of the product and no conversion was observed in the presence of montmorillonite K10 alone. Besides catalyzing the C3-alkylation of indoles with epoxides, $SbCl_3$/montmorillonite K10 catalyzed the reaction of pyrrole with aromatic epoxides to afford the corresponding pyrrole derivatives in good yields (85%–90%). Moreover, the reaction of aliphatic epoxides, glycidyl phenyl ether, and 2-chlorooxirane, with pyrrole were catalyzed by $SbCl_3$/montmorillonite K10, giving the desired products in moderate yields (45% and 51%, respectively).

Alkylation of anisole with cyclohexene is an industrially important reaction. Yadav et al. described an application of a solid acid catalyst $Cs_{2.5}H_{0.5}PW_{12}O_{40}$ (cesium-modified dodecatungstophosphoric acid; designated as Cs-DTP/K-10), for a solventless Friedel-Crafts alkylation of anisole with cyclohexene (Scheme 12.19) [188]. Acid treated clay (K-10) as support with 20% m/m 20% $Cs_{2.5}H_{0.5}PW_{12}O_{40}$, was found to be highly efficient. This reaction leads to the production of 2-cyclohexylanisole and 4-cyclohexylanisole. Notably, the

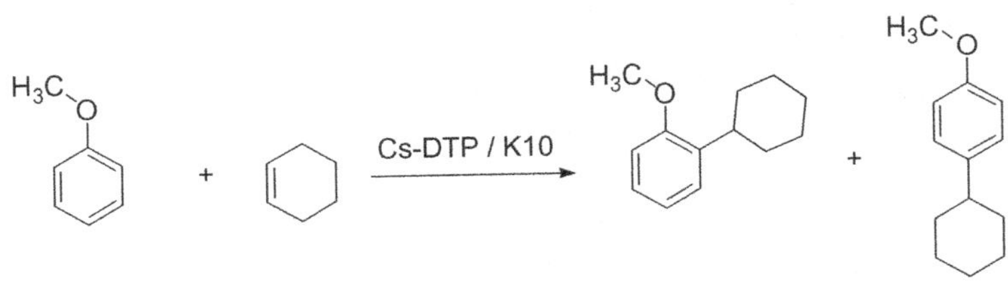

SCHEME 12.19 Alkylation of anisole with cyclohexene catalyzed by $Cs_{2.5}H_{0.5}PW_{12}O_{40}$/K10 (Cs-DTP/K10). *(Reproduced with permission from G.D. Yadav, S.R. More, Appl. Clay Sci. 53 (2011) 254–262, Copyright 2011, Elsevier.)*

reaction was carried out without using solvent. The kinetic model shows that this reaction follows a pseudo first-order mechanism. The apparent activation energy was calculated to be 21.96 kcal/mol. After a fresh run, the catalyst was robust and can be reused for three cycles without loss of activity.

4.5 Other Reactions

The redox isomerization of allylic alcohols catalyzed by transition-metal complexes is a useful and straightforward synthetic route to carbonyl compounds. The mononuclear arene-ruthenium(II) derivatives $[(C_6H_6)RuCl_2(PTA\text{-}Me)]$ (the ionic phosphine ligand 1-methyl-3,5-diaza-1-azonia-7-phosphaadamantane chloride, PTA-Me) have been found to be highly active for the redox isomerization of allylic alcohols; however, the heterogeneous montmorillonite K10-supported $[(C_6H_6)RuCl_2(PTA\text{-}Me)]$ catalyst also showed good activity in the isomerization of the model substrate 1-octen-3-ol (Scheme 12.20) [189]. In addition, it could be easily separated from the reaction media by simple filtration and reused up to 11 consecutive runs with retention of its efficiency.

Methyl mandelate, an important fine chemical in perfumery, was produced through esterification of mandelic acid with methyl alcohol catalyzed by montmorillonite K10 supported dodecatungstophosphoric acid cesium salt (Cs-DTP/K10) (Scheme 12.21) [190]. The Cs-DTP/K10 catalyst showed good activity and selectivity and can also be recycled. Using the same catalyst (Cs-DTP/K10 clay), the esterification of benzoic acid with phenol underwent Fries rearrangement to obtain 2- and 4-hydroxybenzophenones instead of the intermediate phenyl benzoate (Scheme 12.21) [191]. The selectivity of the products depended on the reaction parameters.

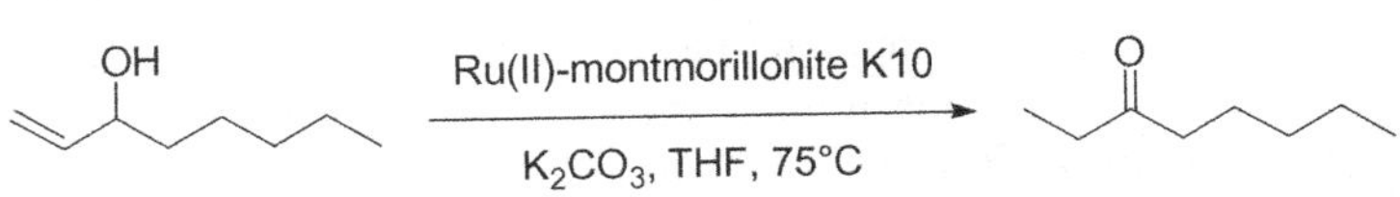

SCHEME 12.20 Isomerization of 1-octen-3-ol into octan-3-one catalyzed by $[(C_6H_6)RuCl_2(PTA\text{-}Me)]$ supported in montmorillonite K10. *(Reproduced with permission from L. Menéndez-Rodríguez, P. Crochet, V. Cadierno, J. Mol. Catal. A: Chem. 366 (2013) 390–399, Copyright 2012, Elsevier.)*

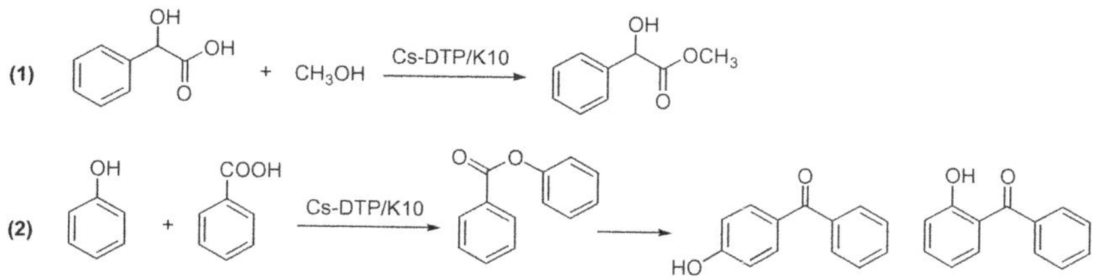

SCHEME 12.21 K10 supported dodecatungstophosphoric acid-Cs salt (Cs-DTP/K10) for esterification of mandelic acid with methyl alcohol. *(Reproduced with permission from G. D. Yadav, R.D. Bhagat, Clean Technol. Environ. Policy 7 (2005) 245–251, with copyright 2005, Springer-Verlag; from G.D. Yadav, G. George, J. Mol. Catal. A Chem. 292 (2008) 54–61, with copyright 2008, Elsevier.)*

Hydroxylation of aryl boronic acid to produce phenols bearing functional groups is an attractive approach to construct more complex molecules because of its harmless nature, stability toward heat, air, moisture, and facile conversion [192,193]. Singh et al. found that a $Cu(OH)_x$-entrapped montmorilonite-KSF catalyzes the ipso-hydroxylation of arylboronic acids under ambient conditions without the requirement of any ligand or base (Scheme 12.22) [192]. This catalyst can be recycled and reused for 10 consecutive cycles without loss of activity.

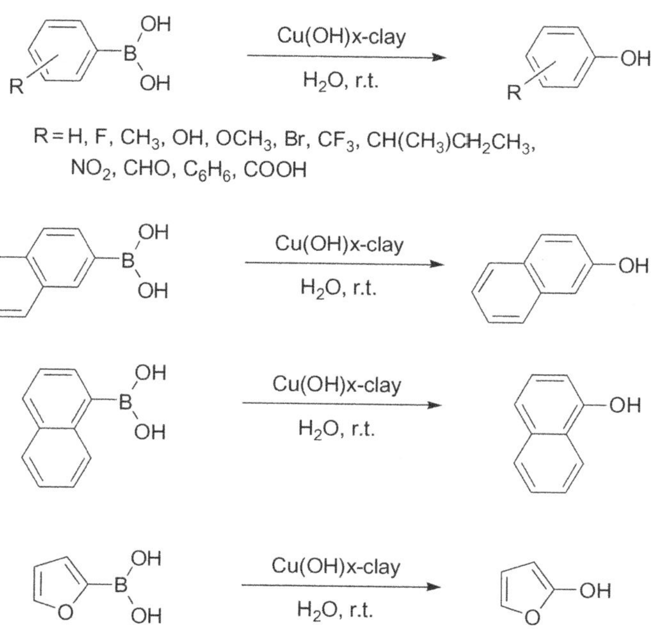

SCHEME 12.22 The ipso-hydroxylation reaction of arylboronic acids over $Cu(OH)_x$-montmorillonite-KSF. *(Adapted with permission from B. Ahmad Dar, P. Bhatti, A.P. Singh, A. Lazar, P.R. Sharma, M. Sharma, B. Singh, Appl. Catal. A: Gen. 466 (2013) 60–67, Copyright 2013, Elsevier.)*

For the dehydrogenation of ethylbenzene to afford styrene a processed Venezuelan smectite clay intercalated by a trinuclear iron complex, $[Fe_3O(OAc)_6 \cdot 3H_2O]^+$ was used [194]. At 410°C for 1 h, ethylbenzene was selectively converted to styrene. This catalyst shows superior activity with a 50% increase in styrene formation as compared to other catalysts.

5 CONCLUDING REMARKS

The recent progress in the ever expanding field of smectite clay-supported heterogeneous catalysts is systematically reviewed in this chapter. Smectite clays can be exploited as efficient catalyst supports due their unique swelling properties and flexible intercalation capacity. They offer an excellent microenvironment in their interlaminar space, which can absorb metal ions, metal complexes, or cationic molecules via ion exchange. The swelling of the interlaminar distance is tunable, the extent of which strongly depends on the layer charge, the exchange agent, and the system environment. From a catalyst design point of view, the catalyst immobilized on the surface of the interlayers can be expected to possess good stability and dispersion, and it can be easily accessible to the substrates. Compared with traditional zeolite catalysts, the larger pore sizes in smectite clays provide opportunities for facilitating the catalytic synthesis and the transformation of larger molecules. Smectite clays offer a good matrix for intercalating metallic catalysts; however, their fragility towards harsh reaction conditions limits their utility to a certain extent. This problem can be solved by clay pillaring and subsequent immobilization of the catalyst. Prospectively, the design of highly efficient smectite clay intercalated metal catalysts remains significantly attractive. This type of catalyst can be a promising alternative in organic synthesis and fine chemistry.

REFERENCES

[1] M.L. Occelli, Ind. Eng. Chem. Prod. Res. Dev. 22 (1983) 553–559.

[2] W. Vermeiren, J.-P. Gilson, Top. Catal. 52 (2009) 1131–1161.

[3] F. Figueras, Catal. Rev. 30 (1988) 457–499.

[4] A.E.C. Collis, I.T. Horváth, Cat. Sci. Technol. 1 (2011) 912–919.

[5] E.D. Park, K.H. Lee, J.S. Lee, Catal. Today 63 (2000) 147–157.

[6] F.T. Sejidov, Y. Mansoori, N. Goodarzi, J. Mol. Catal. A 240 (2005) 186–190.

[7] L.J. Durndell, C.M.A. Parlett, N.S. Hondow, M.A. Isaacs, K. Wilson, A.F. Lee, Sci. Rep. 5 (2015).

[8] N. Linares, A.M. Silvestre-Albero, E. Serrano, J. Silvestre-Albero, J. García-Martinez, Chem. Soc. Rev. 43 (2014) 7681–7717.

[9] J.M. Adams, R.W. McCabe, Clay Sci. 1 (2006) 541–581.

[10] J.M. Adams, K. Martin, R.W. McCabe, J. Incl. Phenom. 5 (1987) 663–674.

[11] J. Billingham, C. Breen, J. Yarwood, Clay Miner. 31 (1996) 513–522.

[12] R.J. White, R. Luque, V.L. Budarin, J.H. Clark, D.J. Macquarrie, Chem. Soc. Rev. 38 (2009) 481–494.

[13] C.H. Zhou, Appl. Clay Sci. 53 (2011) 87–96.
[14] L. Zonghui, S. Guida, Factors Affecting Acidity and Basal Spacing of Cross-Linked Smectites, in: B. Drzaj, S. Hocevar, S. Pejovnik (Eds.), Studies in Surface Science and Catalysis 24, Elsevier, Amsterdam, 1985, pp. 493–500.
[15] J.J. Fripiat, M.I. Cruz-Cumplido, Annu. Rev. Earth Planet. Sci. 2 (1974) 239–256.
[16] J.T. Kloprogge, J. Porous, Mater. Eng. 5 (1998) 5–41.
[17] I.E. Odom, Philos. Trans. R. Soc. Lond. A 311 (1984) 391–409.
[18] R.E. Grim, Clay Mineralogy 4. Structure of Clay Minerals, Smectite Minerals, McGraw-Hill, 1968.
[19] J.T. Kloprogge, S. Komarneni, J.E. Amonette, Clay Clay Miner. 47 (1999) 529–554.
[20] L. Bailey, H.N.W. Lekkerkerkerb, G.C. Maitlandc, Soft Matter 11 (2015) 222–236.
[21] L.J. Michot, I. Bihannic, S. Maddi, S.S. Funari, C. Baravian, P. Levitz, et al., Proc. Natl. Acad. Sci. U. S. A. 103 (2006) 16101–16104.
[22] P.H. Nadeau, D.C. Bain, Clay Clay Miner. 34 (1986) 455–464.
[23] D. Eberl, Clay Clay Miner. 26 (1978) 327–340.
[24] M. Brigatti, E. Galan, B.K.G. Theng, Structure and Mineralogy of Clay Minerals, in: F. Bergaya, B.K.G. Theng, G. Lagaly (Eds.), Handbook of Clay Science, Vol. 1, Elsevier, Amsterdam, 2006, pp. 21–81.
[25] E.A. Emam, ARPN J. Sci. Technol. 3 (2013) 356–375.
[26] J.-H. Cha, M. Kwon, W. Lee, H. Lee, J. Phys. Chem. C 115 (2011) 15655–15660.
[27] D.A. Laird, Layer Charge and Crystalline Swelling of Expanding 2:1 Phyllosilicates (Ph.D. dissertation), Iowa State University, Ames, IA, 1987 (Diss. Abstr. DA87 – 16784).
[28] E.S. Boek, P.V. Coveney, N.T. Skipper, J. Am. Chem. Soc. 117 (1995) 12608–12617.
[29] Z. Király, I. Dékány, Á. Mastalir, M. Bartók, J. Catal. 161 (1996) 401–408.
[30] J. Qian, C.-Y. Guo, Macromolecules 4 (2010) 1–14.
[31] J.M. Adams, T.V. Clapp, E. Clement, Clay Miner. 18 (1983) 411–421.
[32] A. Weiss, Angew. Chem. Int. Ed. 20 (1981) 850–860.
[33] M. Frenkel, Clay Clay Miner. 22 (1974) 435–441.
[34] M.B. McBride, Environmental Chemistry of Soils, Oxford University Press, New York, 1994.
[35] J.M. Fraile, J.I. García, J.A. Mayoral, Chem. Rev. 109 (2009) 360–417.
[36] L.V. Leemput, M.S. Stul, A. Maes, J.B. Uytterhoeven, A. Cremers, Clay Clay Miner. 31 (1983) 261–268.
[37] P. Djomgoue, D. Njopwouo, J. Surf. Eng. Mater. Adv. Technol. 3 (2013) 275–282.
[38] (a) F. Bergaya, G. Lagaly, Appl. Clay Sci. 19 (2001) 1–3. (b) P. Komadel, J. Madejová, Acid activation of clay minerals, in: F. Bergaya, B.K.G. Theng, G. Lagaly (Eds.), Handbook of Clay Science, Vol. 1, Elsevier, Amsterdam, 2006, pp. 263–287. (c) A. Amari, M. Chlendi, A. Gannouni, A. Bellagi, Appl. Clay Sci. 47 (2010) 457–461. (d) P. Komadel, M. Janek, J. Madejová, A. Weekes, C. Breen, J. Chem. Soc. Faraday Trans. 93 (1997) 4207–4210. (e) H. Cetisli, T. Gedikbey, Clay Miner. 25 (1990) 207–215.
[39] (a) C. Breen, J. Madejová, P. Komadel, J. Mater. Chem. 5 (1995) 469–474. (b) C. Breen, J. Madejová, P. Komadel, Appl. Clay Sci. 10 (1995) 219–230. (c) P. Komadel, J. Madejová, M. Janek, W.P. Gates, R.J. Kirkpatrick, J.W. Stucki, Clay Clay Miner. 44 (1996) 228–236. (d) I. Novak, B. Cicel, Clay Clay Miner. 26 (1978) 341–344. (e) C. Breen, Clay Miner. 26 (1991) 473–486. (f) T.-S. Li, A.-X. Li, J. Chem. Soc., Perkin Trans. 1 (1998) 1913–1918.
[40] (a) O. Marvi, L.Z. Fekri, M. Takhti, Russ. J. Gen. Chem. 84 (2014) 1837–1840. (b) J.-Q. Jiang, Z. Zeng, Chemosphere 53 (2003) 53–62.
[41] M. Hirano, K. Ukawa, S. Yakabe, J.H. Clark, T. Morimoto, Synthesis 8 (1997) 858–860.

[42] A. Cornelis, P. Laszlo, Synthesis 10 (1985) 909–919.

[43] (a) C.A. Gorski, M. Aeschbacher, D. Soltermann, A. Voegelin, B. Baeyens, M.M. Fernandes, et al., Environ. Sci. Technol. 46 (2012) 9360–9368. (b) P. Komadei, J. Madejová, J.W. Stucki, Clay Clay Miner. 47 (1999) 458–465.

[44] (a) R.A. Schoonheydt, T.J. Pinnavaia, G. Lagaly, N. Gangas, Pure Appl. Chem. 71 (1991) 2367–2371. (b) S.A. Solin, Annu. Rev. Mater. Sci. 27 (1997) 89–115.

[45] K.N. Bhatt, S.B. Halligudi, J. Mol. Catal. 91 (1994) 187–194.

[46] M.M.T. Khan, S.A. Samad, M.R.H. Siddiqui, G. Ramachandraiah, Polyhedron 10 (1991) 2729–2736.

[47] T.J. Pinnavaia, Clay-Activated Catalytic Hydrogenations: Catalyst Synthesis and Factors Influencing Selectivity, in: P. Laszlo (Ed.), Preparative Chemistry Using Supported Reagents, Academic Press Inc, London, 1987, pp. 484–485.

[48] S. Shimazu, W. Teramoto, T. Uematsu, J. Ion Exch. 4 (1993) 123–129.

[49] E.P. Giannelis, T.J. Pinnavaia, Inorg. Chem. 24 (1985) 3602–3607.

[50] S. Shimazu, T. Hirano, T. Uematsu, Appl. Catal. 34 (1987) 255–261.

[51] T. Mitsudome, K. Nose, T. Mizugaki, K. Jitsukawa, K. Kaneda, Tetrahedron Lett. 49 (2008) 5464–5466.

[52] (a) M.S. Hegde, G. Madras, K.C. Patil, Acc. Chem. Res. 42 (2009) 704–712. (b) M. Flytzani-Stephanopoulos, B.C. Gates, Annu. Rev. Chem. Biomol. Eng. 3 (2012) 545–574.

[53] (a) T.J. Pinnavaia, P.K. Welty, J. Am. Chem. Soc. 97 (1975) 3819–3820. (b) T.J. Pinnavaia, M. Rameswaran, E.D. Dimotakis, E.P. Giannelis, E.G. Rightor, Faraday Discuss. Chem. Soc. 87 (1989) 227–237. (c) S.G. Peng, X.J. Fan, J. Zhang, F.Y. Wang, Appl. Catal. B 115 (2013) 140–141.

[54] M.J. Thomas, Clay, Zeolites and other Microporous Solids for Organic Synthesis, in: R. Scheffold (Ed.), Modern Synthetic Methods, Verlag Salle, Frankfurt am Main, 1989, pp. 259–260.

[55] (a) A. Gil, S.A. Korili, M.A. Vicente, Catal. Rev. Sci. Eng. 50 (2008) 153–221.

[56] (a) M.S. Tzou, T.J. Pinnavaia, Catal. Today 2 (1988) 243–259. (b) A. Gil, L.M. Gandia, M.A. Vicente, Catal. Rev. Sci. Eng. 42 (2000) 145–212. (c) A. Szücs, Z. Király, F. Berger, I. Dékány, Colloids Surf. A Physicochem. Eng. Asp. 139 (1998) 109–118. (d) A. Szűcs, F. Berger, I. Dékány, Colloids Surf. A Physicochem. Eng. Asp. 174 (2000) 387–402.

[57] A. Cornelis, P. Laszlo, Synlett 3 (1994) 155–161.

[58] (a) T.J. Pinnavaia, Science 220 (1983) 365–371. (b) Á. Mastalir, Gy Szöllösi, Z. Király, Zs Rázga, Appl. Clay Sci. 22 (2002) 9–16.

[59] (a) L. Zhou, X. Qi, X. Jiang, Y. Zhou, H. Fu, H. Chen, J. Colloid Interface Sci. 392 (2013) 201–205. (b) G.M. Scheuermann, R. Thomann, R. Muelhaupt, Catal. Lett. 132 (2009) 355–362.

[60] H. Wang, S.X. Deng, Z.R. Shen, J.G. Wang, D.T. Ding, T.H. Chen, Green Chem. 11 (2009) 1499–1502.

[61] B. Török, K. Balázsik, I. Kun, G. Szöllösi, G. Szakonyi, M. Bartók, Stud. Surf. Sci. Catal. 125 (1999) 515–522.

[62] H.A. Patel, H.C. Bajaj, R.V. Jasra, J. Nanopart. Res. 10 (2008) 625–632.

[63] A. Maneshi, In-situ Ethylene Polymerization with Organoclay-Supported Metallocenes for the Preparation of Polyethylene-Clay Nanocomposites (Ph.D thesis), Uni. Waterloo, Waterloo, CA, 2010.

[64] N. Bouazizi, D. Barrimo, S. Nousir, R. Ben Slama, T.C. Shiao, R. Roy, et al., J. Energy Inst. (2016), http://dx.doi.org/10.1016/j.joei.2016.10.002.

[65] C.-C. Chen, P.-L. Kuo, J. Colloid Interface Sci. 293 (2006) 101–107.

[66] I. Dékány, L. Turi, E. Tombácz, J.H. Fendler, Langmuir 11 (1995) 2285–2295.
[67] C. Sievers, Y. Noda, L. Qi, E.M. Albuquerque, R.M. Rioux, S.L. Scott, ACS Catal. 6 (2016) 8286–8307.
[68] M. Haruta, Nature 437 (2005) 1098–1099.
[69] (a) S. Papp, I. Dékány, Colloid Polym. Sci. 281 (2003) 727–737. (b) S. Papp, R. Patakfalvi, I. Dékány, Colloid Polym. Sci. 286 (2008) 3–14. (c) R. Barthos, A. Hegyessy, Z. May, J. Valyon, Catal. Lett. 144 (2014) 702–710. (d) S. Papp, R. Patakfalvi, I. Dékány, Croat. Chem. Acta 80 (2007) 493–502.
[70] T. Mitsudome, K. Nose, K. Mori, T. Mizugaki, K. Ebitani, K. Jitsukawa, et al., Angew. Chem. Int. Ed. 46 (2007) 3288–3290.
[71] A. Orejon, H. Alper, J. Mol. Catal. A 143 (1999) 137–142.
[72] M. Zeng, Y. Wang, Q. Liu, X. Yuan, S. Zuo, R. Feng, et al., ACS Appl. Mater. Interfaces 8 (2016) 33157–33164.
[73] Z. Király, B. Veisz, I. Dékány, Á. Mastalir, Z. Rázga, Chem. Commun. 19 (1999) 1925–1926.
[74] A.V. Martinez, A. Leal-Duaso, J.I. Garcia, J.A. Mayoral, RSC Adv. 5 (2015) 59983–59990.
[75] K. Shimizu, R. Maruyama, S. Komai, T. Kodama, Y. Kitayama, J. Catal. 227 (2005) 202–220.
[76] (a) J.S. Bergman, G.W. Coates, H. Chen, E.P. Giannelis, M.G. Thomas, Chem. Commun. 21 (1999) 2179–2180. (b) A.C. dos Ouros, M.O. de Souzab, H.O. Pastore, J. Braz. Chem. Soc. 25 (2014) 2164–2185.
[77] L. Markó, F. Ungváry, J. Organomet. Chem. 432 (1992) 1–214.
[78] (a) V.B. Jigajinni, R.H. Wightman, Tetrahedron Lett. 23 (1982) 117–120. (b) A.G. Martinez, R.M. Alvarez, M.M. Casadoa, L.R. Subramanian, M.T. Hanack, Tetrahedron 43 (1987) 275–279. (c) P.N. Hawker, Platin. Met. Rev. 39 (1995) 2–8. (d) N. Kimbara, J.-P. Charland, M.F. Wilson, Ind. Eng. Chem. Res. 35 (1996) 3874–3883.
[79] J. Pan, J. Liu, S. Guo, Z. Yang, Catal. Lett. 131 (2009) 179–183.
[80] (a) M. Hiramatsu, T. Machino, K. Mase, M. Hori, H. Kano, J. Nanosci. Nanotechnol. 10 (2010) 4023–4029. (b) F. Cansell, C. Aymonir, J. Supercrit. Fluids 47 (2009) 508–516.
[81] J.B. Harrison, V.E. Berkheiser, G.W. Erdos, J. Catal. 112 (1988) 126–134.
[82] E. Zapp, D. Brondani, I.C. Vieira, C.W. Scheeren, J. Dupont, A.M.J. Barbosa, et al., Sensors Actuators B 155 (2011) 331–339.
[83] M. Bartók, Gy Szöllösi, Á. Mastalir, I. Dékány, J. Mol. Catal. A 139 (1999) 227–234.
[84] C. Claver, E. Fernandez, R. Margalef-Catala, F. Medina, P. Salagre, J.E. Sueiras, J. Catal. 201 (2001) 70–79.
[85] P.K. Saikia, P.P. Sarmah, B.J. Borah, L. Saikia, D.K. Dutta, J. Mol. Catal. A 412 (2016) 27–33.
[86] S. Shimazu, K. Ro, T. Sento, N. Ichikuni, T. Uematsu, J. Mol. Catal. A 107 (1996) 297–303.
[87] R. Ruiz, C. Pesquera, F. González, C. Blanco, Appl. Catal. A 257 (2004) 165–175.
[88] V.L.K. Valli, H. Alper, Chem. Mater. 7 (1995) 359–362.
[89] M. Lenarda, R. Ganzerla, L. Riatto, L. Storaro, J. Mol. Catal. A 187 (2002) 129–134.
[90] (a) C. Philippopoulos, N. Gangas, N. Papayannakos, J. Mater. Sci. Lett. 15 (1996) 1940–1941. (b) S. Morfis, C. Philippopoulos, N. Papayannakos, Appl. Clay Sci. 13 (1998) 203–212.
[91] G. Süss-Fink, J. Organomet. Chem. 751 (2014) 2–19.
[92] R. Berthoud, A. Baudouin, B. Fenet, W. Lukens, K. Pelzer, J.-M. Basset, et al., Chem. Eur. J. 14 (2008) 3523–3526.
[93] (a) M. Womes, T. Cholley, F. Le Peltier, S. Morin, B. Didillon, N. Szydlowski-Schildknecht, Appl. Catal. A 283 (2005) 9–22. (b) B. Chaudret, Top. Organomet. Chem. 16 (2005) 233–259. (c) R.M. Rioux, H. Song, M. Grass, S. Habas, K. Niesz, J.D. Hoefelmeyer, P. Yang, G.A. Somorjai, Top. Catal. 39 (2006) 167–174.

[94] S.H. Joo, J.Y. Park, J.R. Renzas, D.R. Butcher, W. Huang, G.A. Somorjai, Nano Lett. 10 (2010) 2709–2713.

[95] (a) G. Süss-Fink, F.-A. Khan, J. Boudon, V. Spassov, J. Clust. Sci. 20 (2009) 341–353. (b) F.-A. Khan, A. Vallat, G. Süss-Fink, Catal. Commun. 12 (2011) 1428–1431. (c) B. Sun, F.A. Khan, A. Vallat, G. Süss-Fink, Appl. Catal. A 467 (2013) 310–314. (d) B. Sun, D. Carnevale, G. Süss-Fink, J. Organomet. Chem. 821 (2016) 197–205. (e) B. Sun, G. Süss-Fink, J. Organomet. Chem. 812 (2016) 81–86. (f) A. Meister, G. Meister, G. Süss-Fink, J. Mol. Catal. 92 (1994) L123–L126. (g) F.-A. Khan, A. Vallat, G. Süss-Fink, J. Mol. Catal. A 355 (2012) 168–173. (h) G. Süss-Fink, B. Mollwitz, B. Therrien, M. Dadras, G. Laurenczy, A. Meister, G. Meister, J. Clust. Sci. 18 (2007) 87–95. (i) B. Sun, Metallic Ruthenium Nanoparticles Intercalated in Hectorite for Highly Selective Catalytic Hydrogenations [Ph.D thesis], (2015). Retrieved from http://doc.rero.ch/record/257869.

[96] M. Stebler-Röthlisberger, W. Hummel, P.-A. Pittet, H.-B. Bürgi, A. Ludi, A.E. Merbach, Inorg. Chem. 27 (1988) 1358–1363.

[97] F.-A. Khan, Metallic Ruthenium Nanoparticles Derived from Arene Ruthenium Complexes: Synthesis, Characterization and Applications (Ph.D thesis), (2011). Retrieved from https://doc.rero.ch/record/28737.

[98] S. Miao, Z. Liu, B. Han, J. Huang, Z. Sun, J. Zhang, et al., Angew. Chem. Int. Ed. 45 (2005) 266–269.

[99] D. Chen, M. Huang, S. He, S. He, L. Ding, Q. Wang, et al., Appl. Clay Sci. 119 (2016) 109–115.

[100] P.P. Sarmah, D.K. Dutta, Green Chem. 14 (2012) 1086–1093.

[101] P. Upadhyay, V. Srivastava, RSC Adv. 5 (2015) 740–745.

[102] (a) M. Haruta, N. Yamada, T. Kobayashi, S. Iijima, J. Catal. 115 (1989) 301–309. (b) F. Ammari, M. Chenouf, Int. J. Chem. Mol. Nuc. Mater. Metallur. Eng. 9 (2015) 866–869.

[103] (a) L. Prati, M. Rossi, J. Catal. 176 (1998) 552–560. (b) T. Hayashi, K. Tanaka, M. Haruta, J. Catal. 178 (1998) 566–575.

[104] V. Belova, H. Möhwald, D.G. Shchukin, Langmuir 24 (2008) 9747–9753.

[105] V. Belova, D.V. Andreeva, H. Möhwald, D.G. Shchukin, J. Phys, Chem. Rev. 113 (2009) 5381–5389.

[106] M.B. Gawande, A. Goswami, F.-X. Felpin, T. Asefa, X. Huang, R. Silva, et al., Chem. Rev. 116 (2016) 3722–3811.

[107] P.B. Malla, P. Ravindranathan, S. Komarneni, E. Breval, R. Roy, J. Mater. Chem. 2 (1992) 559–565.

[108] B.J. Borah, D. Dutta, P.P. Saikia, N.C. Baruab, D.K. Dutta, Green Chem. 13 (2011) 3453–3460.

[109] K. Bahranowski, M. Gasior, A. Kielski, J. Podobinski, E.M. Serwicka, L.A. Vartikian, et al., Clay Clay Miner. 46 (1998) 89–97.

[110] L. Bergaoui, J.-F. Lambert, H. Suquet, M. Che, J. Phys. Chem. 99 (1995) 2155–2161.

[111] (a) J. Barrault, C. Bouchoule, K. Echachoui, N. Frini-Srasra, M. Trabelsi, F. Bergaya, Appl. Catal. B 15 (1998) 269–274. (b) N. Frini, M. Crespin, M. Trabelsi, D. Messad, H. van Damme, F. Bergaya, Appl. Clay Sci. 12 (1997) 281–292.

[112] R.K. Kukkadapu, L. Kevan, J. Phys. Chem. 93 (1989) 1654–1656.

[113] W. Sleight, Science 208 (1980) 895–900.

[114] A.D.O. Cinneide, J.K.A. Clarke, Catal. Rev. 7 (1972) 213–232.

[115] L.M. Falicov, G.A. Somorjai, Proc. Natl. Acad. Sci. U. S. A. 82 (1985) 2207–2211.

[116] S.J. Tauster, Acc. Chem. Res. 20 (1987) 389–394.

[117] D.W. Goodman, Chem. Rev. 95 (1995) 523–536.

[118] B. Coq, F. Figueras, Coord. Chem. Rev. 178–180 (1998) 1753–1783.

[119] S.M. McClure, M.J. Lundwall, D.W. Goodman, Proc. Natl. Acad. Sci. U. S. A. 108 (2011) 931–936.

[120] A. Bruix, J.A. Rodriguez, P.J. Ramírez, S.D. Senanayake, J. Evans, J.B. Park, et al., J. Am. Chem. Soc. 134 (2012) 8968–8974.

[121] R. Schlögl, Angew. Chem. Int. Ed. 54 (2015) 3465–3520.

[122] P. Rylander, Catalytic Hydrogenation Over Platinum Metals, Academic Press, New York, 1967.

[123] P. Rylander, Catalytic Hydrogenation in Organic Syntheses, Academic Press, New York, 1979.

[124] C.M. Yim, C.L. Pang, D.R. Hermoso, C.M. Dover, C.A. Muryn, F. Maccherozzi, et al., Proc. Natl. Acad. Sci. U. S. A. 112 (2015) 7903–7908.

[125] R.S. Varma, Tetrahedron 58 (2002) 1235–1255.

[126] D.K. Dutta, B.J. Borah, P.P. Sarmah, Catal. Rev. 57 (2015) 257–305.

[127] G. Nagendrappa, Appl. Clay Sci. 53 (2011) 106–138.

[128] T.J. Pinnavaia, R. Raythatha, J.G.S. Lee, L.J. Halloran, J.F. Hoffman, J. Am. Chem. Soc. 101 (1979) 6891–6897.

[129] R.R. Schrock, J.A. Osborn, J. Am. Chem. Soc. 98 (1976) 2134–2143.

[130] C. Copéret, M. Chabanas, R.P. Saint-Arroman, J.-M. Basset, Angew. Chem. Int. Ed. 42 (2003) 156–181.

[131] R. Margalef-Català, P. Salagre, E. Fernández, C. Claver, Catal. Lett. 60 (1999) 121–123.

[132] K. Parida, G.B.B. Varadwaj, S. Sahu, P.C. Sahoo, Ind. Eng. Chem. Res. 50 (2011) 7849–7856.

[133] W.S. Knowles, Angew. Chem. Int. Ed. 41 (2002) 1998–2007.

[134] R. Noyori, Adv. Synth. Catal. 345 (2003) 15–32.

[135] M. Iwai, H. Shoji, S. Shimazu, T. Uematsu, Chem. Lett. 22 (1993) 989–992.

[136] G. Cao, M.E. Garcia, M. Alcala, L.F. Burgess, T.E. Mallouk, J. Am. Chem. Soc. 114 (1992) 7574–7575.

[137] T. Sento, S. Shimazu, N. Ichikuni, T. Uematsu, J. Mol. Catal. 137 (1999) 263–267.

[138] D. Astruc, F. Lu, J.R. Aranzaes, Angew. Chem. Int. Ed. 44 (2005) 7852–7872.

[139] G.A. Somorjai, Introduction to Surface Chemistry and Catalysis, Wiley, New York, 1994.

[140] J.M. Thomas, W.J. Thomas, Principles and Practice of Heterogeneous Catalysis, VCH, Weinheim, 1997.

[141] C. Copéret, A. Comas-Vives, M.P. Conley, D.P. Estes, A. Fedorov, V. Mougel, et al., Chem. Rev. 116 (2016) 323–421.

[142] P.P. Sarmah, D.K. Dutta, Appl. Catal. A Gen. 470 (2014) 355–360.

[143] M.L. Kantam, B.P.C. Rao, B.M. Choudary, B. Sreedhar, Adv. Synth. Catal. 348 (2006) 1970–1976.

[144] T. Subramanian, K. Pitchumani, Cat. Sci. Technol. 2 (2012) 296–300.

[145] D.N. Muraviev, J. Macanas, M. Farre, M. Munoz, S. Alegret, Sensors Actuators B 118 (2006) 408–412.

[146] Y.Y. Xia, T.J. Li, C. Ma, C. Gao, J. Chen, RSC Adv. 4 (2014) 20516–20520.

[147] V.K. Soni, R.K. Sharma, ChemCatChem 8 (2016) 1763–1768.

[148] G.V. Smith, F. Notheisz, Heterogeneous Catalysis in Organic Chemistry, Academic Press, San Diego, 1999.

[149] T. Mallat, A. Baiker, Chem. Rev. 104 (2004) 3037–3058.

[150] V.S. Kshirsagar, S. Vijayanand, H.S. Potdar, P.A. Joy, K.R. Patil, C.V. Rode, Chem. Lett. 37 (2008) 310–311.

[151] V.S. Kshirsagar, A.C. Garade, K.R. Patil, M. Shirai, C.V. Rode, Top. Catal. 52 (2009) 784–788.
[152] F. Wang, G. Yang, W. Zhang, W. Wu, J. Xu, Chem. Commun. 39 (2003) 1172–1173.
[153] M.P.J. Peeters, M. Busio, P. Leijten, Appl. Catal. A Gen. 118 (1994) 51–62.
[154] V.S. Kshirsagar, A.C. Garade, K.R. Patil, R.K. Jha, C.V. Rode, Ind. Eng. Chem. Res. 48 (2009) 9423–9427.
[155] V.S. Kshirsagar, A.C. Garade, R.B. Mane, K.R. Patil, A. Yamaguchi, M. Shirai, et al., Appl. Catal. A Gen. 370 (2009) 16–23.
[156] V.D. Makwana, Y.C. Son, A.R. Howell, S.L. Suib, J. Catal. 210 (2002) 46–52.
[157] C. Doornkamp, V. Ponec, J. Mol. Catal. A 162 (2000) 19–32.
[158] A.C. Garade, N.S. Biradar, S.M. Joshi, V.S. Kshirsagar, R.K. Jha, C.V. Rode, Appl. Clay Sci. 53 (2011) 157–163.
[159] G.R. Krow, Org. React. 1993 (43) (1993) 251–798.
[160] A. Corma, L.T. Nemeth, M. Renz, S. Valencia, Nature 412 (2001) 423–425.
[161] P.K. Saikia, P.P. Sarmah, B.J. Borah, L. Saikia, K. Saikia, D.K. Dutta, Green Chem. 18 (2016) 2843–2850.
[162] H. Sakurai, K. Koga, Y. Iizuka, M. Kiuchi, Appl. Catal. A Gen. 462–463 (2013) 236–246.
[163] K.C. Nicolaou, S.A. Snyder, T. Montagnon, G. Vassilikogiannakis, Angew. Chem. Int. Ed. 41 (2002) 1668–1698.
[164] C. Cativiela, F. Figueras, J.M. Fraile, J.I. Garcia, J.A. Mayoral, L.C. de Ménorval, et al., Appl. Catal. A Gen. 101 (1993) 253–267.
[165] W. Oppolzer, Angew. Chem. Int. Ed. Eng. 23 (1984) 876–889.
[166] W. Oppolzer, Tetrahedron 43 (1987) 1969–2004.
[167] M.R. Dintzner, A.J. Little, M. Pacilli, D.J. Pileggi, Z.R. Osner, T.W. Lyons, Tetrahedron Lett. 48 (2007) 1577–1579.
[168] C. Cativiela, F. Figueras, J.M. Fraile, J.I. García, J.A. Mayoral, Tetrahedron Asymmetry 2 (1991) 953–956.
[169] C. Cativiela, J.M. Fraile, J.I. García, J.A. Mayoral, E. Pires, F. Figueras, et al., Tetrahedron 48 (1992) 6467–6476.
[170] P. Laszlo, J. Luchetti, Tetrahedron Lett. 25 (1984) 1567–1570.
[171] J.M. Fraile, J.I. García, J.A. Mayoral, T. Tarnai, P.J. Alonso, Chem. Commun. 32 (1996) 1981–1982.
[172] G. Sartori, R. Maggi, Chem. Rev. 111 (2011) PR181–PR214.
[173] J.V. McCullagh, K.A. Daggett, J. Chem. Educ. 84 (2007) 1799–1802.
[174] J.K. Groves, Chem. Soc. Rev. 1 (1972) 73–97.
[175] A. Cornélis, A. Gerstmans, P. Laszlo, A. Mathy, I. Zieba, Catal. Lett. 6 (1990) 103–110.
[176] P. Laszlo, M.T. Montaufier, Tetrahedron Lett. 32 (1991) 1561–1564.
[177] A. Cornélis, P. Laszlo, S.F. Wang, Tetrahedron Lett. 34 (1993) 3849–3852.
[178] A. Cornélis, P. Laszlo, S.F. Wang, Catal. Lett. 17 (1993) 63–69.
[179] B.M. Choudary, N.S. Chowdari, M. Lakshmi Kantam, R. Kannan, Tetrahedron Lett. 40 (1999) 2859–2862.
[180] G.A. Oiah, S. Kobayashi, J. Nishimura, J. Am. Chem. Soc. 95 (1973) 564–569.
[181] B.M. Choudary, B.P.C. Rao, N.S. Chowdari, M.L. Kantam, Catal. Commun. 3 (2002) 363–367.
[182] K. Hiyoshizo, O. Takeshi, I. Motoshu, Synlett (1998) 255–256.
[183] B.M. Choudary, N.S. Chowdari, M.L. Kantam, Tetrahedron 56 (2000) 7291–7298.
[184] N.B. Shrigadi, A.B. Shinde, S.D. Samant, Appl. Catal. A Gen. 252 (2003) 23–35.

[185] U. Schwertmann, R.M. Cornell, Iron Oxides in the Laboratory: Preparation and Characterization, VCH, New York, 1991.
[186] N.S. Clarke, P.G. Hall, Langmuir 7 (1991) 672–677.
[187] Y.H. Liu, Q.S. Liu, Z.H. Zhang, Tetrahedron Lett. 50 (2009) 916–921.
[188] G.D. Yadav, S.R. More, Appl. Clay Sci. 53 (2011) 254–262.
[189] L. Menéndez-Rodríguez, P. Crochet, V. Cadierno, J. Mol. Catal. A Chem. 366 (2013) 390–399.
[190] G.D. Yadav, R.D. Bhagat, Clean Techn. Environ. Policy 7 (2005) 245–251.
[191] G.D. Yadav, G. George, J. Mol. Catal. A Chem. 292 (2008) 54–61.
[192] B. Ahmad Dar, P. Bhatti, A.P. Singh, A. Lazar, P.R. Sharma, M. Sharma, et al., Appl. Catal. A Gen. 466 (2013) 60–67.
[193] H. Yang, Y. Li, M. Jiang, J. Wang, H. Fu, Chem. Eur. J. 17 (2011) 5652–5660.
[194] L. Huerta, A. Meyer, E. Choren, Microporous Mesoporous Mater. 57 (2003) 219–227.

FURTHER READING

Y.M. Liu, S.T. Liu, K.Z. Zhu, X.K. Ye, Y. Wu, Appl. Catal. A Gen. 169 (1998) 127–135.
C.V. Rode, M.V. Sonar, J.M. Nadgeri, R.V. Chaudhari, Org. Process. Res. Dev. 8 (2004) 873–878.
S.C. Eyley, Contemp. Org. Synth. 2 (1991) 707–731.
H. Heaney, Contemp. Org. Synth. 2 (1991) 733–752.

Chapter 13

Catalysts Intercalated in Layered Double Hydroxides

Samahe Sadjadi
Iran Polymer and Petrochemical Institute, Tehran, Iran

Contents

1 INTRODUCTION

1.1 Composition and Structure of Layered Double Hydroxides

Layered double hydroxides (LDHs), also known as anionic clays, are a class of synthetic or natural two-dimensional nanostructured layered materials [1,2] with two kinds of metallic cations in the main layers and interlayer domains containing anionic species. The structure of LDHs can be visualized as a CdI_2 type layered hydroxide (e.g., brucite, $Mg(OH)_2$), where some of the divalent cations coordinated octahedrally by hydroxyl groups have been isomorphously substituted by trivalent cations, resulting in positively charged brucite-like sheets. In order to recover the electroneutrality, anions together with water molecules are incorporated between the layers (Fig. 13.1) [3]. The water molecules present in the interlayer of LDHs can form hydrogen bonds to the brucite layers. The hydrogen bonds as well as electrostatic interactions hold the brucite-like sheets bonded. Hydroxyl groups, particularly those

Encapsulated Catalysts. http://dx.doi.org/10.1016/B978-0-12-803836-9.00013-4

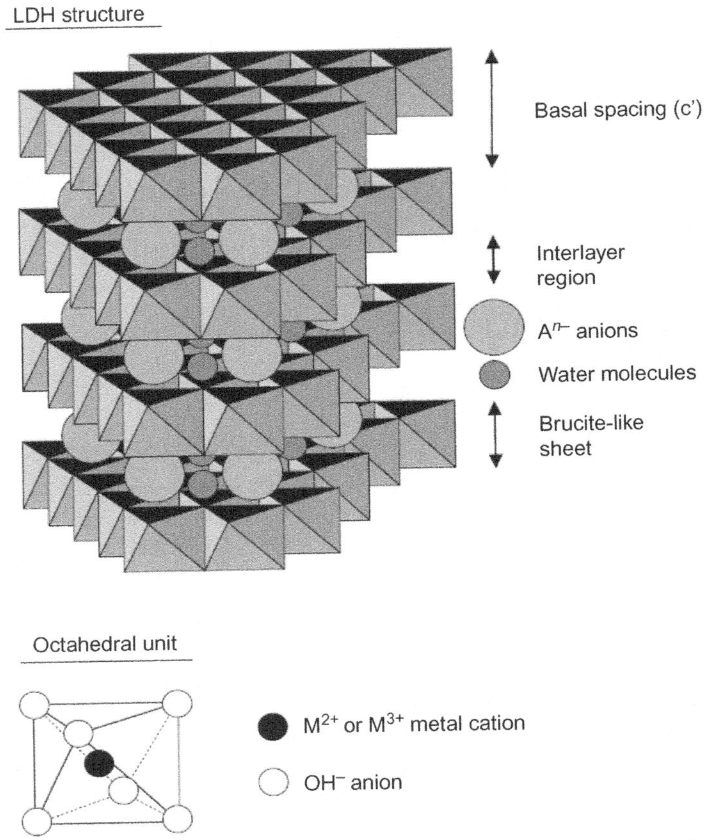

FIG. 13.1 Schematic representation of the LDH structure. *(Reproduced from K.-H. Goh, T.-T. Lim, Z. Dong, Application of layered double hydroxides for removal of oxyanions: A review, Water Res. 42 (2008) 1343–1368, with the permission of Elsevier.)*

bonded to trivalent cations, are strongly polarized and interact with the interlayer anions [4].

The general molecular formula of LDH can be represented by $[M^{2+}_{1-x}M^{3+}{}_x(OH)_2]^{x+}A^{n-}_{x/n}{\cdot}mH_2O$, where M^{2+} and M^{3+} are di- and trivalent metal ions, A^{n-} is the interlayer anion, and x is the fraction of M^{2+}, in other words, $x=M^{2+}/(M^{2+}+M^{3+})$ [5,6].

The divalent metal ions include Mg^{2+}, Zn^{2+}, Co^{2+}, Ni^{2+}, Cu^{2+}, or Mn^{2+}, and trivalent metal ions can be Al^{3+}, Cr^{3+}, Co^{3+}, Fe^{3+}, V^{3+}, Y^{3+}, or Mn^{3+}. In Table 13.1 many of the previously reported binary LDHs are listed. Various anions can be placed in the interlayer of LDHs including conventional inorganic anions such as CO_3^{2-}, NO_3^-, F^-, and Cl^-, coordination compounds, organic anions, biomolecules, and polyoxometalates [3].

TABLE 13.1 Part of Binary LDHs Reported in Patents and Literatures

		M^{2+}, M^{+}												
		Mg	Fe	Co	Ni	Cu	Zn	Ca	Sr	Be	Cd	Pb	Mn	Li
M^{3+}	Al													
	Cr													
	Fe													
	Co													
	Ni													
	Bi													
	Sb													
	Ga													
	In													
	V													
	Cr													
	Y													
	La													
M^{4+}	Ti													
	Sn													
M^{6+}	Mo													

(Reproduced from J. Qu, et al., Mechanochemical approaches to synthesize layered double hydroxides: a review, Appl. Clay 119 (2016) 185–192, with the permission of Elsevier.)

There are different ways for stacking arrangements of the brucite-like sheets. The sheets can be stacked either with two layers per unit cell in hexagonal symmetry (manasseite), with three layers per unit cell in rhombohedral symmetry (hydrotalcite), or in less symmetrical arrangements [2].

One of the interesting structural properties of LDHs is the phenomenon called memory effect, which is the ability to recover the original layered crystalline structure from the calcined form (mixed oxide) upon being subjected to a solution containing suitable anions. The memory effect can be employed for the synthesis and modification of LDHs with different types of intercalating anions, vide infra [3,4].

1.2 Synthesis of LDHs

LDH materials have earned a noticeable interest because of their outstanding features such as acid-base properties, cation-exchange ability, anion exchange capacity, adsorption capacity, and exhibiting memory effect [3,5]. To date, various synthetic protocols including the coprecipitation method, classical ion exchange method, reconstruction method, salt oxide method, sol-gel method, topochemical synthesis, SNAS (separate nucleation and aging steps), electrosynthesis, urea hydrolysis, combustion synthesis, and hydrothermal and microwave treatments have been employed for the preparation of LDHs [7]. He et al. covered various synthetic methods of LDHs in an excellent review article [7]. Moreover, Rives et al. discussed this issue in a book chapter [3]. Because the focus of this chapter is the catalytic species intercalated in LDHs, we discuss here only the most conventional synthetic methods for LDHs. The readers are encouraged to study the abovementioned references for more details on various methods for the syntheses of LDHs.

1.2.1 Coprecipitation Method

Coprecipitation is the most common method for the synthesis of LDHs. In this method appropriate inorganic salts were used as precursors of M^{2+} and M^{3+}. Precipitation is carried out either under a low super saturation condition or a high super saturation condition. Precipitation in a low super saturation condition consists of the slow addition of mixed solutions of divalent and trivalent metal salts with an appropriate ratio into a reactor containing an aqueous solution of the desired interlayer anion. A second solution of an alkali is added into the reactor simultaneously at such a rate to maintain the desired pH for promoting coprecipitation of the two metallic salts. LDHs prepared by this method are usually more crystalline than those synthesized under a high super saturation condition. Moreover, the ratio of divalent to trivalent metals can be easily controlled by adjusting the solution pH under low super saturation coprecipitation. Coprecipitation at a high super saturation condition, however, involves the addition of a mixed salt solution to an alkaline solution containing the desired interlayer anion. Low or high super saturation is followed by a thermal treatment and aging process [2]. It is worth noting that precipitation is influenced by temperature, aging, the nature of the anions, cations, and their ratio and pH [4].

1.2.2 Ion Exchange Method

When the coprecipitation method is limited because of some synthetic problems such as instability of M^{2+} or M^{3+} or anions in an alkaline solution or a more favorable direct reaction between the metal anions and the guest anion, the ion exchange synthetic method (considered by some experts as postsynthesis modification) can be applied. This method is based on exchanging the interlayered anion with a guest under stirring or an ultrasonic-assisted condition to afford specific anion-pillared LDHs [2]. Because there is a strong affinity between the carbonate and LDHs, this method is usually used for carbonate-free LDHs. The anion exchange is affected by the electrostatic interactions between the positively charged layers and interlayer anions. Moreover, the equilibrium constants increase as the ionic radius of the anion decreases, and consequently the exchange is favored for anions with high charge densities [3].

1.2.3 Reconstruction Method

In this method that is based on memory effect, the LDHs containing the desired metal cations are prepared and subsequently calcined to remove the interlayer water molecules, anions, and hydroxyl groups. The obtained mixed metal oxides then are introduced in an aqueous solution containing the desired anions to be intercalated [3].

1.2.4 Hydrothermal Method

Hydrothermal treatment is a useful way to synthesize LDHs with an improved crystallined structure or LDHs with intercalated organic guests that have a low affinity for LDHs. The hydrothermal treatment can also affect the particle size [2,4].

1.3 Applications of LDHs

The unique properties of LHDs such as adjustable element compositions, acid-base properties, ion-exchange ability, adsorption capacity, and memory effect allow a wide range of applications in various fields of science [8]. The basic nature of LDHs has been exploited as an antiacid. The utility of LDHs in some commercial drugs such as Talcid and Almax has been proved [3]. Intercalation of drugs has also been considered for controlled release purposes. LDHs are promising compounds for environmental protection. LDH-assisted removal of organic and inorganic wastes and degradation of environmental contaminants have been reported and reviewed in the literature [2]. Moreover, the development of membrane-like materials using LDHs has been investigated [2]. Another application of LDHs is in material science. They have been used as additives in polymers to improve or tuned the polymers properties such as strength, mechanical properties, and color [3]. One of the most important applications of LDHs is in catalysis. They can be used both as catalyst supports and solid base catalysts. Various reactions such as retro-aldol condensation, aldol condensation, Claisen-Schmidt condensation, Knoevenagel condensation, epoxidation, ketone oxidation to ester, and transestrification processes have been catalyzed by LDHs. Moreover, the catalytic activity of LDHs for promoting reactions under nonconventional green procedures such as ultrasonic and microwave irradiation has been demonstrated [4]. There are some excellent review articles in the literature on the application of LDHs for catalysis [4,6,9]. The scope of the present book chapter is restricted to the LDH-intercalated catalytic species.

1.4 Intercalation in LDHs

Intercalation has been introduced as an efficient way to manipulate magnetic, electronic, chemical, and optical properties of LDHs [6]. To date various anions such as biomolecules, coordination compounds, amino acids, inorganic anions, organic anions, and polyoxometalates have been incorporated between the interlayers of LDHs (Fig. 13.2). There exist some limitations for the intercalation of large molecules in interlayers of pure and unmodified LDHs. Intercalation capacity will be significantly limited if the molecular size and the gallery space do not match.

To enhance the basal spacing of interlayers of LDHs, the intercalation of sterically demanding anions such as biomolecules has been suggested. In this regard, orientation of pillaring species between the interlayer is of great importance.

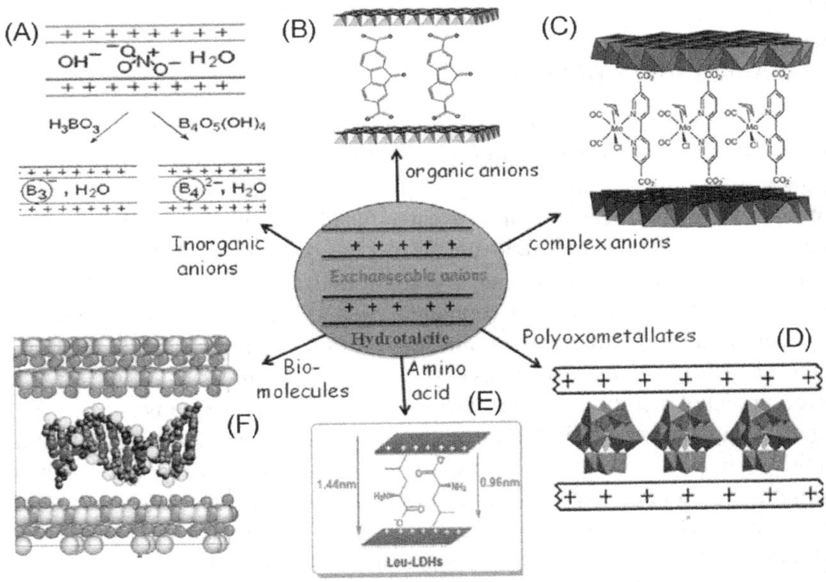

FIG. 13.2 Pictorial representation of various anions intercalated in hydrotalcite materials for various applications. *(Reproduced from T. Baskaran, J. Christopher, A. Sakthivel, Progress on layered hydrotalcite (HT) materials as potential support and catalytic materials. RSC Adv. 5 (2015) 98853–98875, with the permission of the Royal Society of Chemistry.)*

2 CATALYSTS INTERCALATED IN LDH

2.1 Intercalated Complexes

Intercalation of metal-containing anions in the interlayer space of hydrotalcites provides several advantages for an improved use in the case of catalysts; it permits incorporation of, at least, a third metal component in the mixed oxide obtained upon thermal decomposition. In other materials, the interlayer space provides unique reaction conditions, thus permitting some reactions to take place under softer experimental conditions than in its absence [1].

Gago et al. have reported the immobilization of oxomolybdenum species in an LDH pillared by 2,2′-bipyridine-5,5′-dicarboxylate anions for liquid-phase epoxidation of *cis*-cyclooctene, 1-octene and *trans*-2-octene using *tert*-butyl hydroperoxide as an oxygen source [10].

LDH-containing metal complexes of chelating ligands, namely, NTA (nitriloacetate) and EDTA (ethylenediamine tetraacetate), have been prepared either by direct intercalation of metal complexes or indirectly by forming metal complexes between the host layers following intercalation of the ligands [11,12].

LDHs intercalated with halocomplexes, cyanocomplexes, oxocomplexes, and macrocyclic ligand-containing complexes have been reviewed by Rives et al. [1]. In Table 13.2 some recent advances in the field of intercalated complexes in LDHs are tabulated. In the following section some recent examples are discussed in more detail to disclose the advantages of incorporation of catalytic species into interlayers of LDHs.

He et al. designed and synthesized a novel bifunctional catalyst, CoTSPP-Zn_2Al-LDH, for the Baeyer-Villiger oxidation based on intercalation of the

TABLE 13.2 Recent Advances in Complexes Intercalated in LDHs

Reaction	LDH	Complex	Refs.
Oxidation	Ni-Zn	D-Valine-palladium(II)	[13]
Oxidation	Ni-Zn	$[Pd(OH)_4]^{2-}$	[14]
Oxidation	Zn-Al	Ce(III)-complex	[15]
Oxidation	Ca-Al	Mn(II)-amino acid complex(L-cysteine, L-histidine and L-tyrosine)	[16]
Oxidation	Mg-Al	Transition metal sulphonato-Schiff base complexes (Cr, Mn, Fe, Co, Cu)	[17]
Oxidation	Mg-Al	Sulphonato-salen-chromium(III)	[18]
Oxidation	Zn-Al	Metalloporphyrins (MTSPP, M=Co, Fe, and Mn)	[19]
Oxidation	Mg-Al	$CuBDACl_2$[a] anions	[20]

Continued

TABLE 13.2 Recent Advances in Complexes Intercalated in LDHs—cont'd

Reaction	LDH	Complex	Refs.
Oxidation	Zn-Al	Anionic porphyrin $[H_2(TSPP)]^{4-b}$, iron(III) porphyrin $[Fe(TSPP)]^{3-}$, and second generation anionic complexes $[Fe(TDFSPP)]^{3-c}$ and $[Fe(TDCSPP)]^{3-d}$, $[Fe(TSPP)]^{3-}$	[21]
Oxidation	Mg-Al	$[Fe(TDFSPP)]^{3-}$ and $[Fe(TCFSPP)]^{3-e}$	[22]
Oxidation	Mg-Al	Cobalt phthalocyanine tetrasulfonate	[23]
Oxidation	Mg-Al	$[Mn(III)TSPPCl]^{4-f}$	[24]
Epoxidation	Zn-Al	Chiral sulphonato-salen-manganese(III) complex	[25]
Epoxidation	Zn-Al	Sulphonato-Salen-Manganese(III) complex	[26]
Epoxidation	Zn-Al	Chiral sulphonato-salen-manganese(III) complex	[27]
Epoxidation	Zn-Al	Sulphonato-salen complexes containing Mn(III), Co(III), and Fe(III) metal centers	[28]
Epoxidation	Zn-Al	$[Mo(\eta^3\text{-}C_3H_5)Cl(CO)_2(bpdc)^g]^{2-}$	[29]
Epoxidation	Zn-Al	Ti(IV)-Schiff base complex	[12]
Epoxidation	Zn-Al	Mn(TPP)[h] OAc	[30]
Hydrophenylation	Ni-Zn	$[Rh(OH)_6]^{3-}$	[31]
Hydroformylation	Zn-Al	*trans*-$RhCl(CO)(TPPTS)_2$[i]	[32]
Arylation	Zn-Al	Rh-*m*-TPPTC[j] complex	[33]
Polymerization	Mg-Al	Pyridine carboxylate complexes of Mo^{II}	[34]

[a] *H_2 BDA, 2, 2′-bipyridine-5, 5′-dicarboxylic acid.*
[b] *TSPP: [5,10,15,20-tetrakis(4-sulfonatophenyl)porphyrin].*
[c] *[5,10,15,20-tetrakis (2,6-difluorphenyl-3-sulfonatophenyl) porphyrinate] iron(III).*
[d] *5,10,15,20-tetrakis(2,6-dichloro-3-sulfonate phenyl) porphyrin.*
[e] *[5,10,15,20-tetrakis (2-fluoro-6-chlorophenyl-3-sulfonatophenyl) porphyrinate] iron(III).*
[f] *Tetra(p-sulfonatophenyl) porphinatomanganese(III) anions.*
[g] *2,2′-bipyridine-5,5′-dicarboxylate.*
[h] *Meso-tetraphenylporphyrin dianion.*
[i] *Trisodium salt of tri-(m-sulfophenyl)-phosphine*
[j] *Trilithium salt of 3, 3′, 3″-phosphane triylbenzenesulfonic acid.*

tetra (4-sulfonato-phenyl) porphinato-cobalt(II) anion (CoTSPP) into Zn_2Al hydrotalcite [35]. The authors proposed that the reaction promoted by the O_2/aldehyde oxidation system proceeded by oxidation of ketones by the O_2/aldehyde system catalyzed by CoTSPP-Zn_2Al-LDH. The oxidation process started from the oxidation of the aldehyde to form the peracid catalyzed by intercalated CoTSPP, and the produced peracid could readily enable the Baeyer-Villiger oxidation of the ketone catalyzed by the basic material Zn_2Al-LDH (Fig. 13.3). It is worth noting that intercalation of metalloporphyrin in hydrotalcite rendered the catalyst heterogeneous and enhanced its durability. It was demonstrated that the catalyst could promote Baeyer-Villiger oxidation of various ketones at a relatively low temperature, 40°C, and with a low amount of aldehyde to furnish the corresponding products in high yields.

Recently, Palinko et al. reported intercalation of Mn(II)-amino acid (L-cysteine, L-histidine and L-tyrosine) complex anions in CaAl-LDH to afford composite materials with high oxidation catalytic performance [16]. Two procedures were employed for preparing the hybrid systems. One was based on intercalation of the ligand first followed by constructing the complex, and another one consisted of the initial formation of complex and subsequent intercalation among the CaAl-LDH. The authors also modeled the arrangement of complexes between the layers. During building the model, it was revealed that the phenolate oxygen in the tyrosinate and the carboxylate oxygen could not coordinate at the same time due to steric reasons; therefore, the carboxylate group and the amino nitrogen are proposed as coordination sites (Fig. 13.4). To investigate the catalytic activities of the novel catalysts, oxidative transformation of cyclohexane with two oxidants (peracetic acid and in situ formed iodosylbenzene) was selected as a model reaction. The results (Tables 13.3 and 13.4) indicated the superior catalytic performance of the intercalated catalysts compared to bare oxidants and CaAl-LDH. As is obvious in Tables 13.3 and 13.4, in the presence of peracetic acid high selectivity toward epoxide was obtained while superior diol selectivities were observed in the case of iodosylbenzene. The catalysts were reusable, and no complex or ligand leaching was observed.

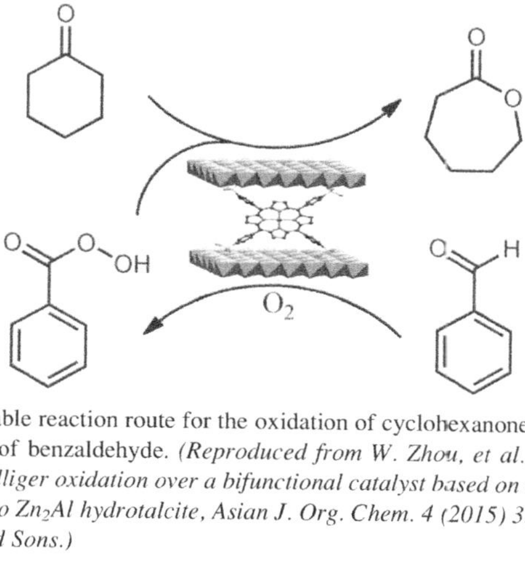

FIG. 13.3 The probable reaction route for the oxidation of cyclohexanone over CoTSPP-Zn_2Al-LDH in the presence of benzaldehyde. *(Reproduced from W. Zhou, et al., Efficient catalysis of the Aerobic Baeyer-Villiger oxidation over a bifunctional catalyst based on cobalt tetraphenylporphyrin intercalated into Zn_2Al hydrotalcite, Asian J. Org. Chem. 4 (2015) 33–37, with the permission of John Wiley and Sons.)*

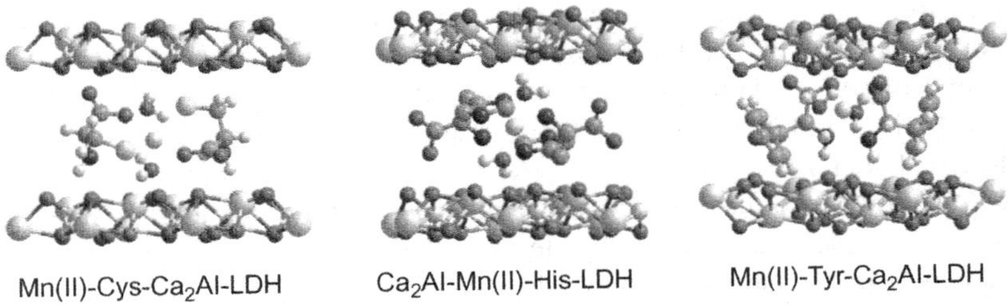

FIG. 13.4 The proposed steric arrangement of the intercalated complexes between the layers of Ca_2Al-LDH, based on XRD and XAS measurements. *(Reproduced from G. Varga, et al., Mn(II)–amino acid complexes intercalated in CaAl-layered double hydroxide—well-characterized, highly efficient, recyclable oxidation catalysts, J. Catal. 335 (2016) 125–134, with the permission of Elsevier.)*

TABLE 13.3 The TOF/Conversion and Selectivity Results of the Oxidation of Cyclohexene After 3 h (Catalyst: 25 mg, Acetone: 10 cm^3, Cyclohexen: 5 mmol, Peracetic Acid: 2.5 mmol, Temperature: 298 K)

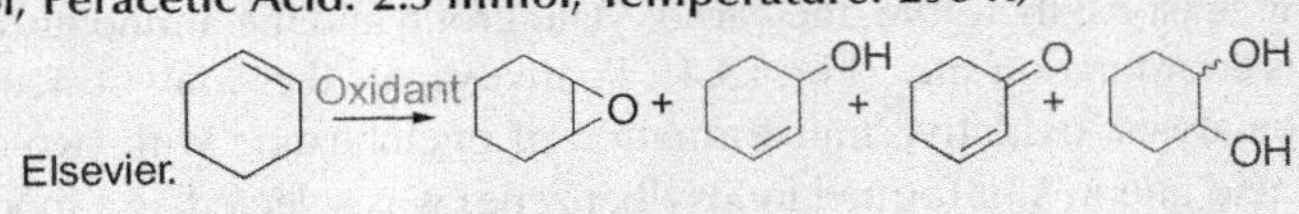

Catalyst	TOF(1/h)/ Conversion (%)	Epoxide (%)	2-Chex-1-ol (%)	2-Chex-1-one (%)	*trans*-Diol (%)
–	nr/21	64	4	2	30
Ca_2Al-LDH	nr/18	63	4	5	28
Mn(II)-Tyr-Ca_2Al-LDH	31/33	79	0	0	21
Mn(II)-Cys-Ca_2Al-LDH	38.5/40	100	0	0	0
Ca_2Al-Mn (II)-His-LDH	100	50/53	0	0	0

(Reproduced from G. Varga, et al., Mn(II)-amino acid complexes intercalated in CaAl-layered double hydroxide—well-characterized, highly efficient, recyclable oxidation catalysts, J. Catal. 335 (2016) 125–134, with the permission of Elsevier.)

Shimazu et al. intercalated catalytic species in Ni-Zn hydroxy double salts (NiZn) and proved their catalytic activities for chemical reactions [13,14,31,36]. As an example, an anionic $[Pd(OH)_4]^{2-}$ species was intercalated into an interlayer of NiZn by the anion exchange method (Fig. 13.5). The catalytic activity of the catalyst, Pd/NiZn (0.02), (number in parentheses indicates Pd loading amount in a unit of mmol g^{-1}) was proved for the oxidation of

TABLE 13.4 The TOF/Conversion and Selectivity Results of the Oxidation of Cyclohexene After 3 h (Catalyst: 25 mg, Aqueous Acetone (5/95% by Volume): 10 cm^3, Cyclohexen: 5 mmol, Diacetoxyiodo) Benzene Acid: 2.5 mmol, Temperature: 298 K)

Catalyst	TOF (1/h)/ conversion (%)	Epoxide (%)	2-Chex-1-ol (%)	2-Chex-1-one (%)	*cis* Diol (%)
–	nr/21	49	28	17	6
Mn(II)-Tyr-Ca_2Al-LDH	28/30	12	0	0	88
Mn(II)-Cys-Ca_2Al-LDH	35.6/37	5	0	0	95
Ca_2Al-Mn (II)-His-LDH	47.2/50	7	0	0	93

(Reproduced from G. Varga, et al., Mn(II)-amino acid complexes intercalated in CaAl-layered double hydroxide—well-characterized, highly efficient, recyclable oxidation catalysts, J. Catal. 335 (2016) 125–134, with the permission of Elsevier.)

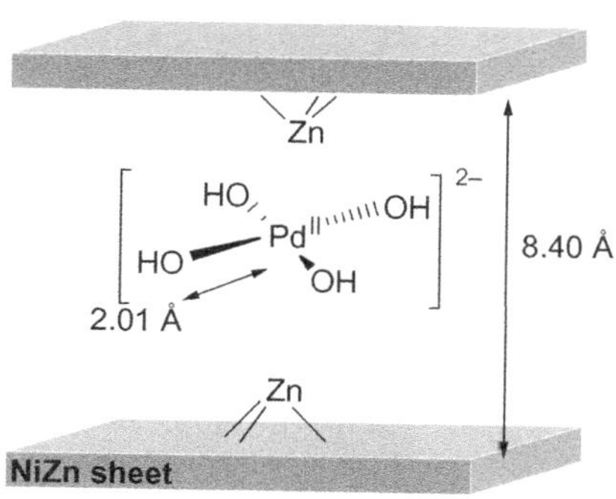

FIG. 13.5 Proposed structure around Pd(II) the center of the Pd/NiZn(0.02) catalyst. The nearest oxygen atoms around the Pd(II) are shown. *(Reproduced from T. Hara, et al., Creation of highly stable monomeric Pd(II) species in an anion-exchangeable hydroxy double salt interlayer: application to aerobic alcohol oxidation under an air atmosphere, Green Chem. 11 (2009) 2034–2040, with the permission of the Royal Society of Chemistry.)*

various alcohols under an air atmosphere. The catalysts could be easily recovered and reused without loss of catalytic performance. The authors believed that the divalent Pd species in the interlayer of NiZn maintained their original monomeric structure during the aerobic alcohol oxidation, due to the strong electrostatic interaction between the NiZn host and anionic Pd(II) species.

In another attempt a novel catalytic system was developed by intercalation of a monomeric hexahydroxy rhodate(III) complex, $[Rh(OH)_6]^{3-}$, into the interlayer of NiZn [31]. The authors investigated the catalytic activity of the catalyst, Rh/NiZn, for the hydrophenylation of internal alkynes with arylboronic acids under mild conditions. The catalytic activity was significantly influenced by the application of olefins as ligands, and the reaction did not proceed in the absence of such ligands. The comparison of the catalytic activity of this catalyst with Rh catalysts under a similar condition established the superior activity of Rh/NiZn (Table 13.5). It is worth noting that the Rh leaching was negligible and the catalyst was reusable. XAFS analysis of fresh and reused catalysts confirmed no change in either Rh oxidation or a coordination environment, suggesting that Rh species in the NiZn matrix remained in a trivalent monomeric state throughout the reaction. That observation was attributed to the strong electrostatic interactions between the anionic Rh species and the NiZn lamellar host.

TABLE 13.5 Effects of Catalysts on Hydrophenylation Between Octa-4-yen and Phenylboronic Acid[a]

Entry	Catalyst	Time (h)	Conv. (%)	Yield (%)[b]
1	Rh/NiZn	1.5	84	71
2	Rh/NiZn	1	62	52
3[c]	Rh/NiZn	1	63	50
4[d]	Rh/NiZn	1	60	48
5[e]	CH_3COO^-/NiZn	5	n.d.	n.d.
6[e]	OH^-/NiZn	5	n.d.	n.d.
7	$Na_3RhCl_6 \cdot nH_2O$	5	54	45
8	$Rh(OH)_3$	5	Trace	Trace
9	Rh_2O_3	5	Trace	Trace
10[f]	Rh(0)/carbon	5	Trace	Trace
11	–	5	n.d.	n.d.

[a]*Rh catalyst (Rh: 0.68 mol%), 1a (0.5 mmol), 2a (1.5 mmol), 1,5-COD (4 eq. relative to Rh), toluene (4 mL), H_2O (0.45 mL), 100°C.*
[b]*Determined by GC analysis using an internal standard technique.*
[c]*Reaction was carried out under an O_2 atmosphere.*
[d]*Under N_2.*
[e]*CH_3COO^-/NiZn (0.1 g) or OH^-/NiZn (0.1 g) was used as a catalyst.*
[f]*5 wt% Rh, purchased from Aldrich.*
(Reproduced from T. Hara, et al., Hydrophenylation of internal alkynes with boronic acids catalysed by a Ni–Zn hydroxy double saltintercalated anionic rhodium(III) complex, Catal. Sci. Technol. 6 (2016) 863–868, with the permission of the Royal Society of Chemistry.)

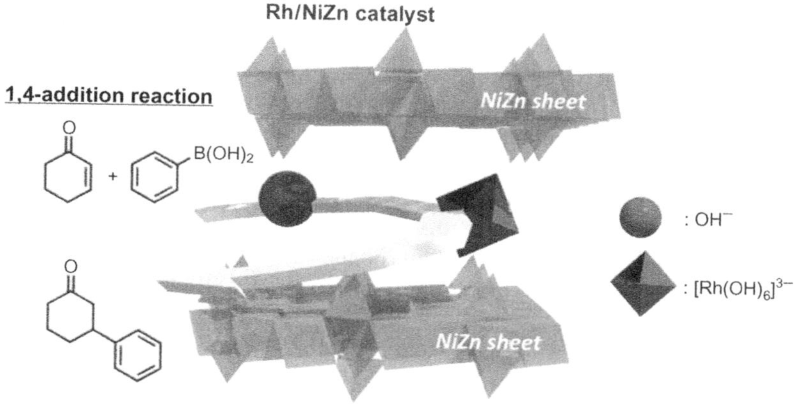

FIG. 13.6 Schematic illustration of the process. *(Reproduced from T. Hara, et al., Efficient 1,4-addition of enones and boronic acids catalyzed by a Ni-Zn hydroxyl double salt-intercalated anionic rhodium(III) complex, ACS Catal. 4 (2014) 4040–4046, with the permission of the American chemical Society.)*

This research group also reported the catalytic activity of Ni-Zn LDH intercalated $[Rh(OH)_6]^{3-}$, Rh/NiZn, for 1,4-addition of a variety of enones and phenylboronic acids to their corresponding β-substituted carbonyl compounds (Fig. 13.6) [37]. The authors believed that the original monomeric trivalent state of the intercalated catalytic species was preserved because of strong electrostatic forces between the LDH and $[Rh(OH)_6]^{3-}$. Comparing the catalytic activity of the novel catalyst with those of alternative Rh catalysts such as Rh^0/carbon, Rh_2O_3, and $Rh(OH)_3$ proved the superior activity of Rh/NiZn under a similar condition. Moreover, the catalyst could be recovered and reused without significant deactivation.

2.2 Intercalated Polyoxometalates

The outstanding features and versatility in chemical composition make LDHs promising candidates for polyoxometalates, PMOs, intercalation. The intercalated PMOs benefit from some advantages such as (1) the host-guest interactions (e.g., electrostatic, van der Waals, or hydrogen bonding) that assure a homogeneous distribution of POMs at the molecular level; (2) suppressed PMO leaching; and (3) improved photo and chemical stability [38]. Because POM intercalated LDHs exhibited more superior application capabilities than individual LDHs or PMOs, much effort has been placed on developing LDH/PMO nanocomposities [39,40]. However, some challenges remain unresolved.

Firstly, during the anion exchange reactions between the LDH precursor and the POM anions, it is possible that the M^{2+}/M^{3+} cations are leached out of the LDHs, which occurs under the neutral to slightly acidic reaction conditions.

Secondly, the control over the final M^{2+}:M^{3+} ratio in the POM pillared LDHs is essential since this ratio determines the charge density of the LDH layers and therefore affects the basal spacing between the intercalated POMs [41]. To synthesize the POM/LDH nanocomposites with different pore size distributions, it is necessary to retain, as much as possible, the desired M^{2+}:M^{3+} ratio from the initial step of synthesis through to the ultimate product. Finally, some POM anions are hydrolytically unstable at weakly acidic to basic pH and the pillaring reactions are accompanied by the co-formation of an impurity phase, which is the M^{2+} rich salt of the POM that deposits on the surface of the LDH crystallites. Deposition of POM salt can block the micropores of LDHs, resulting in low surface areas [39].

Heretofore, various PMOs have been intercalated in several LDHs and used for promoting chemical transformations (Table 13.6). In this context, Song et al. provided an excellent review of polyoxometalate intercalated LDHs in 2014, which covered synthetic approaches as well as various applications of these nanocomposites [39]. Rives et al. also published a comprehensive book chapter on heterogeneous catalysis by polyoxometalate-intercalated LDHs [3]. We strongly encourage the readers to study these references for more details.

Here, some recent examples of LDH-intercalated PMO will be discussed in detail to show the merits of POM/LDH nanocomposites as catalysts.

To circumvent the problems of conventional synthetic approaches for the preparation of POM/LDH nanocomposites, Song et al. developed a novel, simple, and one-pot method for intercalation of Keggin and Dawson POM clusters including $Na_3[PW_{12}O_{40}]\cdot 15H_2O$ (Na_3PW_{12}), $Na_9LaW_{10}O_{36}\cdot 32H_2O$ (Na_9LaW_{10}) and $K_6[P_2W_{18}O_{62}]\cdot 14H_2O$ ($K_6P_2W_{18}$), in the interlayer of tris(hydroxymethyl) aminomethane (Tris)-modified LDH [38]. To this purpose, the LDH precursors and Tris were mixed in aqueous solution to furnish Tris-LDH-CO_3. Subsequently, the ion exchange of PMO with Tris-LDH-CO_3 proceeded (Fig. 13.7). Using Tris-modified LDH, the intercalation could proceed under ambient conditions with no need to degas CO_2. The catalytic activities of the obtained nanocomposites were investigated for degradation of three dyes, methylene blue (MB), rhodamine B (RB), and crystal violet (CV), in the presence of hydrogen peroxide to non- or less toxic anions such as SO_4^{2-}, NO_3^-, $C_2H_3O_2^{2-}$, and CHO^{2-}. The results indicated the superior activity of Tris-LDH-PW_{12}. Additionally, degradation of a mixture of RB, MB, and CV by Tris-LDH-PW_{12} followed the order of CV > MB > RB, which was directly related to the designed accessible area of the interlayer space. It is worth mentioning that the catalyst was reusable and could be recovered and reused with negligible loss of activity.

Recently, Miras, Song, and their coworkers designed and prepared a series of heterogeneous bifunctional catalysts by intercalation of PMO anions $[X_4(H_2O)_2(PW_9O_{34})_2]^{10-}$ (X = Mn, Fe, Co, Ni, Cu, and Zn) into (Tris)-modified LDHs [70]. The catalysts were achieved through a simple ion exchange method and possessed the general formula of Tris-LDH-$X_4(PW_9)_2$

TABLE 13.6 Summary of Reactions Catalyzed by PMO-Intercalated LDHs

Reaction	LDH	POM	Refs.
Oxidant	ZnAl or NiAl	$[SiW_{12}O_{40}]^{4-}$,$[SiW_{11}O_{39}]^{8-}$	[42]
Oxidant	MgAl	$[CoW_{12}O_{40}]^{5-}$	[43]
Oxidant	Zn, Al	$[NaP_5W_{30}O_{110}]^{14-}$	[44]
Oxidant	Mg, Al	α-$[SiW_9O_{37}(Co(H_2O)_3)]^{10-}$	[45]
Oxidant	Mg, Al	$PO_4[WO(O_2)]_4$	[46]
Oxidant	Mg, Al	$[W_7O_{24}]^{6-}$	[47]
Oxidant	Mg, Al	$[WO_4]^{2-}$, $[W_7O_{24}]^{6-}$	[48]
Oxidant	Mg, Al	$[Mo_7O_{24}]^{6-}$	[49]
Oxidant	Mg, Al	$[Mo_7O_{24}]^{6-}$	[48]
Oxidant	Mg, Al	$[V_{10}O_{28}]^{6-}$	[48]
ODH[a]	Mg, Al	$[Mo_7O_{24}]^{6-}$	[50]
ODH	Mg, Al	$[V_{10}O_{28}]^{6-}$	[51]
ODH	Mg, Al	$[V_{10}O_{28}]^{6-}$,$[V_4O_{12}]^{4-}$	[52]
ODH	Mg, Al	$[V_{10}O_{28}]^{6-}$,$[V_4O_{12}]^{4-}$	[53]
ODH	Mg, Al	$[V_{10}O_{28}]^{6-}$	[54]
Reduction	MgAl	$[PW_{12}O_{40}]^{3-}$	[55]
Reduction	MgAl	$[SiW_{12}O_{40}]^{4-}$	[55]
Epoxidation	Zn, Al	$[SiW_{12}O_{40}]^{4-}$	[42]
Epoxidation	Ni, Al	$[SiW_{12}O_{40}]^{4-}$	[42]
Epoxidation	Zn, Al	$[SiW_{11}O_{39}]^{8-}$	[42]
Epoxidation	Mg, Al	$[W_4Nb_2O_{19}]^{4-}$	[56]
Epoxidation	Zn, Al	$[W_4Nb_2O_{19}]^{4-}$	[56]
Epoxidation	Mg, Al	$[H_2W_{12}O_{40}]^{6-}$	[57]
Epoxidation	Zn, Al	$[H_2W_{12}O_{40}]^{6-}$	[57]
Epoxidation	PdMgAl	$[PW_{11}O_{39}]^{7-}$	[58]
Epoxidation	PdZnAl	$[PW_{11}O_{39}]^{7-}$	[58]
Epoxidation	Mg, Al	$[W_7O_{24}]^{6-}$	[59]
Epoxidation	Mg, Al	$C_6H_5PO_3(WO_5)_2$	[59]
Epoxidation	Mg, Al	$C_{12}H_{23}PO_3(WO_5)_2$	[59]
Epoxidation	Mg, Al	$[WZn_3(ZnW_9O_{34})_2]^{12-}$	[60]

Continued

TABLE 13.6 Summary of Reactions Catalyzed by PMO-Intercalated LDHs—cont'd

Reaction	LDH	POM	Refs.
Epoxidation	Mg, Al	$[WCo_3\ (CoW_9O_{34})_2]^{12-}$	[61]
Epoxidation	Mg, Al	$[WZnMn_2\ (ZnW_9O_{34})_2]^{12-}$	[61]
Epoxidation	Mg, Al	$[PW_{11}O_{39}]^{7-}$	[61]
Epoxidation	Zn, Al	$[WCo_3\ (CoW_9O_{34})_2]^{12-}$	[61]
Epoxidation	Zn, Al	$[WZnMn_2\ (ZnW_9O_{34})_2]^{12-}$	[61]
Epoxidation	Zn, Al	$[PW_{11}O_{39}]^{7-}$	[61]
Epoxidation	Mg, Al	$[Mo_7O_{24}]^{6-}$	[57]
Epoxidation	Zn, Al	$[Mo_7O_{24}]^{6-}$	[57]
Epoxidation	Mg, Al	$[Mo_7O_{24}]^{6-}$	[62]
Epoxidation	Zn, Al	$[Mo_7O_{24}]^{6-}$	[62]
Epoxidation	Ni, Al	$[Mo_7O_{24}]^{6-}$	[62]
Epoxidation	Mg, Al	$[V_{10}O_{28}]^{6-}$	[63]
Esterification	Zn, Al	$[PW_{12}O_{40}]^{3-}$	[64]
Esterification	Zn, Al	$[PMo_{12}O_{40}]^{3-}$	[64]
Photo.	MgAl	$[W_7O_{24}]^{6-}$	[65]
Photo.	Zn, Al	$[SiW_{11}O_{39}]^{8-}$	[66]
Photo.	Zn, Al	$[SiW_{11}O_{39}Mn\ (H_2O)]^{6-}$	[66]
Sulfoxidation	Mg, Al	$[WO_4]^{2-}$, $[W_7O_{24}]^{6-}$	[49]
Sulfoxidation	Mg, Al	$[WO_4]^{2-}$, $[W_7O_{24}]^{6-}$	[67]
Sulfoxidation	Mg, Al	$[V_{10}O_{28}]^{6-}$	[49]
Alkoxylation	Cu, Cr	$[Mo_7O_{24}]^{6-}$	[68]
Alkoxylation	Cu, Cr	$[V_2O_7]^{4-}$	[68]
Alkoxylation	Cu, Cr	$[V_{10}O_{28}]^{6-}$	[68]
Alkoxylation	Mg, Al	$[V_{10}O_{28}]^{6-}$	[69]

ODH, *oxidative dehydrogenation;* Photo, *photodegradation.*

(X = Mn, Fe, Co, Ni, Cu, and Zn). The authors studied the catalytic activities of the catalysts for the cascade reaction of oxidation-Knoevenagel condensation between benzyl alcohol and ethyl cyanoacetate to afford benzylidene ethyl cyanoacetate. The plausible mechanism consisted of initial formation of benzaldehyde via benzyl alcohol oxidation that proceeded by PMO clusters in the

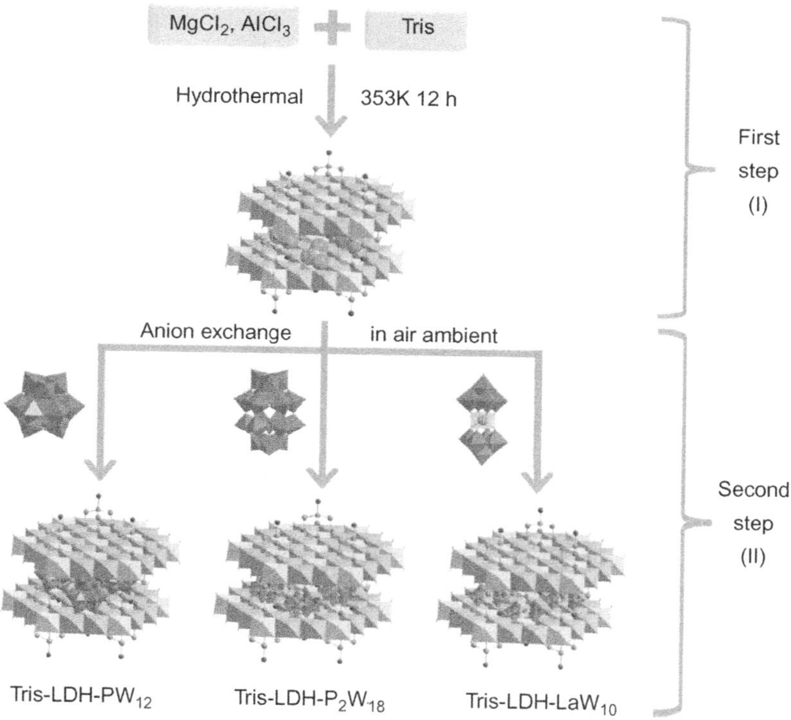

FIG. 13.7 Schematic representation of the synthesis of (I) Tris-modified LDHs and (II) a series of classical POMs intercalated into Tris-modified LDHs. *(Reproduced from Y. Chen, et al., Modular polyoxometalate-layered double hydroxide composites as efficient oxidative catalysts, Chem. Eur. J. 21 (2015) 10812–10820, with the permission of Wiley-VCH Verlag.)*

presence of hydrogen peroxide. In the same time, the basic sites in LDH activated ethyl cyanoacetate and promoted condensation with benzaldehyde to furnish the desired product (Fig. 13.8). The catalysts could be easily recovered and reused for at least ten reaction runs without significant loss of activity or structural integrity.

In another attempt, Song et al. intercalated POM clusters of $[\alpha_2\text{-}P_2W_{17}O_{61}(X{\cdot}OH_2)]^{n-}$ (for $X = Mn^{III}$ and Fe^{III}, denoted as $[P_2W_{17}X]^{7-}$; for $X = Zn^{II}$, Co^{II}, Cu^{II}, and Ni^{II}, denoted as $[P_2W_{17}X]^{8-}$) into LDH to develop novel catalysts, $Mg_3Al\text{-}P_2W_{17}X$ catalysts ($X = Mn^{III}$, Fe^{III}, Zn^{II}, Co^{II}, Cu^{II}, Ni^{II}). The preparation procedure consisted of two steps. Initially, intercalation of suberic acid into $Mg_3Al\text{-}CO_3$ was performed to widen the gallery. Subsequently, PMO was intercalated through ion exchange (Fig. 13.9). Sulfoxidation of a broad range of sulfides in the presence of H_2O_2 as the oxidant was carried out to confirm the catalytic activities of the catalysts. The results established the superior catalytic activity and selectivity of $Mg_3Al\text{-}P_2W_{17}Zn$. The authors believed that the 2-D confinement effect of LDH leads to the uniform and

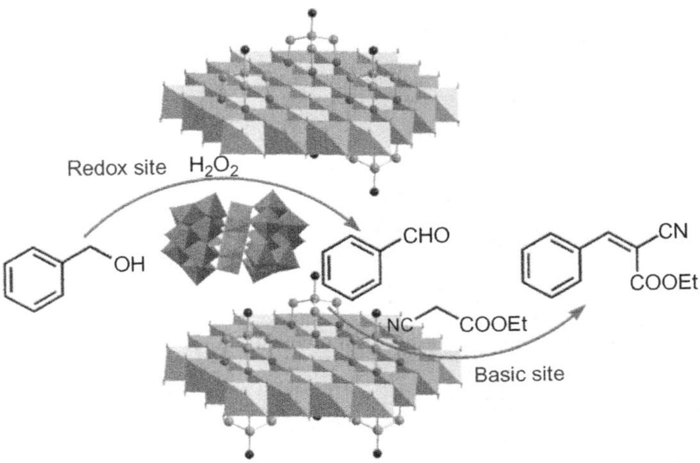

FIG. 13.8 Reaction pathway of the oxidation-Knoevenagel condensation reaction catalyzed by Tris-LDH-$Zn_4(PW_9)_2$. Upper and lower layer (*yellow* and *light blue octahedral* in print versions)=MgO_6 or AlO_6, the octahedral in the middle (*dark blue* and *green octahedral* in print versions)=WO_6 and ZnO_6, spheres (*gray* and *red* in print versions)=Tris. *(Reproduced from K. Liu, et al., Polyoxometalate-intercalated layered double hydroxides as efficient and recyclable bifunctional catalysts for cascade reactions, Chem. Cat. Chem. 8 (2016) 929–937, with the permission of Wiley-VCH Verlag.)*

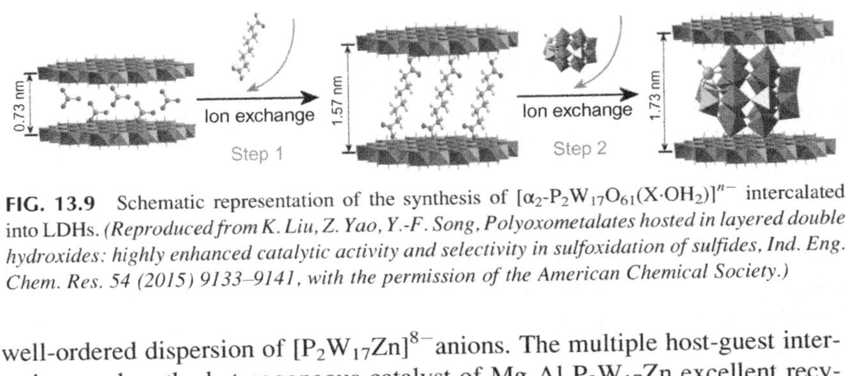

FIG. 13.9 Schematic representation of the synthesis of $[\alpha_2\text{-}P_2W_{17}O_{61}(X\cdot OH_2)]^{n-}$ intercalated into LDHs. *(Reproduced from K. Liu, Z. Yao, Y.-F. Song, Polyoxometalates hosted in layered double hydroxides: highly enhanced catalytic activity and selectivity in sulfoxidation of sulfides, Ind. Eng. Chem. Res. 54 (2015) 9133–9141, with the permission of the American Chemical Society.)*

well-ordered dispersion of $[P_2W_{17}Zn]^{8-}$ anions. The multiple host-guest interactions endow the heterogeneous catalyst of $Mg_3Al\text{-}P_2W_{17}Zn$ excellent recycling performance. Moreover, investigation of leaching ruled out the detectable tungsten in the filtrate [40].

2.3 Encapsulated and Intercalated Nanoparticles

LDHs proved their utilities for developing magnetic composites [71] with a core-shell structure that possesses the magnetization of magnetic materials and multiple functionalities of the LDH materials. Because direct immobilization of hierarchical units onto the magnetic cores is challenging, great effort has been dedicated to circumvent this issue [72]. For instance, the Fe_3O_4 submicrospheres were first coated with a thin carbon layer and then coated with MgAl-LDH [73].

In another example, to develop an active and easily-separable catalyst, the magnetic composite was synthesized through coating ultrafine and uniform Fe_3O_4 (1–3 nm) with Mg-Al LDH. The hydrotalcite-encapsulated core-shell particles were then subjected to an anion exchange with WO_4^{2-} to furnish the tungstate-intercalated, Mg-Al-LDH-encapsulated Fe_3O_4 nanoparticles, WO_4^{2}/LDH@Fe_3O_4 (Fig. 13.10) [74]. The obtained nanocomposite was used as a catalyst for selective oxidation of sulfides to sulfoxides in the presence of H_2O_2 as an oxidant. It was found that the reactions did not proceed under LDH@Fe_3O_4 catalysis while WO_4^{2}/LDH@Fe_3O_4 exhibited high catalytic activity. This observation proved the key role of WO_4^{2-} in catalytic performance. It is worth noting that tungstate leaching was insignificant and the catalyst could be easily recovered magnetically and reused without negligible loss of activity. The catalyst could tolerate the harsh reaction condition because of encapsulation of Fe_3O_4 nanoparticles within LDH.

In this line, another example was introduced by Ganguly and Jain et al. [75]. The authors coated magnetic nanoparticles, Fe_3O_4, with LDHs to develop a nanocomposite, LDH@MNP, on which they immobilized cobalt phthalocyanine (Fig. 13.11). The hybrid catalyst was successfully used for the oxidation of mercaptans to disulfides in the presence of molecular oxygen as an oxidant under an alkaline-free aqueous medium. The novel catalyst exhibited high catalytic activity and could promote the selective oxidation of various mercaptans such as aliphatic, aromatic, and long-chain aliphatic to their corresponding disulfides in near quantitative yields without any evidence for the formation of corresponding sulphonic acid. The magnetic core allowed simple recovery of the

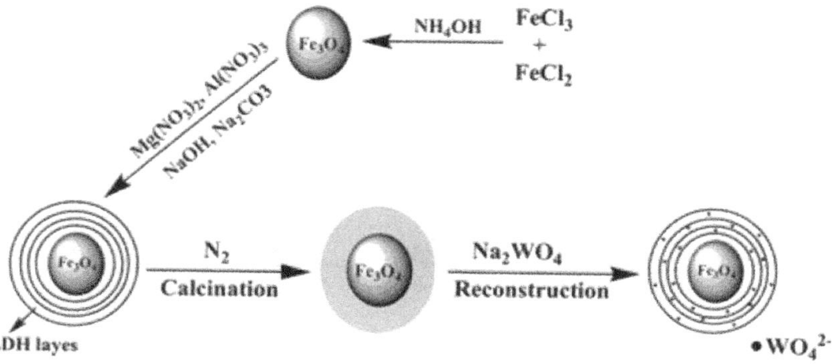

FIG. 13.10 Preparation of $WO_4^{\ 2-}$/LDH@Fe_3O_4. *(Reproduced from C. Yuan, H. Liu, X. Gao, Magnetically recoverable heterogeneous catalyst: tungstate intercalated Mg-Al-layered double hydroxides-encapsulated Fe_3O_4 nanoparticles for highly efficient selective oxidation of sulfides with H_2O_2, Catal. Lett. 144 (2014) 16–21, with the permission of Springer.)*

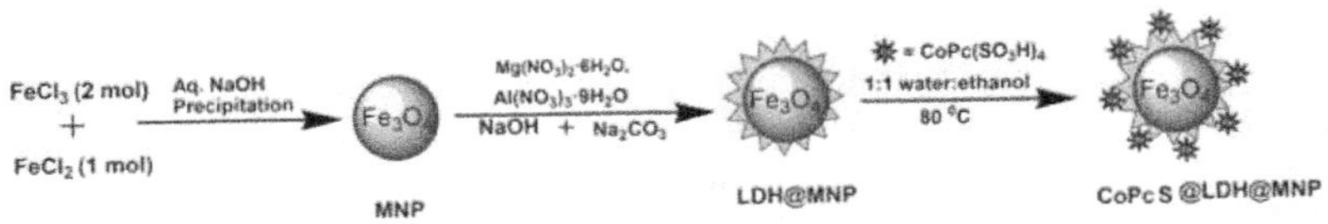

FIG. 13.11 Synthesis of the CoPcS@LDH@MNP catalyst. *(Reproduced from P. Kumar, et al., Magnetic Fe_3O_4@MgAl-LDH composite grafted with cobalt phthalocyanine as an efficient heterogeneous catalyst for the oxidation of mercaptans, J. Mol. Catal. 401 (2015) 48–54, with the permission of Elsevier.)*

catalyst by using an external magnet while the basic sites on the LDH shell provided active sites for the covalent attachment of the tetrasulfonated cobalt phthalocyanine catalyst (CoPcS).

Zhang et al. developed a facile procedure for the synthesis of honeycomb-like hierarchical core-shell Fe_3O_4@MgAl-LDH@Au nanocomposites and used them for the catalytic oxidation of alcohols under atmospheric O_2 without additives [72]. The synthetic procedure included preparing fully negatively charged Fe_3O_4 spheres followed by coating with a layer of oriented grown carbonate-MgAl-LDH to afford Fe_3O_4@MgAl-LDH. Au nanoparticles (~7.0 nm) were subsequently supported by a deposition-precipitation method (DP).

This research group also developed a series of honeycomb-like hierarchical hollow nanostructured magnetic catalysts γ-Fe_2O_3@M_3Al-LDH@$Au_{25\text{-}x}$ (x refers to the mass of Au_{25} nanoclusters in wt%) with nearly atomic precise Au clusters for alcohol oxidation. The catalyst was prepared by fabrication of magnetic supports Fe_3O_4@M_3Al-LDH (M=Ni, Mg or Cu/Mg(0.5/2.5), followed by modified electrostatic adsorption of water-soluble captopril-capped Au_{25} nanoclusters ($Au_{25}Capt_{18}$) and calcinations (Fig. 13.12). In the case of

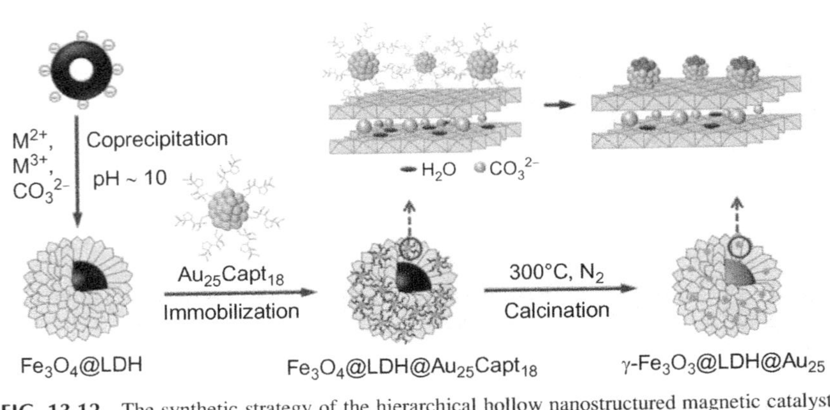

FIG. 13.12 The synthetic strategy of the hierarchical hollow nanostructured magnetic catalysts γ-Fe_2O_3@M_3Al-LDH@$Au_{25\text{-}x}$. *(Reproduced from S. Yin, J. Li, H. Zhang, Hierarchical hollow nanostructured core@shell recyclable catalysts γ-Fe_2O_3@LDH@$Au_{25\text{-}x}$ for highly efficient alcohol oxidation (2016), with the permission of the Royal Society of Chemistry.)*

higher gold loading catalysts, the synergetic interaction between Au_{25} nanoclusters (Au_{25}NCs) and magnetic LDH-based support along with the sintering effect led to slightly aggregated Au_{25}NCs for the series γ-Fe_2O_3@M_3Al-LDH@Au_{25-x} ($x \sim 0.2$) catalysts (ca. 3.0 ± 1.3,3.3 ± 1.2 and 4.2 ± 1.5 nm for Ni-, Mg-, and CuMg-based catalysts, respectively) on calcinations, while the lower gold loading catalysts γ-Fe_2O_3@Ni_3Al-LDH@Au_{25-x} (x = 0.053, 0.11) exhibited nearly atomic precise Au_{25}NCs (*ca.* 1.4 ± 0.3–1.9 ± 0.6 nm) highly dispersed on the surface of shell Ni_3Al-LDH. In comparison to previously reported Fe_3O_4@MgAl-LDH@Au, all γ-Fe_2O_3@M_3Al-LDH@Au_{25-x} catalysts showed much higher catalytic activity for the oxidation of 1-phenylethanol under atmospheric O_2 without basic additives. Among all catalysts, γ-Fe_2O_3@Ni_3Al- LDH@$Au_{25-0.053}$ was the most active catalysts that could be used for the oxidation of a vast range of alcohols under solvent-free conditions. The high catalytic activity of this catalyst was attributed to the nearly atomic precise electron-rich Au_{25}NCs and remarkable Au_{25}NCs-Ni_3Al-LDH-magnetic core three-phase synergetic effect along assisted by abundant Ni-OH sites [76].

Li and Shi and their coworkers present an efficient procedure, flocculation, and in situ reduction method for developing LDHs confined Au colloids [77]. Au nanocolloids were evenly dispersed in an LDH gallery and its particle size was less than 1 nm. The catalytic activity of the composite was studied for low-temperature CO oxidation. The results established that the oxidation reaction could be initiated at a very low temperature (lower than −30°C) and almost a 100% conversion can be achieved at 0°C. The catalyst stability studies show that after a 12 h reaction at 0°C, the CO conversion did not have any apparent decrease (Fig. 13.13).

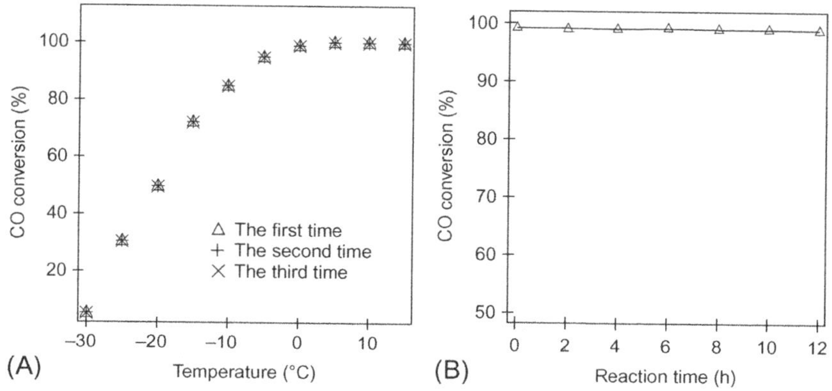

FIG. 13.13 Catalytic performances of LDH confined Au colloidal catalyst for CO oxidation: (A) CO conversion at different temperatures and (B) catalyst stability test (0°C for 12 h). *(Reproduced from L. Li, et al., Layered double hydroxide confined Au colloids: high performance catalysts for low temperature CO oxidation, Catal. Commun. 26 (2012) 15–18, with the permission of Elsevier.)*

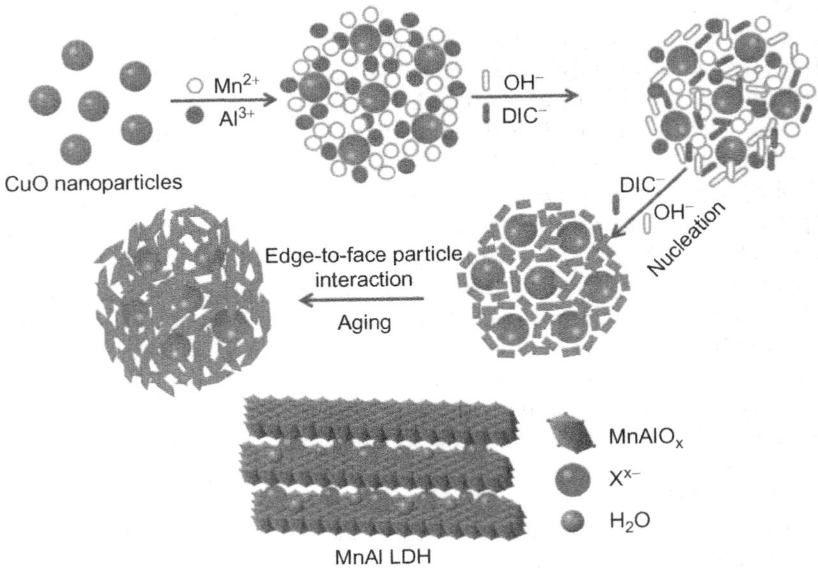

FIG. 13.14 Schematic illustration of the fabrication mechanism of CuO@MnAl NSs. *(Reproduced from M. Asif, et al., Metal oxide intercalated layered double hydroxide nanosphere: withenhanced electrocatalyic activity towards H_2O_2 for biologicalapplications, Sens. Actuator. B 239 (2017) 243–252, with the permission of Elsevier.)*

Recently, novel hybrid nanosphres, CuO@MnAl LDHs, were developed by wrapping CuO nanoparticles in MnAl LDH through simple coprecipitation and a hydrothermal approach (Fig. 13.14) [78]. The utility of the system was investigated by using it as a highly effective electrocatalyst and incorporating it in an electrochemical sensor system in real-time, tracking H_2O_2 secreted from human live cells and detecting ubiquitous H_2O_2 in human serum and urine samples. The results established that the integration of the *n*-type spinel of CuO and *p*-type semiconductive channels of MnAl LDHs can accelerate electron transfer at breakdown voltage of the *p-n* junction. Owing to the synergistic effect of the high surface area of CuO NPs, superb intercalation features of semiconductive MnAl LDHs for encapsulating nanospheres, and their intrinsic p-n junction characteristics, CuO@MnAl NPs have exhibited excellent electrocatalytic activity towards the reduction of H_2O_2.

It is known that noble metal nanoparticles exhibit unusual redox activity by readily accepting electrons from a suitable donor. If such metal particles come in contact with a charged semiconductor nanostructure, they can equilibrate and undergo Fermi-level equilibration. Such electron transfer would greatly enhance the separation of photogenerated electrons and holes and hence improve the photocatalytic efficiency. On the other hand, the utility of LDHs composed of various metals such as Zn, Ni, and Cr has been confirmed for

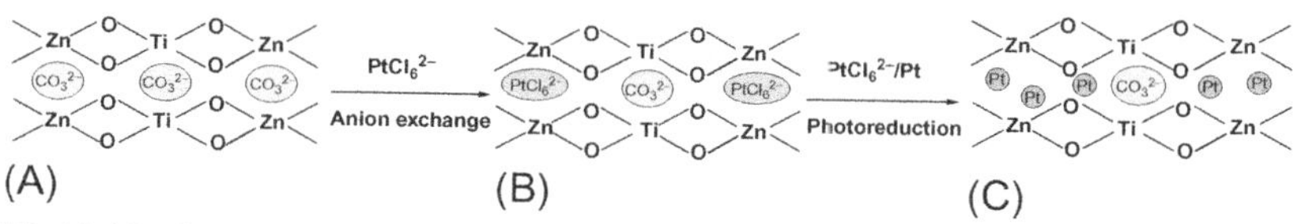

FIG. 13.15 Schematic illustration for synthesis of nano-Pt intercalated Zn-Ti LDHs composite: (A) Zn-Ti LDHs; (B) subsequent exchange of $PtCl_6^{2-}$ with anions of CO_3^{2-} (C) followed by photoreduction to form the assembly of Pt/Zn-Ti LDHs. *(Reproduced from G. Chen, et al., Enhancement photocatalytic degradation of rhodamine B on nanoPtintercalated Zn-Ti layered double hydroxides, Appl. Surf. Sci. 293 (2014) 345–351, with the permission of Elsevier.)*

the photodecomposition of organic compounds [79,80]. It is attributed to the good distribution of metals in the brucite-like layer of LDHs that allows the dye sorption inside the particle and high light absorption on the structure of LDHs in the spectral range. Combining the two abovementioned facts, Luo, Tu, and their coworkers designed a novel composite: nano Pt intercalated-Zn-Ti LDH for photocatalytic degradation of RB. The synthetic procedure of the nanocomposite, depicted in Fig. 13.15, involved three steps: coprecipitation, ionic exchange, and photochemical reduction [81].

Studying the photocatalytic activity of the composite indicated that the combination of Zn-Ti-LDHs and Pt introduced some properties of Pt into photocatalysis such as excellent conductivity and controllability. Remarkable 17-fold enhancement in the degradation of the RB reaction was observed on as-prepared Pt/Zn-Ti LDHs compared with pure Zn-Ti LDHs under simulated sunlight irradiation. The authors attributed the enhanced photocatalytic activity to the high specific surface (111.46 m^2/g), extended photoresponding range, the negative shift in the flat-band potentials, and the high migration efficiency of photoinduced electrons, which may suppress the charge recombination effectively.

2.4 Intercalation in Hydrotalcite and Enhancement of Enantioselectivity

LDHs can also be employed for the asymmetric synthesis. In this regard, LDHs can enhance the enantioselectivty through the confinement effect. Moreover, LDHs can allow the reactions to proceed under psedohomogeneous conditions that can be considered as efficient as homogeneous systems. In an initiative work, pristine α-amino acids as chiral ligands were attached as anions to LDH through electrostatic forces. As seen in Fig. 13.16, α-amino acid was deprotonated, and the proton was replaced by positively charged LDH layers. The ligand was then used for the asymmetric vanadium-catalyzed epoxidation of allylic alcohols.

Asymmetric epoxidation of 2-methyl-cinnamyl alcohol in the presence of $VO(OiPr)_3$ by using intercalated L-glutamate proved the superior enantioselectivity of this catalytic system. To elucidate whether the enhanced enantioselectivity

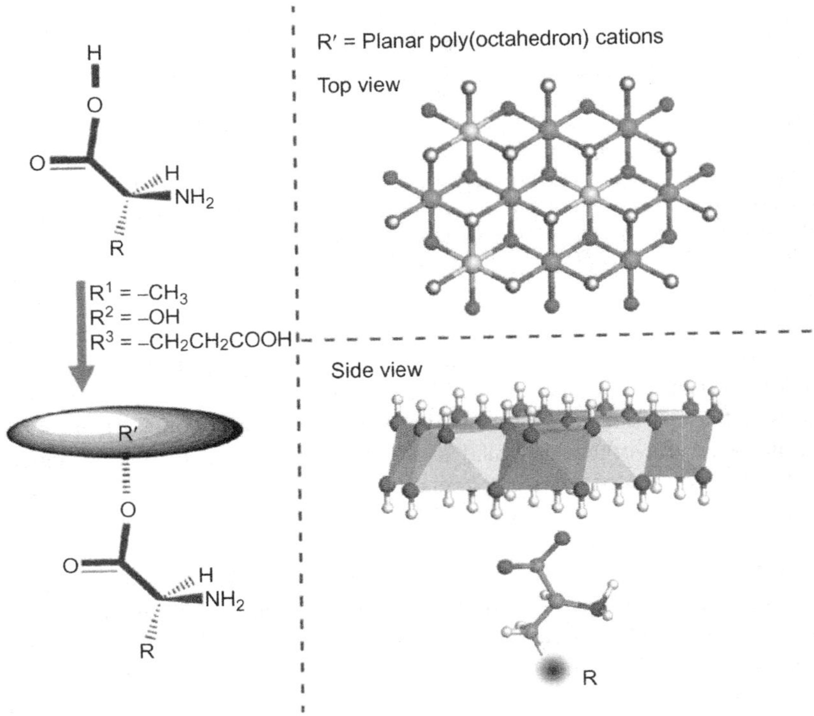

FIG. 13.16 Brucite-like layer as a huge and rigid planar substituent for α-amino acids. *(Reproduced from L.-W. Zhao, et al., Nanosheet-enhanced enantioselectivity in the vanadium-catalyzed asymmetric epoxidation of allylic alcohols, Chem. Eur. J. 18 (2012) 9911–9918, with the permission of Wiley.)*

emerged from the spatial confinement of the interlayer region or the steric/electronic effects of the single layer, the intercalated ligand was delaminated to form exfoliated brucite-like layers. It was proved that the increase in enantioselectivity was due to attachment of α-amino acid to LDH layers and the epoxidation activity was due to the vanadium centers. It was proposed that the LDH layers played the role of a huge planar substituent on a chiral ligand that can boost the chiral induction of the ligand via steric or electronic effects or both. DFT calculations indicated that the LDH layer improved the enantioselectivity by influencing the formation and stability of the catalytic transition states, both in terms of steric resistance and H-bonding interactions [82].

Using the reconstruction approach with the increase in interlayer space of LDHs, Pitchumani et al. intercalated sixteen different amino acids such as arginine, lysine, leucine, isoleucine, phenylalanine, α-aminobutyric acid, aspartic acid, glutamic acid, histidine, alanine, serine, glycine, tyrosine, cystine, and proline into Mg-Al LDHs. It was found that structural features of amino acids, namely, carbon chain length, structure, and physicochemical properties, could affect the intercalation behavior of various amino acids. The obtained hybrids were employed for promoting chemoselective *o*-methylation of phenol and *S*-methylation of thiophenol with dimethyl carbonate (DMC) as a green methylating agent [83]. The amino acids having a large hydrophobic pocket exhibited high catalytic activity with excellent selectivity in *o*-methylation of phenol using DMC as a methylating reagent, and no *C*-methylated products were observed.

The suggested reaction mechanism for chemoselective *o*-methylation of phenol with DMC promoted by Leucine-LDH (Leu-LDHs) is depicted in Fig. 13.17. The intercalation of Leu into anionic LDHs proceeded in such a way that the amine group was exposed and the elongated side chain created a hydrophobic pocket. The reaction proceeded through abstraction of acidic hydrogen from phenol by a free amine to afford a softer phenolate anion, which in turn attacked DMC at the methyl group (softer electrophile), forming a six-membered transition state inside the hydrophobic pocket; this further rearranged to provide anisole and methanol as the end products. It is worth noting that Leu-LDHs could be easily recovered and reused without the loss of catalytic activity.

He et al. developed a heterogeneous Ti (IV) complex through preimmobilization of a chiral L-tartrate ligand between the interlayers of Mg/Al LDHs [84]. Changing the solvent, the authors could control the ligand in a perpendicular-standing (TA-LDH-*p*) or flat-lying arrangement (TA-LDH-*f*) in the interlayer (Fig. 13.18). The authors demonstrated that the orientation of tartrate influenced their coordination to metal centers and consequently their catalytic activity for

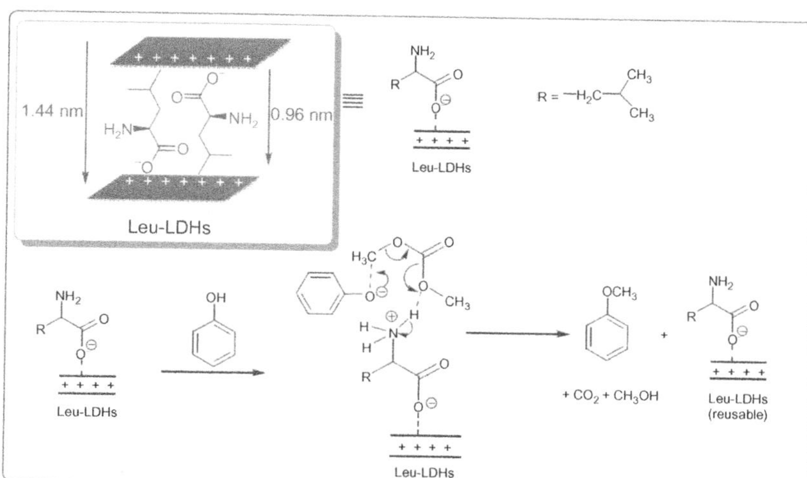

FIG. 13.17 Plausible reaction mechanism for Leu-LDH catalyzed O-methylation of phenol with DMC. *(Reproduced from T. Subramanian, A. Dhakshinamoorthy, K. Pitchumani, Amino acid intercalated layered double hydroxide catalyzed chemoselective methylation of phenols and thiophenols with dimethyl carbonate, Tetrahedron Lett. 54 (2013) 7167–7170, with the permission of Elsevier.)*

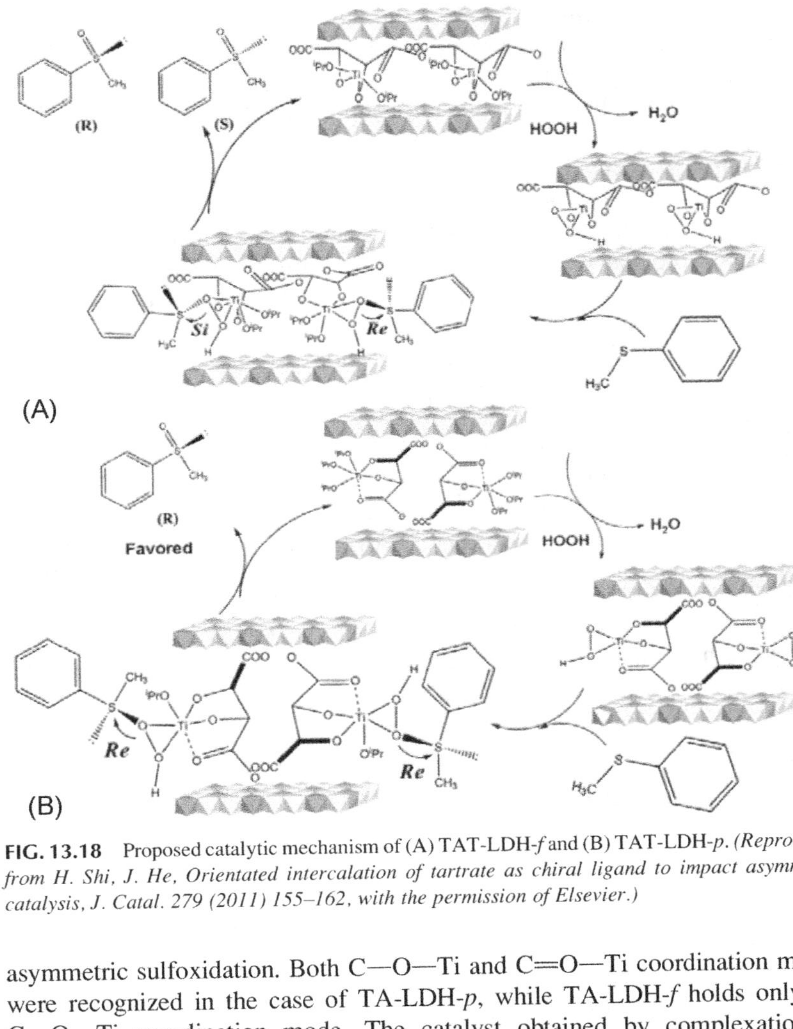

FIG. 13.18 Proposed catalytic mechanism of (A) TAT-LDH-*f* and (B) TAT-LDH-*p*. *(Reproduced from H. Shi, J. He, Orientated intercalation of tartrate as chiral ligand to impact asymmetric catalysis, J. Catal. 279 (2011) 155–162, with the permission of Elsevier.)*

asymmetric sulfoxidation. Both C—O—Ti and C═O—Ti coordination modes were recognized in the case of TA-LDH-*p*, while TA-LDH-*f* holds only the C—O—Ti coordination mode. The catalyst obtained by complexation of LDH-*p* in both modes with the Ti (IV) center led to higher catalytic activity for the asymmetric sulfoxidation. Furthermore, higher enantioselectivity for sulfoxide was observed in this case.

2.5 LDH and Biocatalytic System

Recently, a novel nanobiocatalytic system has been developed by environmentally benign, low-temperature, and organic compound-free synthesis of "Desert Rose Stone"-like Mg-Al-CO_3 LDH nanoflowers through a one-step coprecipitation procedure and biomolecule loading (Fig. 13.19) [85]. In the synthetic procedure, atmospheric carbon dioxide and aluminum sheets were employed as

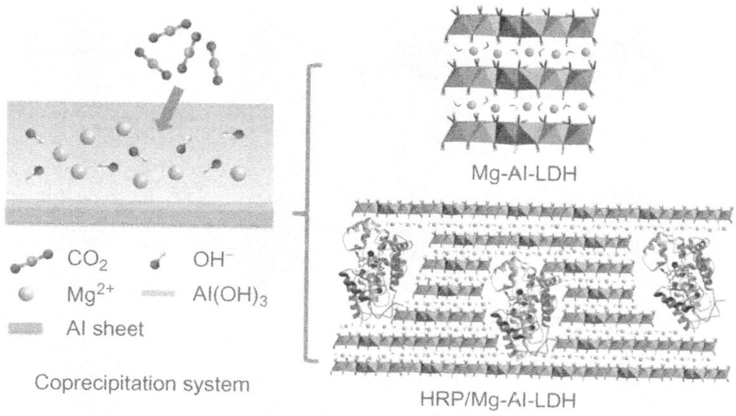

FIG. 13.19 Illustration of the proposed green coprecipitation method for the synthesis of Mg-Al-LDH and HRP/Mg-Al-LDH nanoflowers. *(Reproduced from M. Wang, et al., A green approach to the synthesis of novel "Desert rose stone"-likenanobiocatalytic system with excellentenzyme activity and stability, Sci. Rep. (2014) 4.)*

carbonate and alumina ions respectively. Moreover, horseradish peroxidase (HRP) was selected as the model enzyme. The specific surface area of the in situ-grown 3D hierarchical Mg-Al-LDH nanoflowers was higher than conventional plate-like LDHs.

To study the nanobiocatalytic activity, 3,3′,5,5′-tetramethylbenzidine dihydrochloride hydrate (TMB) and hydrogen peroxide were used as model substrates for HRP. The oxidation of TMB was promoted by HRP in the presence of hydrogen peroxide to afford TMBox. The comparison of the catalytic activity of the HRP/Mg-Al-LDH with bare HRP indicated the superior catalytic reaction velocity of the former. The higher catalytic activity of HRP HRP/Mg-Al-LDH was attributed to the conformation change of the immobilized HRP, which further results in an increase in the proportion of the exposed active sites.

2.6 Intercalated Mesoporous Silica

There are only a few reports regarding preparation of hydrotalcite-SBA composites. However, they are mostly used for CO_2 capture [86] or sorption and rarely for catalytic purposes. In a pioneer work, Sakthivel et al. intercalated an SBA-15 molecular sieve between the interlayer of Mg-Al hydrotalcite [87]. Using various characterization techniques such as XRD, FTIR, and TEM, the authors demonstrated successful intercalation of SBA-15. Interestingly, incorporation of SBA-15 did not result in the collapse of a hydrotalcite structure, and MASBA-HT retained its layered structure even after calcination at 550°C (Fig. 13.20). The obtained hybrid system, MASBA-HT, exhibited high catalytic activity for vapor-phase hydroisomerization of 1-octene. The catalytic activity of the hybrid system was compared with those of pure hydrotalcite (HT)

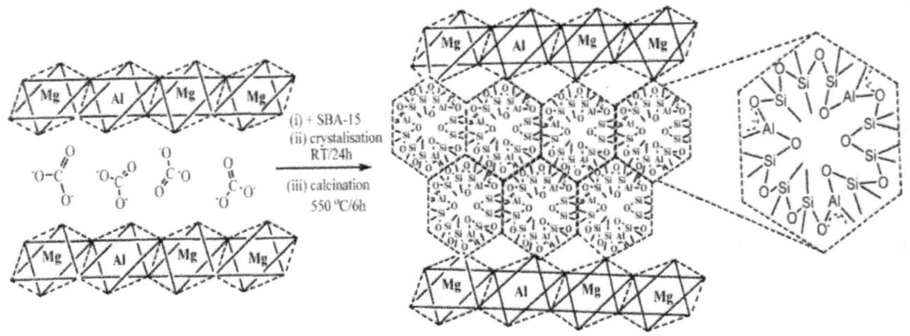

FIG. 13.20 Schematic representation of SBA-15 intercalation on Mg-Al-HT. *(Reproduced from T. Baskaran, et al., SBA-15 intercalated Mg-Al hydrotalcite: an environmental friendly catalyst for hydroisomerization of olefin, Appl. Catal. A. 488 (2014) 119–127, with the permission of Elsevier.)*

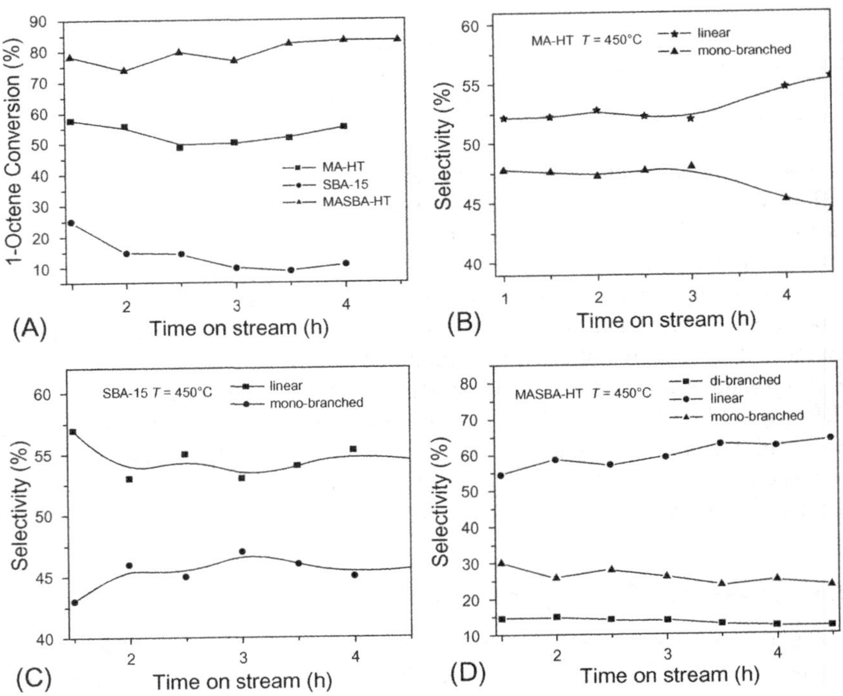

FIG. 13.21 The catalytic performance of MASBA-HT, HT, and SBA-15. Effect of time on stream on different catalysts (A) 1-octene conversion and (B–D) product distribution. *(Reproduced from T. Baskaran, et al., SBA-15 intercalated Mg-Al hydrotalcite: an environmental friendly catalyst for hydroisomerization of olefin, Appl. Catal. A. 488 (2014) 119–127, with the permission of Elsevier.)*

and SBA-15. The results indicated the superior catalytic activity of MASBA-HT, 80%, which can be attributed to dual functionality of this system, in other words, moderate acidity emerged from incorporation of aluminum into the framework of SBA-15 and soluble basicity similar to HT (Fig. 13.21A).

Moreover, the three catalysts exhibited different catalytic performance in terms of product selectivity. For all three catalysts the selectivity of linear to branched isomers remained constant. However, a dibranched isomer was observed just in the case of the MASBA-HT catalyst (Fig. 13.21B). Notably, the catalytic activity remained constant over a longer duration and repeated runs.

CONCLUSION

The interlayer space of LDHs can be considered as an appropriate host for embedding various catalytic species such as PMOs, complexes, nanoparticles, mesoporous silica, and chiral ligands. From various examples discussed in this chapter, it can be concluded that confinement of catalysts through intercalation between interlayers of LDHs can improve the leaching and consequently the reusability of the catalysts. Moreover, the arrangement of catalytic species between LDH layers can affect the catalytic performance. To date, various chemical transformations such as cascade reactions and enantioselective processes were promoted by LDH-intercalated catalysts. Considering the promising results of this class of catalysts, rapid advances in this field are expected in the near future. We hope that this chapter is helpful to push forward the advancement of this research field.

ACKNOWLEDGMENT

The author appreciates the efforts of Mr. Vahid Farzaneh in providing the tables and required articles for covering this chapter.

REFERENCES

[1] V. Rives, A.A. Ulibarri, Layered double hydroxides (LDH) intercalated with metal coordination compounds and oxometalates, Coord. Chem. Rev. 181 (1999) 61–120.

[2] K.-H. Goh, T.-T. Lim, Z. Dong, Application of layered double hydroxides for removal of oxyanions: a review, Water Res. 42 (2008) 1343–1368.

[3] V. Rives, D. Carriazo, C. Martin, Heterogeneous catalysis by polyoxometalate-intercalated layered double hydroxides, in: A. Gil, S. Korili, R. Trujillano, M.A. Vicente (Eds.), Pillared Clays and Related Catalysts, Springer, New York, NY, 2010, , pp. 319–397.

[4] S.N. Basahel, et al., Nanostructured Mg–AL hydrotalcite as catalyst for fine chemical synthesis, J. Nanosci. Nanotechnol. 14 (2014) 1931–1946.

[5] S. Nishimura, A. Takagaki, K. Ebitani, Characterization, synthesis and catalysis of hydrotalcite-related materials for highly efficient materials transformations, Green Chem. 15 (2013) 2026–2042.

[6] T. Baskaran, J. Christopher, A. Sakthivel, Progress on layered hydrotalcite (HT) materials as potential support and catalytic materials, RSC Adv. 5 (2015) 98853–98875.

[7] J. He, et al., Preparation of layered double hydroxides, Struct. Bond. 119 (2005) 89–119.

[8] J. Qu, et al., Mechanochemical approaches to synthesize layered double hydroxides: a review, Appl. Clay Sci. 119 (2016) 185–192.

[9] Z.P. Xu, et al., Catalytic applications of layered double hydroxides and derivatives, Appl. Clay Sci. 53 (2011) 139–150.

[10] S. Gago, et al., Immobilization of oxomolybdenum species in a layered double hydroxide pillared by 2,2'-bipyridine-5,5'-dicarboxylate anions, Inorg. Chem. 43 (2004) 5422–5431.

[11] K.A. Tarasov, D.O. Hare, V.P. Isupov, Solid-state chelation of metal ions by ethylenediaminetetraacetate intercalated in a layered double hydroxide, Inorg. Chem. 42 (2003) 1919–1927.

[12] K.M. Parida, M. Sahoo, S. Singha, A novel approach towards solvent-free epoxidation of cyclohexene by Ti(IV)–schiff base complex-intercalated LDH using H_2O_2 as oxidant, J. Catal. 276 (2010) 161–169.

[13] T. Hara, et al., An anionic D-valine–palladium(II) complex supported on a hydroxyndouble salt with a brønsted basic phosphate anion: application for a heterogeneous catalyst toward aerobic alcohol oxidation, Catal. Sci. Technol. 1 (2011) 1376–1382.

[14] T. Hara, et al., Creation of highly stable monomeric Pd(II) species in an anion-exchangeable hydroxy double salt interlayer: application to aerobic alcohol oxidation under an air atmosphere, Green Chem. 11 (2009) 2034–2040.

[15] S. Singha, M. Sahoo, K.M. Parida, Chemoselective oxidation of primary alcohols catalysed by Ce(iii)-complex intercalated LDH using molecular oxygen at room temperature, Dalton Trans. 40 (44) (2011) 11838–11844.

[16] G. Varga, et al., Mn(II)-amino acid complexes intercalated in CaAl-layered double hydroxide—well-characterized, highly efficient, recyclable oxidation catalysts, J. Catal. 335 (2016) 125–134.

[17] X. Wang, et al., Base-free selective oxidation of glycerol over LDH hosted transition metal complexes using 3% H_2O_2 as oxidant, Catal. Commun. 6 (7) (2016) 101.

[18] X. Wang, et al., Base-free selective oxidation of glycerol with 3% H_2O_2 catalyzed by sulphonato-salen-chromium(III) intercalated LDH, Catal. Commun. 28 (2012) 73–76.

[19] W.Y. Zhou, et al., Highly efficient transformation of alcohol to carbonyl compounds under a hybrid bifunctional catalyst originated from metalloporphyrins and hydrotalcite, J. Catal. 335 (2016) 105–116.

[20] W. Guo, et al., Silylated layered double hydroxide nanosheets prepared by a large-scale synthesis method as hosts for intercalation of metal complexes, Appl. Catal. A 522 (2016) 101–108.

[21] M. Halma, et al., Synthesis, characterization, and catalytic activity of anionic iron(III) porphyrins intercalated into layered double hydroxides, J. Catal. 257 (2) (2008) 233–243.

[22] S. Nakagaki, et al., First insight into catalytic activity of anionic iron porphyrins immobilized on exfoliated layered double hydroxides, J. Colloid Interface Sci. 281 (2) (2005) 417–423.

[23] Z. Sun, et al., A structured catalyst based on cobalt phthalocyanine/calcined Mg-Al hydrotalcite film for the oxidation of mercaptan, Green Chem. 14 (7) (2012) 1909–1916.

[24] Z. Tong, T. Shichi, K. Takagi, Oxidation catalysis of a manganese (III) porphyrin intercalated in layered double hydroxide clays, Mater. Lett. 57 (15) (2003) 2258–2261.

[25] S. Bhattacharjee, J.A. Anderson, Epoxidation by layered double hydroxide-hosted catalysts. Catalyst synthesis and use in the epoxidation of R-(+)-limonene and (−)-α-pinene using molecular oxygen, Catal. Lett. 95 (3–4) (2004) 119–125.

[26] S. Bhattacharjee, J.A. Anderson, Novel chiral sulphonato-salen-manganese (III)-pillared hydrotalcite catalysts for the asymmetric epoxidation of styrenes and cyclic alkenes, Adv. Synth. Catal. 348 (1–2) (2006) 151–158.

[27] S. Bhattacharjee, J.A. Anderson, Synthesis and characterization of novel chiral sulfonato–salen–manganese (III) complex in a zinc–aluminium LDH host, Chem. Commun. 5 (2004) 554–555.

[28] S. Bhattacharjee, T.J. Dines, J.A. Anderson, Comparison of Co with Mn and Fe in LDH-hosted sulfonato-salen catalysts for olefin epoxidation, J. Phys. Chem. C 112 (36) (2008) 14124–14130.

[29] A.C. Gomes, et al., Intercalation of a molybdenum η3-allyl dicarbonyl complex in a layered double hydroxide and catalytic performance in olefin epoxidation, Dalton Trans. 42 (23) (2013) 8231–8240.

[30] Y. Liu, et al., Enhanced catalytic efficiency in the epoxidation of alkenes for manganese complex encapsulated in the hydrophobic interlayer region of layered double hydroxides, Ind. Eng. Chem. Res. 52 (50) (2013) 17821–17828.

[31] T. Hara, et al., Hydrophenylation of internal alkynes with boronic acids catalysed by a Ni–Zn hydroxy double saltintercalated anionic rhodium(III) complex, Catal. Sci. Technol. 6 (2016) 863–868.

[32] M. Wei, et al., Rh-TPPTS intercalated layered double hydroxides as hydroformylation catalyst, AIChE J. 53 (11) (2007) 2916–2924.

[33] F. Neaţu, et al., Arylation of alkynes over hydrotalcite docked Rh-m-TPPTC complex, Catal. Today 247 (2015) 155–162.

[34] M. Vasconcellos-Dias, et al., Pyridine carboxylate complexes of MoII as active catalysts in homogeneous and heterogeneous polymerization, Eur. J. Inorg. Chem. 2007 (18) (2007) 2917–2925.

[35] W. Zhou, et al., Efficient catalysis of the aerobic baeyer–villiger oxidation over a bifunctional catalyst based on cobalt tetraphenylporphyrin intercalated into Zn2Al hydrotalcite, Asian J. Org. Chem. 4 (2015) 33–37.

[36] T. Hara, et al., Epoxidation of cyclic enones with hydrogen peroxide catalysed by alkylcarboxylate-intercalated Ni–Zn mixed basic salts, Catal. Sci. Technol. 5 (2015) 578–583.

[37] T. Hara, et al., Efficient 1,4-addition of enones and boronic acids catalyzed by a Ni-Zn hydroxyl double salt-intercalated anionic rhodium(III) complex, ACS Catal. 4 (2014) 4040–4046.

[38] Y. Chen, et al., Modular polyoxometalate-layered double hydroxide composites as efficient oxidative catalysts, Chem. Eur. J. 21 (2015) 10812–10820.

[39] S. Omwoma, et al., Recent advances on polyoxometalates intercalated layered double hydroxides: from synthetic approaches to functional material applications, Coord. Chem. Rev. 258–259 (2014) 58–71.

[40] K. Liu, Z. Yao, Y.-F. Song, Polyoxometalates hosted in layered double hydroxides: highly enhanced catalytic activity and selectivity in sulfoxidation of sulfides, Ind. Eng. Chem. Res. 54 (2015) 9133–9141.

[41] M.R. Weir, R.A. Kydd, Synthesis of metatungstate pillared layered double hydroxides with variable layer composition. Effect of the Mg:Al ratio on the microporous structure, Microporous Mesoporous Mater. 20 (1988) 339–347.

[42] Y. Watanabe, K. Yamamoto, T. Tatsumi, Epoxidation of alkenes catalyzed by heteropolyoxometalate as pillars in layered double hydroxides, J. Mol. Catal. A Chem. 145 (1) (1999) 281–289.

[43] X. Wei, et al., Tungstocobaltate-pillared layered double hydroxides: preparation, characterization, magnetic and catalytic properties, J. Solid State Chem. 181 (6) (2008) 1292–1297.

[44] E.A. Gardner, et al., Layered double hydroxides pillared by macropolyoxometalates, Appl. Clay Sci. 13 (5) (1998) 479–494.

[45] S.K. Jana, Y. Kubota, T. Tatsumi, Cobalt-substituted polyoxometalate pillared hydrotalcite: synthesis and catalysis in liquid-phase oxidation of cyclohexanol with molecular oxygen, J. Catal. 255 (1) (2008) 40–47.

[46] B.M. Choudary, et al., Tungstate-exchanged Mg-Al-LDH catalyst: an eco-compatible route for the oxidation of sulfides in aqueous medium, J. Chem. Soc. Perkin Trans. 18 (2002) 2069–2074.
[47] V. Hulea, et al., Catalytic oxidation of thiophenes and thioethers with hydrogen peroxide in the presence of W-containing layered double hydroxides, Appl. Catal. A 313 (2) (2006) 200–207.
[48] A. Maciuca, et al., V-, Mo-and W-containing layered double hydroxides as effective catalysts for mild oxidation of thioethers and thiophenes with H2O2, Catal. Today 138 (1) (2008) 33–37.
[49] A. Maciuca, et al., Mild oxidation of tetrahydrothiophene to sulfolane over V-, Mo-and W-containing layered double hydroxides, Appl. Catal. A 338 (1) (2008) 1–8.
[50] P. Mitchell, S.A. Wass, Propane dehydrogenation over molybdenum hydrotalcite catalysts, Appl. Catal. A 225 (1) (2002) 153–165.
[51] R. Dula, et al., Layered double hydroxide-derived vanadium catalysts for oxidative dehydrogenation of propane: influence of interlayer-doping versus layer-doping, Appl. Catal. A 230 (1) (2002) 281–291.
[52] K. Bahranowski, et al., Oxidative dehydrogenation of propane over calcined vanadate-exchanged Mg, Al-layered double hydroxides, Appl. Catal. A 185 (1) (1999) 65–73.
[53] S. Blanco, S. Carrazán, V. Rives, Oxidative dehydrogenation of propane on Mg-V-Al mixed oxides, Appl. Catal. A 342 (1) (2008) 93–98.
[54] A. Węgrzyn, et al., Nanostructured V-containing hydrotalcite-like materials obtained by non-stoichiometric anion exchange as precursors of catalysts for oxidative dehydrogenation of n-butane, Catal. Today 116 (1) (2006) 74–81.
[55] M. Ogawa, S. Asai, Hydrothermal synthesis of layered double hydroxide-deoxycholate intercalation compounds, Chem. Mater. 12 (11) (2000) 3253–3255.
[56] D. Carriazo, et al., Metatungstate and tungstoniobate-containing LDHs: preparation, characterisation and activity in epoxidation of cyclooctene, J. Phys. Chem. Solids 68 (10) (2007) 1872–1880.
[57] E. Gardner, T.J. Pinnavaia, On the nature of selective olefin oxidation catalysts derived from molybdate-and tungstate-intercalated layered double hydroxides, Appl. Catal. A 167 (1) (1998) 65–74.
[58] F. Basile, et al., Synthesis and thermal evolution of hydrotalcite-type compounds containing noble metals, Appl. Clay Sci. 16 (3) (2000) 185–200.
[59] J. Palomeque, F. Figueras, G. Gelbard, Epoxidation with hydrotalcite-intercalated organotungstic complexes, Appl. Catal. A 300 (2) (2006) 100–108.
[60] P. Liu, et al., Direct immobilization of self-assembled polyoxometalate catalyst in layered double hydroxide for heterogeneous epoxidation of olefins, J. Catal. 256 (2) (2008) 345–348.
[61] P. Liu, C. Wang, C. Li, Epoxidation of allylic alcohols on self-assembled polyoxometalates hosted in layered double hydroxides with aqueous H2O2 as oxidant, J. Catal. 262 (1) (2009) 159–168.
[62] D. Carriazo, et al., Hydrotalcites composition as catalysts: preparation and their behavior on epoxidation of two bicycloalkenes, Microporous Mesoporous Mater. 95 (1) (2006) 39–47.
[63] A.L. Villa, et al., A study of V-pillared layered double hydroxides as catalysts for the epoxidation of terpenic unsaturated alcohols, J. Catal. 198 (2) (2001) 223–231.
[64] J. Das, K.M. Parida, Heteropoly acid intercalated Zn/Al HTlc as efficient catalyst for esterification of acetic acid using n-butanol, J. Mol. Catal. A Chem. 264 (1) (2007) 248–254.
[65] Y. Guo, et al., Photocatalytic degradation of aqueous organocholorine pesticide on the layered double hydroxide pillared by paratungstate a ion, $Mg_{12}Al_6(OH)36(W7O_{24}).4H_2O$, Appl. Catal. B 30 (3) (2001) 337–349.

[66] Y. Guo, et al., Preparation and photocatalytic behavior of Zn/Al/W (Mn) mixed oxides via polyoxometalates intercalated layered double hydroxides, Microporous Mesoporous Mater. 56 (2) (2002) 153–162.

[67] A. Maciuca, et al., Catalytic oxidation processes for removing dimethylsulfoxide from wastewater, Chemosphere 68 (2) (2007) 227–233.

[68] F. Malherbe, et al., Alkoxylation reaction catalysed by layered double hydroxides, Appl. Clay Sci. 13 (5) (1998) 451–466.

[69] F. Malherbe, et al., Highly selective synthesis of 2-butoxy ethanol over Mg/Al/V mixed oxides catalysts derived from hydrotalcites, Catal. Lett. 67 (2–4) (2000) 73–79.

[70] K. Liu, et al., Polyoxometalate-intercalated layered double hydroxides as efficient and recyclable bifunctional catalysts for cascade reactions, ChemCatChem 8 (2016) 929–937.

[71] S. Xia, et al., A thermally stable and easily recycled core–shell Fe2O3@CuMgAl catalyst for hydrogenolysis of glycerol, Catal. Sci. Technol. 4 (2014) 912–916.

[72] F. Mi, et al., Facile synthesis of hierarchical core–shell Fe_3O_4@MgAl–LDH@Au as magnetically recyclable catalysts for catalytic oxidation of alcohols, Chem. Commun. 47 (2011) 12804–12806.

[73] D. Pan, et al., Nearly monodispersed core–shell structural Fe_3O_4@DFUR–LDH submicro particles for magnetically controlled drug delivery and release, Chem. Commun. 47 (2011) 908–910.

[74] C. Yuan, H. Liu, X. Gao, Magnetically recoverable heterogeneous catalyst: tungstate intercalated Mg–Al-layered double hydroxides-encapsulated Fe_3O_4 nanoparticles for highly efficient selective oxidation of sulfides with H_2O_2, Catal. Lett. 144 (2014) 16–21.

[75] P. Kumar, et al., Magnetic Fe3O4@MgAl–LDH composite grafted with cobalt phthalocyanine as an efficient heterogeneous catalyst for the oxidation of mercaptans, J. Mol. Catal. 401 (2015) 48–54.

[76] S. Yin, J. Li, H. Zhang, Hierarchical hollow nanostructured core@shell recyclable catalysts γ-Fe_2O_3@LDH@Au_{25-x} for highly efficient alcohol oxidation, Green Chem. 18 (2016) 5900–5914.

[77] L. Li, et al., Layered double hydroxide confined Au colloids: high performance catalysts for low temperature CO oxidation, Catal. Commun. 26 (2012) 15–18.

[78] M. Asif, et al., Metal oxide intercalated layered double hydroxide nanosphere: withenhanced electrocatalyic activity towards H2O2 for biologicalapplications, Sens. Actuator. B 239 (2017) 243–252.

[79] L. Mohapatra, K.M. Parida, Zn-Cr layered double hydroxide: visible light respon-sive photocatalyst for photocatalytic degradation of organic pollutants, Sep. Purif. Technol. 91 (2012) 73–80.

[80] L. Mohapatra, K. Parida, M. Satpathy, Molybdate/tungstate intercalated oxo-bridged Zn/Y LDH for solar light induced photodegradation of organicpollutants, J. Phys. Chem. C 116 (2012) 13063–13070.

[81] G. Chen, et al., Enhancement photocatalytic degradation of rhodamine B on nanoPtintercalated Zn–Ti layered double hydroxides, Appl. Surf. Sci. 293 (2014) 345–351.

[82] L.-W. Zhao, et al., Nanosheet-enhanced enantioselectivity in the vanadium-catalyzed asymmetric epoxidation of allylic alcohols, Chem. Eur. J. 18 (2012) 9911–9918.

[83] T. Subramanian, A. Dhakshinamoorthy, K. Pitchumani, Amino acid intercalated layered double hydroxide catalyzed chemoselective methylation of phenols and thiophenols with dimethyl carbonate, Tetrahedron Lett. 54 (2013) 7167–7170.

[84] H. Shi, J. He, Orientated intercalation of tartrate as chiral ligand to impact asymmetric catalysis, J. Catal. 279 (2011) 155–162.

[85] M. Wang, et al., A green approach to the synthesis of novel "desert rose stone"-likenanobiocatalytic system with excellentenzyme activity and stability, Sci. Rep 4 (2014) 1–8.

[86] C.V. Pramod, et al., Hydrotalcite-SBA-15 composite material for efficient carbondioxide capture, J. CO_2 Util. 12 (2015) 109–115.

[87] T. Baskaran, et al., SBA-15 intercalated Mg–Al hydrotalcite: an environmental friendly catalyst for hydroisomerization of olefin, Appl. Catal. A 488 (2014) 119–127.

Chapter 14

Encapsulated Catalysts for Synthesis of Bulk and Fine Chemicals

Sumeet K. Sharma and Raksh V. Jasra
Reliance Industries Limited, Vadodara, India

Contents

1 INTRODUCTION

Generally, supported homogeneous catalysts are prepared by impregnation of homogeneous catalysts on high surface area supports, anchoring of the catalyst on a support using functional groups, tethering or grafting of a homogeneous complex on a support, and intercalation or encapsulation of homogenous catalysts in the porous/layered inorganic materials. In the intercalation or encapsulation approach, the homogeneous catalyst is sandwiched between the layers or in the pores or channels of a support to avoid or minimize the leaching of a homogeneous catalyst especially for liquid-phase catalytic reactions. In the present article, catalytic applications of encapsulated catalysts are discussed for the synthesis of fine and bulk chemicals.

The metal nanoparticles can be encapsulated in the supports using various techniques such as core-shell, yolk-shell, core-tube, or ship-in-a-bottle techniques [1]. The design, encapsulation methods, and characterization of

Encapsulated Catalysts. http://dx.doi.org/10.1016/B978-0-12-803836-9.00014-6

zeolite encapsulated and clay intercalated metal-Schiff base, porphyrin, phthalocyanine complexes, and their reactivity for biomimetic oxidation reactions are compiled by Bedioui [2]. Recently, synthesis and characterization of transition metal complexes encapsulated in zeolites is reviewed by considering the flexible ligand and template synthesis methods [3]. These methods are used for the synthesis of an encapsulated catalyst in which the metal complex or ligand is trapped in the pores of zeolite. In the present chapter, catalytic activity of encapsulated homogeneous metal complexes in the pores and structure of inorganic materials is described for the synthesis of fine and bulk chemicals.

2 HYDROFORMYLATION OF ALKENES USING ENCAPSULATED RHODIUM COMPLEXES

Hydroformylation (oxo) is a reaction between the double bond of an alkene with a syngas (mixture of hydrogen and carbon monoxide) using homogeneous a rhodium/cobalt-based catalyst to produce aldehydes (*n* and *iso*) as primary products with an additional carbon atom as compared to the alkene (Scheme 14.1).

Commercially, hydroformylation of lower carbon chain length alkenes (C_2-C_5) is carried out using a rhodium complex, $HRh(CO)(PPh_3)_3$ as a homogeneous catalyst in milder reaction conditions (80–100°C, 20–40 pressure). Decomposition of the $HRh(CO)(PPh_3)_3$ complex is the main problem associated with catalyst separation from high boiling point products for hydroformylation of higher carbon number alkenes. For hydroformylation of higher carbon chain alkenes, cobalt carbonyl type complexes are widely used at around 200°C and 200–250 pressure. The yield of linear (*n*) aldehyde is significantly lower upon using cobalt-based catalysts. Typically, cobalt is recovered by a decobalting process in the form of cobalt formate or acetate by adding oxygen and formic acid or acetic acid [4,5]. The drawbacks of higher reaction temperature, pressure, longer reaction time, and lower *n*-aldehyde selectivity as compared to rhodium-based catalysts still persist in cobalt complex catalyzed hydroformylation of alkenes. Therefore, the development of a heterogeneous catalyst is critical for economical hydroformylation of alkenes. This is presently done using two approaches. The first approach involves heterogenization of an $HRh(CO)(PPh_3)_3$ metal complex

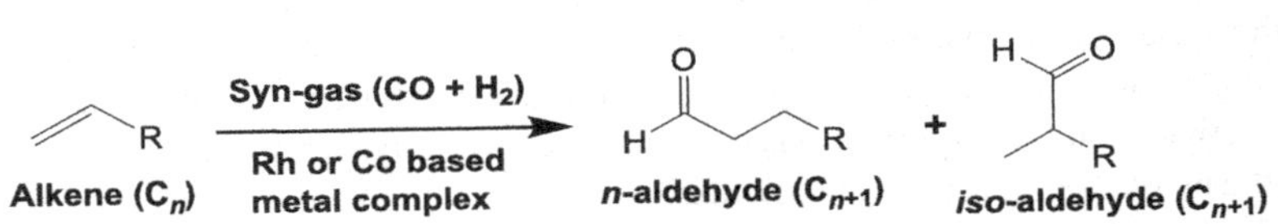

SCHEME 14.1 Hydroformylation (oxo) of alkenes. *(Reproduced from B. Cornils, W.A. Herrmann, Applied Homogeneous Catalysis with Organometallic Complexes, VCH, Weinheim, 1996, with the permission of Wiley-VCH Verlag.)*

using inorganic material as a support, and second method employs a water soluble ligand to separate the rhodium complex from the product mixture.

Impregnation is the commonly used method for the heterogenization of a homogeneous catalyst in which inorganic solid support is mixed with the solution of a homogeneous complex. The resulted slurry is stirred for desired time either at room temperature or at a particular predefined temperature. Generally, silica, alumina, zeolites, and polymers are used as a solid inorganic support to prepare a heterogeneous catalyst. The main problem of the impregnation method is the leaching of rhodium complex from supported catalyst during the catalytic reaction, which leads to reduced catalytic activity in the reusability experiments. Leaching problems can be solved up to a certain extent by anchoring or encapsulation of a homogeneous rhodium complex in a suitable support.

2.1 Rhodium Metal Complex Encapsulated in Zeolites and Zeotype Materials

Zeolites and zeotype (molecular sieves) are very useful inorganic materials in refining, petrochemical, and fine chemical industries due to their high stability, high surface area, shape selectivity, uniform pore size, and varied acidic and basic properties [6–8]. Besides avoiding the leaching of an immobilized metal complex from the pores of zeolites, selectivity to the desired product could be enhanced due to a well-defined and molecularly sized pore structure of the support. Inorganic materials from the M41S family such as MCM-41 (hexagonal), MCM-48 (cubic), and MCM-50 (lamellar) came into the picture due to the larger pore size of these materials [9–12]. A higher surface area (more than 700 m^2/g), pore size in the range of 20–100 Å, and varied pore size distribution made them as versatile potential materials for catalytic applications. To avoid leaching of the homogeneous metal complex from the encapsulated catalyst, the metal complex should be chemically bounded inside the pores of the inorganic material. Apart from the improvement of the recycling of the catalyst, sometimes the support structure also helps to avoid mass transfer limitations during a catalytic reaction, which results in higher selectivity and catalytic activity using an encapsulated catalyst.

To prevent leaching of the transition metal complexes during liquid-phase hydroformylation reaction, Chaudhari et al. encapsulated $HRh(CO)(PPh_3)_3$ complex in the pores of zeolite Na-Y, MCM-41, and MCM-48 supports [13]. The $HRh(CO)(PPh_3)_3$ complex was encapsulated in zeolite Na-Y by the addition of dissolved $HRh(CO)(PPh_3)_3$ complex in ethanol to the zeolite Na-Y gel at 13.1 pH. The slurry was stirred for 1 h followed by 12 h aging at 100°C. After completion of the aging, the encapsulated catalyst was separated by filtration of slurry, washed with deionized water, and dried at 80°C. Similarly, the rhodium metal complex was anchored in the channels of MCM-41 and MCM-48 using respective functionalized support [13]. To functionalize, the supports were stirred in dichloromethane followed by addition of the dichlorodiphenylsilane with continuous stirring [13]. The 3-aminopropyltrimethoxysilane was added dropwise to this slurry at −78°C and stirred for 24 h at 40°C. The functionalized MCM-41 and MCM-48 were obtained after filtration, washing with dry dichloromethane, and vacuum drying. The anchoring of the $HRh(CO)(PPh_3)_3$ complex was carried out by the addition of functionalized support to the solution of $HRh(CO)(PPh_3)_3$ complex in ethanol. The slurry was stirred at room temperature for 16 h followed by filtration and washing with ethanol to remove the free complex from the support; it was then dried in a vacuum. These catalysts were claimed as highly active, stable, and reusable catalysts (six times recycled) for hydroformylation of alkenes [13]. The oxidation state of the rhodium complex determined by XPS spectra was found as Rh(I) in the both fresh and spent catalysts. The hydroformylation of 1-hexene and styrene was also studied using a reusable $HRh(CO)(PPh_3)_3$ encapsulated zeolite Na-Y catalyst to give more than 95% conversion of substrate and *n*/*iso* ratio of aldehyde 3.2 and 2.3, respectively [14].

Encapsulation of rhodium metal in zeolites- A, X, Y, and ZSM-5 was carried out by the ion exchange method in which sodium ions present in zeolite were partly replaced by rhodium ions [15]. The carbonylation of rhodium encapsulated zeolites resulted in the rhodium carbonyl complex that is observed to be active for hydroformylation of alkenes. Higher regioselectivity (*n*/*iso* ratio) of aldehydes was reported for the hydroformylation of 1-hexene using a rhodium encapsulated catalyst as compared to the analogues homogeneous rhodium carbonyl complex. The rhodium encapsulated in zeolite-Y showed higher selectivity to linear aldehyde as compared to zeolite-X as a support. Similarly for hydroformylation of styrene where *iso* aldehyde is an important product, the rhodium-zeolite-Y catalyst showed higher selectivity to branched aldehyde [15].

Davis et al. synthesized heterogeneous catalysts for the hydroformylation reaction using a classical ion-exchange method and the addition of rhodium metal salts during the synthesis of zeolite-A and Na-X [16–20]. The Rh-zeolite A, prepared by the ion-exchange method was used as a catalyst for 1-hexene hydroformylation in liquid and vapor phases. In liquid-phase hydroformylation, Rh-zeolite-A sample showed leaching of rhodium metal and the reaction was catalyzed by leached-out rhodium in the solution [18]. In another study, cationic rhodium complexes, $[Rh(diphosphinite)(cod)]^+$, were prepared by the reaction of $[RhCl(cod)]_2$ with diphosphinite ligands synthesized from (2*S*,4*R*), (2*S*,4*S*)-1-benzyl-4-hydroxy-4-phenyl-2-(1,1-diphenylmethyl)-pyrrolidinylmethanol, and (2*S*,4*R*), (2*S*,4*S*)-1-(3-triethoxysilyl)propylaminocarbonyl-4-hydroxy-4-phenyl-2-(1,1-diphenylmethyl)-pyrrolidinylmethanol [21]. To prepare an encapsulated rhodium complex, the triethoxysilyl group of ligands was covalently coordinated to zeolite-Y by controlled hydrolysis. The catalytic activity of rhodium complex encapsulated catalysts was studied for asymmetric hydroformylation of styrene. These catalysts showed enhanced activity and stability over a prolonged time with the benefits of reusability of the catalyst by filtration for styrene hydroformylation [21].

In another approach to prepare a core-shell-type encapsulated catalyst for hydroformylation reaction, a tailor-made silicalite-1 zeolite membrane of varied thickness was reported to encapsulate the Rh/C catalyst [22]. By changing the thickness of the silicalite-1 membrane, the *n/iso* ratio of aldehydes can be altered due to the differences in the molecular size of the linear and branched aldehydes, which help in faster desorption/diffusion of corresponding aldehyde from the active sites of the catalyst. In a typical synthesis procedure, Rh/C was added to a solution of silicalite-1 precursors used in a molar ratio of H_2O:TPAOH:EtOH:SiO_2 = 6000:24:400:100 at continuous stirring and 70°C for 90 min. The slurry was kept for crystallization of silicalite-1 zeolite at 180°C and 1 rpm for 2 days. After that, the slurry was filtered, washed with distilled water, and dried. The catalytic activity of Rh/C encapsulated silicalite-1 zeolite membrane was evaluated for 1-hexene hydroformylation. Conversion of 1-hexene was obtained up to 96% with *n/iso* ratio of aldehydes 2.3 using an encapsulated catalyst as compared to 0.6 using Rh/C catalyst without encapsulation.

2.2 Encapsulation of Rhodium Complexes in Clays by the Intercalation Technique

The structure of hydrotalcite is similar to brucite wherein sheets are stacked and placed together by hydrogen bonding between the sheets. When magnesium of the brucite sheet is replaced by trivalent cation of a similar radius, a positive change is generated in the sheet that is balanced by the anions located in the interlayer space. Hydrotalcite or layered double hydroxides (LDH) come under the category of anionic clays, represented by the following formula [23]:

$$\left[M^{2+}_{1-x}M^{3+}_{x}(OH)_2\right]^{x+}A^{n-}_{x/n}(1-3x/2)H_2O, (0.20<x<0.33)$$

$$M^{2+} = \text{divalent cation, } Mg^{2+}, Cu^{2+}, Ni^{2+}, Zn^{2+}, Fe^{2+}, \text{etc.}$$

$$M^{3+} = \text{trivalent cation, } Al^{3+}, Ga^{3+}, Ni^{3+}, Fe^{3+}, \text{etc.}$$

$x = M^{3+}/\left[M^{2+}+M^{3+}\right]$ molar ratio, the value of x should be in the range of $0.2 \le x \le 0.33$ to obtain a pure hydrotalcite phase. A = anions present in the interlayer space such as any inorganic anions or metal complexes in anionic form, organic acids, heteropoly acids, etc. These anions are the exchangeable anions that can be replaced from interlayer space of hydrotalcite by a variety of organic and inorganic anions due to anion exchange properties of hydrotalcite, and the obtained inorganic materials can be used either as encapsulated catalysts or for other applications [24–30].

Intercalation of *trans*-RhCl(CO)(TPPTS)$_2$ into Zn-Al LDHs was carried out by Zhang et al.; however, they have studied its application neither as a catalyst nor adsorbent [31]. Thermal stability of intercalated hydrotalcite was increased significantly by intercalation of *trans*-$RhCl(CO)(TPPTS)_2$ complex in the interlayer space of Zn-Al layered double hydroxides. In another study, the $RhCl(TPPTS)_3$ complex was intercalated in Zn-Al LDHs by anion exchange method to prepare a heterogeneous catalyst for hydrogenation of bicyclo[2.2.2] octenes [32]. Sharma et al. studied the intercalation of $HRh(CO)(TPPTS)_3$ complex in the interlayer space of Mg-Al hydrotalcite, and the intercalated complex was used as a multifunctional catalyst for hydroformylation, aldol condensation, and hydrogenation reactions in a single pot [33,34].

The intercalation/encapsulation of the $HRh(CO)(TPPTS)_3$ complex in Mg-Al hydrotalcite was carried out by dissolving a desired amount of $HRh(CO)(TPPTS)_3$ complex in deionized, decarbonated, double distilled water. The hydrotalcite having nitrate anions in the interlayer space or calcined hydrotalcite was added to the solution of $HRh(CO)(TPPTS)_3$ complex under an inert atmosphere. The suspension was stirred at room temperature for 72–96 h. After that, the yellow suspension was filtered and dried in the vacuum to obtain hydrotalcite intercalated $HRh(CO)(TPPTS)_3$ complex. The intercalation of $HRh(CO)(TPPTS)_3$ complex was confirmed by the shifting of peak positions corresponding to (003) and (006) planes towards the lower 2θ degree and broadening of the *d*-spacing value of (003) plane of intercalated hydrotalcite from 8.79 to 16.8 Å [33]. The intercalation of $HRh(CO)(TPPTS)_3$ complex without decomposition was further confirmed by FT-IR and the solid-state ^{31}P-NMR of the intercalated catalyst. The surface area of the intercalated catalyst decreased from 76 m^2/g (for pristine hydrotalcite) to 33 m^2/g. The intercalated catalyst showed excellent catalytic activity for hydroformylation of alkenes in the order of propylene > 1-pentene > 1-hexene > 1-heptene > 1-octene > 1-nonene > 1-decene > 1-undecene > 1-dodecene > 1-tridecene. For hydroformylation of cyclic alkenes, the activity order was found as: cyclohexene > cycloheptene > cyclooctene under identical reaction conditions. The intercalated catalyst was reused five times without significant loss of the activity for 1-hexene hydroformylation. As a multifunctional catalyst, intercalated hydrotalcite showed up to 51% selectivity to C_8 aldehydes from propylene via hydroformylation, aldol condensation, and hydrogenation reactions in a single pot (Scheme 14.2) [34]. Leaching of rhodium was not observed in the ICP analysis of the product mixture and spent catalyst.

2.3 Encapsulation of Rhodium Complexes in Silica

The hexagonal mesoporous silica (HMS) prepared by a neutral templating method has a high surface area and uniform mesoporous cross-linked channels that arrange in a hexagonal-like shape to form a sponge or wormhole-like structure [35–37]. Generally, dodecyl and hexadecyl amines are used as neutral templates for the synthesis of HMS. Alper and Ajjou reported a catalyst prepared by the covalent coordination of rhodium trichloride on HMS for

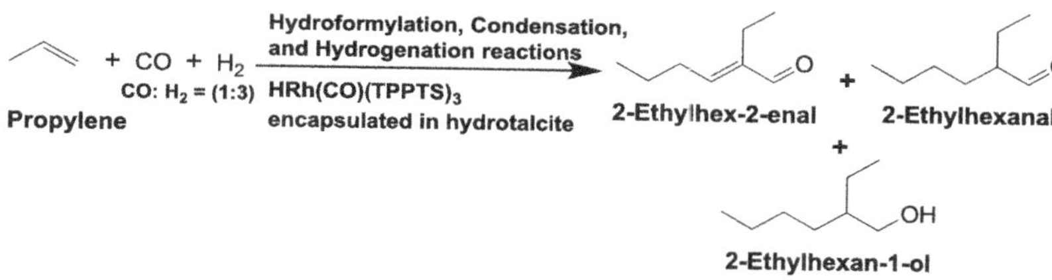

SCHEME 14.2 Single-pot synthesis of C_8 aldehydes and alcohol from propylene using a multifunctional intercalated catalyst. *(Reproduced from S.K. Sharma, R.V. Jasra, Synthesis of 2-ethylhexanal from propylene in a single pot using an ecofriendly multifunctional catalyst synthesized by intercalation of HRhCO(TPPTS)3 complex in the interlayer space of hydrotalcite, Ind. Eng. Chem. Res. 50 (2011) 2815–2821, with permission of the American Chemical Society.)*

hydroformylation of vinyl acetate and other related substrates at 120°C and 40.8 atmosphere of syngas ($CO/H_2 = 2/1$) pressure [38]. The catalyst exhibited excellent activity with higher selectivity for aldehydes at studied reaction conditions. This was the first report on the use of HMS to prepare a heterogeneous hydroformylation catalyst. Later on, Sudheesh et al. modified the synthesis procedure of HMS and encapsulated the $HRh(CO)(PPh_3)_3$ complex during HMS synthesis [39]. The concept behind the encapsulation of the $HRh(CO)(PPh_3)_3$ complex in the organic phase of the template located inside the pores of HMS is to use the HMS pores as nanophase reactors for a hydroformylation reaction. In a typical catalyst preparation method, the desired amount of hexadecylamine was dissolved in an ethanol-water solution followed by the addition of an $HRh(CO)(PPh_3)_3$ complex at constant stirring speed and room temperature. The calculated amount of tetraethyl orthosilicate (TEOS) was added dropwisely to the stirred suspension. After complete addition of TEOS, stirring was continued for an additional 1 h, and then the obtained suspension was kept at room temperature for 18 h. The precipitated solids were filtered, washed with an ethanol-water mixture, and dried in a vacuum to obtain the $HRh(CO)(PPh_3)_3$ complex encapsulated HMS. The encapsulated catalyst was characterized by the P-XRD, ^{31}P CP-MAS NMR, FT-IR, TGA, SEM, and surface area measurements. The P-XRD patterns of HMS encapsulated catalyst showed an intense reflection of the (100) plane at around 2.0 degrees, which is the characteristic peak of mesoporous silica, whereas the ^{31}P CP-MAS NMR spectrum showed the encapsulation of $HRh(CO)(PPh_3)_3$ complex in the pores of HMS without decomposition. The TEM image of the encapsulated catalyst showed a hexagonal array of the HMS mesoporous structure [40]. For hydroformylation of lower carbon chain-length alkene (propylene), almost complete conversion of propylene with 100% aldehyde selectivity (*n*/*iso* ratio = 0.65–0.9) was achieved using an encapsulated catalyst. The catalytic activity was also studied for the hydroformylation of higher carbon chain-length alkenes [39]. The $HRh(CO)(PPh_3)_3$ encapsulated HMS showed 100% conversion of alkenes with *n*/*iso* ratio of aldehydes in the range of 0.7–1 for hydroformylation of linear

C_5-C_{12} alkenes. The $HRh(CO)(PPh_3)_3$ encapsulated HMS was reused six times without any significant loss in the activity for hydroformylation reaction. Kinetic studies for 1-hexene hydroformylation showed a higher rate of reaction at a higher catalyst amount, partial pressure of carbon monooxide, and 1-hexene concentration up to certain range using $HRh(CO)(PPh_3)_3$ encapsulated HMS [41]. With respect to partial pressure of hydrogen, the rate of reaction was observed to be first order. Interestingly the rate of reaction was observed to decrease with the increase in the carbon number of olefin for the hydroformylation of C_5-C_{12} olefins using an encapsulated catalyst [41].

Catalytic activity evaluation of $HRh(CO)(PPh_3)_3$ encapsulated HMS was further extended for the hydroformylation of heterocyclic compounds such as, 2,3-dihydrofuran and 2,5-dihydrofuran (Scheme 14.3) [42].

The optimum conversion of 2,3-dihydrofuran, 98% with 89% aldehyde selectivity and 90% conversion of 2,5-dihydrofuran with 72% selectivity to aldehyde was achieved under similar reaction conditions. The effect of the catalyst weight, substrate concentration, reaction temperature, partial pressure of hydrogen and carbon monoxide on the conversion of substrate and aldehyde selectivity was studied. The catalyst was reused six times without significant loss of the activity for 2,3-dihydrofuran and 2,5-dihydrofuran hydroformylation.

This study was further extended for the encapsulation of rhodium complex containing water-soluble ligand (TPPTS, trisodium triphenylphosphine-3,3′,3″-trisulfonate) in the HMS channels for the hydroformylation reaction [43]. The TEM image of $RhCl(TPPTS)_3$ encapsulated HMS showed the typical wormhole structure of HMS with uniform rhodium dispersion. The catalytic activity of RhCl $(TPPTS)_3$ encapsulated HMS was studied for hydroformylation of vinyl esters

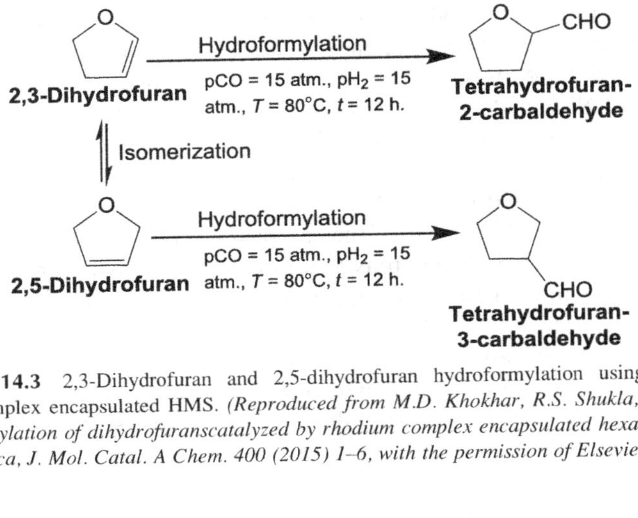

SCHEME 14.3 2,3-Dihydrofuran and 2,5-dihydrofuran hydroformylation using HRh(CO) $(PPh_3)_3$complex encapsulated HMS. *(Reproduced from M.D. Khokhar, R.S. Shukla, R.V. Jasra, Hydroformylation of dihydrofuranscatalyzed by rhodium complex encapsulated hexagonal mesoporous silica, J. Mol. Catal. A Chem. 400 (2015) 1–6, with the permission of Elsevier.)*

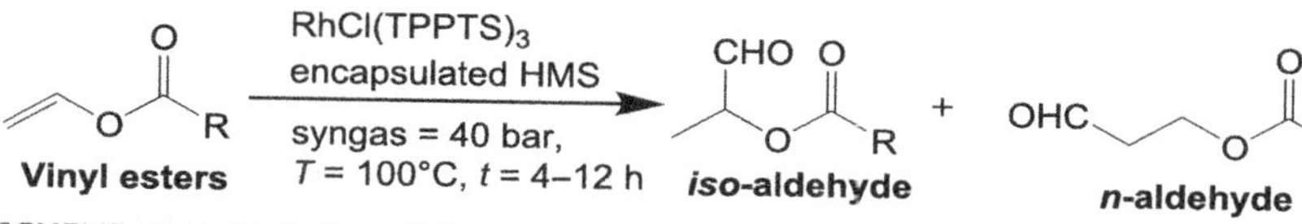

SCHEME 14.4 Hydroformylation of vinyl esters using $RhCl(TPPTS)_3$ encapsulated HMS as a catalyst. *(Reproduced from N. Sudheesh, A.K. Chaturvedi, R.S. Shukla, RhCl(TPPTS)3 encapsulated into the hexagonal mesoporous silica as an efficient heterogeneous catalyst for hydroformylation of vinyl esters, Appl. Catal. A Gen. 409–410 (2011) 99–105, with the permission of Elsevier.)*

(Scheme 14.4; vinyl acetate, vinyl propionate, vinyl butyrate, vinyl benzoate) to achieve 100% conversion of vinyl esters and higher *iso*-aldehyde selectivity in the range of 88%–94% at 40 bar syngas pressure and 100°C reaction temperature.

Apart from the hydroformylation activity, $HRh(CO)(PPh_3)_3$ encapsulated HMS was also reported as a highly active catalyst for the hydroaminomethylation reaction for synthesis of amines from alkenes (Scheme 14.5) [44]. The hydroaminomethylation of various alkenes such as 1-pentene, 1-hexene, and cyclopentene with primary and cyclic amines like hexylamine, morpholine, cyclohexylamine, pyrrolidine, etc. showed 100% alkene conversion with more than 98% selectivity to amines. The catalyst was recycled up to four times. Other encapsulated catalysts, Rh-ETS-10 and Rh-ETS-4, were prepared by rhodium cation exchange in sodium form of titanosilicates ETS-10 and ETS-4 repeatedly for three times. Around 4.8% and 3.2% (by wt.) rhodium was exchanged in ETS-10 and ETS-4, respectively, to prepare rhodium-encapsulated catalysts. The Rh-ETS-10 and Rh-ETS-4 catalysts also exhibited comparable activity to $HRh(CO)(PPh_3)_3$ complex encapsulated HMS for hydroaminomethylation of alkenes [45]. The rhodium carbonyl species formed in-situ may be the active catalytic species for hydroaminomethylation reaction using Rh-ETS-10 and Rh-ETS-4 catalysts.

Other than HMS, sodium bentonite pillared clay [46] and mesoporous FSM-16 silica [47] were also used to prepare the encapsulated hydroformylation catalyst. The rhodium metal was encapsulated in bentonite clay by the ion

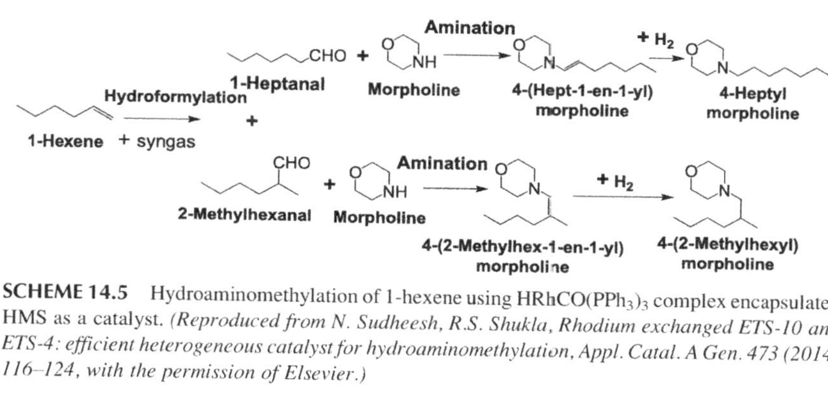

SCHEME 14.5 Hydroaminomethylation of 1-hexene using $HRhCO(PPh_3)_3$ complex encapsulated HMS as a catalyst. *(Reproduced from N. Sudheesh, R.S. Shukla, Rhodium exchanged ETS-10 and ETS-4: efficient heterogeneous catalyst for hydroaminomethylation, Appl. Catal. A Gen. 473 (2014) 116–124, with the permission of Elsevier.)*

exchange method to prepare an Rh-bentonite catalyst (rhodium content 0.46% by weight) for hydroformylation of ethylene and propylene in a fixed bed reactor at atmospheric pressure and 100–220°C temperature range [46]. The encapsulated catalyst showed 78.5% selectivity to *n*-butanol and a constant rate of reaction even at the end of 1 week of an experimental run for hydroformylation of propylene. Hydroformylation of propylene in a continuous flow reactor was studied using in situ $[RhMo_6O_{18}(OH)_6]^{3-}$, $[Rh(COD)_2]^+ - [AlMo_6O_{18}(OH)_6]^{3-}$, and $[Rh(COD)_2]^+$ supported in/on ordered mesoporous silica FSM-16 as a catalyst [47]. The activity of the supported catalyst was compared with $RhCl_3$ impregnated FSM–16 catalyst. A higher selectivity to butanols (more than 98%) was achieved using an $[RhMo_6O_{18}(OH)_6]^{3-}$/FSM-16 catalyst as compared to 73% selectivity using an impregnated $RhCl_3$/FSM-16 catalyst. The addition of molybdenum as a promoter exhibited further increase in the selectivity to butanols. The nanoparticles of rhodium and molybdenum formed in situ in the case of the $[RhMo_6O_{18}(OH)_6]^{3-}$/FSM-16 catalyst could be the probable reason for the enhanced activity for hydrogenation of butanols.

3 ENCAPSULATED CATALYSTS FOR OXIDATION REACTIONS

Varkey and Jacob studied one of the most challenging reactions, oxidation of methane to methanol using copper- and manganese-salen, substituted salen, and octa-azamacrocyclic complex encapsulated zeolite-Y catalysts [48]. The oxidation reaction was carried out in acetonitrile solvent at 0°C, with air as an oxidant and *tert*-butyl hydroperoxide (TBHP) as a promoter. Conversion of methane up to 5.2% with 61.7% selectivity to methanol was achieved in 8 h reaction time. Apart from the salen ligand, the copper- and manganese-saltin and salcyhexen complexes were also encapsulated in the pores of zeolite-Y to prepare a catalyst for oxidation of *p*-xylene [49]. The major products of *p*-xylene oxidation were observed as *p*-toluic acid, toluyl aldehyde, and *p*-toluyl alcohol with 42%–62% conversion of *p*-xylene at 33 bar air pressure, 130°C reaction temperature, and 18 h reaction time.

In another study, the activity of zeolite-Y encapsulated chromium(III), iron(III), bismuth(III), nickel(II), and zinc(II) metal complexes of *N,N'*-bis(salicylidene)propane-1,3-diamine was studied for liquid-phase oxidation of carvacrol (Scheme 14.6) using a hydrogen peroxide oxidant. The conversion of carvacrol was achieved up to 27.6% using the Fe(III) complex of the *N,N'*-bis (salicylidene)propane-1,3-diamine encapsulated zeolite-Y catalyst [50].

The catalytic activity of other zeolite-Y encapsulated metal complexes such as oxovanadium(VI), copper(II), and nickel(II) was evaluated for liquid-phase oxidation of styrene, cyclohexane, and methylphenyl sulfide using a hydrogen peroxide oxidant (Scheme 14.7) [51]. For example, oxovanadium(VI) complex

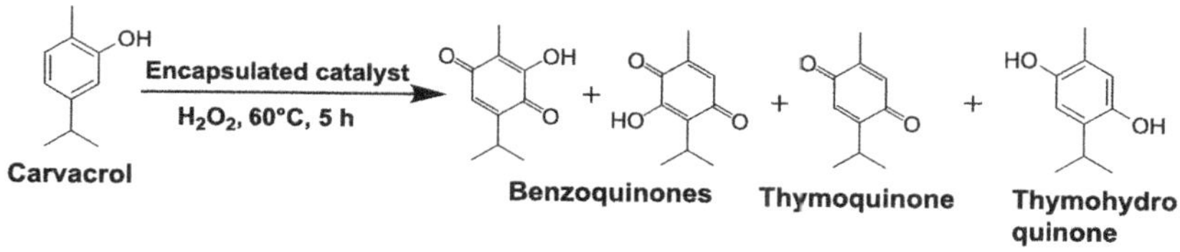

SCHEME 14.6 Oxidation of carvacrol using encapsulated catalysts. *(Reproduced from A. Gunes, O. Bayraktar, S. Yılmaz, Liquid-phase oxidation of carvacrol using zeolite-encapsulated metal complexes, Ind. Eng. Chem. Res. 45 (2006) 54–61, with the permission of American Chemical Society.)*

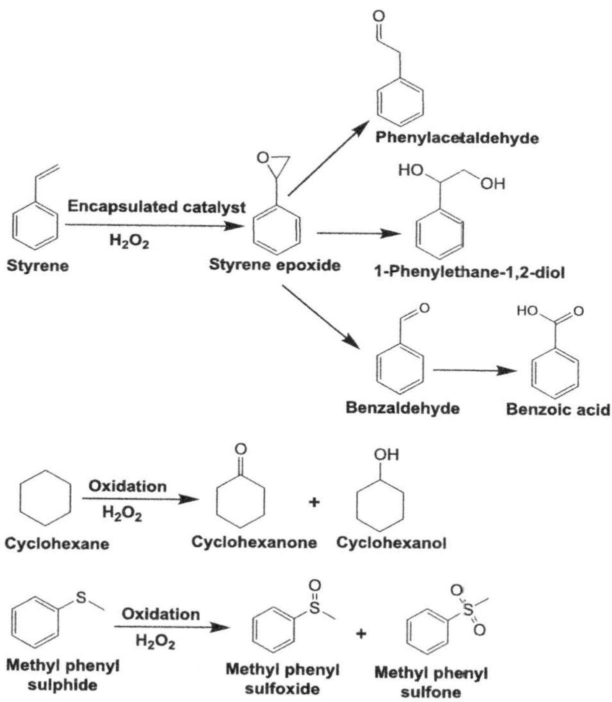

SCHEME 14.7 Styrene, cyclohexane, and methylphenyl sulfide oxidation using encapsulated catalysts. *(Reproduced from M.R. Maurya, A.K. Chandrakar, S. Chand, Zeolite-Y encapsulated metal complexes of oxovanadium(VI), copper(II) and nickel(II) as catalyst for the oxidation of styrene, cyclohexane and methyl phenyl sulphide, J. Mol. Catal. A Chem. 274 (2007) 192–201, with the permission of Elsevier.)*

encapsulated zeolite-Y showed 93.2% conversion of methylphenyl sulfide with 96.9% selectivity to methylphenyl sulfoxide.

Fan et al. synthesized cyclohexanol as a major product with selectivity up to 59% from cyclohexane oxidation using iron and copper salen complex encapsulated zeolite-Y as catalysts and H_2O_2 (30% in aqueous solution) as an oxidant at 60°C in 2 h reaction time [52]. These encapsulated metal complexes in zeolite-Y were also found to be very active catalysts for the oxidation of ethylbenzene. Higher selectivity to acetophenone was achieved from oxidation of ethylbenzene using tri- and tetra-azamacrocyclic Cu(II) complex encapsulated zeolite-Y catalyst compared to the use of homogeneous Cu(II) complexes at 60°C and 50% TBHP in an ethylene dichloride oxidant [53]. The oxidation of other compounds such as 1-octane, cyclohexane, and 2,5-dimethylhexane was also carried out by vanadium picolinate peroxo complex encapsulated zeolite-Y catalyst and an H_2O_2 oxidant [54]. The oxidation of another fine chemical, *cis*-pinane, was also studied using iron and cobalt phthalocyanine encapsulated zeolite-Y [55]. In another report, catechol was synthesized selectively by oxidation of phenol using an H_2O_2 oxidant and Cu(II)bis(picolinato) complex encapsulated zeolite-Y as a catalyst. A slightly higher turnover number (3.4) was observed for Cu(II) complex encapsulated zeolite-Y as compared to the homogeneous $CuCl_2{\cdot}2H_2O$ catalyst (2.6) [56]. The conversion of phenol increased up to 23.6% with 65.2% selectivity to catechol using tri- and tetra-dentate oxovanadium(IV) complex encapsulated zeolite-Y catalyst at 70°C and 6 h reaction time [57]. Apart from oxidation of phenol, zeolite-Y encapsulated vanadium(IV) complexes were observed as active catalysts for the oxidation of benzene, cyclohexene, and styrene using H_2O_2 as an oxidant. For example, 74.6% conversion of cyclohexene with 24.8% selectivity to cyclohexen-1-ol was obtained using vanadium(IV) complex encapsulated zeolite-Y, whereas selectivity to cyclohexen-1-ol increased to 89.7% using Fe-bipyridine complex encapsulated zeolite-Y catalysts [58]. Tanav et al. prepared titanium containing mesoporous silica molecular sieves by partial substitution with titanium as a selective catalyst for the oxidation of 2,6-ditert-butyl phenol and benzene using H_2O_2 as an oxidant [35]. Apart from zeolite-Y, mesoporous Al-MCM-41 (Si/Al = 20) was also used to encapsulate iron(III)-5,10,15,20-tetra-(4-pyridyl)porphyrin complex for a challenging reaction to produce phenol from benzene using hydrogen peroxide oxidant [59].

The catalytic activities of most of the encapsulated catalyst were studied using either H_2O_2 or TBHP as an oxidant. Joseph et al. used air as an oxidant for the oxidation of cyclooctene to cyclooctene epoxide with zeolite-Y encapsulated *cis*-MoO_2(SAL-SH)(DMF) as a catalyst. Higher conversion of cyclooctene (75%) and a nine-time increase in turnover number was reported using zeolite-Y encapsulated catalyst as compared to the homogeneous *cis*-MoO_2(-SAL-SH)(DMF) complex at 60°C [60]. The cobalt(II)-saloph complex encapsulated zeolite-Y that catalyzed β-isophorone oxidation to keto-isophorone also showed encouraging results for the use of air as an oxidant at atmospheric pressure and 25°C (Scheme 14.8) [61]. Among the studied catalysts, higher activity of cobalt(II)Cl-saloph encapsulated catalyst was observed for the oxidation reaction followed by cobalt(II)Br-saloph, cobalt(II)nitro-saloph, and cobalt (II)-saloph encapsulated zeolite-X.

Molecular oxygen was used as an oxidant for partial oxidation of benzyl alcohol and ethylbenzene using cobalt(II), nickel(II), and copper(II) complexes of

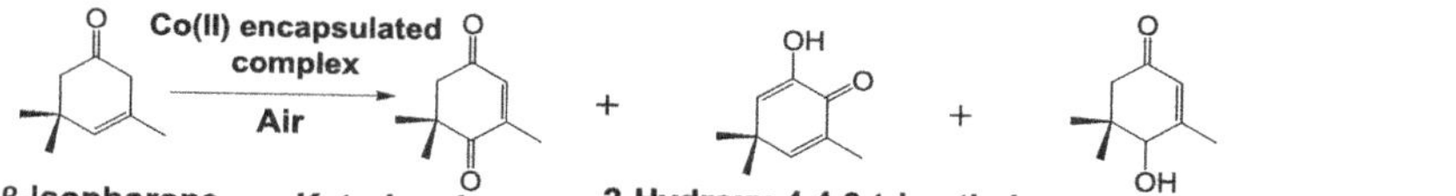

β-Isophorone Keto-isophorone 2-Hydroxy-4,4,6-trimethyl cyclohexa-2,5-dien-1-one 4-Hydroxy-3,5,5-trimethyl cyclohex-2-en-1-one

SCHEME 14.8 Oxidation of β-isophorone using air as an oxidant. *(Reproduced from T. Joseph, S.B. Halligudi, C. Satyanarayan, D.P. Sawant, S. Gopinathan, Oxidation by molecular oxygen using zeolite encapsulated Co(II)saloph complexes, J. Mol. Catal. A Chem. 168 (2001) 87–97, with the permission of Elsevier.)*

dimethylglyoxime and *N,N′*-ethylene-bis(7-methylsalicylideneamine) encapsulated zeolite-Y catalysts [62]. Air as an oxidant showed lower conversion of reactant and lower turnover frequency as compared to H_2O_2 oxidant at 50°C. Felthouse and Murphy reported encapsulation of platinum metal in zeolite-Y, ZSM-5 and mordenite for oxidation of formaldehyde to carbon dioxide at 95°C and 207 kPa partial pressure of oxygen [63]. The molecular oxygen as an oxidant showed 78% selectivity to cyclohexanone with 9.4% conversion of cyclohexane for solvent-free oxidation of cyclohexane using the cobalt encapsulated SBA-15 catalyst at 160°C, 1 MPa oxygen pressure, and 6 h reaction time [64].

Hydroxyapatite, $Ca_{10}(PO_4)_6(OH)_2$, is a useful material as a catalyst support due to its structural stability and ion-exchange properties. The ruthenium was encapsulated in hydroxyapatite using the cation exchange method to prepare an encapsulated catalyst for the oxidation reaction [65]. Further, to prepare a magnetic separable encapsulated catalyst, Mori et al. encapsulated ruthenium metal in the hydroxyapatite by equimolar substitution of calcium, and then magnetic γ-Fe_2O_3 nanocrystallites were dispersed in the ruthenium encapsulated hydroxyapatite [66]. The catalytic activity of the resulting encapsulated catalyst was studied for the synthesis of carbonyl compounds by oxidation of alcohols using molecular oxygen as an oxidant. Conversion of alcohols was achieved in the range of 89%–99% (depends on the reactivity of alcohol) with yield of product in the range of 81%–99% at 90°C. The separation of the catalyst after completion of the reaction is very easy using a magnet due to the magnetic properties of the catalyst. Similarly, a magnetically separable silver nanoparticle containing hydroxyapatite encapsulated γ-Fe_2O_3 was prepared for synthesis of *N*-monoalkylated hydroxylamines from selective oxidation of primary amines [67]. The isolated yield of *N*-monoalkylated hydroxylamines was observed in the range of 75%–85% for oxidation of various primary amines. The catalyst was reused five times for oxidation of phenylethylamine without any significant decrease in the isolated yield of monoalkylated hydroxylamine. Viswanadham et al. studied the ammoxidation of propene using rhenium oxide encapsulated in ZSM-5 and HY zeolites by the chemical vapor deposition method [68]. The catalyst, rhenium oxide encapsulated zeolite ZSM-5, showed higher selectivity (80%) to ammoxidation products.

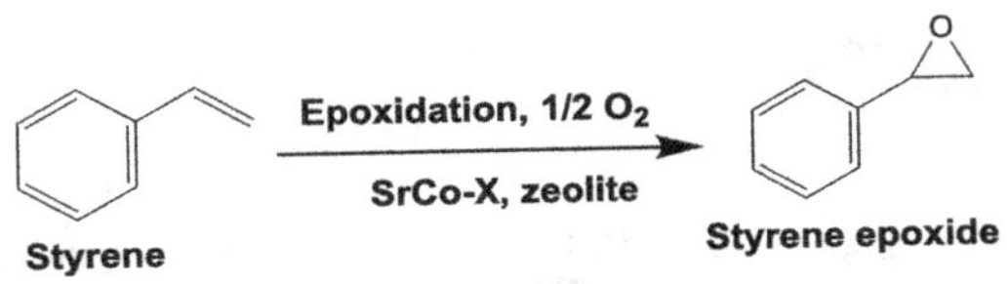

SCHEME 14.9 Epoxidation of styrene. *(Reproduced from J. Sebastian, K.M. Jinka, R.V. Jasra, Effect of alkali and alkaline earth metal ions on the catalytic epoxidation of styrene with molecular oxygen using cobalt(II)-exchanged zeolite X, J. Catal. 244 (2006) 208–218, with the permission of Elsevier.)*

Styrene epoxide, which is conventionally synthesized by epoxidation of styrene using a homogeneous catalyst, is a commercially important fine chemical. To develop a heterogeneous catalyst for styrene epoxidation, cobalt exchanged zeolite-X was prepared for the synthesis of styrene epoxide in the liquid phase using molecular oxygen as an oxidant at atmospheric pressure [69]. Using a cobalt exchanged Na-X catalyst, more than 99% conversion of styrene with 68% selectivity to styrene oxide was obtained (Scheme 14.9).

A significant increase in the styrene selectivity from 68% to 85% with 100% styrene conversion was observed with partial replacement of sodium cations in cobalt exchanged Na-X by alkaline earth metals. The alkaline earth metal containing Co-X catalyst was recycled three times. This study was claimed as the first report on the use of a heterogeneous catalyst for epoxidation of styrene using molecular oxygen as an oxidant. The cobalt exchanged zeolite-X catalyst with alkali metals also showed good activity and reusability for epoxidation of cycloalkenes such as cyclohexene, cycloheptene, *cis*-cyclooctene, and 1,5-cyclooctadiene at 100°C temperature, 200 mg catalyst, 60 psi oxygen pressure, 8 h reaction time, and 40 mL DMF solvent. [70]. The reactivity of cycloalkenes was found in the order of cyclohexene < cycloheptene < *cis*-cyclooctene for the Co-X catalyzed epoxidation reaction. In place of zeolite-X, the large pore zeolite-Y was also used as an inorganic support to prepare a catalyst for the epoxidation reaction. The activity of cobalt-exchanged zeolite-Y with alkali and alkaline metals was studied for epoxidation of α-pinene at 100°C, 20–200 psig oxygen pressure and DMF solvent [71]. The catalyst, NaCsCo-Y, showed 47% conversion of α-pinene with 61% selectivity to epoxide at 80 psig oxygen pressure. Significantly lower styrene conversion and higher styrene epoxide selectivity was reported using cobalt exchanged microporous framework zeolite, SSZ-51, compared to the alkaline earth metal containing Co-X catalyst [72]. In the case of cationic manganese-porphyrin encapsulated titanium-substituted mesoporous molecular sieves, conversion of styrene was obtained in the range of 31%–40% with 85%–90% selectivity to epoxide for epoxidation of styrene with iodosylbenzene [73].

Metal organic frameworks (MOFs) are self-assembled crystalline materials consisting of infinite network structures by linking of organic ligands with metal ions or clusters. MOFs are useful for catalysis due to their high surface

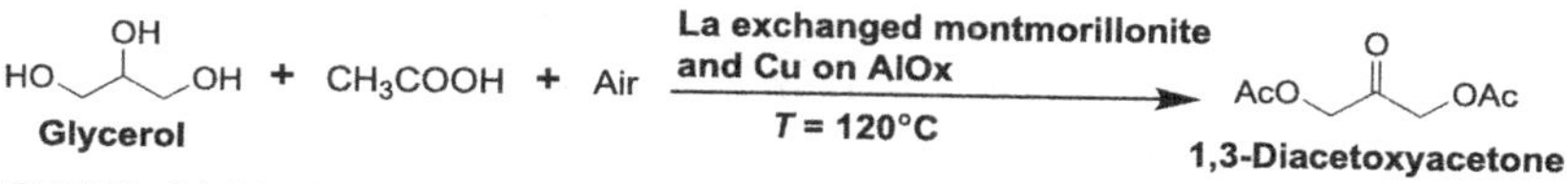

SCHEME 14.10 Synthesis of 1,3-diacetoxyacetone from glycerol. *(Reproduced from T. Mizugaki, R. Arundhathi, T. Mitsudome, K. Jitsukawa, K. Kaneda, Highly efficient and selective transformations of glycerol using reusable heterogeneous catalysts, ACS Sustain. Chem. Eng. 2 (2014) 574–578 with permission of the American Chemical Society.)*

area, high crystallinity, tunable pore size, and functionality. The catalytic activity of MOF depends on the metallic components and organic linker molecules of the framework. Yang et al. encapsulated phosphotungstic acid in Zr-based MOF, UiO-66, for selective oxidation of cyclopentene to glutaraldehyde using a hydrogen peroxide oxidant [74]. Conversion of cyclopentene was achieved up to 94.8% with up to 78.3% yield of glutaraldehyde using 35 wt% encapsulated catalyst. The MOF based on 1,3,5-benzenetricarboxylate and $[M(H_2O)_6]^{2+}$ (M = Mn or Cu) was used to encapsulate the $[Mn(1,10\text{-phenanthroline})_2(H_2O)_2]^{2+}$ complex through hydrogen bonding and π-π interaction with the supramolecular framework of MOF [75]. These encapsulated complexes in MOF showed excellent catalytic activity for phenol and 1-naphthol oxidation using an H_2O_2 as an oxidant. The hydroquinone-to-catechol ratio was obtained up to 3.7 at around 21% conversion of phenol in 9 h reaction time and 35°C temperature.

Glycerol is a low-value commercial byproduct that is produced in a significant amount during biodiesel production. For value addition of glycerol, Mizugaki et al. reported a lanthanum exchanged montmorillonite catalyst for acetylation of glycerol with acetic acid to diacetylglycerols [76]. The 1,3-diacetylglycerol was subsequently oxidized to 1,3-diacetoxyacetone up to the 90% isolated yield using air and copper nanoparticles loaded on AlO_x support (Scheme 14.10).

4 HYDROGENATION REACTIONS USING ENCAPSULATED CATALYSTS

Considering the fact that ruthenium is an active metal for hydrogenation reactions, the ruthenium impregnated catalysts on alumina, silica, etc. have been reported for hydrogenation reactions. The encapsulation of ruthenium metal was reported in the brucite-like sheet of hydrotalcite to prepare an encapsulated catalyst for hydrogenation reaction [77]. All the planes corresponding to the hydrotalcite structure were observed in the P-XRD of ruthenium encapsulated hydrotalcite (Ru-HT) with identical peak positions to pristine hydrotalcite. Ru-HT catalyst was found to give 100% selectivity with 100% conversion of benzene to cyclohexane at 120°C and 60 atmosphere hydrogen pressure in 2 h reaction time. The effect of moisture was not significant on the catalytic activity of Ru-HT catalyst, however, activity decreased considerably with increase of sulfur content in the feed. The encapsulated Ru-HT catalyst was reused four times for hydrogenation of benzene. Another catalyst, ruthenium-salen complex encapsulated zeolite-Y, showed 98.3% conversion of benzene at 60°C, 3.0 MPa hydrogen pressure, and 3 h reaction time for benzene hydrogenation [78]. The ruthenium-salen complex was also encapsulated in SBA-16 support to prepare an active catalyst for hydrogenation of 1-(naphthalene-2-yl)ethanone (98% conversion), 2-methylcyclohexanone (98% conversion), 1-(4-chlorophenyl)ethanone (84% conversion), 1-(4-bromophenyl)ethanone (80% conversion), 1-(4-methoxyphenyl)ethanone (79% conversion), propiophenone (67% conversion), cyclododecanone (60% conversion), (2-hydroxyphenyl) ethanone (20% conversion), and (2-hydroxyphenyl)(phenyl)methanone (20% conversion) to their corresponding alcohols [79].

To prepare the carbon-based catalyst for the hydrogenation reaction, encapsulation of gold nanoparticles in spherical mesoporous carbon was carried out using the core-shell approach for hydrogenation of *p*-nitrophenol to *p*-aminophenol. The catalyst showed higher conversion of *p*-nitrophenol (97%) in 8 min reaction time and reused up to 10 cycles [80]. Recently, platinum nanoparticle-encapsulated in MFI zeolite was prepared for selective synthesis of 4-aminostyrene by hydrogenation of 4-nitrostyrene that was formed in-situ from a mixture of nitrobenzene and dimethylnitrobenzene [81]. Complete conversion of 4-nitrostyrene with 80% selectivity to 4-aminostyrene was reported in 13 h reaction time (Scheme 14.11).

Selective hydrogenation of α,β-unsaturated carbonyl compounds such as cinnamaldehyde is a challenging reaction due to the presence of double bonds at various locations to produce fine chemicals for pharmaceuticals, flavors, and fragrance industries (Scheme 14.12).

The platinum nanoparticle encapsulated mesoporous silica samples were prepared by varying the platinum particle sizes of 8, 4.9, 3.6, and 1.8 nm to evaluate the platinum particle size effect on the activity of the encapsulated catalyst for hydrogenation of cinnamaldehyde [82]. The yield of cinnamyl alcohol

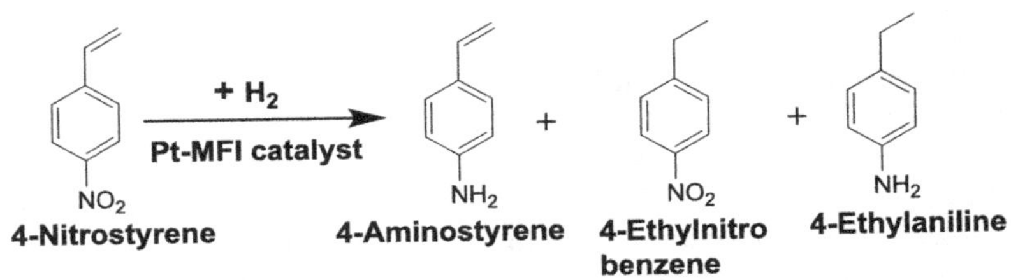

SCHEME 14.11 Synthesis of 4-aminostyrene by hydrogenation of 4-nitrostyrene using Pt encapsulated in MFI zeolite. *(Reproduced from J. Gu, Z. Zhang, P. Hu, L. Ding, N. Xue, L. Peng, et al., Platinum nanoparticles encapsulated in MFI zeolite crystals by a two-step dry gel conversion method as a highly selective hydrogenation catalyst, ACS Catal. 5 (2015) 6893–6901, with the permission of the American Chemical Society.)*

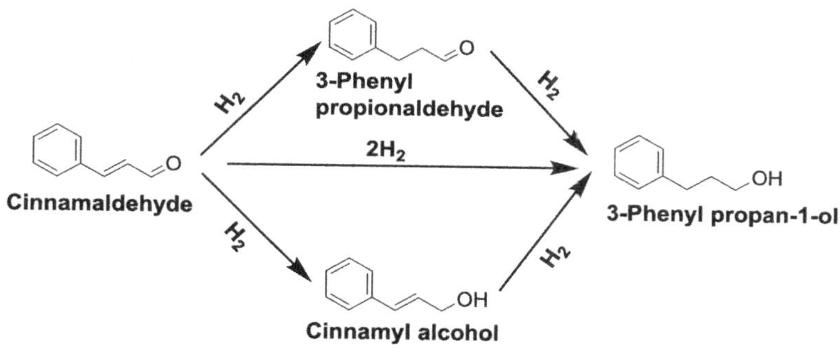

SCHEME 14.12 Hydrogenation of cinnamaldehyde. *(Reproduced from S. Galvagno, G. Capannelli, G. Neri, A. Donato, R. Pietropaolo, Hydrogenation of cinnamaldehyde over Ru/C catalysts: effect of Ru particle size, J. Mol. Catal. 64 (1991) 237–246, with the permission of Elsevier.)*

increased with increase of the particle size of platinum metal in mosoporous silica. Li et al. entrapped platinum nanoparticles of 2–3 nm diameter in the ZSM-5 nanoshells (about 15 nm thickness) to prepare a sinter resistant catalyst for hydrogenation of toluene and mesitylene [83]. A significantly higher rate of hydrogenation reaction (about 100 times) was observed for the Pt-ZSM-5 nanoshell catalyst as compared to conventional Pt-ZSM-5 catalysts. Apart from ruthenium and platinum metals, the nickel metal complex encapsulated catalyst was also reported as an active catalyst for hydrogenation reaction [84]. The hydrogenation of cyclohexene, cyclooctene, 1-hexene, and benzene was studied at 40°C and 60 atmosphere hydrogen pressure using a nickel salen complex encapsulated zeolite-Y catalyst.

5 ENCAPSULATED CATALYSTS FOR ISOMERIZATION AND OTHER REACTIONS

The ruthenium metal, known as an active homogeneous catalyst for double-bond isomerization of perfumery chemicals [85,86], is encapsulated in the hydrotalcite matrix by isomorphic substitution of aluminum metal with ruthenium in the brucite-like sheet [87]. Apart from hydrogenation reaction, Ru-HT is also observed to be a highly active catalyst for isomerization of methyl chavicol, eugenol, allylbenzene, safrole, and 3-carene (Scheme 14.13). Conversion up to 98% with 89% selectivity to the *trans* isomer was obtained in 2 h reaction time. More than 95% substrate conversion was observed within 10 min reaction time, indicating an extremely high activity of the encapsulated catalyst. The Ru-HT catalyst was reused several times for double-bond isomerization of methyl chavicol. Similar to the ruthenium, nickel can also be intercalated in the hydrotalcite. The nickel aluminum hydrotalcite (layered double hydroxide) showed 77% conversion of eugenol and 84% selectivity to *trans*-isoeugenol at 200°C, a substrate to catalyst ratio of 2, 6 h reaction time, and 20 mL DMF

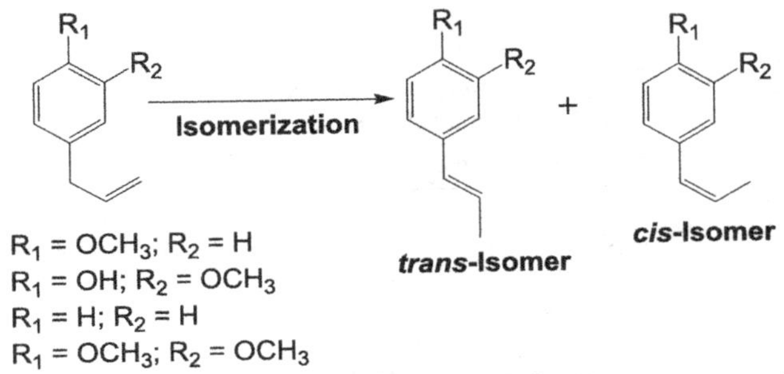

SCHEME 14.13 Isomerization of perfumery chemicals. *(Reproduced from S.K. Sharma, P.A. Parikh, R.V. Jasra, Ruthenium containing hydrotalcite as a solid base catalyst for >C=C< double bond isomerization in perfumery chemicals, J. Mol. Catal. A Chem. 317 (2010) 27–33, with the permission of Elsevier.)*

solvent [88]. For isomerization of estragole, 98% conversion with 85% selectivity to *trans*-anethole was reported using nickel-containing Mg-Al hydrotalcite at a substrate to catalyst ratio of 10, 200°C, 6 h reaction time, and 20 mL DMF solvent [89]. The nickel-containing hydrotalcite exhibited lower activity and selectivity for *trans*-isomer than the Ru-HT for isomerization of perfumery chemicals.

The use of microwave irradiation is also reported for in-situ encapsulation of heteropoly acid (12-tungstophosphoric acid) in the supercages of ultrastable Y (USY) zeolite for the synthesis of 4,4′-dimethyldiphenylmethane by the condensation of toluene and formaldehyde [90]. The activity of encapsulated catalysts was also studied for the synthesis of amines that are intermediate compounds for pharmaceutical industries. The benzylation of aniline was carried out by Dhakshinamoorthy et al., using ZnO nanoparticles encapsulated in K-10 clay [91]. Up to 82.5% selectivity of monobenzylated product with 88% conversion of aniline was achieved in 12 h reaction time at room temperature. The catalyst also showed good activity for benzylation of various substituted anilines and straight-chain amines. Synthesis of propargylamines was carried out by coupling reaction of aldehydes, amines, and alkynes using a copper (I)-Schiff base complex encapsulated zeolite-Y catalyst [92]. The reactivity of various aldehydes, amines, and alkynes (linear carbon chain, aromatic, and functionalized molecules) was studied in detail using a reusable encapsulated catalyst. The electron deficient aryl aldehydes showed a higher yield of product as compared to electron rich aldehydes. The aliphatic amines, piperidine and morpholine, gave higher yields as compared to aromatic amines such as phenyl methylamine under similar reaction conditions. Instead of aldehydes, Shah et al. used ketones as reactants (Strecker reaction) along with amines and trimethylsilyl cyanide (TMSCN) to synthesize α-aminonitriles under solvent-free reaction conditions using Tin-exchanged H-β-zeolite as a catalyst [93]. The yield of α-aminonitriles was achieved up to 93% in 30 min reaction time at room temperature. The Strecker reaction of various aldehydes with amines and trimethylsilyl cyanide (TMSCN) proceeded faster than the ketones. Up to 96% isolated yield of product was achieved in 10 min reaction time. Tin-exchanged H-β-zeolite catalyst was reused five times without significant loss of its activity for reaction of acetophenone, aniline, and TMSCN.

6 ENCAPSULATED CHIRAL COMPLEXES

Asymmetric synthesis is a useful method to produce stereoisomeric compounds for pharmaceutical applications because different enantiomers of molecules are known to have different biological activities. Generally, asymmetric synthesis is carried out by homogenous catalysts for the commercial production of chiral compounds. Immobilization of chiral metal complexes for ease of catalyst separation without compromising with the stereoselectivity for asymmetric synthesis is also attempted by researchers [94]. In some cases, chiral metal complex-encapsulated catalysts showed high enantiomeric excess (*ee*) for the synthesis of stereoisomeric compounds.

Yang et al. demonstrated the enhancement in the hydrolytic kinetic resolution of epoxides using a chiral Co(salen) complex (salen = (*R*,*R*)-*N*,*N*′-bis(3,5-di-tertbutylsalicylidene)-1,2-cyclohexanediamine) encapsulated SBA-16 catalyst [95,96]. Co(salen) complexes can move freely in the nanocages of SBA-16 because there is no covalent bonding between the metal complex and support. Propyltrimethoxysilane was used as a silylating agent to reduce the pore entrance and internal surface modification of SBA-16 support. Using Co(salen) encapsulated SBA-16, conversion of propylene epoxide was achieved up to 50% with 98% *ee* of diol (250 h^{-1} TOF) (Scheme 14.14). For ring opening of epichlorohydrin, a 46% diol yield with lower *ee* (87%) was observed which, may be due to the lower reactivity of epichlorohydrin than that of propylene epoxide. The catalyst was reused ten times for asymmetric ring-opening of propylene oxide.

Achiral Mn(salen) complex encapsulated in cross-linked peptidic phosphonate homochiral support showed low enantioselectivity in the range of 5%–35% *ee* for epoxidation of styrene derivatives, whereas in-situ hydrolysis of epoxides showed *ee* more than 98% to the corresponding diols [97]. Higher enantioselectivity of epoxides was achieved with the chiral Mn(salen)

SCHEME 14.14 Hydrolytic kinetic resolution of propylene epoxide. *(Reproduced from H. Yang, L. Zhang, L. Zhong, Q. Yang, C. Li, Enhanced cooperative activation effect in the hydrolytic kinetic resolution of epoxides on [Co(salen)] catalysts confined in nanocages, Angew. Chem. Int. Ed. 46 (2007) 6861–6865, with the permission of Wiley-VCH Verlag.)*

complex covalently coordinated to MCM-41 and SBA-15 supports modified by 3-aminopropyltriethoxysilane compared to the respective homogeneous Mn(salen) complex for enantioselective epoxidation of alkenes [98,99]. The *ee* was observed in the range of 95%–99% for epoxidation of 2,2-dimethylchromene. The Mn(salen) complex supported on SBA-15 showed higher TOF values as compared to MCM-41 due to lower diffusional resistance in the larger pores of SBA-15 than MCM-41. For asymmetric epoxidation of styrene to styrene epoxide, chiral Mn(salen) complex exchanged montmorillonite clay showed more than 99% yield of styrene epoxide (70% *ee*) in 4 h reaction time [100]. For substrate like 6-nitro-2,2-dimethylchromene, more than 99% *ee* was reported in 4.5 h reaction time using Mn (salen) complex-exchanged montmorillonite clay, whereas lower *ee* (82%–92%) was observed in 12 h for epoxidation of 2,2-dimethylchromene catalyzed by chiral Mn(salen) complex immobilized on pyridine N-oxide-modified MCM-41 [101]. Epoxidation of styrene, indene, 2,2-dimethylchromene showed higher catalytic activity using Mn(salen) complex exchanged montmorillonite clay and SBA-15 as compared to pyridine N-oxide-modified MCM-41 as a support. Synthesis of (*S*)-Levcromakalim, which is an antihypertensive drug, is carried out up to 98% *ee* using chiral dimeric Mn(III)salen complex with the 1*R*, 2*R*-(−)-diaminocyclohexane collar supported silica catalyst [102]. This silica supported catalyst was also reported as an active catalyst for epoxidation of styrene, indene, 2,2-dimethylchromene, and 6-cyano-2,2-dimethylchromene up to 98% *ee*.

Selective oxidation of alkylphenyl sulphides and nerol was studied using Mn-complex encapsulated zeolite Y [103]. The encapsulated catalyst exhibited lower activity than the covalently anchored Mn-complex on ultrastable Y zeolite and MCM-41 supports. The asymmetric Henry reaction of aldehyde and nitromethane was studied using silica and MCM-41 supported chiral lanthanum-lithium-binaphthol complex catalysts that provided up to 84% *ee* to the (*S*)-enantiomer [104].

Asymmetric hydrogenation is an important reaction to produce enantiomers for pharmaceutical applications. Zsigmond et al. encapsulated rhodium complexes, $[Rh(COD)L]^+$, where L = L-prolinamide and *N*-*tert*-butyl-L-prolinamide in zeolite-Y by the ship-in-a-bottle approach, and compared their catalytic activities for hydrogenation of 1-hexene, cyclohexene, and 1-methylcyclohexene [105]. The anchored catalyst showed higher conversion and initial specific activities for hydrogenation of studied substrates. Similarly, for asymmetric hydrogenation of *trans*-2-methylpent-2-enoic acid, higher *ee* (31%) was observed for cationic [Rh(COD)*N*-*tert*-butyl-L-prolinamide] complex anchored on zeolite-Y due to the higher accessibility of the active sites as compared to the encapsulated catalyst (17% *ee*). Only 13% *ee* was obtained for asymmetric hydrogenation of *trans*-2-methylpent-2-enoic acid using a homogeneous cationic [Rh(COD)*N*-*tert*-butyl-L-prolinamide] complex.

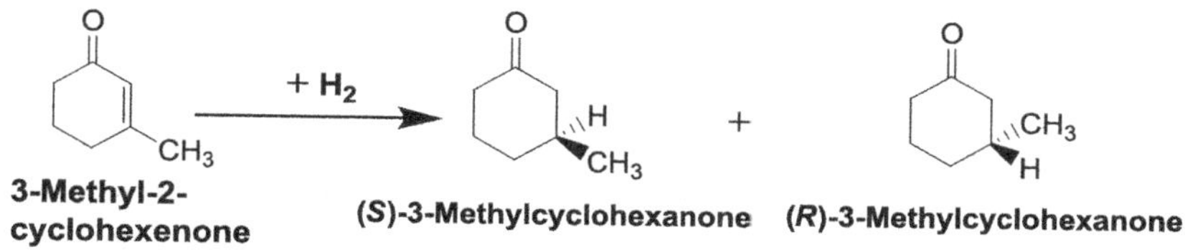

SCHEME 14.15 Hydrogenation of 3-methyl-2-cyclohexenone. *(Reproduced from S. Ernst, E. Fuchs, X. Yang, Enantioselective hydrogenation on zeolite-encapsulated chiral palladium–salen complexes, Microporous Mesoporous Mater. 35–36 (2000, 137–142, with the permission of Elsevier.)*

The palladium chiral salen complex (chiral salen = (*R,R*)-*N′*,*N′*-bis-(3,5-di-tertbutylsalicylidene)-1,2-cyclohexanediamine was encapsulated in zeolites FAU and EMT for the synthesis of cyclooctene by asymmetric hydrogenation of 1,5-cyclooctadiene [106]. For the first time, Ernst et al. encapsulated palladium chiral salen complex in FAU and EMT zeolites. Around 90% yield of cyclooctene was achieved using chiral encapsulated catalysts. The palladium chiral salen complex-encapsulated zeolites FAU and EMT were also used as catalysts for hydrogenation of the prochiral compound, 3-methyl-2-cyclohexenone (Scheme 14.15), to synthesize more than 55% yield of (*R*)-3-methyl-cyclohexanone with the significant lower *ee* (around 7%) [107].

Milo and Neumann adopted an approach using encapsulation of the achiral ruthenium complex in titanium phosphate homochiral peptide-based supports to prepare a supported catalyst for asymmetric hydrogenation of ketones to secondary alcohols [108]. The porous, amorphous support was prepared by using homochiral polypeptides attached to a tripodal core capped with phosphonate moieties. The achiral ruthenium complexes $RuCl_2(R_2PCH_2CH_2NH)_2$(R = Ph, *i*-Pr, or *t*-Bu) were encapsulated in the support during the cross-linking procedure that involves nonhydrolytic condensation of $Ti(O\text{-}i\text{-}Pr)_4$ with phosphonic acid. The *ee* of (*R*)-1-phenylethanol was achieved up to 70% for hydrogenation of acetophenone; however, for hydrogenation of acetophenone derivatives, the *ee* of the corresponding secondary alcohols increased significantly up to 95% with a 42 turnover number. A higher *ee* of (*R*)-1-phenylethanol, up to 75.5%, was reported using ruthenium chiral complex, (1S, 2S)-DPEN-$RuCl_2(TPP)_2$ (DPEN = 1,2-diphenylethylenediamine; TPP = triphenylphosphine) encapsulated in mesopores of SBA-16 silylated with phenyltrimethoxysilane for asymmetric hydrogenation of acetophenone [109]. Another chiral ruthenium complex, $RuCl_2$(*R*-Binap)(dmf)$_n$, was encapsulated in the nanopores of a silica-based support C-FDU-12 using chlorosilane Ph_2Cl_2Si as a silylating agent [110]. In the encapsulated catalyst, ruthenium complex was not covalently coordinated with support; therefore, the ruthenium complex can be moved freely in the nanopores of support during the catalytic reaction (asymmetric hydrogenation of β-keto esters) to achieve higher activity and enantioselectivity.

In conclusion, encapsulated catalysts showed significant potential to substitute the homogeneous catalysts in the lab scale. The further scale-up to pilot level is required to generate activity and reusability data for commercial production of fine chemicals using encapsulated catalysts in place of the homogeneous catalysts presently being used. Deeper understanding of the reaction mechanism and reaction kinetic helps to design a novel, cost effective, and reusable catalyst with enhanced activity for chemical transformation.

REFERENCES

[1] H. Tian, X. Li, L. Zeng, J. Gong, Recent advances on the design of group VIII base-metal catalysts with encapsulated structures, ACS Catal. 5 (2015) 4959–4977.

[2] F. Bedioui, Zeolite-encapsulated and clay-intercalated metal porphyrin, phthalocyanine and Schiff-base complexes as models for biomimetic oxidation catalysts: an overview, Coord. Chem. Rev. 144 (1995) 39–68.

[3] A.H. Ahmed, Zeolite-encapsulated transition metal chelates: synthesis and characterization, Rev. Inorg. Chem. 34 (3) (2014) 153–175.

[4] B. Cornils, W.A. Herrmann, Applied Homogeneous Catalysis With Organometallic Complexes, VCH, Weinheim, 1996.

[5] S.K. Sharma, V.K. Srivastava, R.V. Jasra, Catalytic processes for the hydroformylation of alkenes for producing oxo-products, in: M.A. Siddiqui, R.P. Verma (Eds.), Catalysis in Hydrocarbon Processing and Fertilizer Industry, Lovraj Kumar Memorial Trust, New Delhi, 2005.

[6] D.W. Breck, Zeolite Molecular Sieves: Chemistry and Uses, Wiley, New York, NY, 1974.

[7] A. Corma, From microporous to mesoporous molecular sieve materials and their use in catalysis, Chem. Rev. 97 (1997) 2373–2419.

[8] T.E. Bitterwolf, J.D. Newell, C.T. Carver, R.S. Addleman, J.C. Linehan, G. Fryxell, Rhodium catalysts bound to functionalized mesoporous silica, Inorg. Chim. Acta 357 (2004) 3001–3006.

[9] A. Taguchi, F. Schuth, Ordered mesoporous materials in catalysis, Microporous Mesoporous Mater. 77 (2005) 1–45.

[10] U. Ciesla, F. Schuth, Ordered mesoporous materials, Microporous Mesoporous Mater. 27 (1999) 131–149.

[11] D.T. On, D.D. Giscard, C. Danumah, S. Kaliaguine, Perspectives in catalytic applications of mesostructured materials, Appl. Catal. A Gen. 253 (2003) 545–602.

[12] G. Oye, J. Sjoblom, M. Stocker, Synthesis, characterization and potential applications of new materials in the mesoporous range, Adv. Colloid Interface Sci. 89–90 (2001) 439–466.

[13] K. Mukhopadhyay, A.B. Mandale, R.V. Chaudhari, Encapsulated $HRh(CO)(PPh_3)_3$ in microporous and mesoporous supports: novel heterogeneous catalysts for hydroformylation, Chem. Mater. 15 (2003) 1766–1777.

[14] K. Mukhopadhyay, V.S. Nair, R.V. Chaudhari, Hydroformylation of olefins using encapsulated $HRh(CO)((PPh_3)_3$ in NaY zeolite as a catalyst, Stud. Surf. Sci. Catal. 130 (2000) 2999–3003.

[15] J.M. Andersen, Zeolite-encapsulated rhodium catalysts—the best of both worlds? Platin. Met. Rev. 41 (3) (1997) 132–141.

[16] B.E. Hanson, M.E. Davis, D. Taylor, E. Rode, Intrazeolite rhodium carbonyl and rhodium carbonyl phosphine complexes, Inorg. Chem. 23 (1984) 52–56.

[17] M.E. Davis, P.M. Butler, J.A. Rossin, B.E. Hanson, Hydroformylation of 1-hexene by soluble and zeolite-supported rhodium species, J. Mol. Catal. 31 (1985) 385–395.

[18] R.J. Davis, J.A. Rossin, M.E. Davis, Hydroformylation by rhodium zeolite A catalysts, J. Catal. 98 (1986) 477–486.

[19] M.E. Davis, J. Schnitzer, J.A. Rossin, D. Taylor, B.E. Hanson, Hydroformylation of 1-hexene by soluble and zeolite-supported rhodium species part II, J. Mol. Catal. 39 (1987) 243–259.

[20] M. Nowotny, T. Maschmeyer, B.F.G. Johnson, P. Lahuerta, J.M. Thomas, J.E. Davies, Heterogeneous dinuclear rhodium(II) hydroformylation catalysts-performance, evaluation and silsesquioxane-based chemical modelling, Angew. Chem. Int. Ed. 40 (2001) 955–958.

[21] A. Fuerte, M. Iglesias, F. Sanchez, New chiral diphosphinites: synthesis of Rh complexes. Heterogenisation on zeolites, J. Organomet. Chem. 588 (1999) 186–194.

[22] X. Li, Y. Zhang, M. Meng, G. Yang, X. San, M. Takahashi, N. Tsubaki, Silicalite-1 membrane encapsulated Rh/activated-carbon catalyst for hydroformylation of 1-hexene with high selectivity to normal aldehyde, J. Membr. Sci. 347 (2010) 220–227.

[23] F. Cavani, F. Trifiro, A. Vaccari, Hydrotalcite-type anionic clays: preparation, properties and applications, Catal. Today 11 (1991) 173–301.

[24] A.I. Khan, D. O'Hare, Intercalation chemistry of layered double hydroxides: recent developments and applications, J. Mater. Chem. 12 (2002) 3191–3198.

[25] D.G. Evans, X. Duan, Preparation of layered double hydroxides and their applications as additives in polymers, as precursors to magnetic materials and in biology and medicine, Chem. Commun. (2006) 485–496.

[26] Z.P. Xu, J. Zhang, M.O. Adebajo, H. Zhang, C. Zhou, Catalytic applications of layered double hydroxides and derivatives, Appl. Clay Sci. 53 (2011) 139–150.

[27] S. He, Z. An, M. Wei, D.G. Evans, X. Duan, Layered double hydroxide-based catalysts: nanostructure design and catalytic performance, Chem. Commun. 49 (2013) 5912–5920.

[28] G.F. Li, D.G. Evans, X. Duan, Catalytic applications of layered double hydroxides: recent advances and perspectives, Chem. Soc. Rev. 43 (2014) 7040–7066.

[29] Y. Kuthati, R.K. Kankala, C.H. Lee, Layered double hydroxide nanoparticles for biomedical applications: current status and recent prospects, Appl. Clay Sci. 112–113 (2015) 100–116.

[30] T. Baskaran, J. Christopher, A. Sakthivel, Progress on layered hydrotalcite (HT) materials as potential support and catalytic materials, RSC Adv. 5 (2015) 98853–98875.

[31] X. Zhang, M. Wei, M. Pu, X. Li, H. Chen, D.G. Evans, X. Duan, Preparation and characterization of *trans*-$RhCl(CO)(TPPTS)_2$-intercalated layered double hydroxides, J. Solid State Chem. 178 (2005) 2701–2708.

[32] F. Iosif, V.I. Parvulescu, M.E.P. Bernal, R.J.R. Casero, V. Rives, K. Kranjc, S. Polanc, M. Kocevar, E. Genin, J.P. Genet, V. Michelet, Heterogeneous hydrogenation of bicyclo [2.2.2]octenes on Rh/TPPTS/LDH catalysts, J. Mol. Catal. A Chem. 276 (2007) 34–70.

[33] S.K. Sharma, P.A. Parikh, R.V. Jasra, Hydroformylation of alkenes using heterogeneous catalyst prepared by intercalation of $HRh(CO)(TPPTS)_3$ complex in hydrotalcite, J. Mol. Catal. A Chem. 316 (2010) 153–162.

[34] S.K. Sharma, R.V. Jasra, Synthesis of 2-ethylhexanal from propylene in a single pot using an ecofriendly multifunctional catalyst synthesized by intercalation of $HRhCO(TPPTS)_3$ complex in the interlayer space of hydrotalcite, Ind. Eng. Chem. Res. 50 (2011) 2815–2821.

[35] P.T. Taney, M. Chibwe, T.J. Pinnavaia, Titanium-containing mesoporous molecular sieves for catalytic oxidation of aromatic compounds, Nature 368 (1994) 321–323.

[36] P.T. Taney, T.J. Pinnavaia, A neutral templating route to mesoporous molecular sieves, Science 267 (1995) 865–867.

[37] P.T. Taney, T.J. Pinnavaia, Mesoporous silica molecular sieves prepared by ionic and neutral surfactant templating: a comparison of physical properties, Chem. Mater. 8 (8) (1996) 2068–2079.

[38] A.N. Ajjou, H. Alper, Hydroformylation of olefins catalysed by rhodium trichloride anchored on Ti-HMS, Molecules Online 2 (1998) 53–57.

[39] N. Sudheesh, S.K. Sharma, R.S. Shukla, R.V. Jasra, $HRh(CO)(PPh_3)_3$ encapsulated mesopores of hexagonal mesoporous silica (HMS) acting as nanophase reactors for effective catalytic hydroformylation of olefins, J. Mol. Catal. A Chem. 296 (2008) 61–70.

[40] N. Sudheesh, J.N. Parmar, R.S. Shukla, Hydroformylation of propene heterogeneously catalyzed by $HRh(CO)(PPh_3)_3$ encapsulated in to hexagonal mesoporous silica-parametric variation and mass transfer study, Appl. Catal. A Gen. 415–416 (2012) 124–131.

[41] N. Sudheesh, S.K. Sharma, R.S. Shukla, R.V. Jasra, Investigations on the kinetics of hydroformylation of 1-hexene using $HRh(CO)(PPh_3)_3$ encapsulated hexagonal mesoporous silica as a heterogeneous catalyst, J. Mol. Catal. A Chem. 316 (2010) 23–29.

[42] M.D. Khokhar, R.S. Shukla, R.V. Jasra, Hydroformylation of dihydrofuranscatalyzed by rhodium complex encapsulated hexagonal mesoporous silica, J. Mol. Catal. A Chem. 400 (2015) 1–6.

[43] N. Sudheesh, A.K. Chaturvedi, R.S. Shukla, $RhCl(TPPTS)_3$ encapsulated into the hexagonal mesoporous silica as an efficient heterogeneous catalyst for hydroformylation of vinyl esters, Appl. Catal. A Gen. 409–410 (2011) 99–105.

[44] N. Sudheesh, R.S. Shukla, Rhodium complex encapsulated functionalized hexagonal mesoporous silica for heterogeneous hydroaminomethylation, Appl. Catal. A Gen. 453 (2013) 159–166.

[45] N. Sudheesh, R.S. Shukla, Rhodium exchanged ETS-10 and ETS-4: efficient heterogeneous catalyst for hydroaminomethylation, Appl. Catal. A Gen. 473 (2014) 116–124.

[46] M. Lenarda, R. Ganzerla, L. Storaro, A. Trovarelli, R. Zanoni, J. Kaspar, Vapour phase hydroformylation of ethylene and propene catalyzed by a rhodium-containing aluminium pillared smectite clay, J. Mol. Catal. 72 (1992) 75–84.

[47] Y. Izumi, K. Konishi, M. Tsukahara, D.M. Obaid, K. Aika, Selective butanol synthesis over rhodium – molybdenum catalysts supported in ordered mesoporous silica, vapour phase hydroformylation of ethylene and propene catalyzed by a rhodium-containing aluminum pillared smectite clay aluminum polyoxocations, J. Phys. Chem. C 111 (2007) 10073–10081.

[48] S.P. Varkey, C.R. Jacob, Direct catalytic oxidation of methane to methanol over zeolite encapsulated transition metal complexes, Indian J. Chem. 38A (1999) 320–324.

[49] C.R. Jacob, S.P. Varkey, P. Ratnasamy, Oxidation of para-xylene over zeolite-encapsulated copper and manganese complexes, Appl. Catal. A Gen. 182 (1999) 91–96.

[50] A. Gunes, O. Bayraktar, S. Yılmaz, Liquid-phase oxidation of carvacrol using zeolite-encapsulated metal complexes, Ind. Eng. Chem. Res. 45 (2006) 54–61.

[51] M.R. Maurya, A.K. Chandrakar, S. Chand, Zeolite-Y encapsulated metal complexes of oxovanadium(VI), copper(II) and nickel(II) as catalyst for the oxidation of styrene, cyclohexane and methyl phenyl sulphide, J. Mol. Catal. A Chem. 274 (2007) 192–201.

[52] B. Fan, H. Li, W. Fan, C. Jin, R. Li, Oxidation of cyclohexane over iron and copper salen complexes simultaneously encapsulated in zeolite Y, Appl. Catal. A Gen. 340 (2008) 67–75.

[53] T.H. Bennur, D. Srinivas, S. Sivasanker, Oxidation of ethylbenzene over "neat" and zeolite-Y-encapsulated copper tri- and tetra-aza macrocyclic complexes, J. Mol. Catal. A Chem. 207 (2004) 163–171.

[54] A. Kozlov, A. Kozlova, K. Asakura, Y. Iwasawa, Zeolite-encapsulated vanadium picolinate-peroxo complexes active for catalytic hydrocarbon oxidations, J. Mol. Catal. A Chem. 137 (1999) 223–237.

[55] A.A. Valente, J. Vital, Oxidation of pinane using zeolite encapsulated metal phthalocyanine catalysts, Stud. Surf. Sci. Catal. 108 (1997) 461–468.

[56] K.K. Bania, R.C. Deka, Zeolite-Y encapsulated metal picolinato complexes as catalyst for oxidation of phenol with hydrogen peroxide, J. Phys. Chem. C 117 (2013) 11663–11678.

[57] G. von Willingh, H.S. Abbo, S.J.J. Titinchi, Selective oxidation reactions over tri- and tetradentateoxovanadium(IV) complexes encapsulated in zeolite-Y, Catal. Today 227 (2014) 96–104.

[58] S. Yamaguchi, T. Fukura, K. Takiguchi, C. Fujita, M. Nishibori, Y. Teraoka, H. Yahiro, Selective hydroxylation of cyclohexene over Fe-bipyridine complexes encapsulated into Y-type zeolite under environment-friendly conditions, Catal. Today 242 (2015) 261–267.

[59] H. Nur, H. Hamid, S. Endud, H. Hamdan, Z. Ramli, Iron-porphyrin encapsulated in poly(methacrylic acid) and mesoporous Al-MCM-41 as catalysts in the oxidation of benzene to phenol, Mater. Chem. Phys. 96 (2006) 337–342.

[60] S.N. Rao, K.N. Munshi, N.N. Rao, Catalytic air oxidation of cyclooctene using *cis*-[MoO_2(SAL-SH)(DMF)](SAL-SH = salcylidenesalicyloylhydrazinato-) and its zeolite-Y encapsulated composite, J. Mol. Catal. A Chem. 145 (1999) 203–210.

[61] T. Joseph, S.B. Halligudi, C. Satyanarayan, D.P. Sawant, S. Gopinathan, Oxidation by molecular oxygen using zeolite encapsulated Co(II)saloph complexes, J. Mol. Catal. A Chem. 168 (2001) 87–97.

[62] K.O. Xavier, J. Chacko, K.K.M. Yusuff, Zeolite-encapsulated Co(II), Ni(II) and Cu(II) complexes as catalysts for partial oxidation of benzyl alcohol and ethylbenzene, Appl. Catal. A Gen. 258 (2004) 251–259.

[63] T.R. Felthouse, J.A. Murphy, Zeolite-encapsulated platinum catalysts: preparation, characterization by transmission electron microscopy, and their shape selective behavior toward various nitrogen base poisons during the catalytic oxidation of aqueous formaldehyde, J. Catal. 98 (1986) 411–433.

[64] S.S. Reddy, B.D. Raju, A.H. Padmasri, P.K.S. Prakash, K.S.R. Rao, Novel and efficient cobalt encapsulated SBA-15 catalysts for the selective oxidation of cyclohexane, Catal. Today 141 (2009) 61–65.

[65] K. Yamaguchi, K. Mori, T. Mizugaki, K. Ebitani, K. Kaneda, Creation of a monomeric Ru species on the surface of hydroxyapatite as an efficient heterogeneous catalyst for aerobic alcohol oxidation, J. Am. Chem. Soc. 122 (2000) 7144–7145.

[66] K. Mori, S. Kanai, T. Hara, T. Mizugaki, K. Ebitani, K. Jitsukawa, K. Kaneda, Development of ruthenium-hydroxyapatite-encapsulated superparamagneticγ-Fe_2O_3nanocrystallites as an efficient oxidation catalyst by molecular oxygen, Chem. Mater. 19 (2007) 1249–1256.

[67] L. Mamani, M. Sheykhan, A. Heydari, Nano silver embedded on hydroxyapatite-encapsulated γ-Fe_2O_3: superparamagnetic catalyst for chemoselective oxidation of primary amines to N-monoalkylated hydroxylamines, Appl. Catal. A Gen. 395 (2011) 34–38.

[68] N. Viswanadham, T. Shido, Y. Iwasawa, Performances of rhenium oxide-encapsulated ZSM-5 catalysts in propene selective oxidation/ammoxidation, Appl. Catal. A Gen. 219 (2001) 223–233.

[69] J. Sebastian, K.M. Jinka, R.V. Jasra, Effect of alkali and alkaline earth metal ions on the catalytic epoxidation of styrene with molecular oxygen using cobalt(II)-exchanged zeolite X, J. Catal. 244 (2006) 208–218.

[70] K.M. Jinka, J. Sebastian, R.V. Jasra, Epoxidation of cycloalkenes with cobalt(II)-exchanged zeolite X catalysts using molecular oxygen, J. Mol. Catal. A Chem. 274 (2007) 33–41.

[71] M.V. Patil, M.K. Yadav, R.V. Jasra, Catalytic epoxidation of α-pinene with molecular oxygen using cobalt(II)-exchanged zeolite Y-based heterogeneous catalysts, J. Mol. Catal. A Chem. 277 (2007) 72–80.

[72] K.M. Jinka, S.M. Pai, B.L. Newalkar, N.V. Choudary, R.V. Jasra, Co-SSZ-51 as a catalyst for styrene epoxidation using molecular oxygen, Catal. Commun. 11 (2010) 638–642.

[73] A.K. Rahiman, K. Rajesh, K.S. Bharathi, S. Sreedaran, V. Narayanan, Manganese(III) porphyrin-encapsulated Ti, Si-mesoporous molecular sieves as heterogeneous catalysts for the epoxidation of alkenes, Appl. Catal. A Gen. 314 (2006) 216–225.

[74] X.-L. Yang, L.-M. Qiao, W.-L. Dai, Phosphotungstic acid encapsulated in metal-organic framework UiO-66: an effective catalyst for the selective oxidation of cyclopentene to glutaraldehyde, Microporous Mesoporous Mater. 211 (2015) 73–81.

[75] L.-G. Qiu, A.-J. Xie, Li-De Zhang, encapsulation of catalysts in supramolecular porous frameworks: size- and shape-selective catalytic oxidation of phenols, Adv. Mater. 17 (6) (2005) 689–692.

[76] T. Mizugaki, R. Arundhathi, T. Mitsudome, K. Jitsukawa, K. Kaneda, Highly efficient and selective transformations of glycerol using reusable heterogeneous catalysts, ACS Sustain. Chem. Eng. 2 (2014) 574–578.

[77] S.K. Sharma, K.B. Sidhpuria, R.V. Jasra, Ruthenium containing hydrotalcite as a heterogeneous catalyst for hydrogenation of benzene to cyclohexane, J. Mol. Catal. A Chem. 335 (2011) 65–70.

[78] P. Chen, B. Fan, M. Song, C. Jin, J. Ma, R. Li, Zeolite-encapsulated Ru(III) tetrahydro-schiff base complex: an efficient heterogeneous catalyst for the hydrogenation of benzene under mild conditions, Catal. Commun. 7 (2006) 969–973.

[79] A. Bordoloi, A.P. Amrute, S.B. Halligudi, [Ru(salen)(NO)] complex encapsulated in mesoporous SBA-16 as catalyst for hydrogenation of ketones, Catal. Commun. 10 (2008) 45–48.

[80] I. Nongwe, G. Bepete, A. Shaikjee, V. Ravat, B. Terfassa, R. Meijboom, N.J. Coville, Synthesis of gold encapsulated in spherical carbon capsules with a mesoporous shell structure. A robust catalyst in a nanoreactor, Catal. Commun. 53 (2014) 77–82.

[81] J. Gu, Z. Zhang, P. Hu, L. Ding, N. Xue, L. Peng, X. Guo, M. Lin, W. Ding, Platinum nanoparticles encapsulated in MFI zeolite crystals by a two-step dry gel conversion method as a highly selective hydrogenation catalyst, ACS Catal. 5 (2015) 6893–6901.

[82] (a) S. Galvagno, G. Capannelli, G. Neri, A. Donato, R. Pietropaolo, Hydrogenation of cinnamaldehyde over Ru/C catalysts: effect of Ru particle size, J. Mol. Catal. 64 (1991) 237–246. (b) A.K. Prashar, S. Mayadevi, R.N. Devi, Effect of particle size on selective hydrogenation of cinnamaldehyde by Pt encapsulated in mesoporous silica, Catal. Commun. 28 (2012) 42–46.

[83] S. Li, A. Tuel, F. Meunier, M. Aouine, D. Farrusseng, Platinum nanoparticles entrapped in zeolite nanoshells as active and sintering-resistant arene hydrogenation catalysts, J. Catal. 332 (2015) 25–30.

[84] D. Chatterjee, H.C. Bajaj, A. Das, K. Bhatt, First report on highly efficient alkene hydrogenation catalysed by Ni(salen) complex encapsulated in zeolite, J. Mol. Catal. 92 (1994) L235–L238.

[85] S.K. Sharma, V.K. Srivastava, P.H. Pandya, R.V. Jasra, Solvent-free isomerization of methyl chavicol to *trans*-anethole using transition metal complexes as catalysts, Catal. Commun. 6 (2005) 205–209.

[86] S.K. Sharma, V.K. Srivastava, R.V. Jasra, Selective double bond isomerization of allyl phenyl ethers catalyzed by ruthenium metal complexes, J. Mol. Catal. A Chem. 245 (2006) 200–209.

[87] S.K. Sharma, P.A. Parikh, R.V. Jasra, Ruthenium containing hydrotalcite as a solid base catalyst for $>C=C<$ double bond isomerization in perfumery chemicals, J. Mol. Catal. A Chem. 317 (2010) 27–33.

[88] C.M. Jinesh, Churchil A. Antonyraj, S. Kannan, Isomerization of eugenol and alkenyl aromatics of perfumery interest over Ni-containing layered double hydroxides as solid base catalysts, Catal. Today 141 (2009) 176–181.

[89] D. Kishore, S. Kannan, Catalytic isomerization of estragole to anethole over hydrotalcites and HT-like compounds, J. Mol. Catal. A Chem. 244 (2006) 83–92.

[90] D. Jin, J. Gao, Z. Hou, Y. Guo, X. Lu, Y. Zhu, X. Zheng, Microwave assisted in situ synthesis of USY-encapsulated heteropoly acid (HPW-USY) catalysts, Appl. Catal. A Gen. 352 (2009) 259–264.

[91] A. Dhakshinamoorthy, P. Visuvamithiran, V. Tharmaraj, K. Pitchumani, Clay encapsulated ZnO nanoparticles as efficient catalysts for N-benzylation of amines, Catal. Commun. 16 (2011) 15–19.

[92] H. Naeimi, M. Moradian, Encapsulation of copper(I)-schiff base complex in NaY nanoporosity: an efficient and reusable catalyst in the synthesis of propargylamines via A^3-coupling (aldehyde-amine-alkyne) reactions, Appl. Catal. A Gen. 467 (2013) 400–406.

[93] A.K. Shah, N.H. Khan, G. Sethia, S. Saravanan, R.I. Kureshy, S.H.R. Abdi, H.C. Bajaj, Tin exchanged zeolite as catalyst for direct synthesis of α-amino nitriles under solvent-free conditions, Appl. Catal. A Gen. 419–420 (2012) 22–30.

[94] C. Bianchini, P. Barbaro, Recent aspects of asymmetric catalysis by immobilized chiral metal catalysts, Top. Catal. 19 (2002) 17–32.

[95] H. Yang, L. Zhang, L. Zhong, Q. Yang, C. Li, Enhanced cooperative activation effect in the hydrolytic kinetic resolution of epoxides on [Co(salen)] catalysts confined in nanocages, Angew. Chem. Int. Ed. 46 (2007) 6861–6865.

[96] H. Yang, L. Zhang, W. Su, Q. Yang, C. Li, Asymmetric ring-opening of epoxides on chiral Co(salen) catalyst synthesized in SBA-16 through the "ship in a bottle" strategy, J. Catal. 248 (2007) 204–212.

[97] A. Milo, R. Neumann, An achiral manganese salen catalyst encapsulated in a peptidic phosphonate homochiral solid for the enantioselective formation of diols by consecutive epoxidation and hydration reactions, Chem. Commun. 47 (2011) 2535–2537.

[98] R.I. Kureshy, I. Ahmad, N.H. Khan, S.H.R. Abdi, K. Pathak, R.V. Jasra, Encapsulation of a chiral MnIII(salen) complex in ordered mesoporous silicas: an approach towards heterogenizing asymmetric epoxidation catalysts for non-functionalized alkenes, Tetrahedron Asymmetry 16 (2005) 3562–3569.

[99] R.I. Kureshy, I. Ahmad, N.H. Khan, S.H.R. Abdi, K. Pathak, R.V. Jasra, Chiral Mn(III) salen complexes covalently bonded on modified MCM-41and SBA-15 as efficient catalysts for enantioselective epoxidation of nonfunctionalized alkenes, J. Catal. 238 (2006) 134–141.

[100] R.I. Kureshy, N.H. Khan, S.H.R. Abdi, I. Ahmad, S. Singh, R.V. Jasra, Dicationic chiral Mn(III) salen complex exchanged in the interlayers of montmorillonite clay: a heterogeneous enantioselective catalyst for epoxidation of nonfunctionalized alkenes, J. Catal. 221 (2004) 234–240.

[101] R.I. Kureshy, I. Ahmad, N.H. Khan, S.H.R. Abdi, S. Singh, P.H. Pandia, R.V. Jasra, New immobilized chiral Mn(III) salen complexes on pyridine N-oxide-modified MCM-41 as effective catalysts for epoxidation of nonfunctionalized alkenes, J. Catal. 235 (2005) 28–34.

[102] T. Roy, R.I. Kureshy, N.H. Khan, S.H.R. Abdi, A. Sadhukhan, H.C. Bajaj, Immobilized dimeric chiral Mn(III) salen complex on short channel ordered mesoporous silica as an effective catalyst for the epoxidation of non-functionalized alkenes, Tetrahedron 68 (2012) 6314–6322.

[103] M.J. Alcón, A. Corma, M. Iglesias, F. Sánchez, New Mn(II) and Cu(II) chiral C2-multidentate complexes immobilised in zeolites (USY, MCM41) reusable catalysts for selective oxidation reactions, J. Mol. Catal. A Chem. 194 (2003) 137–152.

[104] A.P. Bhatt, K. Pathak, R.V. Jasra, R.I. Kureshy, N.H. Khan, S.H.R. Abdi, Chiral lanthanum–lithium–binaphthol complex covalently bonded to silica and MCM-41 for enantioselective nitroaldol (Henry) reaction, J. Mol. Catal. A Chem. 244 (2006) 110–117.

[105] Á. Zsigmond, K. Bogár, F. Notheisz, Comparative study of "ship-in-a-bottle" and anchored heterogenized Rh complexes, J. Catal. 213 (2003) 103–108.

[106] S. Ernst, H. Disteldorf, X. Yang, Liquid-phase hydrogenation of 1,5-cyclooctadiene on zeolite-encapsulated noble metal–salen complexes, Microporous Mesoporous Mater. 22 (1998) 457–464.

[107] S. Ernst, E. Fuchs, X. Yang, Enantioselective hydrogenation on zeolite-encapsulated chiral palladium–salen complexes, Microporous Mesoporous Mater. 35–36 (2000) 137–142.

[108] A. Milo, R. Neumann, Achiral ruthenium catalyst encapsulated in titanium phosphonate homochiral peptide-based solids for enantioselective hydrogenation of ketones to secondary alcohols, ACS Catal. 2 (2012) 2531–2536.

[109] J. Liu, B. Fan, D. Liang, X. Shi, R. Li, H. Chen, Asymmetric hydrogenation of aromatic ketones by chiral (1S,2S)-DPEN-Ru(II)$Cl_2$$(TPP)_2$ encapsulated in SBA-16, Catal. Commun. 11 (2010) 373–377.

[110] J. Peng, X. Wang, X. Zhang, S. Bai, Y. Zhao, C. Li, Q. Yang, Asymmetric hydrogenation by $RuCl_2$(*R*-Binap)(dmf)$_n$ encapsulated in silica-based nanoreactors, Catal. Sci. Technol. 5 (2015) 666–672.

Chapter 15

Heterogeneous Catalysts Encapsulated in Inorganic Systems to Enhance Reaction Performances for XTL Processes

Ummuhan Cimenler and John N. Kuhn
University of South Florida, Tampa, FL, United States

Contents

1 INTRODUCTION

In recent years, encapsulation of particles and molecules has received attention. Many studies have been performed on the synthesis of encapsulated structure to increase stability and control the accessibility of the confined species as well as to control the size and shape of materials for applications in catalysis, nanoelectronics, nano-optics, biomedicine, and material sciences [1–3] (Fig. 15.1).

Encapsulated structures can be classified as core-shell (or egg-shell), which is a single noble metal core encapsulated by a shell, yolk-shell, or rattle-type structures. They have a single noble metal core inside a hollow shell, multicore-shell (or pomegranate-like) which is multiple supported or unsupported noble metal cores coated with a metal oxide shell (Fig. 15.2).

Heterogeneous catalysts encapsulated in inorganic materials (such as zeolites and SiO_2) are also a focus of interest for developing inexpensive, selective, and stable catalysts for XTL process steps (e.g., hydrocarbon reforming, water-gas shift (WGS), and Fischer-Tropsch synthesis (FTS)) since interactions between components can lead to improving the properties over the single components. Moreover, encapsulated catalytic structures open the possibility of preserving the catalysts from deactivation and increase the selectivity of the

Encapsulated Catalysts. http://dx.doi.org/10.1016/B978-0-12-803836-9.00015-8

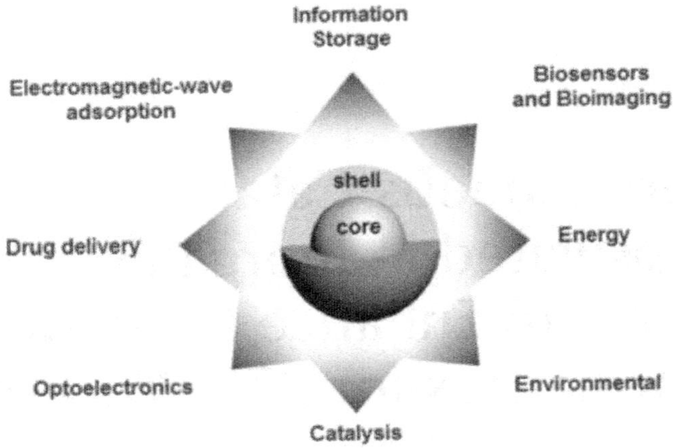

FIG. 15.1 Applications of the encapsulated nanoparticles. *(Reproduced from X. Fang, S. Liu, J. Zang, C. Xu, M.-S. Zheng, Q.-F. Dong, D. Sun, N. Zheng, Precisely controlled resorcinol–formaldehyde resin coating for fabricating core–shell, hollow, and yolk–shell carbon nanostructures, Nanoscale 5 (2013) 6908–6916, with permission from the Royal Society of Chemistry.)*

FIG. 15.2 Schematic illustration of core-shell, hollow, and yolk-shell catalysts. *(Reproduced from X. Fang, S. Liu, J. Zang, C. Xu, M.-S. Zheng, Q.-F. Dong, D. Sun, N. Zheng, Precisely controlled resorcinol–formaldehyde resin coating for fabricating core–shell, hollow, and yolk–shell carbon nanostructures, Nanoscale 5 (2013) 6908–6916, with permission from the Royal Society of Chemistry.)*

catalysts due to the effect of encapsulation. Eggshell, multicore-shell, and yolk-shell structures are the most studied encapsulated structures in the literature as XTL catalysts.

Although the encapsulated structures demonstrate many advantages, there is also a disadvantage for catalytic reactions [4]. The shell layer could lower the diffusion rates of the reactants toward the active cores. Diffusion of molecules through the pores of a zeolite occurs in a regime that is called "configurational diffusion" [5,6]. Configurational diffusion occurs when the pore size is on the order of the molecular size and continuous molecule-wall interaction occurs (Fig. 15.3). Very small changes in pore diameter and molecular dimensions can cause large differences in the diffusion coefficients of the reactants, and this brings the shape selectivity (reactant or product selectivity) property of the zeolites [5,6].

Thus, encapsulated catalysts with a specific size and pore channels can turn this disadvantage to an advantage. For example, using suitable zeolites as a shell layer can protect the core catalyst from deactivation caused by coking because

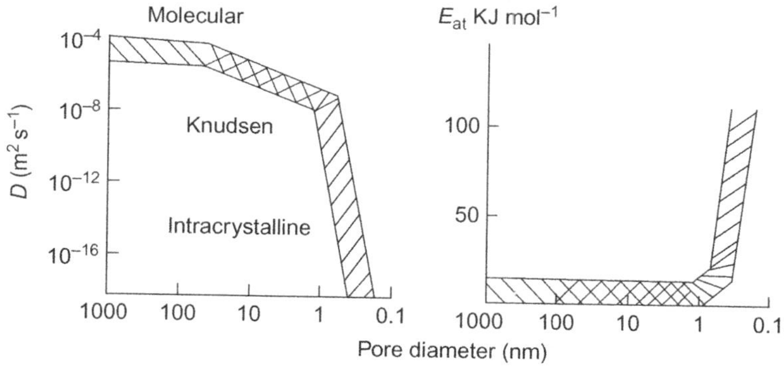

FIG. 15.3 Effect of pore size on the diffusivity and activation energy of diffusion. *(Reproduced from A. Corma, Inorganic solid acids and their use in acid-catalyzed hydrocarbon reactions, Chem. Rev. 95 (1995) 559–614, with permission from the American Chemical Society.)*

zeolites have a reactant selectivity property, which means the microporous character of the zeolites with uniform pore dimensions allows certain molecules to enter the crystals while rejecting others with large molecular sizes [7] (Fig. 15.4A). Thus, the core catalyst can be protected from deactivation by hindering unwanted molecules from accessing the core.

The encapsulated catalysts can also be used as bifunctional catalysts to combine two different active sites in one particle. If the core catalyst that is used for a reaction is encapsulated within a shell membrane that is used as a catalyst for another reaction, the reactants have to first pass through the shell membrane to

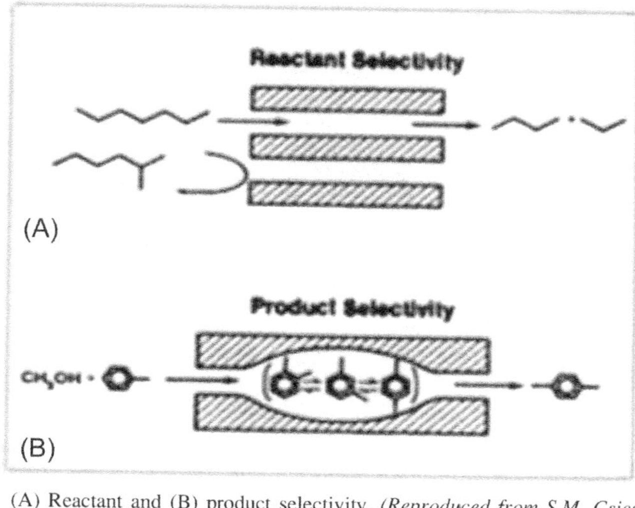

FIG. 15.4 (A) Reactant and (B) product selectivity. *(Reproduced from S.M. Csicsery, Shape-selective catalysis in zeolites, Zeolites 4 (1984) 202–213, with permission from Elsevier.)*

reach the inner core catalyst, where they react to form the intermediate products [8]. After that, intermediate products have to enter the membrane channels, where they are converted into the final products at the active sites in the membrane to leave the catalyst [8]. Hence, the cascade catalytic systems can be designed. There are many studies [8–16] that used core-shell structure catalysts as bifunctional catalysts for FTS in the literature. In these bifunctional structures, the core catalysts were encapsulated with a zeolite membrane shell that is used for isomerization and a cracking reaction since zeolites have acidic properties, varied molecular diffusion rate in their pores, and product selectivity properties (Fig. 15.4). Thus, the products are converted to smaller molecules on the active sites of the shell while diffusing away from the core. In FTS encapsulated catalysts, reactants pass through the zeolite membrane pores and reach the core FT catalyst where hydrocarbons form. When the formed hydrocarbons diffuse through the zeolite pores, they are cracked and isomerized by acidic sites of the zeolite to achieve a synthetic gasoline [12] (Fig. 15.5).

In addition, an encapsulated structure can help to protect the core catalyst from deactivation caused by sintering due to migration and high temperature since the outer shells isolate the catalytically active nanoparticle cores and prevent the possibility of sintering core particles during catalytic reactions, even under harsh reaction conditions [4,17,18] (Fig. 15.6).

The mobility of reactants in the encapsulated catalyst also plays an important role in obtaining high activity. Galanti et al. have recently proposed that the change in the polarity of the shell network, (especially thermosensitive hydrogel shell) alters the mass transport from bulk to the catalyst through a shell

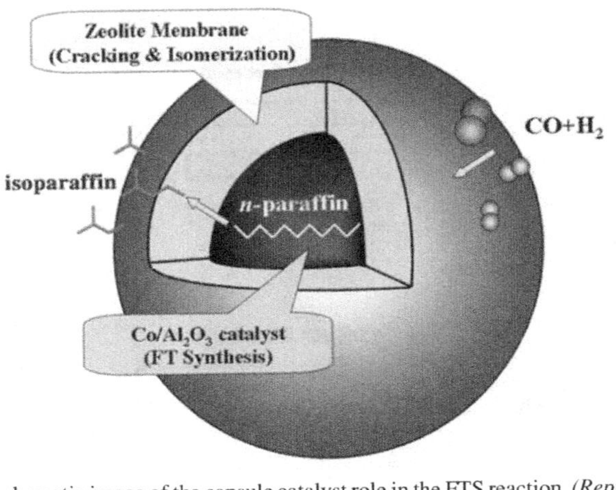

FIG. 15.5 A schematic image of the capsule catalyst role in the FTS reaction. *(Reproduced from J. Bao, J. He, Y. Zhang, Y. Yoneyama, N. Tsubaki, A core/shell catalyst produces a spatially confined effect and shape selectivity in a consecutive reaction, Angew. Chem. 120 (2008) 359–362, with permission from John Wiley and Sons.)*

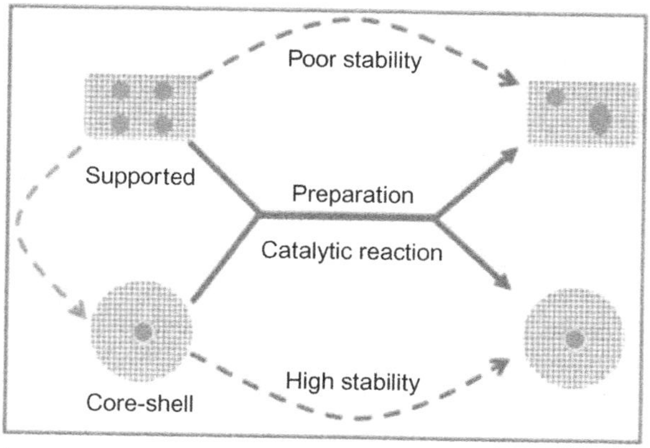

FIG. 15.6 Using encapsulated catalysts as an alternative to conventional supported catalysts for inhibiting the migration and aggregation of small noble metal NPs into large ones during catalytic reaction processes. *(Reproduced from G. Li, Z. Tang, Noble metal nanoparticle@ metal oxide core/yolk–shell nanostructures as catalysts: recent progress and perspective, Nanoscale 6 (2014) 3995–4011, with permission from the Royal Society of Chemistry.)*

medium [19]. Another study demonstrated that the catalytic activity of the encapsulated nanoparticles decreased while the network shrank by increasing the temperature [20]. The reason for the activity decrease was explained as an increased diffusional resistance mass transport within the shrunken network [20] (Fig. 15.7).

According to Galanti et al. [19], the reactants diffuse through a free-energy environment of the shell. This environment impacts the reactants' solvation free energies when entering the shells from bulk, which is regarded as adsorption-

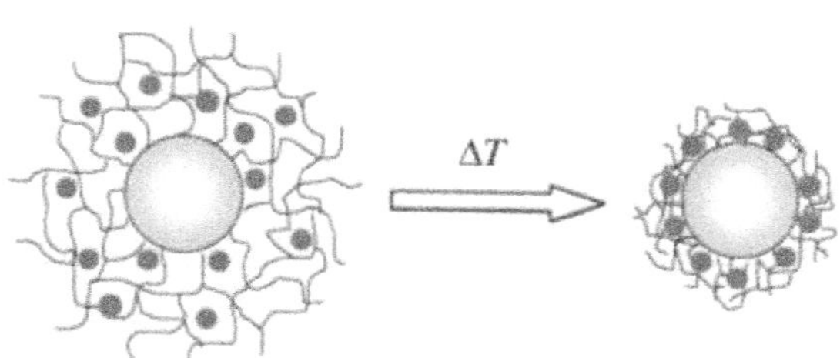

FIG. 15.7 PS-NIPA-Ag composite particles consisting of thermosensitive core-shell particles in which Ag nanoparticles are embedded. The composite particles are suspended in water that swells the thermosensitive network attached to the surface of the core particles. In this state the reagents can diffuse freely to the nanoparticles that act as catalysts. At higher temperatures ($T > 308°C$) the network shrinks and the catalytic activity of the nanoparticles is strongly diminished. *(Reproduced from Y. Lu, Y. Mei, M. Drechsler, M. Ballauff, Thermosensitive core–shell particles as carriers for Ag nanoparticles: modulating the catalytic activity by a phase transition in networks, Angew. Chem. Int. Ed. 45 (2006) 813–816, with permission from John Wiley and Sons.)*

free energy or transfer-free energy. As an example, the free energy of a reactant may be lowered upon entering the network and thus the number of reactant molecules in the network will be increased, so that their increased concentration in the vicinity of the catalyst will lead to a higher reaction rate. In their study, based on this explanation, a theory was developed to calculate the total reaction rate of core-shell catalysts with multiple catalysts embedded in the shell [19].

As explained, there are many advantages of the encapsulated catalysts. In this chapter, studies made in the literature for inorganic encapsulated catalysts for XTL processes will be reviewed and discussed. Catalytic application of the encapsulated catalysts in which the core is a metal and shell is the mesoporous or microporous inorganic material (such as zeolites, SiO_2, CeO_2 etc.) in the reforming, WGS, and FTS steps of the XTL processes will be explained.

2 HYDROCARBON REFORMING

Reforming is a key step of the XTL process to produce syngas (CO and H_2). Syngas can be synthesized using three processes: steam reforming (SR), CO_2 reforming (dry reforming), and oxy-methane reforming (partial oxidation). The combination of steam and oxy reforming can also be used as a fourth process that is known as auto thermal reforming (ATR). In the ATR, steam and oxygen are combined in proportions to give a net $\Delta H \sim 0$ [21]. The equations of SR, CO_2 reforming (dry reforming), oxymethane reforming, and ATR reactions are given below:

Steam reforming

$$CH_4(g) + H_2O(g) \rightarrow CO(g) + 3H_2(g) \tag{15.1}$$

Dry (CO_2) reforming

$$CH_4(g) + CO_2(g) \rightarrow 2CO(g) + 2H_2(g) \tag{15.2}$$

Oxy reforming

$$CH_4(g) + 1/2O_2(g) \rightarrow CO(g) + 2H_2(g) \tag{15.3}$$

Autothermal reforming

$$4CH_4(g) + 2H_2O(g) + O_2(g) \rightarrow 10H_2(g) + 4CO(g) \tag{15.4}$$

Catalysts for SR are transition metals from group VIII such as Ni, Pd, Co, Rh, Ru, Pt, and Ir [22–24]. Similarly, most of the catalysts studied for dry reforming are Ni, Pd, Pt, Ru, Co, Rh, and Ir [25–27]. SiO_2, Al_2O_3, TiO_2, La_2O_3, CaO, MgO, CeO_2, and ZrO_2 are used as oxide supports and K, Mg, Ca, and La have been used as promoters to improve the performance of catalysts [21,27–29].

Catalyst deactivation is one of the biggest challenges in catalytic SR and dry reforming processes. Nickel is inexpensive relative to the other metals and thus it is the most studied catalyst for reforming processes. However, supported Ni catalysts are deactivated by carbon formation, sulfur poisoning, and loss of active sites due to the sintering of nickel particles [30,31]. Ni particle size is an important factor for the activity of SR catalysts since smaller particles will provide a larger surface for the reaction and improve the catalyst activity [32]. Smaller particles have also been reported to be more resistant to carbon formation [32]. Thus, encapsulation of the Ni particles may prevent the possibility of sintering during catalytic reactions caused by migration and high temperature since encapsulation of the Ni catalyst with a stable shell will keep particle size smaller by inhibiting the migration and aggregation and help to protect the catalyst from coking. Because of all these advantages, encapsulation of supported Ni catalysts in an inorganic system for the reforming process has been studied recently to protect the catalysts from deactivation and to improve their performance.

High reaction temperatures can be employed to thermodynamically inhibit coke formation [33]. However, high temperatures increase catalyst sintering, thereby reduce conversion and coke resistance. Thus, to hinder sintering and coking of Ni particles during the dry reforming of methane (DRM), Gould et al. [34] prepared porous alumina coated Ni/Al_2O_3 particles using molecular layer deposition (MLD). The study found that the dry reforming rates for the MLD-coated catalysts remained stable for 108 h, while the uncoated catalyst continuously deactivated during DRM at 973 K (Fig. 15.8).

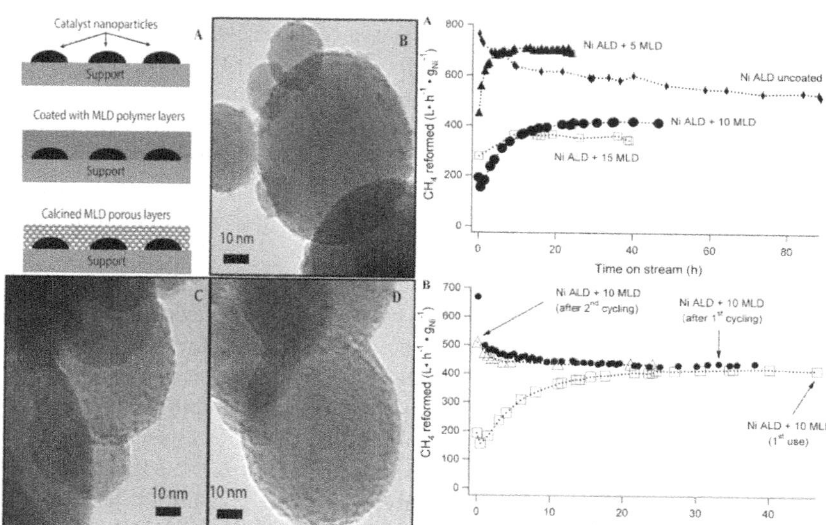

FIG. 15.8 Left figure: (A) Schematic representation of the MLD coating process to produce porous alumina overlayers. (B) Ni nanoparticles (2–5 nm dark particles) deposited on spherical alumina; (C) 5-MLD cycles coating; and (D) 10-MLD cycles coating. Right figure: (A) Dry reforming rates at 973 K for uncoated Ni ALD catalyst and the same catalyst coated with 5, 10, and 15 MLD layers. (B) Effect of cycling on the DRM rates for the 10-MLD catalyst. *(Reproduced from T.D. Gould, A. Izar, A.W. Weimer, J.L. Falconer, J.W. Medlin, Stabilizing Ni catalysts by molecular layer deposition for harsh, dry reforming conditions, ACS Catal. 4 (2014) 2714–2717, with permission from the American Chemical Society.)*

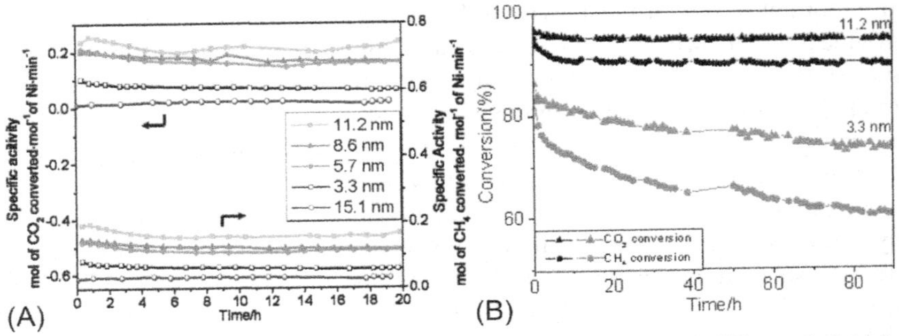

FIG. 15.9 (A) Specific activity for Ni@SiO_2 and Ni-yolk@Ni@SiO_2 with different shell thicknesses. Conditions: 800°C,GHSV = 1440 L g^{-1} cat h^{-1}, Wcat = 0.01 g, CO_2:CH_4:N_2 = 1:1:1. (B) Stability test for Ni@SiO_2 with a 3.3-nm shell thickness and Ni-yolk@Ni@SiO_2 with a 11.2-nm shell thickness. Conditions: 800°C, GHSV = 36,000 mL g^{-1} cat h^{-1}, CO_2:CH_4:N_2 = 1:1:1. *(Reproduced from Z. Li, L. Mo, Y. Kathiraser, S. Kawi, Yolk-satellite-shell structured Ni-yolk@ Ni@SiO_2 nanocomposite: superb catalyst toward methane CO_2 reforming reaction. ACS Catal. 4 (2014) 1526–1536, with permission from the American Chemical Society.)*

In addition to improved stability, encapsulation of the reforming catalysts has been also studied to achieve high activity. Encapsulation can yield large interfaces between materials, which contributes to active metal-support interactions. For instance, Li et al. synthesized a yolk-satellite-shell structured Ni-yolk@Ni@SiO_2 with different shell thicknesses (3.3, 5.7, 11.2, and 15.1 nm) for dry reforming to obtain high catalytic activity and stability [35]. The authors found that a small increase in shell thickness from 3.3 to 5.7 nm resulted in a nearly doubled increase in CH_4-specific activity and a tripled increase in CO_2-specific activity (Fig. 15.9A). The 11.2-nm shell thickness Ni-yolk@Ni@SiO_2 demonstrated the most stable and active material for CH_4 and CO_2 specific activities (Fig. 15.9B).

In another study, Li et al. synthesized Ni@Ni-Mg phyllosilicate core-shell catalysts by the hydrothermal treatment of Ni@SiO_2 nanoparticles to increase the exposed Ni area for reactants [36]. The improved catalytic performance of the DRM reaction is attributed to the high Ni accessibility, strong basicity, and high structural stability of the Ni@Ni-phyllosilicate core-shell catalyst.

Besides the stability and activity, encapsulation of the reforming catalysts is used to improve reactant selectivity. Reactant selectivity is the property of the porous materials, which occurs when only part of the reactant molecules are small enough to diffuse through the pores [9] (Fig. 15.4). For this aim, catalysts are encapsulated with a microporous material such as zeolites since their uniform pore dimensions and microporous characters permit small molecules to enter the crystals while they reject others with large molecular sizes [7]. For instance, Cimenler et al. studied a composite H-β zeolite membrane encapsulated 1.6 wt% Ni/1.2 wt% Mg/$Ce_{0.6}Zr_{0.4}O_2$ SR catalyst and performed methane (CH_4) and toluene (C_7H_8, as a tar model that causes catalyst deactivation) SR to

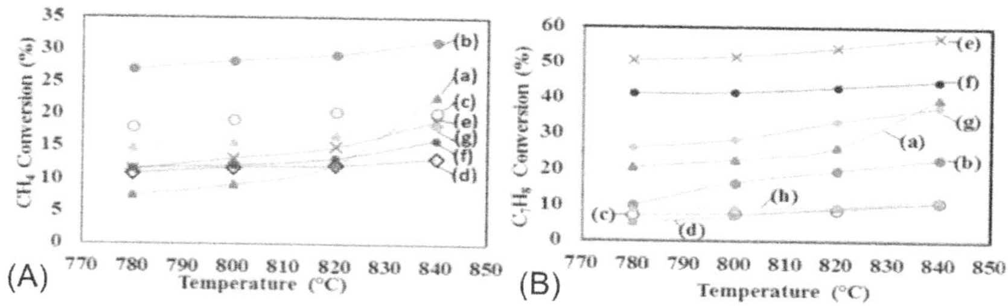

FIG. 15.10 (A) CH_4 steam reforming results. (a) Uncoated SR catalyst, (b) 17.2 mg of SR@β34.3% catalyst, (c) 23.6 mg, and (d) 17.2 mg of SR@β51% catalyst, (e) PM-51% β/SR catalyst, (f) PM-51% SiO_2/SR catalyst, and (g) CH_4-C_7H_8 steam reforming reaction with a molar ratio of $CH_4/C_7H_8/H_2O = 1.44{:}1{:}7$. (B) C_7H_8 steam reforming reaction results. (a) Uncoated SR catalyst, (b) 34.3 wt% H-β-zeolite coated composite catalyst, (c) SR@β51% catalyst, (d) H-β zeolite, (e) PM-51% β/SR catalyst, (f) PM-51% SiO_2/SR catalyst, (g) PM-51% SiC/SR catalyst, and (h) CH_4-C_7H_8 steam reforming reaction on SR@β51%. Conditions were 780–840°C, atmospheric pressure, and a molar ratio of $CH_4/H_2O = 1$ and $C_7H_8/H_2O = 1{:}7$. *(Reproduced from U. Cimenler, B. Joseph, J.N. Kuhn, Effect of zeolite membrane shell thickness on reactant selectivity for hydrocarbon steam reforming using layered catalysts. Energy Fuel 30 (7) (2016) 5300–5308, with permission from the American Chemical Society.)*

investigate the reactant selectivity property of the H-β zeolite shell [37,38]. The authors used a H-β zeolite shell to coat the core SR catalyst and prepared 34.3 and 51 wt% zeolite encapsulated SR catalysts to investigate the zeolite shell thickness on both CH_4 and C_7H_8 conversions [38]. Due to the reactant selectivity of the zeolite membrane, the composite 34.3 wt% H-β-zeolite-coated reforming catalyst demonstrated a decrease in C_7H_8 conversion and an increase in CH_4 conversion due to confined reaction environment and/or Al^{+3} promotion to active sites when compared to the uncoated reforming catalyst [37]. Increasing the zeolite loading from 34.3 to 51 wt% decreased both the CH_4 and C_7H_8 conversions as a result of the increased diffusion limitation [38] (Fig. 15.10).

They also confirmed that the reactions were occurring in the diffusion-limited regime by calculating the Weisz-Prater criterion and effectiveness factor. Diffusion control increased with increasing zeolite loading. Reactant selectivity defined as the ratio of the conversions rates of CH_4 to C_7H_8 increased by 1.5 times with increase of zeolite thickness. Moreover, the authors found that the composite catalysts remained stable during the combined SR for 10 h TOS, but the uncoated SR catalyst deactivated. Thus, they concluded that the zeolite coated SR catalyst can be used practically in industrial processes to reduce deactivation and improve reactant selectivity, which is important to change the product distribution [38].

In another study, Cimenler et al. compared a 51 wt% nonacidic Silicalite-1 zeolite membrane encapsulated 1.6 wt% Ni-1.2 wt% $Mg/Ce_{0.6}Zr_{0.4}O_2$ SR composite catalyst and 51 wt% acidic H-βzeolite membrane encapsulated 1.6 wt% Ni-1.2 wt% $Mg/Ce_{0.6}Zr_{0.4}O_2$ SR composite catalyst to investigate the effect of encapsulation on CH_4 and C_7H_8 SR [39]. The nonacidic Silicalite-1

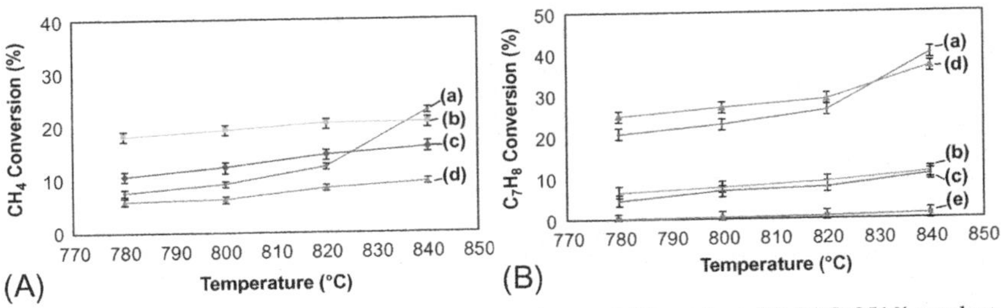

FIG. 15.11 (A) CH_4 steam reforming results. (a) Uncoated SR catalyst, (b) SR@ β51% catalyst, (c) SR@ Sil51% catalyst, (d) PM-51%Sil/SR catalyst. (B) C_7H_8 steam reforming reaction results. (a) Uncoated SR catalyst, (b) SR@ β51% catalyst, (c) SR@ Sil51% catalyst, (d) PM-51% Sil/SR catalyst, and (e) silicalite-1 powder. Conditions were 780–840°C, atmospheric pressure, and molar ratios of $CH_4/H_2O=1$ and $C_7H_8/H_2O=1/7$. *(Reproduced from U. Cimenler, B. Joseph, J.N. Kuhn, Hydrocarbon steam reforming using silicalite-1 zeolite encapsulated Ni-based catalyst, AIChE J. (2016), with permission from John Wiley and Sons.)*

encapsulated catalyst showed ~2%–7% decreases (depending on temperature and reactant) in both CH_4 and C_7H_8 conversions compared to acidic H-β zeolite, confirming that shell acidity did contribute to conversion (Fig. 15.11). Moreover, the authors found that the nonacidic Silicalite-1 encapsulated composite catalyst remained stable during the combined SR for 10 h TOS, but the uncoated SR catalyst deactivated.

3 WATER-GAS SHIFT REACTION

The WGS is a catalytic process to convert CO and H_2O to produce H_2 as shown in reaction (15.5). This process was first reported by Mond and Langer in 1888 [40]. WGS catalysis has an important role in the industry for the production of ammonia, methanol, and hydrogen [41]. This reaction is also used in the XTL process to produce synthetic fuel with the SR of hydrocarbons, to adjust the H_2/CO ratio of the produced gas stream.

$$CO\,(g) + H_2O\,(g) \leftrightarrow CO_2(g) + H_2(g) \quad \Delta H^{\circ}_{298} = -41.2\,\text{kj/mol} \qquad (15.5)$$

To balance between kinetics and thermodynamic limitations, the WGS reaction is usually conducted in two- or three-stage catalytic converters instead of one to make the process more economical [42]. The first stage is a high-temperature step that is operating at 320–450°C, favoring fast CO consumption and minimizing catalyst bed volume. The second stage is a low-temperature step that is operating at 200–240°C to obtain higher conversions [43,44].

The catalyst selection for the WGS reaction depends on the stage that the catalyst will be used. There are two main types of WGS catalysts. The first type is promoted iron oxide catalysts that are high temperature shift catalysts [43,45]. They are active at high temperatures (320–450°C). The second type is copper oxide catalysts, which are low temperature shift catalysts and used at relatively low temperatures (180–260°C) [43,46,47]. Besides these catalysts, precious metal-based catalysts (mainly platinum and gold) have been studied recently for use in fuel cell applications [48–50].

Despite developments on Fe and Cu catalysts, rapid deactivation is a major issue over such catalysts [51,52]. The addition of noble metals, use of different promoters, and improving strong metal-support interaction with porous structures or confining metal into a well-defined channel are some of the methods that are used in the literature to solve deactivation problems [53]. Encapsulating metals with inorganic shells such as SiO_2, Al_2O_3, CeO_2, ZrO_2, Fe_2O_3, and TiO_2 are currently receiving great attention due to their high thermal stability and high level of metal-support interaction [53–55]. For instance, Yeung et al. reported that the microemulsion prepared ceria encapsulated noble metal catalysts that show a comparable or even higher catalytic activity towards the WGS reaction than those catalysts prepared by traditional methods due to the intimate contact with noble metal particle and ceriumoxide [54,56,57]. In another study, Yeung et al. compared the WGS and methanation activities of commercial Cu/ZnO/Al_2O_3, Co-precipitated 2% Pt/ceria, wet impregnated 5% Pt/ceria, ceria encapsulated 5% Pt, and ceria encapsulated 5%Pt-5%Au core-shell catalysts; they found that bimetallic core of Pt and Au in a 1:1 ratio at 5wt% with respect to the ceria shell showed the highest WGS activity with no methanation activity [46]. Saw et al. synthesized bimetallic Ni-Cu core encapsulated by a CeO_2 shell catalyst using the combined method of positive emulsion and self-assembly [53]. They varied the core loadings and compared them with the conventional impregnation method in the WGS. They found that the core-shell catalyst exhibited higher catalytic activity at 500°C than the supported catalyst with the same metal loadings due to a high level of metal-support interaction and small bimetallic Ni-Cu particle size. Wieder et al. synthesized an alumina-supported, Pd@CeO_2, core-shell catalyst having 1 wt% Pd and 9 wt % ceria for the WGS reaction and found that the Pd@CeO_2 catalyst initially exhibited similar WGS rates to that of a conventional Pd/ceria catalyst, but the catalyst deactivated severely over a period of 1 h. According to the authors, the deactivation was caused by the reduction of the Pd@CeO_2/Al_2O_3 catalysts [58]. Cargnello et al. developed a new synthesis method for Pd@CeO_2 core-shell nanoparticles that were easily dispersible in common organic solvents since the dispersibility of core-shell structures with oxide shells is a key property for avoiding formation of aggregates [59]. They demonstrated that the Pd@CeO_2 core-shell impregnated Al_2O_3 support showed significant CO uptake and the Pd is accessible to CO and water, implying that the ceria shell must be porous [59]. In a later study, Cargnello et al. synthesized and compared 1-wt% Pd-CeO_2 catalysts prepared by co-precipitation of Pd nanoparticles with ceria (Pd@CeO_2-CP), by a microemulsion procedure (Pd@CeO_2-ME), and by normal impregnation of Pd salts (Pd/CeO_2-IMP) to test the concept that Pd-CeO_2 catalysts could be more stable for the WGS reaction when the Pd is embedded in CeO_2 [60]. The authors found that Pd@CeO_2-ME sample demonstrated good

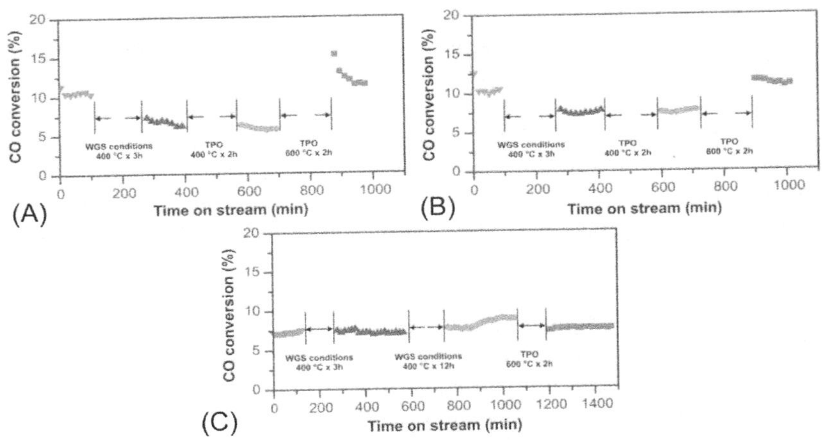

FIG. 15.12 CO conversions at 250°C (A,B) and 300°C (C) under WGS reaction conditions of Pd/CeO_2-IMP (A), Pd@CeO_2-CP (B), and Pd@CeO_2-ME (C) catalyst after different subsequent treatments. *(Reproduced from M. Cargnello, T. Montini, S. Polizzi, N.L. Wieder, R.J. Gorte, M. Graziani, P. Fornasiero, Novel embedded Pd@ CeO_2 catalysts: a way to active and stable catalysts, Dalton Trans. 39 (2010) 2122–2127, with permission from the Royal Society of Chemistry.)*

stability for WGS, suggesting that more effective encapsulation of Pd can limit the sintering of the metal phase and result in stable catalysts under a high-temperature reaction (400°C), as shown in Fig. 15.12.

4 FISCHER-TROPSCH SYNTHESIS

FTS is the last step of the various XTL technologies, and it is employed to synthesize liquid fuel from syngas. Franz Fischer and Hans Tropsch discovered FTS in 1922, converting a mixture of carbon monoxide and hydrogen to hydrocarbons using an iron catalyst [61]. The FTS reaction is a strongly exothermic polymerization reaction ($\Delta H = -165$–204 kJ/mol CO) [62–68]. The details of the FTS reactions are demonstrated in Table 15.1.

The most common Fischer-Tropsch catalysts are Fe, Co, Ni, Rh, and Ru [68–71]. Co and Fe have been mostly used in industry due to their advantages such as low price (compared to Rh and Ru) and lower methane selectivity (compared to Ni-based catalysts) [68,71,72]. However, promoters are required for Fe and Co catalysts to achieve good activity and selectivity. Although hydrocarbons produced by FTS reaction are sulfur, nitrogen, and aromatics-free, which makes the FTS products capable of being desirable fuels, the desired product selectivity of the FTS reaction is low [73]. In order to obtain high selectivity in the FTS reaction, "egg-shell" Co catalysts have been developed by Exxon [74,75] and core-shell type catalysts have been studied widely since then. Using zeolites to encapsulate core FTS catalysts is the most chosen method to increase the desirable product selectivity since zeolites have product selectivity property

TABLE 15.1 Reactions in the Fischer-Tropsch Synthesis (FTS)

Main reactions	
1. Paraffins	$(2n+1)H_2 + nCO \rightarrow C_nH_{2n+2} + nH_2O$ (15.6)
2. Olefins	$2nH_2 + nCO \rightarrow C_nH_{2n} + nH_2O$ (15.7)
3. WGS reaction	$CO + H_2O \leftrightarrow CO_2 + H_2$ (15.8)
Side reactions	
4. Alcohols	$2nH_2 + nCO \rightarrow C_nH_{2n+2}O + (n-1)H_2O$ (15.9)
5. Catalyst oxidation/reduction	(a) $M_xO_y + yH_2 \leftrightarrow yH_2O + Xm$ (15.10) (b) $M_xO_y + yCO \leftrightarrow yCO_2 + xM$ (15.11)
6. Bulk carbide formation	$yC + xM \leftrightarrow M_xC_y$ (15.12)
7. Boudouard reaction	$2CO \rightarrow C + CO_2$ (15.13)

(Reproduced from G.P. Van Der Laan, A. Beenackers, Kinetics and selectivity of the Fischer-Tropsch synthesis: a literature review, Catal. Rev. 41 (1999) 255–318, with permission from Taylor & Francis.)

(Fig. 15.4), varied molecular diffusion rate in their pores, and acidic property. They are also used as hydrocracking and isomerization catalysts in industry due to their acidic property.

In zeolite-encapsulated bifunctional catalysts, the reactants first diffuse the zeolite porous membrane and arrive to FTS sites, which convert the syngas into long-chained linear hydrocarbons. These intermediate products desorb and diffuse through to the acidic sites of the zeolite shell where they undergo further hydrocracking and isomerization to form desired branched hydrocarbons (i.e., gasoline cut) to exit the catalyst (Fig. 15.13) [8,12]. The role of the encapsulated

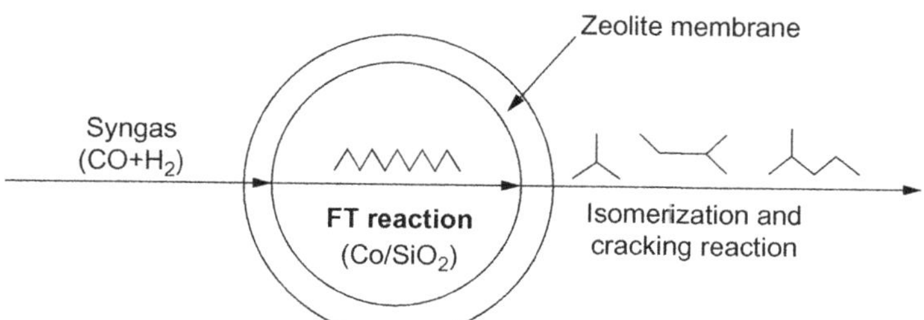

FIG. 15.13 A schematic image of the capsule catalyst role in the FTS reaction. *(Reproduced from J. He, Y. Yoneyama, B. Xu, N. Nishiyama, N. Tsubaki, Designing a capsule catalyst and its application for direct synthesis of middle isoparaffins, Langmuir 21 (2005) 1699–1702, with permission from the American Chemical Society.)*

catalyst structure is to separate two active sites where the different reactions are catalyzed independently, which is crucial for product selectivity. Moreover, the encapsulated bifunctional catalyst is expected to show high selectivity in a consecutive reaction because it significantly improves the collision possibility between intermediates and active sites [8]. The main difference of the encapsulated catalysts and the conventional bifunctional catalysts that are prepared by mixing is that the active sites have random distribution in the conventional bifunctional catalysts, which promotes the cascade reactions randomly. Thus, the intermediate products can leave without reacting at the other active sites, which means low desired product selectivity [8]. Furthermore, shape selectivity for the products also is expected since molecules of different sizes have a different diffusion efficiencies in the membrane channels [8,11]. Thus, a low diffusion rate of the long-chain hydrocarbons forces them to stay in the zeolite membrane layer longer, which means there is a higher chance of isomerization and cracking reaction inside the membrane [12].

There are many studies in the literature about the zeolite-encapsulated FTS catalyst. For instance, Bao et al. studied a core/shell catalyst in which an H-β zeolite membrane was coated directly onto the surface of a Co/Al_2O_3 pellet to increase the selectivity of isoparaffins in the FTS reaction [8]. The authors observed that the formation of C_{12+} hydrocarbons was suppressed completely and the middle isoparaffins became the main products. Yoneyama et al. synthesized Co/SiO_2 core and HZSM-5 zeolite shell FTS catalysts to produce isoparaffins from synthesis gas [16]. The authors obtained low methane selectivity and high isoparaffin selectivity in an FTS reaction using Co/SiO_2/HZSM-5 hybrid catalyst for one-step isoparaffin production from syngas. Li et al. investigated an H-β zeolite-enwrapped Co/Al_2O_3 FTS catalyst and they also obtained extremely high isoparaffin selectivity; the catalyst remained stable for 10 h without deactivation [13]. He et al. wrapped an H-ZSM-5 membrane onto the surface of the Co/SiO_2 pellet to increase the selectivity of the light hydrocarbon in the FTS reaction [12]. They found that long-chain paraffin selectivity reduced upon using an encapsulated catalyst due to secondary isomerization and hydrocracking on the zeolite shell [12]. In another study, He et al. examined the thickness effect of the H-ZSM-5 zeolite shell on the selectivity of the FTS reaction and they found that methane selectivity increased with the zeolite membrane thickness since H_2 diffuses more quickly than CO through the shell [11]. Bao et al. synthesized an H-ZSM-5/fused-iron core-shell catalyst for the synthesis of isoparaffin from syngas [10]. The result of their study showed that the formation of C_{10+} hydrocarbons was suppressed completely and the middle isoparaffins became the main products; they also achieved low methane selectivity. Yang et al. synthesized an H-ZSM-5 zeolite shell on the Co/SiO_2 FTS catalyst using physically adhesive and hydrothermal synthesis methods and compared their catalytic performances with a physically mixed catalyst for the direct synthesis of light isoparaffin from syngas [15]. The authors observed that the selectivity of light isoparaffin obtained by zeolite capsule catalysts increased sharply and the formation of heavy hydrocarbons was suppressed comparing to physically mixed H-ZSM-5 and Co/SiO_2catalyst. Yang et al. synthesized H-ZSM-5 and Silicalite-1 zeolite encapsulated $Cu/ZnO/Al_2O_3$core-shell catalysts for dimethyl ether (DME) production from syngas and compared the activity and selectivity with an uncoated catalyst. The authors found higher DME selectivity but lower activity (CO% conversion) on the core-shell catalysts, which was attributed to the coverage of the active sites by zeolite crystals [14] (Fig. 15.14).

Catalyst	Si/Al[b]	Zeolite (%)	CO conversion (%)	Selectivity (%)		
				MeOH	DME	Others
CZA			44.02	98.38	0.50	1.12
CZA-Z[c]	32	7.60	5.59	3.41	96.59	0.00
CZA-S[d]	221	9.63	30.40	21.43	78.57	0.00
CZA-M[e]	163	10.0	58.07	57.29	40.51	2.20

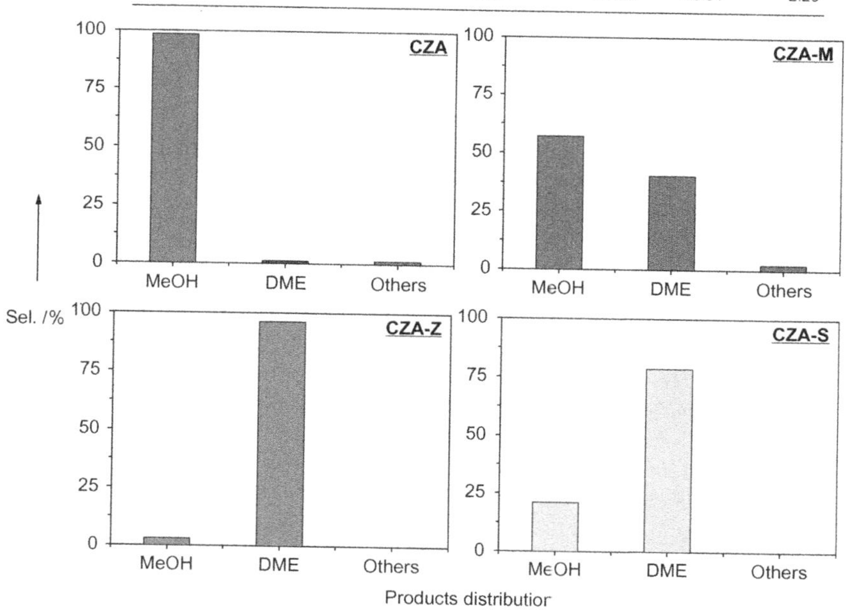

FIG. 15.14 Catalyst catalytic performance in syngas to DME reaction and product distribution of $Cu/ZnO/Al_2O_3$ catalyst (CZA), H-ZSM-5, and silicalite-1 zeolite capsule catalysts (CZA-Z and CZA-S, respectively) and the physically mixed catalyst (CZA-M). *(Reproduced from G. Yang, N. Tsubaki, J. Shamoto, Y. Yoneyama, Y. Zhang, Confinement effect and synergistic function of H-ZSM-5/Cu-ZnO-Al2O3 capsule catalyst for one-step controlled synthesis. J. Am. Chem. Soc. 132 (2010) 8129–8136, with permission from the American Chemical Society.)*

5 CONCLUSION

XTL processes have received an immense deal of interest in recent years. Although these processes are very attractive, active, selective, and stable catalysts for XTL process steps (e.g., reforming, WGS, and FTS) are crucial to achieve affordable synthetic fuel. Encapsulated catalysts (core-shell, yolk-shell, etc.) have been studied widely in the literature over the last several years to increase the stability and activity and product and reactant selectivity for applications in catalysis, especially in reforming, WGS and FTS. Encapsulating metals with inorganic shells such as SiO_2, Al_2O_3, CeO_2, ZrO_2, Fe_2O_3, and TiO_2 are currently receiving great attention due to their high thermal stability and high level of metal-support interaction for the water-gas shift. Coating the Fe or Co catalysts is studied to obtain high product selectivity (especially isoparaffin selectivity) for the FTS step in the XTL processes. The studies in the literature about encapsulated catalysts provided promising results that demonstrated that encapsulation can be used to increase stability and activity, control the reactant selectivity, and obtain products with high desirability in the XTL processes.

REFERENCES

[1] Y. Yin, R.M. Rioux, C.K. Erdonmez, S. Hughes, G.A. Somorjai, A.P. Alivisatos, Formation of hollow nanocrystals through the nanoscale Kirkendall effect, Science 304 (2004) 711–714.

[2] C.-H. Kuo, Y. Tang, L.-Y. Chou, B.T. Sneed, C.N. Brodsky, Z. Zhao, C.-K. Tsung, Yolk–shell nanocrystal@ ZIF-8 nanostructures for gas-phase heterogeneous catalysis with selectivity control, J. Am. Chem. Soc. 134 (2012) 14345–14348.

[3] A. Trovarelli, Catalysis by Ceria and Related Materials, World Scientific, Singapore, 2002.

[4] G. Li, Z. Tang, Noble metal nanoparticle@ metal oxide core/yolk–shell nanostructures as catalysts: recent progress and perspective, Nanoscale 6 (2014) 3995–4011.

[5] J. Wei, Nonlinear phenomena in zeolite diffusion and reaction, Ind. Eng. Chem. Res. 33 (1994) 2467–2472.

[6] J.L. Plawsky, Transport Phenomena Fundamentals, CRC Press, Bosa Roca, 2014.

[7] E.M. Flanigen, J. Jansen, H. van Bekkum, Introduction to Zeolite Science and Practice, Elsevier, 1991.

[8] J. Bao, J. He, Y. Zhang, Y. Yoneyama, N. Tsubaki, A core/shell catalyst produces a spatially confined effect and shape selectivity in a consecutive reaction, Angew. Chem. 120 (2008) 359–362.

[9] S.M. Csicsery, Shape-selective catalysis in zeolites, Zeolites 4 (1984) 202–213.

[10] J. Bao, G. Yang, C. Okada, Y. Yoneyama, N. Tsubaki, H-type zeolite coated iron-based multiple-functional catalyst for direct synthesis of middle isoparaffins from syngas, Appl. Catal. A Gen. 394 (2011) 195–200.

[11] J. He, Z. Liu, Y. Yoneyama, N. Nishiyama, N. Tsubaki, Multiple-functional capsule catalysts: a tailor-made confined reaction environment for the direct synthesis of middle isoparaffins from syngas, Chem. A Eur. J. 12 (2006) 8296–8304.

[12] J. He, Y. Yoneyama, B. Xu, N. Nishiyama, N. Tsubaki, Designing a capsule catalyst and its application for direct synthesis of middle isoparaffins, Langmuir 21 (2005) 1699–1702.

[13] X. Li, J. He, M. Meng, Y. Yoneyama, N. Tsubaki, One-step synthesis of H–β zeolite-enwrapped Co/Al_2O_3 Fischer–Tropsch catalyst with high spatial selectivity, J. Catal. 265 (2009) 26–34.

[14] G. Yang, N. Tsubaki, J. Shamoto, Y. Yoneyama, Y. Zhang, Confinement effect and synergistic function of H-ZSM-5/Cu-ZnO-Al2O3 capsule catalyst for one-step controlled synthesis, J. Am. Chem. Soc. 132 (2010) 8129–8136.

[15] G. Yang, C. Xing, W. Hirohama, Y. Jin, C. Zeng, Y. Suehiro, T. Wang, Y. Yoneyama, N. Tsubaki, Tandem catalytic synthesis of light isoparaffin from syngas via Fischer–Tropsch synthesis by newly developed core–shell-like zeolite capsule catalysts, Catal. Today 215 (2013) 29–35.

[16] Y. Yoneyama, X. San, T. Iwai, N. Tsubaki, One-step synthesis of isoparaffin from synthesis gas using hybrid catalyst with supercritical butane, Energy Fuel 22 (2008) 2873–2876.

[17] M.B. Gawande, A. Goswami, T. Asefa, H. Guo, A.V. Biradar, D.-L. Peng, R. Zboril, R.S. Varma, Core–shell nanoparticles: synthesis and applications in catalysis and electrocatalysis, Chem. Soc. Rev. 44 (2015) 7540–7590.

[18] S.H. Joo, J.Y. Park, C.-K. Tsung, Y. Yamada, P. Yang, G.A. Somorjai, Thermally stable Pt/mesoporous silica core–shell nanocatalysts for high-temperature reactions, Nat. Mater. 8 (2009) 126–131.

[19] M. Galanti, D. Fanelli, S. Angioletti-Uberti, M. Ballauff, J. Dzubiella, F. Piazza, Reaction rate of a composite core–shell nanoreactor with multiple nanocatalysts, Phys. Chem. Chem. Phys. 18 (2016) 20758–20767.

[20] Y. Lu, Y. Mei, M. Drechsler, M. Ballauff, Thermosensitive core–shell particles as carriers for Ag nanoparticles: modulating the catalytic activity by a phase transition in networks, Angew. Chem. Int. Ed. 45 (2006) 813–816.

[21] S.L. Suib, New and future developments in catalysis: catalysis for remediation and environmental concerns, Elsevier, Amsterdam, 2013.

[22] D. Li, Y. Nakagawa, K. Tomishige, Methane reforming to synthesis gas over Ni catalysts modified with noble metals, Appl. Catal. A Gen. 408 (2011) 1–24.

[23] T. Hirai, N.-O. Ikenaga, T. Miyake, T. Suzuki, Production of hydrogen by steam reforming of glycerin on ruthenium catalyst, Energy Fuel 19 (2005) 1761–1762.

[24] E. Kikuchi, S. Tanaka, Y. Yamazaki, Y. Morita, Steam reforming of hydrocarbons on noble metal catalysts (part 1), Bull. Jpn. Petrol. Inst. 16 (1974) 95–98.

[25] J. Rostrupnielsen, J.B. Hansen, CO_2-reforming of methane over transition metals, J. Catal. 144 (1993) 38–49.

[26] P.K. Cheekatamarla, C. Finnerty, Reforming catalysts for hydrogen generation in fuel cell applications, J. Power Sources 160 (2006) 490–499.

[27] M.F. Mark, W.F. Maier, CO_2-reforming of methane on supported Rh and Ir catalysts, J. Catal. 164 (1996) 122–130.

[28] V.S. Guggilla, J. Akyurtlu, A. Akyurtlu, I. Blankson, Steam reforming of n-dodecane over Ru-Ni-based catalysts, Ind. Eng. Chem. Res. 49 (2010) 8164–8173.

[29] K. Sutthiumporn, S. Kawi, Promotional effect of alkaline earth over Ni–La_2O_3 catalyst for CO_2 reforming of CH_4: role of surface oxygen species on H 2 production and carbon suppression, Int. J. Hydrog. Energy 36 (2011) 14435–14446.

[30] J.M. Mayne, A.R. Tadd, K.A. Dahlberg, J.W. Schwank, Influence of thiophene on the isooctane reforming activity of Ni-based catalysts, J. Catal. 271 (2010) 140–152.

[31] J. Sehested, Four challenges for nickel steam-reforming catalysts, Catal. Today 111 (2006) 103–110.

[32] K.O. Christensen, D. Chen, R. Lødeng, A. Holmen, Effect of supports and Ni crystal size on carbon formation and sintering during steam methane reforming, Appl. Catal. A Gen. 314 (2006) 9–22.

[33] J.R. Rostrup-Nielsen, J. Sehested, J.K. Nørskov, Hydrogen and synthesis gas by steam-and CO_2 reforming, Adv. Catal. 47 (2002) 65–139.

[34] T.D. Gould, A. Izar, A.W. Weimer, J.L. Falconer, J.W. Medlin, Stabilizing Ni catalysts by molecular layer deposition for harsh, dry reforming conditions, ACS Catal. 4 (2014) 2714–2717.

[35] Z. Li, L. Mo, Y. Kathiraser, S. Kawi, Yolk–satellite–shell structured Ni–yolk@ Ni@SiO_2 nanocomposite: superb catalyst toward methane CO_2 reforming reaction, ACS Catal. 4 (2014) 1526–1536.

[36] Z. Li, Y. Kathiraser, J. Ashok, U. Oemar, S. Kawi, Simultaneous tuning porosity and basicity of nickel@ nickel–magnesium phyllosilicate core–shell catalysts for CO_2 reforming of CH4, Langmuir 30 (2014) 14694–14705.

[37] U. Cimenler, B. Joseph, J.N. Kuhn, Molecular-size selective H-β zeolite-encapsulated Ce-Zr/Ni-Mg catalysts for steam reforming, Appl. Catal. A Gen. 505 (2015) 494–500.

[38] U. Cimenler, B. Joseph, J.N. Kuhn, Effect of zeolite membrane shell thickness on reactant selectivity for hydrocarbon steam reforming using layered catalysts, Energy Fuel 30 (7) (2016) 5300–5308.

[39] U. Cimenler, B. Joseph, J.N. Kuhn, Hydrocarbon steam reforming using silicalite-1 zeolite encapsulated Ni-based catalyst, AIChE J. 63 (2017) 200–207.

[40] L. Mond, C. Langer. British Patent 12 (1888) 608.

[41] P. Smirniotis, K. Gunugunuri, Water Gas Shift Reaction: Research Developments and Applications, Elsevier, Amsterdam, 2015.

[42] M. Muhler, J.M. Thomas, W.J. Thomas, Principles and Practice of Heterogeneous Catalysis, VCH, Weinheim, 1997. ISBN 3-527-29239-X, Preis: 88,–DM. Berichte der Bunsengesellschaft für physikalische Chemie 101 (1997) 1560–1560.

[43] D. Mendes, A. Mendes, L. Madeira, A. Iulianelli, J. Sousa, A. Basile, The water-gas shift reaction: from conventional catalytic systems to Pd-based membrane reactors—a review, Asia Pac. J. Chem. Eng. 5 (2010) 111–137.

[44] X. Qi, M. Flytzani-Stephanopoulos, Activity and stability of Cu-CeO_2 catalysts in high-temperature water-gas shift for fuel-cell applications, Ind. Eng. Chem. Res. 43 (2004) 3055–3062.

[45] C.-S. Chen, W.-H. Cheng, S.-S. Lin, Study of iron-promoted Cu/SiO_2 catalyst on high temperature reverse water gas shift reaction, Appl. Catal. A Gen. 257 (2004) 97–106.

[46] C.M.Y. Yeung, S.C. Tsang, Some optimization in preparing core-shell Pt–ceria catalysts for water gas shift reaction, J. Mol. Catal. A Chem. 322 (2010) 17–25.

[47] Y. Li, Q. Fu, M. Flytzani-Stephanopoulos, Low-temperature water-gas shift reaction over Cu-and Ni-loaded cerium oxide catalysts, Appl. Catal. B Environ. 27 (2000) 179–191.

[48] C. Ratnasamy, J.P. Wagner, Water gas shift catalysis, Catal. Rev. 51 (2009) 325–440.

[49] Q. Fu, H. Saltsburg, M. Flytzani-Stephanopoulos, Active nonmetallic Au and Pt species on ceria-based water-gas shift catalysts, Science 301 (2003) 935–938.

[50] D. Andreeva, V. Idakiev, T. Tabakova, L. Ilieva, P. Falaras, A. Bourlinos, A. Travlos, Low-temperature water-gas shift reaction over Au/CeO_2 catalysts, Catal. Today 72 (2002) 51–57.

[51] V. Tsipouriari, A. Efstathiou, Z. Zhang, X. Verykios, Reforming of methane with carbon dioxide to synthesis gas over supported Rh catalysts, Catal. Today 21 (1994) 579–587.

[52] C.H. Kim, L.T. Thompson, Deactivation of Au/CeO_x water gas shift catalysts, J. Catal. 230 (2005) 66–74.

[53] E.T. Saw, U. Oemar, M.L. Ang, K. Hidajat, S. Kawi, Highly active and stable bimetallic nickel–copper core–ceria shell catalyst for high-temperature water–gas shift reaction, ChemCatChem 7 (2015) 3358–3367.

[54] C.M. Yeung, K.M.K. Yu, Q.J. Fu, D. Thompsett, M.I. Petch, S.C. Tsang, Engineering Pt in ceria for a maximum metal-support interaction in catalysis, J. Am. Chem. Soc. 127 (2005) 18010–18011.

[55] H. Tian, X. Li, L. Zeng, J. Gong, Recent advances on the design of group viii base-metal catalysts with encapsulated structures, ACS Catal. 5 (2015) 4959–4977.

[56] C.M. Yeung, F. Meunier, R. Burch, D. Thompsett, S.C. Tsang, Comparison of new microemulsion prepared "Pt-in-Ceria" catalyst with conventional "Pt-on-Ceria" catalyst for water-gas shift reaction, J. Phys. Chem. B 110 (2006) 8540–8543.

[57] C.M.Y. Yeung, S.C. Tsang, Noble metal core-ceria shell catalysts for water-gas shift reaction, J. Phys. Chem. C 113 (2009) 6074–6087.

[58] N.L. Wieder, M. Cargnello, K. Bakhmutsky, T. Montini, P. Fornasiero, R.J. Gorte, Study of the water-gas-shift reaction on Pd@ CeO_2/Al_2O_3 core-shell catalysts, J. Phys. Chem. C 115 (2010) 915–919.

[59] M. Cargnello, N.L. Wieder, T. Montini, R.J. Gorte, P. Fornasiero, Synthesis of dispersible Pd@ CeO_2 core-shell nanostructures by self-assembly, J. Am. Chem. Soc. 132 (2009) 1402–1409.

[60] M. Cargnello, T. Montini, S. Polizzi, N.L. Wieder, R.J. Gorte, M. Graziani, P. Fornasiero, Novel embedded Pd@ CeO_2 catalysts: a way to active and stable catalysts, Dalton Trans. 39 (2010) 2122–2127.

[61] F. Fischer, H. Tropsch, Uber die Herstellung synthetischer olgemische (Synthol) durch Aufbau aus Kohlenoxyd und Wasserstoff, Brennst. Chem. 4 (1923) 276–285.

[62] E. de Smit, B.M. Weckhuysen, The renaissance of iron-based Fischer–Tropsch synthesis: on the multifaceted catalyst deactivation behaviour, Chem. Soc. Rev. 37 (2008) 2758–2781.

[63] M. Dry, J. Hoogendoorn, Technology of the Fischer-Tropsch process, Catal. Rev. Sci. Eng. 23 (1981) 265–278.

[64] H. Kölbel, M. Ralek, The Fischer-Tropsch synthesis in the liquid phase, Catal. Rev. Sci. Eng. 21 (1980) 225–274.

[65] V. Rao, G. Stiegel, G. Cinquegrane, R. Srivastava, Iron-based catalysts for slurry-phase Fischer-Tropsch process: technology review, Fuel Process. Technol. 30 (1992) 83–107.

[66] C.N. Satterfield, R.T. Hanlon, S.E. Tung, Z.M. Zou, G.C. Papaefthymiou, Effect of water on the iron-catalyzed Fischer-Tropsch synthesis, Ind. Eng. Chem. Prod. Res. Dev. 25 (1986) 407–414.

[67] R. Srivastava, V. Rao, G. Cinquegrane, G. Stiegel, Catalysts for Fischer-Tropsch, Hydrocarb. Process. 69 (2) (1990).

[68] Q. Zhang, J. Kang, Y. Wang, Development of novel catalysts for Fischer–Tropsch synthesis: tuning the product selectivity, ChemCatChem 2 (2010) 1030–1058.

[69] G.P. Van Der Laan, A. Beenackers, Kinetics and selectivity of the Fischer–Tropsch synthesis: a literature review, Catal. Rev. 41 (1999) 255–318.

[70] A. Dalai, B. Davis, Fischer–Tropsch synthesis: a review of water effects on the performances of unsupported and supported Co catalysts, Appl. Catal. A Gen. 348 (2008) 1–15.

[71] H. Schulz, Short history and present trends of Fischer–Tropsch synthesis, Appl. Catal. A Gen. 186 (1999) 3–12.

[72] M.E. Dry, The Fischer–Tropsch process: 1950–2000, Catal. Today 71 (2002) 227–241.

[73] N. Jiang, G. Yang, X. Zhang, L. Wang, C. Shi, N. Tsubaki, A novel silicalite-1 zeolite shell encapsulated iron-based catalyst for controlling synthesis of light alkenes from syngas, Catal. Commun. 12 (2011) 951–954.

[74] E. Iglesia, S.L. Soled, J.E. Baumgartner, S.C. Reyes, Synthesis and catalytic properties of egg-shell cobalt catalysts for the Fischer-Tropsch synthesis, Top. Catal. 2 (1995) 17–27.

[75] E. Iglesia, H. Vroman, S. Soled, J. Baumgartner, R.A. Fiato, Selective Catalysts and Their Preparation for Catalytic Hydrocarbon Synthesis, Google Patents, 1991.

FURTHER READING

X. Fang, S. Liu, J. Zang, C. Xu, M.-S. Zheng, Q.-F. Dong, D. Sun, N. Zheng, Precisely controlled resorcinol–formaldehyde resin coating for fabricating core–shell, hollow, and yolk–shell carbon nanostructures, Nanoscale 5 (2013) 6908–6916.

A. Corma, Inorganic solid acids and their use in acid-catalyzed hydrocarbon reactions, Chem. Rev. 95 (1995) 559–614.

Index

Note: Page numbers followed by *f* indicate figures, *t* indicate tables, and *s* indicate schemes.

C

D

E

F

I

N

O

P

Made in the USA
Monee, IL
13 September 2026

62977011R00305